Urban Agriculture and Food Systems:

Breakthroughs in Research and Practice

Information Resources Management Association
USA

IGI Global
DISSEMINATOR OF KNOWLEDGE

Published in the United States of America by
IGI Global
Engineering Science Reference (an imprint of IGI Global)
701 E. Chocolate Avenue
Hershey PA, USA 17033
Tel: 717-533-8845
Fax: 717-533-8661
E-mail: cust@igi-global.com
Web site: http://www.igi-global.com

Library of Congress Cataloging-in-Publication Data

Names: Information Resources Management Association, editor.
Title: Urban agriculture and food systems : breakthroughs in research and practice / Information Resources Management Association, editor.
Description: Hershey, PA : Engineering Science Reference, [2019]
Identifiers: LCCN 2018043590| ISBN 9781522580638 (hardcover) | ISBN 9781522580645 (ebook)
Subjects: LCSH: Urban agriculture. | Food security.
Classification: LCC S494.5.U72 U732 2019 | DDC 338.109173/2--dc23 LC record available at https://lccn.loc.gov/2018043590

British Cataloguing in Publication Data
A Cataloguing in Publication record for this book is available from the British Library.

The views expressed in this book are those of the authors, but not necessarily of the publisher.

For electronic access to this publication, please contact: eresources@igi-global.com.

List of Contributors

Table of Contents

Preface

As the world's population continues to increase, it becomes imperative that nutritional needs are being met. Sustaining the populace can become difficult as regions become increasingly more urbanized and rural areas shrink. New strategies and techniques need to be explored to utilize urban areas to their fullest capacity in supporting agricultural development. Additionally, proper food safety, security, and trade must be adhered to for the safety of the public and the prevention of major food crises.

That is why IGI Global is pleased to offer this one-volume comprehensive reference that will empower students, urban developers, professionals, engineers, policymakers, and researchers with a stronger understanding of urban agriculture and food systems.

This compilation is designed to act as a single reference source on conceptual, methodological, and technical aspects, and will provide insight into emerging topics including but not limited to food manufacturing, agricultural technologies, genetically modified crops, poverty reduction, and smart agriculture. The chapters within this publication are sure to provide readers the tools necessary for further research and discovery in their respective industries and/or fields.

Urban Agriculture and Food Systems: Breakthroughs in Research and Practice is organized into seven sections that provide comprehensive coverage of important topics. The sections are:

1. Agricultural Trade;
2. Food Consumption and Demand;
3. Food Safety and Inspection;
4. Food Security;
5. Organic Agriculture and Consumption;
6. Sustainable Agriculture; and
7. Urban Agriculture.

The following paragraphs provide a summary of what to expect from this invaluable reference source:

Section 1, "Agricultural Trade," opens this extensive reference source by highlighting the barriers, implications, and challenges of trading agricultural products and food security concerns. The first chapter in this section, "Trade in Agricultural Products and Food Security Concerns on Emerging Markets: How to Balance Protection and Liberalization," authored by Prof. Vasilii Erokhin from Harbin Engineering University, China, addresses recent food security trends in emerging countries, including BRICS, Eurasian Economic Union (EAEU), and Russia, and assesses the effects of food trade distortions in order to balance protection and liberalization policies for achieving effective levels of food security.

The second chapter in this section, "Implications of Trade Liberalization for Food Security Under the ASEAN-India Strategic Partnership: A Gravity Model Approach," authored by Profs. Ishita Ghosh and Ishita Ghoshal from Symbiosis International University, India, focuses on the agricultural trade between India and the Association of Southeast Asian Nations (ASEAN) under their strategic partnership and regional cooperation. It also provides insights from the results of the analysis to augment the food trade relationship between India and the ASEAN economies.

The third chapter in this section, "The Barriers to Innovation Diffusion: The Case of GM Food in Sri Lanka," authored by Profs. Dilupa Nakandala and Tim Turpin from Western Sydney University, Australia, focuses on barriers to diffusion of innovation and analyzes the case of genetically modified (GM) food diffusion in Sri Lanka using the Rogers' classical model of innovation diffusion. It also examines the potential for GM foods to contribute to the process of social innovation for ameliorating hunger in developing countries through the potential benefits of the introduction of new plant varieties, increased productivity, and reduced costs for pesticides and herbicides.

The fourth chapter in this section, "Agriculture, Trade Liberalization, and Poverty in the ACP Countries," authored by Prof. Theresa Ann Rajack-Talley from the University of Louisville, USA, examines the links between agriculture production and trade, food security, small farmer production, and poverty reduction in developing countries. It also analyzes the impact of the acceleration of trade liberalization and agriculture protectionism on ACP countries.

The fifth and final chapter in this section, "Adaptation of the Russian Food Market to the Contemporary Geopolitical Challenges: Bans vs. Liberalization," authored by Prof. Ivan Ivolga from Stavropol State Agrarian University, Russia, explores the expected influences of trade restrictions on trade in agricultural commodities and assesses the degree of distortion and return effects on domestic food markets. It also addresses the possible effects of trade restrictions between the EU, the USA, Russia, and other countries.

Section 2, "Food Consumption and Demand," includes chapters on emerging trends in food consumption patterns and expenditures and the concept of food porn. The first chapter in this section, "Food Consumption Patterns in Times of Economic Recession," authored by Prof. Glykeria Theodoridou from the Department of Agricultural Economics, Aristotle University of Thessaloniki, Thessaloniki, Greece; Prof. Efthimia Tsakiridou from Aristotle University of Thessaloniki, Thessaloniki, Greece; Prof. Nikos Kalogeras from Wageningen University, Wageningen, The Netherlands; and Prof. Konstantinos Mattas from Aristotle University of Thessaloniki, Thessaloniki, Greece, explores the factors influencing consumers' purchasing behavior in times of crisis and identifies consumers with similar behavior based on their socio-demographic characteristics and their attitudes towards food consumption in times of crisis.

The second chapter in this section, "Food Porn and the Invitation to Gaze: Ephemeral Consumption and the Digital Spectacle," authored by Prof. Yasmin Ibrahim from Queen Mary University of London, UK, explores the notion of food porn with relevance to new media technologies and contemporary digital culture. It posits the notion that the sociality around producing and sharing food in the user generated content (UGC) economy in many ways re-negotiates the unattainable quality that food porn signifies in the mainstream media.

The third and final chapter in this section, "Food Consumption Expenditure and Standard of Living in Romania," authored by Prof. Marian Zaharia from Petroleum-Gas University, Romania and Prof. Rodica-Manuela Gogonea from the Academy of Economic Studies, Romania, highlights the way in which the average food consumption expenditure per inhabitant, per food product categories (cereals and cereal products, meat and meat products, milk and dairy products, vegetables and vegetable products, fruits and fruit products) and beverages (wine and wine products, beer, distilled alcoholic beverages, soft drinks), consumed by populations of different social categories (employee, farmer, unemployed, and pensioner), residential areas, and macro regions is influenced by the average per capita income developments.

Section 3, "Food Safety and Inspection," presents coverage on food safety from consumer perspectives and on food waste management. The first chapter in this section, "Food Safety From Consumer Perspective: Consumer Confidence in Food Safety," authored by Prof. Ivana Domazet from the Institute of Economic Sciences, Serbia and Prof. Nenad Djokić from the University of Novi Sad, Serbia, examines different approaches to food safety and the results of previous research dealing with consumer perspectives of food safety. It also explores how to implement consumer confidence in the food safety scale in Serbian consumer research.

The second chapter in this section, "Food Waste Management and Corporate Social Responsibility in Indian Food Service Industry: A Conceptual Analysis," authored by Prof. Anish Slath from Panjab University, India and Prof. Ashish Nikhanj from the Institute of Hotel Management, India, identifies practices that are being followed in the Indian hospitality industry in the field of food waste management and reviews the policies of food waste management in Indian hotels. It also examines how hotels are managing food waste under current corporate social responsibility practices.

The third and final chapter in this section, "Consumer Co-Operatives and Perceptions of Food Safety: Shaping Markets in Post-Fukushima Japan," authored by Profs. Catherine Burns, Kumiko Katayama, and Robin E. Roberts from Griffith University, Australia, examines the perceptions and purchasing practices of consumer co-operative members located in a small regional city, with particular attention to meanings of food safety following the Fukushima nuclear accident.

Section 4, "Food Security," discusses coverage and research perspectives on food security issues and solutions in various regions around the globe. The first chapter in this section, "Integrating Spatial Technologies in Urban Environments for Food Security: A Vision for Economic, Environmental, and Social Responsibility in South Bend, Indiana," authored by Profs. Edwin Joseph and Elizabeth O'Dea from Indiana University – South Bend, USA, explores South Bend's history and economic development strategies and reviews practices aimed at strengthening food security for the urban poor. The chapter also documents how numerous disparate organizations have been trying to help alleviate urban poverty and hunger and reviews previous strategies used to foster sustainable growth and development.

The second chapter in this section, "The Economic Determinants of Food Security in the MENA Region," authored by Prof. Ahmet Ali Koç from the Department of Economics, Akdeniz University, Antalya, Turkey; Prof. Oznur Ozdamar from the Aydın Faculty of Economics, Department of Econometrics, Adnan Menderes University, Efeler, Turkey; and Prof. Peyman Uysal from the Department of Economics, Akdeniz University, Antalya, Turkey, analyzes the determinants of food insecurity and examines the relationship of important economic, social, and natural factors with an overall food insecurity index that derived by dividing the food imports value to the sum of total exports and the net remittance inflows.

The third chapter in this section, "A System Dynamics Model for Subsistence Farmers' Food Security Resilience in Sub-Saharan Africa," authored by Profs. Benedict Oyo and Billy Mathias Kalema from the Department of Informatics, Tshwane University of Technology, Pretoria, South Africa, investigates the factors that have enabled subsistence farmers to succeed despite the previous shocks and stresses and develop a system dynamics model for sustainable food security based on initiatives exclusive to the farmers. The model is used to examine the question: How can innovative subsistence farmers engage in better livelihood and market orientated production irrespective of external public or donor support?

The fourth chapter in this section, "Rights-Based Approach to Food and Nutrition Security in Nigeria," authored by Prof. Clementina Oluwafunke Ajayi from the Federal University of Technology – Akure, Nigeria and Prof. Kemisola O. Adenegan from the University of Ibadan, Nigeria, focuses on the need for a rights-based approach to food and nutrition security in Nigeria. It creates an awareness of the need to adopt a rights-based approach to food and nutrition security and help define the context of the right to food in Nigeria. It also sets out the roles of right holders, duty bearers, and accountable agents in food and nutrition security and highlights their rights, obligations, and responsibilities, as well as voluntary guidelines and implications for a rights-based approach to food.

The fifth chapter in this section, "Indigenous Food Preservation and Management of Postharvest Losses Among the Akan of Ghana," authored by Prof. Kofi Quan-Baffour from the University of South Africa, South Africa, examines the Akan heritage of indigenous food preservation as a strategy to manage postharvest losses and ensure food security and sustainable livelihoods. It also describes the various indigenous methods of preserving, storing, and managing specific food items to ensure food security and discusses some of the major challenges to indigenous food preservation and storage.

The sixth and final chapter in this section, "Determinant of Food Security on Upland Agriculture Households in Paletwa Township, Chin State of Myanmar," authored by Profs. San Lwin and Supaporn Poungchompu from Khon Kaen University, Khon Kaen, Thailand, examines food security in upland agriculture households in rural areas. It also explores the main goals of food security: sustainable development goals, no poverty, and zero hunger.

Section 5, "Organic Agriculture and Consumption," includes chapters on organic food consumption and the demographic influences involved in purchasing organic food options. The first chapter in this section, "The Role of Private Labels in the Organic Food Market," authored by Profs. Miguel Llorens and Sonia Carcelén from the Universidad Complutense de Madrid, Spain, investigates the role of store brands in the development of the organic food product category. It provides a comprehensive picture of the current status of store brands in the Spanish distribution channels. It also reviews some private label marketing strategies for retailers to implement such as branding, labelling, pricing, merchandising, and promotion.

The second chapter in this section, "Green Consumption: A Study to Understand Consumers' Organic Food Consumption," authored by Profs. Ozan Kaya and Feridun Duman from Mugla Sitki Kocman University, Turkey, reveals the profile of organic food consumers and the factors that affect consumers' motivation for organic food consumption as being one of the most important aspects of green marketing and green consumption.

The third and final chapter in this section, "Moderating Role of Demographics on Attitude Towards Organic Food Purchase Behavior: A Study on Indian Consumers," authored by Prof. Arpita Khare from the Indian Institute of Management, India, examines the role of demographics and attitude on organic food purchase behavior. It looks at the verbal, actual, and affect commitment to buying organic products and at demographic factors such as income, gender, and age.

Section 6, "Sustainable Agriculture," presents coverage on the various aspects of sustainable agriculture. The first chapter in this section, "New Food Industries Toward a New Level of Sustainable Supply: Success Stories, Business Models, and Strategies," authored by Profs. Vittorio D'Aleo, Francesco D'Aleo, and Roberta Bonanno from the University of Messina, Italy, demonstrates that novel food sources can be a substitute for other food that require production processes and thus make a big impact on the environment. It also examines existing business models on the novel food market and the way they operate and create wealth.

The second chapter in this section, "Human Overpopulation and Food Security: Challenges for the Agriculture Sustainability," authored by Profs. Rishikesh Singh and Pratap Srivastava from Banaras Hindu University, India; Prof. Pardeep Singh from the University of Delhi, India; and Profs. Shweta Upadhyay and Akhilesh Singh Raghubanshi from Banaras Hindu University, India, examines issues related to food security such as overpopulation, the impact of agriculture on various natural resources, the nexus of overpopulation-food security-climate change-agriculture, sustainable agriculture, and recommended agricultural practices for meeting future food demand without compromising environmental resources.

The third and final chapter in this section, "Integrating Environment, Food Systems, and Sustainability in Feeding the Growing Population in Developing Countries," authored by Prof. Abiodun Elijah Obayelu from the Federal University of Agriculture – Abeokuta (FUNAAB), Nigeria, provides an overview of the current status of food systems, their interaction with the environment and an assessment of the sustainability of current agricultural practices and their effects on food production.

Section 7, "Urban Agriculture," discusses coverage and research perspectives on the urban farming movement and the challenges and opportunities in urban agriculture. The first chapter in this section, "Urban Farming Movement," authored by Prof. Pierluigi Nicolin from the Politecnico di Milano, Italy, examines the phenomenon surrounding the urban farming movement.

The second chapter in this section, "Developing an Economic Estimation System for Vertical Farms," authored by Profs. Yiming Shao, Tim Heath, and Yan Zhu from the University of Nottingham, Nottingham, UK, presents an integrated knowledge-based software system developed by the authors, called "VFer" (short for "vertical farmer"), to enable better cost and revenue estimations of vertical farms. The chapter examines this highly flexible, customized system and results from several typical vertical farm configurations in three mega-cities (Shanghai, London, and Washington DC) to elucidate the potential economic return of vertical farms.

The third chapter in this section, "Building an Urban Food System Through UDC Food Hubs," authored by Profs. Sabine O'Hara, Dwane Jones, and Harris B. Trobman from the University of the District of Columbia, USA, describes urban food hubs, their locations, and the training, wellness, and leadership opportunities they offer to UDC students and DC residents.

The fourth chapter in this section, "Geospatial Evaluation for Urban Agriculture Land Inventory: Roanoke, Virginia, USA," authored by Prof. Tammy E. Parece from the Department of Social and Behavioral Sciences, Colorado Mesa University, Grand Junction, CO, USA and Prof. James B. Campbell from the Department of Geography, Virginia Tech, Blacksburg, VA, USA, examines how land inventory forms the critical first step in identifying sites for urban agriculture. The authors' analysis greatly expands on prior land inventory strategies, first analyzing land cover to identify all open areas available for siting urban agriculture. Then in GIS, the authors completed a land use suitability analysis, and finally a demographic analysis to assess potential sites for contribution to food security of lower income populations.

The fifth chapter in this section, "Urban Agriculture as Livelihood Strategy: A Case Study in the Southern Fringe of Addis Ababa," authored by Dr. Ashenafi Tilahun Hailie from the Ministry of Construction, Ethiopia, assesses the benefits of urban agriculture in the southern fringe of Addis Ababa. It surveys the general characteristics of urban farmers, identifies the major components of urban agriculture, sorts-out the relatively environmentally friendly and feasible aspects of it, assesses its role as survival strategy, and explores the challenges of urban agriculture.

The sixth and final chapter in this section, "Challenges and Opportunities in Urban Farming: A Study With Reference to Chennai," authored by Prof. V. Hemanth Kumar from Sri Sairam Engineering College, India and Prof. K. Sentamilselvan from SRM Valliammai Engineering College, India, focuses on the challenges and opportunities of urban farming in society. The chapter examines the societal views on urban farming and suggests ways to overcome hindrances and to assist people with utilizing the opportunities of urban farming with its benefits to enrich health for present and future generations.

Although the primary organization of the contents in this work is based on its seven sections, offering a progression of coverage of the important concepts, methodologies, technologies, applications, social issues, and emerging trends, the reader can also identify specific contents by utilizing the extensive indexing system listed at the end.

Section 1
Agricultural Trade

Chapter 1
Trade in Agricultural Products and Food Security Concerns on Emerging Markets:
How to Balance Protection and Liberalization

Vasilii Erokhin
Harbin Engineering University, China

ABSTRACT

Although free trade is a key factor in promoting economic growth, there are markets, which require specific approaches. The food market is one of them due to its vital importance for securing nutritional requirements of the population. Providing the population with food in sufficient quantity and variety is a challenge, which includes a range of issues of food production and dependence on import of the food market. Food security is increasingly influenced by foreign trade policies implemented by national governments. Many of them are re-examining their strategies for dealing with food security concerns and seeking to protect domestic food producers by restricting presence of foreign competitors on the domestic food market. The chapter addresses recent food security trends in emerging countries, including BRICS, Eurasian Economic Union (EAEU), and Russia, assesses the effects of food trade distortions in order to balance protection and liberalization policies for achieving effective levels of food security.

INTRODUCTION

In the early XX century, developed countries came to an understanding of the necessity of foreign trade liberalization for effective development of international trade and easier access to their commodities to foreign markets. Contrariwise, developed countries when practicing efficient protectionist policy were reluctant to abandon completely the protectionist practices and kept securing their domestic markets from foreign competitors (Erokhin, Ivolga, & Heijman, 2014).

DOI: 10.4018/978-1-5225-8063-8.ch001

Principles of trade liberalism are widely accepted by the World Trade Organization (WTO) and various alliances of countries as the state of things most advantageous for everyone, however, it is rather challenging to find the proper balance between protectionism and liberalization especially between developed countries and emerging economies. Too much liberalization could lead to less protection for domestic producers against international corporations, while too much protectionism could create an artificial market regime and stifle competition, the very driver of effectiveness and productivity.

Countries depending on their economic power and position on the global market apply varying policies in terms of the relationship between protectionism and liberalization in the sphere of international trade. Developed countries are within their full economic and political power to implement a wide range of tools to protect their national producers and promote their influence over the "weaker" nations on the global market. Developing economies are still in some or other way in a dependent position on the global market in comparison with developed countries. They supply raw materials, mineral resources, and agricultural commodities and consume products of higher conversion. Evidently, being dependent, developing economies are to a far greater degree concerned that the major part of benefits of trade liberalization goes to the developed states. Shah (2010) claims that often those nations that promote free trade for all, want protectionism for themselves. The major concern shared also by Smith (2005) and Akyüz (2005) is that developed nations are pressing developing nations to abandon any protectionism before their economies are ready to enter the global markets.

That is why, many developing economies continue to protect and regulate various sectors of the domestic market in order to ensure their ability to provide standards for the full range of population, especially those related to agricultural production and food consumption. As of Shah (2010), developing economies consider their food security as a crucial parameter of independence, and even if other countries may be more efficient producers of some agricultural products or food items, a country may decide to grow some of their own and support their farmers.

Those developing countries which have embarked on economic development and reform programs, and have begun to open up their markets and emerge onto the global scene are referred to as emerging economies. Most of the emerging economies are characterized as transitional, meaning they are in the process of moving from a closed economy to an open market economy, such as Russia and other countries of the former Soviet Union. Another criterion for a country to be characterized as emerging is its rate of economic growth. Emerging economies are considered to be fast-growing economies, such as China, India, and other countries of Asia and Latin America.

Emerging countries have played a significant role in increasing global food availability in the past decade (Henneberry & Diaz, 2014). They have increased their agricultural production in order to meet their growing demand, but still, one in eight people in emerging markets suffer from hunger and malnutrition. As of The Swiss Re Group (2013), out of the 850 million people suffering from hunger worldwide, 98% are located in emerging markets. The Asia-Pacific region has the greatest number (528 million), followed by sub-Saharan Africa (237 million).

Food security is a holistic concept that addresses a wide range of dimensions, including availability, access, utilization, and stability of food (Food and Agriculture Organization of the United Nations [FAO], 2006). In terms of availability, in order for a country to achieve food security, the total food supply needs to equal its actual food demand, not only in quantitative but also in a qualitative way (variety, nutrient content, and safety). Food supply involves elements such as production, imports, exports, and also changes in national stocks. Similarly, demand involves the amount of food needed to feed, seed, manufacture, and the amount wasted to post-harvest loss (Henneberry & Diaz, 2014). However, steady population growth, insufficient farm investment, socio-economic instability, and adverse weather events remain key challenges to the world's food supply, often barring people's access to food. Moreover, dietary preferences and nutritional demand are changing and are also putting pressure on food production and availability. Recently, rising food prices have become a major concern due to a combination of complex factors (The Swiss Re Group, 2013).

For the emerging economies, where the economic reforms are still in the progress, market institutions are very fragile, and many people suffer from malnutrition, food security issues are vitally important. For such countries, ensuring food security has become one of the 21st century's greatest challenges. On the one hand, governments try to increase the competitiveness of the national economy by integrating it into the global market, on the other hand, they tend to secure food sovereignty of their countries by implementing a wide range of tools, including hidden measures of protectionism in the sphere of agricultural production and trade in food. In this chapter, the author will investigate the tendencies in trade in agricultural products and food in the countries, which comprise the two alliances of emerging economies, i.e. BRICS (Brazil, Russia, India, China, and South Africa) and EAEU (Eurasian Economic Union, including former countries of the Soviet Union, namely Armenia, Belarus, Kazakhstan, Kyrgyzstan, and Russia). The author will also analyze various policies those countries implement for the purpose of achieving food security levels, with a special focus on extremities – the ban on import of certain categories of agricultural products and food imposed by Russia against some developed countries.

BACKGROUND

The concept of food self-sufficiency means the extent to which a country can satisfy its food needs from its own domestic production (FAO, 1999). The practical application of this concept is that food self-sufficiency means that a country produces a proportion of its own food needs that approaches or exceeds 100% of its food consumption (Clapp, 2015). Ritson (1980) explored the relationship between food security and food self-sufficiency in the case of the UK. He understood food self-sufficiency as a proportion of imported food consumption, while in terms of security, he suggested that an efficient system for the international trading of food was a vital component of world food security and that absolute national self-sufficiency was not commensurate with this. Luan, Cui, and Ferrat (2013) define self-sufficiency as the ability of a region to sustain its own requirement for food. Therefore, self-sufficiency depends on two factors: production and consumption.

While food self-sufficiency is mainly focused on the provision of food security worldwide through the import of food products, their self-sufficiency, or optimum balance between domestic food production and import (Meskhia, 2016), food security is a wider concept. Food security is usually categorized as a non-trade concern within trade policy as it incorporates factors other than those directly relevant to the operation of an international market system for the production and exchange of agricultural goods. FAO (1992) defines food security as physical, social and economic access by all people at all times to sufficient, safe and nutritious food which meets their dietary needs and food preferences for an active and healthy life.

Food security exists when people have access to sufficient amounts of safe and nutritious food and are therefore consuming the food required for normal growth and development, and for an active and healthy life. Food insecurity happens due to the unavailability of food or insufficient purchasing power (FAO, 1992). Thus, FAO's approach to food security is based on two pillars which are the physical availability of food and agricultural products on the domestic market (domestic production plus import) and economic access to adequate supply by all people (purchasing power, food inflation, distribution, etc.).

Physical food availability is the level of saturation of the domestic market with both domestically produced and imported foodstuffs when the population in every locality of the country can purchase them in quantities and assortments no less than the established food consumption rates. Such consumption rates are various foodstuffs in quantities and combinations that meet the modern scientific principles of the optimum nutrition taking into consideration the existing nutritional structure and traditions of the country.

The second pillar, economic access to adequate supply, is a possibility to purchase foodstuffs at prevailing prices in quantities and assortments no less than the established consumption rates supported by the appropriate income level.

According to this two-pillar line of argumentation, food self-sufficiency is just a component part of food security, specifically, physical availability, since it is directly related to a balance between domestic production and imports, not to economic access to supply.

Anderson (1990), who defines food security as an access by all members of a household at all times to enough food for an active, healthy life, has the alike understanding of food security pillars. In his opinion, they are the availability of nutritionally adequate and safe foods and assured the ability to acquire acceptable foods in socially acceptable ways. The implication of such approach is that food insecurity means uncertain (market environment) or limited (administrative regulations) availability of food on the domestic market, or ability to acquire acceptable foods by means of domestic production or import.

The liberalization of international trade has become a significant source of tension in contemporary agricultural change with the incorporation of agriculture into the world trading system (Lee, 2007). In the conditions of globalization, where liberalization of food trade and the reduction of administrative protection of food producers are mandated by the rules of the WTO, many countries have lost a part of their sovereignty over food policies (Lawrence & McMichael, 2012). Some of them (primarily, developing ones) have become food dependent, others managed to benefit from easier access to foreign markets and unified framework of global trade in food. In general, globalization has refocused attention from trade-based food self-sufficiency to availability-based food security.

WTO officials argue that in order to achieve food security nations should encourage international trade in food (Lee, 2007). Trade liberalization can affect the availability of certain foods by removal of barriers to imports, but also to foreign investments in the development of domestic production. Thow and Hawkes (2009) found that availability of processed food has risen in developing countries after foreign direct investment by multinational food companies. Thus, changes in trade policies have facilitated the rising availability and consumption of foods. Dorosh (2004) found the positive contribution of trade liberalization to food security and stabilization of supply and food market in the case of Bangladesh. He concluded that food self-sufficiency was not a prerequisite for food security in the country.

Protection policies, including administrative restrictions of food imports with a view to increasing food self-sufficiency of the country, on the contrary, may cause unnecessary social costs and place food self-sufficiency into conflict with the goals of food security and poverty reduction. Warr (2011) studied the effects of an import ban on rice introduced in Indonesia in an effort to increase food self-sufficiency of the country on this staple food. He concluded that such a policy let to achieve the required level of food self-sufficiency, but only at the expense of reducing food security of the most vulnerable people – the poorest net consumers.

Food security can be severely decreased not by administrative regulations only, but by high food costs, poor availability of healthy food, and low purchasing power of population (Socha, Zahaf, Chambers,

Abraham, & Fiddler, 2012). Samaratunga (2011) studied food security issues in the case of Sri Lanka and concluded that the problem of food security in developing countries rested on many parameters apart from domestic production and import, such as growth and distribution of income, population growth, and food consumption patterns. Power (2005), Sonnino, Faus, and Maggio (2014), Kalugina (2014), Erokhin, Ivolga, and Heijman (2014) support that food insecurity is directly related to low income, not to dependence on imports of foodstuffs.

MAIN FOCUS OF THE CHAPTER

Food Security on Emerging Markets: Major Trends

Indeed, the level of economic development is an essential ingredient of food security. In developed countries, where peoples' incomes are high, food security is also sufficient. In those countries, which grow faster than their neighbors do, economic development raises the incomes of the poor and improves their ability to gain access to food.

From 2005 to 2015, the share of emerging markets in the World GDP grew from 9% to 25%. China itself tripled its own contribution from 5% in 2005 to around 15% in 2015. Economic growth in these emerging countries, in general, has been strongly related to agricultural development (Henneberry & Diaz, 2014). However, the level of economic development and speed of economic growth do not mean food security is achieved automatically. Hunger in the world still exists, especially in those countries which tend to demonstrate higher rates of economic growth than the developed nations, i.e. developing economies. On the fast-growing emerging markets, food insecurity remains rooted in disparities in income, access to food, its availability to all citizens of a country, and its safety and quality.

Henneberry and Diaz (2014) state that food security may be determined by the ability of each country to include the poorest population in the nation's development. This can be achieved by creating the proper infrastructure and providing the poor with new or better income opportunities. Resting on such broad categories as "proper infrastructure" and "better income opportunities", it is rather challenging to measure the progress of a single country in achieving food security. Throughout the years, there have been several approaches to measuring food security within countries with different levels of economic development. One of the most widely accepted tools is the Global Food Security Index (GFSI), which examines food security comprehensively across the three internationally established dimensions of affordability, availability, and quality. GFSI is a scoring model, constructed from 28 indicators (Table 1), that measures the drivers of food security across countries irrespective of the level of their economic development, i.e. developed, developing, emerging, economies in transition, etc.

The overall goal of the study is to assess which countries are most and least vulnerable to food insecurity (The Economist Intelligence Unit, 2016). The overall score for the GFSI is calculated from a simple weighted average of the evaluated indicators and scaled from 0-100, where 100 equals the most favorable (Henneberry & Diaz, 2014). Developed countries of North America and Europe are on the top, led by the United States while developing and emerging economies have lower rankings (Table 2).

Not everyone agrees entirely on which countries are emerging markets. The International Monetary Fund (IMF), Morgan Stanley Capital International (MSCI), Standard and Poor's (S&P), Russell, and Dow Jones each have their lists of emerging markets. There are 16 countries that each organization classifies as emerging markets as of 2016: Brazil, Chile, China, Colombia, Hungary, Indonesia, India, Malaysia,

Table 1. Indicators within the GFSI

Category	Indicator	Content
Affordability	Food consumption as a share of household expenditure	Measures the average percentage of household expenditure that is spent on food
	Proportion of population under global poverty line	Measures the percentage of the population living under $3.10/day in purchasing power parity
	Gross domestic product per capita	Measures individual income and, hence, the ability to afford food in $US purchasing power parity terms
	Agricultural import tariffs	Measures the average applied most favored nation (MFN) tariffs on agricultural imports
	Access to financing for farmers	A qualitative indicator that measures the availability of financing to farmers from the government, multilateral, and private sectors
	Presence of food safety net programs	A qualitative indicator that measures public initiatives to protect the poor from food-related shocks
Availability	Sufficiency of supply	A composite indicator that measures the food availability through the supply of kcal/capita/day and levels of food aid
	Public expenditure on agricultural R&D	Measured as a percentage of agricultural GDP. It is a proxy for agricultural innovation and technology that increases market efficiency and access
	Agricultural infrastructure	Composite indicator that measures the ability to store and transport crops to market
	Volatility of agricultural production	Measured as the standard deviation of the annual growth in agricultural production
	Political instability	Qualitative indicator that measures the presence of general political instability
	Corruption	A measure of the pervasiveness of corruption in a country, by measuring the risk of corruption
	Urban absorption capacity	A measure of a country's capacity to absorb the stresses placed by urbanization and still ensure food security
	Food loss	A measure of post-harvest and pre-consumer food loss as a ratio of the total domestic supply of crops, livestock and fish commodities
Quality and safety	Diet diversification	Measures the share of non-starchy foods in total dietary energy consumption
	Nutritional standards	A composite indicator that measures nutrition governance
	Micronutrient availability	A composite indicator that measures the availability of iron and vitamin A in the food supply
	Protein quality	This indicator measures the grams of quality protein consumed using the methodology of the Protein Digestibility Corrected Amino Acid Score
	Food safety	A composite indicator that measures the enabling environment for food safety

Source: The Economist Intelligence Unit, 2016

Mexico, Peru, Philippines, Poland, Russia, South Africa, Thailand, and Turkey. GFSI rankings for those countries along with the leading developed economies are presented in Table 2.

Over the past five years, the GFSI has shown improvements in food security. Overall global economic growth has led to improvements in the structural areas that are essential to improving people's access to a wide range of affordable, nutritious foods, This is particularly evident in emerging-market countries, which have reached the economic and development threshold necessary to enable them to focus on improving government programmes to enhance food security (The Economist Intelligence Unit, 2016).

Table 2. GFSI rankings of top ten developed countries and selected emerging economies in 2012-2016

Ranking	Country	2012	2013	2014	2015	2016	Average in 2012-2016	5-year change
1	United States	86.1	85.6	86.2	85.8	86.6	86.1	+0.5
2	Ireland	82.1	81.4	81.8	82.3	84.3	82.4	+2.2
3	Singapore	82.5	80.8	83.4	83.0	83.9	82.7	+1.4
4	Australia	81.7	81.0	80.5	81.8	82.6	81.5	+0.9
5	Netherlands	82.9	82.4	82.8	82.2	82.6	82.6	-0.3
6	France	83.5	83.0	81.5	81.6	82.5	82.4	-1.0
7	Germany	81.2	80.9	81.5	81.6	82.5	81.5	+1.3
8	Canada	82.3	81.0	81.4	81.5	81.9	81.6	-0.4
9	United Kingdom	78.8	79.0	79.4	79.3	81.9	79.7	+3.1
10	Sweden	80.7	80.1	80.3	80.0	81.3	80.5	+0.6
…								
24	Chile	70.6	70.8	72.4	72.9	74.4	72.2	+3.8
…								
29	Poland	72.6	71.6	71.7	72.1	72.4	72.1	-0.2
…								
34	Hungary	71.9	71.6	69.6	68.5	69.3	70.2	-2.6
35	Malaysia	66.8	66.6	68.6	68.9	69.0	68.0	+2.2
…								
39	Mexico	66.8	65.9	65.7	67.2	68.1	66.7	+1.3
…								
41	Brazil	67.8	67.7	67.6	66.8	67.6	67.5	-0.2
42	China	62.3	62.3	63.5	64.3	65.5	63.6	+3.2
…								
45	Turkey	62.3	63.0	62.5	64.1	63.6	63.1	+1.3
…								
47	South Africa	60.8	61.0	62.3	62.3	62.9	61.9	+2.1
48	Russia	66.7	65.6	62.8	61.5	62.3	63.8	-4.4
49	Colombia	57.7	57.6	57.8	58.8	61.0	58.6	+3.3
…								
51	Thailand	58.7	59.4	59.5	59.0	59.5	59.2	+0.8
…								
55	Peru	54.7	55.0	55.6	56.9	57.7	56.0	+3.0
…								
71	Indonesia	47.7	48.2	48.0	47.9	50.6	48.5	+2.9
…								
74	Philippines	49.8	48.6	48.8	49.1	49.5	49.2	-0.3
75	India	48.7	48.2	47.9	48.9	49.4	48.6	+0.7

Source: The Economist Intelligence Unit, 2016

Most of the emerging economies succeeded to increase their food security level during 2012-2016, with three countries of Latin America (Chile, Colombia, and Peru) and China gaining the biggest increase. These positive developments have largely been driven by rising incomes in most emerging countries and general improvements in the global economy. Falling food prices have also positively impacted food security (The Economist Intelligence Unit, 2016). Russia was the worst performer with a decline of 4.4 points, and the reasons of that are discussed below.

Production and Trade in Agricultural Products in BRICS

The alliance of five major emerging national economies, Brazil, Russia, India, China, and South Africa, is BRICS. The BRICS members are countries, distinguished by their large, fast-growing economies and growing influence on regional and global affairs. As of 2015, five BRICS countries represent almost 3 billion people, which are 40% of the world population, with a combined nominal GDP of $17 trillion, which is 20% of the world GDP.

Three of these countries, except Brazil, India, and South Africa, entered the WTO in 1995 when the organization was established. Two decades of liberalization of trade in agricultural products and food influenced agricultural production increased imports of food and benefited food security in those countries. Particularly, Brazil has increased its agricultural production in almost three times, while export increased in 5.3 times, and import increased by 53% (Table 3).

India also has been able to ensure a positive trade balance in food and agricultural products during 1995-2015. The share of import in GDP is growing but still remains very low (Table 4).

South Africa is the smallest economy among the five members of BRICS and the most dependent on import of food. Import trade quota is 0.80 in 2014, which means that the country imports almost as much as it produces domestically. However, South Africa keeps its trade balance positive (Table 5).

China entered the WTO in 2002. By 2015, the country has become the leading producer and consumer of food and agricultural products among BRICS countries, with almost $1 trillion worth of agricultural production, and $102 billion of agricultural import. Trade balance in agricultural products has been remaining negative since 2008 (Table 6).

Table 3. Agricultural production and foreign trade in agricultural products in Brazil in 1995-2015, $ million

Indicators	1995*	2000	2005	2010	2011	2012	2013	2014	2015
GDP by kind of economic activity: agriculture, hunting, forestry, and fishing	37,791	30,519	41,435	91,132	113,921	107,736	114,542	111,497	n.a.
Export of agricultural products	13,258	12,808	30,432	60,778	77,389	77,235	82,108	78,845	70,981
Import of agricultural products	5,756	3,665	3,210	8,373	10,244	10,626	11,713	11,224	8,817
Trade balance	7,502	9,143	27,222	52,405	67,145	66,609	70,395	67,621	62,164
Import Trade Quota (Import/ GDP)	0.15	0.12	0.08	0.09	0.09	0.10	0.10	0.10	n.a.

Source: Author's development based on the United Nations Conference on Trade and Development [UNCTAD] (2016)
Note: * - the year of accession to the WTO

Table 4. Agricultural production and foreign trade in agricultural products in India in 1995-2015, $ million

Indicators	1995*	2000	2005	2010	2011	2012	2013	2014	2015
GDP by kind of economic activity: agriculture, hunting, forestry, and fishing	91,022	100,517	145,375	289,880	322,598	312,268	321,028	321,895	n.a.
Export of agricultural products	5,912	5,418	9,001	18,199	27,009	30,534	37,429	36,003	29,954
Import of agricultural products	1,675	3,216	6,332	13,612	16,556	18,399	17,418	19,980	20,777
Trade balance	4,237	2,202	2,669	4,587	10,453	12,135	20,011	16,023	9,177
Import Trade Quota (Import/GDP)	0.02	0.03	0.04	0.05	0.05	0.06	0.05	0.06	n.a.

Source: Author's development based on the UNCTAD (2016)
Note: * - the year of accession to the WTO

Table 5. Agricultural production and foreign trade in agricultural products in South Africa in 1995-2015, $ million

Indicators	1995*	2000	2005	2010	2011	2012	2013	2014	2015
GDP by kind of economic activity: agriculture, hunting, forestry, and fishing	5,499	4,084	6,160	8,961	9,446	8,556	7,608	7,801	n.a.
Export of agricultural products	2,779	2,232	3,990	8,230	9,012	8,742	9,294	9,455	7,977
Import of agricultural products	1,409	1,254	2,451	5,483	7,009	7,448	6,777	6,272	4,896
Trade balance	1,370	978	1,539	2,747	2,003	1,294	2,517	3,183	3,081
Import Trade Quota (Import/GDP)	0.26	0.31	0.40	0.61	0.74	0.87	0.89	0.80	n.a.

Source: Author's development based on the UNCTAD (2016)
Note: * - the year of accession to the WTO

Table 6. Agricultural production and foreign trade in agricultural products in China in 1995-2015, $ million

Indicators	1995	2000	2002*	2005	2010	2011	2012	2013	2014	2015
GDP by kind of economic activity: agriculture, hunting, forestry, and fishing	145,301	180,496	199,759	273,518	598,526	734,708	829,468	919,435	979,224	n.a.
Export of agricultural products	12,299	13,559	16,164	24,635	44,152	54,168	56,318	59,983	63,495	63,236
Import of agricultural products	9,236	9,043	9,891	21,541	59,556	75,455	90,650	98,646	105,263	102,787
Trade balance	3,063	6,273	6,273	3,094	-15,404	-21,287	-34,332	-38,663	-41,768	-39,551
Import Trade Quota (Import/ GDP)	0.06	0.05	0.05	0.08	0.10	0.10	0.11	0.11	0.11	n.a.

Source: Author's development based on the UNCTAD (2016)
Note: * - the year of accession to the WTO

China is an ultimate leader among BRICS countries in agricultural production. The country has been increasing the volume of domestic agricultural production starting from 2003, right after its accession to the WTO (Figure 1).

Due to a huge domestic market, China is not able to export much of its agricultural products. Brazil is the leader in the volume of agricultural export. India is also increasing its export (Figure 2).

China is the most distinct importer among the four countries, with over $100 billion of import in 2015. Other countries are really low, with only India reaching a level of $20 billion (Figure 3).

South Africa, being the smallest economy and food producer among the BRICS countries, is very dependent on import of food, and the dependency is rising. China, India, and Brazil are quite stable in terms of Import/GDP ratio (Figure 4).

The fifth member of BRICS, Russia, is a special case in terms of its recent trade policy and influences on food security. According to the GFSI, Russia was the only country among emerging economies which lost several ranking points in terms of all three dimensions of food security, i.e. affordability, availability, and quality and safety, during 2012-2016. At the same time, Russia is the only emerging country which abandoned trade liberalization pathway and banned imports of a variety of food products from the developed countries, including the EU, the USA, Canada, Australia, and a number of countries. In an effort to make up losses in terms of its influence on global affairs, Russia stresses on regional integration in the form of the emerging alliance of countries which used to be the republics of the Soviet Union and gained independence in 1991. The alliance is the Eurasian Economic Union which originated from the Commonwealth of Independent States in 2000 and was transformed into the EAEU in 2014. Russia is a founding member and a major driving force of this alliance established for the effective promotion of free movement of goods, services, capital, and labor between participating countries (Armenia, Belarus, Kazakhstan, Kyrgyzstan, and Russia). That is why, before the investigation of the Russia's food security case, contemporary trends in trade in food in the macro-region of Eurasia has to be addressed.

Figure 1. Agricultural production of BRICS countries in 1995-2014, $ million
Source: Author's development based on the UNCTAD (2016)

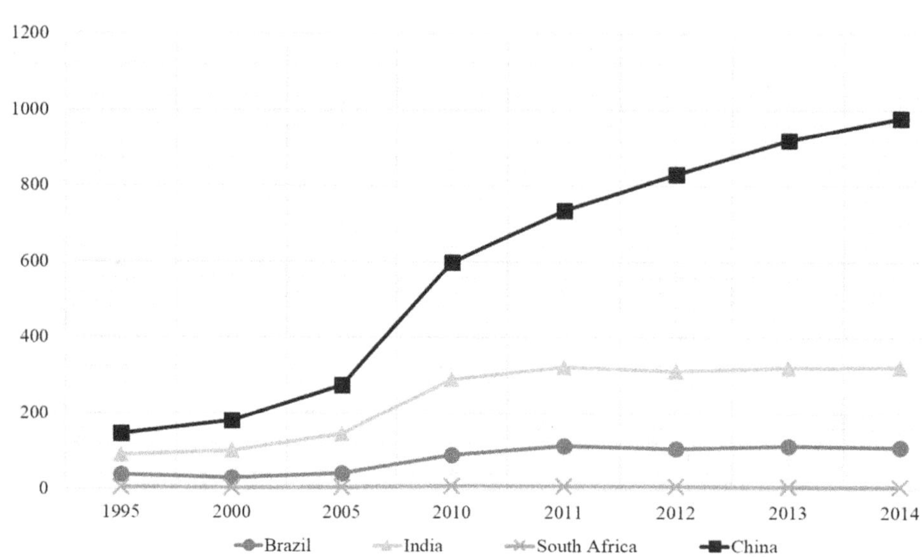

Figure 2. Export of agricultural products and food in BRICS countries in 1995-2015, $ million
Source: Author's development based on the UNCTAD (2016)

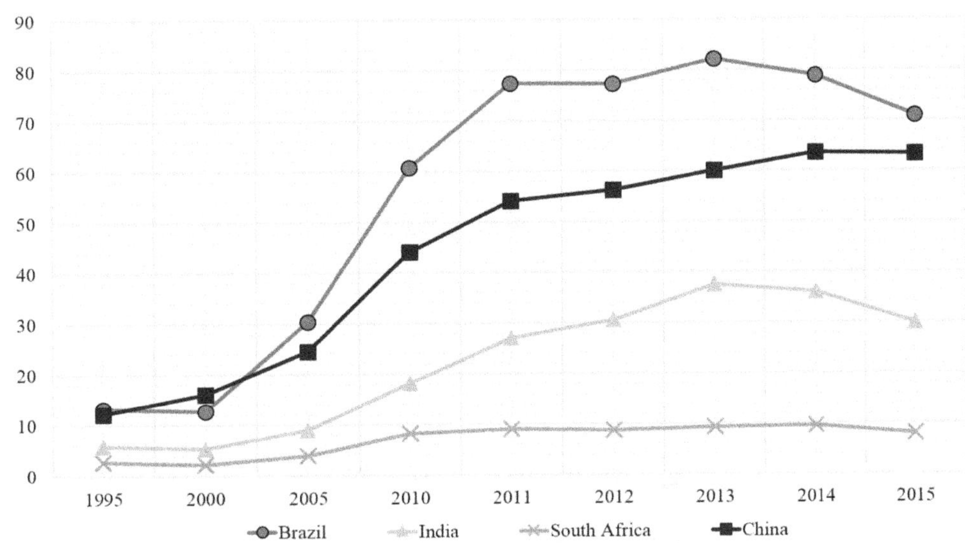

Figure 3. Import of agricultural products and food in BRICS countries in 1995-2015, $ million
Source: Author's development based on the UNCTAD (2016)

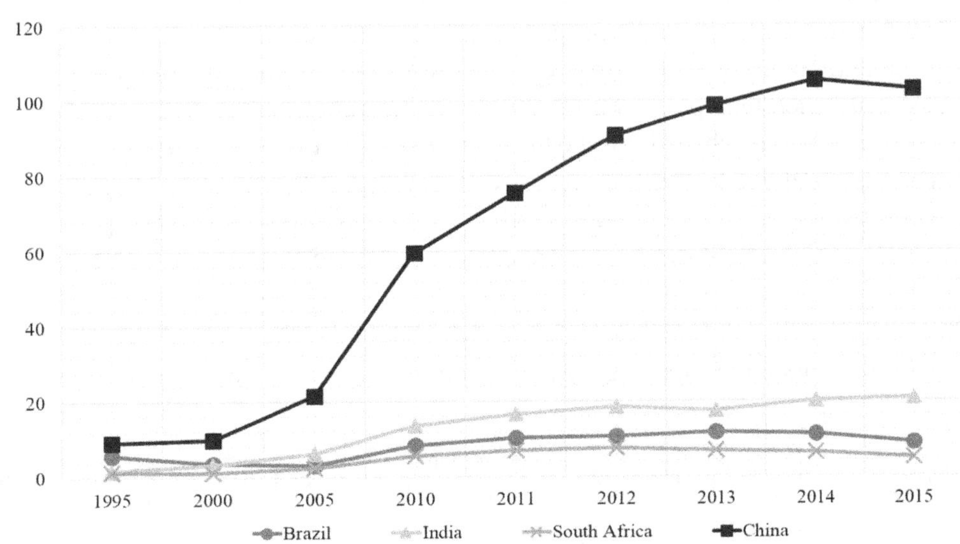

Production and Trade in Agricultural Products in the EAEU

While the EAEU may appear small in terms of its members, it still represents about 183 million people spread amongst the vast territory from the Baltic Sea in the West to the Pacific Ocean in the East. Participating countries are Armenia, Belarus, Kazakhstan, Kyrgyzstan, and Russia. Agricultural production in Armenia is growing, but the country is not self-sufficient. Trade balance in food and agricultural

Figure 4. Import Quota of BRICS countries in 1995-2014
Source: Author's development based on the UNCTAD (2016)

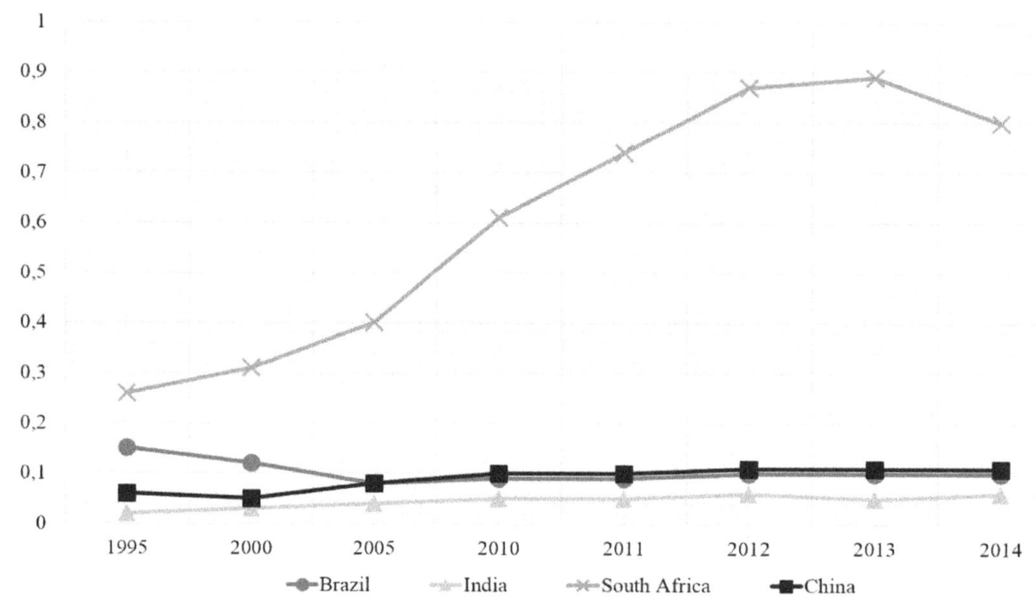

products is negative. Import Quota is 0.37 in 2014 which means that import comprises 37% of the total agricultural production of the country (Table 7).

Belarus is the country which focuses on agricultural production the most among other countries of the EAEU. The volume of agricultural production has been growing rapidly starting from 1995, and especially during the last 2-3 years, and reached almost $6 billion in 2014. Starting from 2009, the Belarussian trade balance in food and agricultural products became positive and remains like that until now. The country substantially increased its export in 2013-2014, when Russia imposed a ban on import of agricultural products from certain countries of the world, and deliveries started to go through Belarus (Table 8).

Table 7. Agricultural production and foreign trade in agricultural commodities in Armenia in 1995-2015, $ million

Indicators	1995	2000	2003*	2005	2010	2011	2012	2013	2014	2015
GDP by kind of economic activity: agriculture, hunting, forestry, and fishing	524	444	605	936	1,574	2,061	1,901	2,050	2,149	n.a.
Export of agricultural products	34	32	84	111	164	235	346	404	408	383
Import of agricultural products	210	210	213	299	662	769	798	830	789	644
Trade balance	-176	-178	-129	-188	-498	-534	-452	-426	-381	-261
Import Trade Quota (Import/ GDP)	0.40	0.47	0.35	0.32	0.42	0.37	0.42	0.40	0.37	n.a.

Source: Author's development based on the UNCTAD (2016)
Note: * - the year of accession to the WTO

Table 8. Agricultural production and foreign trade in agricultural products in Belarus in 1995-2015, $ million

Indicators	1995	2000	2005	2010	2011	2012	2013	2014	2015
GDP by kind of economic activity: agriculture, hunting, forestry, and fishing	2,185	1,277	2,573	5,096	4,950	5,446	5,188	5,935	n.a.
Export of agricultural products	89	495	1,319	3,190	3,792	4,519	5,349	5,333	4,169
Import of agricultural products	665	1,027	1,562	2,748	3,099	3,509	3,972	4,656	4,278
Trade balance	-576	-532	-243	442	693	1,010	1,377	677	-109
Import Trade Quota (Import/GDP)	0.30	0.80	0.61	0.54	0.63	0.64	0.77	0.78	n.a.

Source: Author's development based on the UNCTAD (2016)

Kazakhstan, on the contrary, turned to become a consumer-oriented country – import deliveries of food are growing, the trade balance is negative, and Import Quota is rather high – 0.43 in 2014, which means that 43% of the food on the domestic market is of foreign origin (Table 9).

Kyrgyzstan is more or less in the same position as Kazakhstan – growing deficit in trade in agricultural products and food, growing import quota. The country accessed the WTO in 1999, and the author sees the immediate response of the domestic market in 2000 and onwards – negative trade balance (Table 10).

Belarus and Kazakhstan are two leaders among the EAEU countries on agricultural production. Belarus increases its agricultural production quite rapidly, while Kazakhstan slows down. Agricultural production in Kazakhstan is rather seasonal and very much dependent on climatic conditions and yield of wheat, while Belarus is rather more diversified (Figure 5).

Being the leading food producer, Belarus is also a leading food exporter among the four countries under consideration. Its export is growing steadily, except the decline in 2015, while Kazakhstan's export is fluctuating in parallel with the production of wheat (Figure 6).

The two countries are also leaders in import. Belarus increases its import from the EU, process raw agricultural commodities, or just repacks food products, labels them and then export them to Russia – this is the way how food products of the EU origin come to Russia in circumvention of the food ban, being formally produced in Belarus (Figure 7).

Table 9. Agricultural production and foreign trade in agricultural products in Kazakhstan in 1995-2015, $ million

Indicators	1995	2000	2005	2010	2011	2012	2013	2014	2015*
GDP by kind of economic activity: agriculture, hunting, forestry, and fishing	2,535	1,484	3,639	6,678	9,610	8,920	10,657	9,143	n.a.
Export of agricultural products	518	600	669	1,934	1,826	3,087	2,675	2,605	1,346
Import of agricultural products	374	431	1,135	2,203	3,172	3,823	4,125	3,961	2,709
Trade balance	144	169	-466	-269	-1,346	-736	-1,450	-1,356	-1,363
Import Trade Quota (Import/GDP)	0.15	0.29	0.31	0.33	0.33	0.43	0.39	0.43	n.a.

Source: Author's development based on the UNCTAD (2016)
Note: * - the year of accession to the WTO

Table 10. Agricultural production and foreign trade in agricultural products in Kyrgyzstan in 1995-2015, $ million

Indicators	1995	1999*	2000	2005	2010	2011	2012	2013	2014	2015
GDP by kind of economic activity: agriculture, hunting, forestry, and fishing	607	436	468	701	837	1,207	1,100	1,074	1,098	n.a.
Export of agricultural products	89	92	71	84	340	306	239	213	239	156
Import of agricultural products	96	82	81	166	544	708	780	842	869	567
Trade balance	-7	10	-10	-82	-204	-402	-541	-629	-630	-411
Import Trade Quota (Import/GDP)	0.16	0.19	0.17	0.24	0.65	0.59	0.71	0.78	0.79	n.a.

Source: Author's development based on the UNCTAD (2016)
Note: * - the year of accession to the WTO

Figure 5. Agricultural production in the EAEU in 1995-2014, $ million
Source: Author's development based on the UNCTAD (2016)

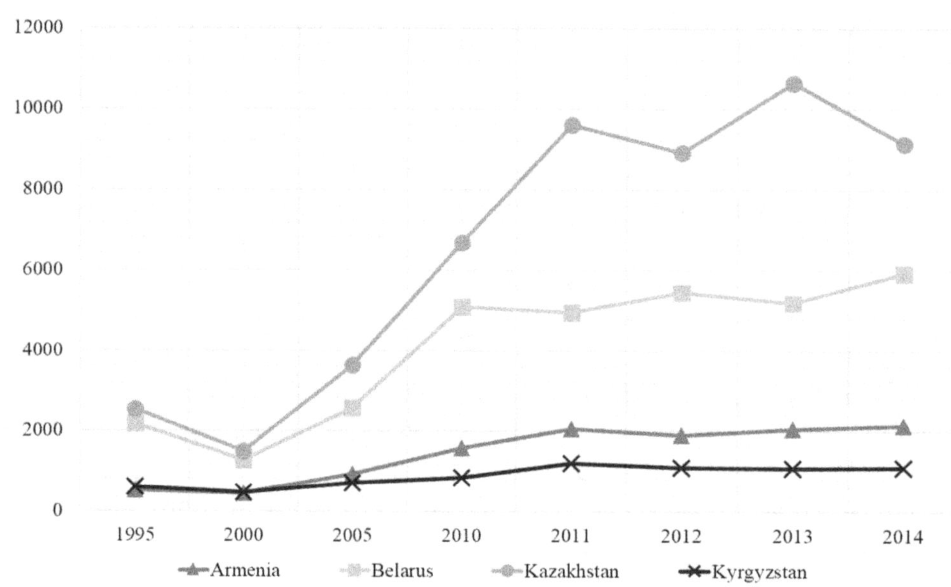

Belarus is the best performer among other EAEU countries in relation to food independence. Being heavily dependent on import deliveries of food products and agricultural commodities in the mid-1990s, the country was able to decrease its dependency (Figure 8).

Protectionism and Food Security Issues: Case of Russia

As we have seen on the example of some EAEU and BRICS countries, domestic production of staple agricultural products and food in many countries, including such big agricultural producers as Russia, fails to meet demand. Such countries have to rely on agricultural imports, leaving them vulnerable to global price fluctuations and affecting their export revenues, which tremendously threat food security of those nations.

Figure 6. Export of agricultural products and food in the EAEU in 1995-2015, $ million
Source: Author's development based on the UNCTAD (2016)

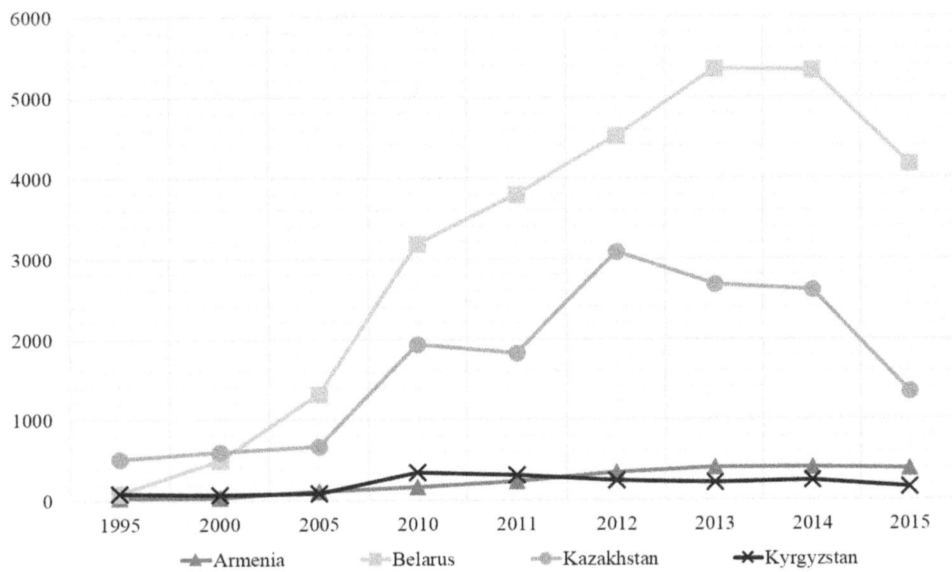

Figure 7. Import of agricultural products and food in the EAEU in 1995-2015, $ million
Source: Author's development based on the UNCTAD (2016)

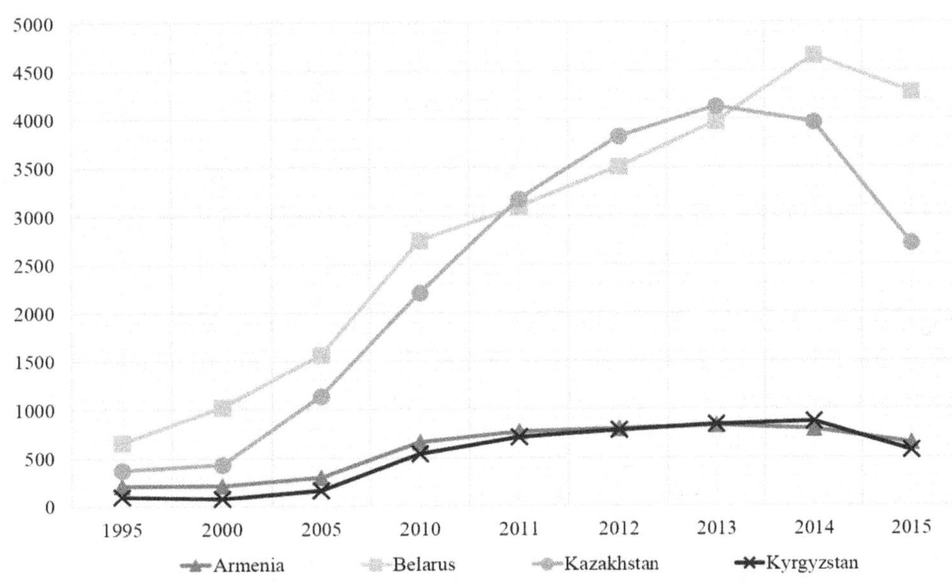

Regardless of the world's biggest territory and high volumes of domestic agricultural production, Russia remains a net importer of agricultural products. Currently, over 41% of Russia's agricultural GDP is produced by imports. Russia received up to 55% of its agricultural imports from the countries it has so far sanctioned, including the EU.

Figure 8. Import Quota of the EAEU countries in 1995-2014
Source: Author's development based on the UNCTAD (2016)

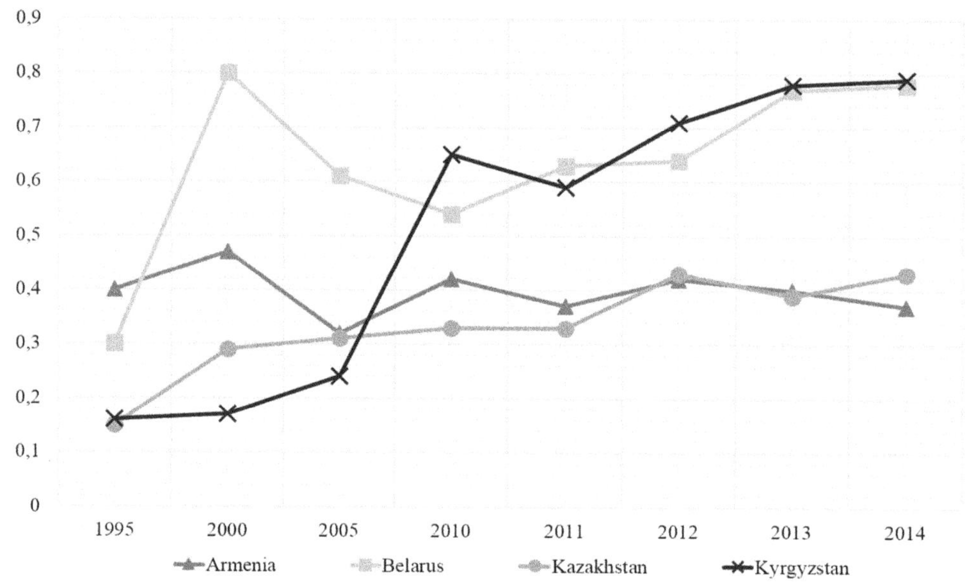

Until recently, Russia was not among the WTO members. Russia's accession to this global trading system in 2012 and membership of its major EAEU and BRICS trade partners in the WTO are considered as a precondition of essential structural changes in interregional and even international trade.

However, trade sanctions imposed on both sides between Russia and the Western countries, demonstrated, that trade liberalization in general and the WTO, in particular, had not much to do with those changes. The real threats for Russian agricultural market and trade are unilateral actions, which completely distort trade patterns and destroy trade links between countries and regions. Trade restrictions threat sustainable development not only because of their direct effects (market volatilities, destruction of trade linkages, etc.) but largely because of the hidden ones (distortion of natural economic order, the establishment of artificial market environment and overprotection).

Considering the retrospectives of the Russia's food market, the author may emphasize the rapid decrease of agricultural production (and its share in the GDP) in the beginning of 1990s, after the fall of the Soviet Union and during the early period of liberal economic reforms (Table 11). That sort of the

Table 11. Major indicators of Russia's agriculture in 1990-2015

Indicators	1990	1995	2005	2010	2011	2012	2013	2014	2015
Share of agricultural production in GDP, %	17.0	6.3	4.5	3.3	3.4	3.4	3.4	3.3	3.5
Agricultural land, million hectares	213.8	209.6	193.0	191.9	191.7	191.9	191.8	191.8	191.8
Rural population, million people	39.5	39.4	38.8	37.5	37.3	37.2	37.1	37.0	38.02
Share of agriculture in employment of population, %	13.2	15.7	10.2	7.9	7.9	7.3	7.0	7.0	7.0

Source: Author's development based on Trading Economics (2016)

liberalization effect is common to most of the economies in transition, when the opening of the domestic market causes an immediate increase of import deliveries, while local producers do not have facilities to respond swiftly to such a change of external competitive environment.

Russia currently provides 1.17% of the world food export and 1.77% of the world food import (UNCTAD, 2016). Over the recent decades, Russia has been a net importer of agricultural commodities and food. The share of the country in the total world food import has been growing since the very fall of the Soviet Union. Food Foreign-Trade-to-GDP ratio (Foreign Trade Quota) has been growing over two decades onwards (Figure 9). The outrun growth of the Food Import-to-GDP ratio (Import Quota) in comparison with the Food Export-to-GDP ratio (Export Quota) should be noted.

Even in 2014, when Russia's domestic food market experienced two tendencies, unfavorable for import (depreciation of food import because of the devaluation of the national currency and import ban, introduced in August 2014), Import Quota spiked and reached 47.1%, while export predictably decreased, as Russian producers reoriented on the domestic market.

The negative trade balance has been the permanent trend of Russia's trade in agricultural products over the past two decades (Figure 10). Falling import and introduced import ban can hardly break the tendency. Simple linear trend model shows the expected $25 billion of negative trade balance by 2020.

Russia's major food import commodities are fruit, meat and meat products, vegetables, and cheese. There is a decrease of import deliveries in 2015 in comparison with 2013 and 2014 as a result of both introduced import ban and depreciation of import because of the devaluation of the Russian ruble (Table 12).

Figure 9. Food export, import, and foreign trade ratios to Russia's GDP (agriculture) in 1995-2014 and forecast for the period until 2020, %
Source: Author's development based on UNCTAD (2016)

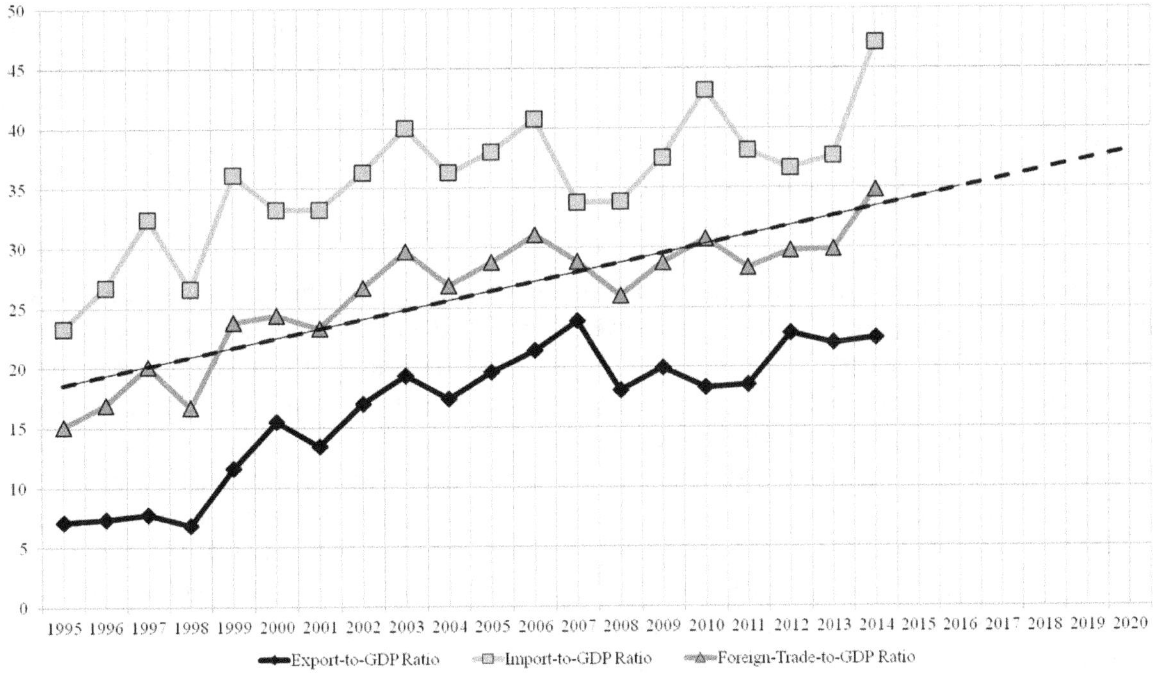

Figure 10. Food export, import, and trade balance of Russia in 1995-2015 and forecast for the period until 2020, %
Source: Author's development based on the UNCTAD (2016)

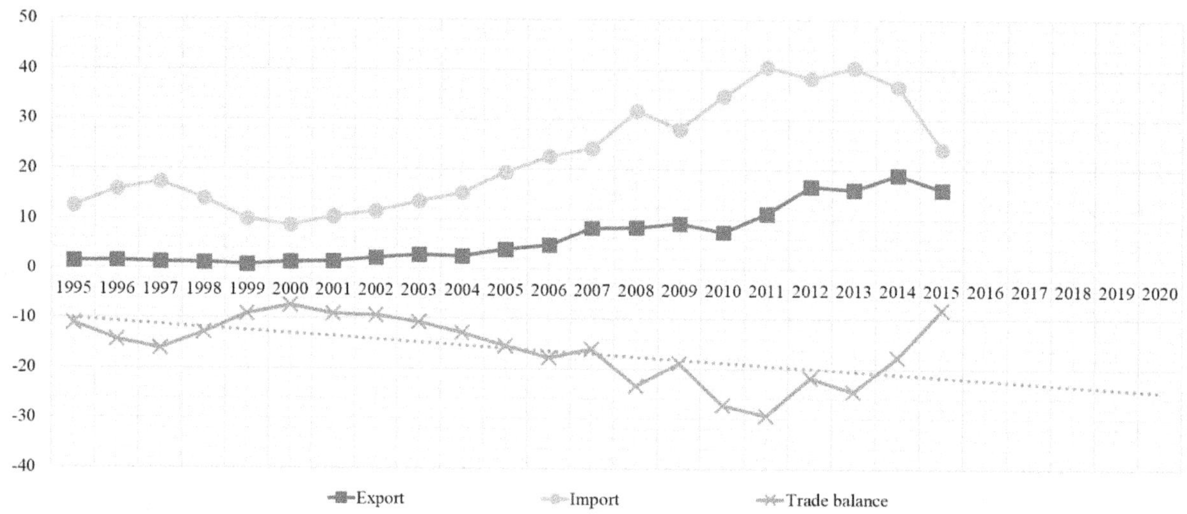

Table 12. Import of selected agricultural and food products to Russia in 1990-2015, $ billion

Product groups	1990	1995	2000	2005	2010	2011	2012	2013	2014	2015
Meat and meat products	0.51	0.62	0.48	1.44	2.82	3.16	3.23	2.94	2.70	1.54
Cheese	0.12	0.21	0.20	0.91	1.93	2.07	2.00	2.33	1.65	0.77
Milk and dairy products	0.19	0.24	0.20	0.40	1.06	0.85	1.00	1.51	1.30	0.81
Fish (fresh, chilled and frozen)	0.09	0.17	0.18	0.98	1.75	2.02	1.98	2.33	1.83	0.92
Vegetables	0.30	0.34	0.28	0.75	2.20	2.88	2.23	2.59	2.60	1.72
Fruit	0.57	0.93	0.72	2.20	5.06	5.71	5.48	5.71	4.71	3.58
Eggs	0.08	0.13	0.41	0.58	0.14	0.18	0.20	0.26	0.33	0.28
Sugar and honey	1.04	1.23	0.96	1.23	1.98	2.44	0.96	0.91	0.84	0.61

Source: Author's development based on the UNCTAD (2016)

Analyzing the structure and volumes of food import to Russia, it is worthwhile to note the high share of the EU – about one-fifth of total import. However, if in 2010 the EU supplied almost 30% of Russia's food import, in 2015 its share decreased down to 21.7% (Table 13). Banned foodstuffs from Europe have been replaced on the Russian food market by deliveries from Asia, Latin America, Africa, and some of the CIS countries, particularly, Belarus and Kazakhstan.

Top ten leading food suppliers of Russia are Belarus, Brazil, Germany and other countries of the EU, South America, and Asia. Geographic structure of Russia's food import is rather more diversified in comparison with export since no country has any serious advantage in its market share.

Today with escalating political conflict between Russia and Ukraine, economic sanctions imposed by the EU and the USA, and Russia's reorientation on Asia and Latin America, it is not clear how further progress can be achieved between Russia and its trade partners.

Table 13. Russia's food import by regions in 1995-2015

Regions	1995		2000		2010		2015	
	$ billion	**share in total, %**	**$ billion**	**share in total, %**	**$ billion**	**share in total, %**	**$ billion**	**share in total, %**
World, total	14.90	100.00	6.98	100.00	33.62	100.00	24.10	100.00
CIS	3.71	24.89	1.67	23.93	3.41	10.14	4.61	19.13
EU	6.76	45.37	1.84	26.36	10.04	29.86	5.23	21.70
North America	1.37	9.19	0.77	11.03	1.77	5.26	0.57	2.37
Latin America	1.00	6.71	0.68	9.74	6.80	20.23	5.27	21.87
Asia	1.67	11.21	0.86	12.32	5.87	17.46	6.65	27.59
Africa	0.22	1.48	0.21	3.01	1.38	4.10	1.38	5.73
Other	0.17	1.15	0.96	13.61	4.35	12.94	0.39	1.62

Source: Author's development based on the UNCTAD (2016)

Russia banned beef, pork, poultry, fish, fruit, vegetables, cheese, milk and other dairy products from the USA, Canada, the EU, Norway, and Australia. Russia imported about €33 billion worth of food products in 2014, of which the banned goods accounted for nearly €7 billion, or 21.2%. Top ten suppliers to Russia include the countries of Latin America, Europe, and Asia. Five out of the top ten are banned, which are the Netherlands, Poland, Norway, France, and the USA (Table 14). The ban hit about 7% of the EU agricultural exports to Russia. For certain EU countries, i.e. Poland, France, and Norway, losses were essential. Russia used to be a large market for Polish fruits, French cheese, and Norwegian fish.

The desire for the ban was that, by banning those products that Russia had become dependent on, the country would start to become a stronger domestic producer of agricultural goods, thus diminishing the need for heavy reliance on imports (Girard, 2015). However, the degree of dependence is higher than official statistics reports.

Wegren (2013) argues that since trade integration is to lead to more food imports, it is unlikely that Russia's targets for domestic food production can or will be met completely. However, when food security is primarily influenced by such direct-acting factor like food import ban, and imports plunge, how does the domestic Russian market become saturated? As a matter of fact, it does not, since consumption levels of staple foods in Russia are below the WHO recommendations. Per capita consumption of milk and dairy products decreased by 4 kg in Russia in 2015 compared to 2014. The same is true for the Russia's market of meat and meat products: despite some increase in production and decrease in imports, the domestic market is not saturated, and consumption of meat is below the WHO recommendation of healthy nutrition.

Due to the decreased purchasing power, nutritional habits of Russian consumers changed from high-fat animal products to starches (Lunze, Yurasova, Idrisov, Gnatienko, & Migliorini, 2015). Russian citizens are being forced to switch to cheaper products and spend more on food. According to the Analytical Center at the Government of the Russian Federation (2016), already from 2014, when the food import ban was launched, the share of food products in the consumer basket in Russia resumed growth and reached 38%. This is much higher than the levels typical for developed countries (10-20%) and even some emerging countries (in Brazil, food costs account for 17.8%). Morozova, Glushakova, and Fadeikina (2016) discovered that Russian people were to redistribute family means in favor of food,

Table 14. Import of selected food and agricultural products to Russia from selected countries in 2013 and 2015, $ million

Commodity group	EU		USA		Norway		Canada		Australia	
	2013	2015	2013	2015	2013	2015	2013	2015	2013	2015
Meat of bovine animals, fresh, chilled or frozen	139.91	0.11	0.87	0.00	0.01	0.01	0.58	0.00	178.04	0.00
Other meat and edible meat offal	1,582.72	10.35	349.48	0.60	0.02	0.02	258.82	0.04	59.70	9.98
Meat, edible meat offal, salted, dried; flours, meals	20.95	1.06	0.01	0.00	0.01	0.00	0.00	0.00	0.00	0.00
Meat, edible meat offal, prepared, preserved	174.73	28.17	4.30	0.00	0.01	0.00	0.04	0.00	0.06	0.00
Milk, cream and milk products (excluding butter, cheese)	303.42	12.46	4.45	0.81	0.54	0.17	0.00	0.00	2.30	0.00
Butter and other fats and oils derived from milk	183.08	0.05	0.02	0.00	0.00	0.00	0.00	0.00	67.14	0.00
Cheese and curd	1,247.67	20.59	0.06	0.03	4.42	0.00	0.00	0.00	10.71	0.25
Fish, fresh (live or dead), chilled or frozen	158.41	11.30	75.39	0.00	1,140.51	7.70	34.25	0.00	0.67	0.00
Fish, dried, salted or in brine; smoked fish	1.68	0.01	0.56	0.00	0.17	0.03	0.01	0.00	0.00	0.00
Fish, aqua. invertebrates, prepared, preserved	65.66	23.31	1.98	0.02	0.25	0.07	0.52	0.12	0.00	0.00
Vegetables	1,008.52	80.54	7.40	1.12	0.09	0.00	2.85	0.11	0.11	0.02
Vegetables, roots, tubers, prepared, preserved	358.18	250.13	3.30	7.87	0.00	0.00	0.01	0.01	0.18	0.05
Fruits and nuts (excluding oil nuts), fresh or dried	1,528.03	88.09	235.31	3.20	0.02	0.00	0.11	0.00	9.82	0.00
Fruit, preserved, and fruit preparations (no juice)	147.40	81.30	10.29	14.97	0.00	0.00	2.67	0.02	0.83	0.01

Source: Author's development based on the UNCTAD (2016)

reducing the consumption level at the same time. The import ban drastically decreased the available variety of foods, leaving shoppers with fewer options at the higher prices. Food prices rose, in some cases increasing more than by 70% (Girard, 2015). Krivonos (2015) also confirms that import restrictions naturally contributed to overall reduced food imports, but also to higher consumer prices and a lower level of consumption in Russia. Due to the recent food ban, Russia is suffering from a critical lack of diversity in its food selection (Girard, 2015). Along with the decrease in consumption, there is a trend that counterfeit products, such as fake cheese, are appearing on the market.

In fact, liberalization of trade in the form of the Russia's accession to the WTO in 2012 did not cause any essential increase in imports of agricultural products and food. The situation is rather unnatural since the removal of barriers is likely to bring more imports to the country (see the examples of BRICS and EAEU countries). In Russia, quite the opposite, food imports went down during 2014-2015 as a result of a series of macroeconomic events overlapping any effects of trade liberalization: reciprocal sanctions between Russia and Western countries, direct food import ban, devaluation of the ruble, food inflation, and erosion of purchasing power. In such conditions, there are no any preconditions for market capacity growth – people do not increase consumption of domestically produced food staples, as Russian government expected when launching the import ban, but decrease it due to shrinking financial capacity.

Therefore, the country is still heavy dependent on imports. Russia is unlikely to achieve domestic self-sufficiency in food, namely meat, milk, and high-value crops in the near future (Petrick, 2015). The ban makes things worse rather than contributes to achieving of food security of the country in terms of availability, affordability, and even quality and safety of food. Russia must look to abandon distorting restrictions and follow the market way of regulation of foreign trade activities if it wants its population to be adequately fed.

SOLUTIONS AND RECOMMENDATIONS

In emerging countries, food security is influenced by two groups of factors. Direct-acting ones are those related to administration of availability of food on the domestic market, including regulation of domestic food production through support and imports through various forms of protectionism, including food import bans. Liberalization of trade and growing openness of domestic market to international competition are preconditions to ensurance of food security. Lower trade barriers reduce domestic food prices, increase the purchasing power of consumers and afford them a greater variety of food products. Domestic markets reach saturation point through openness as domestic agricultural production grows and the remaining deficit is covered by imports. Freer trade can thus help enhance the availability and affordability of food and contribute to a better-balanced diet. Indirect-acting factors are those, which determine consumption behavior on the domestic market, i.e. dietary habits, the level of life, inflation, the purchasing power of population, etc.

From the above reasoning, assessing food security should be undertaken with consideration of both administrative regulations and economic patterns of domestic food market of any country.

FUTURE RESEARCH DIRECTIONS

Despite the food production growth in many emerging countries, their dependence on international trade is strengthening. This is because more open trade and WTO-supported reduction in trade barriers will encourage specialization and exports according to the comparative advantage of each country (The Economist Intelligence Unit, 2016). Emerging countries often lack basic infrastructure, and smaller incomes inhibit access to and affordability of nutritious food. Such countries also have less power to protect themselves and their domestic food markets against competition from the developed countries. Greater progress requires that country-based food-security policies shift from addressing the entire population to targeting the most vulnerable groups. One of the most powerful tools in terms of improving food security is to contain food price inflation. In such cases, as Russia's food import embargo coupled with economic decline, it is important to understand any changes that may have occurred in the food consumption pattern and to observe any nutritional implications of these changes, with a particular focus on the supply of minimum amounts of staple foods required for healthy human life. It is important to determine the actual level of security of domestic food market with an account of the internationally recognized reference nutrient intakes. There should be monitoring of the level of security of certain market over the certain period of time, which allows comparing it with both the official statistics reports and levels of food self-sufficiency and assessing influences of various foreign trade policies on food security.

CONCLUSION

The growth in international agricultural trade has led to increasing interest in the notion of food security (Lee, 2007). As the author's research shows, trade has an important role to play in increasing food self-sufficiency, which is a proportion of imported food in domestic consumption, but also in improving food security. Transforming paradigm of international economic relations causes rising influence of integration and trade liberalization on domestic food markets of developing and emerging countries. Experiences of some of the emerging economies constituting BRICS and the EAEU show direct links between trade integration (accession to the WTO) and growth of dependency of domestic food markets on import. The absence of such effect in the case of Russia is the result of "overlapping" macroeconomic issues, not related to the WTO: sanctions, currency devaluation, decreasing consumer purchasing power, etc. In the mid-term and long-term perspective effects of trade and integration show up. Internal bottlenecks may become accelerators of food security and economic sovereignty problems when trade liberalization goes on.

What has to be done in such a situation? Adaptation of domestic food markets of emerging countries in the conditions of trade integration and liberalization should have two major directions: promotion of external opportunities and overcoming of internal bottlenecks. Most of the emerging economies are dependent on import deliveries of agricultural products and food (food dependency), in addition to the existing (and even growing, in case of Russia) nonconformance of food consumption volumes to international standards (food insecurity), that is why the attention has to paid to the quality of domestic consumption, not only its quantity. Promotion of external opportunities should include the development of food export potential in the conditions of trade integration and liberalization and should be founded on the discovery of comparative advantages and their promotion on the global and domestic food markets.

REFERENCES

Akyüz, Y. (2005). *Trade, Growth and Industrialization: Issues, Experiences and Policy Challenges.* Penang, Malaysia: Third World Network.

Analytical Center at the Government of the Russian Federation. (2016). *The Food Embargo: The Results of 2015.* Moscow: Government of the Russian Federation.

Anderson, S. A. (1990). Core Indicators of Nutritional State for Difficult-to-Sample Populations. *The Journal of Nutrition, 120*, 1557–1600. PMID:2243305

Clapp, J. (2015). *Food Self-Sufficiency and International Trade: A False Dichotomy?* Rome: Food and Agriculture Organization of the United Nations.

Dorosh, P. A. (2004). Trade, Food Aid and Food Security: Evolving Rice and Wheat Markets. *Economic and Political Weekly, 36*(39), 4033–4042.

Erokhin, V., Ivolga, A., & Heijman, W. (2014). Trade Liberalization and State Support of Agriculture: Effects for Developing Countries. *Agricultural Economics*, *60*(11), 524–537.

Food and Agriculture Organization of the United Nations. (1992). *Food, Nutrition, and Agriculture*. Rome: Food and Agriculture Organization of the United Nations.

Food and Agriculture Organization of the United Nations. (1999). *Implications of Economic Policy for Food Security: A Training Manual*. Retrieved February 3, 2017, from http://www. fao.org/docrep/004/x3936e/x3936e03.htm

Food and Agriculture Organization of the United Nations. (2006). *Food Security*. Policy Brief. Retrieved February 1, 2017, from ftp://ftp.fao.org/es/ESA/policybriefs/pb_02.pdf

Girard, M. (2015). *Trade and Transition: Russia's Agricultural Pitfalls and Prospects for the 21st Century*. Orono, ME: University of Maine.

Henneberry, S. R., & Diaz, C. C. (2014). Global Food Security: Emerging Economies and Diverging Food Markets. *Journal of Food Distribution Research*, *45*(1), 29–34.

Kalugina, Z. (2014). Agricultural Policy in Russia: Global Challenges and the Viability of Rural Communities. *International Journal of Sociology of Agriculture and Food*, *1*(21), 115–131.

Krivonos, E. (2015). *Russian Restrictions on Agri-food Imports: Effects on Domestic and World Markets*. Paper presented at the International Conference of Agricultural Economists, Milan.

Lawrence, G., & McMichael, P. (2012). The Question of Food Security. *International Journal of Sociology of Agriculture and Food*, *2*(19), 135–142.

Lee, R. (2007). *Food Security and Food Sovereignty*. Newcastle upon Tyne, UK: Centre for Rural Economy, University of Newcastle upon Tyne.

Luan, Y., Cui, X., & Ferrat, M. (2013). Historical Trends of Food Self-Sufficiency in Africa. *Food Security*, *3*(5), 393–405. doi:10.100712571-013-0260-1

Lunze, K., Yurasova, E., Idrisov, B., Gnatienko, N., & Migliorini, L. (2015). Food Security and Nutrition in the Russian Federation – a Health Policy Analysis. *Global Health Action*, *8*(1), 10. doi:10.3402/gha.v8.27537

Meskhia, I. E. (2016). Food Security Problems in Post Soviet Georgia. *Annals of Agrarian Science*, *2*(14), 46–51. doi:10.1016/j.aasci.2016.03.004

Morozova, E., Glushakova, O., & Fadeikina, N. (2016). Food Consumption as an Indicator of the Quality of Life of the Population in Regions. *Foods and Raw Materials*, *4*(1), 171–180. doi:10.21179/2308-4057-2016-1-171-180

Petrick, M. (2015). *The Prospects for Russian Food Self-Sufficiency*. Paper presented at the International Conference of Agricultural Economists, Milan, Italy.

Power, E. M. (2005). Determinants of Healthy Eating Among Low-Income Canadians. *Canadian Journal of Public Health*, *96*, 537–542. PMID:16042163

Ritson, C. (1980). *Self-sufficiency and Food Security*. Reading, UK: Centre for Agricultural Strategy, University of Reading.

Samaratunga, P. A. (2011). Multiple Facets of Food (In) Security in Sri Lanka: An Input to Food Policy. In S. Mittal & D. Sethi (Eds.), *Policy Options to Achieve Food Security in South Asia* (pp. 51–76). New Delhi: Foundation Books.

Shah, A. (2010). *Deregulation or Protectionism?* Retrieved June 21, 2016, from http://www.globalissues.org/article/43/deregulation-or-protectionism

Smith, J. W. (2005). *Economic Democracy: The Political Struggle for the 21st Century*. Sun City, AZ: Institute for Economic Democracy.

Socha, T., Zahaf, M., Chambers, L., Abraham, R., & Fiddler, T. (2012). Food Security in a Northern First Nations Community: An Exploratory Study on Food Availability and Accessibility. *Journal of Aboriginal Health*, *2*(8), 5–14.

Sonnino, R., Faus, A. M., & Maggio, A. (2014). Sustainable Food Security: An Emerging Research and Policy Agenda. *International Journal of Sociology of Agriculture and Food*, *1*(21), 173–188.

The Economist Intelligence Unit. (2016). *Global Food Security Index 2016. An Annual Measure of the State of Global Food Security*. London: The Economist Intelligence Unit Limited.

The Swiss Re Group. (2013). *New Swiss Re Sigma Study Puts the Spotlight on the Role that Insurance Can Play in Improving Food Security for over 850 Million People Globally*. Retrieved February 1, 2017, from http://www.swissre.com/media/news_releases/nr_20130116_improving_food_security.html

Thow, A. M., & Hawkes, C. (2009). The Implications of Trade Liberalization for Diet and Health: A Case Study from Central America. *Globalization and Health*, *28*(1), 5. doi:10.1186/1744-8603-5-5 PMID:19638196

Trading Economics. (2016). Statistics Database [Data file]. Retrieved from http://ru.tradingeconomics.com/countries

United Nations Conference on Trade and Development. (2016). Statistics Database [Data file]. Retrieved from http://unctad.org/en/Pages/Statistics.aspx

Warr, P. (2011). *Food Security vs. Food Self-Sufficiency: The Indonesian Case*. Canberra: The Australian National University.

Wegren, S. K. (2013). *Return to Putin's Russia. Past Imperfect, Future Uncertain*. Plymouth: Rowman & Littlefield Publishers.

ADDITIONAL READING

Anderson, K., Jha, S., Nelgen, S., & Strutt, A. (2013). Re-examining Policies for Food Security in Asia. *Food Security*, *5*(2), 195–215. doi:10.100712571-012-0237-5

Bairoch, P., & Kozul-Wright, R. (1996). *Globalization Myths: Some Historical Reflections on Integration, Industrialization and Growth in the World Economy*. Geneva: UNCTAD.

Boehringer, C., & Rutherford, T. (1999). *Decomposing General Equilibrium Effects of Policy Intervention in Multi-Regional Trade Models: Method and Sample Application.* Retrieved April 21, 2014, from http://EconPapers.repec.org/ RePEc:zbw:zewdip:5249.

Bokusheva, R., Hockmann, H., & Kumbhakar, S. (2012). Dynamics of Productivity and Technical Efficiency in Russian Agriculture. *European Review of Agriculture Economics*, *39*(4), 611–637. doi:10.1093/erae/jbr059

Bremmer, I. (2015). The New World of Business. *The Fortune*. Retrieved July 23, 2015, from http://fortune.com/2015/01/22/the-new-world-of-business/

Bussière, M., Delle Chiaie, S., & Peltonen, T. A. (2014). Exchange Rate Pass-Through in the Global Economy: The Role of Emerging Market Economies. *IMF Economic Review*, *62*(1), 146–178. doi:10.1057/imfer.2014.5

Calderón, C., & Kubota, M. (2013). Sudden Stops: Are Global and Local Investors Alike? *Journal of International Economics*, *89*(1), 122–142. doi:10.1016/j.jinteco.2012.05.010

Capello, R., & Perucca, G. (2013). Globalization and Growth Patterns in Eastern European Regions: From the Transition Period to the Economic Crisis. Milan: Politecnico di Milano; GRINCOH.

Chang, H.-J., & Nolan, P. (1995). Europe versus Asia: Contrasting Paths to the Reform of Centrally Planned Systems of Political Economy. In H.-J. Chang & P. Nolan (Eds.), *The Transformation of the Communist Economies* (pp. 3–45). London: Macmillan. doi:10.1007/978-1-349-23916-0_1

Dawson, T. C. (2003). *Globalization, the Transition Economies, and the IMF*. Vienna: Joint Vienna Institute.

Devereux, M. (1999). Growth and the Dynamics of Trade Liberalization. *Journal of Economic Dynamics & Control*, *23*(5-6), 773–795. doi:10.1016/S0165-1889(98)00043-8

Erokhin, V., Heijman, W., & Ivolga, A. (2014). Trade Tensions between EU and Russia: Possible Effects on Trade in Agricultural Commodities for Visegrad Countries. *Visegrad Journal on Bioeconomy and Sustainable Development*, *3*(2), 52–57. doi:10.2478/vjbsd-2014-0010

Erokhin, V., & Ivolga, A. (2012). How to Ensure Sustainable Development of Agribusiness in the Conditions of Trade Integration: Russian Approach. *International Journal of Sustainable Economies Management*, *2*(1), 12–23. doi:10.4018/ijsem.2012040102

Estevadeordal, A., Freund, C., & Ornelas, E. (2008). Does Regionalism Affect Trade Liberalization Toward Nonmembers? *The Quarterly Journal of Economics*, *124*(4), 1531–1575. doi:10.1162/qjec.2008.123.4.1531

Gill, I. S., Izvorski, I., Van Eeghen, W., & De Rosa, D. (2014). *Diversified Development: Making the Most of Natural Resources in Eurasia*. Washington, DC: World Bank. doi:10.1596/978-1-4648-0119-8

Henneberry, S. R., & Diaz, C. C. (2015). Food Security Issues: Concepts and the Role of Emerging Markets. In A. Schmitz, P. L. Kennedy, & T. G. Schmitz (Eds.), *Food Security in an Uncertain World* (pp. 63–79). Bingley: Emerald Group Publishing Limited. doi:10.1108/S1574-871520150000015005

Ivolga, A., & Erokhin, V. (2013). Current State and Post-Crisis Developments of EU-CIS Trade: Perspective Tools to Ensure Sustainability. *Visegrad Journal on Bioeconomy and Sustainable Development, 1*, 38–42.

Ivolga, A., Labenko, O., & Ivolga, I. (2013). Vector of the EU and Russia Integration in the Sustainable Development. *Visegrad Journal on Bioeconomy and Sustainable Development, 2*(2), 53–56. doi:10.2478/vjbsd-2013-0010

Khanna, T., & Palepu, K. G. (2010). *Winning in Emerging Markets: A Road Map for Strategy and Execution.* Boston, MA: Harvard Business Press.

Kierzkowski, H. (2000). Challenges of Globalization: The Foreign Trade Restructuring of Transition Economies. *Russian & East European Finance and Trade, 36*(2), 8–41.

Kolapo, I. A., Olufemi, A. J., & Olusola, A. (2012). Globalization and the Transitional Economy: Impacts and Effects on Africa Regional Growth. *European Journal of Globalization and Development Research, 2*(1), 117–126.

Kose, M. A., & Prasad, E. (2010). *Emerging Markets: Resilience and Growth amid Global Turmoil.* Washington, DC: Brookings Institution Press.

McMillan, J., & Naughton, B. (1992). How to Reform a Planned Economy: Lessons from China. *Oxford Review of Economic Policy, 8*(1), 130–143. doi:10.1093/oxrep/8.1.130

Raynolds, L., Murray, D., & Wilkinson, J. (2007). *Fair Trade: The Challenges of Transforming Globalization.* London: Routledge.

Roman-Alcala, A. (2013). *From Food Security to Food Sovereignty?* Retrieved June 22, 2016, from http://civileats.com/2013/05/29/from-food-security-to-food-sovereignty/

Santacreu, A. M. (2015). *The Economic Fundamentals of Emerging Market Volatility*. St. Louis, MO: Federal Reserve Bank of St. Louis.

Schmitz, A., & Meyers, W. H. (Eds.). (2015). *Transition to Agricultural Market Economies. The Future of Kazakhstan, Russia and Ukraine*. Boston, Oxfordshire: CABI. doi:10.1079/9781780645353.0000

Vos, R., & Koparanova, M. (Eds.). (2011). *Globalization and Economic Diversification: Policy Challenges for Economies in Transition.* London: Bloomsbury Academic. doi:10.5040/9781849665742

Windfuhr, M., & Jonsen, J. (2005). *Food Sovereignty: Towards Democracy in Localized Food Systems.* Heidelberg: FIAN ITDG Publishing. doi:10.3362/9781780441160

Wittman, H., Desmarais, A., & Wiebe, N. (2010). *Food Sovereignty: Reconnecting Food, Nature and Community.* Oakland, CA: Food First Books.

Zelenko, O. (2012). *Trade Policy Options of CIS Countries: Twenty Years After.* Rotterdam: Erasmus University.

KEY TERMS AND DEFINITIONS

BRICS: The association of five major emerging countries (Brazil, Russia, India, China, and South Africa), distinguished by their large, fast-growing economies and growing influence on regional and global affairs.

Emerging Economies: The countries, progressing toward becoming advanced, but still lacking market efficiency and sustainable market mechanisms, having greater domestic and external risks due to political instability, infrastructure problems, and currency volatility.

Eurasian Economic Union: The union of Armenia, Belarus, Kazakhstan, Kyrgyzstan, and Russia established in 2014 for effective promotion of free movement of goods, services, capital, and labor between participating countries.

Food Security: A combination of physical food availability on the domestic market by means of domestic production and import, and economic access to sufficient amounts of foodstuffs in accordance with the international standards of adequate food consumption and healthy life.

Food Self-Sufficiency: An extent to which a country can satisfy its food needs from its own domestic production.

Foreign Trade: The system of international commodity-money relations composed of foreign trade activities of all countries worldwide.

Trade Liberalization: A process of reduction or elimination of constraints in the sphere of international trade, including reduction or removal of customs tariffs, import quotas, abolishment of multiple exchange rates, and simplification of administrative requirements to import and export operations.

Trade Protectionism: A set of measures to limit imports or promote exports by putting up barriers to trade that countries use to limit unfair competition from foreign industries.

This research was previously published in Establishing Food Security and Alternatives to International Trade in Emerging Economies edited by Vasily Erokhin, pages 28-54, copyright year 2018 by Business Science Reference (an imprint of IGI Global).

Chapter 2
Implications of Trade Liberalization for Food Security Under the ASEAN–India Strategic Partnership:
A Gravity Model Approach

Ishita Ghosh
Symbiosis International University, India

Ishita Ghoshal
Symbiosis International University, India

ABSTRACT

The objective of the chapter is to understand India's role in providing food security through trade with the ASEAN under the India-ASEAN Strategic Partnership. A dynamic panel has been employed to assess and estimate the said objective, given that India and most of the ASEAN countries are considered emerging markets. Econometric investigation points out that while almost more than half the variables in the augmented gravity model hold good, food imports from India, agricultural labour force, common maritime border, distance from India, food price index, food production index, GDP (size) of the importing country, per capita GDP of the importing economy, and trade openness have statistically significant effect on the food trade from India to the ASEAN. Improving maritime infrastructure and agri-logistics, investing in climate change and water management while augmenting the agri-labour productivity are of paramount importance in order to improve food trade between India and the ASEAN.

INTRODUCTION

The world population is predicted to cross nine billion by the year 2050, resulting in growing concerns to ensure food security (Food and Agriculture Organization of the United Nations [FAO], 2001). Food security has had evolving definitions since the mid-1970s. The latest definition given as per the State of Food Insecurity, 2001, is: "Food security [is] a situation that exists when all people, at all times, have

DOI: 10.4018/978-1-5225-8063-8.ch002

physical, social and economic access to sufficient, safe and nutritious food that meets their dietary needs and food preferences for an active and healthy life" (FAO, 2001).

In the modern context, it is not really a concern if the world is capable of meeting the growing demands to adequately cover calorific and nutritional demands; however, barriers to worldwide access such as political factors, ownership, institutions, poverty and inequality still pose a threat to global supply chains and distribution (Gillson & Fouad, 2016).

Improvements in food security have been noted in almost all the regions of the world and particularly in emerging markets (Emerging Market Economy, n.d.). An emerging market economy describes a nation's economy, that is, progressing toward becoming more advanced, usually by means of rapid growth and industrialization. These countries experience an expanding role both in the world economy and on the political frontier (Emerging Market Economy, n.d.). The Global Food Security Index [GFSI] (The Economist, 2016) mentions that barring Europe, the low-income and lower-middle-income countries have progressed towards ensuring better measures for food security, thereby reducing the global gap between the least and the most food secure nations. Although structural changes in economies have increased the access to a wide range of affordable and nutritious food, put in place more extensive food safety-net programs and expanded crop storage capacity and dietary diversity, food security still remains a concern and challenge in the modern day. Despite investment in infrastructure, most developing and emerging economies are still grossly short on road and port facilities that aid in accessing food. Also, absorbing urban migration is a gradual process – pressures of urbanization have been growing worldwide and agricultural labor is depleting, while there has been a downward trend in the world Gross Domestic Product (GDP) growth. Macroeconomic improvements and stability have been compromised by geopolitical turmoil and climate change (The Economist, 2016).

According to the Asian Development Bank [ADB] (2016), two-thirds of the world's one billion hungry reside in Asia and the Pacific. While food prices are rising, bringing about food shortages and malnutrition in the region's poor, both rural and urban dwellers are forced to spend more than half of their income on food and related goods. The persistent volatility in food price has become an impediment to reaching the Millennium Development Goals, especially the ones related to the mitigation of hunger and poverty (Asian Development Bank, 2016).

The projected population growth that will reside in urban areas of Asia by 2026 is close to 50% of the constantly growing population. This will need comprehensive planning and significant investments to cater to food distribution, storage and marketing facilities to ensure food security. There will also be a heavy requirement of improved land cultivation techniques and labor productivity (Briones, Chavez, Durand-Morat, & Wailes, 2012).

Climate change also plays a vital role in ensuring food security. The Asia-Pacific El Niño has, and the predicted La Niña will cause droughts and hence, reduction in agricultural output. This has affected and will continue to affect inflation rates, poverty, exports, etc., adversely. Prolonged effects have may affect macroeconomic and geopolitical stability (El Niño in Asia, 2016; Tabor, Ginting, & Aji, 2015).

Currently, Asia faces serious water supply crisis. The demand for water in expected to rise by 40% towards 2030. While 80% of the water supply is directed to agricultural production, the water crisis is going to compound the food security issue. Using less water to grow more food alongside sustainable water management will have to be the order of the day.

The ADB (2016) has also projected that domestic food prices will rise in the developing and emerging markets in Asia by 10% between 2020 and 2025. This is likely to push a population of 64 million Asians into poverty, over and above the current status. Extraordinary food price surges, globally increase food insecurity thereby posing great economic difficulties to the region, nations trading in food and agriculture, and governments (Huh & Park, 2013; Jha, Kubo, & Ramaswami, 2013).

As a consequence of the 2007-2008 food price crisis, the Association of Southeast Asian Nations [ASEAN] Summit of 2009 promised to ensure measures towards food security as an urgent and long-term priority. By 2010, the ASEAN Ministry of Agriculture and Forestry [AMAF] fostered economic ties between India, China and the Republic of Korea, as well as, within the regions of the Gulf Cooperation Council [GCC] and the ASEAN+3 (ASEAN plus China, Japan, and South Korea) (Choulamany, n.d.).

It should also be noted that according to Morgan Stanley Capital International (n.d.), India and several nations in ASEAN are emerging economies. The ASEAN being largely agrarian in nature, it is pertinent to examine this trading bloc from the viewpoint of food security and trade.

In this backdrop, the chapter focuses on the agricultural trade between India and ASEAN under their strategic partnership and regional cooperation. The background and problems pertaining to food security under the said alliance are discussed and a gravity equation of trade is employed to empirically identify the variables that may be significant in improving food security in the region through trade with India as well as, variables that impede the bilateral trade in agriculture. The authors hope to provide insights from the results of the analysis to augment the food trade relationship between India and the ASEAN economies.

BACKGROUND

Out of the seven aims and purposes of the ASEAN Declaration (1967), collaboration "for the better utilization of agriculture and industry to raise the living standards of the people", is a key tenet (Overview, n.d.). As mentioned in the introductory section, ASEAN has made ties with India to cope with and cater to food security issues. In 2010, the ASEAN – India Ministerial Meeting on Agriculture and Forestry [AIMAF] and the ASEAN – India Working Group on Agriculture and Forestry [AIWGAF] were put in place to help improve relations in bilateral and sustainable trade in food and agriculture (Choulamany, n.d.). For a while (up to 2011-2012), ASEAN has been India's fourth-largest trading partner after China, the EU, and the USA. Trade between the two regions touched nearly $80 billion in 2014-2015 while the expected potential has been set for $200 billion by 2022. In 2016, India has slid down to the ninth position from the former fourth. Possible reasons for this stagnation and decline could be a change in the government and policies which are yet to take effect. Moreover, the "Make in India" initiative since 2014 has been "inward looking" where the country has essentially concentrated on inward Foreign Direct Investments (FDI) for self-sufficiency and sustainability in manufacturing. The India-ASEAN regional partnership has spanned over two decades now (ASEAN-India Relations, 2016). However, since the execution of the India-ASEAN Free Trade Area (FTA) in 2010, India's exports of garlic, onions, turmeric, wheat, meslin, cane sugar, groundnuts, oilcake/ oilcake meal of soybean, oilcake/oilcake meal of rape/ colza seed, and millets (sorghum and bajra) to ASEAN have registered a sharp increase, year-on-year (Government of India, 2016).

MAIN FOCUS OF THE CHAPTER

Drivers of Food Security and Bilateral Trade in Agriculture: A Literature Review

According to Holt-Giménez (2008), the well-established causes of food crises have been poor weather and climate change, low grain reserves, high oil prices that push up the prices of fertilizers and food transportation, increase in meat consumption worldwide with the rise in global GDP, that has resulted in focusing on the farming of crops for cattle and livestock, thereby decreasing or replacing production of other crops. Also, the new-found need for agrofuels has increased grain prices. Deregulation of stock and commodity markets and the over-enthusiasm to make quick and large amounts of profits have also added to the speculative bubbles in the futures markets. In fact, right after the 2007-2008 subprime crisis, investors sought new avenues to put in money. Rising food prices showed the way to attractive commodity futures investments, which in turn further increased the food prices. When they saw food prices going up, they poured investments into commodities futures, pumping up the price of grains and worsening food price inflation (Holt-Giménez, 2008).

The four main concepts of food sufficiency are given differently by various researchers. According to Reutlinger and Knapp (1980) and Gouel and Jean (2013), food sufficiency is the minimal level of food consumption, while according to Siamwalla and Valdes (1980), food sufficiency is a target level. Barraclough and Utting (1987) have said that food sufficiency is the food adequate to meet nutritional needs. Another way to evaluate food sufficiency is to ensure that the population in question have "enough" food. This can be measured by estimating by how far below the consumption is from the threshold (Lipton & Heald, 1984; Clay, Shaw, & Singer, 1987), while Maxwell (1990) has tried measuring the "intensity" of food insecurity.

The second concern is access to food, which assesses if individuals are able to obtain adequate food. This is motivated by Sen's (1983) concept of "entitlements" of food – that is, "an individual's entitlement is rooted in his/her endowment and is then transformed via production and trade into food or commodities which can be exchanged for food". While the secure access to adequate food is the third concern, the concepts of risk and its avoidance are key to the idea of food security due to various reasons such as variability in crop production and food supply, market and price variability, risks in employment and wages, etc. (Applanaidu, Bakar, & Baharudin, 2014).

It is also important to mention that the temporal dimension in food security has become a vital concept in the latter day policymaking. The World Food Programme (2009) notes that there may be several ways of incorporating the time variable in case of food insecurity – chronic (long term, or persistent for six months), transitory (short-term, or temporary), and cyclical (seasonal variability).

In Malaysia, Mohamed Arshad and Hameed (2010) have inspected the effect of increasing food prices on food security. According to their study, it is the fundamentals like decreasing agricultural production growth leading to falling supply, decreasing stocks of global cereal and rising food demand from the emerging economies around the world.

Aker and Lemtouni (1999) looked into the national level factors and integrated domestic and global markets that affect food security. Macroeconomic variables considered in the study were domestic food production, average annual rainfall, world food prices, gross domestic product, Gini coefficient, and exports of goods and services.

The study of Applanaidu et al. (2014) on Malaysia analyzed the dynamic impact of macroeconomic factors like biodiesel production, exchange rate, government expenditure on rural development, GDP, food price index and population on food security with the vector autoregressive (VAR) method. Variance decomposition analysis shows that biodiesel production, exchange rate and government expenditure on rural development affect food security the most in the tenth year, GDP in the sixth year, and population and exchange rate in the fifth year.

Volatility and rise in food prices have received great attention from the policy makers due to the challenges posed by the dynamics of food price inflation and volatile prices (Lee & Park, 2013). They conducted a panel analysis for 72 countries for the span 2000 to 2011 assessing how food price inflation and volatilities are transmitted globally alongside checking for the impact of several internal and external variables on them. The findings suggest that lagged global food price inflation is significantly associated with domestic food price inflation in Asia. Increasing economic growth rates, shares of food imports and fall in growth rate of food imports lead to a rise in food prices. Higher economic growth rate lowers volatility of food prices. On the other hand, appreciation of the currency, higher political stability, and income of the economy lead to fall in food price inflation.

Among the various economic and econometric concepts to assess food security, the use of the gravity model is gaining pace amid modern day researchers. Very simply put, the model has its origin in the law of universal gravitation in physics, which was developed by Sir Isaac Newton in 1687. The law of universal gravitation describes the gravitational force between two masses in relation to the distance that lies between these two masses. The gravitational constant is an empirically determined value (equation 1)

$$F_{ij} = G * (M_i M_j) / d_{ij}^2 \tag{1}$$

The gravitational force F_{ij} proportionate to the product of the two masses M_i and M_j and inversely proportional to the square of the distance d_{ij}^2 that keeps the two masses apart from each other. The gravitational constant G is an empirically determined value. This relationship is applicable to any context where the modeling of flows or movements is demanded. The gravity model in international economics puts the law of universal gravity into an economic context. Assuming that the trade volume between countries i and j is X_{ij} and furthermore economic masses defined as E_i and E_j, trade flows can be characterized. Thus, the predicted movement of goods or labor between countries i and j would be (equation 2):

$$X_{ij} = (E_i E_j) / \varphi_{ij}^2 \tag{2}$$

In other words, a mass of goods, labor, or other factors of production at origin i (E_i) is attracted by a mass of demand for goods or labor at destination j (E_j), but the potential flow is reduced by the distance between them (φ_{ij}^2).

However, the problem with drawing parallels between trade and physical force of gravity is that there is no set of parameters for which equation (2) will hold exactly for an arbitrary set of observations. Departing from strict analogy, traditional gravity model allows the coefficients of (2) applied to the mass variables and of (3) applied to the bilateral distance to be generated by data to fit statistically inferred the relationship between data on flows and the mass variables and distance. Typically, the stochastic version of the gravity equation has the following form (equation 3):

$$X_{ij} = a_0 E_i^{\alpha 1} E_j^{\alpha 2} \varphi_{ij}^{\alpha 3} \varepsilon_{ij}, \tag{3}$$

where a_0, α_1, α_2, and α_3 are unknown parameters.

In the original version by Tinbergen (1962), as given in (2), the model is expressed in a log-log form, so that the parameters are the elasticity of the trade flow with respect to the explanatory variables. With respect to equation (3), adjacent countries are assumed to have a more intense trade than what distance alone would predict; the adjacency is indicated by the dummy variable N_{ij}, that took the value 1 if the two countries share a common border. Moreover, the equation is augmented with political factors: a dummy variable V_{ij} indicates that goods traded received a preferential treatment if they belonged to some unilateral trade or system of preferences. The strategy of considering the effect of Preferential Trade Agreements (PTA) through the use of dummy variable has been prominent in the literature. Only recently, the alternative strategy of explicitly including the preferential margin guaranteed by the agreement has been taken into account: the authors will come back to this issue in the following. As customary, an independent, identically distributed stochastic term, ε_{ij}, is included. Therefore, the expression would take the form (equation 4):

$$log\,X_{ij} = a_0 + \alpha_1 logE_i + \alpha_2 logE_j + \alpha_3 log\varphi_{ij} + \alpha_4 logN_{ij} + \alpha_5 logV_{ij} + \varepsilon_{ij}, \quad (4)$$

where:

a_0 = constant,
$\alpha_1 \text{log}E_i + \alpha_2 \text{log}E_j$ = economic indicators,
$\alpha_3 \text{log}\varphi_{ij} + \alpha_4 \text{log}N_{ij}$ = distance,
$\alpha_5 \text{log}V_{ij}$ = policy,
ε_{ij} = error term.

Bilateral trade flows are determined by the variables included in the right-hand-side of the gravity equation. This implies a clear direction of causality that runs from income and distance to trade. This direction of causality is, however, theory-driven and based on the assumption that the gravity equation is derived from a microeconomic model where income and tastes for differentiated products are given. Three decades of theoretical work has shown that the gravity equation can be derived from many different – and sometimes competing – trade frameworks (Salvatici, 2013; Starck, 2012; Masudul Haq et al., 2014; Sarker & Jayasinghe, 2007; van Dijk, 2011; Natale, Borrello, & Molotova, 2015; Wang, 2016).

Material and Methods

Given the previous discussion, the following seventeen variables were considered for estimating the gravity model in the current study:

1. Food import from India into the ten ASEAN countries.
2. GDP of India (origin country) in current USD.
3. GDP of the importing country in current USD.
4. GDP per capita of importing country in current USD.
5. Trade as a percentage of GDP to measure trade openness of importing country.
6. Official exchange rate of the importing country (local currency unit (LCU) per USD, period average).

7. Agriculture, value added (percentage of GDP).
8. Depth of the food deficit (kilocalories per person per day).
9. Food production index (2004-2006 = 100).
10. Distance from India.
11. Per capita food production variability.
12. A dummy for the common maritime border.
13. A dummy for Bilateral Trade Agreement (BTA).
14. Food price index for importing country.
15. Quality of port infrastructure, WEF (1 = extremely underdeveloped to 7 = well developed and efficient by international standards).
16. Agricultural labor force as a share of total labor force.
17. Political stability and absence of violence/terrorism.

The gravity equation of trade in this context will help only to determine the variables that play a role in agricultural trade between India and the ASEAN countries. Customs and other import duties and corruption index were dropped from the analysis due to the lack of available data for all the countries in the study.

For the purposes of analysis, the authors have considered the following model using data for the ten ASEAN countries (variables 3-17) and India from 1995 to 2015. Given the nature of the data, the study has utilized panel data methods to estimate the augmented gravity model for deducing the impact of various macroeconomic and policy variables on food import from India to these countries.

$$\begin{aligned}\log(\text{Food Import}_{it}^{\text{India to }i\text{th country}}) &= a + b_1 \log(GDP_t^{\text{India}}) + b_2 \log(GDP_{it}^{\text{Importing Country}}) + b_3 D_{it} \\ &+ b_4 \log(GDP_{it}^{PC}) + b_5 TO_{it} + b_6 ER_{it} + b_7 GDP_{it}^{\text{Agriculture}} + b_8 \text{Food Deficit}_{it} + b_9 F\text{Prod}I_{it} \\ &+ b_{10} \text{Food Prod}n\text{Variability}_{it}^{PC} + b_{11} CMB_i + b_{12} BTA_i + b_{13} FPI_{it} + b_{14} QPI_{it} \\ &+ b_{15} \text{Agri Labour Force}_{it}^{\text{Share of total}} + b_{16} \text{Political Stability}_{it} + \varepsilon_{it}\end{aligned} \tag{5}$$

where:

$\log(\text{Food Import}_{it}^{\text{India to }i\text{th country}})$ is the log of food import from India into the ten ASEAN countries over the years;
$\log(GDP_t^{\text{India}})$ is the log of India's GDP which is the host or origin country;
$\log(GDP_{it}^{\text{Importing Country}})$ is the log of the importing country's GDP;
$\log(GDP_{it}^{PC})$ is the log of per capita GDP of the importing country (to capture the population factor in the model);
D_{it} is the distance of the importing country from India;
TO_{it} is the trade openness of importing country measured by total trade as a share of GDP;
ER_{it} is the exchange rate of the importing country;
$GDP_{it}^{\text{Agriculture}}$ is the share of agricultural production in total GDP of the importing country;
Food Deficit_{it} is the extent of the food deficit that the importing country suffers from;

$F\text{Prod}I_{it}$ is the food production index;

Food Prod$n$ Variability $^{PC}_{it}$ is the per capita food production variability;

CMB_i is the dummy for common maritime borders where $\text{CMB}_i = 1$ if the importing country share a maritime border with India, 0 otherwise;

BTA_i is the dummy for bilateral trade agreement where $BTA_i = 1$ if the importing country has a separate bilateral trade agreement other than the India-ASEAN agreement, 0 otherwise;

FPI_{it} is the food price index;

QPI_{it} is the quality of port infrastructure for the importing country;

Agri Labour Force $^{\text{Share of total}}_{it}$ is the share of labor force involved in agricultural activities in total labor force of the importing country;

Political Stability$_{it}$ is the index for political stability and absence of violence/terrorism.

Data

Panel data for the ten ASEAN countries and India has been used for the period of 1995 to 2015. Data for food import from India into the ten ASEAN countries and share of labor force involved in agricultural activities in total labor force of the importing country have been collected from the United Nations Conference on Trade and Development [UNCTAD] reports; GDP of importing country, per capita GDP, trade openness, exchange rate, food production index, agricultural GDP, quality of port infrastructure have been collected from World Bank Indicators; depth of food deficit (kilocalories per person per day), per capita food production variability, food price index for importing country, political stability and absence of violence/terrorism have been collected from FAO Food Security Indicators. The dummy for BTA was constructed based on information available from Asian Regional Integration Centre of Asian Development Bank. The dummy for CMB has been constructed by the authors based on the current world map. The distances of the importing countries from India have been calculated based on the flying distances available from www.distancefromto.net.

Customs and other import duties data is available with the World Bank, and Corruption Index data is available from Transparency International – The Global Coalition against Corruption website.

Methodology

Panel data estimation techniques have been used to understand the extent and direction of the relationships of the above-mentioned variables with a value of food imported from India by the ASEAN countries. The authors have used panel vector autoregression analysis (VAR) to estimate the gravity model. The fixed effect or random effect methods of estimation were not relied upon since these are single equation analyses which assume explicit exogeneity of the regressor variables. The nature of the economic variables considered for the study does not allow the authors to strictly assume that the regressors are exogenous. VAR analysis allows the authors to treat all the variables as endogenous and then estimate a set of equations instead of a single equation. This gives the more reliable result as the inter-relationships among all the variables in the model are taken into consideration for the estimation process. The following elucidates the detailed analytical methods used for the study:

Panel Data Unit-Root Tests

The authors have used panel data unit root tests to check whether the variables are stationary or not. Several statistical methods (Levin, Lin, & Chu, 2002; Im, Pesaran, & Shin, 2003; Choi, 2001; Breitung, 2000; Hadri, 2000) are constructed to test for unit roots in panel data. Among those, the study has employed Levin-Lin-Chu test and Im-Pesaran-Shin tests to check for stationarity in the variables.

Panel Causality for a Pair of Variables

Granger causality is checked to analyze the causality hypotheses. In the case of a long-run relationship, panel Vector Error Correction Model (VECM) is used while panel VAR method is used in case of no cointegration. The conventional Granger causality tests raise two critical issues for a panel data case, both dealing with the potential heterogeneity of the individual cross sections. The first source of heterogeneity is cross-sectional variation due to the distinctive intercepts; such heterogeneity may be addressed with a fixed effects model (that is, it is controlled by introduction of individual effects in the model). The more crucial case is where heterogeneous slope coefficients should be considered (that is, it should be controlled by introducing individual dimension for regression slopes in the model). The second source of heterogeneity affects the causality relationships. For instance, for some individuals, the introduction of past values may improve the forecast, whereas for others there may be no improvement. Therefore, we should distinguish two subgroups of individuals according to the causality relationships. If this heterogeneity is not considered, the test of causality hypothesis may lead to a fallacious conclusion concerning the relative size of the two subgroups. In a nutshell, the Granger causality for panel data sets should consider the different sources of heterogeneity of the data-generating. A newly suggested theory of Granger Causality in Panel by Hurlin and Venet (2001) is used here to test causality which controls for both sources of heterogeneity. The procedure has three main steps, which are related to the homogeneous non-causality (HNC), homogeneous causality (HC) and heterogeneous non-causality (HENC) hypotheses.

Panel Vector Auto Regression (VAR)

To check for causality in the case of no long-run relationship, panel VAR is used. Following fixed effects model is estimated:

$$\Delta Y_{it} = \sum_{k=1}^{p} a_k \Delta Y_{it-k} + \sum_{k=1}^{p} b_k \Delta X_{it-k} + \lambda_i + \mu_{it}$$

$$\Delta X_{it} = \sum_{k=1}^{p} a_k \Delta X_{it-k} + \sum_{k=1}^{p} b_k \Delta Y_{it-k} + \lambda_i + \mu_{it}$$

where p = lag length, λ_i: country fixed effects – controls the potential heterogeneity of cross sections.

Panel ARDL Approach: Multivariate Analysis

Let us consider the following ARDL process for panel data:

$$y_{it} = \rho_i y_{i(t-1)} + X_{it}\delta_i + \varepsilon_{it}$$

where $i = 1,2,\ldots N$ cross-section units or series, that are observed over periods $t = 1,2,\ldots T$; X_{it} represents the exogenous variables in the model, including any fixed effects or individual trends; ρ_i are the autoregressive coefficients; ε_{it} are the errors which are assumed to be a mutually independent idiosyncratic disturbance.

If $\rho_i < 1$, y_i is said to be weakly (trend) stationary. On the other hand, if $\rho_i = 1$ then y_i contains a unit root. For purposes of testing, there are two assumptions that we can make about the ρ_i. First, one can assume that the persistence parameters are common across cross-sections so that $\rho_i = \rho$ for all i. The Levin-Lin-Chu (LLC), Breitung and Hadri tests all employ this assumption. Alternatively, one can allow ρ_i to vary freely across cross-sections. The Im-Pesaran-Shin (IPS), and Fisher-ADF and Fisher-PP tests are of this form.

After determining the level of integration of all variables, the study proceeds to the application of panel ARDL to estimate the long-run and short-run relationships between regression variables. Autoregressive Distributed Lag (ARDL) models are standard least squares regressions which include lags of both the dependent variable and independent variables as regressors (Greene, 2008). Although ARDL models have been used in econometrics for decades, they have gained popularity in recent years as a method of examining long-run and co-integrating relationships between variables (Pesaran, Shin, & Smith, 1999). In panel settings with individual effects, standard regression estimation of ARDL models is problematic due to bias caused by the correlation between the mean-differenced regressors and the error term. This bias only vanishes for large numbers of observations T, and cannot be corrected by increasing the number of cross-sections N. To address this problem, a number of small T–large N, dynamic panel data GMM estimators have been developed (for example, Arellano and Bond, 1991). In large datasets, these assumptions underlying dynamic GMM are often inappropriate, and the estimator breaks down. In these cases, a popular alternative is the Pooled Mean Group (PMG) estimator of Pesaran et al. (1999). This model takes the cointegration form of the simple ARDL model and adapts it for a panel setting by allowing the intercepts, short-run coefficients, and co-integrating terms to differ across cross-sections.

Analysis

The authors begin by checking for stationarity of the variables. The IPS test has been conducted including a trend for all the panel variables. LLC test requires panel data to be strongly balanced and hence the test could not be run for most of the variables. However, the authors have relied on the LLC test instead of IPS for the variable measuring depth of food deficit since all the observations of the variable had zero value for Singapore as the country does not suffer from food deficit at all and IPS test requires at least six non-zero observations per panel. The dummy variables, distance from India, India's GDP and quality of port infrastructure index (on a scale of 1-7) were excluded from the tests due to the non-panel nature of the variables. The results show that all the variables considered for the tests are stationary in the level form (p-values are lesser than the level of significance) hence they are all integrated of order 0 [I(0)]. The results are presented in Table 1.

Given that neither food import from India nor any of the regressors for the model are integrated of order 1 [I(1)], we cannot establish a co-integrating relationship between the regressors and food import from India that would have shed light on the long run relationship among the said variables. Hence we move on to check for causality among the panel variables running from the factors that are crucial for food import, to food import from India.

The pairwise causality tests were run for various lag lengths. It was seen that given the sample size of the panel data, the maximum lag length that can be considered for testing causality is 9. It is found that the GDP of the importing country has a causal effect on food import from India at lag lengths 1, 2, 3, 6, and 9. This emphasizes the importance of the size of the economy in determining its capacity to import food. Food production index causes food import from India at the 1st lag. The dearth or abundance of domestically produced food products will promote or stunt food import from India. The index for political stability and absence of violence or terrorism is also seen to cause import of food from India at the 3rd lag. A more stable and peaceful economy may pave the way for sustainable policies towards food security of the economy, triggering a change in the food import pattern for the country. However, this process being a long drawn one, the effect of the stability is expected to be seen only after three years.

Table 1. Unit root tests

Variables	IPS with trend	LLC
	p-value	
	Level	Level
Agricultural lab force, total	0.0575*	
Agricultural GDP	0.0445**	
BTA	NA	
CMB	n.a.	
Food deficit	....	0.0002***
Distance from India	n.a.	
FPI	0.0595*	
FProdI	0.0012***	
log(ER)	0.0629*	
log(Food Import from India)	0.0006***	
log(GDP Importing Country)	0.0178**	
log(GDP India)	n.a.	
log(GDP Per Capita)	0.0155**	
Food production variability	0.0408**	
Political stability and absence of violence/terrorism	0.0028***	
QPI	n.a.	
TO	0.0038***	

Source: Authors' development

Note: *, ** and *** indicate statistically significant values at 10%, 5%, and 1% levels of significance, respectively

Food production variability is found to cause food import from India at the 8th lag implying again that a more (or less) dispersed food production will prompt (or curtail) food imports only when the volatility has been experienced over some time and not right away. Agricultural GDP, depth of food deficit and the food price index of the importing economies is found to cause food import from India at a lag length of 9 which leads us to believe that it is only the persistent behaviour of these variables that may take shape through effects on other variables which finally go on to facilitate or impede the food imports from India. The rest of the variables were not found to have any causal effect on food import from India. The results of the causality tests for various lag lengths are presented in Table 2. The statistically insignificant results have not been reported here.

Next, the authors estimate a VAR model to understand the extent and direction of the effects of the chosen variables on food import from India. The estimated model has been reported below:

$$\log(\text{Food Import}_{it}^{\text{India to } i\text{th country}}) = 26.39 - 0.04\log(GDP_t^{\text{India}}) + 2.923log(GDP_{it}^{\text{Importing Country}})$$
$$-2.133\log(GDP_{it}^{PC}) - 0.005D_{it} + 0.006TO_{it} - 0.337ER_{it} - 0.061GDP_{it}^{\text{Agriculture}} + 0.004\text{Food Deficit}_{it}$$
$$+0.021F\text{Prod}I_{it} - 0.012\text{Food Prod}n\text{Variability}_{it}^{PC} - 9.058CMB_i + 1.025BTA_i + 0.215FPI_{it}$$
$$-0.109QPI_{it} - 18.835\text{Agri Labour Force}_{it}^{\text{Share of total}} - 0.019\text{Political Stability}_{it}$$
$$+0.068\log(\text{Food Import}_{i(t-1)}^{\text{India to } i\text{th country}}) - 0.396\log(\text{Food Import}_{i(t-2)}^{\text{India to } i\text{th country}})$$

Table 3 reports the coefficients and the student's t-statistic associated with the variables which allow the authors to establish the direction and value of the impact of a unit change in the regressor variable on the endogenous variable, log of food import from India and the statistical significance of the variables in explaining the variability of the endogenous variable. The critical values of the t-statistic for 10%, 5%, and 1% levels of significance and relevant degrees of freedom were found to be 1.684, 2.021, and

Table 2. Panel causality

Lag Length	Causal Variables	F-statistic	p-value
1	Food production index (2004-2006 = 100)	3.51351	0.0627
	log(GDP Importing Country)	10.6838	0.0013
2	log(GDP Importing Country)	4.21413	0.0163
3	log(GDP Importing Country)	2.79414	0.0418
	Political stability and absence of violence/terrorism	6.1952	0.0007
4, 5, and 7	-	-	-
6	log(GDP Importing Country)	2.15896	0.0497
8	Per capita Food Production variability	1.96164	0.0691
9	Agriculture, value added (% of GDP)	1.89106	0.0608
	Depth of the food deficit (kilocalories per person per day)	1.74314	0.0847
	Domestic food price index	4.05604	0.0019
	log(GDP Importing Country)	1.92637	0.0528

Source: Authors' calculation

Table 3. VAR results

Variables	Co-efficient	t-statistic
log(Food import from India) _ (t-1)	0.068	0.578
log(Food import from India) _ (t-2)	-0.396***	-5.559
Agriculture GDP	-0.061	-1.148
Agriculture lab force, total	-18.835***	-3.33
BTA	1.025	0.52
CMB	-9.058**	-2.134
D	-0.005***	-3.102
Food deficit	0.003	0.637
FPI	0.215*	1.724
FProdI	0.021***	3.501
log(ER)	-0.337	-0.801
log(GDP Importing Country)	2.924***	5.189
log(GDP India)	-0.041	-0.085
log(GDP Per Capita)	-2.133***	-3.935
Per capita Food Production variability	-0.012	-0.675
Political stability and absence of violence/terrorism	-0.019	-0.058
QPI	-0.109	-0.448
TO	0.006**	2.125

Source: Authors' development

Note: *, ** and *** indicate statistically significant values at 10%, 5%, and 1% levels of significance, respectively

2.704, respectively. The authors reject the null hypothesis of no significant effect of the regressor on the regressand when the calculated *t* statistic is greater in absolute value than the critical/tabulated statistic. The results show that the 2nd lag of food imports from India, agricultural labor force, common maritime border, distance from India, food price index, food production index, GDP (size) of the importing country, per capita GDP of the importing economy, and trade openness have statistically significant effect on food imports from India. Lagged food import from India, agricultural labor force, common maritime border, distance from India, and per capita GDP of the importing country affect food import from India in an adverse manner. The rise in the values of these variables will lead to a decline in the flow of food imports to these economies from India. On the other hand, food prices, food production, GDP of the importing country, and trade openness have a positive effect on food imports.

The values of food import from India to the ASEAN countries in the past have a negative impact on the current value of the variable. It may be due to the fact that the country of origin prefers to diversify with respect to trading partners for food items exports and there is a cyclical pattern for inflows of food imports into the importing economy that recurs periodically. It can also be attributed to weather conditions of the importing country – periodic bad weather may prompt the need to import food from India rather than every year.

The rise in agricultural labor force leads to a dip in food imports. It may be because the additional labor units are adding to the food production of the economy making it more self-sufficient and at the same time reducing its dependence on foreign countries for food needs. This also goes on to suggest that the agricultural sector of the ASEAN economies suffer from insufficient labor inputs comparative to capital and land resources. Hence, increase in the agricultural labor force takes the economy closer to the production possibility frontier with greater and more efficient utilization of other factors of production which may have been left comparatively less productive due to lack of human capital.

The existence of a common maritime border reduces food imports from India due to the higher cost of transportation through waterways. The port facilities of the Indian sub-continent are still poor. Also, the high traffic in these very waterways which lead to the common maritime borders for India and the ASEAN countries add to the time and costs of transportation of the traded products.

The distance of the ASEAN economy from India is also a deterring factor for food imports. This result emphasizes the significance of the gravity model considered to estimate the model for the study. The farther apart the economies are, the lesser the chances of them entering into trade relations – this is the essence of the basic gravity model. The obvious reason for this is the increased per unit cost of transportation with the rise in the distance to be covered to trade the products.

Per capita GDP of the importing country adversely affects food imports from India since a rise in the population of the economy will bring along greater demand for food, which in turn will force the economy to rely more on imports than what it used to earlier. Rising population, at the same time, will lead to falling in the per capita GDP. Hence, it may be natural to expect a negative relationship between per capita GDP and food imports.

The rise in food prices (probably because of inefficient production processes with high costs of production) will put the economy in a comparatively disadvantageous position making it more cost efficient to import the necessary food products rather than producing them inside the country. Either absolute or comparative advantage will be at work and push the economy towards importing food at cheaper prices and maximize the welfare of the nation.

The increase in food production may lead to increased food imports if the products produced within the country and the ones imported are not substitutes but complements of each other.

GDP of the importing country reflects the size of the nation and the potential demand of the economy. With higher purchasing power, the economy may choose to opt for more variety of food products and/or better quality of the same food products, necessitating and facilitating food import.

Higher trade openness may be attributable to the overall temperament of the economy towards new, better and/or cheaper food products. It is a characteristic of a more globalized economy which will automatically lead to a rise in the food imports as well.

For the ARDL model, the optimal number of lags for the endogenous variable, that is, food imports from India was found to be 2. Hence, the results are exactly the same as the VAR results reported above. As a result, the authors decided not to report them separately.

SOLUTIONS AND RECOMMENDATIONS

As shown through the analysis above, the augmented gravity model establishes the factors that are significant in determining agricultural trade flows.

Literature indicates, that while India features in the list of top-ten countries trading in agricultural commodities with the ASEAN nations, it still remains at the bottom of that list (ASEAN Statistics,

2015). India has high potential in making the ASEAN nations food secure but probably not through agricultural trade, alone.

The ASEAN nations, by tradition, have been involved in wide-spread extensive intra- and extra-regional trade in agricultural products. Since the ASEAN Economic Community (AEC) propagates "free flow of goods" as its principal tenet, trade facilitation through tariff reduction has been the key focus in this bloc – not only among the ASEAN nations but also with the non-ASEAN trade alliances. While tariff reductions have been on a steady decline (as indicated by the increase in the proportion of agricultural trade), the removal of non-tariff barriers (NTBs) is still not a success story. The quantification of the effects of NTBs and non-tariff measures (NTMs) are not easy. The costs arising from such sources are directly or indirectly captured as "trade costs" under the distance variable in the gravity equation. Although, the above analysis undertaken does not consider tariff rates but it does throw light on some of the reasons why agricultural trade is sluggish between India and the ASEAN. Based on the significant variables, some of the following recommendations can be made.

Since most of India and ASEAN are oil-importing nations, keeping food prices under check is a challenge for both and will have to be achieved by primarily reducing transportation costs – a factor embodied within the distance variable of the gravity equation of trade.

Barring Lao PDR, ASEAN countries and India have large coastlines, with many ports. However, there are large structural differentials in the shipping and maritime infrastructure within ASEAN and also between India and ASEAN. Within the ASEAN economies, Cambodia, Laos, Myanmar, Vietnam (CLMV) fall far behind when compared to other ASEAN countries and India, in terms of cargo volume and shipping tonnage. Predominant ports in ASEAN are located in developing parts of ASEAN and have been growing faster than those located in CLMV or the Mekong sub-region. In the containerized cargo segment, out of world's top 25 ports, seven ports come from ASEAN and only one from India. Singapore Port ranks second in the world in terms of efficiency and infrastructure whereas India's key port, Jawaharlal Nehru Port, is only in the 21st position worldwide. Also, the Jawaharlal Nehru Port is on the western coast of India, which means that the cargo has to travel additionally from west to east. If the eastern ports are made as efficient, India will cut down on trading costs, by far.

Pan-Asian trade integration is possible through better maritime connectivity. Well-connected ports and feeder routes can substantially improve trade between ASEAN and India. While only a few ports are connected directly by shipping lines in this trading bloc, a lot of resources is being allocated to reach ports in Cambodia, Indonesia, Malaysia, Myanmar, and Thailand with whom India's trade has been growing steadily. The existing route through Malacca Strait is heavily congested and also relatively unsafe. As an alternative, economic corridor based multimodal connectivity such as Mekong-India Economic Corridor will connect the Indian coastline with Southeast Asian coastlines in a shorter time and at a lesser cost. Other numerous infrastructural projects also serve to tie India closer to East Asia. India is participating in the United Nations Economic and Social Commission for Asia and the Pacific and the Pacific initiatives for an Asian Highway Network and the Trans-Asian Railway network. Discussions are also proceeding on reopening the World War II-era Stilwell Road linking the India's state of Assam with China's Yunnan province through Myanmar, under the Kaladan Multi-modal Transit Transport project. India should also explore developing the Dawei port and SEZ project in Myanmar under a joint venture. Cooperation in terms of financial credit and infrastructural support would also go a long way in enhancing trade volumes and reducing costs.

Simplifying and standardizing customs and trade logistics will improve the trade competitiveness and integration. Customs Single Window and such like should be implemented urgently.

Poverty reduction and capacity building will also increase labor productivity. Proper healthcare and education augment food utilization and bring about stability in the labor force. India has competencies in research and development (R&D), education and healthcare, which can benefit several ASEAN countries. The nation can categorically help through technology transfers, R&D, augmented human capital by means of providing better healthcare and education and making investments focused specifically towards supporting sectors that will strengthen food security directly or indirectly. This kind of an arrangement will probably involve more of services (and some goods) trade initially, but that, in turn, will pave the way to increasing the pace for agricultural trade and food security between India and the ASEAN economies.

As for managing water crises, climate change and the onslaught of population increase and urbanization, joining hands with each other and finding solutions to integrate as a region, would go a long way for this trading union.

FUTURE RESEARCH DIRECTIONS

This chapter could not include the Corruption Index and import duties due to the lack of adequate data. The inclusion of these variables would have made a better impact on the analysis and findings. Instead of considering total food imports, if the import composition can be bifurcated, more robust results and better insights may have been obtained. Again, if the composition of bilateral investment between India and ASEAN could be disaggregated, then determining how much is apportioned to food and agriculture related infrastructure and development leading to better food security, may have been determined.

CONCLUSION

Emerging countries have played an important role in increasing global food availability in the past decade. While they have increased their agricultural production in order to meet growing demand, the fast increase in food supplies in these markets have raised concerns regarding social and environmental impacts. India and the ASEAN countries may be similar in many ways but as far as economic structure is concerned, these nations are not equals. Growth, infrastructure, and the determinants of the supply-side economy vary widely within this FTA. In order to move onwards sustained food security, measures to augment bilateral trade in agriculture through improved global value chains, port and road infrastructure, better storage and logistics for access to food, reduced tariffs, R&D and knowledge sharing to effectively manage the consequences of climate change will put India and ASEAN in good stead, in the future.

REFERENCES

Aker, J. C., & Lemtouni, A. (1999). A Framework for Assessing Food Security in the Face of Globalization: The Case of Morocco. *Agroalimentaria*, *8*, 13–26.

Applanaidu, S. D., Bakar, N. A., & Baharudin, A. H. (2014). An Econometric Analysis of Food Security and Related Macroeconomic Variables in Malaysia: A Vector Autoregressive Approach (VAR). *UMK Procedia*, *1*, 93–102. doi:10.1016/j.umkpro.2014.07.012

Arellano, M., & Bond, S. (1991). Some Tests of Specification for Panel Data: Monte Carlo Evidence and an Application to Employment Equations. *The Review of Economic Studies*, *58*(2), 277–297. doi:10.2307/2297968

ASEAN-India Relations. (2016). In *ASEAN India Progress & Prosperity*. Retrieved October 17, 2016, from http://www.mea.gov.in/aseanindia/20-years.htm

ASEAN Statistics. (2015). *Top Ten ASEAN Trade Partner Countries/Regions*. Retrieved December 13, 2016, from http://www.asean.org/storage/2016/01/statistic/table20_asof121Dec15.pdf

Asian Development Bank. (2016). *Agriculture and Food Security Issues in Asia and the Pacific*. Retrieved October 6, 2016, from https://www.adb.org/sectors/agriculture/issueshttps://www.adb.org/sectors/agriculture/issues

Barraclough, S., & Utting, P. (1987). *Food Security Trends and Prospects in Latin America*. Notre Dame, IN: The Hellen Kellogg Institute of International Studies.

Breitung, J. (2000). The Local Power of Some Unit Root Tests for Panel Data. In B. H. Baltagi (Ed.), *Nonstationary Panels, Panel Cointegration, and Dynamic Panels* (pp. 161–177). Amsterdam: Elsevier. doi:10.1016/S0731-9053(00)15006-6

Briones, R. M., Chavez, E. C., Durand-Morat, A., & Wailes, E. J. (2012). *Climate Change and Price Volatility: Can We Count on the ASEAN Plus Three Emergency Rice Reserve*. Manila: Asian Development Bank.

Choi, I. (2001). Unit Root Tests for Panel Data. *Journal of International Money and Finance*, *20*(2), 249–272. doi:10.1016/S0261-5606(00)00048-6

Choulamany, X. (n.d.). *Issues and Strategies to Assure Food Security in ASEAN: ASEAN Level Initiative on Food Security* [PowerPoint slides]. Retrieved September 12, 2016, from https://www.oecd.org/tad/events/Issues%20and%20Strategies%20to%20Assure%20Food%20Security%20in%20ASEAN%20-%20%20%20by%20Xaypradeth%20Update.pdf

Clay, E. J., Shaw, D. J., & Singer, H. W. (1987). *Poverty, Development, and Food: Essays in Honour of H.W. Singer on his 75th Birthday*. Houndmills, UK: Macmillan. doi:10.1007/978-1-349-09214-7

El Niño in Asia. (2016). In *The United Nations Office for the Coordination of Humanitarian Affairs*. Retrieved September 11, 2016, from http://www.unocha.org/el-nino-asia

Emerging Market Economy. (n.d.). In *InvestingAnswers*. Retrieved October 2, 2016, from http://www.investinganswers.com/financial-dictionary/world-markets/emerging-market-economy-1518

Food and Agriculture Organization of the United Nations. (2001). *The State of Food Insecurity in the World*. Rome: Food and Agriculture Organization of the United Nations.

Gillson, I., & Fouad, A. (2016). *Trade Policy and Food Security: Improving Access to Food in Developing Countries in the Wake of High World Prices*. Washington, DC: World Bank.

Gouel, C., & Jean, S. (2013). Optimal Food Price Stabilization in a Small Open Developing Country. *The World Bank Economic Review*, *29*(1), 72–101. doi:10.1093/wber/lht018

Government of India. (2016). *State of Indian Agriculture 2015-16*. Retrieved October 10, 2016, from http://agricoop.nic.in/imagedefault/state_agri_1516.pdf

Greene, W. H. (2008). *Econometric Analysis*. New York, NY: Prentice Hall.

Hadri, K. (2000). Testing for Stationarity in Heterogeneous Panel Data. *The Econometrics Journal*, *3*(2), 148–161. doi:10.1111/1368-423X.00043

Holt-Giménez, E. (2008). *The World Food Crisis: What Is behind It and What We Can Do?* Oakland, CA: Institute for Food and Development Policy.

Huh, H., & Park, C. (2013). *Examining the Determinants of Food Prices in Developing Asia*. Manila: Asian Development Bank.

Hurlin, C., & Dumitrescu, E. (2012). *Testing for Granger Non-causality in Heterogeneous Panels*. Retrieved November 17, 2016, from https://halshs.archives-ouvertes.fr/halshs-00224434/document

Hurlin, C., & Venet, B. (2001). *Granger Causality Tests in Panel Data Models with Fixed Coefficients*. Paris: EURIsCO.

Im, K. S., Pesaran, M. H., & Shin, Y. (2003). Testing for Unit Roots in Heterogenous Panels. *Journal of Econometrics*, *115*(1), 53–74. doi:10.1016/S0304-4076(03)00092-7

Jha, S., Kubo, K., & Ramaswami, B. (2013). *International Trade and Risk Sharing in the Global Rice Market: The Impact of Foreign and Domestic Supply Shocks*. Manila: Asian Development Bank.

Lee, H., & Park, C. (2013). *International Transmission of Food Prices and Volatilities: A Panel Analysis*. Manila: Asian Development Bank.

Levin, A., Lin, C. F., & Chu, C.-S. J. (2002). Unit Root Tests in Panel Data: Asymptotic and Finite Sample Properties. *Journal of Applied Econometrics*, *108*(1), 1–22. doi:10.1016/S0304-4076(01)00098-7

Lipton, M., & Heald, C. (1984). *African Food Strategies and the EEC's Role: An Interim Review* (1st ed.). Brussels: Centre for European Policy Studies.

Masudul Haq, M., Kaitibie, S., Huda, S., Vock, N., Coughlan, K., Abdella, K., … Moustafa, A. (2014). *Using a Gravity Approach to Explain Food Imports for Improved Food Security in Qatar*. Paper presented at Qatar Foundation Annual Research Conference, Doha.

Maxwell, S. (1990). Food Security in Developing Countries: Issues and Options for the 1990s. *IDS Bulletin*, *21*(3), 2–13. doi:10.1111/j.1759-5436.1990.mp21003002.x

Mohamed Arshad, F., & Hameed, A.A.A. (2010). Global Food Prices: Implications for Food Security in Malaysia. *Journal of the Consumer Research and Resource Centre*, 21–38.

Morgan Stanley Capital International. (n.d.). *Market Classification*. Retrieved September 6, 2016, from https://www.msci.com/market-classification

Natale, F., Borrello, A., & Molotova, A. (2015). Analysis of the Determinants of International Seafood Trade Using a Gravity Model. *Marine Policy*, *60*, 98–106. doi:10.1016/j.marpol.2015.05.016

Overview. (n.d.). In *Association of Southeast Asian Nations*. Retrieved November 2, 2016, from http://asean.org/asean/about-asean/overview/

Pesaran, M. H., Shin, Y., & Smith, R. J. (1999). *Bounds Testing Approaches to the Analysis of Long-run Relationships*. Cambridge Working Papers in Economics, 9907. Cambridgeshire, UK: University of Cambridge.

Reutlinger, S., & Knapp, K. (1980). *Food Security in Food Deficit Countries* (1st ed.). Washington, DC: World Bank.

Salvatici, L. (2013). *The Gravity Model in International Trade*. Washington, DC: African Growth and Development Policy Modelling Consortium.

Sarker, R., & Jayasinghe, S. (2007). Regional Trade Agreements and Trade in Agri-food Products: Evidence for the European Union from Gravity Modelling Using Disaggregated Data. *Agricultural Economics*, *37*(1), 93–104. doi:10.1111/j.1574-0862.2007.00227.x

Sen, A. (1983). *Poverty and Famines: An Essay on Entitlement and Deprivation*. Oxford, UK: Oxford University Press. doi:10.1093/0198284632.001.0001

Siamwalla, A., & Valdes, A. (1980). Food Insecurity in Developing Countries. *Food Policy*, *5*(4), 258–272. doi:10.1016/0306-9192(80)90055-X

Starck, S. (2012). *The Theoretical Foundation of Gravity Modeling: What are the Developments that have Brought Gravity Modeling into Mainstream Economics?* (Unpublished master dissertation). Copenhagen Business School, Copenhagen, Denmark.

Tabor, S., Ginting, E., & Aji, P. (2015). *Preparing for El Niño: Policy Options*. Manila: Asian Development Bank.

The Economist. (2016). *The Global Food Security Index*. Retrieved September 10, 2016, from http://foodsecurityindex.eiu.com/

Tinbergen, J. (1962). *Shaping the World Economy: Suggestions for an International Economic Policy* (1st ed.). New York, NY: Twentieth Century Fund.

Van Dijk, M. (2011). *African Regional Integration: Implications for Food Security*. Retrieved October 18, 2016, from http://edepot.wur.nl/168845

Wang, J. (2016). Analysis and Comparison of the Factors Influencing Worldwide Four Kinds of Vegetable Oil Trade: Based on Gravity Model. *Modern Economy*, *7*(2), 173–182. doi:10.4236/me.2016.72019

World Food Programme. (2009). *Emergency Food Security Assessments (EFSAs). Technical Guidance Sheet No. 5*. Retrieved September 11, 2016, from http://documents.wfp.org/stellent/groups/public/documents/manual_guide_proced/wfp197295.pdf?_ga=1.244370689.508075766.1479356271

ADDITIONAL READING

Bajpaee, C. (2007). *India Rediscovers East Asia.* Retrieved October 14, 2016, from http://www.atimes.com/atimes/South_Asia/IJ31Df01.html

Engle, R. F., & Granger, C. W. J. (1987). Cointegration and Error Correction: Representation, Estimation, and Testing. *Econometrica, 55*(2), 251–276. doi:10.2307/1913236

Henneberry, S. R., & Carrasco, C. D. (2015). Food Security Issues: Concepts and the Role of Emerging Markets. In A. Schmitz, P. L. Kennedy, & T. G. Schmitz (Eds.), *Food Security in an Uncertain World* (pp. 63–79). Bingley: Emerald Group Publishing Limited. doi:10.1108/S1574-871520150000015005

Johansen, S. (1988). Statistical Analysis of Cointegrating Vectors. *Journal of Economic Dynamics & Control, 12*(2-3), 231–254. doi:10.1016/0165-1889(88)90041-3

Kapila, S. (2016). *India's 'Look East' and 'Act East' Policies Reviewed 2016.* Retrieved October 9, 2016, from http://www.southasiaanalysis.org/node/1924

MacKinnon, J. G., Haug, A., & Michelis, L. (1999). Numerical Distribution Functions of Likelihood Ratio Tests for Cointegration. *Journal of Applied Econometrics, 14*(5), 563–577. doi:10.1002/(SICI)1099-1255(199909/10)14:5<563::AID-JAE530>3.0.CO;2-R

Maddala, G. S., & Wu, S. (1999). A Comparative Study of Unit Root Tests with Panel Data and a New Simple Test. *Oxford Bulletin of Economics and Statistics, 61*(s1), 631–652. doi:10.1111/1468-0084.61.s1.13

Pedroni, P. (1999). Critical Values for Cointegration Tests in Heterogeneous Panels with Multiple Regressors. *Oxford Bulletin of Economics and Statistics, 61*(s1), 653–670. doi:10.1111/1468-0084.61.s1.14

Pedroni, P. (2004). Panel Cointegration: Asymptotic and Finite Sample Properties of Pooled Time Series Tests with an Application to the PPP Hypothesis. *Econometric Theory, 20*(03), 597–625. doi:10.1017/S0266466604203073

Wailes, E., Clarete, R., Briones, R., & Pochara, F. (2012). *Food Security in Asia: The 2007-2008 Food Price Crisis.* Manila: Asian Development Bank.

World Health Organization. (n.d.). *Micronutrients Database.* Retrieved November 1, 2016, from http://www.who.int/vmnis/database/en/

KEY TERMS AND DEFINITIONS

Agriculture: A cultivation of animals, plants and fungi for food, fiber, biofuel, medicinal plants, and other products used to sustain and enhance human life.

Association of Southeast Asian Nations: A regional organization comprising ten Southeast Asian states which promotes intergovernmental cooperation and facilitates economic integration amongst its members. Since its formation on August 8, 1967, by Indonesia, Malaysia, the Philippines, Singapore, and Thailand, the organization's membership has expanded to include Brunei, Cambodia, Laos, Myanmar (Burma), and Vietnam. Its principal aims include accelerating economic growth, social progress, and sociocultural evolution among its members, alongside the protection of regional stability and the provision of a mechanism for member countries to resolve differences peacefully.

Causality: It indicates that one event is the result of the occurrence of the other event, that is, there is a causal relationship between the two events. This is also referred to as cause and effect.

Food Access: It exists when all people, at all times, have physical, social and economic access to sufficient, safe and nutritious food, which meets their dietary needs and food preferences for an active and healthy life.

Food Availability: It refers to the supply of food through production, distribution, and exchange.

Free Trade Area: The geographical expanse encompassing the member countries or a trade bloc who have signed a free-trade agreement. Such agreements involve cooperation between at least two countries to reduce trade barriers – import quotas and tariffs – and to increase trade of goods and services with each other.

Gravity Model: A model used to estimate the amount of interaction between two cities or countries. It is based on Newton's universal law of gravitation, which measured the attraction of two objects based on their mass and distance. The amount of interaction between two cities or countries is proportional to the size of the cities and inversely proportional to their distance squared.

Panel Data: This term is used in econometrics and statistics, for multi-dimensional data frequently involving measurements over time. Time series and cross-sectional data are special cases of panel data that are in one dimension only (one panel member or individual for the former and one-time point for the latter).

Stability: An absence of excessive fluctuations in the macroeconomy. An economy with fairly constant output growth and low and stable inflation would be considered economically stable. An economy with frequent large recessions, a pronounced business cycle, very high or variable inflation, or frequent financial crises would be considered economically unstable.

Trade Integration: A unification of economic policies between different states through the partial or full abolition of tariff and non-tariff restrictions on trade taking place among them prior to their integration. This is meant in turn to lead to lower prices for distributors and consumers with the goal of increasing the level of welfare while leading to an increase of economic productivity of the states.

Trade Liberalization: A removal or reduction of restrictions or barriers on the free exchange of goods between nations. This includes the removal or reduction of tariff obstacles, such as duties and surcharges, and nontariff obstacles, such as licensing rules, quotas, and other requirements. The easing or eradication of these restrictions is often referred to as promoting "free trade".

This research was previously published in Establishing Food Security and Alternatives to International Trade in Emerging Economies edited by Vasily Erokhin, pages 98-118, copyright year 2018 by Business Science Reference (an imprint of IGI Global).

Chapter 3
The Barriers to Innovation Diffusion:
The Case of GM Food in Sri Lanka

Dilupa Nakandala
Western Sydney University, Australia

Tim Turpin
Western Sydney University, Australia

ABSTRACT

Genetically Modified (GM) food has been positioned as a significant innovation with a huge potential for alleviating malnutrition in developing economies. Some potential beneficiaries, however, have been reluctant to accept GM food. Many countries have GM food regulations and some have banned GM organisms. This chapter focuses on barriers to diffusion of innovation and analyses the case of GM food diffusion in Sri Lanka using the Rogers's classical model of innovation diffusion. A complete ban on GM products in 2001 was later relaxed to demand only GM labelling regulations, but GM food has not gained a prominent position in the Sri Lankan market. The attributes of GM food perceived by consumers, the communication system, government responses and broader social expectations have been unfavorable to GM food diffusion. The case of GM food innovation in Sri Lanka demonstrates the very social nature of the process, involving far more than seed producers, growers and related commercial enterprises.

INTRODUCTION

Innovation diffusion is considered as a complex process of introducing change in a social system. The classical model introduced by Rogers (2003) identifies innovation diffusion as a type of communication and a process where innovation is communicated among members of a social system through certain channels. Traditional models identify four dominant elements in this process: an innovation; users; communication channels; and a social system. Importantly, many aspects of the social system including the social structure specifically the effects of norms, the roles of opinion leaders and change agents, decision types and consequences interact with the diffusion process in a complex manner. External influences such

DOI: 10.4018/978-1-5225-8063-8.ch003

as opinion leaders as change agents and surrogate users with the delegated authority to make decisions (Aggarwal & Cha, 1997), government policies and regulations (Feldstein & Glasgow, 2008) and social networks and linkages (Mendel *et al.* 2008) can positively or negatively affect the diffusion process.

This chapter focuses on the process of diffusion of genetically modified (GM) food into the Sri Lankan market. GM food carries the potential to contribute to a process of social innovation for ameliorating hunger in developing countries through the potential benefits of the possible introduction of new plant varieties, increased productivity, and reduced costs for pesticides and herbicides. However, social resistance to GM food has made its wide diffusion difficult (Gaskell et al. 2004). For example, the Sri Lankan government completely banned GM foods in 2001 but the external pressure from foreign governments, multinational GM food corporations and local chambers of commerce delayed the administrative introduction of the ban. Subsequently, the regulation for food Importation & Labelling of GM was enacted in Sri Lanka in 2006. In addition, the National Biotechnology Policy 2009 explicitly required that safety test data manipulations and applications should be fully disclosed and made available for public scrutiny. At present there is apparently no official data on GM organisms neither produced in Sri Lanka nor imported to Sri Lanka; there have not been any food items with a GM label in the Sri Lankan market. It has been reported that some imported food items have been tested for GMOs, but no official results have been released (Hirimuthugodage, 2013).

In this chapter the introduction of GM food in Sri Lanka and the social process surrounding it are analyzed according to Rogers's classical model of innovation diffusion. A key observation is that the barriers for GM food adoption have been of multiple dimensions. Consumers have not been convinced of the relative advantage of GM food due to the scientific uncertainty associated with GM food products. Moreover, the technological complexity of GM technologies is not easily understood by consumers. The general appreciation of natural forms of production, based on traditional agricultural and health systems, stands in contrast to negative perceptions of 'artificial' products. In Sri Lanka there is a strong presence of knowledge stocks in traditional agriculture and health practices that tend to generate skepticism about genetically modified organisms. There is a very real concern that buyer resistance in developed economies will eventually undermine its potential export markets. Government action, partly driven by similar concerns, sought to introduce the ban on GM food types as a protective step (Brankov & Lovre 2013). Not surprisingly the Sri Lankan government action increased uncertainty drawing out media reports emphasizing negative aspects of GM food and GM technologies. As discussed below, it is the breadth and depth of societal input to the process of introducing and accepting (or otherwise) this agricultural process and product innovation that generates an innovation 'bottleneck'. Moreover, as the Sri Lankan case demonstrates, it is not just local social input that is pertinent. Rather, it is an international process that engages with science, global food production chains, governments, the media, producers and consumers.

BACKGROUND

Since the first GM crop entered the market in 1996 there has been considerable debate and controversy about whether and how governments should regulate the introduction of GM crops. On the one hand, seed developers have argued that the introduction of 'nutritionally enhanced crops' carries huge potential for alleviating malnutrition in developing economies (see Anderson, 2011). At the same time some countries, for example, Zambia, Poland, France and Greece, have banned genetically modified organisms (Hirimuthugodage, 2013). The argument that has generally persuaded governments to act in this direction has been concerns that growers and consumers fear potential adverse environmental and health impact (Nielson, 2011; Skevas *et al.*, 2012). Other countries such as the U.S., China and Canada have been more encouraging arguing that consumers in these countries are indeed willing to pay a premium for genetically modified food products (Curtis et al. 2004). In short, although the social perspective varies in different parts of the world there is a normative debate underway between (1) seed developers (innovators); (2) farmers (as consumers and producers); (3) the general population (as consumers); and governments (as regulators). There are also non-governmental organizations and various action groups engaged in one way or another in the debate. Effective and reliable communication channels are important to enable the debate to proceed.

While there is still no clear-cut evidence regarding public health risks or potential transfer of traits to wild plants or non-target organisms much of the opposition to GMO is based on a precautionary principle. For many countries, in both the developed and developing world, policy caution has prevailed. Yet, the innovation process for producing GM seeds and GM crops has grown rapidly over two decades. By 2011, there were estimated to be 159 million hectares under GM cultivation (Hirimuthugodage 2013). While North America accounts for over 70% of sown GM crops, Argentina China and India are rapidly expanding their production. The GM food debate is not just a just a dialogue between the North and the South.

Moreover, nor is it simply a seed producer versus 'cropper' debate moderated through IPR legislation. Although lawsuits in the U.S. between seed producers and farmers have been acrimonious and delivered over USD23 million to Monsanto alone between 2005 and 2012, the evidence appears to show that on average 2/3 of economic return from GM crops are enjoyed by farmers and consumers while 1/3 by the seed developers (Skevas *et al.*, 2012; Klumper & Qaim, 2014). And in Greece, where GM crops are banned a recent survey of growers indicated that a considerable majority (61%) would cultivate the GM seed if the ban were to be lifted.

METHODOLOGY

The method used in this study is based on a literature review. The relevant journal articles and book chapters as well as news articles available online and the articles available on non-commercial websites have been analyzed. GM food has created international controversy. This has partly been due to varying responses from governments, the information made available by seed producers, reports by pressure groups and newspaper articles commenting on the actions of the governments. The international dimension has been extended by WTO and multinational organizations and environmental pressure groups. The keywords of GM food and Sri Lanka, GM ban and Sri Lanka, GM labelling and Sri Lanka were used to identify the literature related to the introduction of GM in Sri Lanka in order to trace the diffusion process over time. The Rogers's (2003) model if innovation diffusion was used as a framework for analysis of the Sri Lankan GM adoption case. In a similar approach perceived attributes of innovation as factors influencing innovation diffusion have been used in studies by Wagner and Walchl (2002) and Islam (2014).

Media reports from Sri Lanka and government responses to GM food serve to document the breadth and depth of 'social negotiation' around the introduction, production, marketing and consumption of GM product. While there is a need to conduct a more comprehensive study using empirical data to analyze the innovation diffusion using primary data from multi-stakeholder perspectives the present study serves to focus attention on the innovation bottleneck. This bottleneck, we argue, draws attention to the intrinsic social nature of innovation.

UNDERSTANDING THE GM INNOVATION BOTTLENECK

The evidence overall suggests that the main barrier to GM seed cultivation is a precautionary political concern by governments regarding public perceptions about health and environmental hazards (Paarlberg, 2002). This is not to argue that government opposition is inappropriate, rather it is simply to recognize where the innovation barrier lies. Future changes are likely to rest on the progress of the 'normative debate' about what is socially and environmentally and medically acceptable. The normative debate, as articulated by Mary Douglas (1982) refers to the process of on-going social debate about 'social responsibility'; about what is 'right' or 'moral'. In short, it is the on-going debate about perceptions of the social world in which we live. The normative debate is a major factor in giving rise to social change. But by its very nature, however, such change can be a slow process. Scientific and social debate about climate change has carried important implications for technological innovation. As a consequence technologies such as electric motor cars that failed to progress through the early twentieth century are now at the forefront of R&D and marketing by auto producers in the early twenty first century. In many countries the production and consumption of GM crops and food products is mediated by a similar social process.

Much of the recent debate about GM crops and food products has moved to increasing the capacity of staple crops to feed and nourish consumers in developing economies (Nielson & Anderson, 2011; Anderon, 2011; Ismail et al., 2013). In this context, resistance to GM crops still remains high. On the one hand, while increased production, lower cost and increased nutritional value may benefit domestic markets producers (and governments) fear that potential exports might be rejected if they contain GM product. The debate surrounding the development of 'golden rice' illustrates the complex interaction between science, the public and state perceptions about health and environmental concerns (Nielson & Anderson, 2011). Yet, globally GM innovation persists and in particular in developing countries. As Hirimuthugodage (2013:1) has provocatively put it: "… GM crops are the fastest adopted crop technology in the history of modern agriculture".

In the Philippines development of flood resistant 'Scuba Rice' developed through the cloning of the Sub1 gene and subsequent marker-assisted backcrossing has led to a variety of flood resistant varieties. In 2009, the Philippine National Seed Industry Council approved the release of 'scuba' rice, known locally as 'Submarino rice', with the Philippine Rice Research Institute (PhilRice) acting as distributor (Ismail et al. 2013). The seed has now been sown extensively in India, Bangladesh and Nepal. In Bangladesh it has been described as a 'magic' or 'wonder' variety (Ismail et al., 2013: 89). In pilot farms in the Philippines, this variety has been observed to survive floods with a good yield and less fertilizer use than before, since farmlands receive nutrients from the silt brought by floods. Critics, however, contest this point. They argue that Submarino rice requires 'a high amount of input of chemical fertilizer and pesticides' and that it is therefore 'not affordable by the majority of poor farmers.' They prefer to endorse alternative growing methods, such as the System of Rice Intensification (see Turpin, et al. forthcoming). Opposition to scuba-rice again illustrates the strength and nature of debate with claim and counter claim from growers and scientists with an interest in the innovation process.

Clearly national regulatory systems are needed. They are needed in order to maintain control over the import GM with appropriate scientific assessment. Ismail et al. commenting on the rapid and successful distribution of scuba rice have drawn attention to the engagement of 'lead farmers' together with the support of scientists and agricultural workers which has 'significantly shortened the time required for evaluation and release of new varieties' (Ismail et al., 2013:91) But, from a consumer perspective, there is also a need for regulation to ensure appropriate labelling requirements. Again, this is an issue of communication channels within the social system that can inform the normative debate.

A robust IPR regime is likely to encourage investment in seed development, however, the rapid distribution and uptake of GM food in developing economies suggests this is not a major barrier. . Ultimately, the viability of GM technology as a domestic or export product will rest on public perceptions. This will require sound and reliable scientific evidence that clearly sets out potential advantages and costs of GM food production (a sound communication channel). To this end there is an important role for local and international research investment as well as national authorities to distribute easily accessible and comprehensive information on the topic.

In a country such as Sri Lanka where, GM is not banned, but there are scientific assessment and labelling requirement on all imported GM product (including seed) the critical issue is information and public perception. Scientific capacity to assess the health and environmental impact of new varieties needs to be in place, but perhaps more importantly assessments need to be communicated in a way that that is readily accessible to the general public. Labelling is one part of the communication channel.

In summary, the literature concerning the uptake of GM food in developing countries suggests that the innovation diffusion bottleneck does not rest on any technological or production barrier, nor is it simply an IPR issue. Rather it is a social issue engaging a complex array of interest groups. Negotiation at and around the bottleneck is mediated through public debate and ongoing discourse. Communication channels that are open and easily accessible are likely to be critical in enabling innovation in this area to proceed. GM food products have been discussed widely in academic and other publications. However, they provide "different and often contradictory interpretations of the issues surrounding the diffusion of genetically engineered products" (Weick & Walchli 2002:267). Sri Lanka as a country which initially introduced a total ban for GM food followed by labelling regulations provides an interesting case for analysis in order to understand what has caused the resistance and rigid responses for adoption of an innovation marketed with an upside potential for developing economies.

GM FOOD: THE SRI LANKAN JOURNEY

Various countries have adopted diverse approaches to respond to the inflow of GM seed and food products. In the early 2000s; labelling rules in the form of voluntary, mandatory, or mandatory above some tolerance level were introduced by many countries. The underlying logic appears to have been one that warns the consumer to exercise some potential caution. According to Phillips & McNeill (2000), 28 countries and the European Union either adopted or had plans to introduce labels for GM food products. Of those countries that undertook some action along these lines, Sri Lanka and Brazil planned to ban GM food products without opting for labelling rules. While Brazil adopted a ban to end in 2001 with a proposal for labelling rules for products containing more than 4% GM content, Sri Lanka banned production or imports of GM products without a clear indication of a potential for a replacement of the ban with labelling rules.

Sri Lanka's Health Ministry imposed import restrictions in 2001 requiring identified categories of food imports to be completely free of GM products. Through the gazette no 1178/18 of 06.04.2001, regulations were introduced to ban the import, manufacture for commercial purposes, transport, storage, distribution, sale and offering for sale any raw or processed food, food ingredient or food additive that has been subjected to any genetic modification or that contains any such ingredients that have been subjected any such genetic modification. It clearly listed 21 types of products that should not be without a certificate from an accredited laboratory or Competent Authority to show that the product does not contain any material or ingredient subjected to genetic modification. The schedule consisted of products such as soybeans, corn/maize, tomato, cheese, potato, yeast, beet sugar and microbiological starter cultures used in food. These regulations were dated to come into effect from 01/05/2001. However, it was immediately retracted by a regulation published in the gazette no 1190/5 of 26.06.2001 and the enforcement was postponed to effect from 01/09/2001. Then an indefinite suspension for the operation of the ban was published in gazette no 1199/23 of 0.08.2001.

The factors that influenced the original decision to ban GM food in Sri Lanka were not explicitly stated. However, according to the South-North Development Monitor (2000), the National Food Advisory Committee had studied the possibility of imposing an import ban due to the lack of precise information about their long-term effects of GM food. The Sri Lankan authorities had in parallel been considering less trade-restrictive options. The swift steps of a total ban by the government were apparently based on what was described as 'scientific uncertainty' and a response to the then growing resistance towards GM food from pressure groups. Following the initial introduction of the ban, the inadequacy of the list of food items was criticized by the Green Party of Sri Lanka; they urged the Ministry of Health to reveal the brand information of GM food items (Cooray 2002). The debate was thus moving into the political domain.

According to the local press reports, the Sri Lankan government's decision to ban genetically modified food generated an explicit critique from US authorities. For example, the government's proposed ban had been heavily criticized by the then Agricultural Counsellor for the US Embassy in New Delhi on the grounds that there was no scientific evidence to justify the government's decision (Seneviratne, 2001). Even though there was evidence challenging these reports the threats from the foreign GM producers became widely known. The Green Party of Sri Lanka drew attention to what they described as attempted lobbying from some foreign diplomatic offices and food exporters for the removal of the GM ban regulation (Cooray 2002). The Centre of Environmental Justice stated that some western governments including the United States, Australia and New Zealand, had threatened the Sri Lankan government that '... if the regulations come into effect it will be taken to the World Trade Organization compliance mechanism. Labelling regulations prepared by the Food Advisory Committee in 2002 were again put on hold due to this pressure' (Withanage 2007:2). The attempts to regulate any GMOs and products has apparently been suspended upon a request by the World Trade Organizations (Baumüller 2003; Krimsky & Gruber 2013). By this time the issues and influencing factors for the diffusion of GM food had become very much an international debate.

The government position in completely banning GM food in 2001 was later clarified in the National Biosafety Framework of Sri Lanka (2005) which stated that the National Policy on Biosafety is guided by the principle that ' ... if there is any perceived threat of serious or irreversible damage, lack of scientific certainty shall not be used as a reason for postponing measures to prevent environmental degradation and health impacts' (Ministry of Environment and Natural Resources 2005: 12). By this time the measures adopted by the Ministry of Health regarding the import of GMOs and products in 2001 were identified as a 'Precautionary Approach' considering the fact that there was no certainty of the safety of GM food types. Considering the limited availability of technology for testing and assessment locally, the shift of responsibility to the importer producer to provide the proof of evidence of no containing GM materials or ingredients subjected genetic modifications appears to have been a pragmatic step chosen by the government.

In 2002, a Special Committee of the Ministry of Health submitted a report on genetically modified food that fed into the draft regulations that were made available in January 2003. These draft regulations intended to bring a procedure of approval for all GM-foods considered on a case by case basis and a system of compulsory labelling for consumer awareness. The National Biosafety Framework (2005) required that labelling of GM products was to be mandatory: ' … all products resulting from modern biotechnology, at all stages of the production process, from raw material to final product' (Ministry of Environment and Natural Resources 2005: 14) are required to have compulsory labelling enabling consumers to make informed food choices for purchasing. In implementing these measures, the framework explicitly drew attention to the Consumer Affairs Authority Act, No 9 of 2003. This Act had provisions that could be utilized for issuing directions to manufacturers requiring them to label all goods with GM ingredients. Such directions could be made formal through publication in the gazette as regulations.

Following an application filed by the Centre for Environmental Justice in 2005, the Sri Lankan Government gazetted GM food labelling legislation (Gazette Ex. Ordinary. 1456/22, 3rd August 2006) (Withanage 2007).The government passed legislation to this effect in section 32 of the Food Act No. 26 of 1980, allowing GM food importation from 2006. The only condition attached to this was that an approval was needed from the Chief Food Authority for importing, storing, transporting, distributing and, selling of ' … any genetically modified organism as food for human consumption; any food containing or consisting of genetically modified organisms; or any food produced from or containing ingredients produced from genetically modified organisms' (p.1A). These regulations for the Control of Import, Labelling and Sale of Genetically Modified Foods came into operation on 1 January 2007. From this date all approved products were required to have a label stating that they consisted of genetically modified ingredients. The information requirement of the label was to refer to any genetically modified food or food ingredients used in the preparation of the good and include the statement 'genetically modified' in conjunction with the name of that food or ingredients used in the preparation or processing aid (p2A). The government move away from an outright ban and toward labeling served to satisfy international demands for free trade. This firmly placed the decision to accept or not accept GM seed or food products onto farmers and consumers. To some extent the government was thus 'let of the hook'. Nevertheless, debate continued amid demands for more scientific evidence and for improved crop production, not only in Sri Lanka, but globally.

The Government Position Between International Seed Producers and Local Growers

Biotechnology Policy (National Science Foundation & National Science and Technology Commission, 2009) identified the need to enhance opportunities for increasing food production through the use of biotechnology in the agricultural sector. It also called for fuller information disclosure related to the industries involved in the use of biotechnology. Their demand was that safety test data for GM organisms and human genetic testing, manipulations and applications should be fully disclosed and made available for public scrutiny (2009:12). In promoting industrial biotechnology, it identified the potential extensive use of GM organisms in bio-refineries producing biofuels and bio-chemicals. In 2011, the National Policy on Biosafety was established and that covered the import of Genetically Modified Organisms.

When the recently suggested Seed and Planting Material Act was published on the website of the Ministry of Agriculture for public comments and feedback, it was read by some as a timely move reflecting the current needs of the Sri Lankan agricultural industry and addressing the need for a compensation system for farmers (Hirimuthugodage 2014). On the other hand others argued against what they saw as the compulsory registration of farmers and the need for the certification of seed and planting. From this perspective the new law was seen as an attempt to open up the local industry for international corporations promoting genetically modified seeds. It was opposed by the Movement for National Land and Agricultural Reform on that basis (Kurukulasooriya 2014; Ho 2015). In response to these public criticisms the Department of Agriculture issued a media release clarifying the scope of the proposed new seed law (2014). The argument presented in this media report was that the fear that the law was essentially an attempt to promote GM seed popularization was unfounded and that the feared potential loss of protection for indigenous seed varieties was simply a product of social resistance towards commercialization of GM technologies within the country.

According to Mohamed and Zoysa (2006), the Tea Research Institute of Sri Lanka was unable to use genetic engineering in their breeding programs due to the negative acceptance of such technologies among local users. Genetic engineering had already been adopted by non-food industries such as rubber. The Rubber Research Institute reports their research involving genetic engineering including multiple projects such as 'developing elite cultivars of rubber through tissue culture and genetic engineering' and 'producing new genotypes to obtain new genetic material for the future trials to develop vigorous, high yielding and disease resistant clones' (2015). In the case of food however, the concern about potential customer resistance stood as a counter weight to potential higher yields.

ANALYSING GM FOOD ADOPTION USING ROGERS'S (2003) INNOVATION DIFFUSION FRAMEWORK

Rogers' classic model of the diffusion of innovation provides a useful framework for analyzing the case of GM food adoption in Sri Lanka. According to Rogers (2003), end users have been at the core of existing models where users make final decisions as to whether or not to adopt an innovation. There are external influencers such as opinion leaders and also the socio-economic factors, but one of the major assumptions of existing models is that the end user makes the final decision on adoption in a social context. Roger (2003) has argued that one of the important variables affecting the rate of adoption of an innovation is the perceived attributes of innovation. Such attributes include: relative advantage; compatibility; complexity; trialability; and observability. Other influential variables include: type of innovation-decision; the nature of communication channels diffusing the innovation at various stages in the innovation decision process; the nature of the social system; and the extent of change agents' promotion efforts in diffusing the innovation (p.232). Rogers's model provides a useful framework to analyse the adoption of GM food in Sri Lanka and allows for some useful insights to emerge that account for the nature and stricture of the innovation bottleneck.

The relative advantage of an innovation refers to whether it is perceived to be better than the product it supersedes (Rogers 2003). In the absence of the weakly convincing ability to produce economic benefits for end users, the perceived benefits of GM food compared to other food types that are produced using traditional methods without using any materials or ingredients subjected to genetic modification have not been strongly positive. Considering the near zero tolerance of health risks or food safety for many consumers, the scientific uncertainty of GM food seems to leave them less convinced of the benefits of GM food. As Krimsky and Gruber (2014) have noted since Americans have been consuming large amounts of GM food does not indicate that GM food is highly nutritional unless the other changes in the crop have not taken place is known and clear. Rogers (2003) describes compatibility of an innovation as the ' … degree to which an innovation is perceived as consistent with the existing values, past experiences, and needs of potential adopters' (:16). When innovation is more compatible with an existing value system, existing knowledge bases and user needs then they are more adoptable by users. Weick and Walchli (2002) have proposed that one of the reasons for the reluctance of European consumers to adopt GM food without caution is based on a European notion of *ethics* in the social system. When the demand for basic food types is not mountainous the influence of the variables such as ethicalness becomes more powerful; Blurred clarity of compatibility of GM food with the existing food as a part of the culture sees diffusion as more difficult. While scientific evidence is essentially available globally (providing there is access to such information) responses at a national level can be variable. These differences may depend on context issues such as food scarcity or abundance. Yet even within an economic bloc such as the EU there are differing national responses.

Sri Lanka has traditionally been an agricultural economy and food has been a part of the Sri Lankan identity. With fresh fruit and vegetables are sold in Sunday markets and only a few supermarket chains, there is a high consumer tendency for natural local food production. The social inclination for living in harmony with the nature and the influence of existing knowledge stocks such as traditional farming practices in agriculture and the Ayurvedic medicine practices in health present the difficulty in absorbing all externally available new knowledge without appropriate social assessment. The high interconnectedness of the social system pushes collective decision making by consumers where the effectiveness of communication systems depends on the capability to change of attitude of mass consumer community towards products. In the case of GM food in Sri Lanka, the media reports have been primarily based on the scientific uncertainty and the risk of negative consequences on health and environment raised by the pressure groups persuading consumers against GM food.

Figure 1. Barriers to diffusion of GM food in Sri Lanka

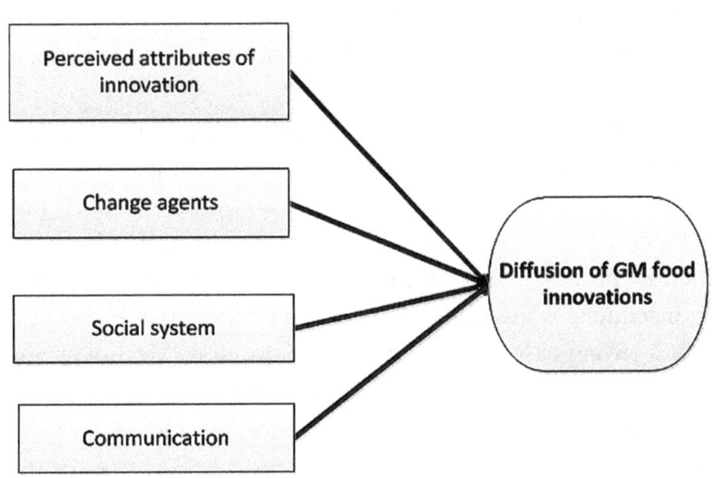

Table 1. Barriers to diffusion of GM food in Sri Lanka

Barriers to Diffusion	Detailed Description of Barriers
Perceived Attributes of Innovation	Scientific uncertainty of effects of GM food consumption Perceived human health risks Uncertainty with long term consequences Environmental risk Low economic benefits for end consumers Benefits are less observable to the end users High complexity of biotechnology applications Low compatibility with Socio-cultural values
Change Agents	Complexity with the food control authority structure The low reliance on local food controlling agencies Lack of food testing technology
Social System	Tendency towards natural food production Existing strong agricultural knowledge
Communication	Media reports demoting the GM food based on scientific uncertainty

The argument put forward for GM Food is that the benefits of new crop varieties with higher yield, increased productivity and less need for pesticides for producers and less environmental pollution through reduced use of chemicals is ultimately a solution to ameliorate hunger from the world (Gaskell t al. 2004). However, the application of bio-technology in agriculture has had many critics. For example, according to Null and Gale (2014), the original foundation for the use of biotechnology in agriculture was based on the idea of making crops immune to toxic chemical spraying. They state that the use of glyphosate and other toxic pesticides and herbicides have increased with the increasing amount of GM food in the market; consequently, the risks to human health, wild life and environment have also been increasing. Similar problems created by the use of glyphosate by farmers in Sri Lanka have been the subject of recent attention. According to Jayasumana *et al.* (2014), a chronic kidney disease of unknown etiology (CKDU) was discovered among rice paddy farmers in the North Central Province of Sri Lanka in mid 1990s. Apparently the disease has spread rapidly to the other farming areas of the country. According to Jayasumana *et al.* glyphosate could destroy the renal tissues of thousands of farmers when it forms complexes with a localized geo environmental factor (hardness) and nephrotoxic metals. In a similar vein, Krimsky and Gruber (2014) have stated that ' …GMO soy contains low level of isoflavnes, GMO canola contains low level of vitamin E and GMO insecticidal rice contains high levels of sucrose, mannitol and glutamic acid than its non-genetically modified counterparts' (need a page number if we are going to quote).

The high degree of complexity associated with biotechnology and low degree of trialability and observability considering the unavailability of scientific evidence of food safety and environmental effects in terms of long term consequences, GM food as an innovation has not been positioned as a powerful innovation convincing the end consumers. As Rogers (2003) has pointed out: diffusion of innovation is an 'uncertainty-reduction process'. Unless the differences between GM food and their non-GM counterparts and their consequences are understood by consumers perceived attributes of GM food by consumers do not present a strong case for a high rate of adoption. Moreover, the claims on high yield crops and less use of pesticides have not been translated into benefits for consumers of GM food. Hence, the superiority of GM food over the food types produced through traditional farming methods have not been visible for end users.

The level of diffusion of an innovation depends on the number of units of decision. As Rogers (2003) explained, when more people are involved in the decision-making process the adoption rate gets lower. He cited a case from the United States where the decision to adopt fluoridation of municipal water supplies is made by a mayor or city manager, the rate of adoption is quicker than when the decision is made collectively by a public referendum. For GM food, decisions of using a product of a technologically complex process are not simple. With the objectives of health and safety of the public, governments tend to impose regulations. As Weick and Walchli (2004) correctly pointed out different countries have different level of trust towards the public institutions. While US agencies involved in regulating GM food products had earned the trust of Americans, the European agencies had not been able to win the trust of Europeans overall. According to Weick and Walchli, the cynicism of the public had been based on the negative experiences such as deadly experience with mad cow disease and contamination of animal food etc.

The food regulatory agencies in Sri Lanka experience resource problems including a lack of technology for food testing and assessment and a lack of qualified human capital with appropriate expertise. In 2013, the government ordered an international milk powder supplier to recall their products from the Sri Lankan market based on the test reports made by the Industrial Technology Institute claiming the samples of Fonterra milk powder from the New Zealand dairy producer Fonterra had the fertilizer additive dicyandiamide, known as DCD in some of the samples. An analytical chemistry expert later reportedly condemned the results of the tests carried out by the Industrial Technology Institute (ITI) claiming that the preferred method for the detection of DCD is Tandem Mass Spectrometry and the methods used by the local institute are in appropriate. ITI maintained that the technique of UV detection used by them for the assessment of DCD presence was well founded (Senanayake 2013). In handling this case, there have been questions on the role played by the Food Advisory Committee (Hettiarachchi et al. 2013). Multi agencies operate in the food control and safety domain including the Food Advisory Committee, Health Ministry and Consumer Affairs Authority, Industrial Technology Institute. Even though this case did not involve genetic engineering, such complexity in handling the issue, multiagency structure and lack of credibility of the food assessment capabilities including the possession of required technologies is likely to leave the public with scant regard for the local food control agencies. In a position paper from Transparency International Sri Lanka (2013) it was argued that there was an urgent need for an independent Food Authority by statute dedicated to food control administration in Sri Lanka.

In summary, as shown in Figure 1 and Table 1, primary barriers to the diffusion of GM food can be multi-dimensional. Although GM food has been positioned as a social innovation to help alleviate hunger from the world there has been little success in defining the major beneficiaries; whether they are seed producers, farmers or consumers. The social process is clearly complex and there is a wide range of interest groups involved arguing the case either for or against, or offering solutions to the issues raised through the international and local debates. For diffusion to progress information requirements need to be better understood along with the complexity of social system including social values and the decision factors that affect decisions about use (Rogers 2003).

DISCUSSION

The entry of GM food into the Sri Lankan market was not supported with necessary instruments for increased consumer awareness to overcome the resistance caused by skepticism. Consequently, the introduction of GM food carried insufficient information to assure consumers that they are safe for consumption. Nor was there assurance for farmers that potential GM crops would be accepted in international markets. The findings of a study by Senarath and Karunagoda (2012) revealed that there was a lack of awareness about GM food among the study participants and there was much interest in knowing more about GM food. Interestingly, they found that the perception of the survey participants towards GM food as strongly negative. While the need for GM food labelling was emphasized among the participants, the implications of labelling are limited as consumers are found to be reading the labels on food items for limited information such as the dates of manufacture and expiry but unlikely to know about the ingredients that indicate the use of GM engineering.

Previous research has indicated that the consumer attitudes are influenced by the specific goals of the use of genetic engineering. For example, if genetic engineering is used for increasing the nutritional content in a food crop (Powell and Griffiths 1993), or when there are concrete befits for human health (Heijs et al. 1993) then public attitudes are likely to change. However for Sri Lankans, GM food is perceived to be risky for human health. Public awareness is likely to increase through media and product exposure. While some have identified that the general public has a little understanding of GM foods there is nevertheless a perception that GM foods as risky to human health (Hirimuthugodage 2013). Others have revealed that there is a growing increase in awareness about GM food among the general public in Sri Lanka. For example, a recent empirical study conducted with respondents from the general public, academics and researchers in Sri Lanka, Sajiwani and Rathnayaka (2014) found that the average awareness of the participants on GM food was high at 96% and the majority of respondents have received information about GM foods from television and radio media.

It has also been found that the attitude towards GM food is influenced by the perceived knowledge about genetic engineering (Frewer et al. 2013). From an empirical study, they found that there have been negative attitudes based on 'unnaturalness' as a determinant of concern as well as ethical concerns of genetic manipulation of animals or human DNA. They also noted that such concerns are more eased with plants or microorganisms. For food related applications, the need of genetic engineering, especially for modifications of the agricultural process or production process has been perceived more than its application for the modification of product characteristics. They explain this behavior by referring to the need for "crops that can grow in dry conditions" and "strawberries that can grow in frosty conditions". Sajiwani and Rathnayaka (2014) have found that the resistance for accepting GM food products by consumers in Sri Lanka has been due to the fear of side effects. They also found that academics, researchers and university students have shown a high level of interest on GM food indicating the access to new knowledge and awareness of new technologies reduce the resistance for GM food, Hence, there is a need for increased awareness among the general public of health risks associated with GM food in order to dilute the fear of long-term health effects from the consumption of GM food.

As with international experiences concerning issues such as climate change, debates continue to flourish for a considerable amount of time. However, over time the argument moves forward. Positions ameliorate, some entrenched views give way and some new advances are made. This is the nature of the 'normative debate'. Innovation diffusion with technologies of this nature depend very much on resolutions to these debates. Information, the media and government regulators are all part of the social process.

Government policies and regulations can positively or negatively affect the diffusion process (Feldstein and Glasgow, 2008). The diffusion of GM technologies in Sri Lanka has been influenced by the government imposing regulations, critiques from environmental and health pressure groups, local industry associations, international trade organizations such as WTO and technology producers from the western world. While the motives of these influencers are distinctly different, end users decisions attitudes are apparently shaped by the government regulatory responses due to lack of awareness of the new technologies and products as well as uncertainty of long term effects. In a recent study by Senarath and Karunagoda (2012), it was found that most of the consumers in Sri Lanka were not aware of GM foods and the majority of them perceived GM foods as risky to human health. Many consumers completely refused buying GM food and some were uncertain whether they would buy GM food or not. A vast majority of the consumers were of the view that GM food should be labeled, although many of them do not read labels on food items all the time. From a different perspective, the influence of perceived benefits could be more significant than the perceived risk. As Gaskell *et al.* (2004) identified, the impact of the absence of perceived benefits of GM food for some consumers forms the basis of rejection where the communication efforts on bringing about public opinion become questionable.

The analysis of the perceived attributes of GM food provides some insight into the barriers of diffusion of GM food from the user perspective. As Roger clearly stated the relative advantage of an innovation is ' … not always very clear-cut or spectacular at least not in the eyes of the intended adopters. They can seldom be very certain that an innovation represents a superior alternative to the previous practice that it might replace' (Rogers 2003:13). The situation with GM food is far from ideal with the associated scientific complexity with genetic engineering and biotechnology and the health risk concerns that have not been proven invalid. Moreover, the presence of a strong knowledge stock in agriculture and traditional health and medical practices as well as the high value of natural production along with food as a strong part of culture form the adoption of GM food which does not have perceived superiority over the existing food products in the market. Beyond the influence of innovation attributes and the social system, the actions of the government confirming their uncertainty of assurance of benefits added with the media reports on the external pressure to the government to revise their decisions might have also caused the market reluctance to adopt GM food even when the regulations are relaxed. The literature identifies user resistance for innovation as two types. Passive resistance for innovation makes most new products fail as they are rejected by consumers due to their resistance to change or the preference towards the existing products before new products are evaluated. Active resistance, however, is based on a negative attitude formed by an unfavorable product evaluation (Talke & Heidenreich 2014). Due to technological complexity and the unavailability of technological resources end users of GM food do not get the opportunity to undertake a comprehensible product evaluation. The dominant barrier formed by the scientific uncertainty strengthened by the actions of multiple governments in being cautious on regulating the entry of GM food into the4 local market has very much limited the diffusion of GM food, even after the regulations have been relaxed.

predisposition to resist innovations

prior to new product evaluation

FUTURE RESEARCH DIRECTIONS

This chapter presents an analysis of the not very successful GM food diffusion in Sri Lanka. It has not aimed to present an all-inclusive retrospective analysis that represents the perspectives of all stakeholders of the diffusion process. An empirical study that collects qualitative data from local consumers to understand the decision factors, from government authorities who determines the food regulations and environmental and from health and environmental pressure groups has the ability to generate more in-depth insights into why GM food has been resisted in Sri Lanka. Moreover, Sri Lanka was not the only country that was reluctant to allow GM food entry into the local market. Brazil and Greece also banned GM food. A cross country study will enable a deeper understanding of the similarity or the diversity of barriers for diffusion of the same innovation in different contexts. Moreover, future work is needed to understand how social innovations can be better positioned to generate greater good in less developed global contexts. Social innovations can occur in all sectors; examples include the micro-credit systems that empower the poor and cheap laptop computers for students in developing economies, cost effective technologies for clean water. Future studies can investigate and compare the barriers to diffusion of different types of innovation. Of course, every innovation comes with some level of uncertainty, the degree of uncertainty of innovations with perceived health risks and environmental risks could be different from other innovations that are more neutral in terms of long-term consequences. Future research can analyze both successful and unsuccessful diffusion of social innovations considering the perceived attributes of social innovation from user perspectives.

CONCLUSION

This chapter aimed to analyze the social process of GM food diffusion in Sri Lanka drawing attention to the barriers of diffusion. Not all innovations are usually adopted or expected to the adopted equally by potential users or user communities. Studying diverse user responses including user resistance of innovations is important for innovation diffusion research knowledge base. Our knowledge on the social resistance for innovations especially when innovations are positioned as benefiting the society is limited. When social innovations are created locally the acceptance of such innovations can be supported by the compatibility with the local systems whereas social innovations positioned as global solutions encounter different responses in different contexts. Some social innovations such as micro-credit financing have been positively accepted whereas GM food and GMOs have been resisted at varying levels in multiple countries. As Dacin *et al.* (2011) have pointed out, it would be interesting to study the scalability of social innovations in order to understand why some social innovations diffuse widely whereas others seem to remain more locally embedded.

The lack of relative advantage of GM food based on the absence of scientific evidence nullifying the claims of long-term health and environmental risks has been the most significant cause of consumer resistance. The complexity of underlying technologies and the scientific uncertainty that cannot be proven and the incompatibility of GM food and GM technologies with traditional knowledge bases and local social system have created a strong barrier for diffusion of GM food. Even though the literature of innovation diffusion identifies different marketing strategies to overcome passive and active resistance for innovation (Heidenreich & Kraemer 2015), GM food adoption requires much more than marketing strategies as innovation itself has not been able to position itself as superior with user benefits relative to the existing products in the market.

Implications of this study are mainly for the producers of innovations. Mitigating social resistance should not be at the end of the innovation process as marketing strategies have limits to increase adoption rates when innovation attributes are not convincing in terms of its superiority over the existing products. Integrating user perspectives in the innovation process with the understanding of what influences user adoption decisions is important. It is important for policymakers to understand the consequences of policy actions in shaping public attitude which calls for responses based on rigorous analyses.

REFERENCES

Aggarwal, P., & Cha, T. (1997). Surrogate buyers and the new product adoption process: A conceptualization and managerial framework. *Journal of Consumer Marketing, 14*(5), 391–400. doi:10.1108/07363769710179419

Anderson, C. (2011, July). The GM Food Potential. *Future Directions International.*

Baumüller, H. (2003). *Domestic import regulations for genetically modified organisms and their compatibility with WTO rules.* Trade Knowledge Network.

Brankov, T. P., & Lovre, K. (2013). WTO Law and Genetically Modified Products. *Challenges for the Global Agricultural Trade Regime After Doha*, 99.

Commission, National Science Foundation & National Science and Technology. (2009). *National Biotechnology Policy.* Colombo: Ministry of Science and Technology of Sri Lanka.

Cooray, P. S. (2002). Health Ministry is Quiet as the Controversy on Genetically Modified Foods Deepens. *Synthesis/Regeneration, 27.*

Curtis, K. R., McCluskey, J. J., & Wahl, T. I. (2004). Consumer Acceptance of Genetically Modified Food Products in the Developing World. *AgBioForum, 7*(1-2), 70-75. Available on the World Wide Web: http://www.agbioforum.org

Dacin, M., Dacin, P. A., & Tracey, P. (2011). Social entrepreneurship: A critique and future directions. *Organization Science, 22*(5), 1203–1213. doi:10.1287/orsc.1100.0620

Douglas, M. (1982). *Essays in the Sociology of Perception.* London: Routledge & Kegan Paul.

Feldstein, A. C., & Glasgow, R. E. (2008). A practical, robust implementation and sustainability model (PRISM). *Joint Commission Journal on Quality and Patient Safety, 34*(4), 228–243. PMID:18468362

Frewer, L. J., van der Lans, I. A., Fischer, A. R. H., Reinders, M. J., Menozzi, D., Zhang, X., ... Zimmermann, K. L. (2013). Public perceptions of agri-food applications of genetic modification–a systematic review and meta-analysis. *Trends in Food Science & Technology, 30*(2), 142–152. doi:10.1016/j.tifs.2013.01.003

Gaskell, G., Allum, N., Wagner, W., Kronberger, N., Torgersen, H., Hampel, J., & Bardes, J. (2004). GM foods and the misperception of risk perception. *Risk Analysis, 24*(1), 185–194. doi:10.1111/j.0272-4332.2004.00421.x PMID:15028010

Heidenreich, Sven, & Kraemer, Tobias. (2015). Innovations—Doomed to Fail? Investigating Strategies to Overcome Passive Innovation Resistance. *Journal of Product Innovation Management.* doi:10.1111/jpim.12273

Heijs, W. J. M., Midden, C. J. H., & Drabbe, R. A. J. (1993). *Biotechnology: Attitudes and influencing factors.* Eindhoven: Eindhoven University of Technology.

Hettiarachchi, K., Jeewandara, S., & Alwis, S. (2013, May 19). Wholesome questions. *The Sunday Times.*

Hirimuthugoda, D. (2014). *Does Sri Lanka Need a New Seed Act?* Retrieved from http://www.ips.lk/talkingeconomics/2014/05/22/does-sri-lanka-need-a-new-seed-act/

Hirimuthugodage, D. (2013). Frankenfoods: A Sri Lankan Perspective of the Inevitable Rise of Genetically Modified Foods. *The Island Online.* Retrieved from www.ik/index.php?page_cat=areticle-details&page=article-details&code_title=71894

Ho, M-W. (2015). CaMV 35S Promoter in GM Seed in Sickened Rats, Transferred into Rat Blood, Liver and Brain Cells. *Science in Society, 65*, 32-37.

Islam, T. (2014). Household level innovation diffusion model of photo-voltaic (PV) solar cells from stated preference data. *Energy Policy*, *65*, 340–350. doi:10.1016/j.enpol.2013.10.004

Ismail, A. M., Singh, U. S., Singh, S., Dar, M. H., & Mackill, D. (2013). The contribution of submergence-tolerant (Sub1) rce varieties to food security in flood-prone rainfed lowland areas in Asia. *Field Crops Research*, *152*, 83–93. doi:10.1016/j.fcr.2013.01.007

Jayasumana, C., Gunatilake, S., & Senanayake, P. (2014). Glyphosate, hard water and nephrotoxic metals: Are they the culprits behind the epidemic of chronic kidney disease of unknown etiology in Sri Lanka? *International Journal of Environmental Research and Public Health*, *11*(2), 2125–2147. doi:10.3390/ijerph110202125 PMID:24562182

Klumper, W., & Qaim, W. (2014). A Meta-analysis of Genetically Modified Crops'. *PLoS ONE*, *9*(11), e111629. doi:10.1371/journal.pone.0111629 PMID:25365303

Krimsky, S., & Gruber, J. (2013). *Genetic explanations: sense and nonsense*. Harvard University Press.

Krimsky, S., & Gruber, J. (2014). *The GMO Deception: What You Need to Know about the Food, Corporations, and Government Agencies Putting Our Families and Our Environment at Risk.* Skyhorse Publishing, Inc.

Kurukulasuriya, L. (2014). *Sri Lankan farmers resist new seed law*. Retrieved from http://newint.org/blog/majority/2014/01/28/sri-lanka-seed-law/

Lanka, Rubber Research Institute of Sri. (2015). *Research and Development*. Retrieved 05/08/2015, 2015, from http://www.rrisl.lk/sub_pags/r&d.html

Mendel, P., Meredith, L. S., Schoenbaum, M., Sherbourne, C. D., & Wells, K. B. (2008). Interventions in organizational and community context: A framework for building evidence on dissemination and implementation in health services research. *Administration and Policy in Mental Health and Mental Health Services Research*, *35*(1-2), 21–37. doi:10.100710488-007-0144-9 PMID:17990095

Ministry of Environment and Natural Resources. (2005). *National Biosafety Framework of Sri Lanka*. Colombo.

Mohamed, M.T.Z., & Zoysa, A.K.N. (2006). Current status and future research focus of tea in Sri Lanka. *The Journal of Agricultural Science*, *2*(2), 32–42.

Nielson, C. P., & Anderson, K. (2011) Genetically Modified Foods, Trade, and Developing Countries: Is Golden Rice Special? *AfBioWorld.* Retrieved from www.agbioworld.org/biotech-info/topics/goldenrice/specialgoldrice.html

Null, G., & Gale, R. (2014). *The Global GMO Food War against Humanity.* Devastating Health and Environmental Impacts Worldwide. Retrieved 29/7/2015, 2015, from http://www.globalresearch.ca/the-global-gmo-food-war-against-humanity-devastating-health-and-environmental-impacts-worldwide/5414669

Paarlberg, R. L. (2002). The real threat of GM crops in poor countries: Consumer and policy resistance to GM foods in rich countries. *Food Policy, 27*(3), 247–250. doi:10.1016/S0306-9192(02)00014-3

Phillips, P.W.B., & McNeill, H. (2000). *A survey of national labeling policies for GM foods.* Academic Press.

Powell, D. A., & Griffiths, M. W. (1993). Public perceptions of agriculturabl iotechnology in Canada. *Food Technology,* 14.

Rogers, E. (2003). *The Diffusion of Innovations.* New York: The Free Press.

Sajiwani, J. W. A., & Rathnayaka, R. M. U. S. K. (2014). Consumer Perception on Genetically Modified Food in Sri Lanka. *Advances in Research, 2*(12).

Senanayake, M. (2013, August 10). DCD in Anchor: conspiracy or genuine error?. *The Island.*

Senarath, S. N., & Karunagoda, R. P. (2012). Consumer Attitude towards Labeling of Genetically Modified Foods in Sri Lanka. *Tropical Agricultural Research, 23*(3), 283–288. doi:10.4038/tar.v23i3.4666

Seneviratne, N. (2001, May 25). US told not to dump untested GM food in Sri Lanka. *The Island.*

Skevas, T., Kikulwe, E. K., Papadopoulou, H., Skevas, I., & Wesseler, J. (2012). 'Do European Union Farmers Reject Genetically Modified Maize? Preferences for Genetically Modified Maize in Greece'. *AgBioForum, 15*(3), 242-256. Available on the World Wide Web: http://www.agbioforum.org

South-North Development Monitor. (2000). *Sri Lanka: Government bans genetically engineered foods.* Author.

Talke, K., & Heidenreich, S. (2014). How to Overcome Pro-Change Bias: Incorporating Passive and Active Innovation Resistance in Innovation Decision Models. *Journal of Product Innovation Management, 31*(5), 894–907. doi:10.1111/jpim.12130

Transparency International. (2013). A single food safety regulatory agency - An imperative for Sri Lanka. Colombo: Author.

Turpin, T., Burgoss, Zhang, & Amaradasa. (2015). *Southeast Asia and Oceania.* UNESCO Science Report.

Weick, C. (2002). Genetically engineered crops and foods: Back to the basics of technology diffusion. *Technology in Society, 24*(3), 265–283. doi:10.1016/S0160-791X(02)00008-8

Withanage, H. (2007, January 1). GM Food labeling assures freedom of choice. *Daily News.*

World Health Organization. (2015). *Food Safety: Frequently asked questions on genetically modified foods.* Retrieved 21/7/2015, 2015, from http://www.who.int/

KEY TERMS AND DEFINITIONS

Compatibility of Innovation: The perception towards the innovation as being consistent with the existing values, past experiences, and needs of potential adopters of the innovation. When an innovation is consistent with the existing system and user values, the resistance for diffusion is low.

Complexity of Innovation: This refers to the degree of complexity of innovation perceived by potential adopters. More specifically, it refers to the degree of difficulty of understanding and using the innovation.

Genetically Modified Food: Genetically modified (GM) foods are foods derived from organisms whose genetic material has been modified in a way that does not occur naturally, e.g. through the introduction of a gene from a different organism (World Health Organization, 2015).

Genetically Modified Organisms: Genetically modified organisms (GMOs) can be defined as organisms (i.e. plants, animals or microorganisms) in which the genetic material (DNA) has been altered in a way that does not occur naturally by mating and/or natural recombination (World Health Organization, 2015).

Innovation: Innovation is a change in products, processes, marketing methods or organizational methods.

Innovation Diffusion: An innovation is adopted by people as a part of a social system is called as diffusion.

Observability of Innovation: The degree to which the results of an innovation are visible to the potential adopters of innovation.

Relative Advantage of Innovation: the benefits of an innovation degree to which an innovation is perceived as better than the idea it supersedes (Rogers 2003: 16).

Trialability of Innovation: The degree to which an innovation may be experimented with on a limited basis (Rogers 2003: 16).

This research was previously published in the Handbook of Research on Driving Competitive Advantage through Sustainable, Lean, and Disruptive Innovation edited by Latif Al-Hakim, Xiaobo Wu, Andy Koronios, and Yongyi Shou, pages 186-203, copyright year 2016 by Business Science Reference (an imprint of IGI Global).

Chapter 4
Agriculture, Trade Liberalization and Poverty in the ACP Countries

Theresa Ann Rajack-Talley
University of Louisville, USA

ABSTRACT

This chapter looks specifically at the impact of the ACP agriculture trade agreements on issues of food production, food security and poverty and how they relate in some ways to each other. It is divided into four distinct but inter-related sections. First, poverty in the ACP regions is briefly described. Secondly, the importance of agriculture, particularly the role of small farming systems to economic growth, employment, food production, food security and poverty reduction in the ACP countries is examined. Thirdly, the impact of the acceleration of trade liberalization and agriculture protectionism on ACP countries is reviewed and analyzed. Here, the trade arrangements and networks in the three ACP regions are described and include international, regional and local agricultural trade and market agreements. Further, the paradoxes and dilemmas that emerge are interrogated and their outcomes discussed. Particular attention is paid to small producers and women. The fourth and final section of the chapter concludes that while the impact of agriculture trade liberalization remains a mixed bag of positive and negative possibilities, one thing that is clear is that while trade liberalization is promoted as a cure for food security and world poverty, the jury is still out on whether it has helped or hindered this endeavour. Trade policies and arrangements do substantially change the lives of people in developing countries, but in very complex ways. Recommendations are made to support the South-South initiatives of the ACP and others as well as enhance regional and domestic agriculture production and trade policies. The author, however, warns that agriculture and trade policies must be linked closer to food security and the healthy diets of local populations.

DOI: 10.4018/978-1-5225-8063-8.ch004

Overcoming poverty is not a gesture of charity. It is an act of justice. – Nelson Mandela

INTRODUCTION

International trade and financial transactions are not new phenomena. From time immemorial, people across the globe have exchanged goods and services. Today, societies have become more specialized and no single country produces all that is needed to sustain itself. Thus, the exchange of goods and services is even more necessary than before. However, the formation of cooperative entities and marketing blocs have made trading more political and convoluted. No longer is the simple use of supply and demand economic principles applicable. Currently, the international movements of goods and services have more to do with decisions made by multinationals and groups of developed countries. According to Koroma and Ford (2006), the rapidly changing structure and patterns of trade are heavily influenced by private sector investment and standards. As a result, freedom to trade at all levels, from the farm gate to the stock market, is now much more complicated and is further removed from the power of producers and/or consumers, who are also finding it increasingly difficult to avoid slipping into the clutches of poverty.

The primary purpose of this chapter is to examine the links between agriculture production and trade, food security, small farmer production, and poverty reduction in developing countries. It specifically analyses the impact of agriculture trade policies and outcomes in select African, Caribbean and Pacific countries known as the ACP group. There are 79 African, Caribbean and Pacific states that form the ACP: 48 are in Sub-Saharan Africa, 16 in the Caribbean region, and 15 in the Pacific. The countries are demographically, geo-physically and economically diverse; 40 are considered less developed countries, 37 small island states (or both), and 15 landlocked developing countries (see table 1). The total ACP region population is approximately 727 million and represents 15% of the total population of all developing countries. Ninety-four percent live in Africa, close to 5% in the Caribbean and just over 1% in the Pacific islands. According to ACP Secretary General, Dr. Patrick Ignatius Gomes, in his address to the UN Summit on Sustainable Development at the 70th United Nations General Assembly (2015), "the domains of the ACP group include: rule of law and good governance; global justice and human security; intra-ACP trade, industrialization and regional integration; building sustainable, resilient and creative economies; as well as financing for development." Further South-South and Triangular Cooperation informs the group's approach to all these domains.

In reality marketing and trade arrangements can sometime result in uneven benefits and the extent to which they assist in poverty reduction remains contested. The chapter infers that agricultural trade policies are often bullied by international trade magnates and are not necessarily the answer to economic growth, food security and poverty reduction in most developing countries. Moreover, trade policies, although still important, do have negative impacts on developing countries including marketing preference erosion, decline of tariffs, and the demand for high level of competitiveness which requires a wide range of mitigating factors. Thus, many developing countries are often confronted with substantial disadvantages in implementing current trade policies that necessitate having substantial resources that they simply do not have (UNCTAD, 2013). Since most of these countries including those member states who belong to the ACP Group are agrarian-based, current agriculture trade policies have direct and indirect impacts on poverty, economic growth and the well-being of the people who live in these countries.

Table 1. The ACP states by region and economic classifications

Least Developed Countries (LDC)			Non-LDCs		
Africa (34)	**Caribbean (1)**	**Pacific (5)**	**Africa (15)**	**Caribbean (15)**	**Pacific (9)**
Angola Benin Burkina Faso Burundi Cape Verde Central African Republic Chad Comoros Dem. Republic of the Congo Djibouti Equatorial Guinea Eritrea Ethiopia Gambia Guinea Guinea-Bissau Lesotho Liberia Madagascar Malawi Mali Mauritania Mozambique Niger Rwanda Sao Tome & Principe Senegal Sierra Leone Sudan Toga Uganda United Rep. of Tanzania Zambia	Haiti	Kiribati Samoa Solomon Islands Tuvalu Vanuatu	Botswana Cameroon Congo Cote d'Ivoire Gabon Ghana Kenya Mauritius Namibia Nigeria Seychelles South Africa Swaziland Zimbabwe	Antigua &Barbuda Bahamas Barbados Belize Cuba Dominican Republic Grenada Guyana Jamaica Saint Kitts & Nevis Saint Lucia Saint Vincent & the Grenadines Suriname Trinidad & Tobago	Fiji Papua New Guinea Tonga Marshall Islands Cook Islands FS of Micronesia Nauru Niue Palau

Table taken from Commodities and Trade Technical paper #8, pg. viii, (2006)
LDC= Less Developed Countries

OVERVIEW OF POVERTY AND AGRICULTURE IN THE ACP COUNTRIES

World Bank estimates indicate that on a global level, the first Millennium Development Goal of cutting the 1990 poverty rate by half in 2015 has already been attained. However, despite the progress made there are still large numbers of people whose standard of living fall at or below the poverty line. Close to 17% or 1 billion people in developing countries live on less than US $1.25 a day, that is, about one in seven of the world's population suffer from hunger (World Bank 2015). Similarly, the United Nations Food and Agriculture Organization (FAO) estimates for 2012-14, indicated that approximately 805 million people were chronically undernourished, i.e., one in every nine people in the world had insufficient food for an active, healthy and productive life.

Progress made towards meeting the Millennium Development Goals internationally has also been uneven. The gap between the wealthiest and poorest countries has widened and the level of hunger and poverty varies among developing regions. For example, although developing countries overall experienced a reduction in poverty, a disproportionate amount (just over 80%) of the world's extremely poor live in South Asia (399 million) and in Sub-Saharan Africa (415 million). Fewer than 50 million of extremely poor people live in Latin America and the Caribbean (LAC), the Middle East, North Africa, Eastern Europe and Central Asia, combined. Furthermore, the vast majority (791 million) of people who live in dire poverty reside in developing countries (WorldBank, 2015). Thirty-nine of the 49 developing countries are members of the ACP groupings where approximately 214 million undernourished individuals can be found in Sub-Saharan Africa[1] and 37 million in Latin America and the Caribbean.

There is also the problem where increasingly larger numbers of families in some of the ACP countries are eating too much of the wrong types of food, resulting in overweight and obesity. For example, while hunger is a major problem in sub-Saharan Africa, the excess intake of calories seriously undermines the health of many in the Caribbean and Pacific regions. Chronic non-communicable diseases, several related to poor nutrition, now account for 57% of deaths in the Caribbean. Further, although undernutrition and iron-deficiency anaemia persists in the region, the increasing incidence of obesity is now the major source of nutrition-related problems. In addition, diseases related to obesity such as diabetes, hypertension and heart conditions are chronic and very expensive to manage placing financial burdens on poor families (PAHO 2014).

A similar story can be told for the Pacific region, where half of the adult population is overweight. Pacific Islanders tend to eat a lot of imported canned meats like spam and corned beef, which has a negative effect on their health. In 2007, eight of the 10 heaviest countries were located in the South Pacific. Nauru, the world's smallest republic with just over 9,000 inhabitants, earned the number one spot with over 90% of their adult population considered obese. At the same time, 38% of Papua New Guineans live below the National Basic Needs Poverty Line, which means 2.7 million people are unable to buy sufficient food and meet basic requirements for housing, clothing, transportation and school fees. Even more alarmingly, 61% of the populace does not have access to safe drinking water (Borgen Project Report 2015).

Ironically, the lack of adequate food, proper nutrition and the resulting high levels of diseases and deaths related to poor nutrition in many ACP countries occur in places where agriculture and food production have historically been core to people's livelihood and dietary needs. Currently, ACP countries need to produce increasing quantities of food to satisfy their growing populations. However, food security is not just about growing more food, but also about high quality items that have the nutritional value for healthy diets. Equally important therefore, is that healthy food items should be accessible and affordable to all sections of the population, including those responsible for producing them. It is therefore imperative that domestic agricultural policies and practices be oriented around these understandings and trade negotiations be sensitive to the needs, capabilities, health and well-being of people.

Role of Agriculture in Livelihoods, Employment, and Food Security

Taken as a whole, ACP countries average significantly lower rates of per capita Gross Domestic Product (GDP) growth than most developing countries. However, with the exception of the Caribbean subgroup, agriculture's share in the GDP of ACP countries is normally higher than in developing countries collectively, and in some ACP countries agricultural products form the larger portion of their total trade. In the case of sub-Saharan Africa, the processing of agricultural products also accounts for two-thirds of manufacturing value-added (CTA 2011). Moreover, agriculture still accounts for more than 60%of total employment in ACP countries compared to around 55% in developing countries jointly. Most of the

ACP countries have also sustained increases in per capita food production, but food production has not kept up with population growth nor the dietary needs of any of the three ACP sub-regions (FAO 2006). Specific to Africa, agricultural production has increased steadily, tripling in value and is almost identical to that of South America, but lower than that of Asia. However, the growth of the sector is uneven and the World Bank estimates that there is a wide spread in agriculture's contribution to GDPs across the various African countries. For example, in Libya, Guinea, Gabon, and others, agriculture accounts for over 45% of the GDPs while in Burundi, Republic of Congo, Liberia, and others, agriculture accounts for under 15% of these countries' GDPs. The average for the continent as a whole is approximately 30-40% (World Bank 2012). More importantly, this sector has always absorbed a large percentage of the adult work force. Agriculture creates most of the jobs in Africa, aside from the oil-producing African countries. Half or more of the working population, an estimated 530 million people or 48% of the total African population work in the agriculture sector. A large proportion of all new entrants to Africa's working population (approximately 330 million) have very little or no option but to turn to agriculture (NEPAD 2013).

The main farming system in Africa is quite similar to that of the Caribbean and the Pacific Islands. Africa's agricultural sector is dominated by small family farms that rely mainly on family labor and women's labor. As a whole, Africa has 33 million farms of less than 2 hectares, accounting for 80% of all farms. However, in the last decade large-scale investment contracts have crept, threatening not only the persistence and support to small farmers but also to the continent's natural resources and biodiversity. These large-scale farming systems are estimated to cover close to 20 million hectares, which represents more than the arable area of South Africa and Zimbabwe combined. While the small family farming systems are unable to expand and enhance their production due to the lack of their own investment capacity and secure land tenure, many African countries have bought into the concept of industrial farming as an avenue to meeting the demands of world trade agreements and the food security of their populations. As such, in many African countries it is easy for foreign groups and multinationals to acquire land and set up operations without adequate policies to protect local production by small producers and the countries' natural resources (NEPAD 2013).

In the last 30 years, Africa's population has doubled overall and tripled in urban areas. The most direct consequence of this exponential population growth is that the continent now has more mouths to feed. Yet, cereal production, one of the main staple for Africa, has been unable to keep pace with population growth, since it has only increased by a factor of 1.8. This gap is even wider for processed products and meat, which are increasingly called for by a larger and larger urban population. In fact, Africa has moved from being self-sufficient in cereal production and other traditional crops in the 1960s to becoming a net importer of basic food items that compete with its own production of meat, dairy products, cereals and oils. In 2011, African countries spent close to US $35 billion in food imports (excluding fish), which amounts to 1.7 times the value of its agriculture exports (Africa Progress Report 2014).

A very similar synopsis is found in the Caribbean Region where agriculture is important for the region's GDP, employment, food production and food security, with these varying from country to country. For example, the 2012 data from the World Bank shows that in the Caribbean, the GDP earned from agriculture ranged as high as 21-13%in some countries (e.g. Guyana, Dominica, Belize) to as low as 1-2%in others (e.g. Barbados, Antigua and Barbuda). As in Africa, the sector is an important source of employment where some 3.3 million individuals are employed as agricultural workers. In the past decade this approximated to 32% of the labor force in 15 countries across the region including Haiti (Bourne 2008).[2] In the last decade farmers made up about 8% of the labor force in the least agrarian country such as Trinidad and Tobago, but comprised as much as 20-30% in countries, e.g. Belize, Guyana, Grenada, Dominica and Jamaica, which are more heavily dependent on agricultural production (Potter et al. 2004). Many others are often employed in jobs attached to the forward and backward linkages in the food sector, especially in agro processing and marketing businesses.

Although the agricultural sector in the Caribbean historically focused on wealth generated from the exportation of raw materials (sugar, cocoa, coffee, etc.) to metropolitan states (and some still do), it also benefitted from a fairly decent domestic food sector. According to Rajack-Talley (2015), the production of food items continue to be carried out primarily by small family farms scattered throughout the landscape of the Region. Although there are variations in size of operations, small producers in the Caribbean are mainly family operated enterprises co-existing alongside larger private and state-owned farms. They have limited access to land and their farms are either owned, rented/leased, shared with other family members, or illegally occupied. The size of farms can vary, but small operations range from ½ - 5 acres (.15-2.0 hectares). They practise mixed farming ; a combination of crop production, livestock husbandry, forestry, artisanal fishery and small scale aquaculture, producing food both for the home and the local markets as well as for the export market. Farmers devote a major part of their time to cultivation using mainly their own labor and that of family members, with little paid help from non-family laborers.

Over the years, agricultural development has suffered so much so, that the Region's food importation bill soars higher every year and was estimated at US $3-6 billion in 2012-2013.[3] With the exception of Belize and Guyana, most of the Caribbean countries are net importers of food, averaging 60% in some cases to as high as 80% in others. Such high levels of food importation compete with cost of local food production and food prices. Furthermore, not only have Caribbean people become more dependent on imported foods but the income needed to purchase these items has dwindled. Although the Region has over the years increased minimum wages, food price inflation has outstripped the minimum wage increases in most countries. According to the Director of the Caribbean Food and Nutrition Institute (CFNI), "soaring food prices present a crisis which has devastating consequences on nutrition, particularly in children" (CFNI Annual Report 2008: v).

Statistics on the agricultural earnings or GDP on the third region of the ACP group, the Pacific Island, are not readily available. What is known is that the agriculture and fisheries (forestry in some of the larger countries) sectors remain the economic mainstay of most Pacific ACP countries (PACPs). Commercialized agriculture does provide a substantial income to these countries based on medium, small and micro levels of production and the US dollar value of commercial market products (see table 2).

The agricultural sectors of the Pacific countries are made up primarily of subsistence or semi-subsistence farmers, producing both food crops and cash crop products in the larger countries (palm oil, cocoa, and coffee). Farming is still a family owned and operated system and many families rely on agriculture as

Table 2. Ranking of commercial agricultural production in PACP countries

Value of Top 5 Commercial Market Products	Country
Medium: Above US$100 million	Papua New Guinea; Fiji
Small: US$10 million-US$ 60 million	Solomon Islands, Vanuatu, Timor-Leste, Kiribati, Tonga.
Micro: Under US$7 million	Federates States of Micronesia, Marshall Islands, Cooks Island, Niue, Nauru, Palau, Tuvalu

Source: CTA (2011), Agrictrade, pg.2

their main source of income. Fisheries, especially tuna, remain an important resource in several of these countries but the processing of fish as an industry remains limiting because it is too capital intensive for most of the people in the Pacific Islands. Timber is another natural resource available to some countries but it is unsustainable and under constant threat, as many of the traditional owners of this resource are unable to protect their interests in commercial negotiations and cannot depend on strong support from their governments (CTA 2011; Malua 2003).

Farming remains a very traditional practice tightly linked to family values and culture. Family farms are considered multi-purpose enterprises whereby women and men have historically provided a self-sufficient food system for households and communities (Farming Matters 2013). However, today imported foods constitute a high percentage of household diets, much like the other two ACP Regions. For example, 90% of the diets in the Marshall Islands are dependent on imported foods and 95% of the diary requirements in Fiji are derived from imported foods, particularly powered milk and cheese (CTA 2011). To break this dependency, the PACPs are actively seeking to promote greater national food self-sufficiency. That is, to produce whatever basic foodstuff can be produced locally (e.g. fruits, vegetables, root crops and small livestock) while importing other food stuff. However, because of the small size, remoteness and geographical fragmentation, many Pacific communities are highly vulnerable and suffer disproportionately from external shocks, such as adverse climatic events or market failures.

Natural disasters, reef degradation and deforestation have caused extensive damage to lives and livelihoods. With the exception of Fiji, in the other islands (Papua New Guinea, Samoa, Solomon Islands, Tonga and Vanuatu), agricultural land is very limited and can do little more than support the local population. Traditional ownership patterns govern land use in the Pacific but are becoming increasingly constrained by the shortage of usable land. As a result, there will always be some dependency upon imported food in the Pacific region in spite of its many negative impacts on local economies (UN-Women 2012). A few Pacific island countries have developed tourism and industry but the lack of infrastructure, especially regular transport services and accommodation, prevent these enterprises from reaching full potential (Bulletin on Asia-Pacific Perspectives 2003/04).

Impact on Women Small Farmers

Throughout the discussions so far, the production of food for families and the domestic market has been discussed as an extremely important part of the agriculture sector. Rajack-Talley and Talley (2000) pointed out that any analysis of the domestic food sector in the Caribbean (as well as in Africa and the Pacific) is essentially an analysis of women's role in food crop production. Women farmers tend to concentrate on local food production for their households, communities, society and region primarily as small producers and/or family farms. Globally, women are responsible for 60-90% of total food production in their respective countries with smaller percentages for exports (FAO 2006). In many countries, women are also responsible for fishing in shallow waters and in coastal lagoons, producing secondary crops, gathering food and firewood, processing, storing and preparing family food, and fetching water for the family (UN-Women 2012).

In many African countries women comprise 70% of the agricultural workers and 60-80% of the labor that produces food for household consumption and sale. Women do all the processing for basic food stuffs and provide 80% of the labor for food storage and transport from farm to village, as well as 60% of the harvesting and marketing activities (NEPAD 2013). Although women play a major role in agriculture, their access to land and other resources is governed by customary laws that discriminate against them.

During the last ten years, many African countries have adopted new land laws in order to strengthen women's land ownership rights. This has helped improve the situation of rural women (Mucavele 2015). According to Dr. Lindiwe Majele Sibanda, Chief Executive Officer and Head of Mission Food, Agriculture and Natural Resources Policy Analysis Network (FANRPAN), "women farmers are the pillars of African agriculture. Yet even as the guardians of food security, they are still marginalized in business relations and have minimal control over access to resources such as land, inputs such as improved seeds and fertilizer, credit and technology" (NEPAD 2011:1).

Women in the Caribbean are also highly visible in the production and marketing of food for domestic, household consumption as well as in the regional export food trade, buying directly at farm-gate and exporting to neighboring islands (FAO 2006; Rajack-Talley 2015). The majority of women are smallholders, with the average farm size being significantly smaller in area than that held by men. Rajack-Talley and Talley (2000) argued that production constraints related to land tenure and other resources impact more heavily on women than on men because women are predominantly in the domestic and/or in the regional sphere of agricultural production and marketing, while men typically are more actively engaged in the production and marketing of traditional agricultural commodities for the international markets (see also FAO 2006; Rajack-Talley 2015).

Similarly, women in the Pacific island countries continue to be prominent in traditional agriculture ensuring the absence of absolute poverty and associated hunger in many homes (Bulletin on Asia-Pacific Perspectives 2004/04). They engage in subsistence farming to feed their immediate families and/or grow cash crops to sell in the market. According to a United Nations study on Entity for Gender Equality and the Empowerment of Women in the Pacific Sub-Region (2012:7), "women's agricultural labor is an indispensable part of food production and consumption practices". Although the ways in which the women participate in agriculture vary by island and local cultural norms, they all contribute in planting, tending, and harvesting crops, as well as in marine activities that support the majority of families throughout the region. The important role of women in agriculture on the Pacific islands is similar to Africa and the Caribbean where women are the nurturers of the family and providers of food on land, while men have been providers from the sea. However, they are disadvantaged by their underrepresentation at all levels of society, especially in the decision-making processes. Customary law established centuries ago by colonial administrative systems prevents women from owning and gaining access to land, credit and other resources in many of the Pacific island countries (Balakrishnan and Fairbairn-Dunlap 2005).

In essence, despite the many discriminatory disadvantages, women across Africa, the Caribbean and the Pacific regions play crucial roles throughout the food chain; from food production to distribution and preparation in the homes. ACP countries have historically tailored their agricultural policies to focus on the export of primary 'raw' products, in response to international trade and domination. Current trade liberalization policies that are export-driven will therefore have gendered effects in production and consumption patterns.

AGRICULTURAL TRADE POLICIES AND AGREEMENTS

There is no doubt that the governments of the ACP countries have attempted to resolve the issue of trying to feed their populations, enhance the agricultural sectors and achieve economic growth. They have over the years attempted to take ownership of their economies by adopting the following strategies: (i) developing policies, structures and trading networks among themselves; (ii) reinvigorating the agricultural (and mineral production) systems and reorienting them towards export markets; and (iii) developing strategies and programs aimed at poverty reduction.

In terms of trade, each region has embarked on several unilateral, bi-lateral and regional agreements as well as extra-regional trade and restrictions, all of which affect food production and movement. In Africa, regional trade arrangements include the West African Economic and Monetary Union (WAEMU), the Regional Integration Facilitation Forum (RIFF) formerly the Cross-Border initiative (CBI), the Southern Africa Customs Union (SACU) and the Central Africa Economic and Monetary Community (CAEMC). Regional trade agreements have also been created under the Southern African Development Community (SAADC), the Common Market for eastern and southern Africa (COMESA), and the Commission for East African Cooperation (EAC), the Indian Ocean Commission (IOC) and the Economic Community of West African States (ECOWAS). With the exception of WAEMU, members of the other regional bodies also belong to one or two other trading blocs[4] From these various configurations, three broad patterns of agriculture reform can be discerned: (1) adherence to market reforms (e.g. maize and fertilizer marketing in East and South Africa); (2) countries that resisted reforms and/or re-imposed controls basically opposing trade liberalization (e.g. maize in Zimbabwe after 1998); and (3) states that maintained control while implementing liberalization (e.g. fertilizer market in Zambia and Ethiopia, coffee in Malawi) (FAO 2003).

Similar regional trading networks and blocs can be found in the Caribbean region. For example, the Community Single Market and Economy (CSME) was established by the Caribbean Community (CARICOM) in response to the evolving international environment in areas specifically trade liberalization and globalization. The idea behind the CSME is to create a safe economic space where goods and services along with the factors of production (labor, capital and in the case of land the right of access for establishment purposes) to move freely and be supported by appropriate institutions, and operate in a coordinated fashion. Similar other trade policies emerged in response to the World Trade Organization's (WTO) Agreement on Agriculture such as the Revised Treaty of Chaguaramus (2001), the Regional Transformation Programme for Agriculture (2002), and the Jagdeo Initiative (2004) to name a few. During the same period that some of the agreements were being promoted, the Caribbean reeled from the loss of privileged markets and tariffs which continued in the mid-2000 as the Doha Sanction Waiver[5] expired. In 2009 the Liliendal Declaration on Agriculture, Climate Change and Food Security, as well as the Regional Food and Nutritional Security (RFNSP) were created to focus on sustainable production, processing, preparation, and the commercialization of affordable, nutritious and high quality Caribbean foods, while preserving the region's natural resources. More recently (2011), the Caribbean like its African partners in the ACP group responded to the EU – CAP (Community Agriculture Policy) by establishing its own CAP that encourages the regional building of broad-based and balanced agricultural sectors that could be internationally competitive. The anticipation that these initiatives would address the Region's issues in the agriculture sectors, and by extension Caribbean countries' fiscal stability and sustainability, food security, hunger, poverty and the protection of the region's natural resources, is still doubtful (Rajack-Talley 2014).

Governments of the Pacific Island countries have also established regional trade agreements and processes. The Pacific Island Countries Trade Agreement (PICTA) is based on having free trade among the participating countries, initially for a 10 year period. However, level of trade between the countries has been low due to geographical and logistical constraints. In fact, a 2010 analysis suggested that intra-regional trade agreements could only make limited contributions to economic development (Scollay 2010). In addition to the intra-regional trade agreement, the Pacific group has had bi-lateral Interim

Economic Partnership Agreements (IEPAs) with the EU and a South Pacific Regional Trade and Economic Cooperation Agreement (SPARTECA). Papua New Guinea has had its own free trade agreement with Australia (PATCRA), and Fiji also established an economic framework agreement with Australia (FATERA). In addition, as a group the PACP countries established free trade with Australia and New Zealand – the PACER Agreement. There are several other agreements negotiated as either individual states and/or as a group with the USA and some Asian countries.

Of all the various extra-regional trade arrangements agreed to by the ACP states, one of the largest is that between the African, Caribbean and Pacific countries and those of the European Union (ACP-EU). As far back as 2000 the ACP-EU agreement was signed into what is known as the Cotonou Agreement which entailed a comprehensive framework for ACP-EU relations. The intent was that the ACP-EU arrangement would help with economic development, regional integration, poverty reduction and the smooth and gradual transition of the ACP into the global economy. To achieve these objectives, the Agreement provided an approach for ending the WTO's compatible trading arrangements between the ACP States and the EU. This was to be accomplished through the conclusion of Economic Partnership Agreements (EPAs) that were bargained during 2002 and 2007. The ACP States negotiated the EPAs as regional groups: CEMAC (Economic and Monetary Community of Central Africa; ECOWAS (Economic Community of West Africa States; ESA (Eastern and Southern Africa); SADC (Southern Africa Development Community); CARIFORUM (Caribbean Forum); and the Pacific Forum.

According to Bilal, Dalleau and Lui (2012: ix), the African, Caribbean and Pacific (ACP) Economic Partnership Agreements (EPAs) with the European Union (EU) resulted in various degrees of progress across the three regions. For example, while in the Caribbean comprehensive agreements have been reached, most African and Pacific countries, particularly the least-developed countries (LDCs), are still negotiating comprehensive regional EPAs.[6] Under an EPA, the EU agreed to grant full and immediate market access to ACP/African countries, however, those ACP members who decide to enter an agreement must remove tariffs on "substantially all the trade" between themselves and the EU "within a reasonable length of time", to comply with Article XXIV of the General Agreement on Tariffs and Trade. However, because governments depend on revenues from trade taxes and custom duties, many ACP countries resisted, because the reduced revenues impacted negatively on social expenditures, meeting the Millennium Development Goals (MDGs) and poverty reduction. These concerns obstructed EPA negotiations and regional integration efforts.

Bertow and Schultheis (2007) argued that the opening of the ACP agricultural markets, following the principle of reciprocity, seems to be obligatory for the EU while causing a lot of difficulties for many of the ACP Group of States. For example, while European farmers are able to enter the world market without export subsidies, the farmers in the 'partner' Africa, Caribbean and Pacific countries are unable to do so especially within the timeframe allotted. This is because of underdeveloped infrastructures and little or no subsidies to the farmers in the ACP group. On the other hand, famers in the EU are supported by subsidies, have access to infrastructure, technologies and other resources, and both the negotiations and pace for implementation of trade is on the EU's terms with tenable outcomes for the 'partner' countries. Goodison (2007) explained that the economic partnership agreement (EPA) negotiations are based on a shift from the Lome Convention's non-reciprocity commitment to a basic regime of free trade between the EU and EPA regions. This involves the liberalization of trade in goods as trade-related areas and services. He further pointed out that while Europe has effectively integrated itself, few African (Caribbean and Pacific) countries have strong regional integration.

Impact of Agriculture Trade on Self-Reliance and Self-Sufficiency

It is difficult to decipher at first glance the advantages and disadvantages of extra-regional agriculture trade agreements and market arrangements for the ACP group and developing countries overall. Both governments and academics remain divided on whether or not agriculture trade liberalization contributes to slow economic growth, the demise of small farming households, and restrictions to achieving food security. Some researchers suggest that trade liberalization is good for growth (Dollar & Kraay 2004; Falvey, Foster, & Greenaway 2012; Sachs & Warner 1995; Summers & Heston 1991), and has generally supported agriculture development and food security (See FAO Report 2003 on Trade Reform and Food Security). Others challenge this perspective (Rodriuez & Rodrik 2000), and some believe that trade liberalization centered on comparative advantage is especially challenging for countries producing agricultural commodities, particularly raw materials and where diversification is difficult. For example, Connell and Soutar (2007) pointed out that while the Windward Islands of the Caribbean have reduced banana production and both Barbados and Fiji experienced contractions in the sugar industry, none have found effective substitutes.

Researchers also point out proposed market arrangements have not worked well for small countries. For example, in the Pacific new regional trade agreements have not been unsuccessful and as a whole international agreements have not recognized the disadvantages of small developing countries, "but continue to stress benefits that are nowhere apparent in the small island states" (Connel & Soutar 2007:41). Some see trade liberalization as having both a negative and positive impact on poverty. For instance, research on household income inequality and poverty using detailed microdata across Brazilian states from 1987 to 2005 found that trade liberalization contributes to growth in poverty and inequality in urban areas but reductions in inequality (possibly poverty) in rural areas (Abarche, Dickerson & Green 2004). It was also observed that both state poverty and inequality decrease with rising export exposure and increase with import penetration. Other researchers concluded that there is not enough data on the actual implementation of agriculture trade liberalization, or on reforms that are only partially implemented in some developing countries (e.g. Sub-Sahara), to reach any conclusive assessment (Milner & Morrissey 1999). What is less refuted, is that there are several direct links between trade liberalization and food production and poverty.

Over a decade ago, in an FAO Report (2003) on trade and food security it was suggested that there were two different approaches in addressing issues of agriculture trade and marketing and food importation and food security. First, there is the *self-reliance* approach whereby a country does not need to produce all the food it needs but to export enough goods and services to earn what is required to purchase all its food needs. The other approach, to be *self-sufficient*, is having the ability to produce all the food needs from domestic agriculture production. In this approach, it is necessary to protect and support local farmers. In reality, not all ACP countries can be self-reliant or self-sufficient, particularly if they have limited export opportunities and access to resources as well as high food needs for their populations. This is particularly true for some of the smaller Pacific and Caribbean territories. On the other hand, the World Bank suggested that with certain conditions in place, Africa's farmers can potentially grow enough food to feed the continent, that is, to become self-sufficient. According to the Bank's 2012 Vice President for Africa, Makhtar Diop:

Africa has the ability to grow and deliver good quality food to put on the dinner tables of the continent's families. However, this potential is not being realized because farmers face more trade barriers in getting their food to market than anywhere else in the world. Too often borders get in the way of getting food to homes and communities which are struggling with too little to eat.

As a general rule, ACP member states should first identify the extent to which their agricultural sector has the potential to meet some, most or all of their food needs - and then develop appropriate agriculture reforms.

Interestingly, discussions about self-reliance and self-sufficiency have been held in the Caribbean for decades. Sir Arthur Lewis (1978a, 1978b) suggested that governments needed to support industrialization and revolutionize domestic agriculture, as well as secure better benefits from international trade. He envisioned that export-led growth could be either detrimental or supportive to the economic development of Caribbean societies, depending on the relationship between the export and domestic sectors. Lewis warned that international trade arrangements should be linked to the specific domestic sectors of a country, including agriculture. Other Caribbean scholars further explained that self-reliance is based on domestic food production as well as local industrial development, to meet the consumption needs of the masses (Beckford 1972; Girvan 1973; Rodney 1974; Thomas 1982). This was offered as an alternative to export-led growth and import substitution. Here, the idea is not to isolate Caribbean economies from the international market but to reduce the external diversion of resources and instead invest in domestic food production and link-industries. For example, Thomas (1982) proposed the re-orientation of agriculture from producing raw materials for the world market, to the production of foodstuffs and needed local industrial inputs. Industry, in its turn, should be required to produce basic consumption goods, in particular those needed to raise agricultural efficiency (see also Amin 1977; Mansour 1979).

Self-sufficiency in food involves the development of the domestic food sector and the ability of the region to produce an adequate amount of the region's most basic food items, such as rice, sugar, citrus, fruit and vegetables, milk, eggs, poultry, fish and meats from small animals. This requires that people have access to suitable agricultural lands and other resources to more than just subsist. Given the current high food importation bills and poor diets of the vast majority of people in the ACP Nation States it is imperative that countries grow more of what they can and promote a change in food culture. This will result in decreased dependency on high levels of imported processed foods and an addiction to fast foods, especially those with high fat, sugar and salt contents. However, a move in this direction necessitates increases in government involvement, greater participation by the people, and the buy-in from the private sector. The success of self-sufficiency is heavily reliant on building the capacity of the people so that they can take advantage of the knowledge, technology and the decision-making processes needed throughout the chain of events (Rajack-Talley 2014).

Both approaches, self-reliance and self-sufficient, are linked to policies that support domestic agriculture with the possibility of one of three outcomes. First, countries can become self-reliant as net food exporters where food security is not an issue. In this case, however, access to foreign market opportunities can be daunting because of the challenges of meeting new and emerging market requirements, especially the more stringent food safety and quality requirements for export products. Additionally, many of the new trade agreements have mandated the removal of guaranteed markets between some ACP countries and the EU. As a result, in reference to the sugar industry (as well as banana and others) in the Caribbean (and other ACP countries), the Chairman of the Sugar Industry Board in Jamaica commented that:

We have moved from a situation of a guaranteed market, guaranteed price, and guaranteed access to one in which we still have guaranteed access but nothing else and we now have to be competitive because that guaranteed access has been expanded to include a number of other countries (Heaven 2010).

Second, countries can become self-sufficient in food meeting local food needs and even engaging in export of some food items. However, to do this requires minimal biases against domestic agriculture and small farm production as well as strong agricultural sector reforms. It is dependent upon the extent to which small producers and the network of family farms in ACP countries can easily access new technologies and other resources that are necessary to increase yields and profitability from food exports, and at the same time lower domestic food prices. If this were to take place, then both rural and urban poverty would be reduced. In the real world however, most foods produced from industrialized countries are sold at subsidized and therefore cheaper prices on the open markets. This common occurrence presents a disadvantage to ACP countries that are aiming for self-sufficiency but who are discouraged by the World Bank and the IMF from subsidizing their own farmers. This practice, along with the mandate of food importation that is tariff-free, creates unfair competition against the small (and some large) domestic producers in the ACP countries. The distortion of world food prices for these reasons is detrimental to the small producers and family farms of developing countries. Consequently, the less efficient farmers who do not have access to technology and other resources are eventually forced out of the domestic and international markets.

Third, some ACP countries may decide that because they are unable to be self-sufficient in local food production aim to generate enough earnings from other agricultural exports to meet food import bills. In this case, it is important to not be dependent on the export of primary commodities with low income elasticity and high price volatility. As Ambassador Derrick Heaven, the former executive chairman of the Sugar Industry Authority (SIA), pointed out in the Jamaica *Gleaner* newspaper, the 2010 divestment of the local sugar factories has not produced the desired result because the new players in the sector are now struggling to keep pace with what is unfolding on the wider global stage. According to Heaven:

We have been weaned on the situation, where for the past several hundred years, we have been making one product - raw sugar, particularly for the mother country (England) and Europe...the situation has changed, but I don't think we have changed as quickly as we should, even though we were aware of the implications.

SUMMARY AND CONCLUSION

There is no doubt that the agriculture sector is important to the 3 regions which comprise the ACP for the following reasons: economic development and employment; food production for domestic use; and foreign exchange earnings through agriculture export production. Nonetheless, most of the ACP countries are experiencing a decline in their agriculture sectors and worse yet are increasingly unable to feed their growing populations. As a result, food import bills in these countries continue to climb. Equally dangerous, is that high levels of food importation are usually accompanied by a fast food culture and the use of unhealthy, highly processed foods – both of which contribute to increases in obesity, hypertension, and other health-related conditions. Consequently, ACP governments are grappling with exorbitant food import bills, loss of employment in the agriculture sector, and a decline in the availability and accessibility of nutritious food that can support healthy and productive populations. Many ACP countries are neither self-sufficient in food production nor self-reliant, that is, having the ability to grow and/or purchase what is required to feed their people. Faced with this situation, international trade policies and market arrangements are often touted as the life line for agriculture growth, food security and poverty reduction for developing countries.

The benefits of agriculture trade liberalization, however, are not straight forward. The impact of these trade policies on food production, food security and food dependency is different for ACP and most developing countries compared to the more developed, industrialized nations. Moreover, there is no 'typical' or 'model' ACP country within and across the three regions. The information and discussions presented in this chapter suggest that individual countries should exercise caution when dealing with trade liberalization policies as no 'one size fits all'. Nevertheless, there are some general concerns about the negative effects of agriculture trade liberalization on the well- being of low-resource households, rural families, small farmers and women farmers in all of the ACP member States.

One glaring feature is that the removal of guaranteed protected markets and subsidies for farmers, as well as the stringent market product mandates (sanitary and phytosanitary measures, standardization and labelling, etc.) has made it difficult for the ACP countries to compete for the same export products on the 'open' and 'free' market. As a result, the agriculture sectors in these countries that are dependent on international trade of 'cash' products find themselves on a downward tailspin as many small producers, farming, rural and working families can no longer generate adequate incomes to support their families. This includes the ability to adequately educate their children, purchase food and other imported items (regardless of cost), afford proper health care and improve their living conditions overall. In other words, the move towards a self-reliant way of life for certain groups in many ACP countries is arduous.

The situation is further compounded because the removal of farm subsidies combined with an increase dependency on imported agriculture inputs, makes it difficult for ACP farmers to grow food for themselves and the domestic markets, at affordable prices. The end result, is that, rather than becoming self-sufficient in food production many of the ACP countries are importing their food, including traditionally grown poultry and other meats, milk and dairy products, some fruit and vegetables, as well as traditionally grown staples (starches, grains, cereals, etc.). So not only do we see local farmers struggling with sales for export crops but also in the production of food for home, community and the domestic market. Currently, many small producers are forced to seek other forms of employment or migrate in search of non-agricultural work. The inability of small producers to maintain their farms also translates to loss of freedom and independence. As farm and rural families migrate, the physical, economic, social and cultural life of entire villages also diminishes.

Equally important, not only are more families unable to grow or purchase adequate food to feed themselves but they are also unable to maintain nutritious diets. Cheaper food does not necessarily mean healthier choices and diets. The quality of food items imported, in particular those that are offered at lower prices tend to have high fats, salt, sugars and carbohydrates. As a result, many ACP countries continue to become dependent on food imports "that contravene their health, nutritional, environmental and cultural norms, including Genetically Modified crops, but without the option of refusing it" (Hurungo 2006:21). Citizens have no control over the price of consumer items, what items should be subsidized by their governments and what foods are imported. Food cultures are changing, health concerns are increasing and the human capabilities of both children and adults are diminishing.

Current trade liberalization strategies are not altogether dissimilar from the old Plantation Model in that they are export production centered. Similarly, they pose serious competition for resources for domestic food production and by extension small producers and women farmers in these countries. As many of the ACP countries adopt policies that support trade liberalization, small producers, especially women farmers are dislocated as land and the support services that should be directed towards local food production are diverted to larger farms producing for the export market. Since men tend to concentrate on cash crop production they are more likely to benefit. Even in cases, Uganda for instance, where the use of women in the production of agriculture goods for export did increase, their access to and control of agricultural resources did not. Neither did the increase in female labor in the export market sector resulted in increased incomes for farming families. Moreover, where some traditional export products (e.g. bananas, sugar, etc.) are 'in trouble' because of the removal of protected markets, there is a tendency for men to shift to growing the same crops as women and compete for limited resources.

There are not many studies on the impact of current agriculture trade liberalization on small farmers and women, but the little evidence that does exist show that there tends to be more harm than benefits. As market operations become more globalized and concentrated at the same time, small producers and women farmers overall are increasingly excluded from decision-making and the advantages of markets (ADB 2013; FAO 2006). International trade agreements, the EU-ACP arrangements in particular, need to be sensitive to the constraints faced by small producers and women farmers in Africa, the Caribbean and the Pacific. Longstanding limitations such as low levels of technical capacity, limited land allocations, credit, extension services, infrastructure and market support continue to inhibit the capacity of small producers and/or women farmers to engage with trade opportunities and switch to higher-return crop or animal production. Worse yet, as more productive resources are reallocated for 'cash crops', the production of food for the domestic market is stymied even though food security and poverty reduction remain top priorities for ACP countries.

On a global level, the ACP group, and developing countries as a whole, are more or less weak partners who are unable to take full advantage of the `intended' benefits from agriculture trade liberalization and 'open' market arrangements. Truth be told, trade policies including the use of tariffs and protectionism have not been very rewarding for African, Caribbean and Pacific countries. While these work for the benefit of developed countries, they do not have the same impact for developing countries and poor people across the globe. To simply follow or respond to the terms of agreement led by groups of select developed countries, multinationals and financial institutions can result in the ACP countries remaining food insecure and income generation from agriculture is stunted. Eventually, some ACP countries may shift from being food producers to becoming reliant on food aid from outside.

Because ACP preferences will continue to be eroded by the WTO, governments should be prepared to adjust to the changes as they expand their own trade. Stronger negotiations are critical so that the agriculture policies and strategies of ACP countries are better integrated into the globalized agriculture and economic trade agreements. According to Koroma and Deep Ford (2006), adjustment assistance should be an accepted dimension of ongoing negotiations by the ACP group on specific products such as rice, sugar, fish, and other items produced largely in these countries. Similarly, it would be advantageous if regional and sub-regional cooperation are strengthened in areas of management, control and surveillance, research and in overall development capabilities. Stepped-up regional economic integration and trade policies, revisiting those that already exist, and addressing issues of natural barriers and transactional costs are all important recommendations. In addition, similar benefits/privileges offered to farmers in developed countries should be given to those in developing countries.

Interestingly, although it is implied that agriculture trade liberalization will have a positive effect on economic growth and poverty, few studies actually examined the distributional effect on both the small farmers and the poor in developing countries, including those of the ACP group. FAO's latest projections indicate that global food production will increase by 60% between 2005/07 and 2050 under its baseline scenario. This represents a downward revision, based on updated data and information, from the 70% increase projected for the same period in 2009 (ESA E Working Paper 2012, No. 12-03). A rise in food production is possible in developing countries, albeit not in all. For example, in Brazil, food production increased from 12.5% in 2003 to 48% in 2008. Likewise, malnutrition declined in Rwanda from 24.5% in 2000 to 15.8% in 2005 (CTA 2011). But this requires that the agriculture in ACP countries to not be driven by international trade liberalization policies, but of meeting the Sustainable Development Goals (SDGs),[7] specifically those related to poverty reduction, ensuring gender equality, economic growth and improving the health and well-being of people. Such gains are more likely to occur from regional cooperation and domestic reforms, rather than from international trade and market arrangements (IFAD 2007).

Nevertheless, it is not recommended or is it practical to simply reverse international trade policies and agreements. However, until the proverbial playing field is levelled, agriculture trade liberalization is not 'fair' nor 'free' and 'open'. The ACP Group of States have proposed South-South Programmes to support food security and nutrition through sustainable food systems and social protection. Governments from these States, while focusing on their local and regional efforts at self-sufficiency and/or self-reliance, must step up their support for this South-South initiative. This is key to the impact of current trade liberalization policies as history has shown us that both wealth and poverty can be derived from agriculture's relationship between the debtor (developing nations) and the creditor (groups of industrialized nations, multinationals and financial institutions).

On the domestic front more attention is required to produce food of high nutritional quality. This requires a range of interventions, including policies that empowers women's decision-making in the sector and providing them with greater access to land, credit, extension services, technology and other resources. Women farmers and small producers are key to both self-reliant and self-sufficient approaches to food production ad food security. They should be encouraged to produce a diversity of grains, roots and tubers, vegetables and fruits, fish and livestock. In doing so, they can provide themselves, and supply local markets, with a diverse diet containing the minerals, vitamins, proteins and calories essential for a healthy life. Raising smallholder productivity is one of the measures being advocated by the Pan-African Nutrition Initiative and the Africa Regional Nutrition Strategy developed by the New Partnership for African Development (NEPAD) and the African Union. Effective implementation of these initiatives in Africa as well as in the Caribbean and the Pacific should improve the nutritional security of the most vulnerable populations in many of the ACP countries. But this calls for a different vision, one that sees the urgent need to create stronger links between agriculture production, food security and health and nutrition. To build stronger partnerships with school feeding projects, local food establishments and cafeterias, with hospitals and other social programs. The end goal is to utilize the agriculture sectors to reduce poverty, hunger, and poor nutrition and to build the capacity, wealth and health of all groups in the ACP countries.

REFERENCES

Abarche, J. B., Dickerson, A., & Green, F. (2004). Trade liberalization and wages in developing countries. *The Economic Journal, 114*(493), 73–96. doi:10.1111/j.0013-0133.2004.00188.x

Africa Progress Report. (2014). *Grain, fish, money, financing Africa's green and blue revolutions*. Author.

Agrictrade. (2011). *Informed analysis, expert opinions- executive brief: Pacific, agric- trade policy debates and developments*. Author.

Amin, S. (1977). Self-reliance and the new international economic order. *Monthly Review (New York, N.Y.)*, *3*(29), 1–21. doi:10.14452/MR-029-03-1977-07_1

Asian Development Bank (ADB). (2013). Gender equality and food security: Women's Empowerment as a Tool against Hunger. Manila, Philippines: Author.

Balakrishnan, R., & Fairbairn-Dunlap, P. (2005). Rural women and food security in Asia and the Pacific: Prospects and paradoxes. UN-FAO. RAP, 2005/30.

Beckford, G. (1972). *Persistent poverty: underdevelopment in plantation economies of the Third World*. Oxford, UK: Oxford University Press.

Bertow, K., & Schultheis, A. (2007). *Impact of EU's agricultural trade policy on smallholders in Africa*. Germanwatch.

Bilal, S., Dalleau, M., & Lui, D. (2012). *Trade liberalization and fiscal adjustments: The case of EPAs in Africa*. European Center for Development Policy Management.

Borgen Project Report. (2015). Retrieved from http://borgenproject.org/9-facts-poverty-south-pacific/

Bulletin on Asia-Pacific Perspectives. (2003/04). *Beyond "subsistence affluence"*. Poverty in Pacific Island Countries.

Caribbean Food and Nutrition Institute. (2008). *Annual Report*. Retrieved from ris.paho.org/xmlui/bitstream/handle/.../AnnualReport08%20CFNI.pdf

Commodities and Trade Technical Paper #8. (2006). *The agricultural dimension of the ACP-EU Economic Partnership Agreements*. United Nations, Food and Agriculture Organization.

Connell, J., & Soutar, L. (2007). Free trade or free fall? Trade liberalization and development in the Pacific and Caribbean. *Social and Economic Studies, 56*(1/2), 41-66.

Dollar, D., & Kraay, A. (2004). Trade growth and poverty. *The Economic Journal, 114*(493), F22–F49.

Falvery, R., Foster, N., & Greenaway, D. (2010). *Trade liberalization, economic crises, and growth*. Bond University.

Falvey, R., Foster, N., & Greenaway, D. (2012). Trade Liberalization, Economic Crises, and Growth. *World Development*, *40*(11), 2177–2193. doi:10.1016/j.worlddev.2012.03.020

Farming Matters. (2013). *Family farming – a way of life*. ILEIA, Center for Learning and Sustainable Agriculture. Retrieved from www.farmingmatters.org

Food and Agriculture Organization. (2003). Trade reforms and food security: Conceptualizing the linkages. United Nations, FAO.

Food and Agriculture Organization. (2006). *Agriculture, trade negotiations and gender.* Rome: United Nations FAO Publications.

Girvan, N. (1973). The development of dependency economics in the Caribbean and Latin America: Review and comparison. *Social and Economic Studies*, *1*(22), 1–33.

Goodison, P. (2007). EU trade policy and the future of Africa's trade relationship with the EU. *Review of African Political Economy*, *34*(112), 247–266. doi:10.1080/03056240701449646

Heaven, D. (2010). *Local sugar facing an uncertain future-divestment not as sweet as was projected.* Retrieved from http://jamaica-gleaner.com/article/news/20151002/local-sugar-facing-uncertain-future-divestment-not-sweet-was-projected

Hurungo, J. (2006). An outline and analysis of EU export subsidies on production of export interest to ESA countries. Academic Press.

International Food Policy Research Institute (IFAD). (2007). *Impact of trade liberalization on agriculture in the Near East and North Africa*. Rome, Italy: Author.

KaringiS. (2008). *Webpate.* Retrieved from http://www.oecd.org/tad/agricultural-policies/39596692.pdf.karingi

Koroma, S., & Deep Ford, J. R. (2006). *The Agricultural dimension of the ACP-EU Economic Partnership Agreements*. Rome, Italy: United Nations, Food and Agriculture Organization.

Lewis, A. (1978a). Growth and Fluctuations 1870-1913. London: Allen & Unwin.

Lewis, A. (1978b). The Evolution of the International Economic Order. Princeton University Press.

Lindiwe, M. S. (2011). *FANRPANO 2011*. Academic Press.

Malua, M. B. (2003). Case Study: The Pacific Island in *Turning Losses into Gains*. SIDS and Multilateral Trade Liberalization in Agriculture. *UN Conference on Trade Development*, (pp. 171-204).

Milner, C., & Morrissey, O. (1999). Measuring trade liberalization in Africa. doi:10.1007/978-1-349-14307-8_3

Morrissey. (Ed.). (n.d.). *Evaluating economic liberalization*. London: MacMillan.

Mucavele, S. (2015). The role of rural women in Agriculture. In *MuGeDe - Women, Gender and Development, Republic of Mozambique - Southern Africa.* Retrieved from http://www.wfo-oma.com/women-in-agriculture/articles/the-role-of-rural-women-in-agriculture.html

NEPAD. (2013). *African agriculture: Transformation and outlook*. Johannesburg, South Africa: NEPAD.

Pan-American Health Organization (PAHO). (2014). *Countries pledge action to reduce child obesity in the Americas*. Author.

Rajack-Talley, T. A. (2014). *Poverty is an economic crime not a moral sin: A history of Globalized food production, foreign dominated economics, trade liberalization and dependency in the Caribbean*. Paper presented at SALISES 15th Annual Conference, St. Augustine, Trinidad.

Rajack-Talley, T. A. (2015). The role of the small farmer, family farms and women in food production, food security, and sustainable development in the Caribbean. In Sustainable Food Production Practices in the Caribbean (vol. 2). Ian Randle Publishers.

Rajack-Talley, T.A., & Talley, C. (2000). The role of farming women in Caribbean agriculture: A historical and contemporary perspective. *Caribbean Journal for Agriculture and Natural Resources, 2*(2), 4-13.

Rodney, W. (1974). *How Europe underdeveloped Africa*. Washington, DC: Howard University Press.

Rodriguez, F., & Rodrik, D. (2000). *Trade policy and economic growth: A skeptic's guide to the cross-national evidence*. Academic Press.

Sachs, J. D., & Warner, A. (1995). Economic reform and the process of integration. Brookings Papers on Economic Activity.

Scollay, R. (2010). *South-South and North-South trade agreements: The Pacific Islands Case*. United Nations University.

Summers, R., & Heston, A. (1991). *The Penn World Table (Mark5): An Expanded Set of International Comparisons, 1950-1987*. NBER Working Paper No. R1562.

Technical Center for Agriculture and Rural Development (CTA). (2011). *Policy brief 3, improving nutrition in AC countries and the role of agriculture*. Retrieved from http://www.cta.int/images/docman-files/policypapers/3_Improving_nutrition_in_ACP_countries.pdf

Thomas, C. Y. (1982). From colony to state capitalism: Alternative paths of development in the Caribbean. *Transition, 5*, 1–20.

UNCTAD. (2013). *Global supply chains: trade and economic policies for developing countries*. Geneva: United Nations.

World Agriculture Towards 2030/2050. (n.d.). *The 2012 revision ESA E Working Paper No. 12-03*. Retrieved from http://www.fao.org/economic/esa/esag/en/

World Bank. (2012). *Press Release*. Retrieved from http://www.worldbank.org/en/news/press-release/2012/10/24/africa-can-feed-itself-earn-billions-avoid-food-crises-unblocking-regional-food-trade

World Bank. (2012). *Little Data book*. Author.

World Bank. (2015). Retrieved from http://www.worldbank.org/en/topic/poverty/overview

ENDNOTES

1 One in every four undernourished people lives on the African continent as a whole.

2 In the Caribbean there is a total of 1.1 million agricultural workers or 17% of the employed labor force, if Haiti is excluded.

3 CARICOM Workshop on Regional Policy for Food and Nutrition Security, Georgetown, Guyana, September 20, 2010.

4 Not all of the African countries however, participate in the regional trade protocols and agreements.

5 The Doha Round is the latest round of trade negotiations among the WTO membership. Its aim is to achieve major reform of the international trading system through the introduction of lower trade barriers and revised trade rules. The work programme covers about 20 areas of trade aimed to improve the trading prospects of developing countries. The Round was officially launched at the WTO's Fourth Ministerial Conference in Doha, Qatar, in November 2001. The *Doha Ministerial Declaration* provided the mandate for the negotiations, including on agriculture, services and an intellectual property topic, which began earlier. In Doha, ministers also approved a decision on how to address the problems developing countries face in implementing the current WTO agreements. The Doha Sanction Waiver expired in 2008 resulting in the loss of tariffs and privilege markets.

6 EPAs represent a fundamental shift in the trading relations between the two parties, from a non-reciprocal preferential trading regime – under which ACP countries could export almost freely to the EU while maintaining their own restrictions on EU imports – to one requiring reciprocity in liberalization, albeit with a certain degree in asymmetry in commitments, in line with rules of the World Trade Organizations (Bilal, Dalleau and Lui, 2012).

7 The Sustainable Development Goals, otherwise known as the Global Goals, build on the Millennium Development Goals (MDGs), eight anti-poverty targets that the world committed to achieving by 2015. The SDGs include a set of 17 Sustainable Development Goals (SDGs) to end poverty, fight inequality and injustice, and tackle climate change by 2030.

This research was previously published in Agricultural Development and Food Security in Developing Nations edited by Wayne G. Ganpat, Ronald Dyer, and Wendy-Ann P. Isaac, pages 1-20, copyright year 2017 by Information Science Reference (an imprint of IGI Global).

Chapter 5
Adaptation of the Russian Food Market to the Contemporary Geopolitical Challenges:
Bans vs. Liberalization

Ivan Ivolga
Stavropol State Agrarian University, Russia

ABSTRACT

Domestic markets of agricultural commodities are increasingly influenced by trade integration and liberalization. Current uncertain political and economic relations in the macro-region of the Commonwealth of Independent States (CIS), Eastern and Western Europe, international tensions around Russia-Ukraine problem, bilateral economic and trade sanctions between Russia and such global economic powers as the USA and the EU transform market patterns and affect agricultural production, rural development and food security in both the CIS and worldwide. The chapter aims to discover the expected influences of such trade restrictions on trade in agricultural commodities, to assess the degree of distortion and return effects on domestic food markets. Chapter specifically addresses possible effects of trade restrictions between the EU, the USA, and other countries from one side, and Russia from the other, particularly imposed ban on agricultural trade. It is concluded with the overview of the expected influences of trade tensions on Russia's domestic food market.

INTRODUCTION

Providing the population with food in sufficient quantity and variety is a challenge, which includes a range of issues of food production, import dependence and export orientation of the food market, solvency and dietary patterns of the population. Domestic production of basic agricultural commodities and food in many countries, including such big agricultural producers as Russia, fails to meet demand. Such countries have to rely on agricultural imports, leaving them vulnerable to global price fluctuations and affecting their export revenues, which tremendously threat food security of those nations.

DOI: 10.4018/978-1-5225-8063-8.ch005

Share of imported agricultural commodities and food on Russia's domestic market exceeds 40% (Federal State Statistics Service of the Russian Federation, 2015). However, in spite of such a high dependency, Russia still decided to ban most imports of Western food products in retaliation for Western trade and economic sanctions.

Such measures will definitely hit World-Russia trade relations. Not only big food producers, processing plants, and retailers, but also a number of small and medium agricultural producers and rural households in trading partner countries are expected to experience losses. Such unilateral actions in the conditions of globalized markets completely distort trade patterns and destroy trade links between countries and regions.

Dedicated to cause losses to the USA and the EU partners, such measures are very much controversial to Russia. Trade protectionism definitely opens up new growth opportunities for domestic producers, but also brings threats to sustainable development of national agriculture and food security (Aslund, 2007).

Regardless of the world's biggest territory and high volumes of domestic agricultural production, Russia remains a net-importer of agricultural commodities. Moreover, its dependency on imports has increased. Currently over 41% of Russia's agricultural GDP is produced by imports, with average EU and USA levels at around 20%. Russia received up to 55% of its agricultural imports from the countries it has so far sanctioned, including the EU. Almost 50% of Russia's meat imports and about 95% of Russia's dairy imports in 2013 came from countries it has now banned, with its biggest suppliers until now being the Netherlands, Germany, Lithuania, Finland, and Poland (Erokhin, 2015a).

Russian farmers are not able to increase their production rapidly, their facilities are outdated; financial resources are limited; quality of their products is low (Bokusheva, Hockmann, & Kumbhakar, 2012; Erokhin & Ivolga, 2012) On the other hand, the ban would definitely backfire by driving up domestic food prices, at least in the short term. Unfortunately, the biggest losers in this will be Russian consumers, who will pay more for their food now as well as in the long run. The ban will hit consumers, along with a number of Russian industries, including food processing plants, shippers and retailers (Erokhin, 2015b). Growth of unemployment and bankruptcies of food processing enterprises are expected.

Until recently, Russia was not among the members of the World Trade Organisation (WTO). Russia's accession to this global trading system in 2012 and membership of some CIS countries in the WTO are considered as preconditions of essential structural changes in interregional and even international trade (Erokhin, 2015a). However, latest trade sanctions imposed on both sides between Russia and Western countries, demonstrated, that WTO had not much to do with those changes. The real threats for Russian agricultural market and trade are unilateral actions, which completely distort trade patterns and destroy trade links between countries and regions. Trade restrictions threat sustainable development not only because of their direct effects (market volatilities, destruction of trade linkages, etc.), but largely because of the hidden ones (distortion of natural economic order, establishment of artificial market environment, overprotection, etc.).

BACKGROUND

The move toward freer trade among both developed and developing countries across the globe has become one of the most essential changes in the world economy since 1980 (Milner, 1999). Liberalization processes increasingly activated in the beginning of the 1990s, when many former socialist countries (republics of the Soviet Union, Central and Eastern European states, and countries of South-East Asia) all chosen to shift from centrally-planned to market economies and therefore liberalized their trade policies.

Dornbusch (1992) defines four major drivers of freer trade regimes during the early stages of economic transition: anti-statism, poor economic performance, information, and evidences of success (pp. 69-70). All economies in transition experienced a rising movement against the beneficial role of the state during the 1990s (anti-statism effect), when trade protectionism was considered as one of the manifestations of an overly intrusive state (Dornbusch, 1992, p. 69). Along with a declining economic performance, growing awareness of the opportunities available in developed countries, and favorable performance of those countries, which adopted outward-oriented policies, anti-statism served to make trade liberalization a general trend of economic policy among the economies in transition.

Notwithstanding certain difficulties in interpreting country experiences during the transition period, many economists agree that liberal trade is important for growth over the long run (World Bank, 2005, pp. 136-137). Srinivasan and Bhagwati (1999) conclude that more outward-oriented countries tend consistently to grow faster than ones that are inward-looking. As of Dollar (2001), countries that opened their domestic markets during the 1990s, experienced faster growth and more poverty reduction.

There are certain benefits of trade liberalization. It allows countries to utilize their competitive advantages and gain profit from greater involvement into international trade. Growing competition from foreign countries spurs domestic producers to increase efficiency and cut production costs. The removal of tariff barriers leads to lower prices for consumers.

Despite such positive effects of trade liberalization, there are many arguments against it, which have to be considered. In case of the economies in transition, the differences are at what stage of development a country should open its market (Spanu, 2003). In the early 1990s, economies in transition undertaken reductions in their protection, at considerable pain and effort, largely with a view to enhancing their own export performance (Anderson et al., 2001). However, as of Rodrik (2000), trade reform is only one element of successful development policy. The entire success of trade liberalisation depends on how flexible an economy is. Trade liberalization requires the development of institutions, but such institutional development can be extremely costly for the economies in transition. Time also matters. Nugent (2002) finds that trade liberalization was successfully implemented in countries where this process was sustained for a long period of time. In the short run, trade liberalisation is often painful and may be even damaging for domestic markets of developing economies.

Although free trade is a key factor in promoting economic growth through international competition and efficient allocation of resources, there are specific markets, which require specific approaches. Food market is one of them due to its vital importance for securing nutritional requirements of the world population. While international trade in recent decades has been gradually liberalized, trade in agricultural products and food has remained among the ones most influenced by international regulations and national policies (Božić, Bogdanov, & Ševarlić, 2011). As of Markovic and Markovic (2014), agricultural protectionism is a part of the agricultural policy of almost every country (p. 424). It is focused on the selection of measures of foreign trade and economic policies to achieve the protection of agricultural sector and domestic food market from foreign competition. In a broader dimension, protectionist policies are aimed at development of sustainable food production, ensurance of food security of a country, and sustainable management of natural resources.

It is worthwhile to note that agricultural protectionism is widely implemented by both developing and developed nations. Moreover, for the EU and the USA it has particular significance in view of protection of domestic food markets and encouragement of exports. Developing countries and economies in transition also implement a variety of measures, which directly restrict imports, stimulate exports, or support domestic agricultural production.

Many governments are re-examining their strategies for dealing with domestic agricultural production and food security concerns (Anderson, Jha, & Nelgen, 2013). As of Josling, Anderson, Schmitz, and Tangerman (2010), developing countries are increasingly providing protection to the import-competing subsector of their agricultural sector (p. 441). In the countries of the Organisation for Economic Co-operation and Development (OECD), almost one third of farmers' income on average is not actually earned in agricultural markets, but rather comes from a range of government subsidies and other support measures that restrict agricultural trade and distort markets (Secretariat of the Convention on Biological Diversity, 2005, p. 3). Effects of such distortions, both positive and negative ones, have to be carefully assessed in order to balance protection and liberalization policies of the economies in transition.

MAIN FOCUS OF THE CHAPTER

Russian Agricultural Market: Retrospectives and Contemporary Situation

Considering the retrospectives of the Russia's food market, the author emphasizes the rapid decrease of agricultural production (and its share in the GDP) in the beginning of the 1990s, after the fall of the Soviet Union (USSR) and during the early period of liberal economic reforms. That sort of the liberalization effect is common to most of the economies in transition, when opening of the domestic market causes an immediate increase of import deliveries, while local producers do not have facilities to respond swiftly on such a change of external competitive environment. There were many reasons for such a situation in Russia in that time: weak financial and economic conditions of the majority of Russian agricultural producers in the 1980s and early 1990s, destruction of production ties under the USSR collapse, high inflation, outdated machineries and technologies, etc. Russia was on the very early stage of market economy, and it was rather hard to compete with foreign producers even on the domestic market.

As a result, the share of agricultural production in GDP slumped from 17.0% in 1990 to 6.3% in 1995, and then continued gradual decrease down to 3.3-3.4% (from the mid-2000s to the present day) (Table 1). Such a level is within the world average range, as in the most developed countries the share of agriculture in GDP oscillates between 1% and 10% (USA: below 1% in 2014).

Table 1. Major indicators of Russia's agriculture in 1990-2014

	1990	1995	2000	2004	2008	2009	2010	2011	2012	2013	2014
Share of agricultural production in GDP, %	17.0	6.3	5.8	4.5	3.6	3.9	3.3	3.4	3.4	3.4	3.3
Agricultural land, million hectares	213.8	209.6	197.0	193.0	191.0	191.8	191.9	191.7	191.9	191.8	191.8
Rural population, million people	39.5	39.4	39.0	38.8	37.9	37.7	37.5	37.3	37.2	37.1	37.0
Share of agriculture in employment of population, %	13.2	15.7	13.0	10.2	8.6	8.4	7.9	7.9	7.3	7.0	7.0

Source: Author's compilation based on the Federal State Statistics Service of the Russian Federation, 2015

In the course of its rapid recovery after financial crisis of 1998, agricultural production in Russia demonstrates far moderate growth rates. In the period of 1999-2012, favorable economic situation in Russia ensured the annual growth of domestic agricultural production by 2.4%. Decrease was registered only in 2010, when the country experienced a severe drought (Ishchukova, 2013).

Agricultural land area in Russia has been decreasing in recent decades. Despite the world's biggest territory, only 7% of land in Russia is suitable for agricultural use. The mass of territories experience unfavorable climatic and environmental conditions for agricultural production. Over the past five years (since 2009), the agricultural land area has decreased by 16.2 million hectares.

About a quarter of Russian population live in rural areas, but rural population is decreasing. The majority of rural territories in Russia experience economic depression and depopulation. Young people in active working age migrate to urban areas seeking better employment opportunities. Fertility rate in rural areas declines at a quicker pace in comparison with urban areas, which, linked with outflow to urban areas, causes ageing of rural population. Number of rural settlements in Russia in 2010 (the latest agricultural census) decreased by 9.2 thousand in comparison with 1989. According to the All-Russian Research Institute of Agricultural Economics, over one third of rural inhabitants consider a migration to urban areas as an opportunity to increase their living standards. Among young people, the share is even bigger – up to a half (Ushachev, 2012).

Such a tendency is not a unique Russia's problem. Similar processes are observed in other countries. In particular, in the USA and the EU over the past five years the share of rural population has declined by 5% (Erokhin, Ivolga, & Heijman, 2014). Lower status value of rural way of life in comparison with an urban one, higher risks of agricultural production in comparison with a wage employment in cities, and absence of significant career and income perspectives force people to seek better opportunities outside the traditional areas of their inhabitation.

Share of agricultural sector in employment of Russia's population is decreasing. Over the past decade, this indicator has declined by more than three percentage points, and by almost one-half from the beginning of the 1990s. The reasons are bankruptcy of local agricultural producers, migration of rural inhabitants to urban areas, decrease of animal production in agricultural GDP. Animal production is a more labour intensive industry in comparison with plant production. Decrease in animal production releases more workforces, then in plant production, furthermore, drives unemployment in related industries.

The above-described situation affects the production-consumption ratio on the domestic food market: production is decreasing along with reduction of employees and migration outflow from rural areas, while demand (effective demand, supported by increasing living standards of urban inhabitants) is growing along with urban population.

The rising demand has to be satisfied, that is why foreign trade balance of agricultural commodities and food has been steadily increasing during 1990-2014 (not only because of growing world food prices, but also by means of expanding import). Since the 1970s, the USSR has been one of the world's biggest importers of grain (mostly forage). In the 1990s, Russia started to import processed products of animal origin instead of operating domestic animal production. Import of meat (including poultry) and meat products reached enormous volumes. Share of food import on the domestic market started to grow (after 24% in the 1980s and following decrease down to 16% by 1990).

As long as market reforms in Russian agriculture assumed wide privatization, volumes of state support were significantly reduced, while transformation of land relations itself caused a bulk of unsolved problems and uncertainties, which negatively affected the effectiveness of agricultural production (Trukhachev, Ivolga, & Lescheva, 2015). Taking into account destruction of existing economic ties between food producers and food consumers because of the USSR fall, territorial changes, introduction of new custom borders, and the fall of agricultural production during 1990-1998 is explainable.

However, in 1998 agricultural production started to grow. There were two preconditions, which had been developed on the domestic food market by that time. Both of them originated from the Russia's financial crisis of 1998. First reason was the significant appreciation of food import because of the Russia's national currency (ruble) devaluation during 1998-1999. Domestic producers managed to capture share of the market and start import substitution with low-cost domestic products. Second reason was related to currency too. Because of devaluation, earnings in foreign currencies became rather attractive and profitable for Russian food producers. Consequently, they started to increase output, aiming at the international food market.

Grain crops, in particular, wheat, have always been and remain the major export commodity of Russian agriculture. Share of grain crops in total agricultural production grew in 1990-2008 up to 20.3% (Table 2).

Considering the ratio between plant production and animal production in the structure of Russia's agricultural GDP, the author emphasizes the priority inversion. If in the late 1980s the bigger part of Russia's agricultural GDP was provided by animal production, starting from 1990 and onwards author observes the increasing share of plant production (Figure 1). As mentioned above, animal production is more labor intensive and less profitable in comparison with plant production.

Along with domestic agricultural production, during the recent decades Russia's food market has also been influenced by price policies. After the fall of the Soviet Union, factor prices started to grow in a more rapid pace, than food prices. Agricultural sector suffered from growing prices on petroleum, oil and lubricants (POL), electric power, transportation and storage of agricultural raw commodities. Consequently, the essential disparity between food prices and factor prices existed, while the profitability of domestic agricultural producers was one of the lowest among Russia's industries.

There was also a negative influence of inflation, which was relatively stabilized (10-20%) only after 2002, while in 1998 it reached 84.4% (Figure 2) (Ishchukova, 2013). For the purpose of a better visual-

Table 2. Structure of Russia's agricultural production in 1990-2014, %

	1990	1995	2000	2004	2008	2009	2010	2011	2012	2013	2014
Total agricultural production	100.0	100.0	100.0	100.0	100.0	100.0	100.0	100.0	100.0	100.0	100.0
Plant production, total, incl.:	36.7	53.1	52.9	51.9	53.1	49.2	45.1	52.2	49.0	52.0	51.0
grain crops	10.3	12.5	16.6	17.5	20.3	14.7	10.5	17.3	13.5	16.7	16.0
industrial crops	2.0	3.3	2.5	4.3	4.5	4.7	5.6	5.5	4.6	5.8	5.7
potatoes	10.8	17.9	14.4	11.7	11.4	12.0	10.7	9.8	10.4	10.0	11.2
vegetables and gourds	6.1	9.9	10.7	9.5	9.1	9.7	10.5	9.9	10.2	9.8	10.0
fruit and berries	1.5	2.8	4.1	4.0	3.6	2.9	2.8	2.7	2.8	2.8	2.8
Livestock production, total, incl.:	63.3	46.9	47.1	48.1	46.9	50.8	54.9	47.8	51.0	48.0	49.0
meat and poultry	21.4	20.2	23.3	23.3	24.5	28.6	30.2	29.5	30.4	29.7	31.4
milk	17.9	18.8	17.0	17.6	16.7	16.2	18.4	15.2	17.9	16.3	16.8
eggs	4.5	4.8	4.7	5.0	4.1	4.1	4.2	4.0	3.9	4.1	4.3
wool	0.9	0.5	0.1	0.1	0.1	0.1	0.1	0.1	0.1	0.1	0.1

Source: Federal State Statistics Service of the Russian Federation, 2015

Figure 1. Structure of Russia's agricultural production (plant and livestock production) in 1990-2014 and forecast for the period until 2020, %
Source: Federal State Statistics Service of the Russian Federation, 2015

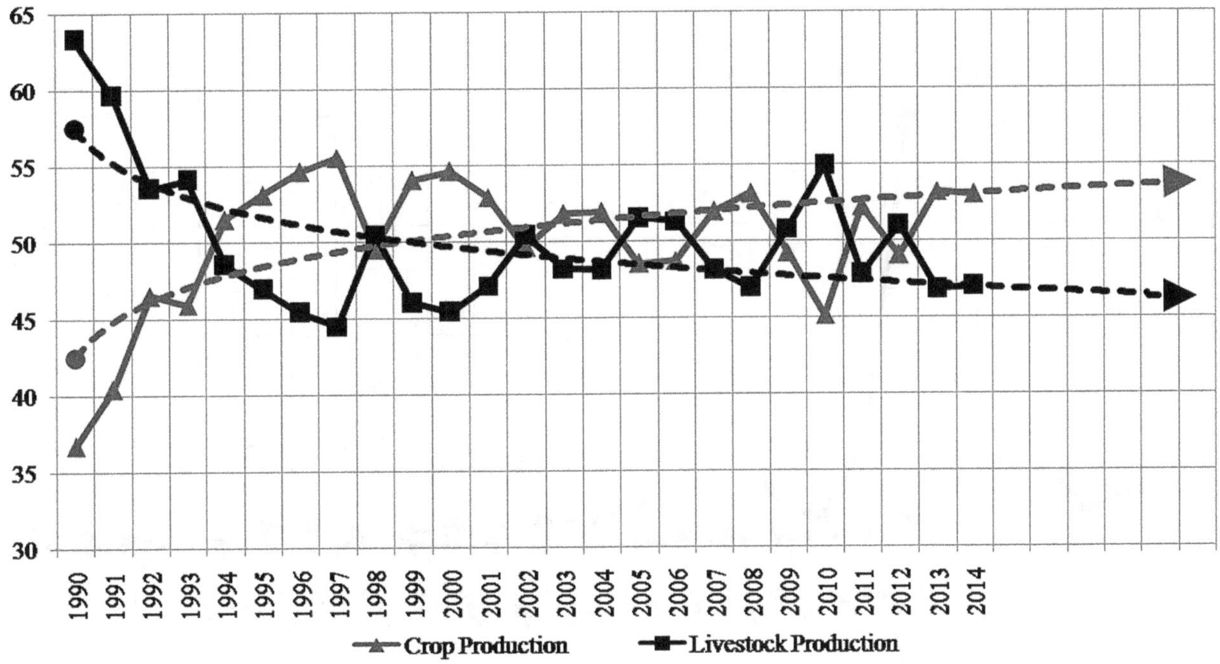

ization of the Figure 2, the author has intentionally excluded Russia's inflation rates for 1991-1994 due to their incomparably higher levels (2500% in 1992, 874% in 1993, and 307% in 1994).

During 1990-1992 food prices in Russia increased fourteen fold (plant production – thirtyfold, animal products – ninefold). Figure 2 shows that food inflation in Russia was caused by domestic factors, not external ones. According to Mikhailov (2011), food inflation in Russia is influenced in a lower extent by the international market in comparison with internal reasons. Petronevich and Kondrashov (2011), and Ishchukova (2013) also provide data on asymmetry of Russia's domestic food prices and the world food price index.

Although Russia's producers' price index has been following the world trend since the beginning of the 2000s and until 2008-2009, global economic recession and deceleration of Russian economy along with external political tensions and bilateral trade sanctions in 2014 caused an essential gap between the world price index and price index of Russian food producers (Figure 3).

World food priced have been growing starting from 2005, the peaks were registered in 2007, 2008, and 2011. In the following years (2012-2014) the prices have been falling. Food price index in December 2014 decreased by 1.7% down to the 2011-2014 minimum. The only commodity, which prices increased, was meat (+8.1%). Price for grain crops slightly increased in the end of 2014 over fears of possible Russia's export restrictions. Nevertheless, Russia introduced the ban on import of agricultural commodities and food from the EU, the USA, Canada, Australia, and Norway.

World food prices have been decreasing since 2011, long before of any bans imposed by Russia. Influencing factors are good yields of recent years in the majority of the world leading food producing countries, amassed reserves, declining transportation costs (because of falling oil prices) and soaring

Figure 2. Annual average inflation in Russia and in the world in 1990-2014, %
Source: Author's development based on the Federal State Statistics Service of the Russian Federation (2015); United Nations Conference on Trade and Development (2015)

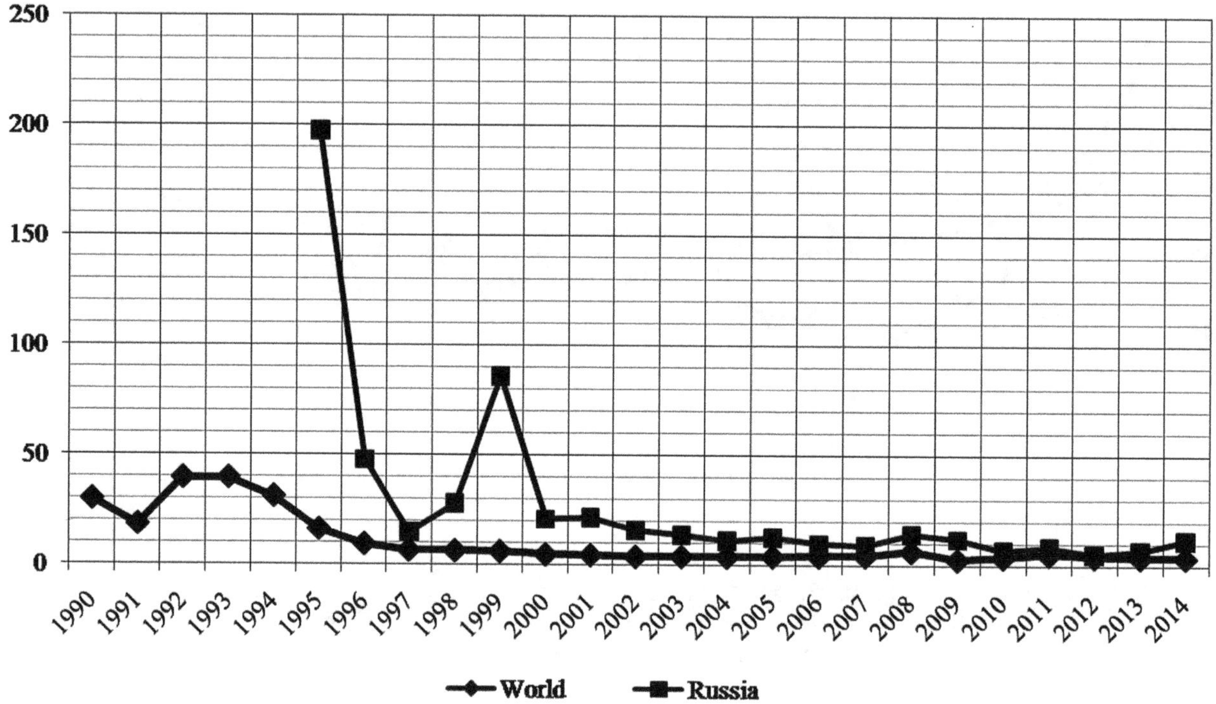

Figure 3. Price indices on agricultural products in Russia and in the world in 1990-2014
Note: basis year is 2002
Source: Author's development based on the Federal State Statistics Service of the Russian Federation (2015); United Nations Conference on Trade and Development (2015)

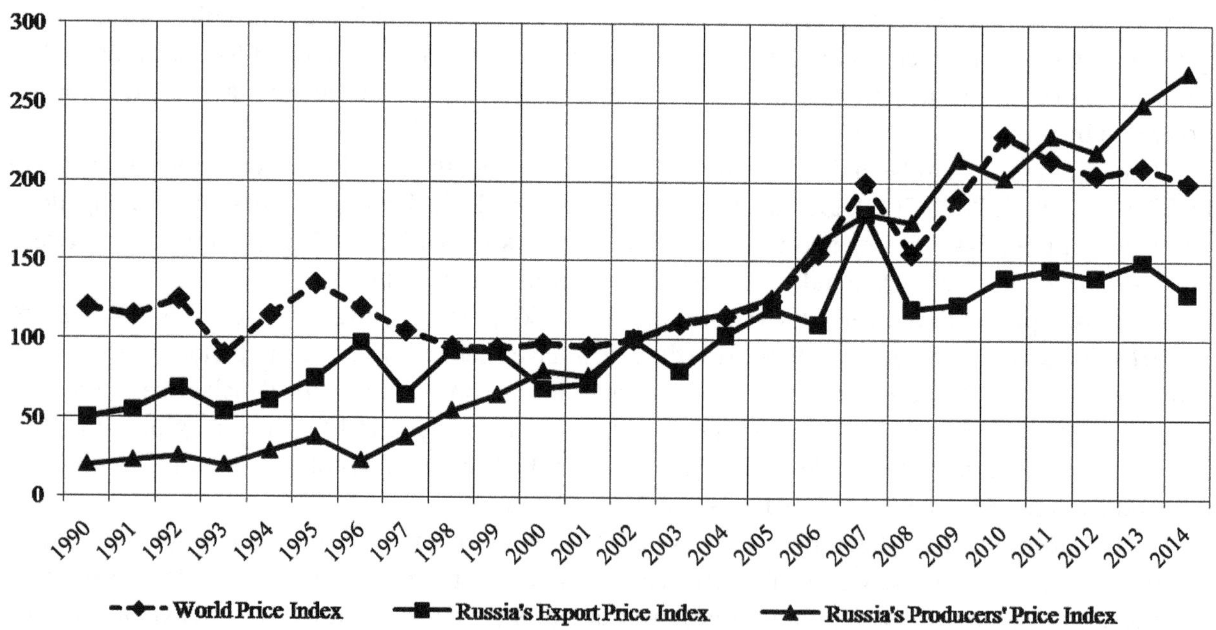

US dollar rate. Russia currently provides about 1.0% of the world food export and 2.7% of the world food import (Table 3), that is why the degree of impact of Russia's actions on the world food market should not be overestimated.

Over the recent decades, Russia has been a net importer of agricultural commodities and food. Moreover, the share of the country in the total world food import has been growing (by 1.3 percentage points) since the very fall of the Soviet Union.

Food Foreign-Trade-to-GDP ratio (Foreign Trade Quota) has been growing over two decades onwards (Figure 4). The outrun growth of the Food Import-to-GDP ratio (Import Quota) in comparison with the Food Export-to-GDP ratio (Export Quota) should be noted. Even in 2014, when Russia's domestic

Table 3. Russia's share on the world food market in 1990-2014, %

	1990	1995	2000	2005	2010	2011	2012	2013	2014	2014 to 1990, percentage points
Share in the world export	0.5	0.4	0.3	0.5	0.5	0.7	1.3	0.9	1.0	0.5
Share in the world import	1.6	2.3	1.7	2.3	2.9	2.8	2.5	2.4	2.7	1.3

Source: United Nations Conference on Trade and Development, 2015

Figure 4. Food export, import and foreign trade ratios to Russia's GDP (agriculture) in 1995-2014 and forecast for the period until 2020, %

Source: Author's development based on the Federal State Statistics Service of the Russian Federation (2015); United Nations Conference on Trade and Development (2015)

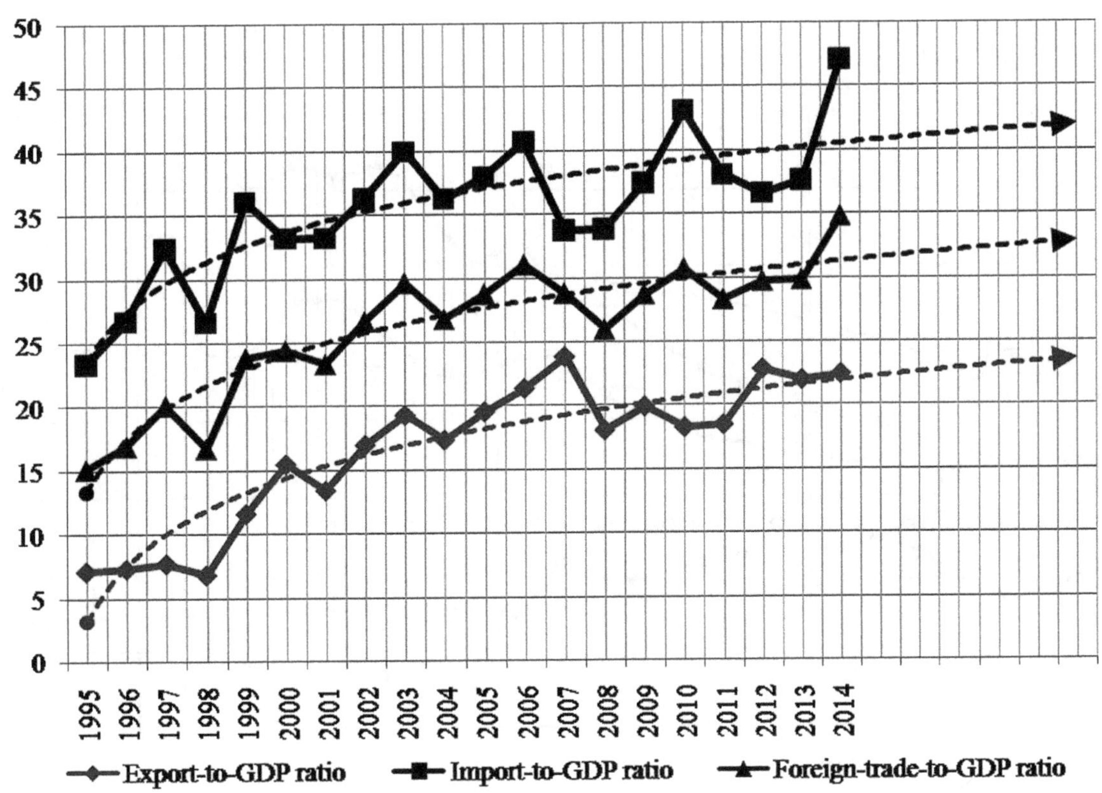

food market experienced two tendencies, unfavorable for import (depreciation of food import because of the devaluation of the national currency and import ban, introduced in August 2014), Import Quota spiked and reached 47.1%, while export predictably decreased, as Russian producers reoriented on the domestic market.

Such a dynamic of Food Import Quota confirms author's previous conclusion on high dependency of Russia's domestic food market on import deliveries. Import of food commodities to Russia in 2014 was $39.7 billion, which is 9.5% less than in 2013. Food export decreased by 26% – from $25.6 billion down to $18.9 billion (Figure 5).

Negative trade balance has been the permanent trend of Russia's trade in agricultural commodities over the past two decades. Falling import and introduced import ban can hardly break the tendency. Simple linear trend model shows the expected $21 billion of negative trade balance by 2020.

Russia's major food import commodities are fruit ($5.26 billion in 2014), meat and meat products ($2.58 billion), vegetables ($2.11 billion), and cheese ($2.09 billion) (Table 4). There is a decrease of import deliveries in 2014 in comparison with 2013 as a result of both introduced import ban and depreciation of import because of devaluation of the Russian ruble.

Russia's food export is rather more homogenous than its import. Its major volume is composed of grain crops, in particular, wheat and barley (Table 5). From the early 2000s, Russia has been among the top world's wheat exporters (up to 14% of global market of wheat). Domestic yield of wheat in 2014 amounted 59 million tons, which is about 8% of the world wheat production. In 2012, wheat provided

Figure 5. Food export, import and trade balance of Russia in 1995-2014 and forecast for the period until 2020, %

Source: Author's development based on the Federal State Statistics Service of the Russian Federation (2015); United Nations Conference on Trade and Development (2015)

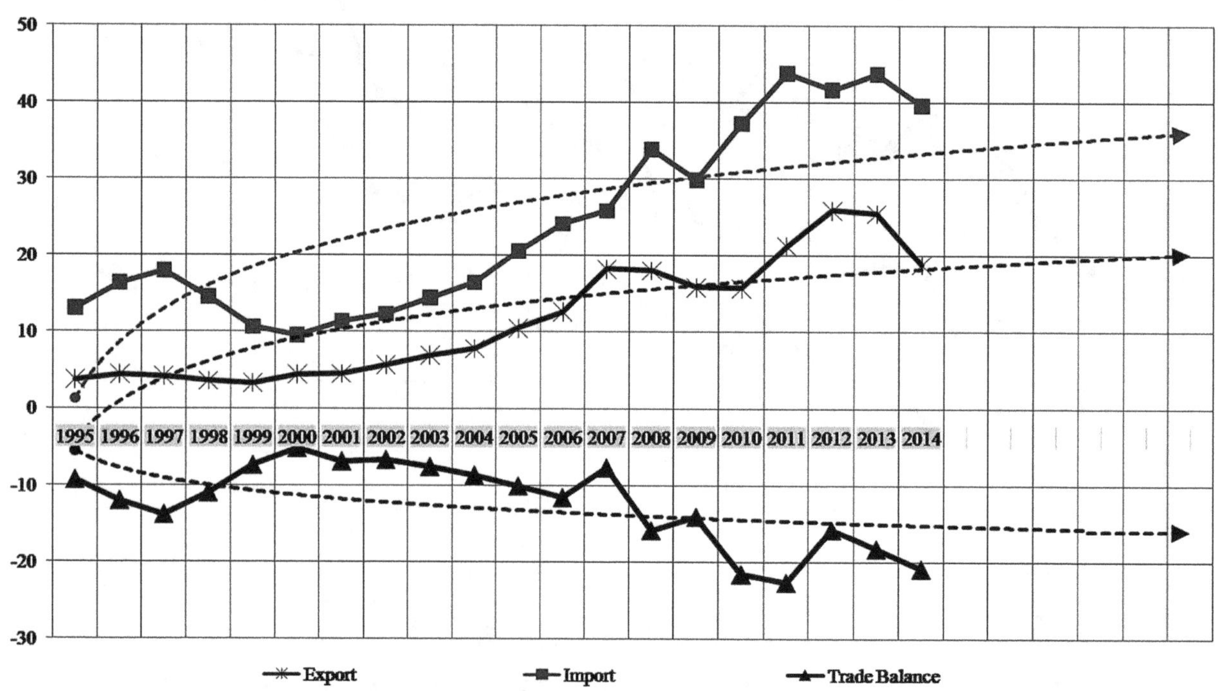

Table 4. Import of selected agricultural and food products to Russia in 1990-2014, $ billion

	1990	1995	2000	2005	2010	2011	2012	2013	2014
Meat and meat products	0.51	0.62	0.48	1.44	2.82	3.16	3.23	2.94	2.58
Cheese	0.12	0.21	0.20	0.91	1.93	2.07	2.00	2.33	2.09
Milk and dairy products	0.19	0.24	0.20	0.40	1.06	0.85	1.00	1.51	1.43
Fish (fresh, chilled and frozen)	0.09	0.17	0.18	0.98	1.75	2.02	1.98	2.33	2.07
Vegetables	0.30	0.34	0.28	0.75	2.20	2.88	2.23	2.59	2.11
Fruit	0.57	0.93	0.72	2.20	5.06	5.71	5.48	5.71	5.26
Eggs	0.08	0.13	0.41	0.58	0.14	0.18	0.20	0.26	0.24
Sugar and honey	1.04	1.23	0.96	1.23	1.98	2.44	0.96	0.91	0.96

Source: United Nations Conference on Trade and Development, 2015

Table 5. Export of selected agricultural and food products from Russia in 1990-2014, $ billion

	1990	1995	2000	2005	2010	2011	2012	2013	2014
Wheat	0.15	0.06	0.04	1.13	2.07	3.67	4.52	3.48	5.25
Barley	0.08	0.03	0.05	0.20	0.20	0.49	0.88	0.55	0.92
Maize	0.02	0.01	0.01	0.01	0.04	0.16	0.57	0.59	0.85
Rice	0.01	0.01	0.01	0.01	0.08	0.09	0.17	0.09	0.10
Milk and dairy products	0.02	0.06	0.09	0.12	0.06	0.07	0.21	0.24	0.26
Fish (fresh, chilled and frozen)	0.06	0.12	0.27	0.39	1.86	2.03	2.12	2.39	2.09
Vegetables	0.01	0.02	0.02	0.05	0.06	0.25	0.34	0.25	0.17
Fruit	0.01	0.01	0.03	0.04	0.01	0.01	0.07	0.05	0.04
Eggs	0.01	0.01	0.01	0.01	0.01	0.01	0.02	0.03	0.01
Sugar and honey	0.05	0.12	0.04	0.04	0.03	0.08	0.10	0.06	0.05

Source: United Nations Conference on Trade and Development, 2015

18.4% of Russia's agricultural export. Such a high dependency on a single commodity is rather dangerous. In case of a bad yield or decrease of world prices domestic wheat producers will suffer losses, which could be hardly compensated by the government.

Reduction of export deliveries in 2014 in comparison with 2013 was caused by both the reorientation of domestic producers on Russia's internal food market and falling world food prices.

During the 1990s, the majority of export deliveries from Russia went to the Western Europe and Asia – approximately in equal proportions. However, by 2010 Asia had become the main consumer of food from Russia, followed by the CIS, Africa, and the EU (Table 6).

Share of Africa in Russia's food export spiked during 1995-2010 from 1.46% in 1995 up to 14.81% in 2010, but had decreased by 2014 (down to 8.05%). The growth was provided by export of wheat and barley to Egypt, Tunisia, Morocco, and Turkey.

Considering the structure of Russia's food export by the biggest consumers, China has a leading position (Figure 6).

Table 6. Russia's food export by regions in 1995-2014

Regions	1995		2000		2010		2014	
	$ billion	share in total, %	$ billion	share in total, %	$ billion	share in total, %	$ billion	share in total, %
World, total	4.12	100.00	1.30	100.00	7.56	100.00	18.90	100.00
CIS	0.68	16.50	0.42	32.31	1.77	23.41	5.35	28.30
EU	1.69	41.02	0.24	18.46	0.78	10.32	3.74	19.78
North America	0.10	2.43	0.02	1.54	0.08	1.06	0.18	0.94
South America	0.01	0.24	0.01	0.01	0.01	0.13	0.01	0.07
Asia	1.29	31.31	0.46	35.38	3.45	45.63	7.09	37.54
Africa	0.06	1.46	0.04	3.08	1.12	14.81	1.52	8.05
Other	0.29	7.04	0.12	9.22	0.35	4.63	1.01	5.32

Source: United Nations Conference on Trade and Development, 2015

Figure 6. Top ten consumers of Russia's agricultural and food products in 2014, share in total Russia's food export, %
Source: United Nations Conference on Trade and Development, 2015

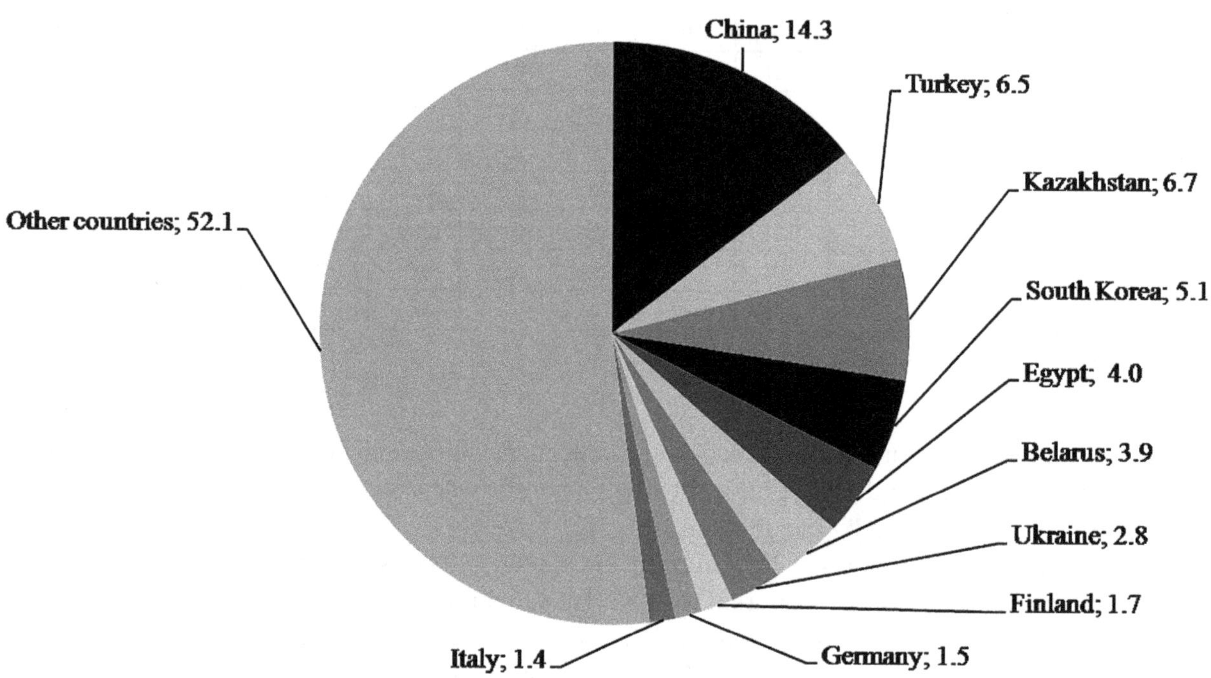

Share of China in Russia's food export is 14.3%. In the late 2000s, it was even higher, but China has been rather successful in development of domestic agricultural production over the past years, which let it to reduce import procurements gradually. Turkey is the second biggest destination of Russia's food export (6.5% in 2014). Over the past years, Turkey has also become the leading supplier of fruit and vegetables to Russia, importing Russia's wheat, sunflower, and barley in return.

Analysing the structure and volumes of food import to Russia, it is worthwhile to note the high share of the EU – about one third of total import (Table 7). In 2014, the EU supplied over 40% of Russia's food import, regardless the introduced import ban.

Among the EU countries, the major trade partner is Germany – over 18% of Russia's food import from the EU. Before the ban, the country supplied to Russia meat and meat products (pork, beef, and poultry), cheese and other food products.

In 2014, the share of South America and Asia in Russia's food import increased. Russia was able to reorient its import procurements to those regions after the ban, introduced in 2014. Shares of South America (in particular, Brazil) and Asia (in particular, China) in Russia's food import increased up to 15.58% and 15.92% correspondingly. Brazil provides up to 85% of Russia's import of raw sugar, up to 45% of beef, and about 40% of pork. Argentina, Uruguay, and Paraguay are other big meat suppliers to Russia.

In the region of Africa, the major food suppliers of Russia are Morocco and Egypt. Two countries together provide about 55% of Russia's food import from Africa.

Top ten leading food suppliers of Russia are Belarus, Brazil, Germany and other countries of the EU, South America, and Asia (Figure 7).

Bans vs. Liberalization: Russia's Geopolitical Actions and Their Influences on Interregional Trade

Today with escalating political conflict between Russia and Ukraine, economic sanctions imposed by the EU and the USA, and Russia's reorientation on Asia and Latin America, it is not clear how further progresses can be achieved between Russia and its trade partners.

Russia banned beef, pork, poultry, fish, fruit, vegetables, cheese, milk and other dairy products from the USA, Canada, the EU, Norway, and Australia for one year (in August 2014) and later (in June 2015) prolonged the ban for till July 2016. Russia imported about €33 billion worth of food products in 2014, of which the banned goods accounted for nearly €7 billion, or 21.2%. Top ten suppliers to Russia include the countries of Latin America, Europe, and Asia. Five out of the top ten are banned, which are

Table 7. Russia's food import by regions in 1995-2014

Regions	1995		2000		2010		2014	
	$ billion	share in total, %	$ billion	share in total, %	$ billion	share in total, %	$ billion	share in total, %
World, total	14.90	100.00	6.98	100.00	33.62	100.00	39.70	100.00
CIS	3.71	24.89	1.67	23.93	3.41	10.14	7.26	18.29
EU	6.76	45.37	1.84	26.36	10.04	29.86	16.03	40.38
North America	1.37	9.19	0.77	11.03	1.77	5.26	1.87	4.72
South America	1.00	6.71	0.68	9.74	6.80	20.23	6.19	15.58
Asia	1.67	11.21	0.86	12.32	5.87	17.46	6.32	15.92
Africa	0.22	1.48	0.21	3.01	1.38	4.10	1.50	3.78
Other	0.17	1.15	0.96	13.61	4.35	12.94	0.53	1.33

Source: United Nations Conference on Trade and Development, 2015

Figure 7. Top ten suppliers of agricultural and food products to Russia in 2014, share in total Russia's food import, %
Source: United Nations Conference on Trade and Development, 2015

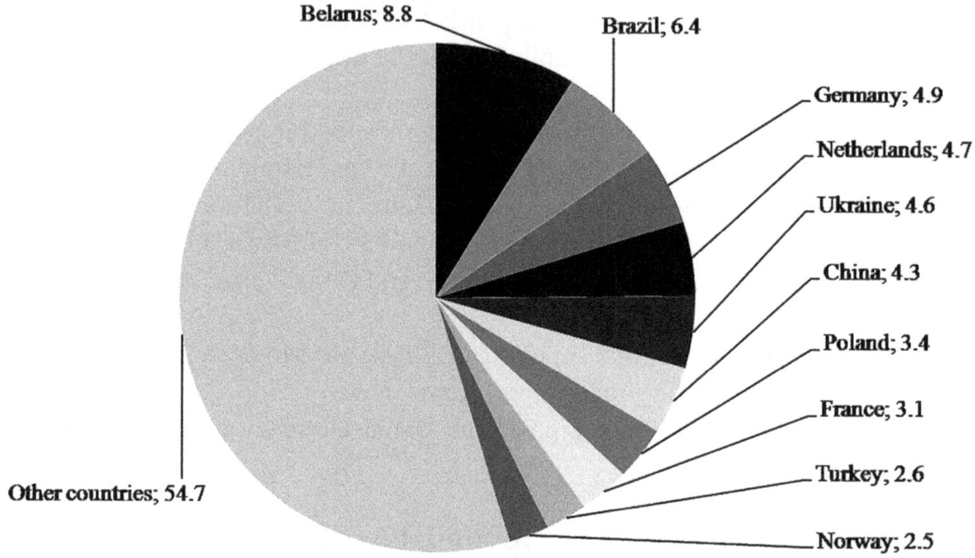

the Netherlands, Poland, Norway, France, and the USA. The ban is expected to hit about 7% of the EU agricultural exports to Russia (Erokhin, Heijman, & Ivolga, 2014). For certain EU countries, i.e. Poland, France, and Norway, losses may be more essential. Russia used to be a large market for Polish fruits, French cheese, and Norwegian fish.

In order to evaluate the expected effects of the ban on Russia-World trade in food and agricultural commodities, it is worthwhile to look at the particular volumes of imports from the EU, the USA, Canada, Australia and Norway to Russia in 2013, 2014, and 1995-2013 (average). Russia banned three kinds of meat and meat products, fish, fruit, vegetables, and all dairy products. As of the UNCTAD classification (United Nations Conference on Trade and Development, 2015), those commodities are included into the following commodity groups (Table 8).

The ban, introduced in August 2014, affected the final import volumes. That is why the author uses 2013 as the basis year for calculation of the potential volume of losses. Using the 1995-2014 data, the author implemented logarithmical forecast of food exports of the EU, the USA, Canada, Australia, and Norway to Russia for the period until 2020. Author assumed the total average exports (Table 8) as the potential losses of the selected countries from Russia's import ban (from 2015 onwards) and correspondingly corrected the logarithmical trend on that value. The resulting graph of the EU food export to Russia with the forecast of the potential losses is presented separately (Figure 8) from the graph of the remaining four countries (Figure 9) in order of a better visualization.

Logarithmical forecast for the period until 2020 shows the potential growth of food export from the EU to Russia up to $16 billion. As the result of both the introduced ban and rapid depreciation of import deliveries in the fourth quarter of 2014, the cutback of export volume down to $11.2 billion is observed. Based on the calculated total average of banned commodity groups ($4.6 billion), the author forecasts further cutback below $9 billion until the ban is not lifted.

Table 8. Import of selected food and agricultural products to Russia from selected countries in 2013-2014, $ million

Commodity group	EU		USA		Norway		Canada		Australia	
	2013	2014	2013	2014	2013	2014	2013	2014	2013	2014
Meat of bovine animals, fresh, chilled or frozen	139.91	165.94	0.87	126.81	0.01	0.01	0.58	5.94	178.04	101.40
Other meat and edible meat offal	1,582.72	792.45	349.48	569.18	0.02	0.02	258.82	321.51	59.70	43.52
Meat, edible meat offal, salted, dried; flours, meals	20.95	16.38	0.01	0.18	0.01	0.01	-	0.88	-	-
Meat, edible meat offal, prepared, preserved	174.73	142.65	4.30	6.58	0.01	0.01	0.04	0.17	0.06	0.21
Milk, cream and milk products (excluding butter, cheese)	303.42	192.33	4.45	1.97	0.54	0.60	-	-	2.30	1.56
Butter and other fats and oils derived from milk	183.08	122.41	0.02	0.02	-	0.27	-	-	67.14	30.36
Cheese and curd	1,247.67	837.93	0.06	0.11	4.42	2.69	-	-	10.71	4.36
Fish, fresh (live or dead), chilled or frozen	158.41	127.89	75.39	58.48	1,140.51	616.52	34.25	20.45	0.67	0.81
Fish, dried, salted or in brine; smoked fish	1.68	1.34	0.56	0.33	0.17	0.05	0.01	0.01	-	-
Fish, aqua. invertebrates, prepared, preserved	65.66	59.79	1.98	1.86	0.25	0.31	0.52	0.21	-	-
Vegetables	1,008.52	730.71	7.40	7.60	0.09	0.04	2.85	4.18	0.11	0.05
Vegetables, roots, tubers, prepared, preserved	358.18	351.50	3.30	5.63	-	-	0.01	0.08	0.18	0.16
Fruits and nuts (excluding oil nuts), fresh or dried	1,528.03	1,143.34	235.31	164.80	0.02	0.01	0.11	0.12	9.82	6.20
Fruit, preserved, and fruit preparations (no juice)	147.40	130.24	10.29	9.59	-	-	2.67	1.87	0.83	0.34
Total	6,920.36	4,814.90	693.42	953.14	1,146.05	620.54	299.86	355.42	329.56	188.97
Total average in 1995-2013	4,643.62		872.90		709.17		244.71		181.81	

Source: Author's own calculation based on the United Nations Conference on Trade and Development (2015)

The similar approach to the assessment of food export cutbacks of the USA, Canada, Australia, and Norway lets us to forecast gaps between potential export volumes with and without import ban.

Shares of food exports, banned by Russia, in overall food exports of the reviewed countries to the world are very small. That is why losses do not seem to be crucial. The worst hit country is expected to be Norway, since Russia's ban is accounted to 5.3% of total Norway's food exports (Table 9).

Bans vs. Liberalization: Controversial Effects for Russia's Domestic Food Market

In Russia, meanwhile, the cost of banned foodstuffs started to rise days after the embargo took effect in early August 2014 (Figure 10).

Figure 8. Dynamics of the EU's food export to Russia in 1995-2014 and estimated cutback in the period until 2020 due to the ban, $ million
Source: Author's own calculation based on the United Nations Conference on Trade and Development (2015)

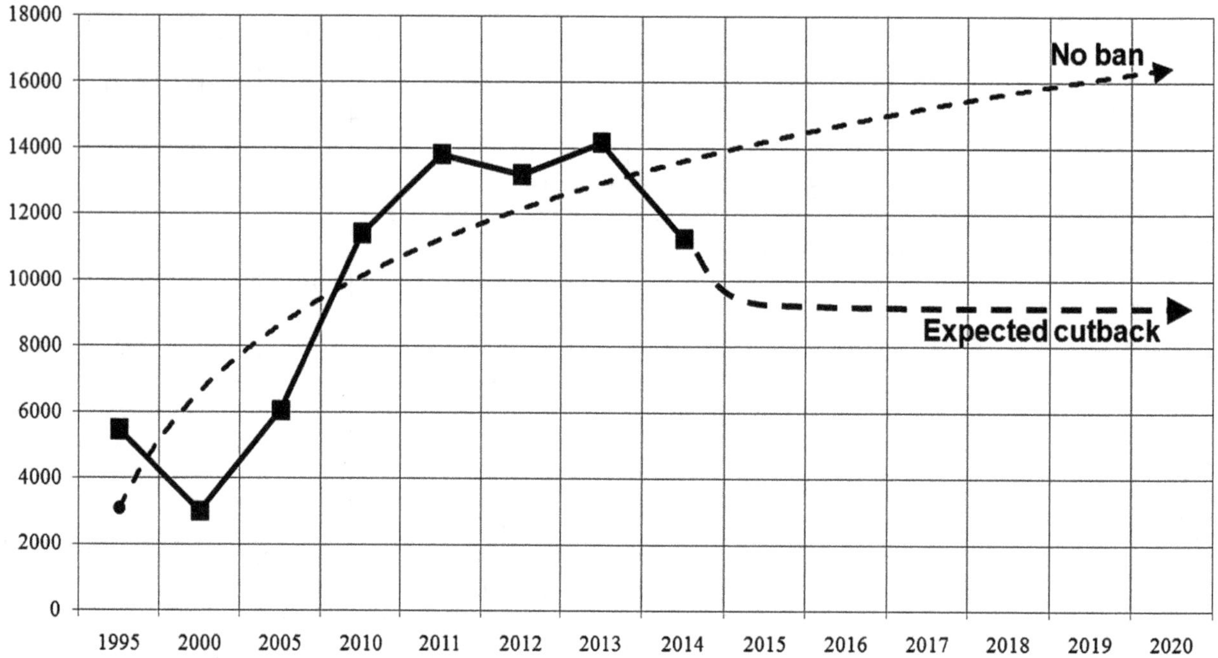

In an effort to downplay the negative effect of the food ban, the Russian government vowed to stem price increases. Russia tried to find alternative suppliers in countries whose goods were not placed under embargo. New suppliers (Turkey, Brazil, Argentina, Vietnam, China, etc.) were not found immediately, and they offered the same goods at higher prices. Ruble depreciation also affected growing costs of import, coming from "new" trade destinations. As a result, import of agricultural commodities and food to Russia decreased in the fourth quarter of 2014 and first quarter of 2015 (Figure 11).

Surprisingly, the volume of domestic food production decreased too, but the downturn trend can be explained by the rapid devaluation of the Russian ruble during the fourth quarter of 2014 and the first quarter of 2015. Expressed in rubles, domestic food production increased by 10% in 2014 in comparison to 2013. In the first quarter of 2015 domestic food production increased by 28.3% in comparison to the first quarter of 2014.

Consequently, the author observes, that trade liberalization (accession to the WTO in 2012) directly influenced neither Russia's external trade in food commodities, nor domestic food production, while import ban had an immediate effect.

In the Russian mass media and official government sources, the introduced ban on food imports is largely presented as a chance for local suppliers to develop. However, most of the analysts doubt that there will be a result in the near future. As of Portansky (2014), complete substitution of imported goods by locally made ones would take a long time. Moreover, it is a waste of resources, which might be utilized more effectively in those industries, where Russia has competitive advantages. Russia is a competitive producer of grain but not of higher-end commodities. Although food production has indeed increased since the imposition of the ban in August 2014, much of it is not cost effective. Russian government

Figure 9. Dynamics of the food export of the USA, Canada, Norway, and Australia to Russia in 1995-2014 and estimated cutback in the period until 2020 due to the ban, $ million
Source: Author's own calculation based on the United Nations Conference on Trade and Development (2015)

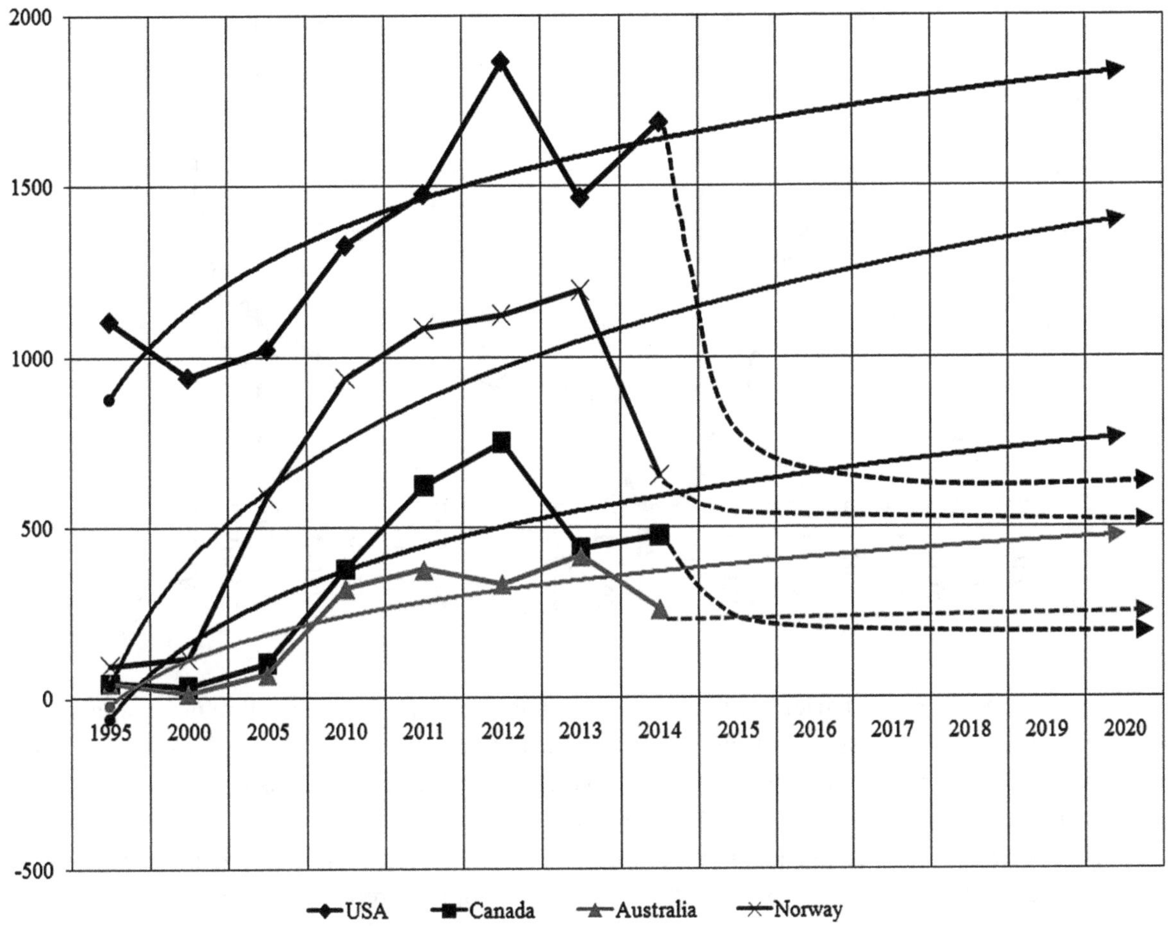

Table 9. Volumes of food exports, banned by Russia, and their shares in the total food exports of the selected countries to Russia and to the world in 2014

Commodity group	EU		USA		Norway		Canada		Australia	
	$ billion	%*	$ billion	%*	$ billion	%*	$ billion	%*	$ billion	%*
Food import, banned by Russia	4.815	-	0.953	-	0.621	-	0.355	-	0.189	-
Total food export to Russia	11.259	42.76	1.687	56.50	0.653	95.00	0.477	74.51	0.261	72.32
Total food export to the world	565.333	0.85	149.684	0.64	11.716	5.30	49.490	0.72	31.984	0.59

* Percentage in total food export of the country
Source: Author's own calculation based on the United Nations Conference on Trade and Development (2015)

Figure 10. Food inflation in Russia in 2010-2015 (quarterly), %
Source: Trading Economics, 2015

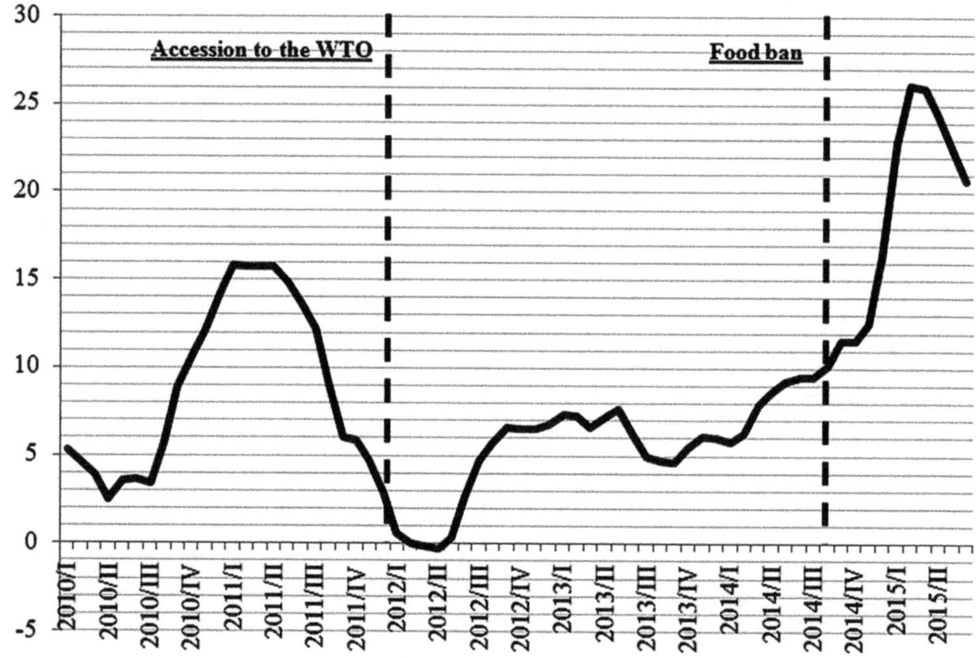

has long subsidized domestic agricultural industries, but agricultural sector is still in need of large-scale investment, which could be essentially cut due to the current economic recession in Russia. As of Stratford Global Intelligence (2015), growing reliance on domestic production that is less cost effective than imports could be a contributing factor in the rising cost of food.

Russia's food production in the near to medium term will not meet domestic demand. Domestic food production is expected to continue rising in 2015, but it will not fully replace lost import volumes. Meat production in Russia is growing, while import is decreasing (Figure 12). For example, domestic pork production in 2015 in Russia is expected to rise up to 2.82 million metric tons (Stratford Global Intelligence, 2015). Still, import volumes are forecast to decrease by 9%, to 0.38 million metric tons, meaning that combined domestic production and imports would not reach the 2013 domestic consumption level of 3.27 million metric tons. Beef, meat, and veal imports are projected to amount to only 0.83 million metric tons in 2015, compared to more than 1.0 million metric tons in 2013. Domestic production is set to rise to 1.4 million metric tons in 2015, up from about 1.38 million metric tons in 2013. Nevertheless, as with pork, production is not expected to make up for the decrease in imports (Stratford Global Intelligence, 2015).

Trade liberalization itself did not affect domestic market of meat and meat products, while embargo immediately distorted competitive environment. As for the meat market, growing domestic production and decreasing import (i.e. food independency) should not be interpreted as categorically positive tendencies. There are serious structural problems observed in the sphere of meat production in Russia (Schetinina, 2015). Growth of output is provided by poultry and pork, while cattle stock has been decreasing since the 1990s (from 36.3 million cattle in 1990 down to 11.1 million cattle in 2012).

Figure 11. Export, import and domestic production of food products in Russia in 2011-2015 (quarterly), $ billion
Source: Federal State Statistics Service of the Russian Federation, 2015

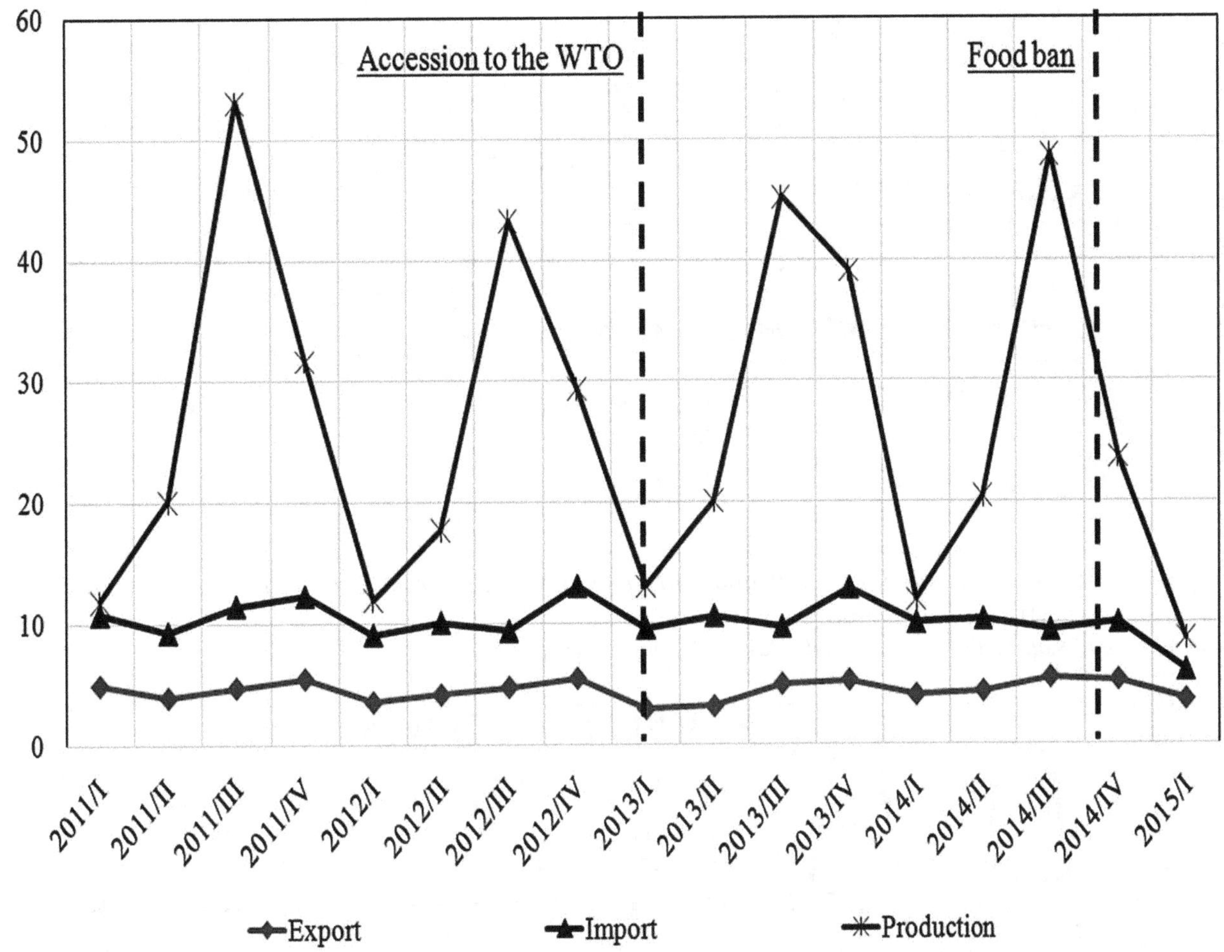

Cattle meat production had not been cost effective before Russia's accession to the WTO in 2012, and stays on like that. Out of the three major segments of meat market (beef, pork, and poultry), beef production has the longest payback period (8-9 years and even longer). The average payback period for poultry is 1-3 years, for pork – 5-7 years (Seninsky, 2015). Because of its unprofitability Russia's meat production is not competitive on the domestic market. Its relative competitiveness has always been supported by the government, including financial support, tariff protection, and such measures, as trade embargoes (both sanitarian and "political" ones, as the recent ban of August 2014). However, having in mind the long-term picture of the supposed Russia's integration into the international market, such unilateral and distorting measures will be limited (or even prohibited) by the Russia's obligations towards the WTO partners. In such conditions, domestic meat production may fall even in those sectors, where Russia has already secured its relative food independency.

While domestic production of meat tends to rise, domestic dairy production decreases (Figure 13).

Figure 12. Domestic production and import of meat (livestock and poultry) in Russia in 2011-2015 (quarterly) and for the period till 2017, thousand tons
Source: Federal State Statistics Service of the Russian Federation, 2015

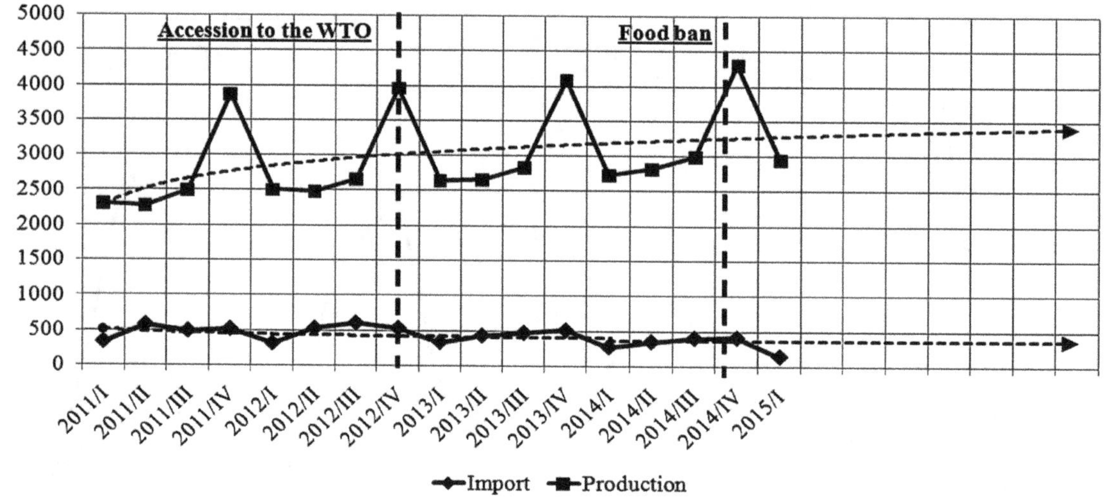

Figure 13. Domestic production of milk and dairy products in Russia in 2011-2015 (quarterly) and for the period till 2017, thousand tons
Source: Federal State Statistics Service of the Russian Federation, 2015

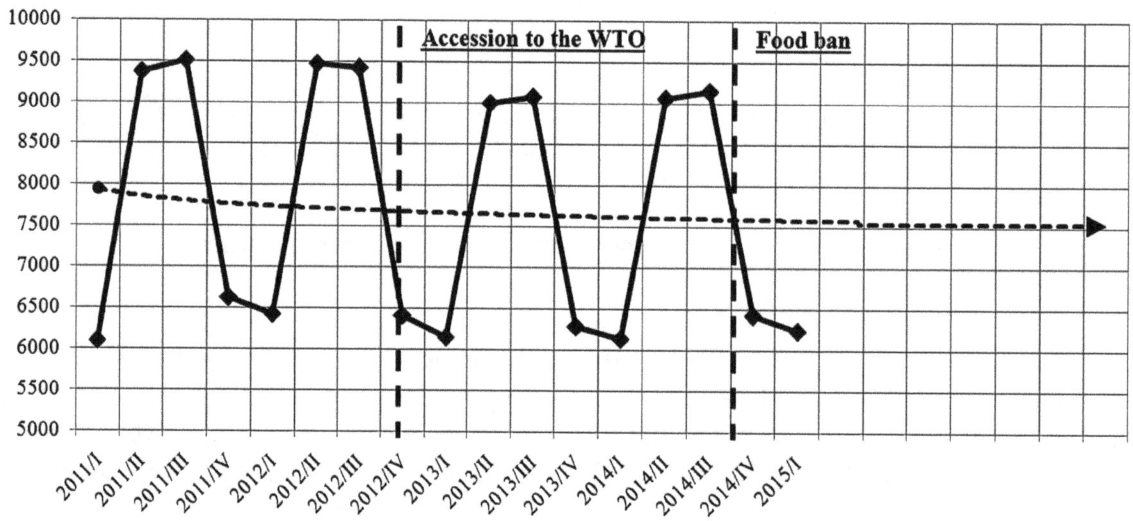

Russia's accession to the WTO weakened tariff protection of most of the domestic agricultural industries, including dairy production. Annual possible increase in productivity of a dairy farm is limited by 15-20% per year, which can be achieved only through the improved animal welfare and technological processes. Dairy farms do not have financial facilities for such an improvement, while growing production costs along with technological backwardness of food processing industries make dairy products non-competitive on the domestic market.

As Russian officials state, the food ban of August 2014 will drive domestic milk production and secure national food independency on dairy products. As of Skove (2015), the government pointed to a rise in total domestic production (3.5% in the first quarter of 2015) and a fall in imports as a proof that the embargo is helping to boost the fortunes of Russian farmers. Output of certain products decreases, which confirms the supremacy of domestic factors over external ones. Dairy farms are not able to react on market changes rapidly; there is always a gap. Producers depend on their cattle population, its productivity, and natural production cycles. That is why expectation of an immediate import substitution as a result of a food embargo is rather a populist affirmation, not reality.

Russia achieves food independency levels neither on meat nor on milk. The official data of the Federal State Statistics Service of the Russian Federation (2015) on the Import-Consumption ratio (presented as "Current Independency" on the Figure 14) shows, that approximately 76-79% of domestic consumption is met by domestic production. However, it is burgeoning poverty, not competitiveness of domestic producers, quality of their products, or food embargo, which is behind such high food independency levels. People tend to consume less. According to the latest Public Opinion Foundation poll (INTERFAX.RU, 2015), 63% of respondents have reconsidered their food consumption behavior since the beginning of 2015, particular have focused on cheaper products (40%), excluded some products from their ration (30%), or reduced food consumption (30%).

Apart from food independency, food security issues have to be considered. Calculation of Russia's real food security levels on meat and dairy products (presented as "Calculated Security" on the Figure 14), conducted on the basis of the World Health Organization (WHO) medical standards of safe food consumption, shows the tremendous gap between current level of "food independency" and real level of "food security".

Food security on milk and dairy products (56.6) means that only 56.6% of safe consumption of dairy products in Russia is provided by domestic production. Security level on meat and meat products is even lower – 53.0%. As import deliveries are severely cut by food embargo and ruble devaluation, and import substitution does not work effectively, domestic market has neither sources of affordable import, nor internal sources of food supply.

According to the Global Food Security Index 2015 (GFSI), published by the Economist Intelligence Unit (2015), "Global food security has made a rapid improvement over the past year. We see this in the increased efficiency of food systems and improvements in the nutritional quality of the food to which populations have access" (p. 4). However, Russia moves in the opposite direction. It has been ranked 43rd among 109 countries, included in the GFSI overall ranking in 2015, 0.3 points and three positions below the 2014 ranking (The Economist Intelligence Unit, 2015).

Aiming at import substitution, Russia's trade policies in fact decrease food security and living standards of population. During the past decade Russia's fears of losing food independency have been associated with the accession to the WTO, easier access of foreign food commodities, and liberalization of domestic market. Liberalization always drives competition in a natural way, while trade bans distort economic environment.

SOLUTIONS AND RECOMMENDATIONS

As the analysis shows, effectiveness of agricultural production is increasingly influenced by foreign economic and trade relations. As Russia has implemented such trade tool as food embargo, it affected

Figure 14. Current food independency and calculated food security on meat and dairy products in 1990-2014, and logarithmical forecast for the period till 2020
Source: Author's own calculation

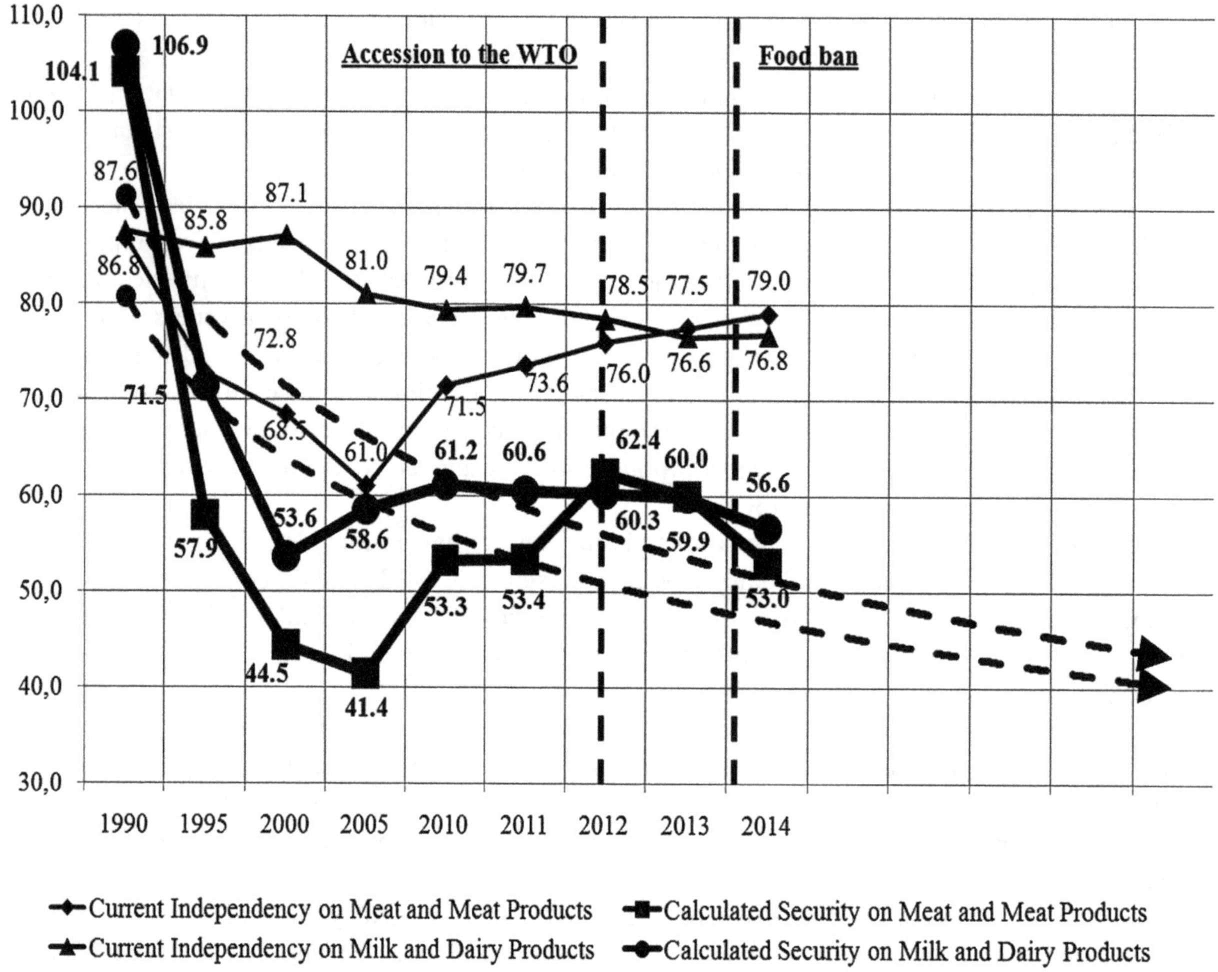

agricultural exports from the countries under the ban (the EU, the USA, Canada, Norway, and Australia), especially those whose trade ties with Russia were the closest (EU and Norway). Economic effects of lower demand from Russia seem to be small, as most countries have relatively limited exposure to exports to Russia in comparison with exports to other countries of the world. The largest impact on the interregional and international trade will occur in the event of a pronounced intensification of the conflict, through worsening global risk sentiment, trade contraction, and energy supply interruptions. For Russia's closest partners (the CIS, East Europe) even a mild intensification of the crisis could have an impact through lower exports of agricultural and food commodities.

Russia itself may turn out to be damaged in a greater extent. On the one hand, food ban is always an opportunity for domestic producers to fill a market gap. However, Russian farmers are not able to increase their output rapidly; their facilities are outdated, financial resources are limited; quality of their products is low. On the other hand, food ban backfires by driving up domestic food prices and lowering food security of the country. Russian consumers have to pay more for their food now as well as in the long run. The ban will hit rural dwellers and small and medium farmers, along with a number of Russian industries, including food processing plants, shippers, and retailers.

There is no any one and only formula to calculate effects of trade restrictions. The very condition, when embargoes, sanctions and counter-sanctions are considered as appropriate tools of international trade policy, is pathological. It means inevitable cut of trade and economic ties with foreign partners, which is self-destructive in the contemporary globalized world. There are no developed and efficient economies, which wall off from the world. Keeping apart from the most developed countries, Russia is endangered to become cut off from the opportunities of efficient modern development. Compulsive import substitution and search of new suppliers and markets require unreasonable expenses, which will result in the loss of competitiveness of national economy.

FUTURE RESEARCH DIRECTIONS

Both trade liberalization and trade bans influence sustainability of agricultural production and rural development. Traditional rural development measures in the developed countries (the EU, the USA, etc.) have tended to centre on agriculture and to concentrate on income redistribution and project support. Situation is different in the economies in transition, where living standards are lower, and the range of economic activities in rural areas is less diverse. There are also wide differences between economies in transition. Author has considered Russia, but every economy in transition has its own peculiarities, reflected in the range of national rural and regional policies being implemented. In general, rural areas in economies in transition are characterized as specialized in agriculture; lacking alternative employment; having poor infrastructure; having developmental problems; lacking or having limited accessibility to basic services; poor investment. That is why shrinkage of domestic market as a result of such trade policies as bans and embargoes definitely aggravates existing problems of rural poverty, unemployment and outflow migration, which have to be investigated, assessed, and solved.

CONCLUSION

In the past decades, despite the widening spread of free trade principles through such organizations as the WTO, trade restrictions have been intensively used by national governments. One of the reasons to protect domestic market is the need to maintain a certain level of national defence, independence, or security. However, as the author discovered in the chapter, in case of Russia's recent food embargo, the desired target is missed. Russia's current protectionist and isolationist policies, accompanied by economic recession, inflation, and ruble devaluation, in fact decrease food security and living standards of population.

Political value of the embargo is minimal. Author's calculations show, that expected losses of the EU, the USA, Canada, and Australia are rather small – below 1% of the total national food exports to the world, which may be easily reoriented to other trade partners in Asia and Latin America, or even supplied domestically.

Russia's domestic food market is in a greater extent influenced by internal factors (low competitiveness of food producers, their financial instability, outdated facilities, infrastructure, lowering effective demand, etc.), than external ones. Author concluded that Russia's accession to the WTO in 2012 and consequent trade liberalization directly influenced neither Russia's external trade in food commodities, nor domestic food production, while import ban had an immediate effect.

Consequently, Russia has to concentrate on solving internal problems and increasing efficiency of domestic agricultural production, not its artificial protection. Available WTO mechanisms, including trade disputes tools, have to be implemented. It will develop the transparent competitive environment and ensure the most effective use of resources. Such measures as embargoes restrict the ultimate beneficial effects of exploiting comparative advantage through free trade, while principle of comparative advantage demonstrates that for the world as a whole free trade leads to a higher level of output and income than trade embargoes. Free trade also enables each nation to obtain a higher level of production and consumption than can be obtained in isolation.

REFERENCES

Anderson, K., Dimaran, B., Francois, J., Hertel, T., Hoekman, B., & Martin, W. (2001). The Cost of Rich (and Poor) Country Protection to Developing Countries. *Journal of African Economies*, *10*(3), 227–257. doi:10.1093/jae/10.3.227

Anderson, K., Jha, S., Nelgen, S., & Strutt, A. (2013). Re-examining Policies for Food Security in Asia. *Food Security*, *5*(2), 195–215. doi:10.100712571-012-0237-5

Aslund, A. (2007). Russia's Accession to the World Trade Organisation. *Eurasian Geography and Economics*, *48*(3), 289–305. doi:10.2747/1538-7216.48.3.289

Bokusheva, R., Hockmann, H., & Kumbhakar, S. (2012). Dynamics of Productivity and Technical Efficiency in Russian Agriculture. *European Review of Agriculture Economics*, *39*(4), 611–637. doi:10.1093/erae/jbr059

Božić, D., Bogdanov, N., & Ševarlić, M. (2011). *Ekonomika Poljoprivrede*. Belgrade: University of Belgrade.

Dollar, D. (2001). *Globalization, Inequality, and Poverty since 1980*. Washington, DC: Development Research Group, World Bank.

Dornbusch, R. (1992). The Case for Trade Liberalization in Developing Countries. *The Journal of Economic Perspectives*, *6*(1), 69–85. doi:10.1257/jep.6.1.69

Erokhin, V. (2015a). Contemporary Reshaping of Eurasian Integration: Russia's Policies and their Implication for the EU and EurAsEC. *Procedia Economics and Finance*, *22*, 402–411. doi:10.1016/S2212-5671(15)00313-5

Erokhin, V. (2015b). Russia's Foreign Trade in Agricultural Commodities in its Transition to Liberalization: a Path to Go Green. In J.-V. Andrei, A. Ion Raluca, & A. Turek Rahoveanu (Eds.), *Green Economic Structures in Modern Business and Society* (pp. 253–273). Hershey, PA: IGI Global. doi:10.4018/978-1-4666-8219-1.ch014

Erokhin, V., Heijman, W., & Ivolga, A. (2014). Trade Tensions between EU and Russia: Possible Effects on Trade in Agricultural Commodities for Visegrad Countries. *Visegrad Journal on Bioeconomy and Sustainable Development, 3*(2), 52–57. doi:10.2478/vjbsd-2014-0010

Erokhin, V., & Ivolga, A. (2012). How to Ensure Sustainable Development of Agribusiness in the Conditions of Trade Integration: Russian Approach. *International Journal of Sustainable Economies Management, 1*(2), 12–23. doi:10.4018/ijsem.2012040102

Erokhin, V., Ivolga, A., & Heijman, W. (2014). Trade Liberalization and State Support of Agriculture: Effects for Developing Countries. *Agricultural Economics, 60*(11), 524–537.

Federal State Statistics Service of the Russian Federation. (2015). *Statistics Database* [Data file]. Available from http://www.gks.ru/wps/wcm/connect/rosstat_main/rosstat/en/main/

INTERFAX.RU. (2015). *Dve treti rossiyan nachali economit na produktakh pitania*. Retrieved June 28, 2015, from https://news.mail.ru/economics/22486768/?frommail=1

Ishchukova, N. (2013). *The Position of the Russian Federation in the International Market of Agricultural and Foodstuff Products* (Doctoral dissertation). Available from the Faculty of Economics and Management, Czech University of Life Sciences Prague.

Josling, T., Anderson, K., Schmitz, A., & Tangerman, S. (2010). Understanding International Trade in Agricultural Products. One Hundred Years of Contributions by Agricultural Economists. *American Journal of Agricultural Economics, 92*(2), 424–446. doi:10.1093/ajae/aaq011

Markovic, I., & Markovic, M. (2014). Agricultural Protectionism of the European Union in the Conditions of International Trade Liberalization. *Economics of Agriculture, 61*(2), 423–440.

Mikhailov, A. (2011). *Kasha iz topora*. Retrieved July 21, 2015, from http://www.gazeta.ru/column/mikhailov/3573933.shtml

Milner, H. V. (1999). The Political Economy of International Trade. *Annual Review of Political Science, 2*(1), 91–114. doi:10.1146/annurev.polisci.2.1.91

Nugent, J. (2002). *Trade Liberalization: Winners and Losers, Success and Failures. Implications for SMEs*. College Park, MD: The IRIS Center at the University of Maryland.

Petronevich, N., & Kondrashov, M. (2011). Zavisimost rossiyskoy inflyatsii ot dinamiki mirovykh tsen na rynke prodovolstvia i nefti. *Economic Policy, 4*, 128–147.

Portansky, A. (2014). Sanctions Pushing to Decline? *Russia in Global Affairs, 3*. Retrieved July 21, 2015, from http://eng.globalaffairs.ru/number/Sanctions-Pushing-to-Decline-16992

Rodrik, D. (2000). *Trade Policy Reform as Institutional Reform*. Cambridge, MA: The Kennedy School, Harvard University.

Schetinina, I. (2015). On Food Security of Russia. *ECO, 1*, 124–138.

Secretariat of the Convention on Biological Diversity. (2005). *The Impact of Trade Liberalization on Agricultural Biological Diversity, Domestic Support Measures and their Effects on Agricultural Biological Diversity*. Montreal: SCBD.

Seninsky, S. (2015). *Osobennosti natsionalnogo importa*. Retrieved July 21, 2015, from http://www.svoboda.org/content/article/26521177.html

Skove, S. (2015). Despite Promises, Food Ban Fails to Deliver. *The Moscow Times*, *5641*, 1–2.

Spanu, V. (2003). *Liberalization of the International Trade and Economic Growth: Implications for both Developed and Developing Countries*. Cambridge, MA: Harvard University.

Srinivasan, T. N., & Bhagwati, J. (1999). *Outward Orientation and Development: Are the Revisionists Right?* New Haven, CT: Yale University Economic Growth Center.

Stratford Global Intelligence. (2015). *How Russia Is Lowering Its Food Prices*. Retrieved July 21, 2015, from https://www.stratfor.com/analysis/how-russia-lowering-its-food-prices

The Economist Intelligence Unit. (2015). *Global Food Security Index* [Data file]. Retrieved October 3, 2015, from http://foodsecurityindex.eiu.com/

Trading Economics. (2015). *Statistics Database* [Data file]. Available from http://ru.tradingeconomics.com/countries

Trukhachev, V., Ivolga, A., & Lescheva, M. (2015). Enhancement of Land Tenure Relations as a Factor of Sustainable Agrocultural Development: Case of Stavropol Krai, Russia. *Sustainability*, *7*(1), 164–179. doi:10.3390u7010164

United Nations Conference on Trade and Development. (2015). *Statistics Database* [Data file]. Available from http://unctad.org/en/Pages/Statistics.aspx

Ushachev, I. (2012). Measures to Secure Competitiveness of Russia's Agricultural Production in the Conditions of its Accession to WTO. *Economics of Agricultural and Processing Enterprises: Academic and Research Journal*, *6*, 1–5.

World Bank. (2005). *Economic Growth in the 1990s. Learning from a Decade of Reform*. Washington, DC: The World Bank.

ADDITIONAL READING

Dıaz-Bonilla, E., & Reca, L. (2000). *Trade and Agroindustrialization in Developing Countries*. Washington, DC: IFPRI Trade and Macroeconomics.

Erokhin, V., & Ivolga, A. (Eds.). (2014). Contemporary Issues of Sustainable Rural Development: International Approaches and Experiences of Eastern Europe and Russia. Stavropol: AGRUS of Stavropol State Agrarian University.

Estevadeordal, A., Freund, C., & Ornelas, E. (2008). Does Regionalism Affect Trade Liberalization Toward Nonmembers? *The Quarterly Journal of Economics*, *124*(4), 1531–1575. doi:10.1162/qjec.2008.123.4.1531

EurActiv. (2014). *Russia's Sanctions Threaten both EU Farmers and Policymakers*. Retrieved August 27, 2014, from http://www.euractiv.com/sections/agriculture-food/russias-sanctions-threaten-both-eu-farmers-and-policymakers-307827

Fock, A., Weingarten, P., Wahl, O., & Prokopiev, M. (2003). Russia's Bilateral Agricultural Trade: First Results of a Partial Equilibrium Analysis. *Russia's Agro Food Sector: Towards Truly Functioning Markets*, 271–297.

Garnik, S., & Sokolnikova, O. (2014). Impact of WTO Accession on Agriculture Sustainable Development in the Russian Federation. *Visegrad Journal on Bioeconomy and Sustainable Development*, *3*(1), 33–38. doi:10.2478/vjbsd-2014-0007

Husabo, E. (2014). Spillovers to Europe from the Crisis in Russia and Ukraine. *Economic Commentaries*, *6*, 1–11.

Ishchukova, N., & Smutka, L. (2013). Comparative Advantage: Products Mapping of the Russian Agricultural Exports. *AGRIS on-line Papers in Economics and Informatics, 3*(V), 13–24.

Kassam, A., Willsher, K., Oltermann, P., Adekoya, R., & Brooks, L. (2014). *Russian Food Embargo Leaves Europe with Glut of Fruit, Pork and Mackerel.* Retrieved March 30, 2015, from http://www.theguardian.com/world/2014/aug/15/ukraine-europe-news

Khusainov, B. (2011). Country Preferences and Competitiveness of Export in Belarus, Kazakhstan and Russia. *Eurasian Economic Integration*, *11*(2), 71–96.

Liapis, P. (2011). *Changing Patterns of Trade in Processed Agricultural Products*. Paris: OECD. doi:10.1787/5kgc3mq19s6d-en

Organisation for Economic Co-operation and Development. (2011). *Evaluation of Agricultural Policy Reforms in the European Union.* Paris: OECD. doi:. doi:10.1787/9789264112124-en

Raynolds, L., Murray, D., & Wilkinson, J. (Eds.). (2007). *Fair Trade: The Challenges of Transforming Globalization*. London: Routledge.

Schmitz, A., Moss, C., Schmitz, T., Furtan, W., & Schmitz, H. (2010). *Agricultural Policy, Agribusiness, and Rent-Seeking Behaviour* (2nd ed.). Toronto: University of Toronto Press.

Tarasov, V. (2012). *Risks and Treats of Competitiveness of Domestic Agriculture and Food Security of Russia at Accession to WTO.* Retrieved July 21, 2015, from http://www.myasoportal.ru/analitika/riski-i-ugrozy-pri-prisoedinenii-rossii-k-vto

Wegren, S. K. (2011). Food Security and Russia's 2010 Drought. *Eurasian Geography and Economics*, *52*(1), 140–156. doi:10.2747/1539-7216.52.1.140

Wehrheim, P., & Wobst, P. (2005). The Economic Role of Russia's Subsistence Agriculture in the Transition Process. *Agricultural Economics*, *33*(1), 91–105. doi:10.1111/j.1574-0862.2005.00136.x

Wittman, H., Desmarais, A., & Wiebe, N. (2010). *Food Sovereignty: Reconnecting Food, Nature and Community*. Halifax, Oakland: Fernwood Publishing and Food First Books.

KEY TERMS AND DEFINITIONS

Adaptation: An adjustment of existing market structures, regulations, and mechanisms to become suitable to new economic conditions.

Economy in Transition: An economy that is changing from central planning to free markets.

Food Independence: Self-sufficiency of a country in agricultural raw materials and food.

Food Market: A medium that allows buyers and sellers of agricultural commodities and food to interact in order to facilitate an exchange; a nominal place where forces of demand and supply operate.

Food Security: Physical and economic access to food that meets people's dietary needs as well as their food preferences.

Trade Ban: A government order that restricts commerce or exchange with a specified country, usually created as a result of unfavorable political or economic circumstances between nations.

Trade Liberalization: The removal or reduction of restrictions or barriers on the free exchange of goods between nations.

This research was previously published in Global Perspectives on Trade Integration and Economies in Transition edited by Vasily Erokhin, pages 185-211, copyright year 2016 by Business Science Reference (an imprint of IGI Global).

Section 2

Food Consumption and Demand

Chapter 6
Food Consumption Patterns in Times of Economic Recession

Glykeria Theodoridou
Aristotle University of Thessaloniki, Greece

Efthimia Tsakiridou
Aristotle University of Thessaloniki, Greece

Nikos Kalogeras
Wageningen University, The Netherlands

Konstantinos Mattas
Aristotle University of Thessaloniki, Greece

ABSTRACT

The aim of this study is to identify the factors influencing consumers' purchasing behaviour for food, in times of crisis. Intercept survey was conducted in a random selected sample consisted of 553 consumers between January and May 2016 in the Prefecture of Thessaloniki. Multivariate data analysis was performed to explore the factors influencing consumers' purchasing behaviour, identify consumers with similar behaviour based on their socio-demographic characteristics and their attitudes towards food consumption in times of crisis. Results demonstrate that the factors affecting consumers' purchasing behaviour are (a) Product's Feature and Natural contents, (b) Economic issues, (c) Identity & Sensory Appeal, (d) Mood, (e) Weight Control & Health and (f) Convenience. Furthermore, cluster analysis was employed to classify consumers with similar buying behaviour towards food, five groups of consumers were identified: (a) neutrals, (b) influenced by psychological issues, (c) influenced by economic issues, (d) low diet/healthy eaters and (e) influenced by availability and easy access.

INTRODUCTION

In the last 60 years, more than 400 economic crises occurred to over 200 countries all over the world, affecting consumer behaviour in several ways (Duquenne and Vlontzos, 2014, Dutt and Padmanabhan, 2009). Health and living conditions of the populations were influenced by an unstable economy. According to Rajmil et al., 2014 and Rico and Blakey, 2012, the impact of crisis in each country depends on

DOI: 10.4018/978-1-5225-8063-8.ch006

the type of recession and the factors affecting it such as duration and intensity, the speed and the types of changes that occur, the period prior to the recession, various measures taken by the governments as an answer to the crisis, and the effect community and family have on every individual.

As the global financial crisis continues to become deeper and deeper, hunger is likely to increase as the poorest people buy less and less products, because of reduced incomes and higher unemployment. While economies continue to fall apart, incomes decrease and people become unemployed, families are cutting back and in some special occasions, they stop eating crucial nourishing food. Some are cutting meals and substituting with less nutritious food (WFP-World Food Programme, 2012).

Increases in the cost of food often lead to changes in the quantity and type of food that is purchased. This may result in a reduction in the quantity of food consumed and/or the substitution of higher priced food for less expensive food which is often less nutritious and of lesser quality. Over a prolonged period, such changes may have negative consequences for nutrition, both through the quantity of food consumed for maintaining energy balance and for the quality of food consumed for maintaining sufficient intakes of proteins, fats and micronutrients such as vitamins, minerals and trace elements (Thompson, 2012).

According to Mansoor and Jalal 2011 and Flatters and Willmott 2009, in a period of economic crisis, consumers' behaviour is described by consumption smoothing at various levels. People are not so willing to pay more for products that can be replaced by cheaper ones. Consumers have redefined what they consider as "necessities" and what are considered "luxuries." This can be verified by the annual research of Boston Consulting Group in 2011, which claims that 73% of the surveyed consumers declared that will purchase only the absolutely necessary goods. Furthermore, the hierarchy of consumers' "values" has changed with the "savings", "health", "value for money", and others to be in the top (Tsourgiannis et al., 2014).

In Greece, economic crisis in 2008, caused a deep recession in its economy. The social consequences of this crisis are very discouraging, with the unemployment rate continuously increasing without knowing until now when and where it will reach its peak. When the crisis started (2008), the official unemployment rate was 14.8%, in June 2011 was 17.2% (Frangos et al., 2012, EL.STAT, 2012), and in September 2014 increased up to 26.1%. The biggest percent of unemployed people exists between 20 and 35 years old, which is almost 50%, while the total number of unemployed people are over one million (Eurostat, 2014).

Consumption to both private and public sectors took a big hit resulting in major reductions which affected every sector of the economy, even the food sector (Karelakis et al., 2013, ELSTAT, 2012). Facing decreased wages up to 30% with average annual household income to be estimated at 13.000 euros and at the same time 20% of people live on 6.500 euros per year (OECD, 2011), increased taxation and recession, low and middle income households have difficulty to even keep the regular accounts (Frangos et al., 2012). Because of the failure to implement working programs through the years, a significant large percent of the population (35,7%, 2015) is living near or below poverty level, with the majority of them to be pensioners, unemployed, immigrants and even former entrepreneurs (KEPE – Centre of Planning and Economic Research, 2012, Vlontzos et al., 2017).

Amid the economic crisis in Greece, the examination of the factors affecting Greek consumers' purchasing attitudes during turbulent economic times is very important because most of our knowledge about consumers' attitudes regarding food consumption (nutritious food and fast food) is mainly derived from studies that have been conducted before the economic crisis period. Also, according to Tsourgiannis et al., 2014, Michalopoulos and Demoussis, 2001, Kupiec and Revell, 1998 and Kotler, 1994, different economic, cultural, demographical, and social characteristics affect consumers' purchasing behaviour in a different way.

According to all the above, the current study examines which of the factors that influence consumers' purchasing behaviour before the economic crisis affect their current purchasing behaviour towards food (nutritious & fast food) during the recent adverse economic conditions. Moreover, the current study aims to identify the factors affecting consumers' purchasing behaviour and classify them into groups with similar behaviour as well as to profile each group of consumers regarding their attitudes towards fast food, their demographic characteristics and the factors that affect them the most, during their shopping.

The paper is organized as follows: first, the methodological background is developed presenting the conceptual model. Next, the research design is presented analyzing the survey procedure, questionnaire design and statistical methodology of the study. Finally, the results are listed followed by concluding remarks and discussion.

METHODOLOGICAL BACKGROUND

Conceptual Model

Choosing food products is a complex procedure of preferences for sensory characteristics, combined with the influence of non-sensory factors, health claims, price, ethical concerns and mood (Fotopoulos et al., 2009, Prescott et al., 2002). Economic, social and cultural factors can concur to the adoption of patterns. Physiological and psychological factors that create consumers' preferences and attitudes towards food, can be differentiated from interpersonal or social factors (Eertmans, 2001).

Many consumer behaviour theories and comprehensive models have been developed the last decades to outline the way people choose food in general. For example, three components that define food choice are personal characteristics, influences and life style (Frust et all, 1996). The special relationship between those three components creates the process that leads to the final choice. But, according to Eertmans et al., 2001, although a model like that describes the complexity of food choice, it is impossible to make predictions about consumers' consumption behaviour towards food (Fotopoulos et al., 2009).

Previous studies suggest that an instrument that could help with the prediction of a general food choice of consumers based on its determinants is the Food Choice Questionnaire (FCQ) (Fotopoulos et al., 2009, Spetoe and Wardle, 1999, Pollard, Spetoe and Wardle, 1998, Spetoe, Polland and Wardle, 1995).

Furthermore, studies have shown that usually consumers would pass through a five stage decision making process (Problem recognition, Information search, Evaluation of alternatives, Purchase decision and Post purchase evaluation) before making their decision to buy food (Stavkova et al., 2008). However, when they buy frequently purchased low-cost items, like food, they usually search less and decide easier (Zander and Hamm, 2011, Carrigan and Attalla, 2001, Kiventz and Simonson, 2000). The general idea across those models was summarized by Howard (1983), as a four stage decision process: Information-Attitude-Intention-Purchase.

However, a simpler view of consumers' behaviour is given by Theory of Reasoned Action (TRA) (Fishbein and Ajzen, 1975). Simply put, according to TRA, consumers examine the consequences of their actions before they act and implicate in a behaviour (Ajzen and Fishbein, 1980). As TRA has been criticized that could not support the relationship between intention and behaviour, Ajzen (1991) and (Gupta and Ogden, 2009) developed the Theory of Planned Behaviour (TPB) model which states that the intention of a person to make an action happen, affects the most the action itself (Tsourgiannis, 2014).

Lastly, the stimulus-response model (or the black box model) examines consumers' behaviour. Specifically, the model shows the interaction between consumer characteristics and responses, decision processes and stimuli. Furthermore, it is related to the black box theory of behaviourism, which focuses in the interaction between stimuli and consumers' attitudes and not in their psychology (Furaiji et al., 2012).

But, all the above-mentioned theories and models of consumer behaviour do not include all the external and internal characteristics that affect consumers purchasing behaviour, but each theory and model uses only some of them.

Therefore, in this study a conceptual model aims to place the key notions in the literature review into an identifiable framework (Figure 1).

In particular, the model tries to investigate the relationship between the factors affecting consumers' purchasing behaviour towards food, both nutritious and fast food, the adoption of a particular purchasing behaviour according to their attitudes towards food consumption during the current economic crisis, the relationship between their demographic characteristics and their purchasing behaviour and the influence of their lower budget for food in their shopping.

RESEARCH DESIGN

Survey Procedure

The research was based on primary data, which were collected using a fully structured questionnaire, supplemented by person to person interviews. The completion of the questionnaires took place in the Prefecture of Thessaloniki, where the period of the survey was from January to May 2016. In the survey, a sample from the corresponding population was included, in order to show that the obtained results could be generalized to some extent in the total population of the Prefecture of Thessaloniki.

Figure 1. Conceptual model

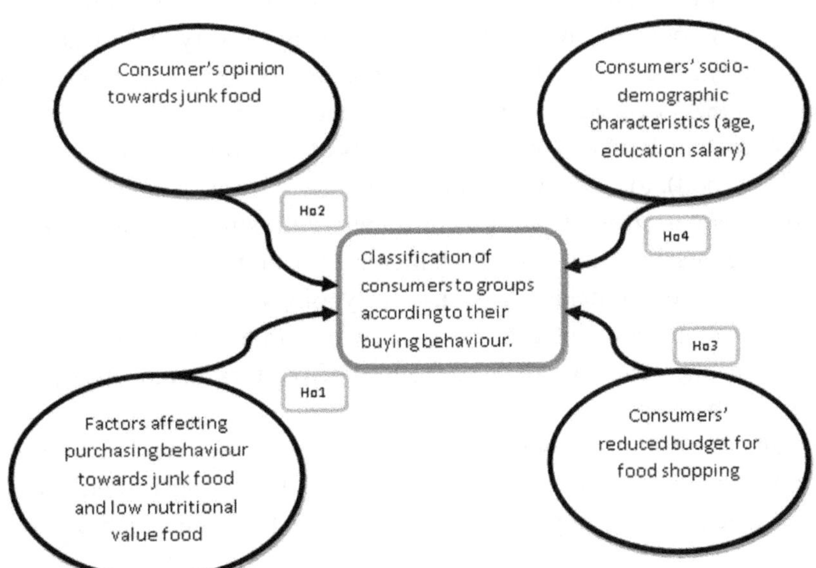

Regarding this research, the available sample data were all the residents of all municipalities and communities of the Prefecture of Thessaloniki. For the entire sample population, the census data of 2011 by the National Statistical Service of Greece (ELSTAT, 2016) were used. In other words, all the above-mentioned people were considered as potential consumers that purchase food, something that made them eligible to participate in the survey. In this study the sampling unit was one person from each household. The sample was selected randomly[1].

Questionnaire Design

Factors that affect consumer behaviour towards food purchasing and also, a tool used in past researches that can be considered having a possible potential to fulfill the aim of predicting general food choice based on its determinants, which is the (FCQ) Food Choice Questionnaire (Steptoe and Wardle, 1999, Pollard et al., 1998, Steptoe et al., 1995), were identified by searching the literature. After that, a pilot survey conducted in December 2015 to 15 consumers of the prefecture of Thessaloniki within three focus groups with in-depth interviews. In the next stage, a questionnaire was designed in order to meet the research objectives and therefore the main survey was conducted between January and May 2016.

The questionnaire was designed in three parts: the first part consists of questions regarding consumers' food habits and characteristics they are looking for, when they are purchasing food. Next, the second part consists of questions that include attitudinal statements in 5 point Likert scale related to consumers' opinions and food consumption under the current economic crisis both for nutritional food and junk food. Lastly, the third part consists of questions regarding consumers' personal information such as age, education, marital status, income, occupation etc.

STATISTICAL METHODOLOGY

In the 553 valid questionnaires collected, multivariate analysis techniques were applied to reveal the key information contained in the responses. The analysis was applied in three stages. First, Confirmatory Analysis (CFA) was used to perform structure validation and to identify the variables that accounted for the maximum amount of variance within the data (components). Bartlett's test of sphericity and measure of sampling adequacy (MSA) was used in order to check the appropriateness of the data. CFA, reduced the key attitude variables, which related to various aspects of consumers' purchasing behaviour towards food (nutritious and fast food), to a smaller set of factors.

Next, Cluster Analysis performed to group consumers with similar patterns of scores into similar clusters based on their purchasing behaviour. Both hierarchical and non-hierarchical methods were used in order to develop a consumers' purchasing behaviour typology.

Discriminant Analysis was performed to assess how accurately the identified key dimensions towards consumers' purchasing behaviour derived from Confirmatory Factor Analysis could predict and discriminate cluster membership.

Lastly, a non-parametric Friedman one way Anova Test performed in order to examine consumers' opinions about junk food and a cross tabulation Chi-Square Analysis performed in order to profile consumers who have a particular purchasing behaviour according to their socio-demographic characteristics.

RESULTS

The general consumer's profile of this study is a woman, aged between thirty-one and forty years old. She graduated from a University or a Technological Institute and she is married without children under eighteen years old. She works full time as a private employee and her monthly family income varies between 1.000 and 1.500 euros.

Confirmatory Factor Analysis, through Principal Component Analysis and Varimax Rotation were conducted to identify the key consumption dimensions. Confirmatory Factor Analysis identified 47 key factors that affect consumers' preferences towards food consumption under the current economic crisis (Table 1).

Table 1. The key factors affecting consumers' behaviour towards food purchasing under the current economic crisis

Key Consumption Dimensions	Variable/ Factors Affecting Purchasing Behaviour	Loading
Product Features & Natural Content	Q.A.3.9 The smallest Packaging	0,796
	Q.A.3.8 To be organic/ Biological	0,788
	Q.A.3.5 Private Label	0,677
	Q.A.3.15 Its Production to be Friendly to the Environment	0,621
	Q.A.3.12 Law Fat	0,579
	Q.A.3.3 Brand	0,569
	Q.A.3.13 Easy to Prepare	0,525
	Q.B.9.2 Contains no additives	0,731
	Q.B.9. 5 Contains natural ingredients	0,692
	Q.B.9.23 Contains no artificial ingredients	0,673
	Q.B.9.22 Contains a lot of vitamins and minerals	0,579
	Q.B.9.32 Has the country of origin clearly marked	0,552
Economic Issues	Q.A.3.7 To be in Offer	0,764
	Q.A.3.2 Price	0,526
	Q.A.3.6 The cheapest Brand/The cheapest Seller	0,737
	Q.B.9.36 Is cheap	0,677
	Q.B.9.35 Can be bought in shops close to where I live or work	0,678
	Q.B.9.11 Is easily available in shops and supermarkets	0,661
	Q.B.9.6 Is not expensive	0,560
	Q.B.9.12 Is good value for money	0,515
Identity & Sensory Appeal	Q.A.3.11 To be Greek/ Not Imported	0,580
	Q.A.3.4 To be Fresh	0,504
	Q.A.3. 10 To be a Local Product	0,504
	Q.A.3. 14Tasty	0,443
	Q.B.9.14 Smells nice	0,653
	Q.B.9.13 Cheers me up	0,587

continued on following page

Table 1. Continued

Key Consumption Dimensions	Variable/ Factors Affecting Purchasing Behaviour	Loading
Mood	Q.B.9.21 Is like the food I ate when I was a child	0,739
	Q.B.9.16 Helps me cope with stress	0,718
	Q.B.9.34 Helps me cope with life	0,713
	Q.B.9.26 Helps me relax	0,704
	Q.B.9.25 Looks nice	0,635
	Q.B.9.18 Has a pleasant texture	0,607
Weight Control & Health	Q.B.9.7 Is low in fat	0,799
	Q.B.9.3 Is low in calories	0,656
	Q.B.9.30 Is good for my skin/teeth/hair/nails etc	0,635
	Q.B.9.29 Keeps me healthy	0,545
	Q.B.9.31 Makes me feel good	0,536
Convenience	Q.B.9.1 Is easy to prepare	0,856
	Q.B.9.28 Takes no time to prepare	0,737
	Q.B.9.15 Can be cooked very simply	0,657
Neighborhood Stores	Q.A.5.3 Neighborhood's Greengrocery Store	0,689
	Q.A.5.5 Neighborhood's Butcher Shop	0,766
	Q.A.5.6 Neighborhood's Fishing Shop	0,769
	Q.A.5.4 Neighborhood's Grocery Store	0,821
Supermarket & Local Market	Q.A.5.8 Peddlers	0,817
	Q.A.5.1 Big supermarket out of town	0,847
	Q.A.5.7 Local Market	0,466

KMO MSA = 0,900, Bartlett tests of Sphericity = 11695,976, P<0,001

In the next stage, hierarchical and non-hierarchical clustering methods were used to develop a typology of the purchasing behaviour of the Greek consumers (Tsourgiannis et al., 2014; Vyncke, 2002; Siardos, 1999, Hair et al., 1998, Helsen and Green, 1991, Harrigan, 1985). There were no outliers, at the 553 consumers that were included in Cluster Analysis.

The cluster analysis identified five groups of consumers that were named according to their purchasing behaviour patterns towards food, during the current economic crisis (Figure 2). These are: (1) the "Neutrals", (2) the "Influenced by Psychological Issues", (3) the "Influenced by Economic Issues", (4) the "Low Diet/Healthy Eaters" and (5) the "Influenced by Availability and Easy Access."

In particular, the "Neutrals" comprise 19.9% of the sample. They declare that they buy food most often from the local markets, but when shopping they are not influenced by any particular factor. Consumers who are "Influenced by psychological Issues" comprise 13.7% of the sample. In the current economic crisis, the consumers who belong to that group are influenced for their buying decisions by factors that are related to their psychology. They buy food that makes them feel good, food which is good for their skin/ teeth/ hair, etc., and most of the times food reminds them what they ate when they were children. They are also influenced by the attractiveness of the food. They want to have a nice packaging, a pleasant texture and to look nice. The consumers who belong to the "Influenced by Economic Issues" group

Figure 2. Groups of consumers according to their buying behaviour

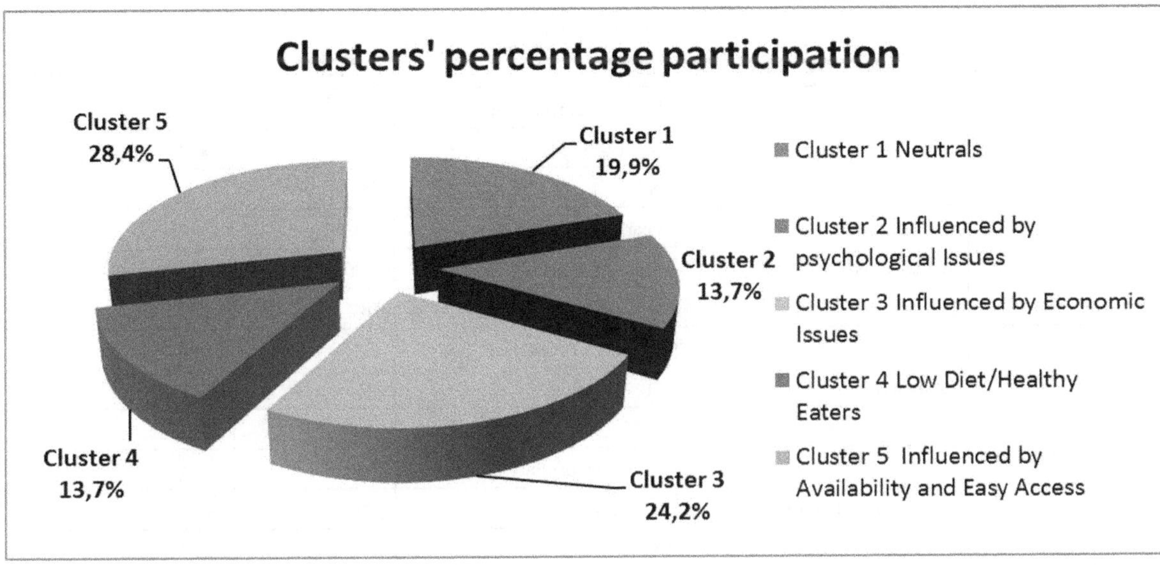

**For a more accurate representation see the electronic version.*

comprise 24.3% of the sample and prefer food which is domestically produced in Greece. They prefer the local products and production methods and they want to contribute to the national economy. Also they are very influenced by factors related to the price of the products. They purchase food that is the cheapest or is in offer or that is privately labeled because they are very influenced by the economic crisis. The "Low diet/Healthy Eaters" comprise 13.7% of the sample. They declare food according to how it makes them feel, helps them relax or helps them to cope with stress. Moreover, they are influenced by the ingredients of food, the calories, if it is low fat, organic, etc. They are very little influenced by hedonism and psychological issues, whilst they are not influenced by economic factors at all. Last but not least, the "Influenced by availability and easy access" comprise 28.4% of the sample and they are influenced by the distance they have to cover to find the food they want and by the simplicity of cooking it. It is a group of consumers that pay attention to factors that "make their life easy" and they are not influenced by price, health protection and psychological issues.

The summary of the cross-validation classification derived from the Discriminant Analysis is shown in Table 2.

The forty seven consumption dimensions derived from Confirmatory Factor Analysis, could accurately predict and discriminate consumers' group membership. Factors that affect consumers' food purchasing behaviour are the same with those before the economic crisis as presented in the literature review. But under the current economic situation, the exceptions subject to the fact that low fat food and generally healthy diet purchases became a trend in Greece the last decade. Consumers search for food that has the country of origin clearly marked and they prefer to be greek and not imported. And unexpectedly, consumers have no interest for the quality of food due to the economic crisis.

In the next stage, the non-parametric Friedman one way Anova Test was performed for each of the five groups of consumers in order to examine their opinions about junk food (Figure 3). All of the five groups of consumers consume junk food, some of them occasionally and some of them more often. In particular, the "Neutrals" mainly order fast food when they do not have time or mood to cook, and they

Table 2. Cross validation classification according to discriminant analysis

Correlations				
Approximately 90% of the Cases (SAMPLE)			**Predicted**	**Cluster Analysis**
0 - 10% Sample	Predicted	Pearson Correlation	1	,808**
		Sig. (2-tailed)		,000
		N	58	58
	Actual Classification	Pearson Correlation	,808**	1
		Sig. (2-tailed)	,000	
		N	58	58
1 - 90% Sample	Predicted	Pearson Correlation	1	,860**
		Sig. (2-tailed)		,000
		N	495	495
	Actual Classification	Pearson Correlation	,860**	1
		Sig. (2-tailed)	,000	
		N	495	495

**. Correlation is significant at the 0.01 level (2-tailed).

Figure 3. Consumers' opinion towards fast food consumption

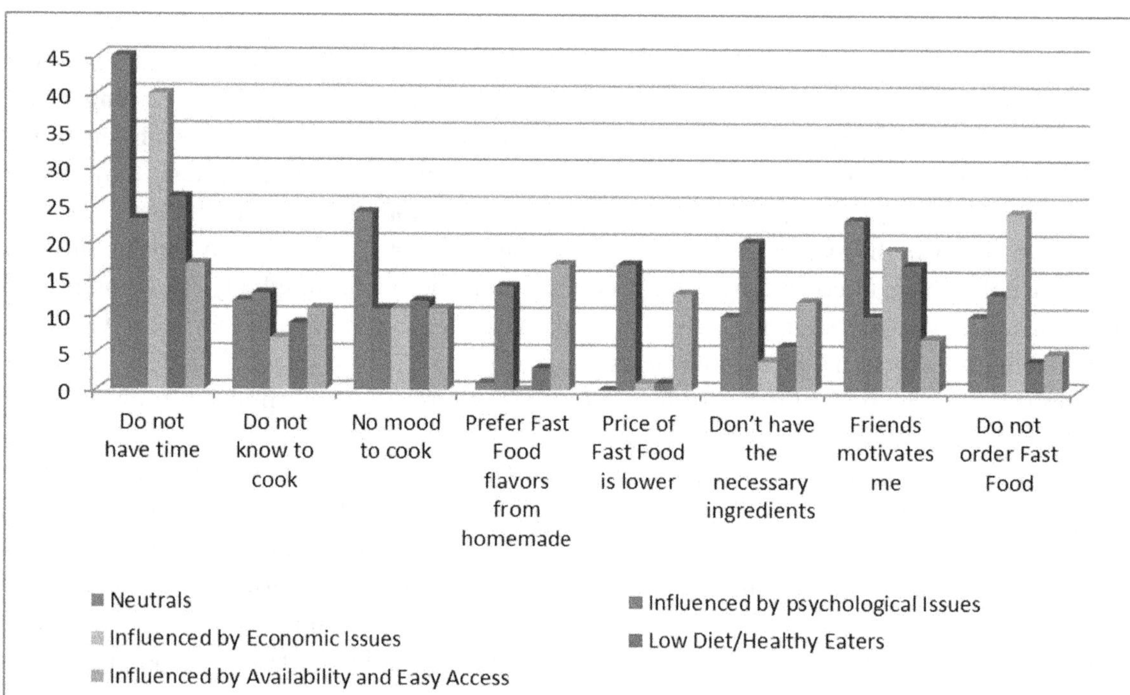

**For a more accurate representation see the electronic version.*

are motivated by friends and company to eat fast food. "Consumers influenced by Economic Issues" do not order fast food, except for the fact when they do not have time to cook. They do not think that the flavors of junk food are better from homemade food and also that fast food's price is lower than nutritional food. On the contrary, those influenced by availability and easy access and those influenced by psychological issues believe that the prices of junk food are lower than nutritional food and also they prefer the flavors of fast food. These factors, including the times that they do not have the necessary ingredients to cook, are the main reasons to eat fast food. Lastly, the "Low Diet/ Healthy Eaters" order fast food because they do not know how to cook and because they are not in a mood to cook. But mostly, because there are motivated by their friends.

Lastly, Cross Tabulation Chi-Square Analysis was performed, profiling each consumers' group regarding their reduced budget of food shopping (Table 3). 410 from 553 (74.1%) consumers declare that they have reduced their food shopping budget during the current economic crisis.

Most of the "Influenced by Availability & Easy Access" and "Influenced by Economic Issues" consumers declared that they have reduced their food shopping budget between 0%-25%. Most of the "Low Diet / Healthy Eaters" reduced their budget between 26%-50%. The consumers "Influenced by psychological Issues" declare that they reduced their budget even near to 70%. Last, the consumers who belong in the "Neutrals" differentiate. Some of them reduced their budget between 0%-25% and others went up to 70%.

Also, a cross tabulation chi-square analysis was performed for each group of consumers in order to develop the profile of consumers who have a particular purchasing behaviour regarding some of their personal characteristics.

As Table 4 indicates the "Influenced by the Availability & Easy Access" consumers mostly are singles, they have graduated from University or Technical Institution and they are private employees. "The low Diet/ Healthy Eaters" are also singles, they have graduated from University or Technical Institution and some of them have a Masters or a PhD degree and they are private employees. Consumers "Influenced by Economic Issues are married or divorced, they have graduated from primary and secondary schools and they are public employees or retired. Furthermore, consumers "Influenced by Psychological Issues" have graduated from middle education and are public employees or self – employed. Lastly, the "Neutrals" mostly live with a partner, are graduates from University or a Technological Institution, a large percentage of them have a Masters or a PhD degree and are mainly unemployed.

Table 3. Profile of each consumers' group regarding their reduced budget

Reduced Budget (%)		Clusters					
		"Neutrals"	"Influenced by Psychological Issues"	"Influenced by Economic Issues"	"Low Diet/ Healthy Eaters"	"Influenced by Availability and Easy Access"	Total
In what % you had reduced your budget	0 - 25%	18,5%	14,6%	23,7%	10,4%	32,8%	100%
	26 - 50%	21,5%	21,5%	16,1%	12,9%	28,0%	100%
	51 - 70%	22,2%	55,6%	22,2%	0,0%	0,0%	100%
Total		19,3%	17,1%	22,0%	10,7%	31,0%	100%

$X^2 = 17.010$, df=8, $p<0,05$

Table 4. Profile of each group of consumers regarding their socio-demographic characteristics

Socio - Demographic Characteristics		Clusters					
		"Neutral"	"Influenced by Psychological Issues"	"Influenced by Economic Issues"	"Low Diet/ Healthy Eaters"	"Influenced by Availability and Easy Access"	
Marital Status	Married	19,0%	18,3%	28,0%	9,7%	25,1%	x^2= 77.837 df=16 p<0,001
	Single	17,9%	6,6%	20,3%	15,6%	39,6%	
	Divorced	8,3%	10,7%	41,7%	0,0%	8,3%	
	Widowed	5,0%	25,0%	12,5%	12,5%	0,0%	
	Living with a partner	33,3%	9,5%	16,7%	35,7%	4,8%	
Education	Primary Education	2,9%	0,0%	57,1%	0,0%	0,0%	x^2= 91.543 df=20 p<0,001
	Middle Education	12,9%	28,6%	28,6%	0,0%	0,0%	
	Secondary Education	27,7%	20,3%	35,4%	12,7%	13,9%	
	Private Educational Center/ Technical School	25,3%	21,3%	16,0%	1,3%	36,0%	
	University / Technological Institution	33,6%	13,2%	20,9%	13,9%	38,5%	
	Master or PhD degree	30,4%	5,4%	27,7%	24,1%	12,5%	
Employment	Public Empoyee	10,0%	22,0%	28,0%	10,0%	30,0%	x^2= 56.296 df=28 p<0,001
	Private Employee	19,5%	14,3%	21,8%	17,3%	27,1%	
	Self - Employed	17,6%	12,1%	22,0%	11,0%	37,4%	
	Retired	4,8%	10,3%	37,9%	0,0%	6,9%	
	Housewife	11,1%	10,5%	26,3%	5,3%	36,8%	
	Student	21,7%	0,0%	21,7%	21,7%	34,8%	
	Unemployed	40,0%	0,0%	25,0%	15,0%	20,0%	
	Other	40,0%	0,0%	40,0%	20,0%	0,0%	

DISCUSSION AND CONCLUSION

This research contributes to the understanding of consumers' purchasing behaviour towards food (nutritious and fast food) under the current economic crisis.

The major findings of this study indicate that there is a significant association between the adoption of a purchasing behaviour and the factors that affect consumers to purchase food.

Moreover, our analysis found that compared to the period prior the economic crisis, Greek consumers seem to attach high importance to more or less the same factors that affect their purchasing behaviour towards food under the current economic crisis. However, the results show that consumers under those turbulent times switch to private label products and buy products in smaller packages. Moreover, in depth analysis helped us to identify that nowadays significant number of consumers switch to cheaper brands. Also, they want to purchase food that has clearly marked the country of origin, and they prefer not to be foreign brands. They observe the prices of food and looking for promotions and noteworthy is that under the current economic crisis they do not care about the quality of food, they just prefer the cheapest one.

The findings from cross tabulation Chi-Square Analysis indicate that economic crisis cause changes in consumers purchasing behaviour and their spending habits. That occurs, because consumers have reduced their budget for food shopping mostly by 0%-25%, but sometimes even up to 70%.

Additionally, the Friedman one way Anova Test shown that under the current economic crisis, consumers still purchase fast food, but only under certain circumstances such as when they do not have time or mood to cook, or they are motivated by friends to eat outside. Also, nowadays they prefer to substitute a meal with friends in a restaurant by fast food retailers and ready meals because of their lower cost.

According to Confirmatory Factor Analysis, consumers still try to have a balanced diet even though nowadays they pay no attention to the quality of a product. Also, they started to eat more at home rather out at taverns or restaurants, they prefer to shop in large supermarkets and local markets, switched from the finest product to economy range and they look for promotion and offers. These findings also identified that for a numerous number of consumers, price has the highest priority over quality.

In the light of these findings, it can be said that income is the most important factor which shape consumers' food purchasing behaviour. The findings of this study indicate that consumers buy the cheapest brand or from the cheapest retailer, in reduced quantities and avoid purchasing unnecessary products like snacks. The results show that consumers clearly prefer to buy products that have clearly marked the country of origin and they do not prefer imported products.

Overall, the findings of the present study confirm that consumers purchasing behaviour towards food is closely connected to the culture and the identity of the population where they are produced, they carry a strong symbolic value. But under the current economic crisis consumers have changed their hierarchy in the factors that affect them when they do their shopping.

Nevertheless, there are some limitations in this study. First that the study even it was a large scaled study, it took place only in the prefecture of Thessaloniki, so it would be valuable to enlarge the sample size to other cities or to rural areas. Another is the natural limitation that the survey conducted on the spot and may not represent the true beliefs and attitudes of consumers purchasing behaviour. It would be good if we could examine overtime since natural circumstances change all over time. Finally, the study was conducted during a particular face of the current economic crisis and it would be interesting to examine the factors that affect consumers' purchasing behaviour throughout all the faces of the economic crisis.

REFERENCES

Ajzen, I. (1991). The theory of Planned Behaviour. *Organizational Behavior and Human Decision Processes*, *50*(2), 179–211. doi:10.1016/0749-5978(91)90020-T

Ajzen, I., & Fishbein, M. (1980). *Understanding attitudes and predicting social behaviour*. Englewood Cliffs, N.J.: Prentice Hall.

Carrigan, M., & Attalia, A. (2001). The myth of the ethical consumer – do ethics matter in purchase behavior? *Journal of Consumer Marketing*, *18*(7), 560–578. doi:10.1108/07363760110410263

Cryssohoidis, G., Krystallis, A., & Perreas, P. (2007). Ethnocentric beliefs and country-of-origin (COO)-effect: Impact of country, product and product attributes on Greek consumers evaluation of food products. *European Journal of Marketing*, *41*(11/12), 1518–1544. doi:10.1108/03090560710821288

Duquenne, M.-N., & Vlontzos, G. (2014). The impact of the Greek crisis on the consumers behaviour: Some initial evidences? *British Food Journal*, *116*(6), 890–903. doi:10.1108/BFJ-11-2012-0279

Dutt, P. & Padmanabhan, V. (2012). When to push the panic button? Impact of currency crisis on consumer behaviours.

Eertmans, A., Baeyens, F., & Van den Bergh, O. (2001). Food likes and their relative importance in human eating behaviour: Review and preliminary suggestions for health promotion. *Health Education Research*, *16*(4), 4430456. doi:10.1093/her/16.4.443 PMID:11525391

Eetermans, A., Victoir, A., Notelaers, G., Vansant, G., & Van de Bergh, O. (2006). The Food Choice Questionnaire: Factorial invariant over western urban populations? *Food Quality and Preference*, *17*(5), 344–352. doi:10.1016/j.foodqual.2005.03.016

EL.STAT (2012). Monthly bulletins. *Hellenic Statistical Service*, October 2009-May 2011, July 2012.

EL. STAT (2016), http://www.statistics.gr/el/2011-census-pop-hous

Eurostat. (2014). Unemployment Statistics. Retrieved from http://epp.eurostat.ec.europa.eu/statistics_explained/index.php/Unemployment_statistics

Fishbein, M., & Ajzen, I. (1975). *Belief, Attitude, Intention and Behaviour* (5th ed.). New York: The Dryden Press.

Flatters, P., & Willmott, M. (2009). Understanding the post-recession consumer. *Harvard Business Review*, *7*(8), 106–112.

Food and Agriculture Organization. The double burden of malnutrition: Case studies from six developing countries. FAO: Food and nutrition paper No84, Rome 2006. 2006

Fotopoulos, Ch., Krystallis, A., Vassallo, M., & Pagiaslis, A. (2009). Food Choice Questionnaire (FCQ) revisited. Suggestions for the development of an enhanced general food motivation model. *Appetite*, *52*(1), 199–208. doi:10.1016/j.appet.2008.09.014 PMID:18929606

Frangos, C., Frangos, C., Sotiropoulos, I., Orfanos, V., Toudas, K., & Gkika, E. (2012, July 4 - 6). The Effects of the Greek Economic Crisis on Eating Habits and Psychological Attitudes of Young People: A Sample Survey among Greek University Students. In *Proceedings of the World Congress on Engineering*, London, UK.

Frust, T., Connors, M., & Bisogbi, C. (1996). Food choice: A conceptual model of the process. *Appetite*, *26*(3), 247–266. doi:10.1006/appe.1996.0019 PMID:8800481

Furaiji, F., Katuszynska, M., & Wawrzyniak, A. (2012). an empirical study of the factors influencing consumer behaviour in the electric appliances market. *Contemporary Economics*, *6*(3), 76–86. doi:10.5709/ce.1897-9254.52

Gupta, S., & Ogden, D. T. (2009). To buy or not to buy? A social dilemma perspective on green buying. *Journal of Consumer Marketing*, *26*(6), 376–391. doi:10.1108/07363760910988201

Hair, J. F., Anderson, R. E., Tatham, R. L., & Black, W. C. (1998). *Multivariate data analysis*. New Jersey: Prentice Hall Inc.

Harringan, K. R. (1985). An application of Clustering for Strategic Group Analysis. *Strategic Management Journal*, *6*(1), 55–73. doi:10.1002mj.4250060105

Helsen, K., & Green, P. E. (1991). A computational study of Replicated Clustering with an Application to market segmentation. *Decision Sciences*, *22*(5), 1124–1141. doi:10.1111/j.1540-5915.1991.tb01910.x

Howard, J. A. (1983). Marketing theory of the firm. *Journal of Marketing*, *47*(4), 90–100. doi:10.2307/1251402

Karelakis, C., Abas, Z., Galanopoulos, K., & Polymeros, K. (2013). *Positive effects of the Greek economic crisis on livestock farmer behaviour*. Agronomy for Sustainable Development; doi:10.1007/S13593-013-0136-y

KEPE – Centre of Planning and Economic Research. (2012). Greek Economic Outlook.

Kivetz, R., & Simonson, I. (2000). The effects of incomplete information on consumer choice. *JMR, Journal of Marketing Research*, *37*(4), 427–448. doi:10.1509/jmkr.37.4.427.18796

Kotler, P. (1994), Marketing management – analysis, planning, implementation, and control. Englewood Cliffs: Prentice Hall Inc.

Kupiec, B., & Revell, B. (1998). Speciality and artisanal cheeses today: The product and the consumer. *British Food Journal*, *100*(5), 236–243. doi:10.1108/00070709810221454

Mansoor, D., & Jalal, A. (2011). The Global Business Crisis and Consumer Behaviour: Kingdom of Bahrain as a Case Study. *International Journal of Business and Management*, *6*(1), 102–115.

Michalopoulos, V. G., & Demoussis, M. P. (2001). Greek household consumption of food away from home: A microeconometric approach. *European Review of Agriculture Economics*, *28*(4), 421–432. doi:10.1093/erae/28.4.421

OECD. (2011). Income. Retrieved from http://www.oecdbetterlifeindex.org/topics/income/

Pollard, T. M., Steptoe, A., & And Wardle, J. (1998). Motives underlying healthy eating: Using the Food Choice Questionnaire to explain variation in dietary intake. *Journal of Biosocial Science*, *30*(2), 165–179. doi:10.1017/S0021932098001655 PMID:9746823

Prescott, J., Young, O., ONeil, L., Yau, N., & Stevens, R. (2002). Motives for food choice: A comparison for consumers from Japan, Taiwan, Malaysia and New Zealand. *Food Quality and Preference*, *13*(7-8), 489–495. doi:10.1016/S0950-3293(02)00010-1

Rajmil, L., Fernandez de Sanmamed, M.-J., Choonara, I., Faresjo, T., Hjern, A., Kozyrskyj, A., ... Taylor-Robinson, D. (2014). Impact of the 2008 Economic and Financial Crisis on Child Health: A Systematic Review. *International Journal of Environmental Research and Public Health*, *11*(6), 6528–6546. doi:10.3390/ijerph110606528 PMID:25019121

Rico, A.; Blakey, E. (2012). Crisis económica y desigualdades en salud [Economic crisis and health inequalities]. In *Colección de Minitemas Online de la Escuela Nacional de Sanidad*. Madrid, Spain: Instituto de Salud Carlos III.

Siardos, G. (1999). *Methods of Multivariate Statistical Analysis. Examination of the association between variables. Part One*. Thessaloniki: Ziti Publications.

Stavkova, J., Stejskal, L., & Toufarova, Z. (2008). Factors influencing consumer behavior. *Agricultural Economics*, *54*(6), 276–284.

Steptoe, A., Pollard, T. M., & Wardle, J. (1995). Development of a measure of the motives underlying the selection of food: The Food Choice Questionnaire. *Appetite*, *25*(3), 267–284. doi:10.1006/appe.1995.0061 PMID:8746966

Steptoe, A., & Wardle, J. (1999). Motivational factors as mediators of socio-economic variation in dietary intake patterns. *Psychology & Health*, *14*(3), 391–402. doi:10.1080/08870449908407336

Thompson, B. (2012). *Impact of the Financial and Economic Crisis on Nutrition – Policy and Programme Responses*. FAO, Nutrition and Consumer Protection Division.

Tsourgiannis, L., Karasavoglou, A., Tsourgiannis, A. Ch., Florou, G., Theodosiou, T., & Valsamidis, S. (2014). Factors affecting consumers in Greece to buy during the economic crisis period food produced domestically in Greece. *Procedia Economics and Finance*, *9*, 439–455. doi:10.1016/S2212-5671(14)00046-X

Vlontzos, G., Duquenne, M. N., & Niavis, S. (2017). *Is Binge drinking prevalent in Greece after the emergence of the economic crisis? Assessment of this idea using the Theory of Plant Behavior* (Vol. 3, p. 3). Beverages.

Vyncke, P. (2002). From attitudes, interest, opinions, to value aesthetic styles, life visions and media preferences. *European Journal of Communication*, *17*(4), 445–463. doi:10.1177/0267323102017004030l

World Food Programme (WFP). (2012). Global food crisis in depth. Retrieved from https://www.wfp.org/global-food-crisis-in-depth

Zander, K., & Hamm, U. (2011). Information search behaviour and its determinants: The case of ethical attributes of organic food. *International Journal of Consumer Studies*, *36*(3), 307–316. doi:10.1111/j.1470-6431.2011.00998.x

ENDNOTE

1 The sample was set at 553 people, for a confidence level of 95% (a = 0.05) with the margin of acceptable error at 5%, according to the equations from Raosoft on determining the random sample size: $x = Z(c/100)^2 r(100-r)$, $n = N\,x/((N-1)E^2 + x)$, $E = \mathrm{Sqrt}[(N - n)x/n(N-1)]$. Where N is the population size, r is the fraction of responses that you are interested in, and Z(c/100) is the critical value for the confidence level c.

This research was previously published in the International Journal of Food and Beverage Manufacturing and Business Models (IJFBMBM), 2(1); edited by Constantin Zopounidis and George Baourakis, pages 56-69, copyright year 2017 by IGI Publishing (an imprint of IGI Global).

Chapter 7
Food Porn and the Invitation to Gaze:
Ephemeral Consumption and the Digital Spectacle

Yasmin Ibrahim
Queen Mary University of London, UK

ABSTRACT

In the digital world, notions of intimacy, communion and sharing are increasingly enacted through new media technologies and social practices which emerge around them. These technologies with the ability to upload, download and disseminate content to select audiences or to a wider public provide opportunities for the creation of new forms of rituals which authenticate and diarise everyday experiences. Consumption cultures in many ways celebrate the notion of the exhibit and the spectacle inviting gaze through everyday objects and rituals. Food as a vital part of culture, identity, belonging, and meaning making celebrates both the everyday and the invitation to renew connections through food as a universal subject of appeal. Food imagery as a form of transacted materiality online offers familiarity, comfort, co-presence but above all a common elemental literacy where food transcends cultural barriers, offering a universal pull towards a commodity which is ephemeral yet preserved through the click economy. Food is symbolic of human solidarity, sociality and sharing and equally of difference creating a spectacle and platform for conversations, conventions, connections, and vicarious consumption. Food images symbolise connection at a distance through everyday material culture and practices.

INTRODUCTION

What is this pervasive act of capturing our meals via our mobile and smart phones and sharing it with wider audiences through personal networks and messaging services? Why are we making a moment of the ordinary? This pervasive food sharing through mobile images is part of our digital literacy today. As an inimical part of digital culture, food images dominate as a universal symbol of social sharing and social gaze, and above all, human solidarity. Food as a subject of digital capture carries social resonance

DOI: 10.4018/978-1-5225-8063-8.ch007

and sociability. As new behavioural and social conventions convene around the production and dissemination of food images online, food as a material yet ephemeral object ingested by the body, amenable to edible iterations as well as decomposition, has become a major subject of material objectification signifying new ways to connect, frame social relations and to negotiate our private and public realms. Food once transformed into a digital image online is often offered for public consumption, transaction and dissemination inviting the gaze of family, friends and strangers. This paper explores the notion of food porn with relevance to new media technologies and contemporary digital culture. It posits the notion that the sociality around producing and sharing food in the user generated content (UGC) economy in many ways re-negotiates the unattainable quality that food porn signifies in the mainstream media. Amateur interpretations of food porn not only democratise it as a social activity but imbue new social conventions around the capture and sharing of these images.

The term 'food porn' is increasingly used to describe the act of styling and capturing food on mobile gadgets, eliciting an invitation to gaze and vicariously consume, and to tag images of food through digital platforms. The mundane and ordinary food is attributed a spectacle in this economy premising food as the message and the medium. The term 'food porn' of course predates these ubiquitous practices of capture today. Used synonymously with 'gastro porn,' it alludes to the fetishisation of food and its coalescence with desire by styling culinary offerings through the vantage point of the camera lens to be consumed by hungry publics. This food is meant to be consumed by sight and other senses (well removed from just ingesting it), evoking our hidden desires while highlighting its unattainability. Its pornographic quality removes food from the mundane and ordinary, elevating it to the level of the pornographic. This paper specifically addresses the imaging of food through new mobile technologies whilst acknowledging the wider ecology of media and marketing phenomena which have socialised us into accepting food images and the practices of imaging food as a resonant part of digital culture in our contemporary reality.

Food throughout history has been a symbol of sharing and sociality, envy and avarice, decadence and depravity, pride and repugnance and equally a resource of commonality as well as difference. Food has an elemental quality of tapping into our emotional reservoir, our nostalgic memories and inner psyche to elicit comfort, familiarity, aversion, desire, greed, degrees of pleasure and displeasure. Food has the ability to trigger forms of affect and ubiquitous food imaging through digital platforms transforms food into a digital commodity, bringing with it a plethora of social and symbolic meanings through the acts of image capture, upload, dissemination as well as archiving. But 'food is extraordinary in its ordinariness, exceptional in the extent to which we treat it as mundane, and outstanding as a focus for the study of consumption' (Marshall 2005: 69). Food serves to confirm membership as well as to set people apart. Food is connected to rituals, symbols and belief systems. It is associated with myth; the sacred and taboos. Food functions in social relationships; in terms of ethnicity, race, nationality, class, individuality and gender.

With increasing appropriation of technologies into our daily lives we don't often stop to question new forms of behaviours, conventions and habits which emerge with mediated communication. As we adopt new rituals and integrate these into our daily lives, we often struggle to completely comprehend why these new rituals become habitual and unquestioning moments of life's patterns and rhythms in contemporary culture. Imaging food and serving these to others via images is a widely accepted phenomenon today. The home photography situated the everyday through imagery. Richard Chalfen (1987) in writing about the 'home mode' and Kodak Culture highlighted the important role of photography in domestic life, where social practices involving family norms, traditions and values can be expressed and sustained. The everyday and its material practices became a site to enact and construct culture. This Kodak culture, which domesticated the everyday through imagery and food, is part of this visual literacy. With the convergence of technologies and the incorporation of recording features on the mobile phone, food becomes a site for multiple iterations from autobiography and memory making to self-representation.

In restaurants, homes and in the streets, food as a commodity is captured, shared and archived, providing a medium for human connection, communication; public and private interactions while contributing to a repository of food images on image sharing platforms online. The domestication of food images through the smartphone and the ability to consume food through screen cultures marks a shift in production values of food porn produced by expert stylists and photographers for the cultivated audience and those produced by the smartphone flaneurs, marking the division between production and consumption in the traditional media economy compared with the prosumer economy. The amateur prosumer equipped with her mobile technology can style and capture food through her sense of aesthetics and disseminate it to a potentially global audience. This smartphone food porn democratises imaging practices around food as a cultural artefact, while domesticating food imagery through pervasive technology as form of social exchange. The distance between the amateur and the professional is not obliterated but enables a mainstreaming of food photography in our cultural practices through the smartphone and mobile gadgets. With the late twentieth century's explosion of imaging and visualising technologies, everyday life has become visual culture (Lister and Wells). In this arena of hypermedia and multimedia age of visuality, 'seeing is much more than believing. It is not just a part of everyday life, it is everyday life' (Mizroeff 1999: 1).

There is a vast cultural economy around food online. From food blogging, recipe sharing, gastro tourism narratives, and food reviews to cataloguing 365 days of meals. Food imagery and meals can mediate the temporal and spatial, thus occupying a primacy in our cultural imagination and interactivity. Images are a primary form of content creation today. But what does it mean to capture something for our personal repository and image archive before we consume it? What is this urge to offer these images of food for vicarious consumption? What is it about food images which offer shared moments of communion between friends and strangers? The explanations reside at many levels and the deconstruction of these offer insights into the therapeutic nature of both food images as a cultural bridge but also social media platform as a means for sharing the banal and the mundane, and how these interactions display social relations as well as one's negotiation of self and life in the private and public realms (Ibrahim 2015). Functioning at the level of spiritual and emotional connection, ubiquitous food imaging and sharing remain an intrinsic part of digital culture today.

FOOD AS PART OF OUR EVERYDAY CONSTRUCTION

Food is more than just a fuel for bodies. It is a site of consumption, pleasure, morality and renewed aesthetics. As Krishnendu Ray (2007) asserts the very visibility of something that was mundane, trivial, and habitual is somewhat embarrassing but the source of discomfort for some may be the dissolution of boundary between the life world and the art world. Kerry Chamberlain (2004: 468) concurs that food permeates our relationships, infiltrating our language, our images, and that metaphors of food surround us. Food is entwined with popular culture, saturating our visual and sensory modes through television, magazines, newspapers, specialist food blogs and recipe sharing cites (Chamberlain 2004; Lupton, 1996). These provide voyeuristic escapes into food and cuisines where money or calories are of no consequence (O'Neill 2003). The re-fashioning of food or gasto porn then refers to the ways in which TV-programs and cookbooks often portray food as perfect, using pornographic features and visual effects (Zukin & Maguire, 2004). Richard Magee (2007) contends that food pornography takes the form of glossily lush photographs of voluptuous and sinfully rich desserts, or of fantasy recipes and lifestyle images that are 'so removed from real life that they cannot be used except as vicarious experience'. The concept of food porn then describes a movement away from motifs of aesthetics and trespasses some kind of threshold towards the obscene and unattainable; rupturing a gap and breaching the lucidity in judging what is *real* and *illusion*.

The term *food porn* is thus not just about the *content* or *way* food is represented, but rather addresses a much more complex construct of human experience (Mitchell 2014). The term gastro porn may be attributed to Alexander Cockburn's 1977 New York Review of Books when he criticised the pornographic character of the unattainable and picture perfect dishes in connection to the insatiable desires these elicited (Cockburn, 1977). McBride (2010) suggests that the term first appeared in 1979 when Michael Jacobson as the co-found for the Centre for Science in the Public Interest used food porn to refer to unhealthy foods in the Center's newsletter. According to Jacobson he "coined the term to connote a food that was so sensationally out of bounds of what a food should be that it deserved to be considered 'pornographic' (cf. Mcbride 2010). Mcbride (2010) employs the term food porn to denote 'watching others cook on television or gazing at unattainable dishes in glossy magazines without actually cooking oneself'. Signe Rousseau (2013) contends that food porn occupies a contested space between constructions of 'legitimate' and 'illegitimate' desires when it comes to eating.

Frederick Kaufman, in likening sex porn to gastro porn observes that 'like sex porn, gastro porn addresses the most basic human needs and functions, idealizing and degrading them at the same time' (cf. Ray 2007). The sex and food corollary is reiterated in Foucault's writing. Michel Foucault (1983) observed that food in the early Christian history was an issue of primacy, just like sex has become in the modern era, where its significance is created through a complex set of restrictions and discursive practices. Food has dominated as a subject of major preoccupation throughout history. Roland Barthes (1972: 22) in discussing the relationship between food, national identity and imperialism was conscious of the centrality of food to other forms of social behaviour;

Today we might say all of them: activity, works, sports, effort, leisure, celebration – every one of these situations is expressed through food. We might almost say that this 'polysemia' of food characterizes modernity. (Barthes 1972: 25)

The fetishisation of food as well as the cultural practices convening around the visual consumption of food have become important in locating the everyday. Raymond Williams (1993) decried that 'culture is ordinary,' in attempting to locate culture as something lived in and common place rather than a work of art removed from social and material practices. Williams (1993: 11) in defining culture as a set of expressive practices implicated not just print, cinema and television, but also employed it to cover a range of new material entities which emerged in post-war Britain including canned foods, contraceptives, aspirin and baby Austins. Culture is then reproduced in multiple sites. Michel de Certeau *et al.*'s *Practice of Everyday Life living and cooking* (De Certeau *et al.*, 1998:3) assert that 'culinary virtuosities' establish 'multiple relationships between enjoyment and manipulation'. In terms of food visuality, food is not just consumed orally and ingested, but consumed through other senses and a sense of play and desire.

Food, when dislodged from the kitchen and devoid of its nutritive or taste qualities enters a realm of the performative; valorising the visual above all else and situating it within the performative (Magee 2007). Barbara Kirshenblatt-Gimblett (1999) argues that food that is dissociated from eating bypasses the nose and mouth. Such food may well be subjected to extreme visual, and for that matter tactile and verbal, elaboration. Our eyes let us 'taste' food at a distance by activating the sense memories of taste and smell. Television shows and cookbooks invite us to eat with our eyes without either cooking or eating oneself. Kirshenblatt-Gimblett reminds us that the wondrous confectionery presented at the conclusion of Renaissance banquets, while technically edible, might never be eaten, though it might be enthusiastically applauded. She contends that these monumental events cultivate the visuality and appearance rather than the flavour, hence situating appearance as enduring compared to the eating of food, which can be ephemeral. The dominance of visuality or appearance is deliberate, for it emphasises a legible (edible) visual language of emblems and signs. Here the visual is the performative. Similarly, Barbara Wheaton in *Savoring the Past* notes that, in the edible allegorical tableaus that were medieval banquets, visual effects, rarity of ingredients, opulence, and sequence of events were more important than the dishes, ingredients, preparation techniques, or flavours. Richard Magee points out that both food pornography and sexual pornography are primarily focused on food or sex as a performance, and, like all performance, are designed as a voyeuristic exercise.

While considerable attention has been accorded to food as image or symbol, food studies is somewhat muted on the topic of food as a performance medium. Allen Weiss (2002) in drawing attention to a minority stance within Western aesthetics that allows for the definition of food as art contends that the Enlightenment concord has been thwarted to expand the boundaries of form and taste. The new postmodern aesthetic premises 'the disappearance of the boundaries between the arts, disrupting the border between art and craft, repositioning classic hierarchies of form and taste.' Weiss challenges with the question; 'if fashion can become a fine art today, why not cooking?'

This variant of food porn through the smartphone invariably offers a performance medium. In digital cultures the performative and visual elements of food porn enter a complex arrangement where it is conjoined with the creation of self-identity, community, invitation to gaze, memory making, and as a form of social capital to transact and exchange. In the digital economy, the unattainable quality of food porn is mediated and made mainstream for the amateur food flaneur. Amateur food images, rather than emphasising its unattainability, hinge on the individual aesthetics and exchange of image as a form of social capital to elicit validation and to seek attention and bring moment to the everyday. The imaging of food elongates the performative, making food inedible and imbuing it with non-perishable qualities. Imaging food accords performance to the everyday while transforming the mundane into commodities for sharing. Imaging food via smartphones and mobile gadgets enables the formation of ownership to an image and intimacy which food porn of the glossy magazines and television may not engender. In so doing they don't eradicate desire or the notion of the unattainable, where one man's food is offered to another for vicarious consumption but it mediates distance and closeness to this porn through a sense of ownership. With image sharing sites allowing for dissemination, display and curation of images, the processes of creation and ownership online allows for the perception of 'this porn is mine'. The flaneur gaze of the Internet means that images can be culled from the wider web and displayed in what one might consider a personalised space online, while sustaining vicarious consumption. Desire is still a manifest part of the food porn but it can reside in the personalised spaces which we curate and publish online.

RITUALS OF THE INTERNET

The Internet as a virtual space is not disembedded or dichotomised from the real world. It is mediated through social and cultural norms of our lived world. But it is equally a space that can produce cultures in its own right (Hine 2000). Our ability to enact new forms of rituals online and to seek communion, validation, therapy or to commemorate and preserve memories through image sharing and uploads has been well documented (Ibrahim 2011; 2015). The emergence of converged mobile technologies which enable publishing and social sharing while diarising the everyday has given rise to a whole host of cultural activities around image sharing and new media visualities online. Imaging the everyday by capturing our daily routines including meals has become an important part of new media rituals and validation of the self. According to a report by Pew entitled 'Photo and Video Sharing Grow Online' (Duggan 2013), 54% of adult users post original photos and videos online that they themselves have created. Another 47% of adult Internet users take videos or photos they found online and repost them on sites designed for sharing images with many people. According to Webstagram which ranks Instagram images by popularity, food images retain popular resonance through multiple tags including 'foodporn', 'yummy', 'yumgasm' and 'dessert'. For instance #food figured as number 25 in the top 100 tags on Instagram in 2014.

Marking life's everydayness and paces of routine through images, communicating these everyday images and inviting the gaze of others authenticates everyday life experiences; capturing the ephemerality of life while archiving and displaying the banal and the unusual. Food and meal times as symbolic imagery of our everyday (or the temporality of life) provide a means of communication and social sharing of a familiarity. They also serve as spatial and temporal devices to create a sense of space and a notion of lifetime online. This image literacy and reciprocity of exchange is part of digital culture. Food has a universal quality of resonance and food images carry this aura of desire, familiarity, comfort, spirituality as well as a sense of offering and a moment of display. The glimpse into one's private moment inevitably conveys an intimacy, but with social media platforms the intimate and public moments occupy a hybrid space where the private and public can co-exist. Food imagery provides an acceptable space of social and cultural exchange where the private and public can merge, forming hybrid spaces of material consumption, exchange and co-presence. These new rituals enacted through new mobile technologies and social media provide a means for shared and collective meaning making while renewing social connections.

The rich sociological literature on rituals confirms its symbolic, functional and therapeutic elements for humanity and society. Durkheimian interpretation of ritual emphasise the creation of social order and strengthening of group solidarity. Emile Durkheim (1976) accorded rituals a seminal role in enacting and sustaining the 'sacred' and in constructing a communally-shared social reality. Victor Turner (1969) conceptualised rituals as being both creative and equally subversive attributing it with 'social drama,' which can challenge social hierarchies. In contrast, Irving Goffman (1967) stresses rituals' ability to capture interactions and the performative aspects of everyday life. Rituals subsumes both the institutional and formal but also the informal and personal, providing a means for collective meaning making accounting for shared experiences as well as individual identity (Helland 2013:26; Rook 1985). Defying static manifestations to embrace the movements of time and space, rituals can reflect the changing patterns of society. As 'repeated and patterned forms of communication, rituals enable us to attach ourselves to 'our surrounding media-related world' (Semulia 2013: 9).

These new rituals online of imaging food have both symbolic and instrumental value of marking everyday moments like meal times and staples, but equally veritable feasts and culinary spectacles which celebrate the occasion, providing a means to commemorate and remember while creating a sense of shared and or individual identities. Several authors (e.g. Sahlin, 1974; Lupton. 2003) argue that gift giving is a way of validating social relations. Furthermore, gift giving might not only relate to the giving of physical objects but might also (or perhaps even more so) relate to more symbolic acts and practices and therefore, food image are part of gift giving, the purpose of which is to validate social relations. New rituals online such as imaging food before consumption (or during consumption) have instrumental and symbolic qualities where the invitation to gaze by others authenticates life's experiences while enabling the formation of everyday and special memories offered through the process of transacting the food image. These rituals have their therapeutic function in terms of social sharing and reciprocity, but they are equally intertextual with the wider world of consumerism representing material consumption as a site of desire, meaning making, performance and engagement with capital.

FOOD IN MEDIA AND CONSUMER CULTURE

Throughout history we have been socialised into food imagery whether in art history, food advertising editorials, magazine features, specialised cookbooks or personal photography. Food once transformed as an image or commodity signifies a permanence, defying its perishable and destructible qualities. Imaging provides an ethereality; an ability to transcend time and space and to be accorded different cultural significance by new audiences who consume it, appealing to their senses and sensitivities in different ways. With the invention of mobile photographic equipment, particularly the Kodak camera, food has remained a subject of rites of passages, of events as well as the everyday, marking the moments and routines of life within the domestic and beyond.

With print culture, broadcasting, advertising and branding, food imagery entered a realm of consumer capitalism denoting the construction of desire and its circulation through mass media platforms. The selling and packaging of food to the masses drew on our nostalgic bonds with food while recruiting consumers for advertisers. Food porn in the hands of capital sought to create a consumption community through desire for new offerings on the market. Food advertising came through our TV sets and magazines, offering the ease of assemblage to the homemakers, enticing new audiences while renewing our intrinsic ties with food in domestic settings. The increasing incorporation of leisure readings in newspapers and proliferation of lifestyle and women's magazine also socialised us into accepting stylised food imagery as part of a new consumer culture which packaged fantasy, escapism and romance into everyday fast moving consumer products. Advertising of food took it to new heights pledging desire and enticement through the lens, which could yield pleasure and yearning en masse.

The everyday staples of bread, butter, milk and coffee could be re-fashioned through the co-optation of photography which worked to lure mass audiences to mass production and into the hands of capital. The emergence of mass advertising of consumer products through mass media enabled food imagery to be accepted as part of the consumerist landscape to fashion instant desire like instant coffee. We were socialised into food porn as part of our consumer culture through mass media even before the advent of satellite television, which could assuage niche interests. The increasing tabloidisation of the public sphere and broadcasting also saw a proliferation of cookery shows, food channels and celebrity chefs, making food a ubiquitous subject of human consumption and of decadence and a symbol of high living where the concerns were about offering creativity and variety through food rather than issues of starvation, austerity or famine. With the tele-visualisation and tabloidisation of food, food imagery entered an era of the spectacle where it can be 'groomed' for consumption through the vantage point of the lens but removed from the immediate pleasure of consuming and ingesting food. Food porn online did not emerge from a vacuum – its prominence on the digital platforms today harks back to a wider media and capitalist socialisation which leveraged on our spiritual connection with food as a symbol of culture, survival, sustenance and solidarity.

With increasing globalisation of the world, food is increasingly a part of popular culture and popular imagination. Claude Fischler contends that what we have seen in fast food is not the "menace of Americanization" or a threat to traditional French cuisine, but an essentially benign reflection of the global circulation of culinary cultures. The globalisation of coffee chains and conveyor belt sushi joints symbolises both the increasing diversification of cuisines in cosmopolitan cultures and equally the dominance of standardised offerings through the Mcdonaldization of the globe. In the process, food acquires a popular global currency and can symbolise a multitude of meanings from the McDonaldization of society, the increasing curiosity and acceptance of world cuisines to the exoticisation of the ordinary meal.

FOOD, UGC AND PROSUMERISM

With the incorporation of Kodak culture and home photography, food retains a primacy as a subject of photography. With the convergence of technologies in the digital age and the incorporation of the digital photographic tools into mobile telephony the mobile phone becomes an extension of the body and memory, archiving and imaging events and rituals of the everyday. The body embedded with technology such as the smart phone is a space of double articulation in terms of functioning as a repository for memory. Beyond the cognitive ability of the brain to store visual memory and images, mobile technology such as the smart phone facilitates the production of mobile repositories, including cloud computing which can produce and carry images while on the move. Food provides a means to document and memorialise the everyday and the eventful. In the social media economy cultural food production also signifies the production of the self and diarisation of everyday life. New rituals around food consumption including the uploading of the food image before consumption, integrates seamlessly into a UGC economy where content creation and human behaviour fuse together to manifest new social conventions and rituals.

UGC is integrated into many commercial platforms and with the proliferation of social networking sites, image sharing and video archives, UGC has in parts entered a monetised economy while feeding this notion of consumer sovereignty and user empowerment online. This user-centred economy of publishing online through its emphasis on interactivity and online conversations is seen as celebrating a prosumerism, blurring the boundaries between production and consumption. In tandem with this, mobile telephony today enables people to record and publish images while on the move. In this digital economy where people can produce and consume incessantly, the image plays an important role in communication and in enabling connection both in the everyday and in marking the unusual. Web 2.0 celebrates the coexistence of a multitude of forms including images both mobile and static. In expressing everyday flows of life and in establishing co-presence as well as endorsing human communion, images are seen as a vital part of this connectivity. The virtual world yields to our desire to see ourselves in it and equally to look into the lives of others. Food provides a means of intimacy into a private realm while being a symbol of collective consumption.

The prosumer economy online has appropriated food photography as a well-established cultural convention of capture and exchange. This is evident where the Internet is saturated with features to improve techniques of imaging food with a smartphone. Conde Nast Publishers in its Traveler magazine ran a feature in 2014 on 'Food Photography Tips for Instagram and Smartphone,' dispensing technical advice on how to get the right lighting and vantage point while staying true to the protagonist, food. Similar articles have featured in Epicurious.com (2015) and the Guardian which ran a feature entitled, ' How to Photograph Food with your Phone' in 2012. A huge litany of similar articles exist online providing amateurs with tips to mainstream food photography with mobile phones. The Internet craves content and food imagery is a vital part of this economy.

People face the burden as well as the freedom to construct their own identities in postmodernity (Giddens, 1991; Christensen, 2008; Baudrillard, 2002), and new media technologies and the image economy are increasingly implicated in this. Consumption in modern (mass) consumer society can be highly individualistic and even narcissistic. While the sensual pleasures of eating are completely individualised, eating is a highly social activity and regulated by the community (Marshall 2005: 71) and food porn becomes a site of both personal consumption and collective desire. Holt (1995, 1998), refers to the way in which consumers use consumption objects to classify themselves in relation to relevant others. Miller and Edwards (2007) similarly explore online photo-sharing practices with sites such as Flickr and Facebook in relation to how different types of people manage public and private boundaries in this space. The coalescing of the public and private realms of consumption and notions of space make food porn amenable to multiple iterations. Like the selfie, food is part of this transaction culture and memory creation. What one consumes and experiences is archived and communicated through food. The selfie and the culture of social networking sites are driven through the uploading of content. The commodification of self and food are entwined online. The construction of the self is renewed through the constant engagement in these sites. Food as an everyday object of materiality and as a part of one's personal domain of consumption offers a means to create a personal connection while providing a common theme of interest for others to gaze.

The increasing number of image sharing sites and the communities which have emerged around gazing, curating, tagging, liking and retweeting, and re-blogging images makes food porn a tool of cultural production and interaction online; equally one of technique and aesthetics. The digital economy is made for forms of public and quasi-public sharing, inviting the gaze of the strange and familiar while enabling the self to be consumed and constructed through this image economy. According to a Pew Report in 2012, photos and videos have become a key social currency online (Lee et al. 2012). The rise of social media, especially YouTube and newer services like Pinterest, Instagram, and Tumblr have made curating activities easier because they are organized for easy image and video-sharing. Beyond content creation, the Internet and social networks drive on a circulation economy constantly merging content in unlikely places. Food porn can be part of disparate narratives and conversations online and equally it can convene dedicated niche communities.

Food as a subject of human connection can transcend cultural boundaries and is often considered a product of cultural diplomacy in terms of tourism and national identity. Food images become tools to personify self, communicate intimacy and to connect to a wider audience of similar and disparate interests. Our intimacy with food means it is a site for self-representation but also a means to leave narratives of social history and autobiographies in a transient mortal world. The Internet's virtuality is in many ways is inerasable and enduring while being amenable to wider circulation and search. Our connection to the image binds us to the mortal world in complex ways. Food, then, is a medium and metaphor for this representation and the preservation of the self through the image economy. Digital cultures are undeniably shaped by the culture of the physical world, which reinforces our need to imagine ourselves beyond our situated mortality; the virtual world provides a means to leave traces of your intimate portrayals where you can both avow and disavow ownership to content you produce. The image economy, particularly the intimate images of creation, tap into this need to preserve ourselves beyond our transient world. The ubiquity of new practices and conventions in many ways reassert this, and the imaging of food is no exception; representing both the need to prolong desire and life through the symbolic metaphor of food, recognising it as an ephemeral entity, but seeking to prolong and immortalise it through the image.

CONCLUSION

Amateur food porn online democratised by the smartphone and convergence technologies appropriates many cultural qualities from the food porn crafted by the experts. It relocates desire as something that can be generated through UGC mediating both distance to food porn and its unattainability through ownership and creation of food images. In the process, it provides a mechanism to create social capital and currency online and to produce these to document the everyday and to renew connections with a wider community. Smartphone food porn democratises the production of desire and in doing so taps into representation of self and identity while being intertextual with our consumerist world and consumption cultures. At the heart of virtual life is our quest for self-presentation beyond our mortal existence. The image economy assuages our need to seek immortality through the virtual world. Food then becomes the medium and message.

REFERENCES

Barthes, R. (1972). *'Ornamental Cookery,' Mythologies* (p. 78). New York: Hill and Wang.

Barthes, R. (1979). Towards a psycosociology of contemporary food consumption. In R. Forster & O. Rarum (Eds.), *Food and drink in history: Selections from the annals economics, societies, civilisations*. Baltimore: John Hopkins University Press. (Original work published 1961)

Baudrillard, J. (2002). The Precession of Simulacra. In G. D. Meenakshi & D. M. Kellner (Eds.), *Media and Cultural Studies: Keyworks* (pp. 521–549). Malden, MA and Oxford: Blackwell Publishers.

Chalfen, R. (1987). *Snapshot Versions of Life*. Bowling Green, Ohio: Popular Press.

Chamberlain, K. (2004). Food and Health: Expanding the Agenda for Health Psychology. *Journal of Health Psychology*, *9*(4), 467–481. doi:10.1177/1359105304044030 PMID:15231050

Christensen, C. L. (2008). Livsstil som TV-underholdning. *Journal of Media and Communication Research*, *45*, 23–36.

Cockburn, A. (1977, December). Gastro-Porn. *The New York Review of Books*, *8*. www.nybooks.com/articles/8309

Culture and practice. *Proceedings of CHI'07*. San Jose, CA. ACM Press.

De Certeau, M., Giard, L., & Mayol, P. (1998). The Practice of Everyday Life, Volume 2: Living and Cooking. University of Minnesota Press: Minneapolis.

Duggan, M. (2013). *Video and Photo Sharing Grow Online.* Pew Research Centre, http://www.pewinternet.org/2013/10/28/photo-and-video-sharing-grow-online/

Durkheim, E. (1976). *'The Elementary Forms of Religious Life'. K.Feilds Glencoe.* Free Press.

Fischler, C. (1988). Food, Self and Identity. *Social Sciences Information. Information Sur les Sciences Sociales*, *27*(2), 275–292. doi:10.1177/053901888027002005

Foucault, M. (1983). *'On the Genealogy of Ethics: An Overview of Work in Progress,' in Hubert Dreyfus and Paul Rabinow, Michel Foucault: Beyond Structuralism and Hermeneutics*. Chicago: University of Chicago Press.

(2007, Winter). Gastronomica. *The Journal of Critical Food Studies*, *7*(1), 50–63.

Giddens, A. (1991). *Modernity and Self-Identity*. Stanford: Stanford University Press.

Goffman, E. (1959). *The Presentation of Self in Everyday Life*. Edinburgh: Social Sciences Research Centre.

Goffman, E. (1967). *Interaction Ritual: Essays on Face-to-Face Behaviour*. New Jersey: Transaction Publishers.

Helland, C. (2013). Ritual. In H. Campbell (Ed.), *Digital Religion: Understanding Religious Practice in New Media Worlds*. Routeledge, New York.

Hine, C. (2000). *Virtual Ethnography*. Thousand Oaks, CA: Sage.

Holt, D. B. (1995). How consumers consume: A typology of consumption practices. *The Journal of Consumer Research*, *22*(1), 1–16. doi:10.1086/209431

Holt, D. B. (1998). Does cultural capital structure American consumption. *The Journal of Consumer Research*, *25*(1), 1–25. doi:10.1086/209523

How to Photograph Food with your Phone. (2012, 15 July). *The Guardian*, http://www.theguardian.com/lifeandstyle/2012/jul/15/how-to-photograph-food-smartphone

Ibrahim, Y. (2011). The non-stop 'capture': The politics of looking in postmodernity. *The Poster*, *1*(2), 167–185. doi:10.1386/post.1.2.167_1

Ibrahim, Y. (2015). Instagramming life: banal imaging and the poetics of the everyday. *Journal of Media Practice* , 1-13.

Jensen, P. (2008). The International Extent and Elasticity of Lifestyle Television. *Mediekultur*, *45*, 24–45.

Kaufman, F. (2005). Debbie Does Salad: The Food Network at the Frontiers of Pornography. *Harper's*, (October): 55–60.

Kirshenblatt-Gimblett, B. (1999). Playing to the senses, Food as Performance Medium'. *Performance Research*, *4*(1), 1–30. doi:10.1080/13528165.1999.10871639

Lee, R., Rainier, J., & Purcell, K. (2012). Photos and Videos as Social Currency Online. *Pew Research Centre*, http://www.pewinternet.org/2012/09/13/photos-and-videos-as-social-currency-online/

Lister, M., & Wells, L. (2001). Seeing Beyond Belief: Cultural Studies as an approach to studying the visual. In Handbook of visual analysis, pp. 61-91.

Lupton, D. (1996). *Food, the Body and the Self.* Thousand Oaks, CA: Sage.

Magee, R. M. (2007). Food Puritanism and Food Pornography: The Gourmet Semiotics of Martha and Nigella. *Americana: The Journal of American Popular Culture, 6*(2), at www.americanpopularculture.com/journal/articles/ fall_2007/magee.htm

Marshall, D. (2005). Food as ritual, routine or convention. *Consumption Markets & Culture, 8*(1), 69–85. doi:10.1080/10253860500069042

McBride, A. E. (2010, Winter). "Food porn." Gastronomica. *The Journal of Food and Culture, 10*(1), 38–46.

Miller, A. D., & Edwards, W. K. (2007). Give and take: A study of consumer photo-sharing

Mirzoeff, N. (1999). *An Introduction to Visual Culture*. London, New York: Routledge.

Mitchell, R. D. (Ed.). (2014). *Food Design on the Edge: Proceedings of the International Food Design Conference and Studio 2014,* 2-4 July 2014, Dunedin, New Zealand. Otago Polytechnic, Dunedin.

O'Neill, M. (2003). Food porn. *Columbia Journalism Review, 5*, 38–45.

Ray, K. (2007). Domesticating cuisine: food and aesthetics on American television.

Raymond, W. (1985). *Keywords: A Vocabulary of Culture and Society*. Oxford: Oxford University Press.

Rook, D. (1985). The ritual dimension of consumer behaviour. *Journal of Consumer Behaviour, 12*(December), 251–264. doi:10.1086/208514

Rousseau, S. (2014). Food "Porn" in Media. In *Encyclopedia of Food and Agricultural Ethics* (pp. 1–8). Netherlands: Springer.

RoutledgeNew York.

Sahlins, M. (1976). *Culture and Practical Reason*. Chicago, IL: Univ. Chicago Press.

Semulia, J. (2013). *Media and Ritual: Death*. Community and Everyday Life.

Snapchat: The Ultimate Guide to Mobile Food Photography, From Lighting to Editing. (2015, 9 Jan). *Epicurious.com*. http://www.epicurious.com/expert-advice/the-ultimate-guide-to-mobile-food-photography-from-lighting-to-editing-article

Sturken, M., & Cartwright, L. (2001). *Practices of Looking: An Introduction to Visual Culture*. Oxford: Oxford University Press.

Turner, V. (1969). *The Ritual Process: Structure and Anti-Structure*. Ithaca: Cornell University Press.

Weiss, A. (2002). *Feast and Folly: Cuisine, Intoxication, and the Poetics of the Sublime Albany*. State University of New York Press.

Wheaton, B. (1983). *Savoring the Past: The French Kitchen and Table from 1300 to 1789 New York.* Simon & Schuster.

Zukin, S., & Maguire, J. (2004). Consumers and Consumption. *Annual Review of Sociology*, *30*(1), 173–197. doi:10.1146/annurev.soc.30.012703.110553

This research was previously published in the International Journal of E-Politics (IJEP), 6(3); edited by Mehdi Khosrow-Pour, D.B.A., pages 1-12, copyright year 2015 by IGI Publishing (an imprint of IGI Global).

Chapter 8
Food Consumption Expenditure and Standard of Living in Romania

Marian Zaharia
Petroleum-Gas University, Romania

Rodica-Manuela Gogonea
Academy of Economic Studies, Romania

ABSTRACT

Food consumption and its structure is significantly influenced by standard of living. But in addition, there are differences in the level and structure of food consumption, determined by characteristics and local habits. Given these considerations, in this chapter, starting from the mean consumption expenditure by detailed level, and the overall structure of consumption expenditure by detailed COICOP level (01-12 - Individual consumption expenditure of households), are analyzed the mean consumption expenditure per inhabitant, types of household, degrees of urbanization, and macro regions of Romania. Also are analyzed the correlations between the evolution of living standards and evolutions structure of consumption expenditure by type of household.

INTRODUCTION

The standard of living of the individual and/or family is in bi-univocal correspondence with the revenue available to them. Put face to face, their preferences and income lead to the formation of demand for goods and services and ultimately to their consumption patterns.

Given the diversity of products (goods and services) that falls into the consumption pattern of families, any study on their consumption expenditure implies the existence of certain classifications to ensure unity and comparability. One of the most widely used and unanimously recognized classification systems is Classification of Individual Consumption according to Purpose (COICOP), published by the United Nations Statistics Division. It divides the consumption expenditure incurred by individual in three groups: expenditure for households, for non-profit institutions and for general government. These are further structured in divisions, groups and classes.

DOI: 10.4018/978-1-5225-8063-8.ch008

In COICOP, food consumption expenditure is structured in two divisions of household expenditure: Food and non-alcoholic beverages (01) and Alcoholic beverages, tobacco and narcotics (02).

The standard of living of individuals and households is quite heterogeneous. This feature is given not only by the structure of population in terms of income levels, but also by significant differences between social categories and between residential areas (village, town). These distinctions are also propagated on the structure of food expenditures.

In Romania, the structure of food consumption patterns and food consumption expenditure are implicitly influenced by local and regional traditions and cultural particularities, which lead to models that are specific for developing regions and even counties within their structure.

Although transnational retail chains are expanding throughout the country, a process which brings uniformity to supply, impacting the demand, local patterns of food consumption have and will have a significant share, especially as these particularities contribute to tourism development in those respective areas and to revenue growth of local communities.

Also, the food consumption expenditure structures of the main social categories in Romania have a number of features determined on the one hand, by great differences between their average income (between employees and retirees or between employed and unemployed) and on the other hand, by the significant differences between rural and urban environments, and not least, by the differences between age groups (pensioners are citizens over 60 years of age, which makes their consumption patterns differ significantly from those of other social categories, even in similar conditions of income and residential area).

This chapter aims to highlight the way in which the average food consumption expenditure per inhabitant, per food product categories (cereals and cereal products, meat and meat products, milk and dairy products, vegetables and vegetable products, fruits and fruit products) and beverages (wine and wine products, beer, distilled alcoholic beverages, soft drinks), consumed by population of different social categories (employee, farmer, unemployed and pensioner), residential areas and macro regions, is influenced by the average per capita income developments.

BACKGROUND: CHARACTERISTICS AND FACTORS OF FOOD CONSUMPTION EXPENDITURE

The increase of economic and social development levels lead to larger or smaller changes of the structure of consumption expenditures by category of goods and services, as well as within them. According to Engel's Law, the largest shares of total expenditures of poor households are allocated to food consumption. In the case of developed countries the stability of the expenditure structure of consumption is higher than in developing countries and significantly different from the expenditure structure of average household income in underdeveloped countries.

In a report of the United States Department of Agriculture (Muhammad, Seale, Meade & Regmi, 2011) which updates the international evidence of food consumption patterns using 2005 International Comparison Program (ICP) data, analyzing the incomes and expenditures from 144 countries, shows that in the countries with low income, 48.5% of it is used for food, beverages and tobacco, and in the countries with high income, the share is only 20,4% (2.37 times less), the ratio being inversely for medical and health expenditure share (8.9% in the countries with high income, compared to 4.5% in countries with low income) and for recreation (9.5% in the countries with high income, compared to 4.5% in countries with low income).

An Italian study made during 1989-2004 (Barigozzi, Alessi, Capasso & Fagiolo, 2012) highlighted the structure stability of expenditure categories, but showed heterogeneity on goods categories. Another study, conducted in the Czech Republic (Hindls, Hronová, Sixta & Vltavská, 2014) for a much longer period (1970-2012), highlighted the decrease of the share spent on food products and the increase of housing expenditure, as household income increased.

In Romania (Babucea & Bălăcescu, 2011), in the first 20 years after the fall of communism, people tried to maintain the level of consumption at the expense of savings, although in the early years of the analyzed period, social confrontations have considerably reduced the pace of economic reforms. The period preceding the economic crisis has led in Romania to the decrease of food expenditure in the context of total household expenditures, and to the increase of expenditures for housing and endowment of goods, health, transport and communications (Cismaş, Miculescu & Oţil, 2010). Also, an analysis of the evolution of household consumption expenditures at the local level, including the economic crisis period, is conducted by Ghiţă and Manea (2013) by applying the Relative distance ranking method and the Real ranks method.

The structure of household consumption expenditures was partially influenced by the European integration process. Analyzing the convergence of new member states towards the old member states using the structure of food expenditure patterns, Liobikienė & Mandravickaitė (2013) showed that during 1995-2007 in the Czech Republic, Estonia, Lithuania and Slovenia, the convergence process was more pronounced than in Latvia and Poland.

Food expenditure patterns of households are complies with Engel's Law also. This fact is underlined by Muhammad, Seale, Meade & Regmi (2011) in their report. They show that in countries with low income, the food pattern budget share for cereals is 23.3% and 18.1% for fruits and vegetables, compared to only 8.6% for cereals and 9.8% for fruits and vegetables, in the countries with high income.

Food expenditures are the result of confrontation between individual preferences and limited resources, a result of supply and demand of the food markets in which the individual acts as an agent. The expansion of transnational retail chains led to the modernization of the food marketing sector and to an increasing volume of purchased food products, which leads to convergence in the agro-food sector (Frazao, Meade & Regmi, 2008).

Household expenditure patterns for food are strongly influenced by family income and the number of their members (Jacobson, 2010, Dudek, 2011) and also by a number of factors such as marital status, gender, place of residence, employment status (Kostakis, 2014). A low level of family income leads to a decreased access to healthy food. This is not only a characteristic of underdeveloped countries or regions, but can also occur in civilized countries during economic crises (Constandache, Condrea, & Stanciu, 2014). Through its long-term impact on public health, food security is a priority area for research. A series of papers () analyzes on the one hand, how individuals and households are concerned with avoiding dietary risk (Baker, 2003), avoiding it at all stages of the distribution chain (Rabonti, 2010, MacLaren, 2006), and on the other hand, issues and concerns for cooperation between national and international institutions, regarding regulation on food safety and health protection at European and international level (Toma-Bianov, 2013, Hassanein, 2011, Buzby & Mitchell, 2006, Ng & Salin, 2012).

Declining birth rates and an aging population are also influential factors of expenditure patterns (Anders, 2010), causing significant changes in expenditures for products with a high sugar content, meat and meat products. Another study (Drichoutis, Nayga & Lazaridis, 2012) that starts from the observation that the population aged 50 and over is an important demographic group in Europe, analyzes using instrumental variable methods, the effects of food consumption among the older population, regardless of home expenditures.

The existence and growth of vegetarian population determines food expenditure patterns which are totally different from those of non-vegetarians. Guillementte & Cranfield (2012) consider that the evolution of vegetarian diet effects on food expenditure may be associated with those of organic food consumption, both being increasing consumption patterns.

An issue that constantly draws the attention is that of inequality and welfare in food consumption expenditure. Its scale, depending on the development level, varies from country to country and is enhanced by the features of the residential area of the individual and his family (Haq, 2009) and also by the social category to which they belong. In the study of inequality in consumption expenditure, conducted on households in the Indian state of Uttar Pradesch (Pandey, 2013), it is highlighted the fact that the main factor of inequality is the education level of the head of household, followed by the size of the household and residential area, stressing the need for increasing public expenses for technical and high level education to reduce inequality.

METHODS AND METHODOLOGY

The research objective consists in assessing trends, over time (2001-2014) and space (on residence areas and at macro regional level), of average monthly expenditures for 13 food products, in order to establish their place and role in Romania's economic growth process.

Achieving the established objective required several analyses regarding the evolution of average monthly expenditures, on 13 food products, social categories, areas of residence and macro-regions.

The research methodology consisted in combining theoretical elements, as guiding principles, with several methods and techniques (ANOVA), as well as ordering processes, development indicators (average indicators) and processing information taken from official data provided by INSSE.

The situation of food purchase expenditures during 2001-2014 is the result of information gathered by conducting the Household Budget Survey (HBS). An analysis of this type, made on expenditure, is important in the context of interdependence between the standard of living and economic growth. It should be noted that from all the listed categories, only four were introduced in the analysis (employee, farmer, unemployed and pensioner), considered as most relevant for achieving the established purpose.

The coordinates of consumption for the main agrifood products bought in Romania were determined in relation to 13 categories, considered most significant of all 59 identified in the information provided by Romanian National Institute of Statistics (NIS).

The main analyzed categories of food and beverage products were: bread and bakery products, fresh meat, meat products, fish and fish products, milk, cheese and cream, eggs, oil, sugar, vegetables and tinned vegetables in fresh vegetables equivalent, fruit, mineral water and other soft drinks, and alcoholic beverages.

Determining the average expenditure for food product purchase by social categories, related to both residence areas (urban, rural) and macro regions (Macro_1 with the North-West and Central regions; Macro_2 with the North-East and South-East regions; Macro_3 with the South-Muntenia and Bucharest-Ilfov regions; Macro_4 with the South-West Oltenia and West regions) is important in the conducted analysis, to highlight the occurring differences and the ways to meet the needs, as a result of food consumption particularities.

The average monthly expenditure on food consumption per residence area was analyzed from 2001 to 2014.

The necessity to detail and clarify the expenditure situation at a territorial level determined NIS to present the average monthly food consumption expenditure on macro regions since 2008, which led to a regional analysis on a shorter period, of just 7 years.

FOOD CONSUPMTION EXPENDITURE PER INHABITANT BY SOCIAL CATEGORY AND AREA

Monthly average food consumption expenditure per inhabitant by social category

The evolutions of the average expenses per capita for buying food and beverage products from one year to another are the result of the influence of many factors, among which social category of consumers, residential area, geographic region of origin, level of education, and so on.

From all the six social categories for which information was provided to the NIS, they were selected for analysis only four considered most significant: employee, farmer, unemployed and pensioner. Total monthly average expenditure per capita for food and beverage, in Romania, during the analyzed period, in current prices, record an increasing continuous trend (Figure 1) from a minimum of 45.60 lei per capita, in 2001, to 195.35 lei per capita, in 2014, with an average rate of 11.84% yearly, which represents an average yearly increase of 11.52 lei per capita.

Figure 1. Evolution of total monthly average expenditures per capita by the social categories (current prices)

Source: own construction based on http://statistici.insse.ro/shop/index.jsp?page=tempo3&lang=en&ind=BUF114I Total consumption for purchase agro-foods products and non-food at main class goods and main class social, by area - monthly averages per person

In the period 2001-2014 each social category had the same increase trend for the entire period of analysis, but with different rhythms from one category to another. The highest monthly average expenditure per capita for food and beverage were registered at employees who, from a minimum of 55.80 lei in 2001 reach in 2014 to 221.82 lei; in the period 2001-2014, this category had the highest monthly average per capita of 152.49 lei and the most significant yearly absolute increase of monthly average expenditure for food and beverage of 12.77 lei per employee.

The farmer category, although recorded the lowest level of monthly average expenditure per capita for food and beverage consumption (thanks to the close interdependence between the environment in which they operates and food) comprised between 20.81 lei in 2001, and 110 lei in 2014 has the highest annual growth rate (by 13.66%) compared to the other three categories; but the absolute value of the yearly average increase per capita of 6.86 lei ranks farmers on last place.

The monthly average expenditure situation of food and beverage consumption per capita for unemployed category in the period 2001 – 2014 is similar to that of farmers. From a monthly average expenditure for food consumption per capita of 35.42 lei registered in 2001, in 2014 is reached to 150.79 lei. The yearly average expenditure was 104.4 lei per unemployed, and the yearly average increase of monthly average expenditure for food consumption was 11.79%, which represents a yearly absolute average increase of 8.87 lei per unemployed.

The evolution for pensioner category is similar and close to the total evolution, seeming that an average level of balance between the other categories; for the period 2001-2014, the monthly average expenditure for food consumption per capita was 133.08 lei with a yearly absolute average increase of 12.35 lei per pensioner and a rate of 12.28% per year.

In food product categories, bread and bakery products represent for Romanian a basic foodstuff. This fact is highlighted by the monthly average expenditure for bread and bakery products per capita on 2001- 2014 period, which is the highest of all social categories in relation to all the others 13 foodstuff (21.95 lei per unemployed ; 21.73 lei per employee ; 21.62 lei per pensioner; 16.65 lei per farmer). Although farmer category recorded the lowest average value compared to the others social categories, however, the increase of expenditure with this type of product was significant, rate of yearly increase being 11.84% yearly, compared with 9.38% yearly at pensioners, 8.53% yearly at unemployed or with 8.11% yearly at employees.

Another food included in the diet is meat. It appears in analysis through three categories: fresh meat, meat products, and fish, fish products and canned fish. An image of the monthly average expenditure per capita for meat, in the period 2001-2014, from this three categories and by social category is illustrate in Figure 2.

The values of monthly average expenditure per capita, in the analyzed period, shows that the employees are the ones who spend the most on fresh meat consumption (25.17 lei), followed by pensioners (20.23 lei), unemployed (16.71 lei) and farmers (7.95 lei).

But, social and economic changes, in past 20 years have brought significant alterations in the standard of living and farmers' mentality so, from a monthly average expenditure for fresh meat by 1.73 lei per farmer, in 2001, in 2014 were registered 13.16 lei per farmer, which, from this point of view, this category ranks first, compared the other three analyzed.

The average rates of growth of monthly expenditure for fresh meat corresponding to pensioners and the unemployed are nearly alike. The pensioners, from a monthly average expenditure with fresh meat of 7.14 lei per capita in 2001, will get to expend 33.18 lei per capita in 2014, with growth of 12.54%

Figure 2. Monthly average expenditure per capita, on types of meat, by social categories, in the period 2001-2014 (current prices)
Source: own construction based on http://statistici.insse.ro/shop/index.jsp?page=tempo3&lang=en&ind=BUF114I. Total consumption for purchase agro-foods products and non-food at main class goods and main class social, by area - monthly averages per person

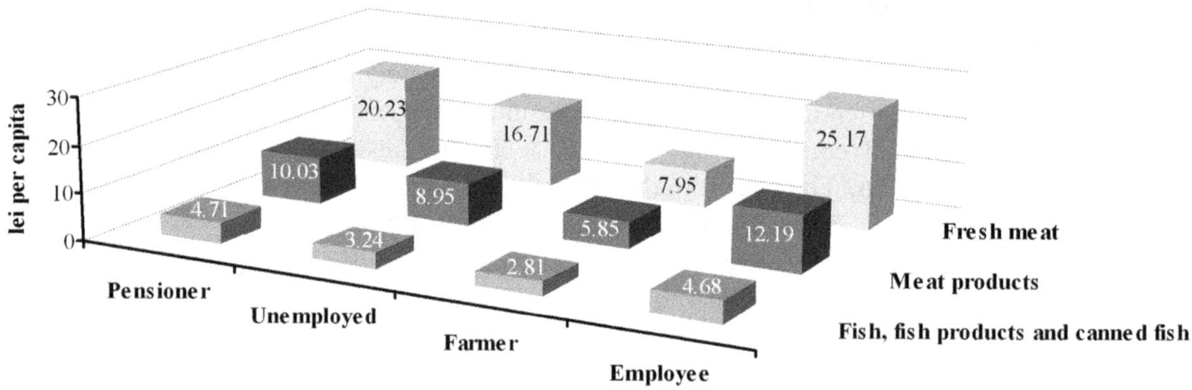

yearly, while an unemployed from 5.21lei per capita in 2001 will reach to expend 23.24 lei per capita, in 2014, the average rate of growth being of 12.19% yearly.

In value terms, the largest average growth is registered by the employees whose monthly average expense with fresh meat per capita increases by 2.07 lei yearly, followed by the pensioners with 2.00 lei yearly, the lowest absolute average increase being registered at the unemployed with 1.39 lei yearly, and farmers with 0.88 yearly.

The monthly average expenditure per capita for meat products had the same trend as the trend of fresh meat. In 2001, they were 4.81 lei per employee, 3.16 lei per pensioner, 2.58 lei per unemployed, and 1.43 lei per farmer. Between 2001-2014 the yearly average increase rates of monthly average expenditure per capita for meat products were 11.06% at employees, 13.39% at pensioners, 13.08% for unemployed, and 15.93% for farmers, which shows a slowly convergence process. However hierarchy is maintained, so that in 2014, the expenditures, at current prices, were: 18.81lei per employee, 16.19 lei per pensioner, 12.75 lei per unemployed, and 9.77 lei per farmer

The situation changes in the case of consumption expenditure of fish, fish products and canned fish. Pensioners and employees expend the most to buy fish, fish products and canned fish, as evidenced by monthly average expenditure during the period analyzed by 4.71 lei per pensioner and 4.68 lei per employee. This product is slightly preferred by unemployed and farmers, the average expense of only 3.24 lei per unemployed and 2.81 lei per farmer. In relative values, the highest increase rate of yearly average expenditure for the purchase of fish, fish products and canned fish (16.42%) was registered at unemployed, followed small difference by pensioners (16.39%), employees (16.29%) and farmers (16.09%).

Milk, eggs and cheese and cream are the food of animal origin, for which the average monthly expenditure at current prices, were under 12 lei per capita. The average values of spending for these products during the period 2001-2014, leaded to the following hierarchy of social categories (Figure 3) employee, pensioner, unemployed and farmer.

Employees are social categories that spend the most in comparison with other social categories, for the purchase of milk and cheese and cream and eggs for consumption. On the second and third places are pensioners and unemployed; pensioners ranks second, and the unemployed ranks third, in terms of

Figure 3. Monthly average expenditure per capita, on eggs, cheese and cream, and milk, by social categories, in the period 2001-2014 (current prices)
Source: own construction based on http://statistici.insse.ro/shop/index.jsp?page=tempo3&lang=en&ind=BUF114I Total consumption for purchase agro-foods products and non-food at main class goods and main class social, by area - monthly averages per person

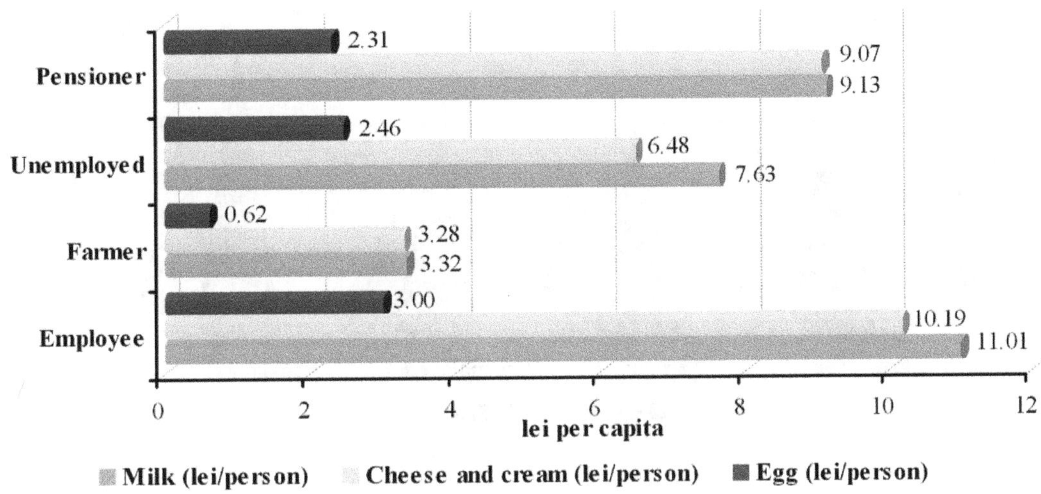

monthly average expenditure for milk and cheese and cream, while for eggs second place is occupied by unemployed, and pensioners ranks third.

Annual rates of evolution of monthly average expenditure for milk per capita, in current prices, were 15.74% at farmers, 15.15% at unemployed, 13.73% at pensioners, 13.48% at employees; in 2014, the average monthly expenditure for milk were: 11.01 lei per employee, 3.32 lei per pensioner 7.63% per unemployed, and only 9.13 lei per farmer.

A similar situation, in terms of yearly growth rates of monthly average expenditure per capita, expressed in current prices, there is at the cheese and cream, which fits between 17.36% for farmers, and 13.81% for employees. In 2014, current prices for cheese and cream are spent on average 10.19 lei per employee, 3.28 lei per pensioner, 6.48 lei per unemployed, and 9.07 lei per farmer.

Regarding the average monthly expenditure for eggs per capita, they increased from 1.67 lei in 2001 to 4.06 lei in 2014 at employees (7.07% yearly), from 1.14 lei to 3.41 lei at pensioners (8.79% yearly), from 1.07 lei to 3.03 lei at unemployed (8.34% yearly), and from 0.25 lei to 0.9 lei at farmers (10.36% yearly).

Along with milk, cheese and cream, and eggs, an important place in nutrition are granted oil and sugar consumption (Figure 4). Evolution of the monthly average consumption expenditure of oil tends to increase in each social category included in the analysis.

The social category which expends the most for the two products is pensioner. Thus, their expenditure for sugar, have evolved at 1.34 lei per capita in 2001 to 3.65 lei per capita in 2014, the average of period being 2.93 lei per capita, and for oil at current prices, its monthly average expenditure has increased from 2.34 lei per capita in 2001 to 5.69 lei per capita in 2014, the average of analyzed period being 4.89 lei per capita.

In the case of employees, average monthly expense for oil increased from 2.10 lei per capita, in 2001, to 4.86 lei per capita, in 2014, the monthly average for the period, being 4.30 lei per capita. For sugar consumption, average monthly expenditure have increased from 1.13 lei per capita in 2001, to 2.87 lei

Figure 4. Monthly average expenditure per capita, on oil and sugar, by social categories, in the period 2001-2014 (current prices)

Source: own construction based on http://statistici.insse.ro/shop/index.jsp?page=tempo3&lang=en&ind=BUF114I Total consumption for purchase agro-foods products and non-food at main class goods and main class social, by area - monthly averages per person

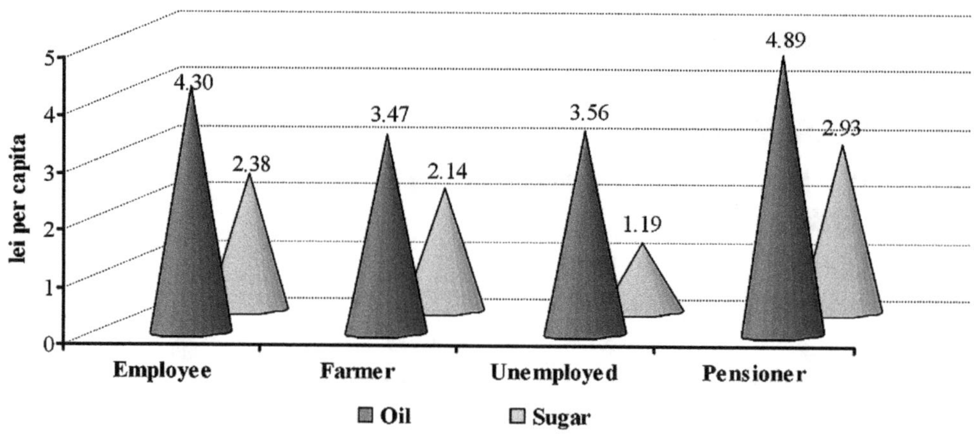

per capita in 2014. In the period 2001 to 2014, the employees expend monthly on average 2.38 lei per capita, which place this social category, second after pensioners for sugar consumption.

As with other categories of foods, the farmers and unemployed share third and fourth, in terms of monthly average expenditures for sugar and oil. The unemployed ranks third, for oil consumption, with an average monthly expenditure of 3.77 lei per capita, and fourth for sugar consumption, with a monthly average expenditure of 2.02 lei per capita. In current prices, the average monthly expenditure per capita of unemployed had increased from 1.66 lei in 2001, to 3.98 lei in 2014 for oil, respectively, from 0.87 lei in 2001 to 2.17 lei in 2014 for sugar.

The monthly average expenditures of farmers, for oil had increased from 1.49 lei per capita in 2001, to 4.03 lei per capita in 2014. Over the entire analyzed period, their monthly average expenses for oil were 3.47 lei per capita, which placed farmers last in oil consumption. In terms of consumption of sugar, farmers were ranked third with an average monthly expenditure of 2.14 lei per capita. In the analyzed period, the monthly average expenditure for sugar have increased from 0.89 lei per capita in 2001, to 2.48 lei per capita in 2014.

In the period 2001-2014, yearly average increase rates of expenditure for sugar, in current prices in Romania were 7.85%, and by social categories of 7.43% at employees, 9.34% at farmers, 7.28% at unemployed, and 8.01% at pensioners. Regarding expenditures for oil, overall yearly average rate was 6.94%, while by social categories were 6.67% at employee, 7.95% at farmer, 6.96% at unemployed, and 7.07% at pensioner.

Daily in food, vegetables and canned vegetables and / or fruit are included, which gives these categories significant importance in all food products. The monthly average expenditures per capita for these product categories have had an upward trend. Both the average expenditures per capita for vegetables and canned vegetables, and for fruit category, determined for entire period analyzed, lead to the same hierarchy of social categories (Figure 5).

So, on the first place are employees who expend the most, both for buying canned vegetables and vegetables, and for fruit consumption. Their monthly average expenses for vegetables and canned veg-

Figure 5. Monthly average expenditure per capita for vegetables and canned vegetables and for fruit by types of social categories, in the period 2001-2014 (current prices)
Source: own construction based on http://statistici.insse.ro/shop/index.jsp?page=tempo3&lang=en&ind=BUF114I Total consumption for purchase agro-foods products and non-food at main class goods and main class social, by area - monthly averages per person

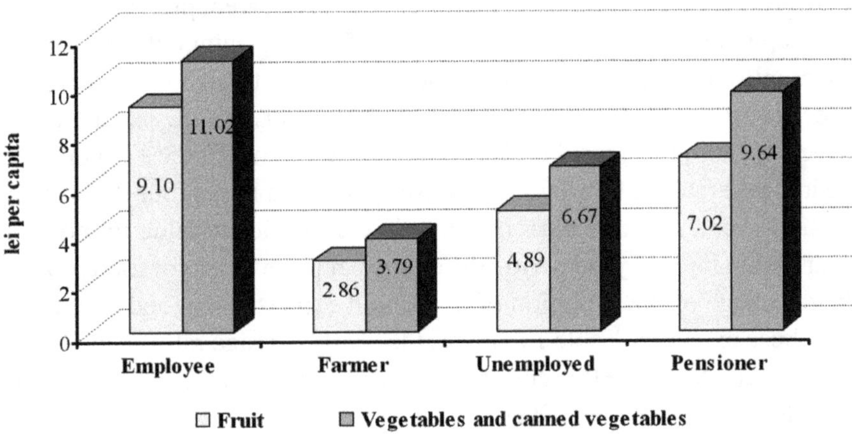

etables have evolved between 3.87 lei per capita in 2001, and 16.14 lei per capita in 2014. The fruits expenditures evolved between 2.56 lei per capita in 2001 and 14.16 lei per capita in 2014.

The monthly average expenses of pensioners for vegetables and canned vegetables have evolved between 3.23 lei per capita in 2001, and 14.93 lei per capita in 2014. For fruit, their spending evolved between 1.69 lei per capita in 2001 and 12.05 lei per capita 2014. On average over the analyzed period, pensioners have expended 7.02 lei monthly per capita, for fruit, and 9.64 lei per capita for vegetables and canned vegetables, which puts them second in the consumption of these products.

The monthly average expenditure per capita of the unemployed, for vegetables and canned vegetables, and fruits are about half the monthly average expenditure of employees. They evolved between 2.17 lei per capita in 2001 and 10.40 lei per capita in 2014, to vegetables and canned vegetables, and between 1.25 lei per capita in 2001, and 7.51 lei per capita in 2014 to fruits, classifying the unemployed third in the consumption of these products.

As regards expenditures of farmers for vegetables and canned vegetables, and for fruits, in 2001 they had very low values (below 1 lei per month). However, in 2014, their monthly average expenditure per capita reached 5.72 lei, for vegetables and canned vegetables, and 5.28 lei for fruit.

In the period 2001-2014, yearly average increase rates of expenditure for vegetables and canned vegetables expenditures, in current prices, in Romania were 12.21%, and by social categories of 11.61% at employee, 14.53% at farmer, 12.81% at unemployed, and 12.50% at pensioner. As regards expenditure for fruits, globally, the yearly average rate was 15.26%, and by social categories of 14.06% at employee, 21.06% at farmer, 14.79% at unemployed, and 16.31% at pensioner.

The monthly average expenditure on consumption of mineral water and other soft drinks in 2014 compared to 2001, they increased at current prices by 8.9 times at farmers (from 0.36 to 3.23 lei per capita), by 7.6 times at pensioners (from 0.8 to 6.08 lei per capita), by 6.5 times, at unemployed (from 0.65 to 4.22 lei per capita) and by 5.1 times for employees (from 1.59 to 8.06 lei per capita). By social categories, the yearly average rates were of 13.30% at employee, 18.39% at farmer, 15.48% at unemployed, and 16.88% at pensioner. In terms of average expenditure for water and other soft drinks, per capita, in the entire analyzed period (Figure 6), employees are first, followed by pensioners, unemployed and farmers.

In the period 2001-2014, the monthly average expenses for alcoholic drinks increased approximately in the same proportions to the four social categories, respectively 4.2 times at employees, 4.3 times at unemployed, 4.4 times at pensioners, and 4.8 times at farmers. In current prices, they have increased from 1.76 to 7.37 lei per capita at employees, from 1.09 to 5.25 lei per capita at farmers, from 0.92 to 3.98 lei per capita, at unemployed, and from 1.37 to 6.0 lei per capita at pensioners. Yearly average increase rate of spending for alcoholic drinks, in analyzed period were 11.95%. By social categories, it was of 11.65% at employee, 12.82% at farmer, 11.93% at unemployed, and 12.03% at pensioner.

Yearly average increase rates of household expenses per capita in the four social categories analyzed in the period 2001-2014, calculated in current prices have significant value in most cases (over 10%). However, given that during the analyzed period the consumer price index for food products has increased continuously, a better image of the evolution of household expenses for food and drink is given by the real growth rates of those expenses. Table 1 displays the yearly average rates of evolution of the monthly average expenditure per capita for the thirteen categories of food and beverages, determined both current prices and comparable prices (2001 = 100%).

To highlight the influence of social category on monthly average expenditure per capita for food and beverage was tested the hypotheses:

H_0: The social category did not significantly influence the monthly average expenditure per capita incurred for the purchase of food and beverages.

H_1: The social category has a significant influence on the monthly average expenditure per capita incurred for the purchase of food and beverages.

Figure 6. Monthly average expenditure per capita, for mineral water and other soft drinks, and alcoholic drinks by types of social categories, in the period 2001-2014 (current prices)

Source: own construction based on http://statistici.insse.ro/shop/index.jsp?page=tempo3&lang=en&ind=BUF114I Total consumption for purchase agro-foods products and non-food at main class goods and main class social, by area - monthly averages per person

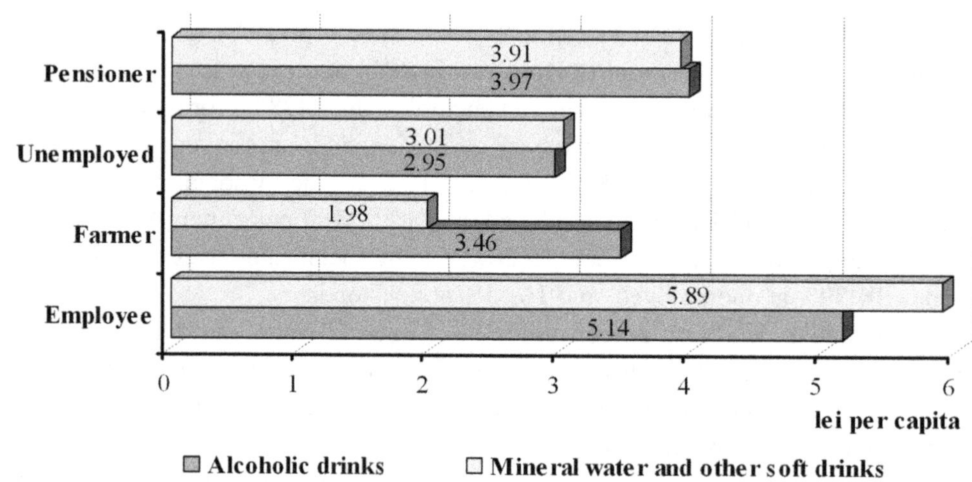

Table 1. Average annual increase rates of monthly expenditures per capita for food and beverages in the period 2001-2014

Category of products	Employee		Farmer		Unemployed		Pensioner	
	R_{CRN} (%)	R_{CMP} (%)	R_{CRN} (%)	R_{CMP} (%)	R_{CRN} (%)	R_{CMP} (%)	R_{CRN} (%)	R_{CMP} (%)
Bread and bakery products	8.11	2.06	11.84	5.79	8.53	2.48	9.38	3.33
Fresh meat	10.45	4.40	16.89	10.84	12.19	6.14	12.54	6.49
Meat products	11.06	5.01	15.93	9.88	13.08	7.03	13.39	7.34
Fish, fish products and canned fish	16.29	10.24	16.09	10.04	16.42	10.37	16.39	10.34
Milk	13.48	7.43	15.74	9.69	15.15	9.10	13.73	7.68
Cheese and cream	13.81	7.76	17.36	11.31	14.89	8.84	14.73	8.68
Eggs	7.07	1.02	10.63	4.58	8.34	2.29	8.79	2.74
Sugar	7.43	1.38	9.34	3.29	7.28	1.23	8.01	1.96
Oil	6.67	0.62	7.95	1.90	6.96	0.91	7.07	1.02
Fruit	14.06	8.01	21.06	15.01	14.79	8.74	16.31	10.26
Vegetables and canned vegetables	11.61	5.56	14.53	8.48	12.81	6.76	12.50	6.45
Mineral water and other soft drinks	13.30	7.25	18.39	12.34	15.48	9.43	16.88	10.83
Alcoholic drinks	11.65	5.60	12.82	6.77	11.93	5.88	12.03	5.98

R_{CRN} – yearly average expenditure rate calculated at current prices
R_{CMP} – yearly average expenditure rate calculated in comparable prices
Source: own calculations based on http://statistici.insse.ro/shop/index.jsp?page=tempo3&lang=en&ind=BUF114I Total consumption for purchase agro-foods products and non-food at main class goods and main class social, by area - monthly averages per person

The results obtained for significance threshold $\alpha = 0.05$ (a probability of 95%) are presented in Table 2. Fisher test critical values in this case ($F_{0.05;\,3,\,52}$) is 3.8286. The condition of accepting the null hypothesis (H_0) is $F < 3.8286$ or $P\_value > 0.05$.

For bread and bakery product, taking into account that $F = 1.574 < F_{critic} = 3.8286$, respectively $P\_value = 0.206569 > \alpha = 0.05$ H_0 is accepted, this means that the social category did not significantly influence the monthly average expenditure per capita for this categories of food. For twelve other products, the null hypothesis (H_0) is rejected, and consequently the alternative hypothesis (H_1) is accepted.

In conclusion, the social category has a significant influence on the monthly average expenditure per capita incurred for the purchase of food and beverages.

Evolutions of Average Food Consumption Expenditure per Inhabitant by Residential Area

The residential area has a significant influence on the size and structure of consumption expenditure for food and beverages fact confirmed by analysis of the evolution of the average food consumption expenditure per capita in urban and rural areas during 2001-2014.

The analysis of the average monthly expenditure by food consumption the 13 food categories, by categories of households by occupational status of the household head, both urban and rural areas as, covered the period 2001-2014.

Table 2. Test results of null hypothesis regarding to the significance of social category on the monthly average expenditure per capita for food and beverages

Category of food and beverages	F	P-value	Accepted hypothesis
Bread and bakery products	1.57400	0.206569	H_0
Fresh meat	12.91600	2.01E-06	H_1
Meat products	7.89500	0.000196	H_1
Fish, fish products and canned fish	11.50850	6.74E-06	H_1
Milk	10.10322	2.38E-05	H_1
Cheese and cream	10.08432	2.42E-05	H_1
Eggs	26.15950	3.57E-10	H_1
Sugar	3.92886	0.013321	H_1
Oil	3.82857	0.014931	H_1
Fruit	10.29897	1.99E-05	H_1
Vegetables and canned vegetables	10.29897	1.99E-05	H_1
Mineral water and other soft drinks	11.79125	5.26E-06	H_1
Alcoholic drinks	11.79125	5.26E-06	H_1

Source: own calculations based on http://statistici.insse.ro/shop/index.jsp?page=tempo3&lang=en&ind=BUF114I Total consumption for purchase agro-foods products and non-food at main class goods and main class social, by area - monthly averages per person

In Romania, the average monthly expenditures for the purchase of food by residence areas have a continuous upward trend from year to year; both urban and rural (Figure 7). In urban areas, the average monthly expenditure for the purchase of food products increased from 58.67 lei per capita in 2001 to 234.70 lei per capita in 2014, the yearly average increase of monthly expenditure being 13.85 lei per capita, respectively, a relative increase by 11.40%.

Figure 7. Evolution of the monthly average expenditure for the purchase of food and beverages, by area of residence (current prices)
Source: own construction based on http://statistici.insse.ro/shop/index.jsp?page=tempo3&lang=en&ind=BUF114I Total consumption for purchase agro-foods products and non-food at main class goods and main class social, by area - monthly averages per person

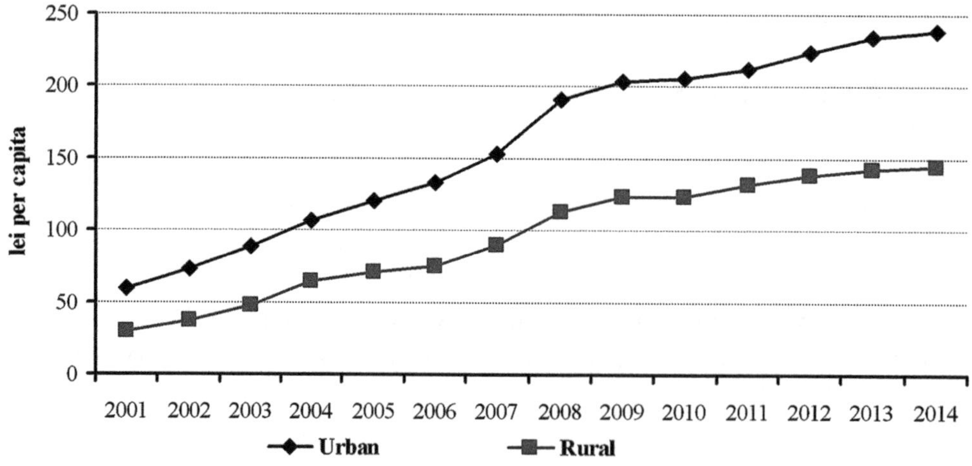

The monthly average expenditure for the purchase of food by rural residents is nearly half of urban residents' expenditure. They evolved from 29.99 lei per capita in 2001 to 144.76 lei per capita in 2014. The average monthly expenditure for food products in the period 2001-2014 were 95.67 lei per capita, the yearly average increase of monthly expenditure being 8.83 lei per capita, which represents an average relative increase of 12.87% yearly. This difference in expenses is determined, among other by the fact that the countryside has a close link with food production, especially the raw material needed for this type of production, the productions obtained in the households of rural inhabitants providing a significant share of their individual consumption.

Analyzing the annual average expenditure in the period 2001-2014 (Figure 8), result that food and beverages monthly expenditure of urban employees (168 lei per capita) and those of rural areas (111 lei per capita), are higher compared to the spending of the other social categories. The yearly average of monthly expenditure of unemployed and farmer per capita, both urban areas and rural areas (83.74 lei for unemployed and 72 28 lei for farmer) are quite close to those of employees. Farmers from urban and the rural areas spend the least with procuring food, fact highlighted by the low values of yearly average of monthly expenditure per capita of 72.39 lei in urban area and of 72.28 lei in rural area.

In terms of the intensity of evolutions of yearly expenditures for food and beverages per capita, the average absolute most significant increases are registered by pensioners from urban area, with 14.76 lei (a rate of 11.50%) and, at short distance, by employees with 14.19 lei (a rate of 11.29%). In the case of rural areas, the average expenditure recorded significant increases for the same social categories only that, compared with those recorded at urban level, the increases are lower. Thus, during 2001-2014, the yearly average expenditures per capita for food and beverages, in rural areas had the most significant growth, with 9.89 lei per pensioner (a relative increase by 13.26% yearly), respectively, with 9.49 lei per employee (a relative increase by 11.67% yearly).

Unemployed social category had, in the analyzed period, an average yearly increase of expenditures for food and beverages by 9.83 lei per capita in urban areas, with average rate of11.56% and 7.02 lei in rural areas with average rate of 12.28%.

Figure 8. The evolution of yearly average of monthly expenditure per capita for the purchase of food and beverages by area of residence and social groups, in the period 2001-2014 (current prices)
Source: own construction based on http://statistici.insse.ro/shop/index.jsp?page=tempo3&lang=en&ind=BUF114I Total consumption for purchase agro-foods products and non-food at main class goods and main class social, by area - monthly averages per person

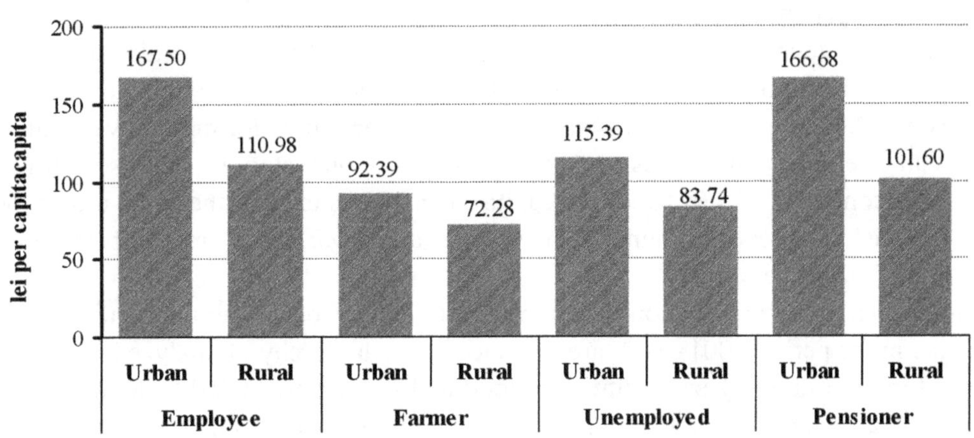

The lowest increase was recorded in both areas of residence at farmer social category, for which the average monthly expenditure destined for food has increased yearly by an average of 8.17 lei per capita in urban areas (an average relative increase of 12.61%) and 6.78 lei per capita in rural areas (an average relative increase of 13.71%).

The evolutions of monthly average expenditure per capita for food and beverages have different weightings and evolutions for analyzed product categories according to area of residence (Table 3) and occupational status of the household head.

In urban areas, from point of view of average monthly expenditure per capita in 2001, the analyzed product categories were ranked so: fresh meat (19.12%), bread and bakery products (16.64%), meat products (7.91%), vegetables and canned vegetables (7.53%), milk (6.36), cheese and cream (6.02%), fruit (4.50%), oil (3.78%), eggs (3.36%), alcoholic drinks (2.68%), mineral water and other soft drinks (2.45), sugar (1.99%), and fish, fish products and canned fish (1.96%).

Fresh meat is the food category for which the urban population expends the most. In the period 2001-2014 the average annual expenditure per capita for its purchase was 27.56 lei (Table 3), the average annual increase of these expenses is 2.31 lei per capita (an average gain relative 10.53). Of the four social category, pensioners monthly expend the most (28.85 lei per capita), followed by employees (27.56 lei per capita), unemployed (18.51 lei per capita) and farmers (11.67 lei per capita).

Bread and bakery products represent the second category of food to which urban population expends much compared with other expenditure for food, of the total of 13 analyzed. Monthly average expense along the analyzed period was 21.43 lei per capita, and its yearly average absolute increase was of 1.27 lei per capita At this category of food are very small differences between the expenditure of the four social categories. They fall between 22.07 lei per capita, to pensioners, and 20.57 lei per capita, to farmers. The monthly average expenditure on bread and bakery products, in the period 2001-2014, increases with a yearly average of 1.27 lei per capita.

For five product categories (meat products, vegetables and canned vegetables, milk, cheese and cream and fruit) monthly average expenses are lower than in the first two presented above, forming a group of food products whose average expenditures, over the entire analyzed period, ranges between 9.67 lei per capita, for fruit and 13.08 lei per capita for meat products. In the case of these five product categories, two social categories occupy the first two places: employees and pensioners.

Employees spend most on consumption of meat products and fruit, the monthly average expenditure being of 14.27 lei per capita, respectively, of 10.41 lei per capita. Pensioners are the ones who spend the most on milk (12.85 lei per capita), cheese and cream (12.39 lei per capita), and vegetables and canned vegetables (14.24 lei per capita).

For the other six categories of products (oil, eggs, alcoholic drinks, mineral water and other soft drinks, sugar, and fish, fish products and canned fish), the average monthly expenditure in the analyzed period are between 2.44 lei per capital at sugar, and 5.72 lei per capita, for mineral water and other soft drinks. By social categories, employees expend monthly more on water and other soft drinks (6.61 lei per capita), and alcoholic drinks (5.50 lei per capita). Pensioners expend the more money on the consumption of fish, fish products and canned fish, oil, eggs and sugar, the average expenditure per capita between 2.44 and 4.92 lei per capita.

The changes that have occurred in consumption patterns of the urban inhabitants and the modifications of their income, in the period 2001-2014, are reflected in the hierarchy of analyzed product categories by their weight in total monthly expenditure for food and beverages. In 2014, the ranking was: fresh meat (17.27%), bread and bakery products (11.00%), meat products (8.18%), vegetables and canned veg-

Table 3. Mean indicators of trends in monthly average expenditure per capita on food and beverages category, in current prices, by residence areas, in the period 2001-2014 [lei per capita]

Category of products	Residence area	Monthly average expenditure per year		Monthly average expenditure per period	Yearly average absolute increase	Yearly average rate [%]
		Minimum (2001)	Maximum (2014)			
Bread and bakery products	Urban	9.76	28.32*	21.43	1.27	39.07
	Rural	7.20	28.45*	20.40	1.48	54.24
Fresh meat	Urban	11.22	41.22	27.56	2.31	54.30
	Rural	2.96	19.68	11.66	1.29	88.04
Meat products	Urban	4.64	19.52	13.08	1.14	61.43
	Rural	2.21	12.23	7.67	0.77	76.88
Fish, fish products and canned fish	Urban	1.15	8.32	4.92	0.55	93.41
	Rural	0.88	6.09	3.54	0.40	90.57
Milk	Urban	3.73	18.80	12.17	1.16	71.45
	Rural	1.38	8.93	5.26	0.58	86.35
Cheese and cream	Urban	3.53	18.92	11.32	1.18	75.00
	Rural	1.33	9.46	5.24	0.63	92.32
Fruit	Urban	2.64	15.40	9.67	0.98	80.01
	Rural	0.87	7.29	4.22	0.49	103.11
Vegetables and canned vegetables	Urban	4.42	19.31	12.76	1.15	63.48
	Rural	1.51	7.39	4.95	0.45	69.78
Mineral water and other soft drinks	Urban	1.44	8.08	5.72	0.51	77.70
	Rural	0.58	4.42	2.91	0.30	96.79
Alcoholic drinks	Urban	1.57	7.31*	4.79	0.43	66.29
	Rural	1.32	5.21	3.76	0.30	58.04
Eggs	Urban	1.97	5.41**	3.63	0.24	36.95
	Rural	0.34	1.38**	0.88	0.08	58.35
Sugar	Urban	1.17	3.59**	2.44	0.14	35.80
	Rural	1.13	3.88**	2.59	0.16	41.92
Oil	Urban	2.22	6.23	4.54	0.22	32.04
	Rural	1.9	5.76	4.12	0.22	35.91

Note: * the year 2013; ** the year 2012

Source: own calculations based on http://statistici.insse.ro/shop/index.jsp?page=tempo3&lang=en&ind=BUF114I Total consumption for purchase agro-foods products and non-food at main class goods and main class social, by area - monthly averages per person

etables (8.09%), cheese and cream (7.93%), milk (7.88%), fruit (6.45%), fish, fish products and canned fish (3.49%), mineral water and other soft drinks (3.39%), alcoholic drinks (3.03%), oil (2.14%), eggs (2.12%), and sugar (1.23%).

In rural areas, from point of view of average monthly expenditure per capita in 2001, the analyzed product categories were ranked so: bread and bakery products (24.01%), fresh meat (9.87%), meat products (7.37%), oil (6.34%), vegetables and canned vegetables (5.04%), milk (4.60%), cheese and cream

(4.43%), alcoholic drinks (4.40%), sugar (3.77%), fish, fish products and canned fish (2.93%), fruit (2.90%), mineral water and other soft drinks (1.93%), and eggs (1.13%). This ranking is significantly influenced by food and beverage resulting from its own production in rural areas and that does not appear explicitly in the monthly expenditure for food and beverages.

In terms of the monthly average expenditure carried out by rural residents, in the period 2001-2014, food and beverage categories can be grouped into four categories.

The first category, the dominant expenditure category, includes fresh bread and bakery products and meat. The total average expenditure level for bread and bakery products are of 20.40 lei per capita and for fresh meat of 11.60 lei per capita. Both for bread and bakery products, and for fresh meat, the highest average costs belong to employees (22.85 lei per capita, respectively 14.68 lei per capita) and pensioners (21.23 lei per capita, respectively 12.18 lei per capita)

The second group comprises three food categories: the first category which is closely related to the first group, and which follow it naturally, contain meat products, with average monthly expenditure of 7.67 lei per capita, and a yearly average absolute increase of 0.77 lei per capita. Also, as part of the Romanian cuisine is the consumption from the other two products, milk, and cheese and cream, for which the rural residents expend annually per capita an average of 5.26 lei, respectively 5.24 lei. Employees living in rural areas are the ones who expend more for these products, the monthly average expenditure being of 9.43 lei per capita, for meat products, of 6.55 lei per capita, for milk, and of 6.31 lei per capita, for cheese and cream.

The third group includes foods with the values of the monthly average expenditure comprised between 4.95 lei per capita for vegetables and canned vegetables and 3.54 lei per capita fish, fish products and canned fish. This group also includes, fruit; oil and alcoholic drinks. Employees are the category that dominates the ranking of monthly average expenditure per capita for vegetables and canned vegetables (5.77 lei), fruit (5.47 lei) and alcoholic drinks (4.15 lei), while for the other two categories pensioners are first, expending an average of 4.11 lei per capita for fish, fish products and canned fish and 4.66 lei per capita, for oil.

The last group includes categories of foods for that, during 2001-2014, rural residents have expended monthly in average, below 3 lei per capita. While the average monthly expenditure for mineral water and other soft drinks was 2.91 lei per capita and for sugar of 2.59 lei per capita, the yearly average expense for eggs is insignificant, fact which shows that in very many rural households are raising birds. Living conditions, the inherited traditions from rural areas determines the inhabitants to not be very interested in purchasing these food categories, fact evidenced not only the low values of the average monthly expenditure per capita, but also the average absolute increases recorded by each of them.

During the last 14 years, the monthly average expenditure of rural residents increased in all categories of food and beverages with values between 1.01 lei, for eggs, and 19.22 lei for bread and bakery products. On the other hand, there were changes in the share of average expenditure by product category in the total average monthly expenditure. Thus, of the 13 product categories for five of them (bread and bakery products, alcoholic drinks, eggs, sugar and oil) fell its expenditure percentages in the total monthly average monthly expenditure per capita with values between 5.76 percentage points for bread and bakery products and 0.2 percentage points for oil.

In 2014, from point of view of monthly average expenditure per capita, the analyzed product categories were ranked so: bread and bakery products (18.25%), fresh meat (13.59%), meat products (8.45%), cheese and cream (6.53%), milk (6.17%), vegetables and canned vegetables (5.11%), fruit (5.04%), fish,

fish products and canned fish (4.21%), alcoholic drinks (3.30%), oil (3.30%), mineral water and other soft drinks (3.05%), sugar (2.23%), and eggs (0.93%).

To highlight the influence of residential area on monthly average expenditure per capita for food and beverage was tested two categories of hypotheses. The first refers to the influence of residential area on the absolute value of the monthly average expenditure per capita. The hypotheses are:

H_0: The residential area did not significantly influence the monthly average expenditure per capita incurred for the purchase of food and beverages.

H_1: The residential area has a significant influence on the monthly average expenditure per capita incurred for the purchase of food and beverages.

The results obtained for significance threshold $\alpha = 0.05$ (a probability of 95%) are presented in Table 4. Fisher test critical values in this case ($F_{0.05,1,26}$) is 4.225. The condition of accepting the null hypothesis (H_0) is F < 4.225 or P_value > 0.05.

At total, taking into account that $F = 10.456 > F_{critic} = 4.225$, respectively $P\_value = 0.003 < \alpha = 0.05$ H_0 is rejected and H_1 is accepted, which meant that the monthly average expenditure per capita for food and beverages differ significantly depending on the area of residence. The conditions that favor these differences were highlighted above.

Table 4. Test results of null hypothesis regarding to the significance of residence areas on the monthly average expenditure per capita for food and beverages

Category of Food and Beverages	First Category of Hypothesis			Second Category of Hypothesis		
	F	P-Value	Accepted Hypothesis	F	P-Value	Accepted Hypothesis
Total	10.456	0.003	H_1	-	-	-
Bread and bakery products	0.144	0.707	H_0	139.746	0.000	H_1
Fresh meat	24.644	0.000	H_1	146.622	0.000	H_1
Meat products	10.910	0.003	H_1	2.213	0.149	H_0
Fish, fish products and canned fish	2.675	0.114	H_0	15.108	0.001	H_1
Milk	17.919	0.000	H_1	86.309	0.000	H_1
Cheese and cream	14.776	0.001	H_1	43.044	0.000	H_1
Fruit	16.187	0.000	H_1	3.479	0.073	H_0
Vegetables and canned vegetables	27.866	0.000	H_1	382.958	0.000	H_1
Mineral water and other soft drinks	13.356	0.001	H_1	10.157	0.004	H_1
Alcoholic drinks	2.601	0.119	H_0	91.989	0.000	H_1
Eggs	80.861	0.000	H_1	153.737	0.000	H_1
Sugar	0.209	0.651	H_0	76.113	0.000	H_1
Oil	0.676	0.418	H_0	17.554	0.000	H_1

Source: own calculations based on http://statistici.insse.ro/shop/index.jsp?page=tempo3&lang=en&ind=BUF114I Total consumption for purchase agro-foods products and non-food at main class goods and main class social, by area - monthly averages per person

In terms of analyzed product categories, in five of them (bread and bakery product, fish, fish products and canned fish, alcoholic drinks, sugar and oil) is accepted the null hypothesis (H_0) and it follows that the residential area of the consumer does not significantly influence its expenses for these products.

The second category of hypotheses refers to the influence of area of residence on the share of expenditures for food in the monthly average expenditure per capita for food and beverage. The hypotheses are:

H_0: The residential area did not significantly influence the share of expenditures for food in the monthly average expenses per capita incurred for the purchase of food and beverages.

H_1: The residential area has a significant influence on the share of expenditures for food in the monthly average expenses per capita incurred for the purchase of food and beverages.

The results are shown in Table 4 and emphasizes that, with the exception of two products (meat products and fruit) null hypothesis is rejected and the alternative hypothesis is accepted. Consequently, area of residence has a significant influence on the share of expenditures for food in the monthly average expenses incurred for the purchase of food and beverages.

ASPECTS OF FOOD CONSUPTION EXPENDITURE AT TERRITORIAL LEVEL

In the period 2008-2014 the monthly average expenditure at current prices, for the thirteen categories of foods and drinks, was a process of continuous growth from year to year (Figure 9) in each region

So, Macro_3 ranks first in terms of monthly average expenses for food and beverage, which varies between 183.88 lei per capita in 2008, and 219.02 lei per capita in 2014, for the entire period, the

Figure 9. Evolution of the monthly average expenditure of the purchase of food and beverage, by macro regions, from 2008-2014 (current prices)

Source: own construction based on http://statistici.insse.ro/shop/index.jsp?page=tempo3&lang=ro&ind=BUF107K Total expenditure per person of main household categories, by macro regions and development regions

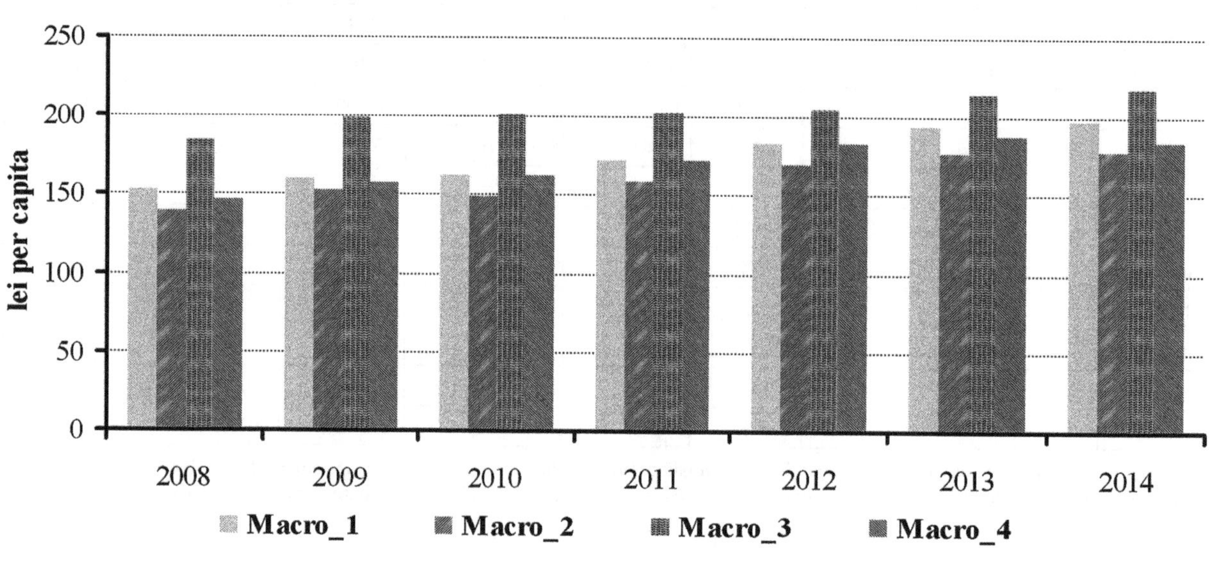

monthly average expenditure being of 203.96 lei per capita. The yearly average increase rate of the monthly expenditure for food and beverages was 2.96%, which in absolute terms represents an annual increase of 5.86 lei per capita.

In Macro_1, the monthly average expenses for these products ranged between 152.54 lei per capita in 2008, and 197.56 lei per capita in 2014, with an absolute yearly average increment of 7.50 lei per capita, and a rate of 4.40%, which gives him second place between the other macro regions.

An increase in monthly average expenditure for food products included in the analysis, quite close to that of Macro_3, was registered by Macro_4. In Macro_4 the monthly average expenditures for food and beverages were between 146.92 lei per capita in 2008 and 184.96 lei per capita in 2014. The yearly average increase of the monthly expenditures for food and beverages was 6.34 lei per capita (a yearly average rate of 3.91%).

Last place was occupied by Macro_2. The monthly average expenditure for the period 2008-2014 were between 139.21 lei per capita in 2008 and 178.39 lei per capita in 2014, with an yearly average increase of 6.53 lei per capita, and with a yearly average rate of. 4.22%.

Although the data series of monthly average expenditure per capita for food and beverages are very short (have only seven values for each macro) for an overview of how the characteristics of macro-regions influenced their evolution over the period 2008 - 2014 were tested the following hypotheses:

H_0: There are no significant differences between macro regions regarding the monthly average expenditure per capita for food and beverages

H_1: There are significant differences between macro regions regarding the monthly average expenditure per capita for food and beverages

Table 5. Test results of null hypothesis regarding to the significance of social category on the monthly average expenditure per capita for food and beverages

Category of Food and Beverages	F	P-Value	Accepted Hypothesis
Bread and bakery products	5.45546	0.005276	H_1
Fresh meat	10.69984	0.000118	H_1
Meat products	10.30786	0.000151	H_1
Fish, fish products and canned fish	25.53598	1.21E-07	H_1
Milk	24.59595	1.69E-07	H_1
Cheese and cream	11.72789	6.3E-05	H_1
Eggs	5.40399	0.005511	H_1
Sugar	1.09089	0.371996	H_0
Oil	0.12991	0.941374	H_0
Fruit	34.82578	6.62E-09	H_1
Vegetables and canned vegetables	71.76027	3.98E-12	H_1
Mineral water and other soft drinks	172.35570	2.29E-16	H_1
Alcoholic drinks	23.96358	2.13E-07	H_1

Source: own construction based on http://statistici.insse.ro/shop/index.jsp?page=tempo3&lang=ro&ind=BUF107K Total expenditure per person of main household categories, by macro regions and development regions

The results obtained for significance threshold $\alpha = 0.05$ are presented in Table 5. Fisher test critical values in this case ($F_{0.05; 3.24}$) is 3.0088. The condition of accepting the null hypothesis (H_0) is F<3.0088 or P_value > 0.05.

Of the thirteen categories of foods and beverages null hypothesis is only accepted for two: sugar (*P_value* = 0.371996 > α = 0.05), and oil (*P_value* = 0.941374 > α = 0.05). For all other eleven products H_0 is rejected and H_1 is accepted, which meant that the monthly average expenditure per capita differ significantly between the fore macro regions of Romania.

Extending the analysis of monthly average expenditure per capita for food and beverages recorded the macro level, by social categories provides significant additional information.

By social categories (Figure 10), the employees had the highest monthly average expenditure per capita for food and beverages, calculated for the entire period. These range from 185.3 lei per capita in Macro_2 to 227.42 lei per capita in Macro_3. Average annual increases of these expenditures were between 5.71 lei per capita in Macro_3 (a yearly average rate of 2.60%) and 7.74 lei per capita in Macro_4 (a yearly average rate of 4.01%).

Pensioner is the social category which is ranked second in terms of monthly average expenditure for food and beverages. They are between 166.35 lei per capita in Macro_4, and 202.05 lei per capita in Macro_3, with a yearly average increases between 9.50 lei per capita in Macro_ 2 and 7.40 lei per capita in Macro_4.

Unemployed in 2008-2012, had expended monthly on average between a maximum 152.64 lei per capita, in Macro_3, and a minimum of 132.91 lei per capita, in Macr_1, the yearly average increases ranging from 3.54 lei per capita in Macro_2, and 8.77 lei per capita in Macro_4.

Figure 10. Evolution of monthly average expenditure for the purchase of food by the macro and social categories in the period 2008-2014

Source: own construction based on http://statistici.insse.ro/shop/index.jsp?page=tempo3&lang=ro&ind=BUF107K Total expenditure per person of main household categories, by macro regions and development regions

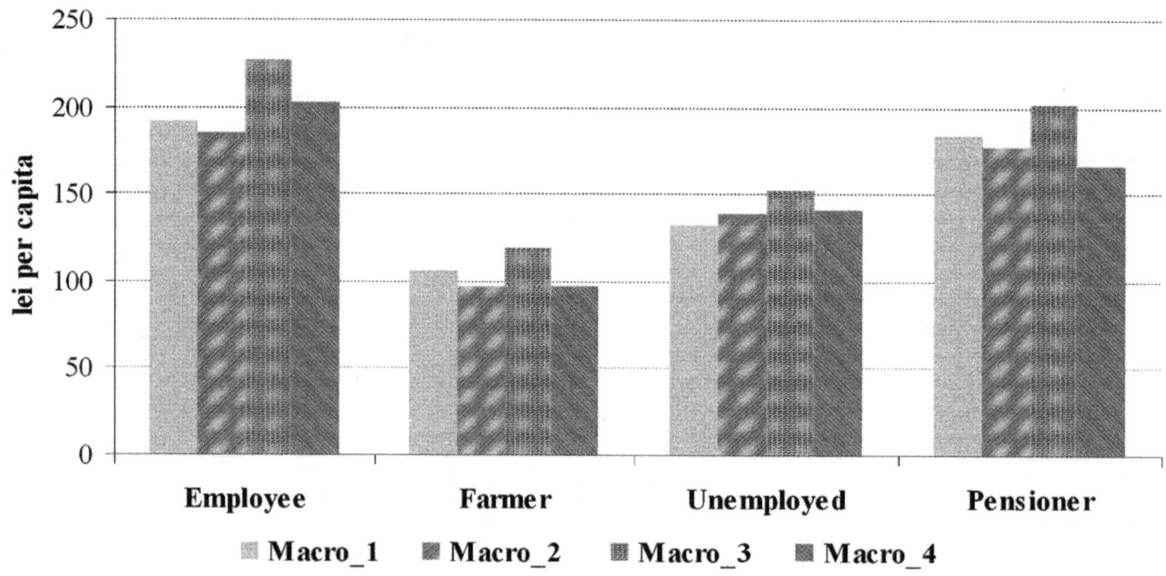

As with the analysis of average expenditure for consumption of thirteen food by residence areas, last place, in terms of categories of households, by occupational status of the household head, is occupied by farmers due to the close link with the place of residence, which in many cases, it may be the necessary basis for food production, reducing the expenditure for their procurement from market.

Regarding the farmers, in the analyzed period, they expended per month, on average, for food and beverages between 96.25 lei per capita, in Macro_2, and 118.98 lei per capita in Macro_3, the yearly average increases ranging between a maximum of 5.63 lei per capita (an yearly average rate of 5.69%) in Macro_1, and a minimum of 2.97 lei per capita (an yearly average rate of 2.62%) in Macro_3.

The evolutions of the monthly average expenditure per capita in current prices, performed in the four macro regions of Romania, presented above, provides only an overview of the place occupied by each in this respect. A summary of the results recorded on the thirteen product categories are shown in Table 6.

Macro_1's level, the highest monthly average expenditure values were recorded in two product categories: bread and bakery products and sugar. Pensioners are those who spend most for both food products. On average, in the analyzed period, a pensioner has spent 30.26 lei per month for bread and bakery products, and 4.13 lei for sugar.

Also, residents of Macro_1 ranks second in terms of monthly average expenditure for six categories of products: fresh meat, meat products, milk, fruit, alcoholic drinks and eggs. For these, the average monthly expenditures are between 3.22 lei per capita for eggs and 25.66 lei per capita for fresh meat. Those who spend the most, for these six food categories are employee. They had expended monthly an average of 3.69 lei per capita for eggs and 29.01 lei per capita for fresh meat.

Table 6. The average monthly expenditure by food categories, end by macro-regions and place of each macro-region for the period 2008-2012 (current prices)

Categories of food and beverages products	Monthly expenditure (lei per capita)				Place			
	MR1	MR2	MR3	MR4	MR1	MR2	MR3	MR4
Bread and bakery products	29.06	24.72	27.62	28.08	I	IV	III	II
Fresh meat	25.66	25.66	32.63	26.42	II	I	II	III
Meat products	15.02	12.04	14.82	15.91	II	IV	III	I
Fish, fish products and canned fish	3.86	6.96	7.95	5.93	III	II	I	IV
Milk	13.02	10.57	15.73	11.95	II	IV	I	III
Cheese and cream	10.70	11.29	15.42	11.12	IV	II	I	III
Fruit	9.52	9.50	12.87	8.57	II	III	I	IV
Vegetables and canned vegetables	10.76	11.52	16.54	10.31	III	II	I	IV
Mineral water and other soft drinks	6.23	4.49	7.70	6.38	III	IV	I	II
Alcoholic drinks	6.21	5.08	6.56	4.89	II	III	I	IV
Eggs	3.22	2.66	3.40	2.82	II	III	I	IV
Sugar	3.45	2.99	3.10	3.42	I	IV	III	II
Oil	5.37	5.56	5.46	5.43	IV	I	II	III

Note: MR1 – Macro_1; MR2 – Macro_2; MR3 – Macro_3; MR4 – Macro_4

Source: own construction based on http://statistici.insse.ro/shop/index.jsp?page=tempo3&lang=ro&ind=BUF107K Total expenditure per person of main household categories, by macro regions and development regions

Macro_1, ranks third in terms of average monthly expenditure per capita for three products: fish, fish products and canned fish, canned vegetables and vegetables, and mineral water and other soft drinks. Also the employees are those expend most to purchase them. The monthly average expenditures are: 4.56 lei per employee for fish, fish products and canned fish, 7.42 lei per employee for mineral water and other soft drinks, and 12.54 lei per employee for vegetables and canned vegetables.

Regarding the consumption of cheese and cream and oil, Macro_1 ranks last. In this macro region those who expend the most per month for cheese and cream are employee (12.32 lei per capita), and those who spend the most per month for oil are pensioner (6.22 lei per capita).

Macro_2 is in first place in terms of monthly average expenditure for fresh meat and for oil. For fresh meat, employee expend monthly most (31.47 lei per capita), yearly average increase per capita being 1.70 lei, while the yearly average rate is 5.58%. For the oil, ones who expend the most are pensioners (as in Macro_1), the average monthly expenditure being 6.53 lei per capita.

For product categories fish, fish products and canned fish, cheese and cream, vegetables and canned vegetables, Macro_2 is second. Pensioners have recorded the highest values of the monthly average expenditures for fish, fish products and canned fish, of 8.36 lei per pensioner (a yearly average increase of 0.77 lei per pensioner, and a yearly average rate of 9.82%), as well as cheese and cream of 13.76 lei per pensioner (a yearly average increase of 1.33 lei per pensioner, and a yearly average rate of 10.29%). For vegetables and canned vegetables those who spend the most are employee (13.64 lei per employee, with an average annual increase of 0.40 lei per employee and a yearly average rate 2.98%).

Macro_2 ranks third in terms of the monthly average expenditures per capita for fruit, alcoholic drinks and eggs. The highest yearly average expenditure for fruit (11.73 lei per capita) and eggs (3.43 lei per capita) were incurred by employees. At alcoholic drinks expenditure, pensioners are in first place with an average expenditure of 5.39 lei per capita.

At five categories of foods, Macro_2 is in last place: These are bread and bakery products, meat products, milk, mineral water and other soft drinks, and sugar. Within this macro region, employee have the highest monthly average expenses for bread and bakery products (26.27 lei per capita), meat products (14.49 lei per capita), milk (12.63 lei per capita) and mineral water and other soft drinks (6.19 lei per capita). For sugar, the social category which spends monthly, on average, most is the pensioners (3.66 lei per capita).

Macro_3 includes the development regions South Muntenia and Bucharest-Ilfov which includes the city of Bucharest where the income per capita is significantly higher than in other areas. Due to this feature, in terms of monthly average expenditure for food and beverages, Macro_3 is first in eight product categories: fish, fish products and canned fish, milk, cheese and cream, fruit, vegetables and canned vegetables, mineral water and other soft drinks, alcoholic drinks, and eggs. For all these categories of products, employee expenses have the highest yearly increase with values ranging between 8.61 lei their per capita for fish, fish products and canned fish, and 18.50 lei per capita for milk.

At the monthly average expenditures per capita for fresh meat and for oil, Macro_3 has second place. For fresh neat, employees expend more (37.98 lei per capita) and, for oil, pensioners are those had the most expenditure for consumption of oil (6.02 lei per capita). And finally, at monthly average expenditures per capita for bread and bakery products, and meat products, Macro_3 ranks third, employees expending monthly most for meat products (17.10 lei per capita), while pensioners spend most for bread and bakery products (28.84 lei per capita), and for sugar (3.55 lei per capita).

Macro_4 is ranked first in terms of monthly average expenditure per capita for meat products (employee spending monthly on average of 18.99 lei per capita for this product) and second at expenditures for bread and bakery, mineral water and other soft drinks and for sugar. The highest monthly average expenses for bread and bakery products, mineral water and other soft drinks are of the employees, (30.35 lei per capita, respectively, 7.83 lei per capita), while for sugar, the biggest expense are of the pensioners, which expend monthly, an average, 3.80 lei per capita.

Fresh meat, milk, cheese and cream, and oil are food categories at that the monthly average expenditures per capita, ranks Macro_4 on third place. Of these categories of foods, employee expend on average 33.38 lei per capita for fresh meat, 15.56 lei per, capita for milk, and 13.36 lei per capita for cheese and cream. In case of monthly average expenditures for oil, pensioners are in first place with an average 5.83 lei per capita

In terms of average monthly expenditure for fish, fish products and canned fish, fruit, vegetables and canned vegetables, alcoholic drinks, and eggs, highest monthly average expenditures, in Macro_4, were registered at the pensioners (6.38 lei per capita for fish, canned fish and fish products) and employees (11.16 lei per capita for fruit, vegetables 12.83 lei per capita and canned vegetables, 5.74 lei per capita for alcoholic drinks and 3.77 lei per capita at the eggs).

In Romania, of the thirteen categories of foods, in terms of monthly average expenditure per capita, bread and bakery products and fresh meat, there are located on the top two places in all four macro regions. On the last places there are the products: fish, fish products and canned fish, sugar and eggs. In terms of volume of the monthly average expenditure for food and beverages, employee and pensioner categories are in most product categories, the first places in all four macro-regions in Romania.

FUTURE RESEARCH DIRECTIONS

The analyzes performed on evolutions of the monthly average expenditure per capita of Romanian citizens, for food and beverages by area of residence, macro regions and social categories, as well as the conclusions resulting from them highlight economic and social situation of households in Romania in the past two decades . This is a starting point for decision making, and strategies to improve quality of life. In this connection, a future research direction is to analyze the links between evolutions and consumer patterns of citizens and indicators of living standards and quality of life.

Also, to get a better image on final consumption of households, the poverty measurement and the level of social exclusion, with their political and economic implications, the analysis of the level and structure of expenditure for the purchase of food should include an analysis of the household size, number of children included in the household, and level of education of household head.

The results of these analyzes could be used for various types of studies and evaluations to identify factors and directions of action for increasing living standards, impacting on policies regarding regional, structural, sectoral and cohesion issues.

We believe that one of the most important directions of research is to identify the influence of programs of food consumption of the population of Romania as a result of income and living standards of their on nutritional requirements of a healthy diet, the Incidence of their income on nutrition-related diseases and default on general health of Romanian.

CONCLUSION

Results of the changes manifested during the transition to a market economy after 1990, have been noticed especially after the year 2000, when the economic growth led to changes in income, expenditure and consumption. The mutual interdependence between revenue, expenditure and consumption, has gradually determined a change in the living standard, with significant impact on economic growth but also on attitudes concerning future needs, thus implicitly changing market demand.

In this context it was established that the changes of monthly average expenditure for the 13 food categories included in the analysis (bread and bakery products, fresh meat and meat products, fish, fish products and canned fish, milk, cheese and cream, fruit, vegetables and canned vegetables, mineral water and other soft drinks, alcoholic drinks, eggs, sugar, oil), from year to year, from 2001 to 2014, experienced a continuous growth, both at the country level and in relation to the four social categories studied (employee, farmer, unemployed, pensioner).

Considering that the highest revenues are those of employees, the analysis concludes that they rank first in terms of consumption of studied food products, a rank that is expressed through the highest average monthly expenditure. The average monthly food expenditure made by pensioners, leads to medium values or average increases, sometimes quite close to those of employees, ranking them on second place. It should be emphasized that these expenses are not always the result of increased food consumption, but are frequently represented by the love and care they have for their children and grandchildren, who they support for an increased standard of living.

The low income of unemployed people determines lower average monthly expenditures for food mentioned in the analysis, in comparison to other household members who have a different employment status.

Farmers represent a category which is found to occupy the last position in almost all rankings regarding mean values for average monthly expenditure for food consumption, which is explained in terms of their interdependence with the various production sources of raw materials, necessary for food production.

In the analysis carried out either by residence areas or by macro regions, this classification of social categories in relation to annual average values of monthly average expenditure, knows fairly insignificant changes from one product to another, in the sense of place exchange between employee and pensioner, or only between unemployed and farmer.

Following the analysis of average monthly expenditure, conducted on the 13 food products and on the two residence areas, it has been established that, in general, mean values determined for urban areas are higher than those in rural areas, considering that part of the rural population's food consumption is provided by their own food production, thus leading to a reduced consumption of purchased products and a low food expenditure.

Whether the urban/rural area, the employee, pensioner, unemployed or the farmer, the biggest expenses are reported for fresh meat and bread and bakery products, two traditional categories in Romanian nutrition, and the lowest expenses are reported for eggs and sugar consumption. The same food products are placed at the beginning and end of the ranking achieved as a result of the analysis conducted on macro regions and social categories. Expenditures for other food categories record various changes leading to ranking modifications, in relation to particularities of residence areas or macro regions.

The results obtained after applying the ANOVA variance analysis, lead to the conclusion that the area of residence significantly influences food and beverage expenses, so the average monthly expenditure per capita obviously differs, depending on the area of residence. However, the residence area of the consumer does not significantly influence his expenses for bread and bakery products, fish, fish products and canned fish, alcoholic drinks, sugar and oil.

Regarding the influence of the residence area on the share of expenditures for a food product, in the average monthly expenditure per capita, made to purchase food and drinks, it was established that this influence is significant, except for two product categories: meat products and fruit.

Another aspect of the analysis results can be highlighted by clarifying the high values of average food expenses, which are not always the result of a high consumption of these products, but especially the result of high prices at which they are sold.

The trend assessment in time (2001-2014) and space (on both residence areas and at macro regional level) of the average monthly expenditure for the 13 food products mentioned in the analysis, led to the conclusion that the costs needed for purchasing food products are the result of a strong link between living standards and economic growth.

An increase in food expenditure can be seen through the diversification of product quality, consumption and as a result of income growth, which would imply an increase of living standards, with direct implications in the creation of GDP and economic growth.

Also, the economic growth process, manifested in Romania, led to a rise in living standards and to higher expenses on food. This increase is not the result of an increasing need, but rather a response to a mentality that changed the people's way of life. Thus, many households producing raw materials and food for their own consumption have been converted into land for recreation purposes, causing a rise in costs for food. In addition, an increase in food expenses often produces negative effects on the environment or causes a reduction in costs for the purchase of other goods and services, thereby causing a decrease in the production or sales of other industries, with negative impact on the economic growth process.

The process of food consumption spending growth must be in harmony with environmental balance and development activities in other areas.

REFERENCES

Anders, S. M. (2010), Ageing and Consumption – The Impact of Demographic Change on Food Expenditure Patterns. *European Association of Agricultural Economists, 115th Joint EAAE/AAEA Seminar*. Retrieved from http://purl.umn.edu/116442

Babucea, A. G., & Balacescu, A. (2011). Statistical Analysis of the Dynamics of Income and Consumption Expenditure of Households in the 1990 - 2010 Period. *Annals - Economy Series from Constantin Brancusi University. Targu Jiu, 1*, 9–16.

Baker, G.A. (2003). Food Safety and Fear: Factors Affecting Consumer response to Food Safety Risk. *International Food and Agribusiness Management Review, 6*(1).

Buzby, J., & Mitchell, L. (2006). Private, National, and International Food-Safety Standards. *Journal of Food Distribution Research, 37*(1), 1–6.

Cismaş, L., Miculescu, A., & Oţil, M. (2010). Income and expenditure of households in Romania. *Annals of University of Craiova - Economic Sciences Series*, 1(38), 18-26.

Constandache, M., Condrea, E. & Stanciu, A.C. (2014). Romanian Food Consumption and its Effects on Population's Health Condition. *SEA - Practical Application of Science, 3*, 135-145.

Drichoutis, A., Nayga, R., & Lazaridis, P. (2012). Food away from home expenditures and obesity among older Europeans: Are there gender differences? *Empirical Economics, 42*(3), 1051–1078. doi:10.100700181-011-0453-4

Dudek, H. (2011). Quantitative analysis of the household's expenditure for food. *Problems of World Agriculture/Problemy Rolnictwa Światowego, 26*(3), 23-30.

Frazao, E., Meade, B. G. S., & Regmi, A. (2008). Converging Patterns in Global Food Consumption and Food Delivery Systems. *Amber Waves., 1*, 22–29.

Ghita, S., & Manea, D. (2013). An Analysis on Consumption Expenditure of Households at Territorial Level, Using Multi-criteria Ranking Methods. Study Case: Romania. *Ovidius University Annals. Economic Sciences Series, 13*(1), 325–329.

Guillementte, A. R., & Cranfield, J. A. L. (2012). Food Expenditures: The Effect of a Vegetarian Diet and Organic Foods. *Agricultural and Applied Economics Association*. Retrieved from http://purl.umn.edu/123391

Haq, R. (2009). Inequality and Welfare by Food Expenditure Components. *Pakistan Development Review, 48*(4), 755–768.

Hassanein, N. (2011). Matters of scale and the politics of the Food Safety Modernization Act. *Agriculture and Human Values, 28*(4), 577–581. doi:10.100710460-011-9338-6

Hindls, R., Hronová, S., & Sixta, J., & Vltavská. (2014). Household Consumption Expenditures between 1970 and 2012. *Politicka Ekonomie*, (6), 725–748. doi:10.18267/j.polek.979

Jacobson, D., Mavrikiou, P. M., & Minas, C. (2010). Household size, income and expenditure on food: The case of Cyprus. *Journal of Behavioral and Experimental Economics, 39*(2), 319–328. doi:10.1016/j.socec.2009.12.009

Kostakis, I. (2014). The Determinants of Households' Food Consumption in Greece. *International Journal of Food and Agricultural Economics, 2*(2), 17–28.

Liobikienė, G., & Mandravickaitė, J. (2013). Convergence of new members of the EU: Changes in household consumption expenditure structure regarding environmental impact during the prosperous period. *Environment, Development and Sustainability, 15*(2), 407–427. doi:10.100710668-012-9386-0

MacLaren, D. (2006). On Aspects of Food Safety and International Trade. *The Singapore Economic Review, 51*(2), 135–145. doi:10.1142/S0217590806002317

Matteo Barigozzi, M., Alessi, L., Capasso, M., & Fagiolo, G. (2012). The distribution of household consumption-expenditure budget shares. *Structural Change and Economic Dynamics, 23*(1), 69–91. doi:10.1016/j.strueco.2011.09.003

Muhammad, A., Seale, J. L., Meade, B. G. S., & Regmi, A. (2011). *International Evidence on Food Consumption Patterns: An Update Using 2005 International Comparison Program Data. TB-1929.* U.S. Dept. of Agriculture, Econ. Res. Serv. March 2011. Revised February 2013. http://purl.umn.edu/184306

Ng, D. W., & Salin, V. (2012). An Institutional Approach to the Examination of Food Safety. *International Food and Agribusiness Management Review, 15*(2), 21–46.

NIS. (n.d.). *National Institute of Statistics*. Retrieved from http://www.insse.ro/cms/en

Pandey, A. (2013). Inequality in consumption expenditure at household level in Uttar Pradesh: Regression Based Inequality Decomposition (RBID). *Journal of Global Economy, 9*(2), 135–152.

Rabontu, C. (2010). Food safety and its role in the evolution of food trade. *Annals - Economy Series, Constantin Brancusi University. Târgu Jiu, 2*, 161–172.

Toma-Bianov, A. (2013). The International and European Regime of Health Protection and Food Safety. *Curentul Juridic, The Juridical Current. Le Courant Juridique, 52*, 110–120.

ADDITIONAL READING

Brown, M., & Lee, J. Y. (2008). Impacts of the distribution of households across income groups in a differential demand system for orange juice. *Empirical Economics, Springer, 34*(3), 567–584. doi:10.100700181-007-0136-3

Browning, M., Crossley, T., & Winter, J. (2014). The Measurement of Household Consumption Expenditures. *Annual Review of Economics, 6*(1), 475–501. doi:10.1146/annurev-economics-080213-041247

Burda, A. (2010). Mutations in Food Consumption in Romania, AMIDST European Integration. *Knowledge horizons*. 2(3), 77-83

Costache, M., Condrea, E., & Stanciu, E. (2014). Romanian Food Consumption And Its Effects On Population's Health Condition, *SEA - Practical Application of Science. Volume II, 1*(3), 135–145.

Cranfield, J. A. L., Preckel, P. V., Eales, J. S., & Hertel, T. W. (2002). Estimating consumer demands across the development spectrum: Maximum likelihood estimates of an implicit direct additivity model. *Journal of Development Economics, Elsevier., 68*(2), 289–307. doi:10.1016/S0304-3878(02)00014-7

Danu, M. C. (2014). The Impact Of Risk Of Poverty On Segments Of The Market In Romania. *Studies and Scientific Researches. Economics Edition, 19*, 27–37.

Dubois, P., Rachel Griffith, R., & Nevo, A. (2014). Do Prices and Attributes Explain International Differences in Food Purchases? *American Economic Review. American Economic Association, 104*(3), 832–867.

Dudek, H, (2011). Quantitative analysis of the household's expenditure for food. *Problems of World Agriculture*. 11(26) Part 3, 23-30.

Ekaterina Rozhkovskaya, E. (2008). An Econometric Model for Analysis and Forecasting of Final Consumption Expenditure Components in the Republic of Belarus: Conceptual and Methodological Approaches, Estimation Results. *Applied Economics*, *12*(4), 27–41.

Fleissig, A. R., & Whitney, G. (2013). Virtual prices and the impact of house rationing in Belgium on consumer choices. *Explorations in Economic History, Elsevier.*, *50*(2), 308–315. doi:10.1016/j.eeh.2012.11.001

Fogarasi, J. (2008). Hungarian and Romanian Agri-Food Trade in the European Union. *Management*, *3*(1), 3–13.

Fousekis, P. & Lazaridis, P. (2001). Food Expenditure Patterns of the Urban and the Rural Households in Greece. A Kernel Regression Analysis. *Agricultural Economics Review*, 2001(1), 5-20.

Gabriella Donatiello, G., D'Orazio, M., Frattarola, D., Rizzi, A., Scanu, M.& Spaziani, M. (2014). Statistical Matching of Income and Consumption Expenditures. *International Journal of Economic Sciences*, 2014(3), 50-65.

Giovanni, A. C. (1994). Poverty, Food Consumption, and Nutrition During the Transition to the Market Economy in Eastern Europe. *The American Economic Review*, *84*(2), 297–302.

Gogonea, R. M., & Zaharia, M. (2013). Statistical Analysis of Evolutions of Demand and Supply of Agro-touristic Boarding Houses in 2000-2012 Period. *Annals of the "Stefan cel Mare" University of Suceava. Fascicle of Faculty of Economics and Public Administration.*, *2*(18), 7–16.

Gundersen, C., & Ribar, D. (2011). Food Insecurity and Insufficiency at low Levels of Food Expenditures. *Review of Income and Wealth*, *57*(4), 704–726. doi:10.1111/j.1475-4991.2011.00471.x

Haq, R., & Arshid, N. (2009). Inequality and Welfare by Food Expenditure Components. *Pakistan Development Review*, *48*(4), II, 755–768.

Hubbard, C., Szigeti, J., & Podruzsik, S. (2010). Distributional Impacts of Food Price Changes on Consumer Welfare in Hungary and Romania following EU Accession. *European Association of Agricultural Economists*,116th Seminar, October 27-30, Parma, Italy, http://purl.umn.edu/95036

Jae, M. K., Ryu, J. S., & Abdel-Ghany, M. (2000). Family characteristics and convenience food expenditure in urban Korea. *Journal of Consumer Studies & Home Economics*, *24*(4), 252–256. doi:10.1046/j.1365-2737.2000.00156.x

Kenneth, C. W., & Dongling, C. (2010). Affluence and Food: A Simple Way to Infer Incomes. *American Journal of Agricultural Economics. Agricultural and Applied Economics Association.*, *92*(4), 909–926.

Kenneth, C. W., & Grace, G. (2012). Quality, quantity, spending and prices. *European Economic Review, Elsevier.*, *56*(7), 1376–1391. doi:10.1016/j.euroecorev.2012.06.006

Lazaridis, P. (2000). Decomposition of Food Expenditure Inequality: An Application of the Extended Gini Coefficient to Greek Micro-Data. *Social Indicators Research*, *52*(2), 179–193. doi:10.1023/A:1007131116344

Macours, K., & Swinnen, J. F. M. (2000). Causes of Output Decline in Economic Transition: The Case of Central and Eastern European Agriculture. *Journal of Comparative Economics, Elsevier, 28*(1), 172–206. doi:10.1006/jcec.1999.1643

Manole, V., & Istudor, N. (2011). Food Security in Romania. *Ovidius University Annals. Economic Sciences Series, XI*(1), 1193–1201.

Militino, A. F., Goicoa, T., & Ugarte, M. D. (2012). Estimating the percentage of food expenditure in small areas using bias-corrected P-spline based estimators. *Computational Statistics & Data Analysis, 56*(10), 2934–2948. doi:10.1016/j.csda.2012.01.009

Noll, H. H., & Weick, S. (2015). Consumption expenditures and subjective well-being: Empirical evidence from Germany. *International Review of Economics, 62*(2), 101–119. doi:10.100712232-014-0219-3

Oliviera, V. (1999). Domestic Food Assistance Expenditures Drop Again. *Food Review: The Magazine of Food Economics, 22*(3), 30–31.

Prasad, A. R. (2005). Engel's Law and Expenditure Elasticities of Specific Food Items in India: 1950-51 to 1999-2000. *Journal of Global Economy, 1*(4), 233–242.

Regmi, A., Takeshima, H., & Unnevehr, L. J. (2008). Convergence in Food Demand and Delivery: Do Middle-Income Countries Follow High-Income Trends? *Journal of Food Distribution Research, 39*(1), 116–122.

Savu, M., & Voicu, O. L. (2015). The Romanian Consumer under the Influence of the Global Crisis. *Annals of the „Constantin Brâncuşi" University of Târgu Jiu. Economy Series, 1*(I), 293–297.

Stanciu, M. (2010). Consumption of Romanian population in last two decades. *Life Quality, XXI*(3-4), 251–273.

Toderoiu, F. (2012). The Romanian Agri-Food Economy – Performance Reductive Effects After Five Years of EU Membership. *Agricultural Economics and Rural Development, 9*(1), 25–45.

Verma, M., Hertel, T., & Preckel, P. V. (2011). Predicting within country household food expenditure variation using international cross-section estimates. *Economics Letters, 113*(3), 218–220. doi:10.1016/j.econlet.2011.07.004

Winters, L. A. (2005). The European Agricultural Trade Policies and Poverty. *European Review of Agriculture Economics, 32*(3), 319–346. doi:10.1093/eurrag/jbi016

This research was previously published in Food Science, Production, and Engineering in Contemporary Economies edited by Andrei Jean-Vasile, pages 245-274, copyright year 2016 by Information Science Reference (an imprint of IGI Global).

Section 3

Food Safety and Inspection

Chapter 9
Food Safety From Consumer Perspective:
Consumer Confidence in Food Safety

Ivana Domazet
Institute of Economic Sciences, Serbia

Nenad Djokić
University of Novi Sad, Serbia

ABSTRACT

Consumer perception is one of the aspects from which food safety can be considered. This view of food safety, named in the literature as subjective safety, does not necessarily correspond to objective food safety and is also considered as not being easily influenced by education of consumers. Having that in mind, the first goal of this chapter is to examine different approaches to food safety and the results of previous research dealing with consumer perspective of food safety. Besides giving broad literature review of the topic, the goal of this chapter is to implement Consumer confidence in food safety scale in Serbian consumer research. The validity, reliability, and one-dimensionality of this scale, which consists of optimism and pessimism (two dimensions of confidence), are tested and determinants of different rating on it are examined. The research from 2016 includes 300 participants in the main survey. Finally, starting from profiling consumers with different levels of general confidence in food safety, managerial recommendations are provided.

INTRODUCTION

The importance given to food safety and quality can be understood as a part of competition among actors in agribusiness based upon adding value rather than relying only on efficiency and quality control, what is caused by a number of food scares in the focus of public attention, segments of the general public paying greater attention to manners of producing food at the farm and processing level as well as by consumers becoming more fragmented in their food choices, especially in developed countries, according to Grunert (2005, pp. 369–391).

DOI: 10.4018/978-1-5225-8063-8.ch009

Ritson and Mai (1998) give a broad interpretation of food safety, which includes nutritional quality of the diet and concerns about novel foods, as well as chemical and microbiological safety, then what binds these characteristics together from an economics perspective is that they all reflect a potentially adverse impact on an individual as a consequence of the consumption of food. Hereby, such impact is structured of two components: hazard, which represent the severity of the impact, and risk, which considers the probability of the hazard occurring. Hazards can be classified into sources of risk: microbiological, chemical, and technological (Yeung & Morris, 2001).

Ritson and Mai (1998) identify four manners for potential market economy "failures" in providing the optimum degree of food safety: asymmetry in the knowledge of risks, aspects of food safety (public goods), social costs of food safety, and the divergence between objective scientific evidence and consumer perception of food safety risks. When it comes to consumer behavior in regard to food safety risk, Verbeke, Frewer, Scholderer, & De Brabander (2007) point out that they tend to systematically overestimate risks in connection to the technical probability of harm occurring, while to underestimate in a large extent other risks. Despite distinguishing broad categories of risks, they do not find enough differences between various types of risk within a particular food group. Even in conditions of existence of information focused on reducing food safety uncertainty, consumers do not process this information. It is in accordance with differentiation of objective food safety (assessment of the risk of consuming a certain food by scientists and food experts) and subjective food safety (in the mind of the consumer), their mutual deviation and failures of attempts to educate consumers in regard to these questions (Grunert, 2005). It is also in accordance with Yeung and Morris (2001) pointing out consumer risk perception of food safety not being largely determined by the hazard per se, but more with its social and psychological characteristics. Finally, this is similar to suggestions of Frewer, De Jonge, & Van Kleef (2010) that evaluation performed by consumers regarding the severity of different risks appear actually differ from valuations given by persons with knowledge about a specific hazard domain.

Having all this in mind, there are several goals of this chapter:

1. Examine different approaches to food safety and the results of previous research dealing with consumer perspective of food safety,
2. Implement consumer confidence in food safety scale, developed by de jonge et al. (2008a), in serbian consumer research,
3. Provide managerial implications (when considering the results of the research conducted in serbia, as well as considering similar researches abroad).

The first goal is implemented in the Background section of the chapter, the second in the Main Focus section, and the third in Solutions and Recommendations and Conclusions sections.

BACKGROUND

As for the consumer relation to food safety, De Jonge et al. (2004) argue that in the conditions characterized by the absence of any incidents in regard to food products, food safety will be "taken for granted" by consumers. Ritson and Mai (1998) stress that even if food safety was easily identifiable when existing the inverse relationship between it and some positive attribute (e.g. taste, appearance, price), consumers are not seeking for 100% safety. Grunert (2005) points out that a way in which food safety affects consumer food choice is different from those of the other dimensions of quality (taste, health, convenience, and process). When considering the consumer purchase decision-making process, it is dominant in two conditions: when major safety problems are perceived and in regard to certain production technologies. Thereby, food safety perceptions are described as a "sleeping giant" – they do not influence quality perceptions in regular conditions, but this influence becomes substantial in the conditions of crisis.

The relation between food quality and safety is also studied by Van Rijswijk and Frewer (2008). The research was conducted in four European countries: Germany, France, Italy, and Spain and included 163 consumers interviewed in qualitative semi-structured interviews. These interviews yielded in consumers' own definitions of food quality and safety, as well as their relations. The greatest number of consumers sees food quality and food safety as interlinked concepts and there are overlapping of definitions of the two concepts. They pay more attention to food quality during the purchase. Hereby, from the consumer perspective, traceability was not only in connection with food safety but with food quality as well.

By the confidence in food safety, De Jonge et al. (2004) understand the general expectation of consumers that food products will not be the cause of any harm for their health or for the environment. The authors identify five potential determinants of consumer confidence in food safety: consumer trust in regulatory institutions and actors in the food chain, the occurrence of food safety incidents, media coverage of food risks, consumer recall of food safety incidents, and consumer confidence in the safety of particular product groups. As for the general confidence in food safety, the authors argue that it is likely to have both a cognitive and an affective dimension. They also identify possible behavioral consequences of changes in general consumer confidence in food safety: information search, brand choice, retail choice, product substitution, and reduced consumption.

Yeung and Morris (2001) propose a conceptual model of consumer food purchase relating to food safety. The combination of perceptions of the degree of exposure to a hazard and the consequences of such exposure influence the perception of food safety risk. Bearing in mind the nature of food safety risk, risk perception (strengthened by uncertain characteristics of food safety risks) shows a tendency to weight the perceived potential severity of unhealthy food more in comparison to the probability of exposure (Zubovic & Domazet, 2009).

Verbeke et al. (2007) explore the reasons for consumer behavior in relation to food safety. They first identify the gap between lay and expert opinions and find its causes in a so-called perception filter related to this bias. Secondly, they explain pathway from risk to crisis in relation to the theory of social amplification of risk. Three factors are identified, providing insight into the problem of why some relatively minor risks frequently become crucial for evolution of risk into a crisis: fright factors (individual's perception of the seriousness of a risk), panic elements (related to the nature of the risk itself, whether it is universal, new, believable, uncertain, etc.), and societal triggers (e.g. media information). Thirdly, the authors make an opposition between lifestyle and technological hazards, whereas optimistic biases are more expressed for lifestyle. Finally, they explain the reluctance to information processing in the context of potential irrelevance or usefulness of information to consumers since failing in targeting particular needs or interests.

Frewer et al. (2010) research similar topics. They point out the importance of adding additional criteria to judging risks to classical factors (level of probability and degree of possible harm): the degree to which a hazard is unknown and the degree of dread experienced by the individual evaluating a particular hazard. In consideration to individuals' perception of risk, there should be mentioned a psychological phenomenon of optimistic bias or unreal optimism. This phenomenon is related to judgment that negative events are less likely to happen to the person making the judgment than to an average member of society. When explaining often-mentioned opposition to emerging food technologies, the authors stress that this opposition does not have to come only from evaluations of the potential personal benefits and health effects, but also from moral concerns, beliefs, and values.

These authors stress the need for deeper understanding individual responses to food issues, especially in the situations in which habits, emotions, and information processing heuristics have an effect on consumer decision-making. Hereby, they notice the connection of individual differences and consumer perceptions of food safety, not only in regard to socio-demographic characteristics (gender, age, education level, number of children in the household, etc.) with lack of consistent results, but also in regard to other variables – personal experience of food poisoning (which reduces optimistic bias), general openness to worry and stress, being neophobic, as well as the locus of control ("a measure of attribution of control of external events which potentially affect people's health status").

Verbeke et al. (2007) also suggest strategies for restoring consumer confidence in food safety: establishment of traceability and labeling, segmented communication approaches, and public involvement in risk management decision-making.

De Jonge et al. (2004) report results regarding a nationally representative sample of Dutch consumers from 2002, whereas data collection was performed in three "waves" (January, May, and December) with around 500 respondents (aged 16 or above) participating in each. Hereby, out of all respondents, 85% considered that, when it comes to food safety, there was more than sufficient attention being paid, as well as actions being taken. As for their perception of the responsibility for the safety of food, 43% considered Government as primarily responsible, 27% perceived manufacturers, and 26% farmers in this role, while only 3% stated consideration of supermarkets as primarily responsible for food safety.

In the same research, consumer confidence in food safety was measured by four items: "I think the quality of food keeps improving", "In recent years my confidence in food products has decreased" (recorded in the analysis), "I am worried about the safety of food products" (recorded in the analysis), and "Food products have never been as safe as nowadays". When analyzing the mean rating of statements, several conclusions were reached. The mean for all statements by all respondents was 3.1, on a scale from 1 (low confidence) to 5 (high confidence), but there were significant differences across gender and age segments. Hereby, men were more confident in the safety of food than women. However, when it comes to age, there were differences in several directions regarding consideration of food safety by different groups. When evaluating the statement "I think the quality of food keeps improving", the youngest age group (16-24) accepted it the most, while the age group 35-49 was the most pessimistic concerning it. Considering the second and the third statement, there was a negative relation between age and their evaluation. However, the oldest respondents accepted the most that "Food products have never been as safe as nowadays", while this was strongly rejected by the middle age groups (25-34 and 35-49).

When asking the same respondents about their primary concerns with regard to food, 16% of them had no concern, whereas more men gave this answer in comparison to women, while there were no differences across age groups when answering. The five most common concerns were: additives, unnatural ingredients, genetic manipulation, lack of information on the product label, and meat. During the evaluation of their confidence in 18 different product categories, 42% of respondents reported no concerns. Confidence was especially high for agricultural products (cheese, dairy, and fresh vegetables) while especially low for the products that have been processed extensively (ready-to-eat products and meals), as well as for the products that have a chemical character (vitamin supplements, sport, and energy drinks). This implied that consumers were more concerned regarding appliance of new technologies in food production than regarding products produced without extensive processing.

De Jonge et al. (2008b) conducted another research in the Netherlands – this time among Dutch consumers responsible for the daily grocery shopping. The authors used data from December 2003 that were collected among 3,669 consumers out of 4,437 consumers belonging to a panel of Dutch consumers. The aim of the research was to evaluate the level in which general consumer confidence in food safety (represented by optimism and pessimism) is caused by their trust in general, trust in specific actors (food manufacturers, the Government, farmers, and retailers), specific trust dimensions (the extent to which consumers perceived actors to take care of public well-being, perceived competence, and perceived openness), as well as specific combinations between actors and trust dimensions. Hereby, separate regression analyses were performed for two dimensions of consumer confidence in food safety – optimism and pessimism, while having in mind that there were identified four actors and three trust dimensions, twelve variables were used within the model as independent variables.

The results have shown that although consumer general trust strengthened their optimism and diminished their pessimism when it came to food safety in general, the strength of such relation varied in the context of the actor being trusted and the trust dimension taken into consideration. Generally, consumer confidence in the safety of food is most strongly influenced by their trust in food manufacturers. The extent of that influence is much larger in comparison to the influence of the Government, farmers, and retailers.

When considering trust dimensions in regression analysis where optimism was the dependent variable, it was concluded that the extent to which consumers perceived actors to take care of public well-being is the most important dimension. Besides, there were significant interaction effects for perceived openness and perceived competence dimensions in the relation of trusting in food manufacturers and the Government. However, while for food manufacturers there was positive interaction effect with perceived competence and negative interaction effect with perceived openness, the results for the Government are opposite since perceived competence was less important and perceived openness more important. When considering trust dimensions in a regression where pessimism was the dependent variable, the greatest difference from the previous analysis, was that the extent to which consumers perceived actors to take care of public well-being is not the most important dimension. All other results could be understood as "the mirror image of the results from the model with optimism as the dependent variable".

Table 1 presents shortly some of the main findings from the previously listed literature.

When it comes to Serbian researches, according to the authors' knowledge, there was no research regarding consumer view on food safety with the implementation of instruments developed abroad. However, having in mind the possible connection of food quality and safety from a consumer perspective, and the results of some previous research of organic and functional food, it can be expected that food safety is an important issue for Serbian consumers. Among such researches can be listed the one of Grubor and Djokic (2016) on a segment of Serbian consumers who prefer organic food. These respondents among other characteristics strongly value the importance of a diet for their health and believe that a disease can be prevented by proper diet. There is also the research of Grubor, Djokic, Milicevic, & Kovac-Znidersic (2014), which identifies segments of consumers preferring food with "health claims", including functional food.

Researching similar topics in domestic conditions as well as in other developing countries can be significantly from two perspectives. Firstly, it can be useful for food producers to try to differentiate and target premium segments. Secondly, if considering exporting food products, researching foreign markets concerning food safety from a consumer perspective is very important.

Table 1. Food safety from consumer perspective

Author(s)	Findings
Grunert (2005)	There is a difference between objective food safety (assessment of the risk of consuming a certain food by scientists and food experts) and subjective food safety (in the mind of the consumer).
Frewer et al. (2010)	A valuation performed by consumers regarding the severity of different risks appear actually differ from valuations given by persons with knowledge about a specific hazard domain.
Yeung and Morris (2001)	- Consumer risk perception of food safety is not being largely determined by the hazard per se, but more with its social and psychological characteristics. - Combination of perceptions of the degree of exposure to a hazard and the consequences of such exposure influence the perception of food safety risk.
Verbeke et al. (2007)	- Consumers tend to systematically overestimate risks in connection to the technical probability of harm occurring, while to underestimate in a large extent other risks. - Three factors are identified, providing insight into the problem of why some relatively minor risks frequently become crucial for evolution of risk into a crisis: fright factors (individual's perception of the seriousness of a risk), panic elements (related to the nature of the risk itself, whether it is universal, new, believable, uncertain, etc.), and societal triggers (e.g. media information).
Frewer et al. (2010)	The authors point out the importance of adding additional criteria to judging risks to classical factors (level of probability and degree of possible harm): the degree to which a hazard is unknown and the degree of dread experienced by the individual evaluating a particular hazard.
De Jonge et al. (2004)	The authors identify five potential determinants of consumer confidence in food safety: consumer trust in regulatory institutions and actors in the food chain, the occurrence of food safety incidents, media coverage of food risks, consumer recall of food safety incidents, and consumer confidence in the safety of particular product groups.

Source: Authors' compilation

In the following sections of the chapter, the results of primary research are presented considering the implementation of consumer confidence in food safety scale in Serbia.

MAIN FOCUS OF THE CHAPTER

Research Methodology

Data collection was carried out in May 2016. The respondents took up to 15 minutes on the average to complete the questionnaires. The questionnaire consisted of two parts. The first part was Consumer confidence in food safety scale, developed by De Jonge et al. (2008a). Hereby, all items were rated on seven-point Likert scales from "strongly disagree" to "strongly agree". During the translation of questionnaire into Serbian, back-translation procedure was applied.

The second part of the questionnaire was related to respondents' socio-demographic characteristics. These characteristics were taken into account having in mind that being implicitly present in the behavior theories in relation to consumers' food choices: the Theory of Reasoned Action and its later modification – the Theory of Planned Behaviour Theory (Aertsens, Verbeke, Mondelaers, & Van Huylenbroeck, 2009). Included socio-demographic characteristics were: gender, age, education, marital status, the presence of children in the family, household size, self-assessed household income (from 1 – the lowest mark, to 5 – the highest mark), and place of residence. The levels of these variables that are nominal and ordinal are presented in Table 2.

Table 2. Description of the sample

Characteristics of the sample		Measures	Data
Gender	Male	Percentage	38.0
	Female		62.0
Marital status	Married		59.0
	Single		41.0
Presence of children in family	Families with no children		47.3
	Families with children		52.7
Education	Finished secondary school		54.3
	Student		17.0
	Finished college or faculty		28.7
Place of residence	Urban		59.7
	Rural		40.3
Age		M (SD)	38.44 (13.906)
Household size			2.84 (1.396)
Self-assessed household income			2.87 (0.816)

Note: M – mean; SD – standard deviation.
Source: Authors' development

Within the research method, structured interviews by means of personal communication with the respondents with the use of questionnaire were conducted. The respondents were chosen from the population aged between 18 and 65 years. In an attempt to reach average Serbian consumer, they were approached near green markets, large shopping centers, and universities. The interviews were performed simultaneously in several big Serbian cities: Belgrade, Novi Sad, Subotica, Kragujevac, Kraljevo, and Niš. The interviewed persons were those who were in charge for performing at least 50% of buying food for their households. Convenience sampling method was implemented.

After formal and logical analysis of questionnaires and elimination of questionnaires containing answers of respondents scoring low or high on all items, 300 questionnaires were included in final analysis. The socio-demographic characteristics of those respondents are summarized in Table 1.

On the first stage, reliability, validity, and unidimensionality of the questionnaires were tested according to the recommendations by Churchill (1979), Anderson and Gerbing (1982), Danes and Mann (1984), and Steenkamp and van Trijp (1991). In operationalisation, this considered that Cronbach's alpha, Cronbach's alpha if item deleted, and Corrected item-total correlation were calculated – separately for two dimensions of confidence – optimism and pessimism. On the next stage, exploratory factor analysis by maximum likelihood factor analysis method and Promax rotation method were conducted for testing validity and included all the question of the scale. Prior to Exploratory factor analysis, adequacy of conducting was tested by Kaiser-Meyer-Olkin (KMO) test and Bartlett's Test of Sphericity. Besides, confirmatory factor analysis with maximum likelihood method was used within each of two dimensions of confidence.

Next, after omission of certain items in accordance with the results of the previous tests, two dimensions of Consumer confidence in food safety scale were analyzed in the context of respondents' socio-demographic characteristics. Hereby, both optimism and pessimism were presented by means of ratings of all respondents in regard to the scales. When testing whether there were statistically significant differences between men and women, married and single respondents, families with and without children, people living in urban and rural areas, Independent samples t-test was implemented in relation to the ratings. On the other hand, when testing whether there were statistical significant relations between consumer education and respondents' ratings of these scales, Spearman rank correlation coefficient was used, while when establishing whether there were statistical significant relations between consumer age, household size, and self-assessed household income with respondents' ratings of scales, Pearson correlation coefficient was implemented.

On the third stage, the influence of all researched socio-demographic characteristics of the respondents (gender, age, education, marital status, presence of children in family, household size, self-assessed household income, and place of residence) on both dimensions of consumer confidence in food safety (optimism and pessimism) was analysed by using multiple regression with General Linear Model. The dependent variables (consumer optimism and pessimism) were presented by means of ratings of all respondents in regard to the items remained in the scales after testing their reliability, validity, and unidimensionality. The results were observed at the level of average age, average household size, and average self-assessed household income. Coding was performed, so that the value of 0 was assigned to the following levels of respondents' socio-demographic characteristics: female, single, living in family with no children, finished secondary school, and living in rural area, while the value of 1 was assigned to the following levels of the characteristics: male, married, living in family with children, student or finished faculty or college, and living in urban area.

Data were processed in the SPSS and AMOS statistical packages.

Reliability, Validity, and Unidimensionality of Scales

The results of the analysis showed that the values of Cronbach's alpha, both for the Optimism dimension of the Consumer confidence in food safety scale (0.932) and the Pessimism dimension of the Consumer confidence in food safety scale (0.866) were above 0.7 suggested by Nunnally (1978). Hereby, although removal of an item "As a result of the occurrence of the food safety incidents, I am suspicious about certain food products" could increase the value of Cronbach's alpha for Pessimism dimension of the Consumer confidence in food safety scale (from 0.866 to 0.901), the option was kept in the questionnaire. Besides, all the values of Corrected item-total correlation were above the threshold of 0.3 (Nunnally & Bernstein, 1994).

The result of Kaiser-Meyer-Olkin test of 0.842 was above 0.6 (Kaiser, 1970, 1974), and the level of significance of the Barthlett's sphericity test of 0.000 suggested adequacy of conducting exploratory factor analysis. Two factors with Eigenvalue above 1 (Kaiser, 1960) were identified. Identified factors explained 77.496% of the total variance.

When analyzing the Patern matrix, it was noticed that the first four items belonging to the Optimism dimension of the Consumer confidence in food safety scale loaded on the first factor, while the other three items belonging to the Pessimism dimension of the Consumer confidence in food safety scale loaded on the second factor. Hereby, although all the loadings on appropriate factor were higher than 0.35, which is the minimum recommended for a sample of 250 respondents, and 0.30, which is the minimum recommended for a sample of 350 respondents (Hair, Tatham, Anderson, & Black, 1998). The item with the weakest loading "Generally, food products are safe" was excluded from further analysis. When comparing two factor loadings of each item, it was identified that all the differences were above 0.2 (Matsunaga, 2010), what suggested no cross-loadings. Previous results of testing reliability and validity of the scale are presented in Table 3.

Following Hair et al. (1998), all items were retained when considering that their factor loadings in confirmatory factor analysis were higher than the minimum recommended for that sample size, but the last item was omitted from further analysis because its R^2 was lower than 0.5 (Table 4).

After removal of the named item and correlating residuals' covariances, according to modification indices, the results of confirmatory factor analysis showed acceptable model fit (compared to thresholds given by Hu and Bentler (1999)) regarding most of the indicators (Table 5).

Even when considering indicators that suggest that model fit is not acceptable, one can have in mind that probability of p-value being significant increases as the sample size increases, no matter what the model-specific features are (Hooper, Coughlan, & Mullen, 2008).

Table 3. Reliability and validity analysis of the scale

Scale	Cronbach's alpha	Items	Cronbach's alpha if item deleted	Corrected item-total correlation	F1	F2
Optimism	0.932	I am optimistic about the safety of food products.	0.919	0.816	**0.899**	0.064
		I am confident that food products are safe.	0.899	0.876	**1.052**	0.179
		I am satisfied with the safety of food products.	0.896	0.884	**0.834**	-0.128
		Generally, food products are safe.	0.928	0.786	**0.569**	-0.451
Pessimism	0.866	I worry about the safety of food.	0.718	0.839	0.025	**0.980**
		I feel uncomfortable regarding the safety of food.	0.788	0.770	0.020	**0.860**
		As a result of the occurrence of the food safety incidents, I am suspicious about certain food products.	0.901	0.638	0.080	**0.714**

Source: Authors' development

Table 4. Unidimensionality analysis of the scale

Scale	Items	Factor loadings	R^2
Optimism	I am optimistic about the safety of food products	0.961	0.923
	I am confident that food products are safe	0.890	0.792
	I am satisfied with the safety of food products	0.955	0.913
Pessimism	I worry about the safety of food	0.978	0.956
	I feel uncomfortable regarding the safety of food	0.840	0.705
	As a result of the occurrence of the food safety incidents, I am suspicious about certain food products	0.668	0.446

Source: Authors' development

Table 5. Results of confirmatory factor analysis – model fit

	Chi-square / df	**p – value for the model**	**CFI**	**GFI**	**AGFI**	**SRMR**	**RMSEA**	**PCLOSE**
Consumer confidence in food safety scale	4.498	0.000	0.977	0.959	0.905	0.0499	0.108	0.02
Thresholds (Hu and Bentler, 1999)	< 3 good; < 5 sometimes permissible	> 0.05	> 0.95 great; > 0.9 traditional; > 0.8 sometimes permissible	> 0.95	> 0.8	< 0.09	< 0.05 good; 0.05-0.10 moderate > 0.10 bad	> 0.05

Source: Authors' development

Consumer Confidence in Food Safety and Their Socio-Demographic Characteristics

When considering dimensions of the Consumer confidence in food safety scale at the total level, it can be seen that average optimism is 4.08 (standard deviation is 1.45), while average pessimism is 3.52 (standard deviation is 1.76). Hereby, there is moderate, negative, and statistically significant correlation among them: $r = (-0.467)$; $n = 300$; $p = 0.000 < 0.05$.

In Table 6, both optimism and pessimism regarding the consumer confidence in food safety are observed in the context of socio-demographic characteristics of respondents. Hereby, besides the results of appropriate tests, when considering nominal and ordinal variables, for each of their levels the averages for both dimensions are given.

When considering gender, there are no statistically significant differences for both optimism and pessimism in regard to the consumer confidence in food safety. As for the presence of children in the family, there are also no statistically significant differences for food safety optimism and pessimism between the families with and without children. The same conclusion can be reached regarding the place of residence since consumers living in urban areas are not statistically different from consumers living in rural areas when it comes to their optimism and pessimism related to food safety. However, during the analysis of marital status, it can be concluded that there is no statistically significant difference between married and single respondents in relation to optimism. However, there are statistically significant differences between these two segments in relation to pessimism. Married people statistically significantly differ from the single ones by being more pessimistic about food safety.

There is no statistically significant relation between the level of education of consumers and their rating optimism scale considering food safety, as well as between level of education and rating pessimism scale. Also, there is no statistically significant relation between rating optimism scale in food safety and respondents' age, the size of their households, and their self-assessed household income. However, when considering rating pessimism scale in food safety, there are statistically significant relations with the size of households their self-assessed household income of the respondents. These significant relations confirm that pessimism in food safety increases along with the increase of consumers' household size, while decreases along with the increase of self-assessed household income.

Table 6. Consumer confidence in food safety scale and socio-demographic characteristics

Characteristics of the sample		Measure	Optimism	Results of the tests	Pessimism	Results of the tests
Gender	Male	Percent	4.06	t(df = 298) = -0.217, p = 0.828 > 0.05	3.27	t(df = 298) = -1.840, p = 0.067 > 0.05
	Female		4.09		3.66	
Marital status	Married		4.08	t(df = 298) = -0.051, p = 0.960 > 0.05	3.81	t(df = 298) = 3.346, p = 0.001 < 0.05
	Single		4.08		3.10	
Presence of children in family	Families with no children		4.20	t(df = 298) = 1.557, p = 0.121 > 0.05	3.62	t(df = 298) = 1.032, p = 0.303 > 0.05
	Families with children		3.94		3.40	
Education	Finished secondary school		4.13	p = -0.047, n = 300, p = 0.412 > 0.05	3.56	p = -0.035, n = 300, p = 0.546 > 0.05
	Student		4.09		3.65	
	Finished college / faculty		3.98		3.37	
Place of residence	Urban		4.15	t(df = 298) = 1.096, p = 0.274 > 0.05	3.41	t(df = 298) = -1.307, p = 0.192 > 0.05
	Rural		3.97		3.68	
Age			r = 0.087, n = 300, p = 0.134 > 0.05		r = -0.005, n = 300, p = 0.934 > 0.05	
Household size			r = -0.022, n = 300, p = 0.700 > 0.05		r = 0.295, n = 300, p = 0.000 < 0.05	
Self-assessed household income			r = 0.016, n = 300, p = 0.783 > 0.05		r = -0.187, n = 300, p = 0.001 < 0.05	

Source: Authors' development

Results of the General Linear Model Application

By implementing multiple regression with the General Linear Model procedure, the results for the dependent variable of the Consumer optimism regarding confidence in food safety (OPT) can be expressed by a regression equation as follows:

OPT = 3.893 – 0.085 * gender + 0.006 * age – 0.084 * marital status + 0.324 * presence of children in family – 0.077 * household size – 0.029 * education + 0.022 * self-assessed household income + 0.186 * place of residence

Table 7 shows the values of parameters for the dependent variable of the Consumer optimism regarding confidence in food safety at the level of average age, average household size, and average self-assessed household income.

It may have sense to interpret the above equation only in the context of coding stated in the procedures section: 0 is assigned to female, single, living in family with no children, finished secondary school, living in rural area, while 1 is assigned to male, married, living in family with children, student or finished faculty or college, living in urban area. Hereby, there is no any socio-demographic characteristic of respondents that statistically significantly influences Consumer optimism regarding confidence in

Table 7. Values of parameters for the dependent variable of the Consumer optimism regarding confidence in food safety

Parameter	B	Std. Error	t	Sig.	95% Confidence Interval	
					Lower Bound	Upper Bound
Intercept	3.893	0.266	14.641	0.000	3.369	4.416
gender	-0.085	0.186	-0.457	0.648	-0.451	0.281
age	0.006	0.007	0.793	0.428	-0.009	0.021
marital status	-0.084	0.220	-0.383	0.702	-0.517	0.349
presence of children in family	0.324	0.230	1.409	0.160	-0.129	0.776
household size	-0.077	0.085	-0.909	0.364	-0.245	0.090
education	-0.029	0.189	-0.152	0.879	-0.401	0.343
self-assessed household income	0.022	0.116	0.188	0.851	-0.206	0.250
place of residence	0.186	0.176	1.056	0.292	-0.160	0.531

Source: Authors' development

food safety. However, it can be stated that at the level of average age, average household size, average self-assessed household income, and under other initial conditions (all other levels except the discussed coded with 0), in the units of seven-point Likert scale, men are less optimistic regarding food safety in relation to women (by 0.085), married people are less optimistic regarding food safety in relation to single ones by (0.084), people living in families with children are more optimistic regarding food safety in comparison to people living in families with no children by (0.324), students or respondents that graduated faculty or college are less optimistic regarding food safety compared to the respondents who graduated secondary school by (0.029), and people living in urban areas are more optimistic regarding food safety in relation to people living in rural areas by (0.186).

Besides, when it comes to age, it can be seen that increase in age by one year in relation to average age (at other initial conditions) increases Consumer optimism regarding confidence in food safety by 0.006 on average on a seven-point Likert scale. As regards household size, it can be stated that the increase by one family member in relation to average household size (at other initial conditions) decreases Consumer optimism regarding confidence in food safety by 0.077 on average on a seven-point Likert scale. When it comes to self-assessed household income, it can be found that the growth by 1 on a five-point Likert scale in comparison with an average grade, at other initial conditions, increases the Consumer optimism regarding confidence in food safety by 0.022 on a seven-point Likert scale.

The level of the Consumer optimism regarding confidence in food safety can be considered as a combination of levels of respondents' socio-demographic characteristics (gender, education, marital status, presence of children in family, and place of residence) at the level of average age, average household size, and average self-assessed household income (Table 8).

At the level of average age, average household size, and average self-assessed household income, Consumer optimism regarding confidence in food safety is mostly manifested by females, single respondents, people living in families with children, respondents who finished secondary school and living in urban places (4.40).

Table 8. Consumer optimism regarding confidence in food safety (decreasing values) and socio-demographic characteristic

Gender	Marital status	Presence of children in family	Education	Place of residence	Consumer optimism regarding confidence in food safety
female	single	with children	graduated secondary school	urban	4.40
female	single	with children	student or graduate faculty or college	urban	4.37
female	married	with children	graduated secondary school	urban	4.32
male	single	with children	graduated secondary school	urban	4.32
female	married	with children	student or graduate faculty or college	urban	4.29
male	single	with children	student or graduate faculty or college	urban	4.29
male	married	with children	graduated secondary school	urban	4.23
female	single	with children	graduated secondary school	rural	4.22
male	married	with children	student or graduate faculty or college	urban	4.20
female	single	with children	student or graduate faculty or college	rural	4.19
female	married	with children	graduated secondary school	rural	4.13
male	single	with children	graduated secondary school	rural	4.13
female	married	with children	student or graduate faculty or college	rural	4.10
male	single	with children	student or graduate faculty or college	rural	4.10
female	single	no children	graduated secondary school	urban	4.08
female	single	no children	student or graduate faculty or college	urban	4.05
male	married	with children	graduated secondary school	rural	4.05
male	married	with children	student or graduate faculty or college	rural	4.02
female	married	no children	graduated secondary school	urban	3.99
male	single	no children	graduated secondary school	urban	3.99
female	married	no children	student or graduate faculty or college	urban	3.97
male	single	no children	student or graduate faculty or college	urban	3.96
male	married	no children	graduated secondary school	urban	3.91
female	single	no children	graduated secondary school	rural	3.89
male	married	no children	student or graduate faculty or college	urban	3.88
female	single	no children	student or graduate faculty or college	rural	3.86
female	married	no children	graduated secondary school	rural	3.81
male	single	no children	graduated secondary school	rural	3.81
female	married	no children	student or graduate faculty or college	rural	3.78
male	single	no children	student or graduate faculty or college	rural	3.78
male	married	no children	graduated secondary school	rural	3.72
male	married	no children	student or graduate faculty or college	rural	3.69

Source: Authors' development

Accordingly to the previous results, by implementing multiple regression with the General Linear Model procedure, the results for the dependent variable of the Consumer pessimism regarding confidence in food safety (PES) can be expressed by a regression equation as follows:

PES = 3.496 – 0.024 * gender – 0.006 * age + 0.290 * marital status – 0.353 * presence of children in family + 0.370 * household size + 0.202 * education – 0.171 * self-assessed household income – 0.079 * place of residence.

The values of parameters for the dependent variable of the Consumer pessimism regarding confidence in food safety at the level of average age, average household size, and average self-assessed household income are presented in Table 9.

It is reasonable to interpret the above equation, as well as the previous one, only in the context of coding stated in the procedures section. Hereby, the only socio-demographic characteristic of respondents that statistically significantly influences the Consumer pessimism regarding confidence in food safety is household size in a manner that its increase by one member in relation to average household size (at other initial conditions) increases the Consumer pessimism regarding confidence in food safety by 0.370 on average on a seven-point Likert scale.

However, when analysing other characteristics, it can be stated that at the level of average age, average household size, average self-assessed household income, and under other initial conditions, in the units of seven-point Likert scale, men are less pessimistic regarding food safety in relation to women (by 0.024), married people are more pessimistic regarding food safety in relation to single ones (by 0.290), people living in families with children are less pessimistic regarding food safety in comparison to people living in families with no children (by 0.353), students or respondents that graduated faculty or college are more pessimistic regarding food safety than respondents who graduated secondary school (by 0.202), and people living in urban areas are less pessimistic regarding food safety in relation to people living in rural areas (by 0.079).

Besides, when it comes to age, it can be concluded that increase in age by one year in relation to average age (at other initial conditions) decreases the Consumer pessimism regarding confidence in

Table 9. Values of parameters for the dependent variable of the Consumer pessimism regarding confidence in food safety

Parameter	B	Std. Error	t	Sig.	95% Confidence Interval	
					Lower Bound	Upper Bound
Intercept	3.496	0.306	11.433	0.000	2.894	4.098
gender	-0.024	0.214	-0.112	0.911	-0.445	0.397
age	-0.006	0.009	-0.715	0.475	-0.023	0.011
marital status	0.290	0.253	1.146	0.253	-0.208	0.788
presence of children in family	-0.353	0.264	-1.337	0.182	-0.873	0.167
household size	0.370	0.098	3.790	0.000	0.178	0.563
education	0.202	0.217	0.929	0.354	-0.226	0.630
self-assessed household income	-0.171	0.133	-1.283	0.200	-0.434	0.091
place of residence	-0.079	0.202	-0.392	0.696	-0.477	0.318

Source: Authors' development

food safety by 0.006 on average on a seven-point Likert scale. When analyzing self-assessed household income, it can be seen that its growth by 1 on a five-point Likert scale in comparison with an average grade, at other initial conditions, decrease the Consumer pessimism regarding confidence in food safety by 0.079 on a seven-point Likert scale.

The level of Consumer pessimism regarding confidence in food safety can also be considered as a combination of levels of respondents' socio-demographic characteristics (gender, education, marital status, presence of children in family, and place of residence) at the level of average age, average household size, and average self-assessed household income (Table 10).

At the level of average age, average household size, and average self-assessed household income, the Consumer pessimism regarding confidence in food safety is mostly manifested by females, married respondents, people living in families with no children, students or people graduated faculty or college and living in rural areas (3.99).

SOLUTIONS AND RECOMMENDATIONS

When analysing the results of research showing average optimism being 4.08, while average pessimism 3.52, in the context of seven-point Likert scale used within this research, it can be seen that in both cases respondents are in average not sure whether they have reasons to be optimistic or pessimistic (in the second case this mark is between "Disagree" and "Neither agree nor disagree"). These results resemble the results of the research by De Jonge et al. (2004), by which the mean for all statements considering consumer confidence in food safety at the level of all respondents was 3.1 on a scale from 1 (low confidence) to 5 (high confidence).

When considering the results in the context of socio-demographic characteristics of the respondents, it can be concluded that married people statistically significantly differ from single ones by being more pessimistic in regard to food safety. Pessimism in food safety significantly increases along with the increase in the size of households, while significantly decreases with the increase of self-assessed household income. These results are also in accordance with De Jonge et al. (2004), where there are significant differences across socio-demographic segments in regard to consumer confidence in food safety. Besides, the results are in accordance with Frewer et al. (2010) who notice the connection of individual differences and consumer perceptions of food safety in regard to socio-demographic characteristics (gender, age, education level, and a number of children in the household). Finally, differences existing when it comes to one of the dimensions of the Consumer confidence in food safety are in accordance with De Jonge et al. (2008b) when stating that such differences suggest that optimism and pessimism are distinct concepts that should not be considered as two end poles of a unidimensional concept.

When it comes to the application of the General Linear Model and consideration of influences of the researched socio-demographic characteristics of respondents on two dimensions of the Consumer confidence in food safety (optimism and pessimism), it is concluded that the only statistically significant influence is the one of household size on the Consumer pessimism in a manner that increases in household size influences increase of pessimism. Such results furthermore support discussion stated above, although some differences occur as a consequence of different statistical approaches.

Besides the managerial implications of the results of the research, whereas optimism in regard to food safety should be increased and pessimism decreased, in the latter case, there can be identified the

Table 10. Consumer pessimism regarding confidence in food safety (decreasing values) and socio-demographic characteristics

Gender	Marital status	Presence of children in family	Education	Place of residence	Consumer optimism regarding confidence in food safety
female	married	no children	student or graduate faculty or college	rural	3.99
male	married	no children	student or graduate faculty or college	rural	3.96
female	married	no children	student or graduate faculty or college	urban	3.91
male	married	no children	student or graduate faculty or college	urban	3.89
female	married	no children	graduated secondary school	rural	3.79
male	married	no children	graduated secondary school	rural	3.76
female	married	no children	graduated secondary school	urban	3.71
female	single	no children	student or graduate faculty or college	rural	3.70
male	married	no children	graduated secondary school	urban	3.68
male	single	no children	student or graduate faculty or college	rural	3.67
female	married	with children	student or graduate faculty or college	rural	3.63
female	single	no children	student or graduate faculty or college	urban	3.62
male	married	with children	student or graduate faculty or college	rural	3.61
male	single	no children	student or graduate faculty or college	urban	3.59
female	married	with children	student or graduate faculty or college	urban	3.56
male	married	with children	student or graduate faculty or college	urban	3.53
female	single	no children	graduated secondary school	rural	3.50
male	single	no children	graduated secondary school	rural	3.47
female	married	with children	graduated secondary school	rural	3.43
female	single	no children	graduated secondary school	urban	3.42
male	married	with children	graduated secondary school	rural	3.41
male	single	no children	graduated secondary school	urban	3.39
female	married	with children	graduated secondary school	urban	3.35
female	single	with children	student or graduate faculty or college	rural	3.34
male	married	with children	graduated secondary school	urban	3.33
male	single	with children	student or graduate faculty or college	rural	3.32
female	single	with children	student or graduate faculty or college	urban	3.27
male	single	with children	student or graduate faculty or college	urban	3.24
female	single	with children	graduated secondary school	rural	3.14
male	single	with children	graduated secondary school	rural	3.12
female	single	with children	graduated secondary school	urban	3.06
male	single	with children	graduated secondary school	urban	3.04

Source: Authors' development

managerial implication coming from the researches listed in the literature review, i.e. establishment of traceability and labelling, segmented communication approaches, and public involvement in risk management decision-making.

FUTURE RESEARCH DIRECTIONS

Future researches could include a more representative sample, connection to certain product categories as well as including additional variables in relation to food safety. These researches could also explore in detail the relation of food safety and quality in the perception of domestic consumers and their relation to organic and functional food.

CONCLUSION

The topic of this chapter is related to the divergence between objective scientific evidence of food safety and consumer perception of food safety risks, i.e. differentiation of objective food safety (assessment of a risk of consuming a certain food) and subjective food safety (existing in consumer's mind). The chapter listed the authors who emphasized consumers' behavior regarding food safety stating that they systematically overestimated risks in relation to the technical probability of harm occurring, while underestimated other risks. Despite distinguishing broad categories of risks, the authors have not found enough differences between various types of risk within a particular food group. Even in the conditions of existence of information focused on reducing food safety uncertainty, consumers do not process this information. In this regard, there was given a review that consumer risk perception of food safety was not largely determined by the hazard per se, but more by its social and psychological characteristics, i.e. that valuation performed by consumers regarding the severity of different risks actually differed from valuations given by the persons with knowledge about a specific hazard domain.

The chapter also seeks in the literature for the explanation of such consumer behavior. Hereby, there were cited the authors stating that the causes of the gap between expert opinions were found in a so-called perception filter. They explained pathway from risk to crisis in relation to the theory of social amplification of risk by identifying three factors, providing insight into the problem of why some relatively minor risks frequently become crucial for the evolution of risk into a crisis: fright factors, panic elements, and societal triggers.

Besides giving a broad description of the researches regarding consumer relation to food safety conducted abroad, the chapter presents the results of the first application of the Consumer confidence in food safety scale in Serbia. The results show that average optimism is 4.08, while average pessimism is 3.52, in the context of the seven-point Likert scale used within this research. The results are considered in the context of socio-demographic characteristics of the respondents firstly by using tests of differences and relations and secondly by using multiple regression by applying General Linear Model. In the first case, it is shown that married people statistically significantly differ from single ones by being more pessimistic in regard to food safety. Pessimism in food safety significantly increases along with the increase of household size, while significantly decreases along with the increase of self-assessed household income of consumers. Secondly, in the application of General Linear Model, it is concluded that the only statistically significant influence is the one of household size on pessimism in a manner that increases in household size cause an increase of pessimism.

The chapter included managerial implications from both primary and secondary research.

ACKNOWLEDGMENT

This chapter is written as a part of the research projects III47009 (European Integrations and Social and Economic Changes in Serbian Economy on the Way to the EU) and OI179015 (Challenges and Prospects of Structural Changes in Serbia: Strategic Directions for Economic Development and Harmonisation with EU Requirements), financed by the Ministry of Education, Science, and Technological Development of the Republic of Serbia.

REFERENCES

Aertsens, J., Verbeke, W., Mondelaers, K., & Van Huylenbroeck, G. (2009). Personal Determinants of Organic Food Consumption: A Review. *British Food Journal*, *111*(10), 1140–1167. doi:10.1108/00070700910992961

Anderson, J. C., & Gerbing, D. W. (1982). Some Methods for Respecifying Measurement Models to Obtain Unidimensional Construct Measurement. *JMR, Journal of Marketing Research*, *19*(4), 453–460. doi:10.2307/3151719

Churchill, G. A. Jr. (1979). A Paradigm for Developing Better Measures of Marketing Constructs. *JMR, Journal of Marketing Research*, *16*(1), 64–73. doi:10.2307/3150876

Danes, J. E., & Mann, O. K. (1984). Unidimensional Measurement and Structural Equation Models with Latent Variables. *Journal of Business Research*, *12*(3), 337–351. doi:10.1016/0148-2963(84)90016-X

De Jonge, J., Frewer, L., Van Trijp, H., Renes, R. J., De Wit, W., & Timmers, J. (2004). Monitoring Consumer Confidence in Food Safety: An Exploratory Study. *British Food Journal*, *106*(10/11), 837–849. doi:10.1108/00070700410561423

De Jonge, J., Van Trijp, J. C. M., Van der Lans, I. A., Renes, R. J., & Frewer, L. J. (2008a). *A Scale for Consumer Confidence in the Safety of Food.* Discussion Paper no. 44. Wageningen: Mansholt Graduate School of Social Sciences.

De Jonge, J., Van Trijp, J. C. M., Van der Lans, I. A., Renes, R. J., & Frewer, L. J. (2008b). How Trust in Institutions and Organizations Builds General Consumer Confidence in the Safety of Food: A Decomposition of Effects. *Appetite*, *51*(2), 311–317. doi:10.1016/j.appet.2008.03.008 PMID:18450326

Frewer, L. J., De Jonge, J., & Van Kleef, E. (2010). Consumer Perceptions of Food Safety. In B. P. Mansourian, A. Wojtczak, & B. McA. Sayers (Eds.), *Medical Sciences* (pp. 244–264). Oxford, UK: EOLSS Publishers.

Grubor, A., & Djokic, N. (2016). Organic Food Consumer Profile in the Republic of Serbia. *British Food Journal, 118*(1), 164–182. doi:10.1108/BFJ-06-2015-0225

Grubor, A., Djokic, N., Milicevic, N., & Kovac-Znidersic, R. (2014). Acceptance of Food with "Health Claims" in Serbia: Consumers' Profile Research and Recommendations for Promotion. *Custos e Agronegócio Online, 10*(4), 199–220.

Grunert, K. G. (2005). Food Quality and Safety: Consumer Perception and Demand. *European Review of Agriculture Economics, 32*(3), 369–391. doi:10.1093/eurrag/jbi011

Hair, J. H., Tatham, R. L., Anderson, R. E., & Black, W. (1998). *Multivariate Data Analysis.* Upper Saddle River, NJ: Prentice Hall.

Hooper, D., Coughlan, J., & Mullen, M. (2008). Structural Equation Modelling: Guidelines for Determining Model Fit. *Electronic Journal of Business Research Methods, 6*(1), 53–60.

Hu, L., & Bentler, P.M. (1999). Cutoff Criteria for Fit Indexes in Covariance Structure Analysis: Conventional Criteria versus New Alternatives. *Structural Equation Modelling: A Multidisciplinary Journal, 6*(1), 1–55.

Kaiser, H. F. (1960). The Application of Electronic Computers to Factors Analysis. *Educational and Psychological Measurement, 20*(1), 141–151. doi:10.1177/001316446002000116

Kaiser, H. F. (1970). A Second Generation Little Jiffy. *Psychometrika, 35*(4), 401–415. doi:10.1007/BF02291817

Kaiser, H. F. (1974). An Index of Factorial Simplicity. *Psychometrika, 39*(1), 31–36. doi:10.1007/BF02291575

Matsunaga, M. (2010). How to Factor-Analyse your Data Right: Do's Dont's, and How-To's. *International Journal of Psychological Research, 3*(1), 97–110.

Nunnally, J. C. (1978). *Psychometric Theory*. New York, NY: McGraw-Hill.

Nunnally, J. C., & Bernstein, I. H. (1994). *Psychometric Theory*. New York, NY: McGraw-Hill.

Ritson, C., & Mai, L. W. (1998). The Economics of Food Safety. *Nutrition & Food Science, 98*(5), 253–259. doi:10.1108/00346659810224163

Steenkamp, J. E. M., & van Trijp, H. C. M. (1991). The Use of LISREL in Validating Marketing Constructs. *International Journal of Research in Marketing, 8*(4), 283–299. doi:10.1016/0167-8116(91)90027-5

Van Rijswijk, W., & Frewer, L. J. (2008). Consumer Perceptions of Food Quality and Safety and Their Relation to Traceability. *British Food Journal, 110*(10), 1034–1046. doi:10.1108/00070700810906642

Verbeke, W., Frewer, L. J., Scholderer, J., & De Brabander, H. F. (2007). Why Consumers Behave as They Do with Respect to Food Safety and Risk Information. *Analytica Chimica Acta, 586*(1/2), 2–7. doi:10.1016/j.aca.2006.07.065 PMID:17386689

Yeung, R. M. W., & Morris, J. (2001). Food Safety Risk – Consumer Perception and Purchase Behavior. *British Food Journal, 103*(3), 170–186. doi:10.1108/00070700110386728

Zubovic, J., & Domazet, I. (2009). Organic Food Market Development: The Case of Serbia. In J.-P. Guichard (Ed.), *Challenges of Economic Sciences in the 21st Century* (pp. 179–189). Belgrade: Institute of Economic Sciences.

ADDITIONAL READING

Alhakami, A. S., & Slovic, P. (1994). A Psychological Study of the Inverse Relationship between Perceived Risk and Perceived Benefit. *Risk Analysis*, *14*(6), 1085–1096. doi:10.1111/j.1539-6924.1994.tb00080.x PMID:7846317

Anderson, E. S., Winett, R. A., & Wojcik, J. R. (2000). Social-Cognitive Determinants of Nutrition Behavior among Supermarket Food Shoppers: A Structural Equation Analysis. *Health Psychology*, *19*(5), 479–486. doi:10.1037/0278-6133.19.5.479 PMID:11007156

Barnett, J., & Breakwell, G. M. (2001). Risk Perception and Experience: Hazard Personality Profiles and Individual Differences. *Risk Analysis*, *21*(1), 171–177. doi:10.1111/0272-4332.211099 PMID:11332545

Baron, J., Hershey, J. C., & Kunreuther, H. (2000). Determinants of Priority for Risk Reduction: The Role of Worry. *Risk Analysis*, *20*(4), 413–427. doi:10.1111/0272-4332.204041 PMID:11051067

Berg, L. (2004). Trust in Food in the Age of Mad Cow Disease: A Comparative Study of Consumers Evaluation of Food Safety in Belgium, Britain and Norway. *Appetite*, *42*(1), 21–32. doi:10.1016/S0195-6663(03)00112-0 PMID:15036780

Dosman, D. M., Adamowicz, W. L., & Hrudey, S. E. (2001). Socioeconomic Determinants of Health- and Food Safety-Related Risk Perceptions. *Risk Analysis*, *21*(2), 307–317. doi:10.1111/0272-4332.212113 PMID:11414539

Eiser, R. J., Miles, S., & Frewer, L. J. (2002). Trust, Perceived Risk, and Attitudes toward Food Technologies. *Journal of Applied Social Psychology*, *32*(11), 2423–2433. doi:10.1111/j.1559-1816.2002.tb01871.x

Fischhoff, B., Slovic, P., Lichtenstein, S., Read, S., & Combs, B. (1978). How Safe is Safe Enough? A Psychometric Study of Attitudes towards Technological Risks and Benefits. *Policy Sciences*, *9*(2), 127–152. doi:10.1007/BF00143739

Frewer, L. J., Howard, C., Hedderley, D., & Shepherd, R. (1996). What Determines Trust in Information about Food-Related Risks? Underlying Psychological Constructs. *Risk Analysis*, *16*(4), 473–485. doi:10.1111/j.1539-6924.1996.tb01094.x PMID:8819340

Frewer, L. J., Miles, S., & Marsh, R. (2002). The Media and Genetically Modified Foods: Evidence in Support of Social Amplification of Risk. *Risk Analysis*, *22*(4), 701–711. doi:10.1111/0272-4332.00062 PMID:12224744

Frewer, L. J., & Salter, B. (2002). Public Attitudes, Scientific Advice and the Politics of Regulatory Policy: The Case of BSE. *Science & Public Policy*, *29*(2), 137–145. doi:10.3152/147154302781781092

Frewer, L. J., & Salter, B. (2003). The Changing Governance of Biotechnology: The Politics of Public Trust in the Agri-Food Sector. *Applied Biotechnology, Food Science and Policy*, *1*(4), 1–8.

Green, J. M., Draper, A. K., & Dowler, E. A. (2003). Short Cuts to Safety: Risk and Rules of Thumb in Accounts of Food Choice. *Health Risk & Society*, *5*(1), 33–52. doi:10.1080/1369857031000065998

Miles, S., Brennan, M., Kuznesof, S., Ness, M., Ritson, C., & Frewer, L. J. (2004). Public Worry about Specific Food Safety Issues. *British Food Journal*, *106*(1), 9–22. doi:10.1108/00070700410515172

Slovic, P. (1987). Perception of Risk. *Science*, *236*(4799), 280–285. doi:10.1126cience.3563507 PMID:3563507

KEY TERMS AND DEFINITIONS

Consumer Confidence: Confidence which consumers have in terms of certain phenomenon or product (food they consume), based on their own or other consumers' experience.

Different Levels of General Confidence: Differences in the degree of confidence which consumers have in certain products depending on their own experience, gender, age, education, and other social, economic, demographic, psychological, and other characteristics.

Dimensions of Confidence: Dimensions of confidence which range from fully positive (optimism) to utterly negative (pessimism).

Food Safety Scale: Measuring scale which shows the degree of confidence of consumers in safety of certain food.

Functional Food: Food which biological characteristics have a positive impact on human health and certain physiological functions in order to provide necessary energy and nutrients, thus preventing certain diseases and improving physical condition.

Organic Food: Food produced in a natural way, without genetically modified or chemical additives, pesticides, and chemical fertilizers.

Profiling Consumers: Defining narrow segments of producers based on similarities in food consumption, while the difference between the defined segments is significant and refers to attitudes, values, and perception of food consumption.

This research was previously published in Establishing Food Security and Alternatives to International Trade in Emerging Economies edited by Vasily Erokhin, pages 316-336, copyright year 2018 by Business Science Reference (an imprint of IGI Global).

Chapter 10
Food Waste Management and Corporate Social Responsibility in Indian Food Service Industry:
A Conceptual Analysis

Anish Slath
Panjab University, India

Ashish Nikhanj
Institute of Hotel Management, India

ABSTRACT

India is a land of cultural diversity, every year thousands of tourist travel to India for enjoying the rich heritage of the country. The growth of tourism and hospitality sector has been phenomenal in the recent years. Due to this growth the inflow of tourists has increased. More tourists mean more waste. It is estimated that on an average an in house guest generates one kg of waste every day which is quite alarming figure. Hospitality Industry is an ever growing industry. In hospitality industry, managing generation of daily operational waste is a perpetual challenge. Every year a hotel spends huge amount of money in managing waste keeping in mind the health and safety of the staff the hotels need separate space to store and sort waste. As a practice the waste that is generated in the industry goes to landfill, but the current study is focused on understanding the use of food waste under CSR practices.

INTRODUCTION

Today, hospitality industry has become a colossal phenomenon at global level accounting for GDP, employment generation, social up-liftment, betterment of society and giving a significant boost to tourism. Hospitality is a part and parcel of everyday life. It is present in almost all the fields' weather it is a hospital or a bank. In today's modern world and in tough competition the customer has a lot of choice, but the edge is gained by the companies or industries that are giving the best hospitality to its customers.

DOI: 10.4018/978-1-5225-8063-8.ch010

Hospitality industry recently became a silver lining in the cloud of the world economy, the income of middle-class has suddenly surged up due to which the trend of eating out has changed in short time of this decade. Further the tourism industry has started offering holiday packages which are economical to a low budget tourist also the travel companies are offering travel on EMI thus the travel and tourism industry has seen a boost as a person who could not afford to travel for tourism purpose before is exploring different holiday packages which are falling in his/her budgets.

Way-back in 1970s India had international hotel chains like Holiday Inn The Marriott, Intercontinental which were preferred choice of global travellers. India made significant pace in this field by launching its own domestic hotel brands of international competency. Today India has more National hotel brands than international brands. Hotel IHC, Oberoi, and the ITC group have become a global icon and contributed more to the hospitality industry in India than any other brands. Hospitality sector of India is progressing at a very fast pace. It is one of the most profitable industries which is contributing to the economy of the country. Hotels provide world class services to the guests. One of the important aspects of hospitality industry is that it has managed to attract foreign clients like Accor Group, Starwood hotels, Marriott; Cabana hotels etc. in the past few years and some of the domestic companies which are leading the Indian hospitality sector are ITC Hotels, Bharat Hotels, Indian Hotels Company (IHCL).

India has the potential to become the number one tourist destination in the world with the demand growing at 10.1 per cent per annum, the World Travel and Tourism Council (WTTC) has predicted. The WTO (World Travel Organisation) predicts that India will receive 25 million tourists by year 2015. According to FHRAI the occupancy level of India shifted from 59.3% in 2011/12 to 57.8% in 2012/2013. In 2013/14 the occupancy level progressed to 58.9%. In 2013/14, branded hotel supply surpassed the 100,000 room's milestone. Nationwide existing supply recorded 103,855 rooms as of March 31, 2014, an increase of around 10.2% (9,600 branded rooms) over the previous year. Amid the main cities of our nation Mumbai has the topmost positions in terms of branded room inventory (13,022) in India, shadowed at close quarters by Delhi (12,025) (excluding Gurgaon, NOIDA, Greater NOIDA), Bengaluru (9,877), Chennai (7,105). In a survey conducted in India it was found that, Taj Hotels, Resorts & Palaces had a total of 12,000 rooms in various of its brands, ITC Hotels followed close by, Carlson Rezidor, Starwood Hotels, Marriott International in that order. (Hospitality, Biz India;, 2014).

In 2013 the international tourist's arrival was at 6.9 million and the number went up by 3.54 million in between January to June 2014. Thus an increase of 5.9% was seen when compared with previous year. The top three international source markets for India continued to be USA (15.6%) followed 1 by UK (11.62%) and Bangladesh (7.53%). Further, he suggested that scheme of tourist on arrival visa which may be extended from 11 to 180 countries and a relaxation of mandatory 60 day gap for frequent foreign travellers to India will boost the inbound travel in near future. Domestic travel expenditures generates approximately 80.7% of Direct tourism GDP ($1,145 million) marking an boost of nearly 10% In the year 2012 (Khanna, 2014).

As per the above statistics the hospitality sector is growing with a fast pace. With this growth there is every possibility of generation of more food waste as more guests national and international are travelling across the country to explore India. According to Global Hunger Index, India is ranked 63 out of 78 hungriest countries (Biswas & Tortajada, 2012) which is an alarming figure. Every day tons and millions of food waste is generated by the hotels but there is no policy or practice in place which directs the hotels to manage this waste properly. If all the food waste that is generated by the hospitality sector can be gathered and distributed among the needy, no person in India will go hungry and our rank will improve accordingly. Therefore it is high time to study the proper and effective use of food waste management practices that are followed by the hospitality sector.

OBJECTIVES OF THE STUDY

Not much of literature is available on food waste management practices in hospitality Industry in India. When resources are found they majorly talk about solid waste management and no focus has been laid down on food waste management in these reports or articles. Therefore the objectives of the current work is to

1. Identify practices that are being followed in Indian hospitality industry in the field of food waste management.
2. Review the policies of food waste management in Indian hotels after an extensive review of related literature
3. Find how the hotels are managing this food waste under CSR practices.
4. Understand the Challenges faced for food waste management.
5. Finally recommended some measures of effective food waste management.

METHODOLOGY

The present study has been based on conceptual review of food waste management patterns, challenges faced in food waste management would and different policies in relation to food waste management. In fact, the main focus of the paper is to develop a conceptual framework for the further empirical researches in the less explored area of hospitality management.

CONCEPTUAL ANALYSIS OF LITERATURE REVIEW

"It is estimated that 30–50% (or 1.2–2 billion tones) of all food produced on the planet is lost before reaching a human stomach" (Fox, 2013). Between 2025 and 2050 world population will increase by 20% to reach 8 billion inhabitants and by 2050 the overall population will be 9.5 billion. It must be noted that Asia and Africa will be contributing about 97% in the population growth and it is expected that after 2025 Asia will have more than two thirds of the world population.

Moreover there would be a remarkable increase in the GDP/c in developing countries by 2025 the world production will be doubled by 2050 it will be double of that 2025. Thus this growth in population and economy will lead to exponential increase of waste volumes (Mavropoulos).

All around the world the food is being wasted and thus the nutrition of the food also is wasted but it goes unnoticed that with the waste of food the resources required to produce these foods like water, land energy are also being wasted. As the nations will develop, the per capita increase in food intake will also increase. Thus it becomes important to manage the food waste so that there is optimum use of the key resource used in the production of food.

With the increase in technology the modern agriculture has become easier. There are numerous ways by which a farmer can improve the agricultural use of the land. Out of 10Gha usable land surface as of now globally 4.9Gha of land is being used for food production. Because of the increase in the population more land would be required for growing food thus it will lead to an unstable ecosystem. Therefore it becomes important to control the food waste as minimum food wastage may lead to use of effective use of agricultural land Water is the main resource that is required for the food production over the past decades the concept of use of fresh water has increased rapidly. As of now per year 8.3 trillion m3 of water is being used by humans of whom 70% is being used by agricultural sector and it is expected that this use will increase in the coming years. How much water would be required for food production will totally depend on how the food is being produced? The demand may increase to 10-13 trillion m3 annually by midcentury which will be 2.5 to 3.5 times more than the total fresh water used by humans every day (Fox, 2013).

It is estimated that a guest generates about 1 Kilograms of food waste per night when he/she stays in a hotel .Now imagine if we multiply this number with the number of rooms in a hotel. For optimum utilization of waste generated in the hotels during building stage a lot of space is allocated to public and guest areas the left over space is used for disposing of waste (Patricklalonde, 2013).

Food production areas in the food service industry generate the major chunk of waste which can be organic (food waste peels, bones egg shells etc.) or inorganic (packaging materials tins, card boards cleaning equipments, etc.) in nature, the housekeeping department Is a secondary source from where cleaning materials, chemicals and left over amenities etc. Are generated more over the horticulture department also generates chemical as well as organic waste, further more back of the house departments like maintenance accounts sales, Hr contribute to in organic waste like toners, stationery etc. and in front of the house it is not only the guest rooms in a hotel from where waste originates, public areas like swimming pools public washrooms, banquets, conference rooms etc. are also significant contributors towards waste generation (Arunima, ABHIJIT, & Rubiya, 2012).

In an independent study conducted by WRAP, (Worldwide Responsible Accredited Production (WRAP), an accreditation agency) in the U.K. The report estimated that over 3.4 million tons (mt) of mixed waste is produced by food service organizations (hotels, pubs, restaurants and quick service restaurants) each year. Of this, 1.6 mt (48 per cent) is recycled, reused or composted, while almost 1.5 mt (43 per cent) is thrown away, mainly to landfill. Furthermore, 600,000 tonnes of food waste was disposed of in 2009, two-thirds of which (400,000 tonnes) could have been eaten if it had been better portioned, managed, stored and/or prepared. This has led to increased recycling rates, but land filling waste remains an issue, as 70 per cent of the mixed waste could be recycled, further Hotels very often say that they waste very little food as the plates come back clean but in actual it is not true. As per the study done by WRAP the food waste comes from a variety of sources. The study reveals that 45% food is wastes in food preparation area, 34% food is wasted by customers and 21% of food goes waste by spoilage. (WRAP, 2009) In hotels there are various sources from where food waste is generated which can be seen as in Figure 1.

A study was conducted by the university of florida for hotel waste audit which showed that each guest room contributes from "½ to 28 ½ pound per day" much of which comes from F&B Department, the study estimated that a large hotel may generate 8 tons of waste per day out of which 50% is recyclable (University Of Florida, 2014).

As GHG (green house gases) and climate change damages worth $390 Billion, besides water wasted worth, $172 Billion Cleared forests and eroded land worth, $73 Billion, Biodiversity and pollinators lost worth, $32 Billion, in addition we should add the social cause,$1 Trillion for food that never added

Figure 1. Food waste in hotels by type
Source: http://www.greenhotelier.org/know-how-guides/reducing-and-managing-food-

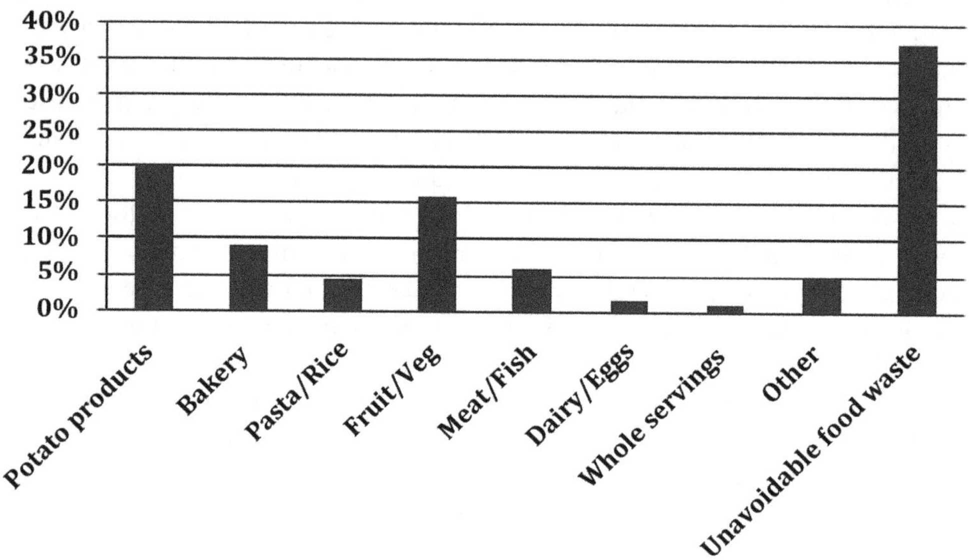

a bit of nutrition to humans, above and beyond the cost of, $153 Billion cost of human health because of the use of pesticides and chemicals, Loss of lively hood as natural resources become more scarce and war induced to get better lands and pressure on resources is waged, $696 Billion and there are many more that cannot be quantified e.g. The cost of wetlands lost (Food and Agriculture Organization of the United Nations, 2014).

"If you don't measure it you don't Manage it" (Lehner, 2013) it is rightly said. It is a prime concern that large amount of food waste goes un-accounted, if a waste management policy is present which measure every ounce of food that goes waste it is every possibility that in future the food wastage can be reduced. Organisations will have enough information with them which will help them identify the various places from where food waste is generated and how much food is wasted from these areas. Thus the figures will help the hospitality industry to reduce the waste up to optimal.

SITUATION OF FOOD WASTE IN INDIA

India is a land of different cultures, festivals and values. According to Hindu religion, food is a gift from God and should be treated with great respect. Despite of all the preaching's and teaching INDIANS waste as much food as the whole of United Kingdom consumes. Up to 40% of the food that is produced in India goes waste says UNDP. According to the agriculture ministry, Rs. 50,000 crore worth of food produced is wasted every year in the country. (Singh, Kanhaiya). It is projected that by 2050 India will generate 150 million tons of waste (Chintan Environmental Research and Action Group).

In India the agriculture farming has seen improvement every year the growth rates of crops has increased, in the year 2005-2006; 208 million tons of food was produced and in the year 2013-2014 an estimated 263 million tons of food was produced. India requires 225-230 million tons of food per year, therefore looking into the India's population growth, food production is clearly not the main issue

(Biswas & Tortajada, 2012). India has always been an agrarian country the food farmers grow enough food surplus to feed the nation the problem is not of food inadequacy the main problem is of food wastage FCI admits that. Approximately 1,94,502 metric tons of food grain was wasted in between 2005 and March 2013 due to various reasons (Hindu, 2014).

At present India has 6,300 cold storage facilities which are spread across the nation. The total food holding capacity of these units is 30.11 million metric tonnes, moreover India actually needs more storage units to stop food waste the current storage units can only store half the amount of food that is produced rest goes to waste. It is estimated that the countries cold storage units should be able to store 61 million metric tons of food which is double the amount that India can store currently. In order to cope up the demand of storing current production levels of fruits and vegetables, by 2015-2016 an investment of Rs 55,000 crore has to be done on making more storage units. Due to lack of adequate cold storage units India is throwing away fresh produce worth Rs, 13,300 crore every year. (Bhosale, 2013) .It is estimated that 21 million tons of wheat is spoilt due to inadequate storage facilities this is equivalent to entire production of food in Australia (Williams, 2011).

This inadequate storage of food and improper management by the government has inflated the prices of fruits. These are sobering figures when one considers that almost 80 million people in Indian cities lived below the poverty line in 2009/10 and 221 million people in India are acutely malnourished (Jackie Fitzgerald, 2013).

Social gatherings in India have become a major source of food wastage. Weddings are ceremonial functions in India where maximum wastage of food can be seen. Approximately 40,000 Marriages are held in Jaipur every year and food waste from these marriages account for more than 25-30% if this wastage is collected it can serve 5000 more marriages. At Bangalore, annually 943 tonnes of food wasted during weddings is enough to serve 2.6 crore people a normal meal and at an average cost of 40/- per meal the food worth Rs 339 crore is wasted. Wastage is more with buffet (22%) than served meals (20%). In Bhubaneswar food waste contributes to 26.63% of the city waste which is directly thrown into the bins and In Jaipur 7500tonnes of food is purchased every day out of which 835 tonnes of food is wasted every day (Agarwal, Vivek S.& Nag, Ambika;, 2013).

Annually 943 tonnes of food is wasted in weddings in Bangalore this waste is sufficient enough to serve 2.6 crore people a normal meal. If calculated in figure the food worth 339 crore is wasted. The food wastage coming out from buffet parties accounts to 22% than the meals that are served. Bhubaneswar throws 26.63% of food waste directly into waste bins (Agarwal, Vivek S.; Nag, Ambika, 2013). Nations Capital New Delhi generates 8000 metric tons of waste per day which is the largest in the country and by 2020 this waste is expected to increase by 23,000 tons. (Chintan Environmental Research and Action Group).

According to the study conducted in Haryana state the pie chart shown in Figure 2 explains the type of waste generated by the hotels. It is very clear from the pie chart that 60% waste generated is solid in nature i.e. non-biodegradable, 30% is wet waste i.e. degradable and 10% is Gaseous Waste (Malik, Sandeep; Kumar, Sanjeev;, 2012).

A study was conducted by Mr. Amar Nath in New Delhi on 8 hotels to find out the waste generation done by them. As per the findings some alarming figures have come out. It is very clear from the study that hotels generate maximum amount of food waste as compared to other inorganic waste. On an average a hotel generates 280.6 tons of waste per year and if combined all the hotels in Delhi would generate

Figure 2. Amount and type of waste generated by hotels
Source: http://prj.co.in/setup/business/paper69.pdf

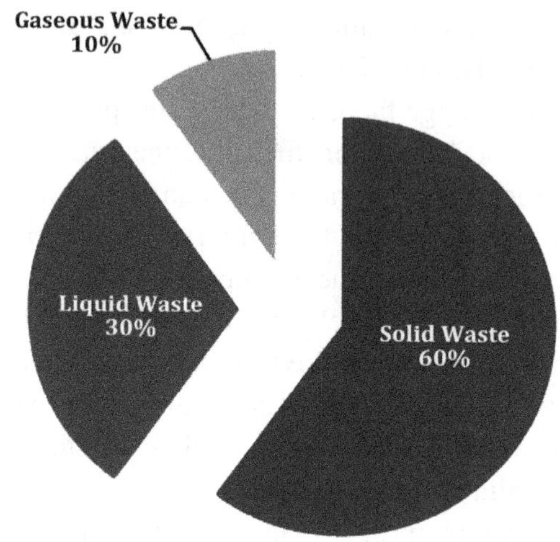

about 26400.60 ton of waste per year. As per Table 1, it can be seen that these 8 hotels in totality generate 3825.36 kg of waste per day and in totality 1,39,6256.40 Kg of waste per year. These figures tell us that hotels are major contributors of generating food waste (Nath Amar, 2014).

Recently in India the disposable income of the families has gone high due to which even a smallest of the function is being celebrated in the hotels. Hotels have got no other option than to provide services and food to their guests. Due to the increased number in the functions taking place in hotels the generation of food waste has also gone high.

It is a national shame that despite of being the largest producer of milk and second largest producer of fruits and vegetables, India is the world's biggest food waster. As a result the cost of fruits and vegetables is twice as it should be and milk costs 50% more than it should be.

Table 1.

Name of the Hotel	Average Organic Waste Per Day in KG	Total Organic Waste Per Year in KG
Hotel La-Shangri-la New Delhi	1559.16	569093.40
Hilton Hotel New Delhi	291.00	106215
Country Inn Hotel New Delhi	900.00	328500
Park Plaza New Delhi	121.00	44165
Jaypee Vasant Continental New Delhi	493.00	179945
Jaypee Green Golf and Spa Resorts New Delhi	117.60	42924
Jaypee Siddharth Hotel	112.60	41099
Lemon Tree Hotel	231.00	84315
Total	3825.36	1396256.40

Source: http://www.jiit.ac.in/uploads/ThesisAmanath.pdf

FOOD WASTE MANAGEMENT POLICY OF INDIA

In a daring move the government in Delhi has ordered hotels to donate any waste food that is edible or untouched to local food banks and allowing changes in menu by allowing guests to order half portions (IBN LIVE, 2013).

IFBN in India is trying to tie up with community groups, NGOs and government agencies so that the food can be forwarded to the needy (Alvi, 2012).

Mumbai's dabbawalas have also taken a huge step and initiative of their own conscience simply by letting its uses put a share sticker on their "DABBA" (tiffin) while bringing it back so that those tiffins that have leftovers be shard and feed the hungry destitute children in their local areas like in the Crawford market station in Mumbai dabbawalas feed an average of 20 children every day (Bhattacharya, 2013).

The hotel worked to accurately forecast business levels to reduce food waste. Additionally, the hotel donates untouched whole-fruit items daily to a local charity called the Food Link Foundation. The hotel also gives used food waste to two local organizations to be used as fertilizer and fish feed, and used cooking oil goes to local company Dynamic Progress for conversion to biodiesel (Green Building and Design Magazine, 2014).

Most of the hotels of the Carlson Rezidor, Marriott International and the Fairmont chain of United States distributes undefiled food from exhibits and trolleys (Baker, 2014) Buffets in restaurants are the main source of food wastage which is used to as staff food or if in excess it is used to feed people in slum as a part of the CSR initiative (Agarwal, 2015).

Tata Subsidary Indian Hotels Limited Clearly outlines Its Policy on Food waste to be composted wherever hotels would have space (Tata group, 2010).

CSR AND FOOD WASTE MANAGEMENT BY HOSPITALITY INDUSTRY

Each year millions and tons of food leftovers are thrown in the garbage. It is a pain to see all the edible food going into trash. What can be done with this food? How can it be utilised are some of the questions that often crosses minds of people associated with Hospitality Industry. In this section of the paper the researcher has tried to identify what the hospitality industry is doing with the leftover food.

Hotel Lalit Ashok, Kumara Krupa Road Bangalore is an initiator of green practices. This hotel follows all the norms of waste and energy management. After Segregation of waste the biodegradable waste is composted in vermicompost onsite and non biodegradable waste is segregated and recycled. ITC Gardenia, Vittal Mallya Road Bangalore is another example of using food waste effectively. It has an onsite organic waste converter which converts the food waste into kitchen manure which is later used in the kitchen garden of the hotel. Hotel Leela Palace, Old Airport Road Bangalore donates its food waste to the poultry farm in Marathahalli for feed (R. S. Rohella et. al).

BLOCKADES TO FOOD WASTE MANAGEMENT

Hereby saying "diversion" we mean how much of food waste that has so much energy and potential not be sent to landfills and further problems for our shrinking resource chain as outlined in the EPA's Food Waste Recovery Hierarchy (Figure 3).

Figure 3. Food waste Recovery Hierarchy from most preferable to least preferable
Source: http://www.epa.gov.in

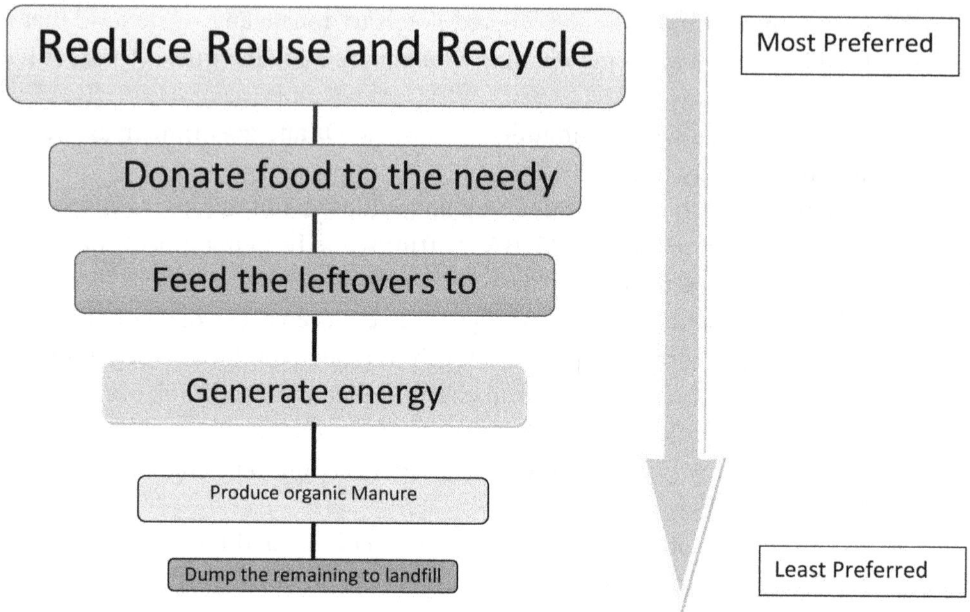

We must first reduce the amount of trash by firstly reducing what we can in terms of waste generation somehow even after reduction we still are left over with food waste we should try to donate as much as possible (in case of a country like US one could even avail tax benefits) on the other hand it is important to understand the blockades to this necessary diversion it might be as simple as lack of facility or infrastructure, funds, education, or lack of Samaritan laws.

Some of the Barriers to effective food waste management are as follows:

1. Leftover food is a perishable commodity and it needs to be handled with care and precision and has to be stored at a proper temperature which hotels think of as a liability. Hotels do not believe in storing leftover food rather it goes into the dustbins.
2. In sufficient cold storage in the food banks is the next hurdle that is faced. due to Insufficient *space and refrigeration,* Food Banks cannot accept large quantities of food left over thus a major chunk of food waste goes to landfills.
3. The cost for transporting food waste is very high it is always a question who will bear this cost the food bank or the hotels. Further laws pertaining to transfer of food waste are stringent which again is a hurdle in managing food waste.
4. A semi processed food or the food which is not good in taste but good to eat generally goes into landfill as the food manufacturer is afraid to donate this food because the hotel doesn't want be party to the damage to livestock after eating this semi processed /mis formulated food.
5. Lack of govt policies that protect good doers in case of legal liability.
6. Some amount of confusion has been added to the trivia of Good food past saleable date in case of use by sell by dates.
7. Apart from the other concerns there is a concern that what food banks might or might not accept.

8. Inability on part of caterer to donate food without prior approval of the party involved. Here a legislation is required to give teeth to catering units to take care of their own CSR.
9. Apart from legal there are some financial implications as well for example the decision to write off at the end of a financial year.
10. Justifying fair market value in case of unconventional donations (especially when there are tax benefits).
11. A very different and potent blockade is the blockade of company policy in this scenario more food is wasted as the company is not able to donate it in the absence of a clear policy directive (example of hotel not donating food but composting it for legal or financial matters.
12. Educating the employees becomes very important in the implementation of CSR in the case of food and that includes in the case of our country segregation of waste in the catering industry and not just reducing workload by disposing it off, only good branded hotels and QSR chains are implementing this segregation of waste policy that becomes a milestone in implementation of CSR in this regard.
13. Human psychology also plays as a blockade in the implementation coupled with the above point as humans do not change easily but consistent reminders and education would help.

SUGGESTIVE CSR FOOD WASTE PRACTICES BY HOSPITALITY INDUSTRY

Hospitality Industry is the major source of food waste generation. What happens to the millions of tons of food waste that is generated by the industry, does it go to landfill? Is it donated? Or is it used to generate energy? These are the questions that often cross someone's mind when talking of food waste initiative in the hotel industry.

According to the researcher's some of the best practices that may be useful in tackling food waste are as given below:

1. Managing of food waste on the site: one of the practices that a hotel can follow is preventing the food waste on the source. At every stage of food preparation if just simple steps are followed it might lead to overall prevention in food waste.
    a. The first stage of any food production cycle is purchasing. A hotel should see that it does not over order the raw materials required for food preparation at any cost.
    b. Effective storing of food items is very essential in reducing food waste. It should be ensured that the food items being stored in hotels are properly labelled for expiry dates. The system of FIFO and LIFO should be practiced. There should be a separate storage units for dry and perishable food products. To increase the shelf life of foods vacuum packaging can also be considered.
    c. The next area in a hotel from where food waste is generated the maximum is its kitchen. A chef of the hotel should ensure that proper food cutting techniques are being followed in the kitchen there should be no over trimming of food peels and maximum yield is taken out from the food stuffs. Only prepare food when required or ordered to avoid food wastage by over preparation. If pre cuts of meat and fish are ordered they can help reduce the problem of meat trimmings, as disposal of meat trimmings is difficult. Standard recipe should be followed to avoid food wastage and adjust the portion size of the food that goes on the plate.

d. Training the manpower to effectively dispose of the food waste is an important aspect in any food waste management practice.

2. Implementation of Samaritan Law that to this date does not exist in our country and proves to yet a big hurdle in the CSR activities concerning wasted food.
3. **Awards Achievements and Recognitions:** Every year in the hospitality industry we gives away awards for best hotels for their green practices but no award is given for the hotels that are following the practice of donating food waste to the charities or piglet farms. The hospitality industry should also look into this aspect of recognition which is also a motivator for a hotel and as a result the hotels will start following the practice of effective food waste management under CSR practices.
4. **Benefits in Taxes:** If it is encouraged by the government it is every possibility that whole of the hospitality industry will follow effective food waste management plans.
5. Food Banks: It is high time to have more food banks that are operational across the country. These food banks are doing a tremendous work on collecting and distribution of food waste.
6. **Government Support and Policies:** None of the practices or policy can be successful if the government does not give a helping hand. There should be a proper support from the government for the people who are using food left over's for good purpose. Various campaigns should be run by govt. in which the issue of food wastage should be covered.
7. **Food Waste Disposal Audit:** Each year hotel does different kinds of audit like HR audit, Financial Audit etc., but none practice the audit of Food waste disposal. If a hotel follows the practice of food waste audit *remember if you can measure only then you can reduce.*
8. **Generating Energy from Food Waste:** Food waste produces methane gas which is harmful if released in the air, causing 21 times more green house effect than co2, but if this gas can be trapped it can be used as an essential fuel to light up the gas ranges of the kitchen.
9. **Donating Food Left Over:** Every possible effort should be made by the hotels to donate the food that is untouched and healthy for human consumption.

CONCLUSION

It can be observed from the above literature that hospitality industry in totality is doing something or the other to effectively use the food waste management under CSR practices but very few hotels in India are following this practice most of the hotels in India send the food waste to landfills rather than to donating the food waste to the needy or to live-stocks. If we are able to reduce our food waste we would be creating profit vis a vis CSR. Due to India's subtropical temperatures and rapid fruit spoilage, and insufficient cold storage units the food is getting wasted every year. Apart from this basic lack of a Samaritan law in India that saves the cost of legal hassles after food donation seems to be one of the primary concerns of food wastage as the brand image of hotels plays an important part in its survival. Any undue legal hassles or defibrillation because of food borne illnesses caused by inappropriate food handling Fed to the needy could land the hotel chains in trouble. Not only this even a small FNB outlets would not dare to step into the shady world of reusing food. It was observed from the study that disposing food was much easier than donating it the best that hotels did was to create compost of that food waste only if the hotel had land available for that.

The government should step in and give proper support to the hospitality industry for managing food waste. Necessary training programs should be run by government to inform help the hospitality industry understand the importance of managing food waste as it is not only the food that is being wasted along with it also goes waste the resources that were used for making food. The study has given some alarming figures in which it can be seen that how much waste is being generated by the hotels. It is high time for the Industry and the policy makers to join hands and tackle this menace of food waste going to landfill. The study is totally conceptual thus generalising it would not be possible an empirical study in this area will give further insights of food waste management as CSR practices in India.

REFERENCES

Agarwal, & Nag. (2013). Sustainable Food Waste Prevention Strategies to Achieve Food Security in India. *International Journal of Agriculture and Food Science Technology*, 189-194.

Alvi, N. (2012, June 9). *Times of India Delhi*. Retrieved May 29, 2015, from http://timesofindia.indiatimes.com/city/delhi/Dont-throw-away-party-leftovers/articleshow/13943807.cms: http://timesofindia.indiatimes.com

Amar, N. (2014). *Profitability and sustainability from waste management practices in hotels and its impact on environment*. New Delhi: Jaypee Business School.

Arunima, K. E., Abhijit, R., & Rubiya, N. (2012, January 8). *Hotel Waste Management Rain Drops*. Retrieved May 25, 2015, from https://businessimpactenvironment.wordpress.com/2012/01/08/hotel-waste-management-rain-drops/: https://businessimpactenvironment.wordpress.com

Baker, C. (2014). *Waste Management World a welcome sign: Hotels adopt reuse and recycling*. Retrieved May 29, 2015, from http://www.waste-management-world.com/articles/print/volume-7/issue-7/features/a-welcome-sign-hotels-adopt-reuse-and-recycling.html: http://www.waste-management-world.com

Bhattacharya, S. (2013, June 9). *The National, World*. Retrieved May 29, 2015, from http://www.thenational.ae/news/world/south-asia/mumbais-dabbawalas-begin-deliveries-to-hungry-children: http://www.thenational.ae

Bhosale, J. (2013, November 28). *The Economic Times*. Retrieved May 27, 2015, from http://articles.economictimes.indiatimes.com/2013-11-28/news/44547246_1_cold-storage-facilities-wastage-cold-chain-infrastructure: http://articles.economictimes.indiatimes.com

Biswas, A. K., & Tortajada, C. (2012, August 12). *World Economic Forum*. Retrieved May 26, 2015, from https://agenda.weforum.org/2014/08/india-perishable-food-waste-population-growth/: https://agenda.weforum.or

Chintan Environmental Research and Action Group. (n.d.). Retrieved May 26, 2015, from http://www.chintan-india.org/documents/fact_sheets/chintan_waste_tales_fact_sheet.pdf: http://www.chintan-india.org

Fitzgerald, J. (2013, September 3). *Ricardo-AEA*. Retrieved May 26, 2015, from http://www.ricardo-aea.com/cms/will-india-s-food-waste-help-fuel-their-economic-growth/#.VWVg9BGJjVI: http://www.ricardo-aea.com

Food and Agriculture Organization of the United Nations. (2014). *Food wastage foot print -youtube*. Viale Delle Terme Di Caracalla.

Fox, T. (2013). *Global Food: Waste Not, Want Not*. London: Institute of Mechanical Engineers.

Green Building and Design Magazine. (2014, May/June). *Green Building and Design Magazine*. Retrieved May 29, 2015, from http://gbdmagazine.com/2014/27-intercontinental-hong-kong/: http://gbdmagazine.com

Hindu, T. (2014, February 12). *The Hindu*. Retrieved May 26, 2015, from http://www.thehindu.com/news/national/fci-admits-194-lakh-mt-foodgrain-wasted-between-200513/article5680994.ece: http://www.thehindu.com

Hospitality, B. I. (2014, September 9). *Hospitality BIZ*. Retrieved May April 20, 2015, from http://www.hospitalitybizindia.com/detailNews.aspx?aid=20171&sid=41: http://www.hospitalitybizindia.com

IBN Live. (2013, August 13). *IBN live*. Retrieved MAY 3, 2015, from http://www.ibnlive.com/news/india/govt-directs-5-star-hotels-to-donate-leftover-food-offer-half-portions-to-guests-631019.html: http://www.ibnlive.com

Khanna, A. (2014). *Hotels in India Trends and Opportunities*. New Delhi: HVS.

Lehner, P. (2013, March 02). *Recipe for cutting food waste*. Academic Press.

Malik, & Kumar. (2012). Management of hotel waste: A case study of small hotels of Haryana state. *Arth Prabhand: A Journal of Economics and Management*, 43-55.

Mavropoulos, A. (n.d.). *Waste Management World*. Retrieved May 24, 2015, from http://www.waste-management-world.com/articles/print/volume-11/issue-2/features/waste-management-2030.html: http://www.waste-management-world.com

Patricklalonde. (2013, April 17). *Stop Food Waste*. Retrieved May 25, 2015, from https://hospitalityfood-waste.wordpress.com/2013/04/17/food-waste-in-the-hospitality-industry/: https://hospitalityfoodwaste.wordpress.com

Rohella, R. S., Panda, S. K., & Das, P. S. (n.d.). *Hotel Food Scraps Go to the Animals-Reduces Disposal Cost and Saves Environment*. Retrieved June 2, 2015, from http://pubs.sciepub.com/env/3/2/3/: http://pubs.sciepub.com

Singh, K. (n.d.). *The CSR Journal Sustainability News for India*. Retrieved May 26, 2015, from http://thecsrjournal.in/food-wastage-in-india-a-serious-concern/: http://thecsrjournal.in

Tata Group. (2010, April). *Green Practises An Indian Hotel Company Initiative*. Retrieved May 29, 2015, from http://www.tata.co.in/company/articlesinside/vzAtfIWNGNo=/TLYVr3YPkMU=: http://www.tata.co.in

University of Florida. (2014). *Managing waste in hotel and motel operations*. Retrieved From: http://infohouse.p2ric.org/ref/13/12170.pdf

Williams, D. (2011, August). *Institute of Mechanical Engineers*. Retrieved May 26, 2015, from http://www.imeche.org/knowledge/industries/energy-environment-and-sustainability/news/WorldFood: http://www.imeche.org

WRAP. (2009). Retrieved from http://www.wrap.org.uk/content/composition-waste-disposed-uk-hospitality-industry-1

WRAP. (n.d.). *Overview of waste in the hospitality and food service sector*. Retrieved May 26, 2015, from http://www.wrap.org.uk/content/overview-waste-hospitality-and-food-service-sector: http://www.wrap.org.uk

This research was previously published in Corporate Social Responsibility in the Hospitality and Tourism Industry edited by Lipika Kaur Guliani and Syed Ahmad Rizwan, pages 92-105, copyright year 2016 by Business Science Reference (an imprint of IGI Global).

Chapter 11
Consumer Co-Operatives and Perceptions of Food Safety:
Shaping Markets in Post-Fukushima Japan

Catherine Burns
Griffith University, Australia

Kumiko Katayama
Griffith University, Australia

Robin E Roberts
Griffith University, Australia

ABSTRACT

This paper examines consumer co-operatives and members' perceptions of food safety. Japan is an ideal place to study given it is undoubtedly the 'best example of a successful consumer co-operative sector in the postwar period' (Birchall, 2002, p. 79). While some co-operatives have evolved into a considerable political force, not all consumer co-operatives are as large or as politically active. This study qualitatively explores the views of the members of two small, less politically active co-operatives in Tokushima. Of particular relevance are the types of produce being consumed by members, and why (and how) purchasing behaviour has been shaped by food safety concerns, post-Fukushima.

INTRODUCTION

Japan has been described as the 'land of co-operatives' (Thompson, 2008). According to the International Cooperative Alliance (2012), the presence of Japanese co-operatives in daily life is unsurpassed by any other country, and their economic impact is substantial. Within Japan's co-operative sector, 74% are consumer co-operatives, constituting 80.4% of combined annual sales. Japanese consumer co-operatives consist of more than 27 million members, comprising 49% of total households (Kurimoto, 2015) and an annual turnover of US$25 billion[1] in 2014 (JCCU 2015). Specialist in socio-economics and the global co-operative movement, Johnston Birchall (2002, p. 72) thus claims, 'The Japanese co-operative move-

DOI: 10.4018/978-1-5225-8063-8.ch011

ment is, on several criteria, the most successful in the world.' The significance of Japan's consumer co-operatives also extends beyond the domestic market. They are significant to global suppliers because, as overall third largest retailer (Thompson, 2008), they are both a competitive force in the Japanese retail food sector and a distribution network for imported products. As such they have demonstrated strength in preventing importation of foods that did not meet specified standards (even though compliant with World Health Organisation standards) (Jussaume, 1991, p. 33).

While Japanese co-operatives boomed in conjunction with Japan's economic success in the 1980s, the proceeding economic downturn, misnamed Japan's 'lost decade' according to Metzler (2012), prompted predictions of decline (Jussaume, 1991; Ashkenazi and Jacob, 2000; Jussaume, Hisano, and Taniguchi, 2000; Parker, 2005). Yet these predictions have not been realised. Membership has expanded and revenue has continued to increase, albeit at modest rates of approximately two to three per cent per year. A major reason for the continued growth in Japanese consumer co-operatives is concern for food safety and security. Numerous food risks resulting from early industrialisation, imported foods, and domestic scandals or disasters (most recently the Fukushima nuclear reactor disaster in 2011), have reinforced perceptions of a 'Precarious Japan', the title of Japan scholar, Ann Allison's most recent book (2013). These factors have drawn attention to the problems of food safety and security. Consumers have become acutely aware of the fact that 'Japan is the world's largest importer of food and approximately 60% of Japanese calorie intake comes from imported food' (Tanaka, 2008, p. 574).

The strength and uniqueness of Japanese consumer co-operatives has resulted in an emerging body of literature, but surprisingly little attention has been paid to the motivations and views of members, despite the fact that they own these businesses (highlighted in Birchall, 2011) title and preferred term of 'Member-Owned Business' rather than 'co-operative'). Much of the existing literature also focuses on a select few very large, politically active co-operatives, which have been treated as exemplars. The purpose of this study is to explore these gaps to garner implications for food marketers. We examine the perceptions and purchasing practices of consumer co-operative members located in a small regional city, with particular attention to meanings of food safety following the Fukushima nuclear accident.

The first section of the paper provides a focused literature review on Japanese co-operatives, food safety, and concepts and policies informing the localisation of food consumption in Japan. This is followed by an outline of the research methodology then a presentation of qualitative findings and implications.

LITERATURE REVIEW

The Rise of Japanese Consumer Co-Operatives

The history of Japanese co-operatives is now well documented (Jussaume, Hisano, & Taniguchi, 2000; Saito, 2010; Vacek, 2011) so only a few salient points will be outlined to provide background context for our analysis of contemporary issues. Japan's first consumer co-operative was formed in 1879, inspired by Great Britain's Rochdale Equitable Pioneer Society (Rochdale Cooperative) established in 1844. However, in unique contrast to European co-operatives (which were largely driven by the demands of civil society), Japan's early co-operatives were founded within the elite and middle classes (Saito 2010; Vacek, 2011).

Japan's post-war co-operatives began to reflect a stronger civil society, formed in response to food safety concerns and pollution scandals accompanying Japan's rapid industrialisation and urbanisation from the early 1950s. With increasing affluence and an industrial relations structure privileging white-collar male workers, emerged a new urban phenomenon: full-time or 'professional housewives' (Vogel, 1978). There is widespread recognition that it was this group of women who formed the backbone of consumer co-operatives from the 1960s. There is now a small body of feminist scholarship on Japanese co-operatives, but most of the literature focuses on the Seikatsu (Lifestyle) Club Consumer Co-operative (SCCC), a uniquely political, urban-based consumer co-operative with its headquarters in Tōkyō (Gelb and Estevez-Abe, 1998; Lam, 1999; LeBlanc, 1999; Grubel, 2004; Jo, 2011; Kimura, 2013). However, little is known about regional consumer co-operatives and those less politically engaged. This study therefore aims to explore the consumer activities and motivations underlying membership in two co-operatives in regional Japan.

Co-operatives range from large multi-function prefectural organisations, with shops the size of large department stores, to *han*: small but very significant groups of four to seven households, who place a weekly collective order to be delivered to a local co-operative distribution centre or house. The *han* joint-buying strategy is unique among global co-operatives (Birchall 2002), providing the basis for the emergence, growth and business success of Japanese co-operatives. The chief purpose of joint-buying is to achieve economies of scale, lowering costs and maximising freshness of produce. While this is attractive to consumers and thus an 'effective marketing strategy' (Jussaume, 1991, p. 30) for the co-operatives, the reduced storage requirement and delivery cost is a major economic advantage.

While cost is important, the popularity and clean image of the *han* system is also due to its focus on *sanchoku*: the delivery of primarily fresh eggs, milk, vegetables and fruit purchased directly from farmers. Through *sanchoku* emerged a discourse associating proximity with food safety.

From a business point of view, Birchall (2002, p. 81) suggests that to a large extent, Japan's co-operative innovations solve the universal problem of 'how to restructure into larger organisations without losing touch with…members.' *Han* organisation and leadership structure facilitate a two-way flow of information and feedback between households and the Japan Consumer Co-operative Union (JCCU).

Much of the research on Japanese consumer co-operatives attributes their success to flexible marketing strategies (e.g. through the *han)* that help them adapt to changing socio-economic environments (Jussaume, 1991; Jussaume, Hisano, and Taniguchi, 2000). Vacek (2011) and Saito (2010) note that the development of high quality, low priced 'co-op' branded goods in the 1960s made a substantial contribution to co-operatives' growing popularity and reputation, in the wake of increasing consumer prices and concerns about food safety (discussed below). Annual growth rates for co-operative products (the majority of which were 'co-op brand') increased from 22% in the period 1966-68, to 45% in 1971 (Vacek, 2011, p. 1058). Throughout the 1980s, sales from co-operatives doubled and sales from *han* increased more than six-fold (Saito, 2010, p. 69). In another astute shift to accommodate increasing consumer demand, co-operatives were the first retailer to sell organic produce in Japan. However, some suggest that because of more recent competition from supermarkets and other retailers selling organic foods, co-operatives may have lost their advantage: in the early 1990s, co-operative membership began to level off (Maclachlan, 2002; Parker, 2005; Reiher, 2012).

Another significant factor in the lack of growth in Japanese co-operative membership is women's increasing engagement in the paid workforce, limiting their ability to commit to *han* leadership and general involvement. It has also been suggested that the decline of the *han* was due to increasing individualism in Japanese society and a diminishing sense of community (Jussaume, 1991). Co-operatives responded by shifting towards individual delivery. Thus, while *han* joint-buying had fallen by 27%, and store sales decreased by 9%, individual delivery rose from 22% to 46% between 2000 and 2005 (Saito, 2010, pp. 128-29). This shift may reflect changing motivations for joining co-operatives.

Although the original motive to establish consumer co-operatives was more competitive prices, the literature confirms that this is no longer the case. JCCU (2015) surveys show the most common reason for joining a co-operative is to buy safe produce, particularly foods with minimal use of petrochemicals. Saito (2010) also found that the younger generation have begun to emphasise convenience in their co-operative experience.

Food Safety and Security

Like most other countries, Japan has experienced numerous food scandals and the importance of food security has increased with trade liberalisation. Food security demands a degree of self-sufficiency in food production, usually measured in terms of per capita daily caloric intake. As already noted, Japan ranks among the lowest in the world. International negotiations, such as the GATT Uruguay Round Agreement, have had a profound impact on Japan in terms of increased agricultural imports (Kimura 2011) and this is likely to continue under the new Trans-Pacific Partnership. This is problematic as imported food carries a risk of contamination and failure to meet domestic food standards, exacerbating safety concerns.

Academic interest in food safety and security in Japan has increased with the opening of Japan's markets. Numerous food safety scandals, prompting citizen campaigns and political responses, are now well documented (for an overview see Jussaume et al., 2000). Etched into Japan's recent historical memory, and foundational to many consumer campaigns, are two court cases involving industrial pollution. The cases involved deaths from cadmium contaminated rice and mercury poisoned fish in the 1950-60s. Frank Upham (1987) provides a comprehensive analysis of the cases. Other scholars have highlighted the significance of the 1955 case of Morinaga milk powder contaminated with arsenic which resulted in 12,000 victims of poisoning, including many babies and young children (Assmann, 2013). This case served as a catalyst for the formation of Seikatsu Club Co-operative (Maclachlan, 2002; Rosenberger 2009). Consumer concerns were subsequently heightened by numerous other food safety scandals and issues involving mislabelled food products, chemical additives, and the use of pesticides and antibiotics (Maclachlan, 2002; Mah, 2010; Reiher, 2012; Matsuo, 2013; Kurimoto, 2015).

In Japan, like most countries, social anxiety becomes particularly acute when scandals involve imported foods. Fernandez-Ferrin et al. (2015) point out that while it is now well established in the marketing literature that generally consumers' food judgements are influenced by a product's country-of-origin, 'consumer ethnocentrism' (moral appropriateness of buying imported foods) and 'consumer animosity' (anger towards a specific country and its products) are useful concepts in better understanding purchasing behaviour. For instance, the media generated 'food terror' (Rosenberger, 2009) following the 2008 case of poisoned dumplings imported into Japan from China provides evidence of consumer animosity (see also Kimura, 2011). Our study aims to draw on these two concepts to further understand why co-operative members choose to avoid certain imported foods.

The most recent 2011 Fukushima nuclear reactor disaster has prompted wide critiques of the government's monitoring mechanisms, information delivery and adequacy of its food radiation standards (Assmann, 2013; Bachev and Ito, 2013; Frid, 2011; Kimura and Katano, 2014). Kimura's (2013) work documents the social confusion and anxiety that escalated as the Japanese public realised that domestic radiation standards were ad hoc, and that the government's 'provisional standards' appeared improvised and inadequate. Moreover, consumers received mixed messages: while sales of leafy vegetables, milk and beef from the area were soon banned, the government attempted to diffuse escalating consumer anxiety by making repeated statements that 'consumption of contaminated food 'will not pose an immediate threat to health'' (Kimura, 2013, p. 12). Over the next few years, the Japanese government (and media) endeavoured to minimise concern and encourage support for Fukushima and neighbouring economies by implementing radiation limits more stringent than international standards (Bachev and Ito, 2013) and by alleging consumer panic was based on harmful rumours (*fuhyō higai*). However, many farmers remain reluctant to return to the Fukushima area (Bachev and Ito 2013) and more critical consumers continue to seek independent sources of information and avoid agricultural products from the contaminated areas in the Northeast (Kimura, 2013). Existing literature is yet to identify the broader socio-economic impact of the Fukushima disaster in regional areas beyond the Northeast.

Another major event that shaped food safety governance and prompted a new discourse on food safety was the domestic outbreak of bovine spongiform encephalopathy (BSE), or mad cow disease, in 2001. Tanaka (2012) argues that the public outcry prompted the emergence of a new discourse: *shoku no anzen anshin* (food safety). She explains:

Anzen... signifies technical, or 'objective,' aspects of food safety while anshin... stresses a social and psychological, or 'subjective,' sense of security that arises from one's confidence and trust in the social and material environments in which one finds himself or herself. To secure both anzen and anshin, the Japanese government enacted the 2003 Food Safety Basic Law and established the Food Safety Commission (Tanaka 2012, 568).

Tanaka reveals through her analysis of Cabinet discussions that legislators agreed the parameters of risk assessment should reflect scientific and technical aspects of food safety (*anzen*) as well as psychological and emotional aspects of food safety (*anshin*). The discussion also suggested some legislators recognised an opportunity to reinforce *anshin,* or peace of mind concerning food, by enhancing localised food systems and trust through developing consumers' 'face-to-face relationships' with producers. This association of local (domestic) and safe, in contrast to foreign (imported) and risk, is now the focus of a number of closely related studies on Japan's 'eat local movement' (*chisan chishō*) and Japan's food education (*shokuiku*) policies (Rosenberger, 2009; Mah, 2010; Kimura, 2011; Reiher, 2012).

Eat Local Movement: Chisan Chisho and Shokuiku

The 'eat local movement' and the embedded concepts of *chisan chishō* and *shokuiku* are crucial to understanding food safety and security responses in Japan and the consumer co-operative landscape. There is limited scholarship exploring the concept of *chisan-chishō*, translated as 'locally produced-locally consumed' (Nishiyama et al., 2007; Usui, 2007; Kimura and Nishiyama, 2008; Rausch, 2015). *Chisan chishō* is both a concept and set of practices focused on the localisation of food. It includes both distribution and consumption, encapsulated in the purchasing practice of *sanchoku. Chisan chishō* emerged as a grassroots movement responding to perceptions of food crisis, including the long-term decline of both rural communities and agriculture, and to more recent food scandals that resulted in a loss of consumer trust in imported foods (Kimura and Nishiyama, 2008).

Kimura and Nishiyama (2008) outline five typical claims made by the *chisan chishō* movement. First, local food equates geographical proximity with food safety and trust because consumers establish a direct affinity with the face of the farmer Secondly, *chisan chishō* links local food with improved self-sufficiency and thus local economic growth. Third, geographical proximity is associated with environmental sustainability due to the reduced ecological costs involved in transport and energy use. Fourth, local food supports a unique traditional food culture that reinforces family and community. Finally, local food constitutes a healthier diet in contrast to westernised, fast and/or mass-produced food.

One might expect that *chisan chishō* was a citizen-based response to the negative social and economic impacts of the globalised industrial agri-food business. However, while *chisan chishō* was initially established on this basis, it did not develop into an anti-globalisation movement. In contrast, the popularisation of *chisan chishō* has been the result of its incorporation into national policy to address food security, evident in the Revitalization Plan for Food and Agriculture (2002). It remains a largely government directed program in association with the Japan Agricultural Cooperative (JAC). Although its name may suggest otherwise, JAC is primarily concerned with the promotion of large-scale industrial farming and almost all farmers are members. As Kimura and Nishiyama (2008, p. 55) note, the 'Localization of food consumption and production would go against its long-held tenet of modernization and mechanization.'

Consumer co-operatives embody the overall concept of *chisan chishō* and are therefore supported by government. *Chisan chishō* activities fall into three areas: raising general community awareness, marketing, and school education (Kimura and Nishiyama, 2008, p. 53). It is here that the nexus between *chisan chishō* and *shokuiku* is evident.

Shokuiku translates directly as 'food education.' While the concept is not new, it was reinvigorated in the context of emerging concerns of two main food-related 'crises': diet-related health problems (e.g. hypertension, diabetes and obesity) and widespread concern about declining food safety and quality. This concern is focused on the over-consumption of Western foods, which have long been linked to diet-related health problems (Takeda 2008; Kimura 2011). The debates have attracted considerable scholarly attention. Through a close examination of the policy background, Mah (2010) argues that food safety and security (*anshin* and *anzen*) concerns drove the promulgation of a new 'food regime', beginning with the establishment of the Committee of Food Education Promotion in 2003, followed by the Food Education Basic Law, enacted in 2005. In addition, under the national 2006 Food Education Promotion Basic Plan, local governments became required to implement their own food education promotional plans (Kimura, 2011; Shokuiku Promotion Policies, 2014). The Food Education Law defines *shokuiku* as the 'acquisition of knowledge about food and nutrition, as well as the ability to make appropriate decisions through practical experience with food, with the aim of developing people's ability to live on a healthy diet' (MAFF, 2016).

The limited English language literature on *shokuiku* reflects a largely critical view of the *shokuiku* definition, the Food Education Law, and its associated policies and campaigns. All agree that the food regime clearly links food safety with food security. The Law and the associated campaign promote the image of domestic food as safe and encourage its consumption, especially rice (Articles 7 and 8; see also Reiher, 2012). The discursive link between rice, nationalism and identity has a long and established history in Japan (Ohnuki-Tierney, 1993), and many scholars note its basis in ethnocentric practices and perceptions. Rosenberger (2009, p. 254) similarly draws attention to the moral undertone of the *shokuiku* campaign with its emphasis on teaching children to appreciate Japanese and local regional food in particular, and the media's re-inscription of 'food as morally meaningful in terms of nation, native agricultural land, family safety, and individual health.' Similarly, the Food Education Law's reinforcement of anachronistic images of Japanese society, especially in relation to notions of family and gender relations has also been critiqued (Kimura, 2011; Reiher, 2012).

A number of scholars note that the Food Education Law shifts responsibility onto individuals and families to make 'right' food choices in terms of nutritional balance and eating habits (e.g., eating with family and not skipping breakfast) while ignoring the responsibilities of food corporations, particularly supermarkets and chain-stores (Kimura, 2011; Reiher, 2012). Kimura (2011) highlights two shortcomings of *shokuiku* policies. She argues that Japan's privatisation of the training of food educators through a certification process has resulted in the over-simplification of social and nutritional issues, including apportioning blame for poor diet and ill-health solely on an individual's inadequate knowledge and skills. She is also critical of the policy-making process, which excluded representatives from various citizen and consumer groups and consumer co-operatives, thus marginalising debates around social issues such as genetically modified organisms. Finally, Reiher (2012) notes that while the Food Education Law promotes knowledge regarding food safety, it does not provide for practices that increase food safety, particularly improved governance, higher standards or more accurate labelling of produce. This is of particular concern in the post-Fukushima environment.

It is evident that concern for food safety, and resulting changes in legislation are driven broadly by the Japanese community and, more specifically, consumer-led. However, once these concerns are integrated into policy, there is a tendency to undermine social critique by marginalising the critical voice. Existing literature focuses on the relative uniqueness and success of Japanese co-operatives and their campaigns to address environmental and food safety issues. However, little research has been carried out on how the (assumed) member concerns regarding food safety shapes purchasing behaviour. Our research seeks to fill this gap.

RESEARCH AIMS

Our research examines the purchasing behaviours of a specific set of concerned consumers after the Fukushima incident. In particular, we were interested in qualitatively exploring the views of consumer co-operative members to better understand their perceptions of food safety and impact of this on their purchasing practices. This cohort was selected on the basis that members of food co-operatives are likely to be well informed and relatively concerned about food in relation to health and well-being, and the overall safety of produce. In addition, although there is some research examining the radiation-affected areas around Fukushima (Koyama, 2013), there is a paucity of information in more distant areas. Tokushima is not only distant from Fukushima, but being a food producing agricultural area, we expected that people who reside there would likely support farmers Japan-wide.

Our research was driven by four major aims:

1. to determine the values and/or attributes that people associate with food;
2. to establish whether (and the extent to which) recent food purchasing has changed as a result of perceived food risks from Fukushima,
3. to determine the sources of information contributing to perceptions of food risk, and
4. to explore the co-operative members' views on the concepts of *chisan chishō* and its relevance in their purchasing behaviours.

With reference to the first research aim, we expect that the product's country of origin is an important factor at the point of purchase and that knowledge about food risks (e.g., contamination, additives and diseases) is linked with safety. The second research aims rest on the expectation that food co-operative members' purchasing behaviours would change as a result of the Fukushima incident. In particular, despite the Japanese government promoting overall support for farmers and produce from potentially radiation-affected areas, we believe that co-operative members would be less likely to purchase produce from these areas. Our third aim was to better understand the source of interviewed consumers' information, the level of trust they placed in this information, and the extent to which this contributed to their perceptions of food safety. Our final aim was to better understand the level of importance placed on the concept and practices of *chisan chishō* (and within this, the related idea of *shokuiku*) and how consumers incorporated these into their purchasing behaviours.

METHOD

Background Information and Location

Shikoku Island is the smallest of four major Japanese islands, constituting 3.1% of Japan's total population (Ministry of Internal Affairs and Communications, 2010). Tokushima is one of four prefectures in Shikoku, and its capital, Tokushima City, has a population of 264,568 (Ministry of Internal Affairs and Communications, 2010); it is a prosperous city, that has been a centre of industry, administration, economic activity, cultural and educational facilities. Tokushima City has two consumer co-operatives: *Co-op Shizenha Tokushima* and *Tokushima Seikyō*.[2] Tokushima City was selected as a viable research site given it is both rural and agricultural, and because the second author has established links and networks in the community, facilitating access and interviews.

Co-op Shizenha Tokushima (hereafter *Shizenha*) was originally accredited in 1991 and, as of late 2015, has approximately 15,000 members (personal communication with Director, 2015). *Shizenha* translates as 'Co-op Natural' and as its name suggests, the co-operative prioritises environmental sustainability and high levels of food safety. *Shizenha* reiterates the views of the Ministry of Agriculture, Forestry and Fisheries [MAFF] (see Kimura, 2011; Assmann, 2015a; 2015b) that too many imported foods with unknown safety risks are readily available in Japan. Food security and support of Japanese, and particularly local, farmers are thus central *Shizenha* principles. Organic farming is promoted and their food safety standards are stricter than the national guidelines. Their members have access to detailed information on food safety, including the name and location of farmers and food producers, and the use and levels of specific chemical additives and pesticides. *Shizenha* thus caters to consumers who have an especially strong interest in food safety.

Tokushima Seikyō (hereafter TS) was established in 1984 and is much larger than *Shizenha* with just over 100,000 members, constituting more than 30% of Tokushima households (Tokushima Seikyō, 2015). TS was established on the general principles of consumer co-operatives, in particular, improving members' livelihoods culturally and financially. Therefore, the co-operative aims to reflect members' opinions and requests for particular products and to provide them at a reasonable price. Unlike *Shizenha*, TS operates for less specific purposes. For instance, while TS's products have all cleared national food safety standards and most of their products are Japanese, they are not aligned with principles such as only selling organic or Japanese products.

Sampling

The data collection phase involved the second author undertaking qualitative interviews with co-operative members in Tokushima. The second author drew on her connection to a not-for-profit organisation, '*Kosodateshien* Network Tokushima' (Parenting Support Network Tokushima), also known as 'K-Net', which provides services and support for parents and employed the snowball sampling technique. This strategy is deemed the most relevant for accessing participants with certain attributes, characteristics, or those difficult to locate (Berg, 2007). The founder of K-Net agreed to an interview and to assist in meeting other potential interviewees. In December 2014 and January 2015, the second author interviewed 22 members of *Seikyō* and/or *Shizenha,* all of whom were women: eight from TS, ten from *Shizenha* and four were members of both. There was a distribution of age: four were between 25 and 34, eleven participants between 35 and 44, two between 45 and 54, and five were over 55 years of age. The participants were involved in a range of occupations: seven identified as housewives, six were employed in childcare, three were clerical workers, three identified in professional roles, two owned and operated small businesses, and one was retired.

Two supplementary interviews were undertaken with the respective Directors of *Co-op Shizenha Tokushima* and *Tokushima Seikyō.* This approach was used to draw on their experiences and expertise in the socio-political landscapes of co-operatives. The Directors were both male and had been members of their co-operatives for a significant period (20 and 30 years).

Interview Schedule

The semi-structured interview schedule asked questions in three major sections. The first asked participants about their thoughts on food safety, and eating for health and general well-being. The second section focused on co-operative membership, including reasons for membership, the interviewee's level of participation in the co-operative and their views on the socio-political functions of the co-operative. The third section asked about the participants' knowledge on *chisan chishō* and the related concept of *shokuiku.*

The interview schedule for the Directors asked the same range of questions, plus three supplementary sections. The first asked questions surrounding the relationship between co-operatives and 'community', the second on co-operative principles and governance and the third on how the Directors themselves personally interacted with the co-operative.

We acknowledge that while the sampling strategy lends itself to provide rich and detailed data from the perspectives of a well-informed cohort, the views expressed will not be representative of all those with co-operative memberships, nor of all those who reside in Tokushima. As a result, the findings are not generalizable.

FINDINGS

Broadly, the interviews found that the reasons for (and advantages of) co-op membership and purchasing behaviour were determined by attitudes towards food safety (including the idea of health and well-being), convenience, and *chisan chishō.*

Of the 22 co-operative members, the length of membership ranged from one to 30 years. Confirming Saito's (2010) findings, the majority of interviewees spoke of two key reasons behind their membership: food safety (and related to this, an element of trust in purchasing food items from the co-operatives), and the convenience of home delivery. Interviewee five said: *'Home delivery is convenient. I am working full-time and I have no time for shopping. They also have safe food. I heard the co-op has strict standards with pesticides.'* Similarly, Interviewee 6 stated, *'My friend recommended the co-op to me. I heard their food is safe... they are strict with monitoring chemicals and radiation. I've become more aware of those since having a child.'*

A few respondents, while highlighting food safety and convenience, also noted that they agreed with the philosophy behind the co-operative. For example, Interviewee 21 stated: *'Shizenha advocates environmental conservation. They believe eating Japanese vegetables and farming ourselves results in protecting the environment. They protect and advocate Japanese produce.'*

Benefits of Co-Operative Membership

The women reinforced that food safety, convenience and a better selection of produce than supermarkets were the primary advantages of membership. As noted by Interviewee 9: *'The benefit of being a member is that I can obtain information in relation to food safety, such as radiation contamination and GMO.'* Similarly, Interviewee 15 stated that the convenience was crucial, but also the co-operative offered food with *'reduced or no pesticides and organic. This makes the choice of safe food easy.'* Interviewee 4 stated that she *'learned a lot about food from Shizenha. Being a member enabled me to get information that I wouldn't have known about otherwise. They also give us opportunities to meet the growers.'* This same Interviewee also said, *'I buy vegetables from Shizenha because I believe their produce is safe... they tell us where they were grown. I only buy vegetables from a local supermarket nearby when I run out. Shizenha's delivery is only once a week.'* The women felt that their co-operative membership provided them with trustworthy knowledge about safe food, and the ability to purchase it in a convenient way.

While both co-op directors' views echoed those of the female consumers, they also emphasised the co-operatives' values and community, a focus unsurprising given their roles. The Director of TS explained:

The biggest reason for me to be a member is that I identify with the philosophy of co-operatives. The next is convenience. Our delivery staff go out to deliver members' orders...they are appreciated in many ways, for example, they find old people who suffer from dementia and are wandering around. We had meetings with a representative in the local community and agreed to a neighbourhood watch.

This comment highlights the philosophy and community-oriented support in which the co-operative engages.

Place of Purchase

Many women sourced their fresh produce according to convenience; buying from the co-op and having it delivered required planning. While co-operatives were a major place of purchase for fresh fruit and vegetables, so too were farmer's markets and local supermarkets. As Interviewee 16 stated:

I often forget to place an order with Shizenha. I have to order by 10am Wednesday. When I forget or run out of vegetables, I buy them from a local vegetable shop. Also, I made friends with a contract farmer of Shizenha. When I buy vegetables from them, I have to order by phone and pick them up on the weekend... Their vegetables are just picked and fresh.

The major reason behind frequenting farmer's markets was cost. As one woman stated: *'Compared to farmer's markets, food from the co-op with reduced or no pesticides, or organic vegetables, are still dear'* (Interviewee 9). The women highlighted the necessity of having various sources to purchase their food items.

Purchasing Factors

When asked about purchasing factors, the women emphasised both buying local produce and food safety. Interviewee 22 explained, 'I try to eat local produce. The closer the place where it was grown, the better the produce is.' In a more political voice, Interviewee 2 said:

I think it is important that local power should be strong. We should be able to produce safe and healthy food locally, so I buy local produce to support local farmers. Even if I heard that local farmers use more pesticides on produce in Tokushima, I would still buy them.

Not only was local produce important, but also many women said that they prefer to buy produce explicitly from Japan, believing that it is safer and better than imported produce. For instance, as Interviewee 1 said, 'I don't believe Japanese produce is 100% safe, but at least Japanese produce is better compared to imported produce… I can trust local produce because the details of the produce are clearly written.'

The perception of higher levels of food safety from Japanese produce was prevalent. Interviewee 5 felt that 'Japanese produce is generally safer. I think they use more chemicals overseas.' Another woman said, 'I trust Japanese farmers in how they grow vegetables. The characteristics of Japanese people wouldn't allow excess use of chemicals. They keep to the rules' (Interviewee 8). Post-harvest chemicals were discussed as one of the major reasons for preferring national produce. As Interviewee 11 stated, 'I don't buy imported vegetables because I am concerned about post-harvest chemicals.' Similarly, Interviewee 9 said, 'It is hard to trace how overseas vegetables were grown and harvested [and the use of] post-harvest chemicals.' Further, Interviewee 18 added, 'I learned the danger of imported produce at seminars organised by the co-op.' Two women specifically stated they would not buy food imported from China. For example, Interviewee 12 noted, 'There are scandals over Chinese produce like using cardboard boxes to make pork buns. If the food contains anything from China, I don't buy it even for Co-op food.' Overall, the women demonstrated well-educated views about food safety and were aligned with supporting nationally produced foods.

Supporting the Economy

In addition to linking food safety with local produce, establishing a face-to-face relationship with farmers and supporting the national and local economy were important to many women. For instance:

At Shizenha, they display the photos of farmers who produce vegetables and their messages like 'please support us!' This makes me feel close to them and I should support them. I do try to buy local produce. (Interviewee 11)

Many women emphasised that the way co-operatives specify the origin of the produce is important. For instance, Interviewee 9 said,

As for the Shizenha delivery, their description is very thorough and clear, such as the use of pesticides. Freshness as well. They put 'Product of Tokushima' rather than just saying 'Product of Japan.' That's good.

Supporting the local and national economy was an idea that clearly emerged from our data, linking in with the idea of *chisan chisho* (discussed below).

Food Risks and Purchasing Behaviour

Despite the majority of women wanting to support local and national produce, many stressed that the Fukushima disaster impacted on where they purchased their fruits and vegetables. For instance, Interviewee 9 remarked: *'I changed my purchasing behaviour and I still don't buy food from Fukushima or the Tōhoku region at all.'* Interviewee 21 also explained, '*Shizenha actually sells vegetables from all over Japan... especially organic vegetables. I still check where they're produced and prefer to buy vegetables from the western side of Japan... Even for wines, I don't want grapes from the places near Fukushima.'* Similarly, Interviewee 9 said:

I only buy Japanese produce on principle. I do have concerns about food from Fukushima, but I know I shouldn't say it openly especially in front of my children. They learn quickly and they repeat what they hear. It could be insulting to those in Fukushima... and I don't want to be too negative.

Geographical proximity and distance were major influences on purchasing decisions. As Interviewee 11 said, *'I don't buy produce from Kantō and Tōhoku in general. I feel bad, but... I set my own areas that I don't buy from: Fukushima, Tochigi, Ibaragi, Iwate, Nagano, Chiba and Shizuoka.'* Similarly, Interviewee 21 stated, 'I buy rice from Kyushu [local]. I prefer vegetables from the western side of Japan.'

The majority of women echoed this sentiment; only a few stated otherwise. For instance, Interviewee 2 said, 'I know I should support the people in Fukushima, but I don't want my children to eat produce containing any trace of radiation. If I were on my own, I would eat it.' Interviewee 15 explained: 'I don't go out of my way to buy food from Fukushima in order to support the region. But I eat out and the vegetables or food may be from that region.' Yet, Interviewee 7 confessed, 'I cared at first, but now I've kind of forgotten about it.' This latter viewpoint was an anomaly. All other participants were generally careful about where their fresh fruit and vegetables were sourced.

Understandings (sometimes quite sophisticated) about food risks among participants, including the already-noted impact of Fukushima, were prevalent. The interviewer highlighted a number of concerns as options for the participants to discuss. These included genetically modified organisms (GMOs), food additives, radiation, pesticides, chemical contaminations, cleanliness / contamination, and diseases caused by foods (e.g., BSE). In response, Interviewee 9 stated:

I am most concerned with GMO because we don't know how much GMO is contained in foods. We can't see it. The data has just started to come out and not enough information is available. I suspect there are lots of foods that contain GMO… [Shizenha's] bread, seasonings and soya sauce are made out of GMO free soya beans.

The *Shizenha* Director also expressed concern about a number of food safety risks, in particular, diseases caused by food additives. In stark contrast, and reflecting perhaps the different principles of the two co-operatives, the Director from TS explained that:

Other than radiation contamination, I don't have many concerns about food available in Japan. I am aware that there are issues in relation to food safety such as additives or GMO, but food distributed around Japan is relatively safe in my view.

Nevertheless, both Directors were in clear agreement that food safety issues, of one kind or another, influenced purchasing behaviour. This was also reflected in the participants' comments. While many women referred to their pregnancy (or feeding infants) as a catalyst for paying more attention to food safety, it was apparent that co-operative membership also had a strong influence on their food safety awareness.

Sources and Trustworthiness of Information About Food

The majority of women stated that their knowledge about food, including safety, was a result of being a member of the co-operative. Both Interviewees 16 and 21 explained that co-operative flyers on food issues were a crucial part of their understanding and they regarded these as trustworthy. Additional sources of information on food included friends and family, the Internet and television. Interviewee 2 explained, *'I often talk to my friends about which food is better. There is definitely more quantity [of information about food now], but not necessarily quality. Some information has no scientific proof… I don't believe things from just one source. I look into various sources and make my own judgement.'*

Most women were very confident in the trustworthiness of co-operative information about food. Interviewee 21 was firm in this view: *'I believe the information provided by Shizenha is factual and true.'* Similarly, Interviewee 9 stated that, *'I generally trust information from Shizenha… I don't trust 100% on so-called 'organic' produce from elsewhere… at some farmer's markets, an old man there wrote by hand 'organic' on his produce. I don't trust it.'* Other women were also sceptical of information from outside of the co-operatives.

Particularly low levels of trust were evident amongst the women regarding information from the government, especially in relation to the Fukushima incident and the resulting radiation. For example, as Interviewee 11 said:

I didn't trust this information at all. When the Fukushima incident happened, I researched intensively to find out what was going on… I [also] read information on a German website… Seafood around that area is out of the question… mushrooms also tend to accumulate contamination.

When asked about levels of trust in government information, Interviewee 12 said: *'I just can't easily remove the feeling of distrust, even if the government says the food there [in the areas affected by the*

Fukushima incident] is safe.' Further, Interviewee 6 made an important point: *'I don't trust the information released by the government. I don't think the government is telling us the truth. If the government had told us the truth, everyone would have panicked.'* Interviewee 15 had low levels of trust regarding information from any source and her comment highlights the importance of face-to-face relationships in evaluating trust:

The number or level of contamination is not actually the issue. If someone I know gives me their produce, I will eat it even if the produce may contain pesticides. I don't eat produce from Hokkaido on principle, but if someone I know sends something to me, I will eat it. It's all to do with trust in people. I don't know anyone in the government in person, therefore I can't trust them.

Overall, it was evident that the women felt a greater level of trust regarding information from the co-operatives.

Chisan Chishō

Many of the ideas discussed above echo the idea of *chisan chishō*. This concept (and practice) is about the localisation of food distribution and consumption (Kimura, 2008). Regarding their views on *chisan chishō*, all interviewees were interested and responded positively to the idea. The above-mentioned themes of local supply and economy, food safety, and health and well-being were all tied into *chisan chishō* and discussed. For example, as Interviewee 18 states, *'I can see the faces of farmers. It is more efficient to eat the produce from where it is produced [and more] delicious.'* Interviewee 6 said, *'Chisan chishō is a good idea. We should eat local produce for Tokushima's sake.'* Interviewee 21's view was quite strongly associated with the socio-political aspiration and community aspect of *chisan chishō*: *'What we can do locally, we should do locally. It is important to build connections between people within a local community to care for each other and protect the environment as a result.' Chisan chishō* is important because *'you can eat fresh produce in season. It supports local farmers [and] food mileage is better'* (Interviewee 22). Likewise, Interviewee 4 expressed: *'I feel we can gain more than the nutrition contained in the food itself when we eat food grown in the place we live.'*

The idea that *chisan chishō* is crucial for the overall health of the Japanese agricultural economy was also evident: *'From the view of supporting local economy, I will support chisan chishō and buy local produce if there is a choice'* (Interviewee 2); *'It is important for our local community to be able to gain financially from agriculture'* (Interviewee 12); and *'Chisan chishō is good for the local economy. I feel a sense of impending crisis for local farmers'* (Interviewee 8).

The majority of women believed that their co-operative ideologically and practically supported *chisan chishō*. As Interviewee 4 said, *'Shizenha promotes Japanese produce and that leads to support the idea of chisan chishō. They promote the idea of eating what we produce, where it is produced.'* Interviewee 9 felt that *chisan chishō* is:

...important to grow and consume produce locally. It is the way it should be to sustain the environment. I think Shizenha tries to contribute to chisan chishō... I don't think the government does, they aren't doing enough. I suspect their motive is something different.

Another woman said, *'I feel really close to the local produce because I know who grows it and how they do it'* (Interviewee 21). Overall, the *chisan chishō* concept was well regarded and practised by many of the women.

Shokuiku

As previously noted, the *shokuiku* campaign emphasises the acquisition of knowledge about food and nutrition, and establishing 'correct' eating habits. This concept is bound within *chisan chishō.* The majority of women said that they practiced what they believed to be elements of *shokuiku,* predominantly in terms of eating a variety of foods, and overall, a balanced diet. However, given the ambiguity of the *shokuiku* concept, the women spoke of different points. For example, Interviewee 15 said that for her, *'Shokuiku means eating food happily together.'* Interviewee 4 said that she and her children practice *shokuiku* in the following way: *'I always check where the product comes from and I talk with my children. When we can't find out where it comes from, we don't buy it.'* Interviewee 2 said that for her *shokuiku* meant: *'having a balanced diet, eating breakfast, being aware of the nutritional chart, eating traditional food and cooking food rather than buying ready-made meals.'* It was clear that there was variation in the views of what *shokuiku* meant to the interviewees.

Some women spoke of the national *shokuiku* agenda and the inclusion of it within the education curriculum. Interviewee 9 spoke of the *shokuiku* education at school as providing necessary information about food:

I think teaching the children where food comes from and showing them how certain foods were grown or caught is good. If we don't eat food, it goes rotten and goes back into the earth. Shokuiku teaches them the natural cycle of food.

There were however a few women who were not convinced. For example, Interviewee 11 said, *'I don't completely agree with what the Ministry of Education and Science is suggesting as shokuiku.'* Similarly, Interviewee 22 was sceptical:

To me developing a special curriculum for shokuiku in schools is not real education. Rather, talking about food with your family is the basic thing to do... I don't oppose shokuiku, but I don't completely agree with the way it has been delivered. It should be done at a family level not a national level.

Overall, there was significant uncertainty about what exactly *shokuiku* is, but it was considered a necessary and desired concept and practice.

DISCUSSION

There is nothing new about the study of nutrition and consumer products among co-operative members. Saito (2010, 65-66) notes the organisation of these types of activities in the 1960s, at a time of rapid development of urban residential zones populated primarily by emerging middle-class nuclear families, in which 'professional housewives' dominated. However, there is a paucity of current research exploring the views of co-operative members. Our research seeks to fill this gap by examining the views of co-operative members living in a rural and agricultural area.

This paper explored the meaning of co-operative membership in a rural agricultural community following a period of intense social concern and debate. It has shown that the co-operative members were particularly concerned with the trustworthiness of their government's response and availability of information on radiation following the Fukushima Daiichi nuclear disaster. This study has shown that the threat to food safety was clearly apparent to co-operative members, which is unsurprising given the long history in which these co-operatives have battled to strengthen Japan's food safety standards, and advocated consumer rights in food safety and environmental protection.

Our research illustrated that co-operative members associated food safety primarily with local produce and exhibited a corresponding aversion to imported foods due to their perception of increased risk. Consumer ethnocentrism, rather than consumer animosity, was more evident among participants. Moreover, while food safety was implicit in the appreciation of local produce for some members, for others, supporting the local economy was more important. Food security, in terms of self-sufficiency, trumped food safety. Nevertheless, in all cases, co-operative members stated that their food purchasing behaviour had changed as a result of perceived food risks from Fukushima. This agreement in opinion was tempered, however, by variations in the extent to which avoidance of potentially contaminated food by different members continued five years on from the accident. Our study also revealed a profound distrust of government information on food safety issues, amongst most members. The need to locate alternative information was thus evident and, for many members, the co-operative was viewed as a trustworthy source. Our findings also indicated clear support and participation in *chisan chishō*.

Consistent with existing literature, we found that the women joined food co-operatives for access to 'safe' food, and the convenience of home delivery. Our study also clarified that the perceived convenience of co-operative membership included the provision of information on food safety, which was seen by many as a useful saving of independent research time. As expected, women correlated food safety with local and national produce. The literature tells us that food nationalism takes a particular form in Japan; the tensions between food security and food safety, described as 'alimentary xenophobia' (McVeigh, 2001) by U.S. officials and industry sources were evident in our findings. In support of this view, we found that the majority of participants preferred supporting Japanese and, in particular, local produce. Obata (2012) reiterates the much-touted fact that food self-sufficiency has dramatically declined in Japan (from 73% in 1965 to 41% in 2008) and that imported food is filling the gap between supply and demand. Although there are other factors contributing to this decline in Japanese agriculture, changes in diet and reliance on cheap, imported food have exacerbated the negative spiral (Obata, 2012). It appears this is one reason why many women viewed imported foods either sceptically, or avoided them altogether. However, the tendency to associate local as positive and foreign as negative, as noted by Kimura and Nishiyama (2008), was particularly evident in the findings. This resonates with the work of anthropologist and Japan specialist, Brian McVeigh (2001) who argues that nationalism is 'implicated in the mundane practices of everyday life, and like other hegemonic ideologies, it garners its strength from its invisibility.' We certainly noted that some of the women almost blindly agreed with the idea of the superiority of national produce over imported produce.

McVeigh (2001) examines the construction and mobilisation of nationalism in everyday life rather than political discourse. McVeigh argues that although the most prominent and visible representations of nationalism (such as prime ministerial visits to the Yasukuni shrine, which houses convicted war criminals) provoke both domestic and international consternation, nationalism attains most power where it is integrated into everyday life, such as in innocuous events or practices that are generally overlooked. The linkage between food and nationalism is not new to studies of Japan. In the eighteenth century, Motoori Norinaga (1730-1801) constructed a narrative of a sacred Japanese identity based on the supreme quality of its rice, and there is considerable scholarship on more recent incarnations of food nationalism (Ohnuki-Tierney, 1993; Ashkenazi and Jacob, 2000; McVeigh, 2006; Rath, 2010; Hook, Mason, and O'Shea, 2015; Ichijo and Ranta, 2016).

Where nationalism is heightened, ethnocentrism is generally prevalent. The relatively unquestioned preference for Japanese food, regarded as better by participants, was a clear finding. Although understandings of the concept of *chisan chishō,* and particularly the policy of *shokuiku,* seemed somewhat ambiguous, we observed a close alignment between daily habits, purchasing behaviour, and the strength of nationalistic sentiment.

To date, the literature about co-operatives has not emerged from regional agricultural areas, and although there have been a few studies of Japanese co-operatives since Fukushima, there has been little consideration of the relevance of the 'eat local movement' post-Fukushima. While both the co-operative members and Directors alike strongly believed in the importance of *chisan chishō,* and felt it important for the local economy, our study reveals an uncomfortable ambiguity and guilt provoked by the concept of *shokuiku* and the associated drive to support Fukushima farmers. Our study supports and provides evidence for McVeigh's (2001) assertion that daily habits strengthen a governmental government's nationalist agenda.

CONCLUSION

Consistent with past research, we found strong evidence to suggest food safety is a primary concern among co-operative members. While our exploratory study cannot generalise, Japan's strong membership in consumer co-operatives suggests a considerable market for food that is understood as safe and environmentally sustainable.

Our study suggests that the consumer-producer, face-to-face relationship marketing strategy has had an effective impact on purchasing decisions, particularly in an environment of heighted concerns about food safety and security. This has implications for food marketers who may benefit from considering the importance of personalising information about food producers, in addition to the provision of technical information, to address both the *anzen* and *anshin* aspects of Japanese perceptions of food safety.

Studies have confirmed that consumer perception of products is shaped by information on country of origin. It is clear that two distinct concepts are useful tools in understanding consumer decision-making: consumer ethnocentrism and consumer animosity (Fernandez-Ferrin et al. 2015). In this study, most interviewees expressed reluctance, but did not rule out the possibility of buying imported products from other countries (consumer ethnocentrism), however a few expressed strong animosity towards, and even an adamant refusal to buy, Chinese products following a range of food safety scandals. Further research is this area would assist marketers to better understand consumer decision-making in Japan when purchasing domestic, imported and 'safe' foods.

REFERENCES

Allison, A. (2013). *Precarious Japan*. Duke University Press. doi:10.1215/9780822377245

Ashkenazi, M., & Jacob, J. (2000). *The Essence of Japanese Cuisine: An Essay on Food and Culture*. Surrey: Curzon Press.

Assmann, S. (2015a). Agriculture in Japan: Free Trade and the Need for Reform. In S. A. Abingdon (Ed.), *Sustainability in Contemporary Rural Japan: Challenges and Opportunities*. Oxford, New York: Routledge.

Assmann, S. (2015b). The Remaking of a National Cuisine: The Food Education Campaign in Japan. In J. Farrer (Ed.), *The Globalization of Asian Cuisines: Transnational Networks and Culinary Contact Zones* (pp. 165–186). New York: Palgrave Macmillan. doi:10.1057/9781137514080_9

Bachev, H., & Ito, F. (2013). Fukushima Nuclear Disaster: Implications for Japanese Agriculture and Food Chains. *International Journal of Food and Agricultural Economics*, *2*(1), 95–120.

Berg, B. (2007). *Qualitative Research Methods for the Social Sciences* (6th ed.). Boston, MA: Allyn and Bacon.

Birchall, J. (2002). *The New Mutualism in Public Policy*. Oxford, New York: Routledge.

Birchall, J. (2011). A 'Member-Owned Business' Approach to the Classification of Co-operatives and Mutuals. *Journal of Co-operative Studies*, *4*(2), 4–15.

Buchler, S., Smith, K., & Lawrence, G. (2010). Food Risks, Old and New Demographic Characteristics and Perceptions of Food Additives, Regulation and Contamination in Australia. *Journal of Sociology (Melbourne, Vic.)*, *46*(4), 353–374. doi:10.1177/1440783310384449

Citizens Radiation Measurement Station (CRMS). (n. d.). Radiation Measurement by Citizens: Movement of Citizens' Radioactivity Measuring Station. Retrieved from http://fukushimaontheglobe.com/citizens-movement/radiation-measurement-by-citizens#sthash.puPlEezy.dpuf

Fernandez-Ferrin, P., Bande-Vilela, B., Klein, J., & Luisa del Rio-Araujo, M. (2015). Consumer Ethnocentricism and Consumer Animosity: Antecedents and Consequences. *International Journal of Emerging Markets*, *10*(1), 73–88. doi:10.1108/IJOEM-11-2011-0102

Frid, M. (2011). Food Safety: Addressing Radiation in Japan's Northeast After 3.11. *The Asia-Pacific Journal*, *9*(3), 1–9.

Gelb, J., & Estevez-Abe, M. (1989). Political Women in Japan: A Case Study of the Seikatsusha Network Movement. *Social Science Japan Journal*, *1*(2), 263–279. doi:10.1093sjj/1.2.263

Grubel, R. (2004). Internationalization in Everyday Life: The Role of Consumer Co-ops in Japan. *Kwansei Gakuin University School of Sociology Journal*, *96*, 137–148.

Ichijo, A., & Ranta, R. (2016). *Food, National Identity and Nationalism: From Everyday to Global Politics*. Basingstoke, Hampshire: Palgrave Macmillan. doi:10.1057/9781137483133

International Co-operative Alliance. (2013). Japan's new co-operative alliance gets active. Retrieved from http://ica.coop/en/media/news/japans-new-co-operative-alliance-gets-active

International Co-operative Alliance (ICA) Committee on Consumer Cooperation for Asia and the Pacific. (2012). Japan. Retrieved from http://www.ccw.coop/images/stories/Members_report/asia_2012_06.pdf

Japan Consumer Cooperative Union (JCCU). (2015). Co-op 2014 Facts and Figures March 21, 2014 to March 20, 2015. Retrieved from http://jccu.coop/eng/public/pdf/ff_2014.pdf)

Jo, I. (2011). The Empowerment of the Women in the Kanagawa Seikatsu Club Movement [PhD diss.]. University of London.

Jussaume, R. (1991). The growing importance of food safety to Japanese consumers and its implications for United States farmers. *American Journal of Alternative Agriculture*, *6*(1), 29–33. doi:10.1017/S0889189300003775

Jussaume, R., Hisano, S., & Taniguchi, Y. (2000). Food Safety in Modern Japan. *Japanstudien. Jahrbuch des Deutschen Instituts für Japanstudien der Philipp Franz von Siebold Stiftung*, 12, 211-228. http://www.contemporary-japan.org/back_issues/japanstudien_12_essen/dij-jb_12-jussaume.pdf

Kimura, A. (2011). Food Education as Food Literacy: Privatized and Gendered Food Knowledge in Contemporary Japan. *Agriculture and Human Values*, *28*(4), 465–482. doi:10.100710460-010-9286-6

Kimura, A. (2011). Nationalism, Patriarchy, and Moralism: The Government-led Food Reform in Contemporary Japan. *Food & Foodways*, *19*(3), 201–227. doi:10.1080/07409710.2011.599778

Kimura, A. (2012). Feminist Heuristics: Transforming the Foundation of Food Quality and Safety Assurance Systems. *Rural Sociology*, *77*(2), 203–224. doi:10.1111/j.1549-0831.2012.00075.x

Kimura, A. (2013). Standards as Hybrid Forum: Comparison of the Post-Fukushima Radiation Standards by a Consumer Cooperative, the Private Sector, and the Japanese Government. *International Journal of Sociology of Agriculture and Food*, *20*(1), 11–29.

Kimura, A., & Katano, Y. (2014). Farming after the Fukushima Accident: A Feminist Political Ecology Analysis of Organic Agriculture. *Journal of Rural Studies*, *34*, 108–116. doi:10.1016/j.jrurstud.2013.12.006

Kimura, A., & Nishiyama, M. (2008). The Chisan-Chisho Movement: Japanese Local Food Movement and its Challenges. *Agriculture and Human Values*, *25*(1), 49–64. doi:10.100710460-007-9077-x

Koyama, R. (2013). Rebuilding of Agriculture and Forestry Fishery from Harmful Rumor. *Journal of Rural Planning Association*, *31*(4), 572–575. doi:10.2750/arp.31.572

Kurimoto, A. (2015, November 8-10). Co-operatives and Trade Unions: From Occasional Partners to Builders of Solidarity-based Society. *Paper presented at the ICA/ILO Research Conference*, Antalya. Retrieved from http://ccr.ica.coop/sites/ccr.ica.coop/files/attachments/2.1Akira%20Kurimoto.pdf

Lam, P. (1999). *Green Politics in Japan*. London: Routledge.

LeBlanc, R. (1999). *Bicycle Citizens: The Political World of the Japanese Housewife*. Berkeley: University of California Press.

Maclachlan, P. L. (2002). *Consumer Politics in Postwar Japan: The Institutional Boundaries of Citizen Activism*. Columbia University Press.

Mah, C. (2010). Shokuiku Governing Food and Public Health in Contemporary Japan. *Journal of Sociology (Melbourne, Vic.)*, *46*(4), 393–412. doi:10.1177/1440783310384455

Matsuki, Y. (2002, July 7-12). Agri-Food Supply Chain Development by Various Chain Leaders: Case Studies in Japan and New Concept of High Nature Value. *Paper presented at 13th International Farm Management Congress*. Wageningen, The Netherlands.

Matsuo, M. (2013). Restructuring Japanese Food Safety Governance: Institutional Changes and Future Challenges. *European Food and Feed Law Review*, *4*, 250–258.

McVeigh, B. (2001). Postwar Japan's 'Hard' and 'Soft' Nationalism. Japan Policy Research Institute (Working Paper No. 73).

McVeigh, B. (2006). *Nationalisms of Japan: Managing and Mystifying Identity*. Oxford: Rowman and Littlefield Publishers.

Metzler, M. (2012). Introduction: Japan at an Inflection Point. *Asia Pacific Business Review*, *18*(2), 135–147. doi:10.1080/13602381.2011.561650

Ministry of Agriculture, Forestry and Fishery (MAFF). (2013). Kanzei Teppaishita Bāi No Keizaikōka Ni Tsuite No Seihu Tōitsu Shisan [Government Unified Estimates of the Economic Effects in the Case of Tariff Elimination]. Retrieved from http://www.maff.go.jp/j/kanbo/saisei/pdf/01_cao.pdf

Ministry of Agriculture, Forestry and Fishery. (2016). What is 'Shokuiku (Food Education)'? http://www.maff.go.jp/e/pdf/shokuiku.pdf

Ministry of Education, Culture, Sports, Science and Technology. (2009). Gakkō kyūshoku jisshi kijun no sekō ni tsuite [Enforcement of practical standard on School Supply Lunch (Notice)]. Retrieved from http://www.mext.go.jp/b_menu/hakusho/nc/1283812.htm

Ministry of Internal Affairs and Communications. (2010). Summary of Population Census of Japan. Retrieved from http://www.stat.go.jp/data/kokusei/2010/kihon1/pdf/gaiyou1.pdf

Nishiyama, M., Shimoura, S., Kurihara, S., Maruyama, A., Hirose, M., & Matsuda, T. (2007). The Analysis of Consumers' Interests for Construction of Local Agri-food System. *Japanese Journal of Farm Management*, *45*(2), 141–146.

Obata, N. (2012). A Study on Reduction of Food Wastes and Sustainability. *Food System and Agriculture Policy Science*, *19*(3), 249–267.

Ohnuki-Tierney, E. (1993). *Rice as Self: Japanese Identities through Time*. Princeton: Princeton University Press.

Parker, G. (2005). Sustainable Food? Teikei, Co-operatives and Food Citizenship in Japan and the UK4 (Working Papers in Real Estate & Planning 11/05). Henley Business School, University of Reading.

Parker, G. (2014). Social Innovation in Local Food in Japan: Choku-bai-jo Markets and Teikei Cooperative Practices (Working Papers in Real Estate & Planning 08/14). Henley Business School, University of Reading.

Rath, E. (2010). *Food and Fantasy in Early Modern Japan*. Berkeley: University of California Press. doi:10.1525/california/9780520262270.001.0001

Reiher, C. (2012). Food Pedagogies in Japan: From the Implementation of the Basic Law on Food Education to Fukushima. *Australian Journal of Adult Learning*, *52*(3), 507–531. Retrieved from http://search.proquest.com.libraryproxy.griffith.edu.au/docview/1315889794?accountid=14543

Rosenberger, N. (2009). Global Food Terror in Japan: Media Shaping Risk Perception, the Nation, and Women. *Ecology of Food and Nutrition*, *48*(4), 237–262. doi:10.1080/03670240903001100 PMID:21883068

Saito, Y. (2010). A Brief Chronicle of the Modern Japanese Consumer Movement. In *Japanese Consumer Co-operative Union* (ed. and trans. T. Suzuki). Retrieved from http://jccu.coop/eng/aboutus/pdf/a_brief_chronicle.pdf

Shokuiku. (2014). Promotion policies: FY2014 (White Paper on Shokuiku). Retrieved from http://www8.cao.go.jp/syokuiku/data/whitepaper/2014/en/pdf/part2-1.pdf

Takeda, H. (2008). Delicious Food in a Beautiful Country: Nationhood and Nationalism in Discourses on Food in Japan. *Studies in Ethnicities and Nationalism*, *8*(1), 5–30. doi:10.1111/j.1754-9469.2008.00001.x

Tanaka, K. (2008). Seven Samurai to Protect our Food: The Reform of the Food Safety Regulatory System in Japan after the BSE Crisis of 2001. *Agriculture and Human Values*, *25*(4), 567–580. doi:10.100710460-008-9152-y

Thompson, D. (2008). Japan: Land of Cooperatives. *Cooperative Grocer Magazine*, 135, March-April. Retrieved from http://www.grocer.coop/articles/japan-land-cooperatives

Tokushima Seikyō. (2015). Dai31kai Sōdaikai Giansho [The 31st General Meeting of Representatives Report]. Tokushima: Tokushima Seikyō.

Upham, F. (1987). *Law and Social Change in Postwar Japan*. Massachusetts: Harvard University Press.

Usui, T. (2007). Fast Food and Slow Food: A Theoretical Introduction to Contemporary Agrifood Problems. [Kibi International University School of Sociological Studies Research Bulletin]. *Kibi Kokusai Daigaku Shakai-gakubu Kenkyu Kiyo*, *17*, 15–22.

Vacek, G. (2011). Japan. In J. Brazda & R. Schediwy (Ed.), *A Time of Crises Consumer Co-operatives and their Problems around 1990*. Retrieved from https://www.google.com.au/url?sa=t&rct=j&q=&esrc=s&source=web&cd=1&ved=0ahUKEwiNqYDCpOzNAhUHp5QKHYmFAvgQFggdMAA&url=https%3A%2F%2Fgenos.univie.ac.at%2Ffileadmin%2Fuser_upload%2Fgenossenschaftswesen%2FGenos%2Fconsum.pdf&usg=AFQjCNH8DxoTwu-aGrr6UIyf8Zsv2QYSLQ&sig2=9RSIdec71UW8_KJPlY29og

Vogel, S. (1978). Professional Housewife: The Career of Urban Middle Class Women in Japan. *Japan Interpreter*, *12*(1), 16–43.

ENDNOTES

1 The yen-U.S. dollar conversion was based on the exchange rate (TTS) at 10 June 2016: US$0.94 = 100yen.

2 *Seikyō* is an abbreviation of Seikatsu Kyōdō Kumiai meaning Consumers' Co-operative.

This research was previously published in the International Journal of Public and Private Perspectives on Healthcare, Culture, and the Environment (IJPPPHCE), 1(1); edited by Mika Markus Merviö, pages 51-69, copyright year 2017 by IGI Publishing (an imprint of IGI Global).

Section 4
Food Security

Chapter 12
Integrating Spatial Technologies in Urban Environments for Food Security:
A Vision for Economic, Environmental, and Social Responsibility in South Bend, Indiana

Edwin Joseph
Indiana University – South Bend, USA

Elizabeth O'Dea
Indiana University – South Bend, USA

ABSTRACT

Food security for the urban poor has been an important topic for both developed and developing countries over the last 15 years. Although South Bend Indiana is a city in a developed country, declining economic circumstances have caused the city to show significant urban decay somewhat similar to some cities in developing countries. In this chapter, we explore South Bend's history and economic development strategies, and review practices aimed at strengthening food security for the urban poor. The chapter documents how numerous disparate organizations have been trying to help alleviate urban poverty and hunger, and reviews previous strategies used to foster sustainable growth and development. The integration of spatial technologies will become a key factor for promoting community social networks, participatory planning, and collaboration. The case is presented for the use of Geographic Information Systems (GIS) and associated technologies to help organizations, community leaders, local organizations, city planners, higher education institutions and the urban poor, work together to alleviate poverty and malnutrition through networking and sustainable urban agriculture.

DOI: 10.4018/978-1-5225-8063-8.ch012

INTRODUCTION

The World Food Summit of 1996 defined food security as "when all people, at all times, have physical, social and economic access to sufficient, safe and nutritious food which meets their dietary needs and food preferences for an active and healthy life". Food insecurity is defined as, "a household-level economic and social condition of limited or uncertain access to adequate food" (World Health Organization, 2015). The United States Department of Agriculture (USDA)'s definition of food security is, "access by all people at all times to enough food for an active, healthy life" (USDA, 2015).

ECONOMIC PROFILE: SOUTH BEND (2015)

The city of South Bend is considered the economic and cultural hub of the greater seven-county region known as Michiana, Indiana in the United States of America. The Chamber of Commerce of St. Joseph County, Indiana defines Michiana as St. Joseph County and "counties that contribute at least 500 inbound commuting workers to St. Joseph County daily." Those counties include Elkhart, La Porte, Marshall, St. Joseph, and Starke in Indiana, and Berrien and Cass in Michigan (Wikipedia, 2016). South Bend is the fourth largest city in the state, with a population of 101,190 (Indiana Demographics, 2016).

The 2015 unemployment rate in St Joseph County was 4.8%. The top three employers came from education [Notre Dame University], health care [Beacon Health System], and K-12 Public Schools [South Bend Community School Corporation] (South Bend Region Economic Development, 2016). There are concerns over the ecological unsustainability within some areas of the city. There is also growing inequality among the urban poor. The big issue is to identify possible remedies for inner city poverty in South Bend, knowing that the gap in the wealth among American households is more extreme than at any time since the Great Depression (Matthews, 2014). Seeking solutions for these problems may take many forms. One way is to explore pathways towards economic growth through urban sustainable agriculture. It is also instructive to review trends that are currently occurring in developing countries.

POVERTY AND SUSTAINABLE DEVELOPMENT WORLDWIDE

There seems to be much programmatic interest about living standards and health among neighborhoods in developing countries, but very little in developed countries. Researchers have indicated that population growth will be concentrated in cities and towns by 2030 (Montgomery & Hewett, 2005). This statistic is supported by the United Nations Department of Economic and Social Affairs world population and distribution research data, which shows that 66% of the world's population would be urban, with North America currently having 82%.

At the 2012 Rio+20 meetings, political leaders acknowledged that there were many challenges to sustainable development, and reiterated that many of the world's poor depend on rapidly disappearing and fragile biodiverse ecosystems (Chappell, et al., 2013). The United Nations department of Economic and Social Affairs stated that there are more than 1 billion people living in extreme poverty and income inequality worldwide. As a result, inadequate and unsustainable consumption and production patterns exist, but at huge economic and social costs (DESA, 2013).

URBAN FOOD SYSTEMS

The term "food system" describes the complex and interconnected activities of agricultural food production, processing, marketing, consumption, and disposal. It is an emerging discipline that brings together scientists and humanists to help address the cultural, social, ecological, physical, ethical and political aspects of food (Berman, 2011). According to Berman (2011), there is no single accepted definition for local food systems, but it is often defined as an area within a 400-mile radius of food origin. The scale of urban agriculture can vary from less than an acre of land worked by a handful of people to several acres.

Simply put, urban agriculture is growing or producing food in a city or heavily populated town or municipality. Producers vary from individuals, friends, non-profits, neighborhood groups and so forth. Reasons for production also vary from sales to restaurants, farmers' markets and donations to local food kitchens and churches. Urban agriculture increases access to locally grown food, and reinforces the importance of community and culture. In short, it is the front line of the food system (Greensgrow, 2016)

Urban Poverty in America: A Brief Overview

In his book about the causes of urban poverty in 1987, Harvard Professor, William Julius Wilson argued that changes in the structure of the U.S. economy were the primary drivers of increased social and economic dislocation of the urban poor. As the American economy shifted from predominantly manufacturing to service-driven, the urban underclass was severely impacted. According to Wilson, joblessness, low socioeconomic status, social isolation, impoverished neighborhoods, low education, lack of community safeguards, resources, and support were typical features of African-American urban communities. It wasn't a cultural problem, but a problem of isolation from community and necessary resources. Discrimination facilitated the persistence of an urban "underclass." (Wilson, 1987).

Urban Sustainable Agriculture: Examples Across the US

Agriculture in the United States has been in social and economic turmoil for over two decades while Urban Sustainable Agriculture and Community Supported Agriculture (CSA) has been gaining in popularity. We can trace CSAs from the 1960s in Japan, when women who were frustrated with the pesticide laden, process prepared, imported food formed an alliance with farmers to supplement their food needs from locally grown, nutritious, organic food. The alliance was known as Teikei, or partnership—a more detailed/literal Japanese translation is "food with the farmer's face on it." Teikei therefore was born out of the rejection of conventional agriculture, on grounds of social, environmental, and economic justice in addition to a simple desire for fresh, quality food (Schnell, 2007) (Henderson & Van En, 2007). The concept eventually made its way to the United States.

Farmers, consumers, and activists in the sustainable agriculture community in Pennsylvania learned to exploit ruptures in the conventional food system and developed new forms of food provisioning in local economies. The partnership has aided in local economic development and poverty alleviation, and demonstrated that this alternative model of food production and consumption has encouraged the rise of urban gardens all over the nation (Trauger, 2009).

In May 2000, Greenpoint Manufacturing and Design Center (GMDC) embarked on a $6 million development of The Greenpoint Wood Exchange—a rehabilitation of an 80,000 square foot former bowling alley building at 810 Humboldt Street Brooklyn New York, and built a rooftop garden. The rooftop accommodates a 15,000 square-foot hydroponic greenhouse that produces over 100 tons of food per year and supplies year-round produce to nearby grocery stores and restaurants (GMDC, 2015).

The Will Allen organization, Growing Power, designed a system of greenhouses and farm locations around Milwaukee that divert and compost more than 44 million pounds of food waste per year from nearby households, beer breweries, coffee shops, and even corporations like Walmart. The organization also produces one million pounds of organically-grown food each year (Kumar, 2015).

In 2011, Noah Link and Alex Bryan built an urban farm in Detroit Michigan known as Food Field that offers a CSA (Community-Supported Agriculture) service. Their goal is to create an alternative to the corporate food system. Local residents enjoy the sustainable produce in weekly CSA boxes containing salad greens, mulberries, and other food items. Food Field now has an aquaponics system for raising Catfish and Blue Gill. The farm also raises chickens and ducks for eggs (FoodTank.com, 2013).

The Distributed Urban Farming Initiative (DUFI) in Bryant Texas transformed vacant lots to produce healthy food. The goal is to build gardens in empty spaces, inspire Bryan residents to eat healthy food, and drive entrepreneurship and tourism. The initiative has been able to successfully grow broccoli, cauliflower, cabbage, and lettuce on raised beds and pallet gardens. DUFI has been working hard to reconnect the Bryan community with the food system (DUFI, 2015).

The Chicago Council on Global Affairs published a report highlighting the importance of urban food security, and offers an action plan for addressing urban food security and nutrition (Lee, 2013).

What is known about many urban centers in the US is that healthy, safe, nutritious food at affordable prices is difficult to find (The Chicago Council on Global Affairs, 2013). In many ways, the City of South Bend Indiana is no different from other cities in its quest to feed its growing population. Although over the past few years there have been efforts to address poverty and malnutrition through community gardens, much more collaborative work is needed to make these efforts sustainable.

URBAN SUSTAINABLE AGRICULTURE IN SOUTH BEND, INDIANA

A report published by the American Planning Association (APA), provides guidance for supporting and expanding urban agriculture. Like Chicago, Cleveland, Detroit, Missouri, Milwaukee, Philadelphia, among others, South Bend is an old industrial city, and possess acres of vacant abandoned city lots. These areas are potential centers for urban agriculture initiatives (Hodgson, Campbell, & Bailkey, 2011). Currently, the most popular form of urban agriculture in the city is community gardening, which contributes significantly to developing social connections, capacity building, and community empowerment. Our investigation seeks to find out how much collaboration exists to support important social networks to help in closing the gap in the wealth among households, and in the satisfaction of food needs.

There is a growing awareness about the role of urban sustainable agriculture in the context of food security and poverty alleviation in South Bend Indiana; however, its implementation is very informal. The local government verbally supports the grassroots efforts of the urban agricultural movement in South Bend. However, the city has yet to incorporate its support into the City Plan (Radican, August 2010). This makes urban agriculture vulnerable, and may also jeopardize its sustainability. Nonetheless, urban sustainable agriculture in the city seems to be steadily increasing. There is also a growing demand for a greener city, social justice, job security, and increased wildlife restoration.

Currently, there is a public listing of active community gardens and food services for individuals and families at risk in South Bend. The original listing included contact persons, telephone numbers, and emails (these have been removed from the Table 1). In order to view the distribution of the gardens and food services within a Geographic Information System (GIS), we geocoded the addresses by assigning geographic coordinates to each location (latitude/longitude). This made it convenient to view other spatial entities such as roads, parks and schools, to understand trends, patterns and distribution in relation to each other.

Table 1. Distribution of community gardens in South Bend IN

Food Entity	Address	Zip	Latitude	Longitude	Description
Corvilla - Bulla	18443 Bulla Road	46637	41.7043715	-86.2273586	Garden
Corvilla - Roelke	1318 E Roelke Dr	46614	41.617287	-86.233396	Garden
Corvilla Office Unity Garden	3231 Sugar Maple Business St.	46628	41.720718	-86.2975079	Garden
El Campito Unity Garden	1024 Thomas St	46601	41.6741565	-86.2676142	Garden
Garden of St Therese	915 N Olive Street	46628	41.6873507	-86.2858679	Garden
Haircrafters Unity Garden	602 Lincoln Way East	46601	41.6702888	-86.2449995	Garden
IHC Unity Garden	1901 W. Western Ave	46619	41.6731319	-86.2800307	Garden
IUSB Unity Garden	1700 Mishawaka Ave	46615	41.6800027	-86.2401604	Garden
Jewish Federation	3202 Shalom Way	46615	41.678144	-86.205174	Garden
JJC Unity Garden	1001 S. Michigan St	46601	41.6640819	-86.2522822	Garden
Kroc Center Unity Garden	900 W. Western Ave.	46601	41.671611	-86.2644002	Garden
LaSalle Square	3701 Prast Blvd	46628	41.6863911	-86.3027314	Garden
Madison Square Unity Garden	1230 E Madison Street	46617	41.6806026	-86.2337549	Garden
McKinley School	228 E McKinley Ave	46615	41.6804295	-86.1814233	Garden
Monroe Park	608 Carroll St.	46601	41.6693708	-86.2492739	Garden
Morningside Unity Garden	333 W Colfax Ave.	46601	41.6781427	-86.2571334	Garden
Muessel Primary Center	1020 Blaine Ave.	46628	41.6883679	-86.2696123	Garden
Ravina Park Unity Garden	1511 Dale Ave	46613	41.6596109	-86.239353	Garden
Red Cross Refuge Unity Garden	3220 E Jefferson Blvd	46615	41.6724718	-86.2039691	Garden
Robinson Center	921 North Eddy Street	46617	41.6880438	-86.2370864	Garden
Safe Station Unity Garden	1322 LincolnWay East	46614	41.6597928	-86.2316855	Garden
Smallest Unity Garden	528 E Colfax Ave	46617	41.6776081	-86.245224	Garden
St Peter's UCC Unity Garden	915 N Ironwood Street	46615	41.687646	-86.2184267	Garden
St Vincents Unity Garden	520 Crescent Ave	46617	41.6838757	-86.2460485	Garden
Unity in God Garden	646 LaPorte Ave	46628	41.6831259	-86.2739902	Garden
Zion Unity Garden	224 E. Jefferson Blvd	46617	41.6731176	-86.1806002	Garden
Notre Dame Ctr.	1045 W. Washington St.	46601	41.6762414	-86.2673326	Garden
Beacon Triangle Unity Garden	1345 N Michigan Street	46617	41.6924393	-86.2528915	Garden
Eddy St Eberhart Enviro Garden	317 S. Eddy Street	46615	41.6730586	-86.2372438	Garden
Walnut Street	421 S. Walnut Street	46619	41.6716617	-86.2726509	Garden

continued on following page

Table 1. Continued

Food Entity	Address	Zip	Latitude	Longitude	Description
Walnut Street	423 S. Walnut Street	46619	41.6716618	-86.2726732	Garden
Hear Our Tears	2421 Bertrand St	46628	41.6815726	-86.2860361	Garden
Hope Ministries Rescue Mission	432 S Lafayette Blvd	46634	41.6712481	-86.2555248	Soup Kitchen
Center for the Homeless	813 S. Michigan Street	46601	41.6659525	-86.2527165	Soup Kitchen
St. Vincent de Paul	3408 Ardmore Trail	46628	41.6885417	-86.2990941	Food Asst.
Berean Community Service	601 W Colfax Ave	46601	41.677674	-86.2600018	Food Pantry
Food Bank Of Northern Indiana	702 S Chapin St	46601	41.6679634	-86.2625769	Food Bank
Community Food Pantry	702 S Chapin St	46601	41.6671644	-86.263539	Food Pantry
Downtown Soup Kitchen	333 N Main St	46601	41.6800648	-86.2546771	Soup Kitchen
Northeast Neighborhood Center	803 N. Notre Dame Avenue	46617	41.685896	-86.2401508	Food Pantry
Catholic Charities Food Pantry	1817 Miami St.	46613	41.6554923	-86.2372697	Food Pantry
La Casa De Amistad	746 S Meade St	46619	41.6667298	-86.28753	Food Pantry
River Park United Methodist	920 S 23rd St	46615	41.6653659	-86.2157277	Food Pantry
Little Flower Catholic Church	5419 N Ironwood Drive	46635	41.6972864	-86.2186849	Food Pantry
Needs for Neighbors	54301 Maple Lane	46635	41.7041949	-86.2159955	Food Pantry
Christ Temple Apostolic	55454 N Lexington Ave.	46628	41.6869748	-86.3148238	Food Pantry
Share Your Blessing	3001 W. Cleveland Rd.	46628	41.7250602	-86.2932771	Food Pantry
Clay United Methodist Church	17646 Cleveland Rd.	46637	41.7229223	-86.2758369	Food Pantry

A review was undertaken of other aspects of urban agriculture that are being used to raise awareness, and inform neighborhoods, decision makers, and planners about the different types of urban agriculture that have the potential to increase household incomes, promote healthy lifestyles, and generate a distribution chain through strong community networks. The importance of GIS cannot be understated since it can transform the way people network, communicate, and collaborate to make decisions—in this case, improve the standard of living for the urban poor. The significance of GIS technology to boost ongoing initiatives by key players and stakeholders in South Bend to improve the livelihoods of the urban poor must be recognized and factored into solutions.

HISTORY AND ITS EFFECTS ON AFRICAN-AMERICANS IN SOUTH BEND

African-American historical narratives not only document the lives of enslaved people, but offer invaluable perspectives of first-hand experience and mistreatment of people of color throughout an important period of American history. After the Civil War, former slaves continued documenting their experiences to ensure social and economic progress as affirmed by the nation's leaders. The African-American History narrative in South Bend mirrors some of the characteristics of post-war slavery, but is unique in many ways. For this reason, this should not be a truncated, incomplete, or untold story. It is a narrative that has the "potential to evoke pride and add a level of complexity" to South Bend's black heritage and history (Andrews & Adams, 2016).

Despite its rich history, little is known about the African-American experience in South Bend, or the efforts that were made to improve the quality of life within the black community; and enhance educational and economic opportunities. The South Bend History Museum documents the life of John Charles Bryant—a descendant of one of South Bend's first African American families with portraits dating back to the 1800s. For Bryant, portraits of his grandparents and great-grand parents have special meaning because they tell the story of how against all odds the Bryant-Powell family were able to escape the south, and settle in Indiana.

John Charles Bryant's grandfather worked at Studebaker, his father worked the tables at the Old LaSalle Hotel, his mother Georgia Bryant, was an organist for 42 years at the Olivet AME church, South Bend's first established African American church. This and other narratives are intertwined with the fabric of American History .

A brief review of South Bend's history and economy reveals that like most of America, African Americans experienced a measure of discrimination in the workplace and in the community. South Bend was a hub of manufacturing fueled by the presence of the Studebaker car company and other significant industries. The 1883 South Bend City Directory lists several African American men employed by local industries such as Wilson Brothers Shirt Factory, Birdsell Manufacturing Company, Singer Sewing Machine Company and the F.P. Box Company, The Oliver Chilled Plow Works (The History Museum, 2016).

Between 1870 and 1910, the African American population increased from 73 to 640; and in 1903 when Booker T. Washington visited South Bend, he found additional opportunities for employment for African Americans at Studebaker. Unfortunately, employment for African-Americans became difficult after the end of World War I when they had to compete with European immigrants for jobs. In the early 1920s the population of South Bend Indiana was 70,983, occupying a land area of 15.4 square miles, and an average population density of 4,609 per square mile. The city was ranked 100 out of the 100 largest urban places in America (U.S. Bureau of Census, 1998).

The population of African-Americans in South Bend was estimated around 1,269 during the 1920s, and increased to 3,431 in 1930 (Robinson, 2015). Studebaker employed more than 400 African-Americans, and at times up to 700. This was considered unique for the period, as several other local companies did not employ African Americans until the 1940s (WNDU, 2010). However, Studebaker underwent financial problems and the plant ceased production in 1963. As a result, the city lost about one-third of its population, and experienced severe growth and economic problems.

During World War II, South Bend became part of the wartime employment boom. Many African-Americans were employed as factory workers, but by the early 1950s, employment for African-Americans was again on the downturn in South Bend. In spite of the unemployment challenges, the 1950s Great Migration of African-Americans from the South caused an increase in the African-American population. And by the mid-1950s, the African-American population exceeded the foreign-born population (South Bend Tribune, 2015).

Based on archived reports, there was a concerted effort to end racial policies in South Bend. For example, in 1963, South Bend Mayor Frank J. Bruggner appointed a 34-man Bi-racial Committee to study the problems of racial discrimination in South Bend. The committee met with community leaders to help eliminate, by voluntary measures, the cause of racial tension (Mayor's Biracial Committee Progress Report, 1963). South Bend also passed an ordinance (4091) known as the South Bend Fair

Employment Practice Ordinance which prohibited discrimination in employment because of race, color etc. (City of South Bend Common Council, 1965).

In 1968, the South Bend Human Relations and Fair Employment Practices Commission endorsed open housing, thus raising the possibility of an open housing ordinance at the county level (The Reformer Newspaper Vol. 1, No. 29, 1968). In spite of these efforts, there were many examples of discrimination in the sale, rental, and financing of private housing in South Bend. One important example is the public hearing on March 19, 1963, in Notre Dame Law School Auditorium (Citizen's Fair Housing Committee of South Bend-Mishawaka, 1963)

South Bend Today

In 2011, Newsweek put South Bend on its list of America's Dying Cities, underscoring the 2.5% decline in the number of young people over the previous decade. (Newsweek, 2011). In 2014 South Bend was listed among the least happy regions in the United States by a Harvard and University of British Columbia study, taking the number 3 spot (UBC News, 2014). This was based on a survey of 300,000 people in 2009. The 2013 Gallup-Healthways Well-Being Index also lists South Bend as one of the "most unhappy" regions in the nation. The degree of unhappiness was based on a number of factors such as physical and emotional health, healthy behaviors, work environment, social and community factors, financial security and access to necessities like food, shelter and health care (South Bend Tribune, 2016).

Moreover, economists have established volumes of literature on the relationship between food insecurity and household incomes (classified by the poverty line). Researchers have concluded that food insecurity has increased dramatically in recent years, and is one of the most important nutrition-related public health issues in the U.S. (Gundersen, Kreider, & Pepper, 2011)

In 2015, South Bend's poverty rate stood at 23.9%, with almost 70% of students in the public school system eligible for free or reduced-cost lunch. South Bend also has double the national average rate of families living in poverty (Buttigieg, 2015). Crime risk statistics show South Bend to be above the US average for personal crime risk, murder risk, rape risk, robbery risk, property crime risk, burglary risk, larceny risk, and automotive theft risk. Assault risk, however, is lower than the US average (Point2Homes, 2016). The mayor of South Bend acknowledged that there were many excellent organizations working on problems, but needed partnership with the city.

In spite of these troubling statistics, the city of South Bend was ranked third in the nation for America's Best Complete Streets Policies of 2015 with a score of 98.2/100. This is an integrated transportation system that supports safe travel for pedestrians, cyclists, transit riders, motorists and others, within the city. The policy is one milestone encouraged in the Mayors' Challenge, a U.S. Department of Transportation initiative in which South Bend is a participant, committed to improving sidewalks, bike lanes, public transportation stops, frequent and safe crossing opportunities, median islands, accessible pedestrian signals, curb extensions, narrower travel lanes, and roundabouts (Smart Growth America, 2016).

A recent article by the South Bend Tribune discussed how it may be possible to develop, attract and keep talent so that South Bend area can compete in the modern economy. The Tribune acknowledges that at its heart, the region is still a manufacturing center, and manufacturing accounts for more than one out of every four jobs. (South Bend Tribune, 2014). Young, talented people are looking for quality of life, and will choose a community with a rich lifestyle over a good job elsewhere. The city knows that young people are more likely to locate near "livable" communities with bike trails, public transportation, and vibrant downtowns among others. South Bend agreed to work together with institutions of higher learning to include significant grass-roots involvement in visioning such a future.

As we fast-forward to the year 2016, we believe it is important to review a 30-year old problem, and carefully examine the issue of how food and other essential needs can be better satisfied among the urban poor in South Bend Indiana, and how we can use Geographic Information Systems Technology (GIS) to analyze proximities, adjacencies, intersections, and containment issues to help solve urban poverty.

For example, one may wish to know how close a community garden might be to a landfill in an urban area; or city planners may want to identify roads that pass within a certain distance of a city park. These are proximity issues. Questions such as, what is the shortest street network route to all community gardens are important proximity questions for understanding distance relationships, and are applicable for both developing and developed countries. Adjacency is generally about objects that are zero distance away from each other—in short, they share at least one common boundary.

An intersection analysis calculates the geometric intersection of features on a landscape. A simple analysis could be which river or stream crosses a major road—at what specific location; or which trail passes through a particular parcel in the city. The analysis does not limit itself to linear objects, but may extend to point on point analysis. In other words, where are the houses (point location) that have gardens on the lot (another point location) - that is considered an intersection of two points. Containment analysis answers questions such as which gardens are within which neighborhoods. These are very important spatial questions that we can query on-the-fly, and use for decision-making related to urban poverty.

Facing the Realities in South Bend: There Is Hope

Many organizations work towards the goal of a sustainable South Bend. They include the Food Bank of Northern Indiana, Unity Gardens, Young Professionals Against Poverty (YPAP), City of South Bend Division of Sustainability, Center for a Sustainable Future, Links Incorporated (South Bend Chapter), The Civil Rights Heritage Center, United Way, 100 Black Men, and a few others can be located using the University of Notre Dame's link http://green.nd.edu/sustainability-in-action/community/.

These disparate groups tend to work independently, but would be more effective if there are stronger collaborations. For example, the Food Bank of Northern Indiana works in partnership with the community to feed the hungry, increase awareness of the effects of hunger and lead programs designed to alleviate hunger (Food Bank of Northern Indiana, 2016). The Food Bank describes that the rate of hunger insecurity in Northern Indiana as a silent epidemic, and sees the challenge for the city of South Bend as addressing food and nutritional security in a sustainable way given very high poverty levels (Foodbank of Northern Indiana, 2015). Their programs include Healthy Choices Market, Senior Bag Program, Mobile Food Pantry program, and Food 4 Kids.

The Unity Gardens, a 501(c)3 Non-profit in South Bend, is committed to improving community health by increasing accessibility to fruits and vegetables by developing a local food system. The goal of Unity Gardens is community building through a "framework of gardens". The gardens would not only provide healthy food for those in need, but also educate the community on how to grow, and utilize healthy food. The overall mission is to improve community health which includes access to locally grown fruits and vegetables, education on food preparation and consumption, increasing social capital, and opportunities for the disadvantaged (Unity Gardens, 2009).

Young Professionals Against Poverty (YPAP) was founded in 2006 by Steve Camilleri and Matt Money. They wanted to raise awareness about poverty in South Bend, and brought young professionals together to help organize events and raise money geared towards improving the quality of life. Steve Camilleri is also the Executive Director of the Center for the Homeless which has been endorsed by the US Department of Housing and Urban Development (HUD). The Center has provided more than 700,000 safe nights and over 1.5 million meals to more than 44,000 men, women, and children since opening in December 1988. The focus is on providing life-changing services to their guests (Center for the Homeless, 1988).

South Bend's chapter of Links Incorporated is an organization whose mission is to promote and engage in educational, civic, and intercultural activities in order to enrich the lives, and to work together toward achieving community goals. The organization's vision is to eliminate disparities by reducing barriers to services through advocacy, education, and service. Their goal is to empower communities and increase the number of sustainable and measurable programs; expand collaborative partnerships; and identify communities with greatest needs (The Links Incorporated, 2015).

The Civil Rights Heritage Center (CRHC) was established on the campus of Indiana University South Bend for the study and documentation of civil rights history. The Center engages in scholarly inquiry, student research, and community education, and is dedicated to creating partnerships with the state and local communities and offer leadership in creative solutions for 21st-century problems. (Civil Rights Heritage Center, 2015). The center sponsors group discussions related to equity and social justice, wages and income, access to healthy food and barriers to employment, among others.

In 2016 the center hosted a number of significant events. Some included community lectures on asset building, and civil rights. Other events were partnerships, for example, the center partnered with the University of Notre Dame to host a Wikipedia Edit-A-Thon to teach people how to become a Wikipedia editor and use proper guidelines for writing articles and finding resources for researching and writing about South Bend's history.

The Center also celebrated the history and experience of African Americans in South Bend through stories from the Black LGBT experience, and explored the significance of open-access, digital resources using The Black Press Research Collective (BPRC). It hosted a scanning event to preserve and share photographs and documents that tell about black history and the black experience. Guests were encouraged to bring artifacts to the Civil Rights Heritage for scanning and inclusion in Michiana Memory, the free, online archive (http://michianamemory.sjcpl.org/). In celebration of Women's History Month, the Civil Rights Heritage Center part sponsored a public talk on the topic "A Band of Sisters: Fighting for Employment Equality from the Assembly Line"—the story of a small group of blue-collar women who, in a class action lawsuit, uncovered endemic gender discrimination in South Bend.

South Bend Reconnecting People Through Sustainable Programs

Recently the city of South Bend has been working in conjunction with the Environmental Protection Agency (EPA) and participating in their program called the Food Recovery Program (FRP), which helps prevent food waste by diverting leftover food from local restaurants to impoverished people in need (Rivest, 2015). There is also renewed interest in traditional food systems in South Bend. In an attempt to reconnect land and people through sustainable gardening, Indiana University South Bend (IUSB) is currently conducting research on urban sustainable agriculture. IUSB has developed agricultural courses focusing on sustainable local food production techniques, and is well on the way to developing a Sustainability Showcase—an area on campus dedicated to demonstrating urban sustainable practices.

The Showcase would contain a plaza/entryway featuring native plants, a zero impact home with the capacity to continue to showcase new home building design and products. It would also serve as a site for constructing Tiny Homes, and invite community participation for the regular construction of this new concept to increase accessibility to low cost housing in South Bend. The showcase will be open to area schools, neighborhood groups, non-profits, and so forth; all working together to solve social, economic, and environmental problems in the city. It will also serve as an open living laboratory to compare best practices for urban gardening. The goal is to share information and strengthen community relationships and networking.

TECHNOLOGY: A PATHWAY TO ECONOMIC RECOVERY IN SOUTH BEND

We believe that introducing new technology may be an appropriate pathway to economic recovery and progress in South Bend. Information and spatial technologies have the potential to provide a platform for people with similar interests to solve problems in the community. These technologies can provide a technical framework for food production and act as a communications platform even in low-income communities. New technologies can be separated into two categories—sustaining, and disruptive. A sustaining technology relies on incremental improvements to an already established technology, while a disruptive technology may not yet have a proven practical application, but has the potential to significantly alter the way organizations operate (Christensen, 1997).

Geographic Information Systems (GIS) technology falls into the category of a disruptive technology. It is a location-based technology that has made people more aware of the dynamic world in which we live. It has improved our ability to interpret our environment, and understand spatial interactions. Therefore, by nature, GIS technology is adaptive, and suited for dynamic, fast-paced innovations. Many entrepreneurs view GIS as a survival imperative that creates revenue opportunities. Some companies see it as a tool to expand their markets, new products or services. Dynamic leaders are curious, and welcome a challenging future, and are constantly looking for institutional changes that adapt to "urgent" disruptive technologies. GIS technology, therefore, encourages a growth mindset—essentially welcoming change instead of fearing it. Yet, GIS technology has not been fully embraced and exploited for solving urban design and environmental problems in South Bend. Integrating GIS with urban sustainable land use planning that includes gardening is crucial for realizing the aim of food security in South Bend Indiana. The city is able to explore and exploit the potential of GIS technologies not only for monitoring the urban infrastructure, but also as an information technology which when shared at the grassroots level through participatory planning, has the potential to promote sustainable gardening and responsible land use. No other technology can provide a powerful combination of digital, mobile, social, geographic and video innovations as tools for investigating different land use options as with GIS. In the case of South Bend Indiana, existing community collaborations provide a solid network of private organizations and individuals, community leaders, government institutions, and educators, to develop urban food networks and related economic activities that fit local realities.

GIS in South Bend, Indiana

Developed in the late 1960s, very few planning departments could afford a GIS because of the prohibitive cost of hardware and the limited capabilities of the software. But by the 1980s, GIS technology was considered the formalized computer-based information system for effective decision-making in urban planning (Han & Kim, 1989). South Bend's GIS department has been using ArcGIS software since 1999 to manage geographic data, and generate maps for different city projects. The city uses GIS to provide services such as; hot spot mapping of crime, tracking of basement flooding, utility inventory and similar others. The city also makes data and information available via an on-line map server (ESRI, 2011). Currently, the system provides basic information about property within the city, and maps include aerial photographs (orthophotos), parcel boundaries containing parcel identification number, property address, and property owner information (City of South Bend Indiana, 2016).

Among the geographic layers that are available for download from the City of South Bend's website are:

- Demographic Data
- Vacant and abandoned properties
- Street centerlines
- Public facilities
- Social services location
- Parks
- Businesses

These information layers are valuable for environmental research, public safety, marketing, etc. (Government Technology, 2016). It is now possible for organizations in the city to invest their time to perform tasks that were previously impractical or impossible without the current datasets. For example, modeling, predictions, and analysis can now be done through dynamic map overlays of physical, cultural, and economic features within the city landscape. GIS technology will enhance public participation, promote greater collaboration among public agencies, and would also be useful to city planners and educators who focus on geographic literacy/education and GIS mapping.

In the quest to acquire data for spatial analysis, several problems were encountered. Important metadata for datasets were missing or incomplete. Online map servers run by the city allowed minimum analyses. Data export was also limited to graphic formats. When GIS data was available for download, they were usually from other sources such as the Census Bureau which, because of scale, did not satisfy the desired unit of analysis.

EVALUATING POPULATION AND INCOME DISTRIBUTION IN SOUTH BEND

South Bend is the largest city in St Joseph County, Indiana, and occupies 457.9 square miles. South Bend's population distribution per square mile was 584.4; with a total population of 101,190 in 2014. This is a 0.2% rise in population from 2010 (STATSINDIANA, 2015). Table 2 shows that the 2015 South Bend population dropped to 101,168 with a race distribution of 60% white, 26% African American, 12% Hispanic or Latino. These three races account for more than 98% of the population. Source shows small variations in population distribution[1].

Table 2. 2015 South Bend population by race

Race	Population	% of Total
Total Population	***101,168***	***100***
White	61,199	60
Black or African American	26,906	26
Hispanic or Latino	13,116	12
Asian	1,318	1
American Indian	478	Below 1
Three or more races	313	Below 1

Source: suburbanstats.org

The 2015 median age in South Bend was 33, with average income for people within the age group 25-44 being $36,620 (Suburban Stats, 2016). Family income distribution when compared with the State of Indiana and the U.S show that above $50,000 family income in South Bend is lower than the State of Indiana, and the U.S (USA.com, 2016). However, when we compare family income below $50,000, South Bend is higher (see Figure 1).

Poverty by Zip Codes

Sixteen zip codes are included in the city of south Bend; 46601, 46613, 46614, 46615, 46616, 46617, 46619, 46624, 46626, 46628, 46634, 46635, 46637, 46660, 46680, 46699 with some zip code routes intersecting the city boundary (zipcode.org, 2016). Population, race, and income characteristics for Whites, Blacks, and Hispanics (the largest three groups) are presented in Table 3 below, for each zip code[2]. Zipcodes 46601, 46613, 46616, 46617, and 46619 show the highest poverty rates.

Figure 1. Family income distribution

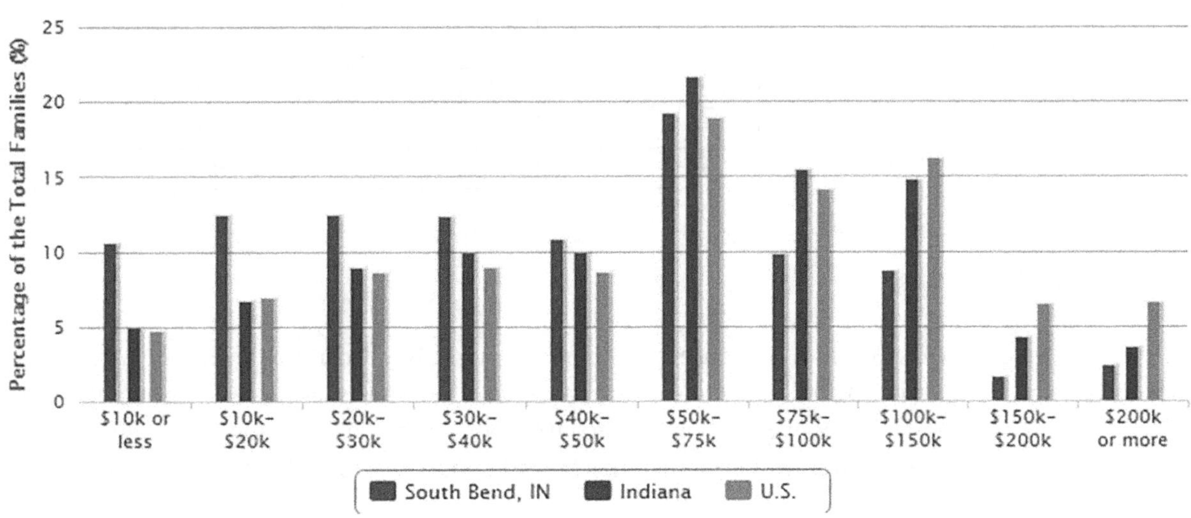

*Table 3. Poverty level data was obtained from zipatlas.com * (NA=Not Available)*

Zipcode	Average House Value USD$	Median Age	Income per H/hold USD$	% Below Poverty Level	% White Pop.	% Black Pop.	% Hispanic Pop.	% of Total Pop.
46601	70,300	33.1	17,434	39.98	43.25	48.49	7.29	92
46613	48,300	30.10	30,717	25.46	64.05	22.21	14.18	100
46614	93,300	41.80	43,160	8.06	91.33	4.89	2.42	99
46615	70,200	34.10	33,557	15.65	82.87	9.43	4.03	13
46616	62,800	32.80	33,449	26.82	66.49	24.95	6.91	98
46617	85,900	34.30	39,517	21.84	67.28	26.52	3.81	98
46619	60,100	33.80	33,463	20.09	59.72	24.86	18.48	103
46624	N/A	N/A	N/A	N/A	N/A	N/A	N/A	N/A
46626	N/A	N/A	N/A	N/A	N/A	N/A	N/A	N/A
46628	69,900	36.80	36,951	17.96	63.67	30.64	4.44	99
46634	N/A	N/A	N/A	N/A	N/A	N/A	N/A	N/A
46635	105,400	44.70	54,296	4.69	90.93	4.49	2.24	98
46637	95,700	37.80	43,225	13.82	89.06	5.45	2.61	97
46660	N/A	N/A	N/A	N/A	N/A	N/A	N/A	N/A
46680	N/A	N/A	N/A	N/A	N/A	N/A	N/A	N/A
46699	N/A	N/A	N/A	N/A	N/A	N/A	N/A	N/A
46616	62,800	32.80	33,449	26.82	3,940	1,583	460	402
46617	85,900	34.30	39,517	21.84	7,010	2,338	610	626
46619	60,100	33.80	33,463	20.09	12,250	5,789	5,911	4,284
46624	N/A	N/A	N/A	N/A	N/A	N/A	N/A	N/A
46626	N/A	N/A	N/A	N/A	N/A	N/A	N/A	N/A
46628	69,900	36.80	36,951	17.96	16,079	8,678	1,975	1,821
46634	N/A	N/A	N/A	N/A	N/A	N/A	N/A	N/A
46635	105,400	44.70	54,296	4.69	5,617	800	318	429
46637	95,700	37.80	43,225	13.82	13,345	1,402	627	933
46660	N/A	N/A	N/A	N/A	N/A	N/A	N/A	N/A
46680	N/A	N/A	N/A	N/A	N/A	N/A	N/A	N/A
46699	N/A	N/A	N/A	N/A	N/A	N/A	N/A	N/A

AARP Livability Index

The AARP (American Association of Retired Persons) Public Policy Institute developed a Livability Index to measure community livability. The scores are based on affordability and access to housing, access to live, work and play, safe and convenient options for transportation, clean air and water, health, civic and social involvement, opportunities for inclusion. The 2015 livability index[3] for South Bend Indiana,

Figure 2. Zipcode map of South Bend Indiana

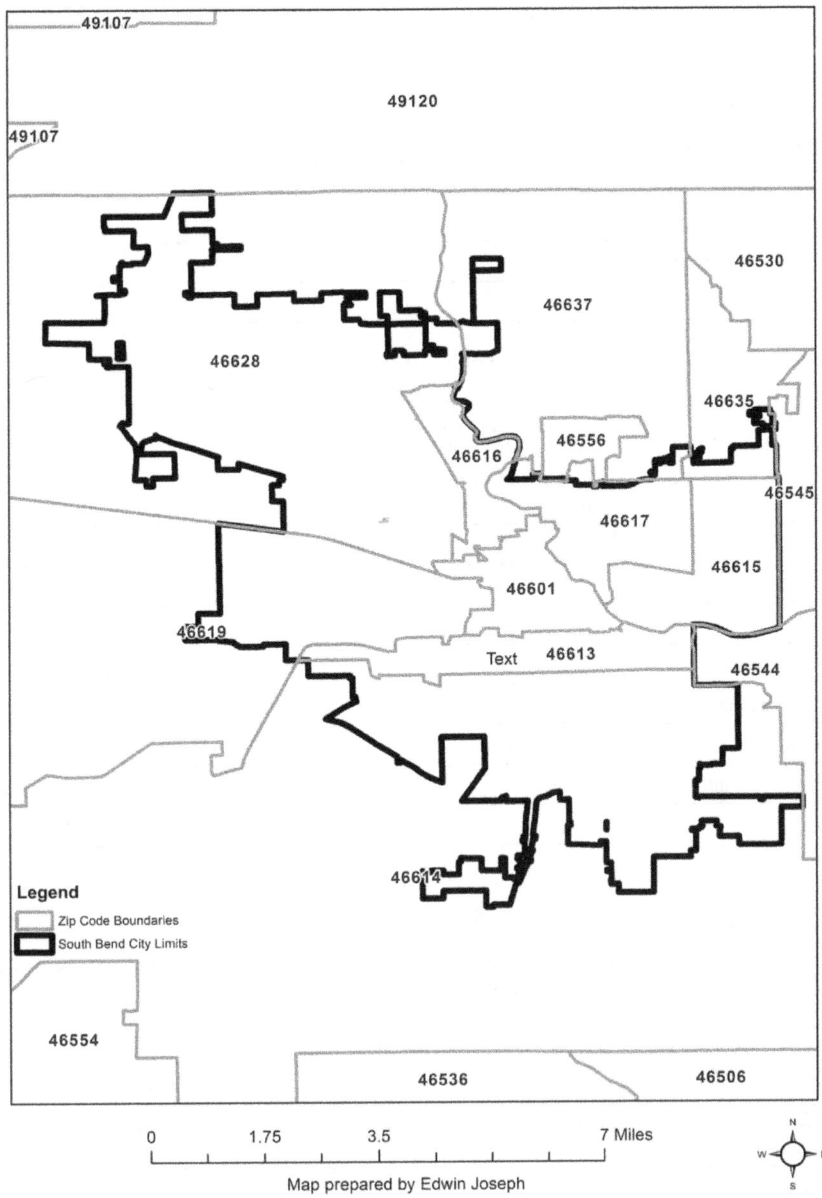

was 51 out of 100p (AARP, 2016). Areavibes.com has also developed livability indexes based on their opinion of important metrics. They include amenities, cost of living, crime rates, education, employment, housing and weather. The individual category scores for AARP are presented in Tables 4 – 17.

Areavibes.com Livability Index

Areavibes.com (2016) assigns South Bend a livability index of 71. This was derived from a distribution and number of important amenities such as grocery stores, other food and drink establishments, shopping

Table 4. AARP livability index for South Bend Indiana

Category	Index
Housing—affordability and access	58
Neighborhood—access to life, work, and play	49
Transportation—safe and convenient options	55
Environment—clean air and water	50
Health—prevention, access, and quality	48
Engagement—civic and social involvement	49
Opportunity—inclusion and possibilities	47
Livability Index	**51**

Table 5. AARP livability index for zip code 46601

Category	Index
Housing—affordability and access	69
Neighborhood—access to life, work, and play	52
Transportation—safe and convenient options	58
Environment—clean air and water	49
Health—prevention, access, and quality	44
Engagement—civic and social involvement	53
Opportunity—inclusion and possibilities	44
Livability Index	**53**

Table 6. AARP livability index for zip code 46613

Category	Index
Housing—affordability and access	**57**
Neighborhood—access to life, work, and play	48
Transportation—safe and convenient options	58
Environment—clean air and water	51
Health—prevention, access, and quality	44
Engagement—civic and social involvement	51
Opportunity—inclusion and possibilities	44
Livability Index	**53**

malls, coffee shops, schools, parks, libraries, entertainment, fitness. The spatial distribution for these amenities is shown in Figures 3 – 12.

The City of South Bend Department of Community and Economic Development, with assistance from the Area Plan Commission of St. Joseph County, gathered data from the 2010 census and other recent census products and compiled maps for displaying population, race, housing units, and income. This

Table 7. AARP livability index for zip code 46614

Category	Index
Housing—affordability and access	55
Neighborhood—access to life, work, and play	42
Transportation—safe and convenient options	47
Environment—clean air and water	45
Health—prevention, access, and quality	50
Engagement—civic and social involvement	50
Opportunity—inclusion and possibilities	46
Livability Index	**48**

Table 8. AARP Livability Index for zip code 46615

Category	Index
Housing—affordability and access	58
Neighborhood—access to life, work, and play	52
Transportation—safe and convenient options	57
Environment—clean air and water	52
Health—prevention, access, and quality	53
Engagement—civic and social involvement	44
Opportunity—inclusion and possibilities	50
Livability Index	**52**

Table 9. AARP Livability Index for zip code 46616

Category	Index
Housing—affordability and access	58
Neighborhood—access to life, work, and play	47
Transportation—safe and convenient options	58
Environment—clean air and water	51
Health—prevention, access, and quality	50
Engagement—civic and social involvement	45
Opportunity—inclusion and possibilities	46
Livability Index	**51**

community profile provides information about demographic trends in St. Joseph County. Much of this data is presented at the census tract level. The data shows that St. Joseph County became more racially diverse with a decrease in the white population. The number of residents who are African American, another race, or of two or more races all increased in absolute and percentage terms. Minority groups

Table 10. AARP livability index for zip code 46617

Category	Index
Housing—affordability and access	56
Neighborhood—access to life, work, and play	48
Transportation—safe and convenient options	57
Environment—clean air and water	48
Health—prevention, access, and quality	49
Engagement—civic and social involvement	49
Opportunity—inclusion and possibilities	47
Livability Index	**51**

Table 11. AARP livability index for zip code 46619

Category	Index
Housing—affordability and access	59
Neighborhood—access to life, work, and play	46
Transportation—safe and convenient options	49
Environment—clean air and water	48
Health—prevention, access, and quality	47
Engagement—civic and social involvement	46
Opportunity—inclusion and possibilities	48
Livability Index	**49**

Table 12. AARP livability index for zip code 46624, 46626, 46628 & 46634

Category	Index
Housing—affordability and access	69
Neighborhood—access to life, work, and play	52
Transportation—safe and convenient options	58
Environment—clean air and water	49
Health—prevention, access, and quality	44
Engagement—civic and social involvement	53
Opportunity—inclusion and possibilities	44
Livability Index	**53**

remained concentrated within South Bend, where the vacancy rate was relatively high—particularly on the near south and west sides of town. (Department of Community & Economic Development, 2011)

A recent review of the status of the urban poor in South Bend (based on 2000-2013 census block group data) revealed that 16.7% of residents had incomes below the poverty level in 2013 while that for the whole state was 9.5% (CDW Services, 2015).

Table 13. AARP livability index for zip code 46635

Category	Index
Housing—affordability and access	54
Neighborhood—access to life, work, and play	46
Transportation—safe and convenient options	44
Environment—clean air and water	52
Health—prevention, access, and quality	53
Engagement—civic and social involvement	53
Opportunity—inclusion and possibilities	44
Livability Index	**49**

Table 14. AARP livability index for zip code 46637

Category	Index
Housing—affordability and access	63
Neighborhood—access to life, work, and play	42
Transportation—safe and convenient options	45
Environment—clean air and water	49
Health—prevention, access, and quality	53
Engagement—civic and social involvement	50
Opportunity—inclusion and possibilities	41
Livability Index	**49**

Table 15. AARP livability index for zip code 46660

Category	Index
Housing—affordability and access	58
Neighborhood—access to life, work, and play	52
Transportation—safe and convenient options	57
Environment—clean air and water	52
Health—prevention, access, and quality	53
Engagement—civic and social involvement	44
Opportunity—inclusion and possibilities	50
Livability Index	**52**

The spatial distribution of poverty shown at a larger scale identifies 4 census blocks with the highest poverty rates (Figures 13 – 16). These blocks are located either south or west of the city center.

Given the spatial distribution of poverty in South Bend, it is important to understand the use of GIS technology to seek solutions for solving environmental, social, and economic problems. Some critical areas range from public safety, to health, and to transportation (Government Technology, 2016). South

Table 16. AARP livability index for zip code 46680

Category	Index
Housing—affordability and access	63
Neighborhood—access to life, work, and play	42
Transportation—safe and convenient options	45
Environment—clean air and water	49
Health—prevention, access, and quality	53
Engagement—civic and social involvement	50
Opportunity—inclusion and possibilities	41
Livability Index	**49**

Table 17. AARP livability index for the State of Indiana

Category	Index
Housing—affordability and access	55
Neighborhood—access to life, work, and play	45
Transportation—safe and convenient options	48
Environment—clean air and water	60
Health—prevention, access, and quality	41
Engagement—civic and social involvement	47
Opportunity—inclusion and possibilities	62
Livability Index	**51**

Bend city planners have invested much time in regulating land use, monitoring infrastructure, and traffic management among other things. And like many other cities, GIS is now used for performing spatial analyses. Planners also realize that the spatial and temporal aspects of GIS not only allow for modeling and predictions, but also useful for in-depth analysis.

HOW GIS CAN HELP URBAN AGRICULTURE AND INCREASE FOOD SECURITY IN SOUTH BEND

There are many strategic advantages for integrating GIS in Sustainable Urban Agriculture. GIS is a location-based system and is ideal for environmental monitoring, flood management, soil classification, pest and disease management, strategic planning, among others. For the average gardener, planning is generally some combination of guesswork and luck, but integrating GIS allows for more spatial organization. Factors that can affect urban agriculture are sunlight, shade, soil moisture content, pH, soil type, artificial materials, dust and lead; all of which can affect the quality and quantity of crops.

Figure 3. Location of grocery stores
Grocery Stores—20 locations

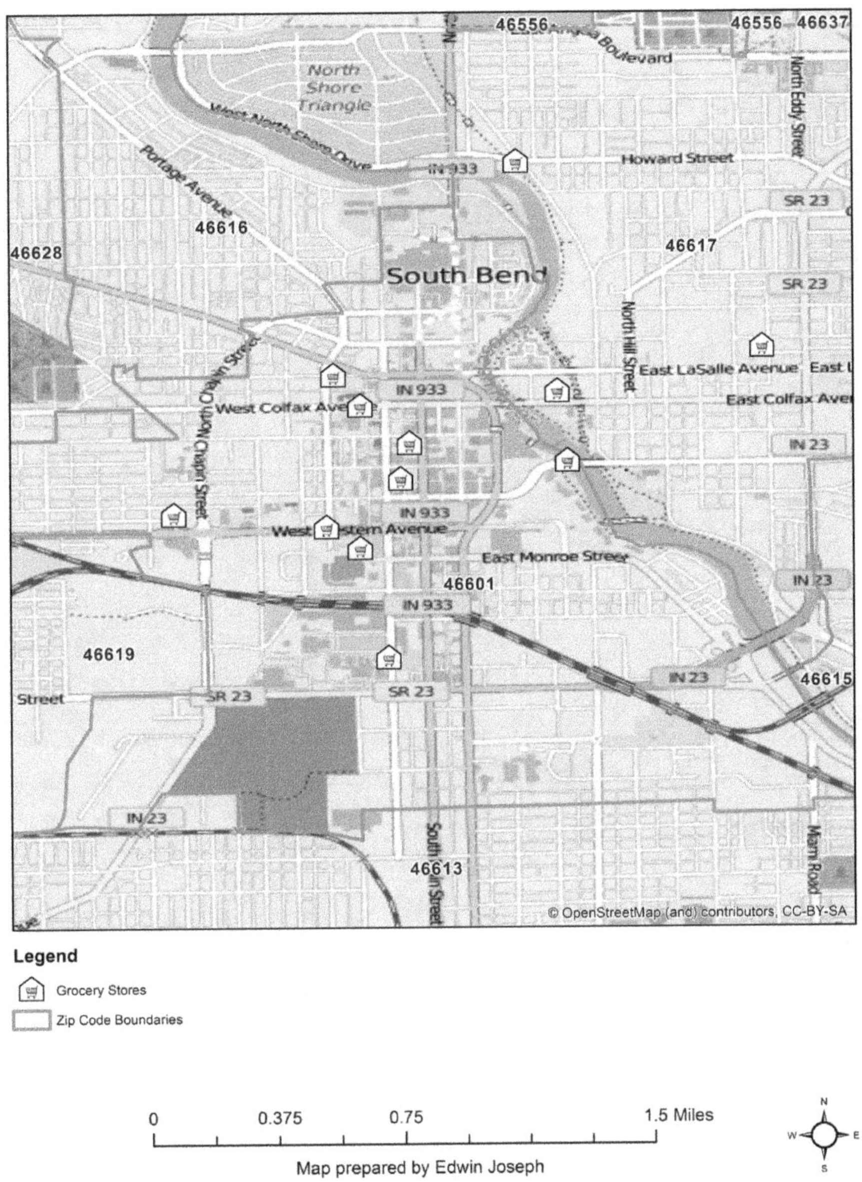

A combination of GIS, GPS and Multimedia mapping will provide continuous data for sustainable food production in the city. Surveying new areas to plant crops, or examining soil and air quality will be an enormous challenge without spatial technologies. These technologies lend themselves to crowd sourcing projects, and greatly assist in managing food production and distribution. The World Food Program, the division of the United Nations recognizes the importance of GIS in food production and is one of the biggest users of GIS data (ArcNews, 2014). The program recognizes the importance of safeguarding

Figure 4. Locations of food & drink establishments
Food and Drink—20 locations

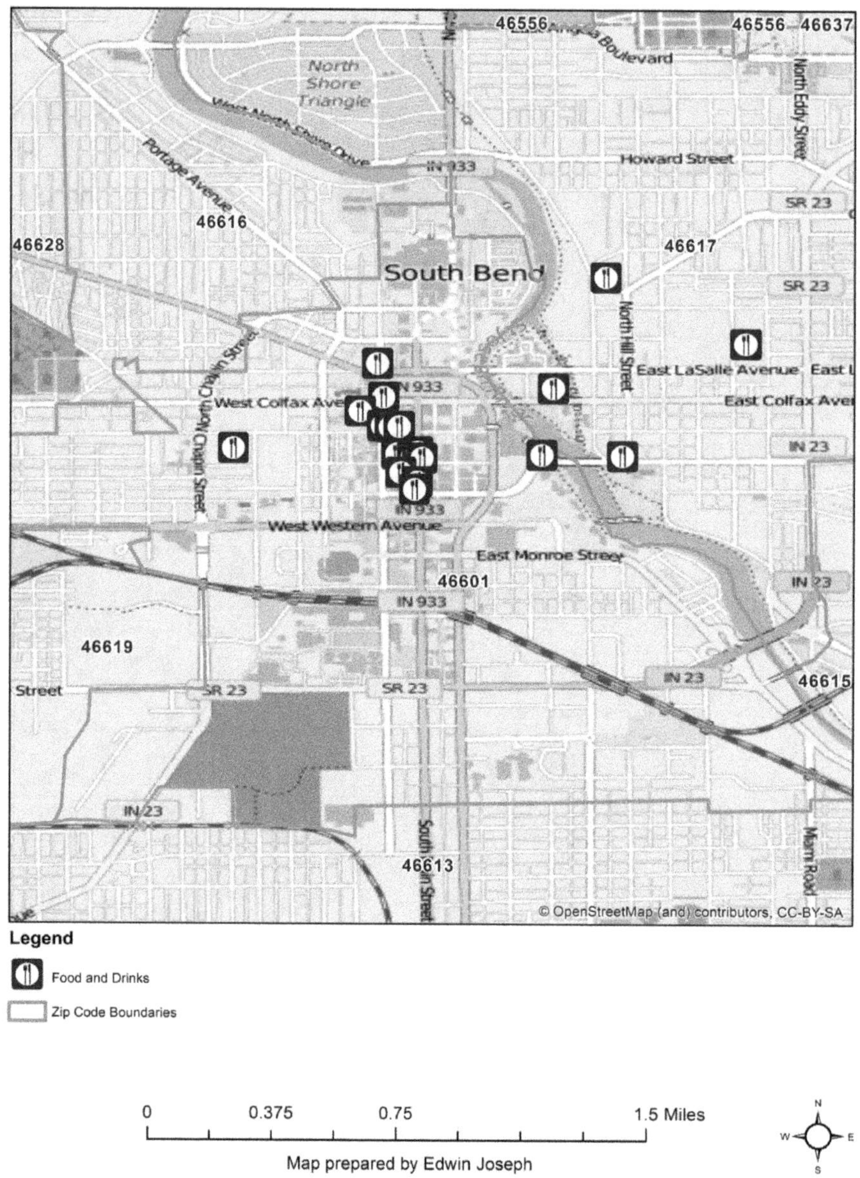

food sources, and engages communities in activities such as land use planning, biodiversity education, and soil fertility management. The World Food Program also monitors changes in the environment, and uses remote sensing to identify disaster prone regions.

In South Bend it is possible to use drone mapping technology to capture high resolution imagery of urban gardens to monitor production. The addition of high resolution orthoimagery (photographic maps) provides information necessary to adapt sustainable agriculture practices. Orthoimagery from drone

Figure 5. Shopping locations
Shopping—20 locations

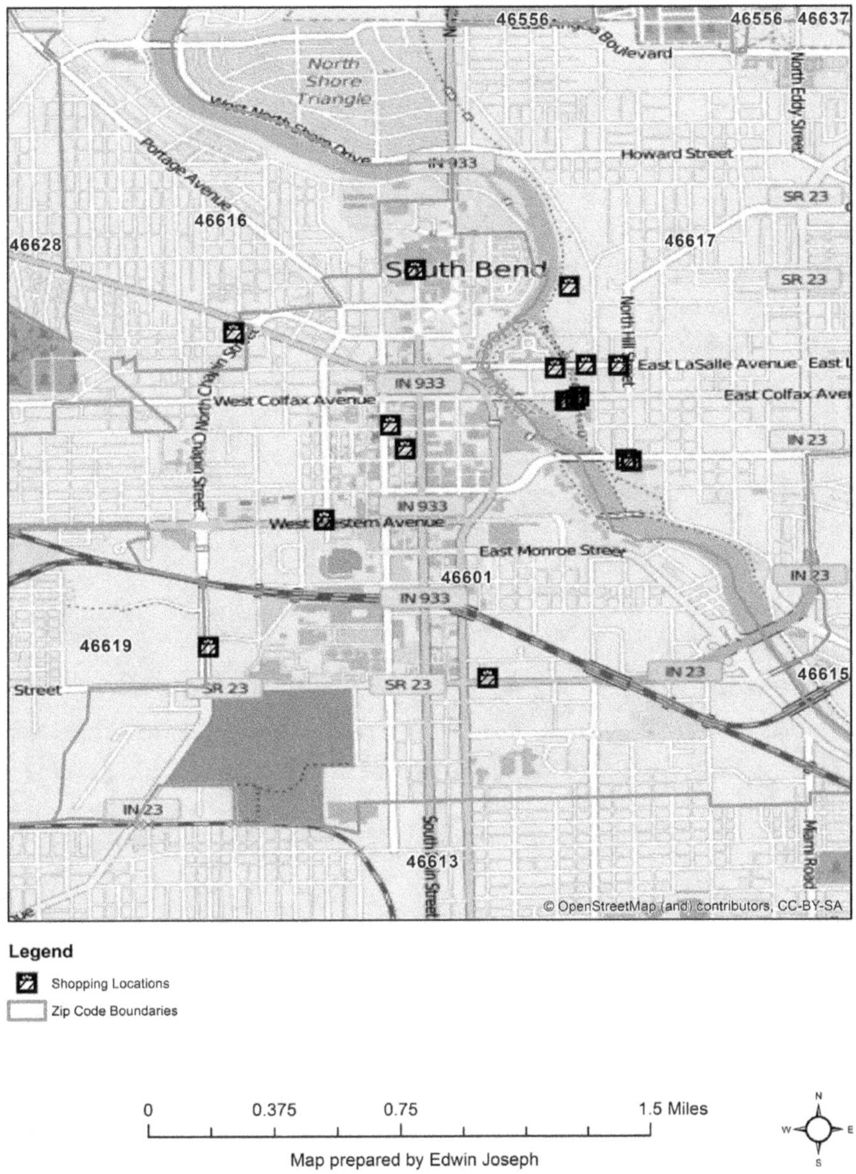

mapping creates 3-D models of urban spaces that are critical for landscape analysis. Thus, generating models for potential gardens will ensure better planning for future generations.

Urban Sustainable Agriculture has the potential to become a dynamic economic sector in South Bend. City planners need to establish a proper program for developing the social, economic and ecological framework for urban agriculture. They must work closely with the urban poor, and as partners, develop a sustainable and resilient city that is socially inclusive, safe, and environmentally-healthy. The chal-

Figure 6. Locations of cafe
Coffee—7 locations

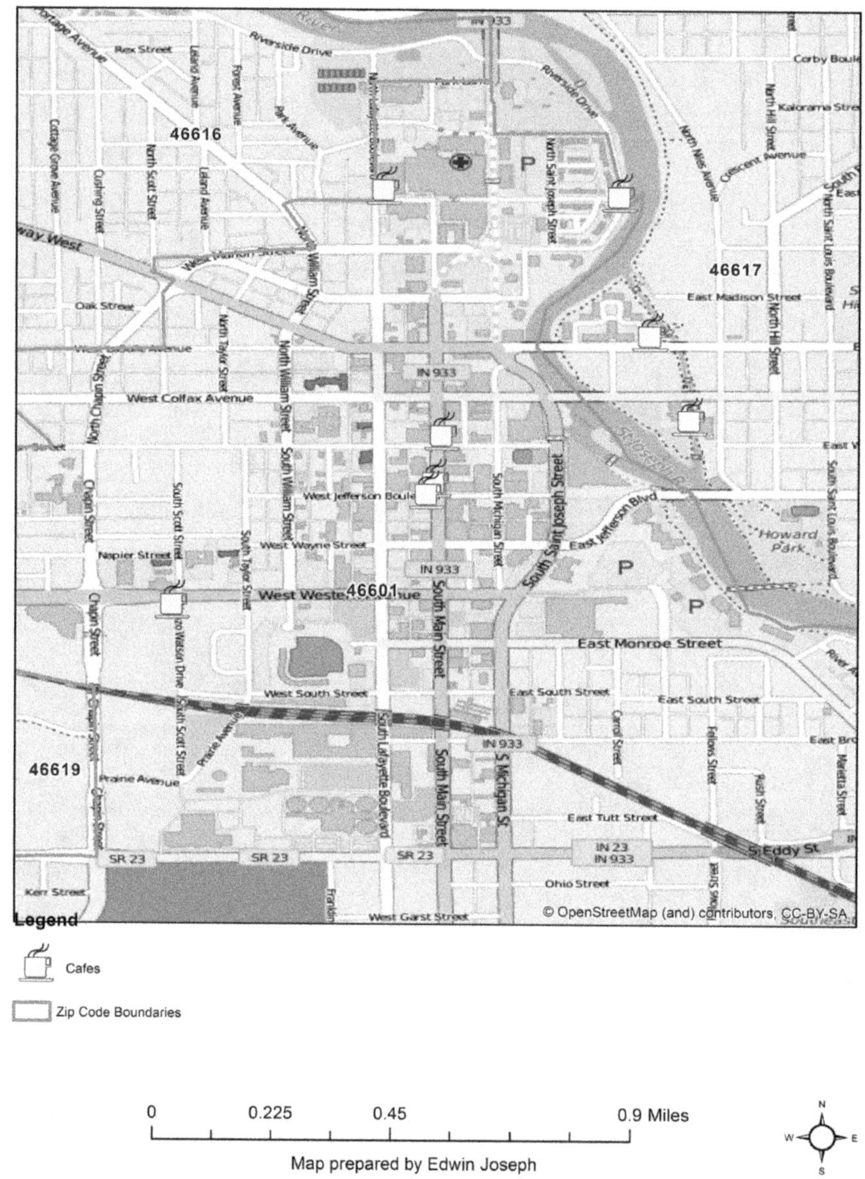

lenge for South Bend is putting together a conducive policy for urban agriculture, and urban land use. The city also has to provide easy access to vacant open urban spaces and partner with higher education institutions to develop formal urban garden programs that include technical workshops, participatory action research, and specific skills field training. Potential urban gardeners should not only have access to technical advice, but also credit (in the case where they seek agri-business opportunities within the city) for activities such as urban beekeeping, hydroponics, aquaponics, egg production and small stock (rabbits) production.

Figure 7. Location of schools
Schools—20 locations

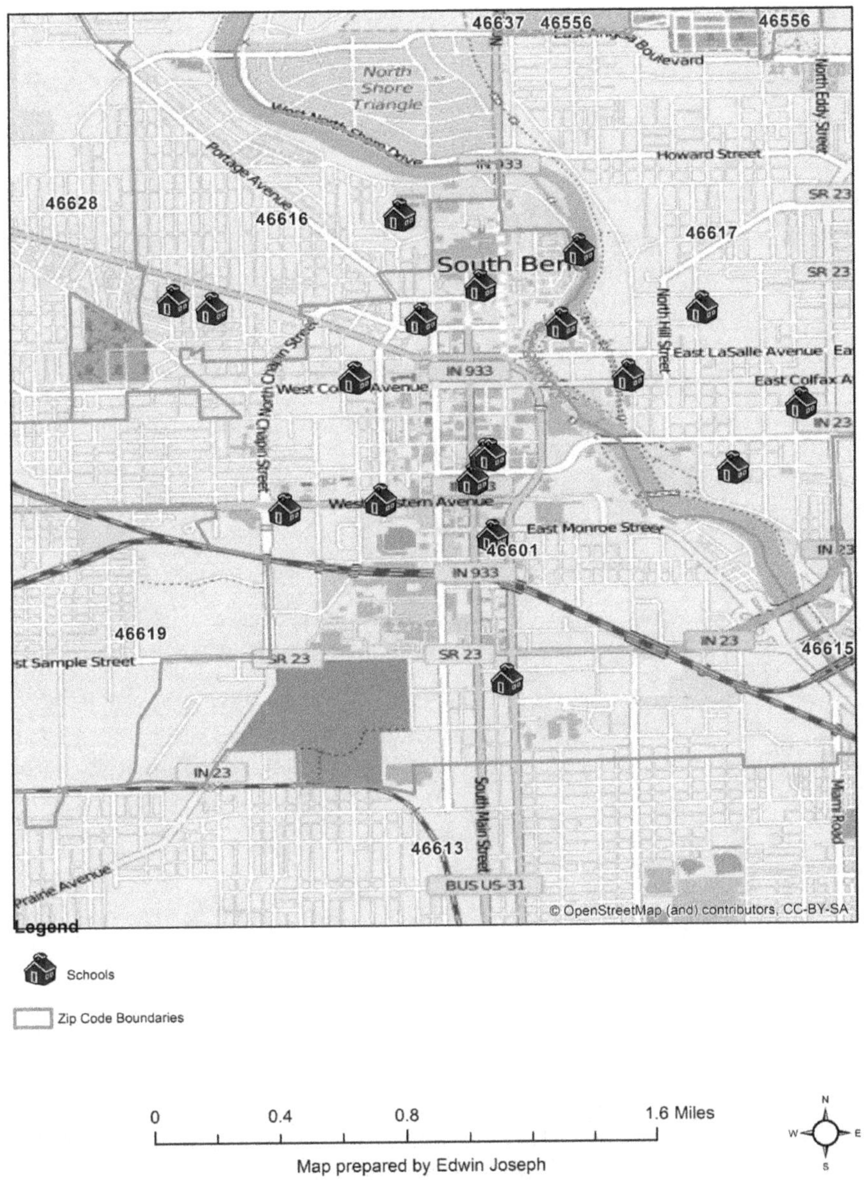

NEW DIRECTIONS FOR A FUTURE SOUTH BEND

At the 2016 State of the City address, Mayor Pete Buttigieg declared that South Bend was at the midpoint of a decade that would go down in history as the most transformative period for the city in modern times (Catanzarite, 2016). The Mayor reported that in 2015, four hundred and thirty-three houses were repaired through a combination of code enforcement and strong community partnerships. The Mayor identified

Figure 8. Distribution of parks
Parks—11 locations

community partners, for example, social service providers, neighborhood representatives, respected pastors, nonprofit stakeholders, police officers and local, county, and federal law enforcement officials working toward stopping violence.

The 2016 address also promised funding ($700 million) to improve the community, empower people, and allow individuals to live the life of their choosing. The Mayor stated that investing in people and

Figure 9. Distribution of libraries
Libraries—11 locations

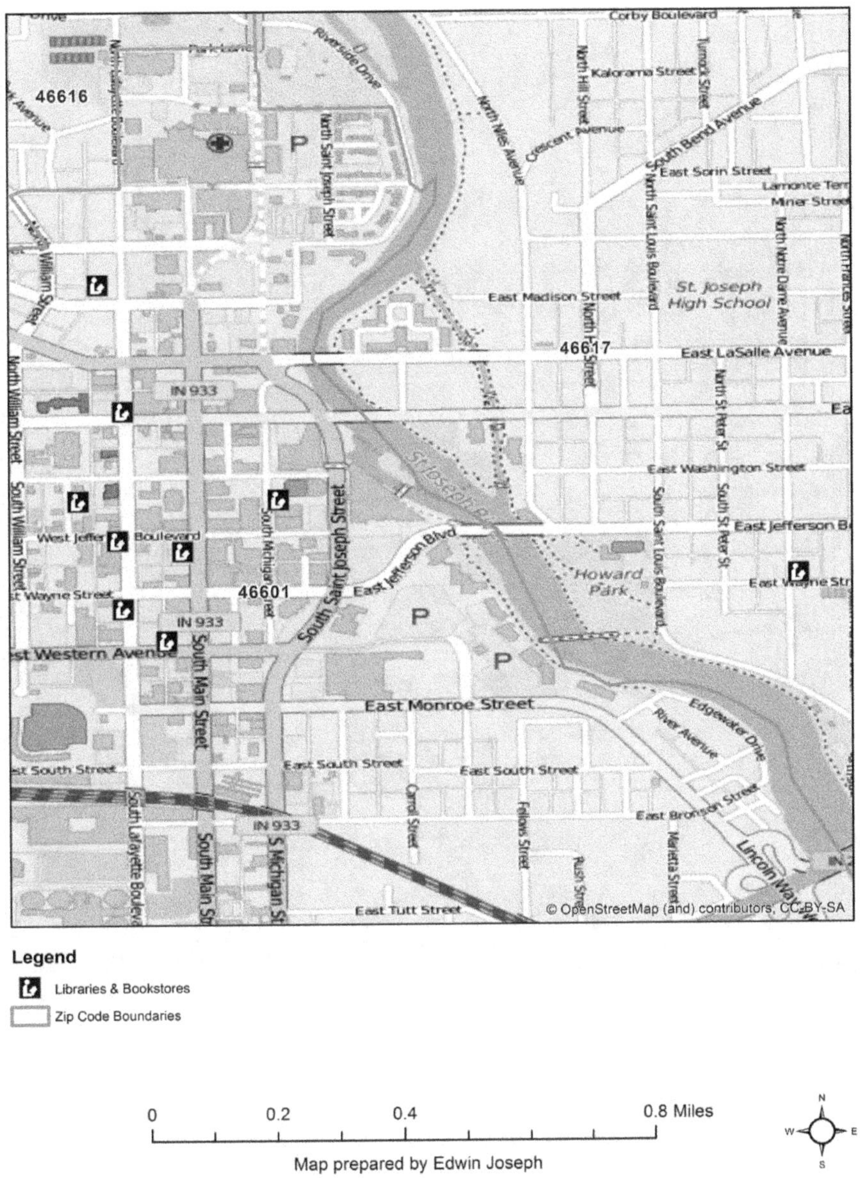

places meant investing in neighborhoods, and outlined the quality of life enhancements such as curbs and sidewalks, ramps, intersections, and wheelchair access completed. The city also invested in parks and recreation, and planted approximately 821 trees in 2015. A common theme in the address was the importance of including diverse community voices and perspectives, and focusing on unlimited imagination. In some ways, the mayor encourages imagination and sees it as a precursor to innovation.

Figure 10. Entertainment locations
Entertainment—11 locations

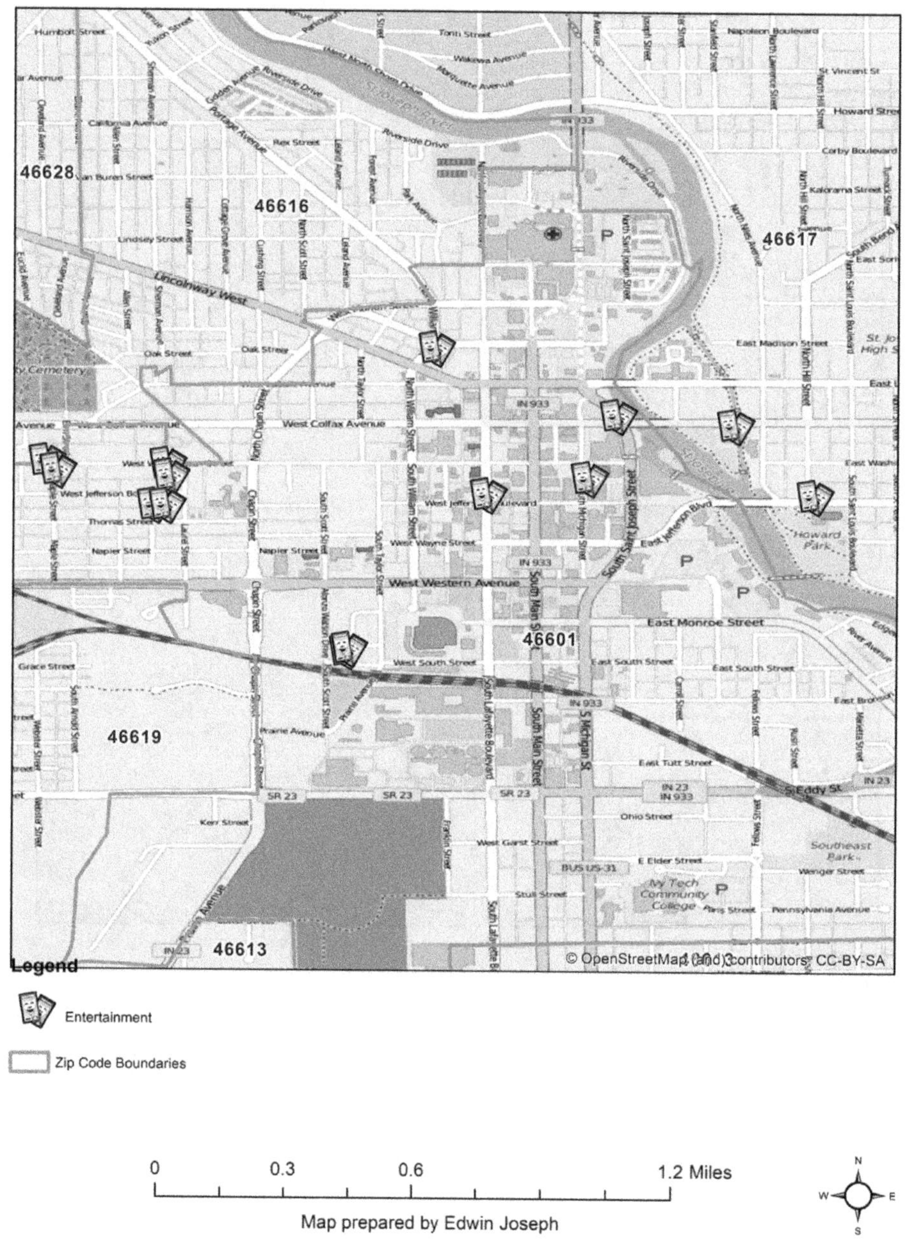

The city's interest in gardening was apparent in the Mayor's address when he spoke about his summer 2015 tour of a vegetable garden with a hundred 4th-graders. He also publicly thanked the leadership of the St. Joseph County Health Department and the Reducing Obesity Coalition of St. Joseph County because South Bend was recognized by First Lady Michelle Obama as the first city in Indiana to meet all five goals of the "Let's Move!" initiative for healthy kids.

Figure 11. Distribution of fitness centers
Fitness—6 locations

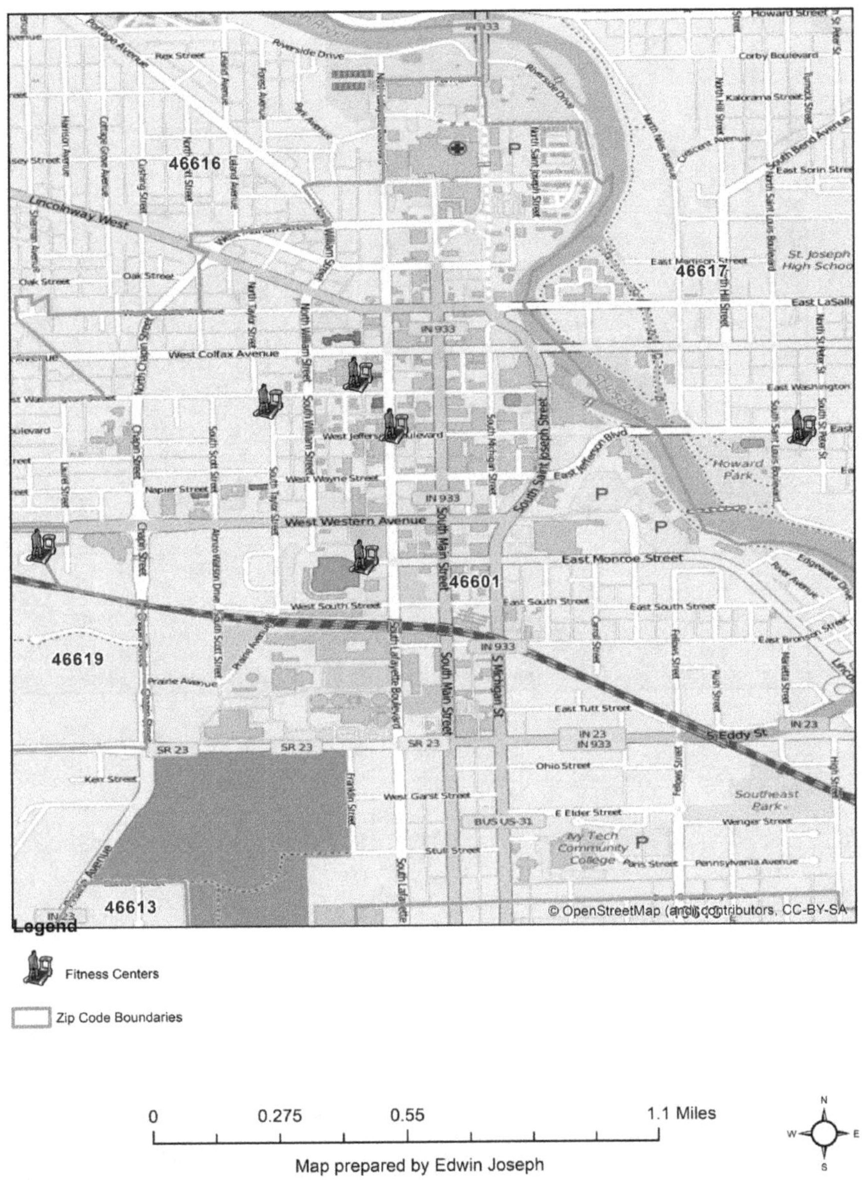

The city passed an innovative food truck ordinance, and created an online tracker that shows on a dynamic map where the food trucks are actually located, and is now working on an online tool to visualize vacant lots, so residents and potential investors or nonprofits could find ways to research and evaluate vacant land. These two initiatives demonstrate the need for expanding the use of GIS because the obvious benefits of GIS is the location factor. Knowing where things are, and how they change in space and time can be viewed in real time. This, however, requires dedicated programming, time, and effort.

Figure 12. Distribution of all amenities
All amenities

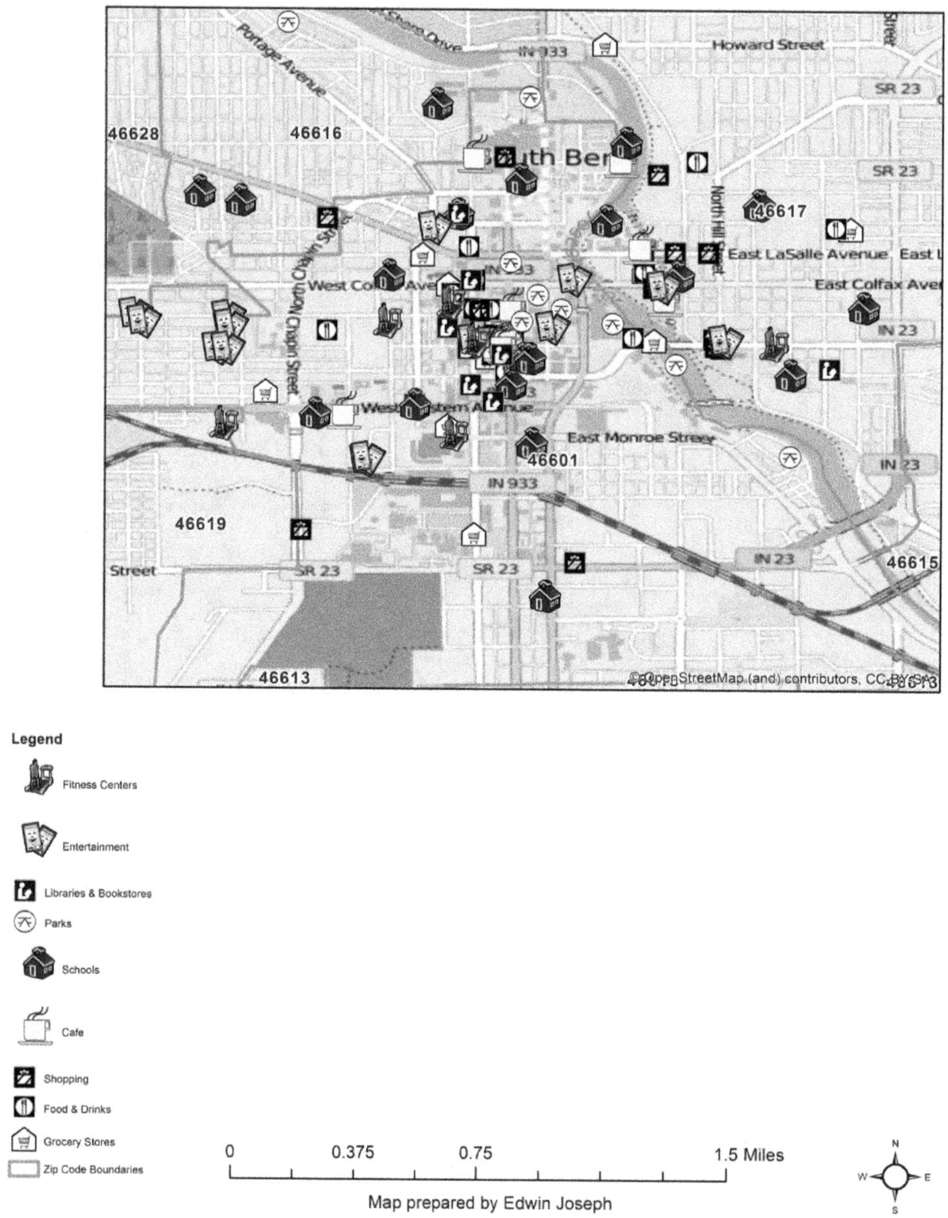

DISCUSSION AND CONCLUSION

GIS and GPS technology are indispensable but neglected tools. These spatial technologies have the potential to generate partnerships among food producers, volunteers, government and other institutions. Developing geographic awareness through collaborative mapping projects will help foster learning communities that can work together to develop sustainable urban garden initiatives—and may include

Figure 13. Residents with income below the poverty level (74.5%): Located roughly along West Sample St., & South Michigan St., zip code 46601 & 46613 (74.1%-76% poverty).

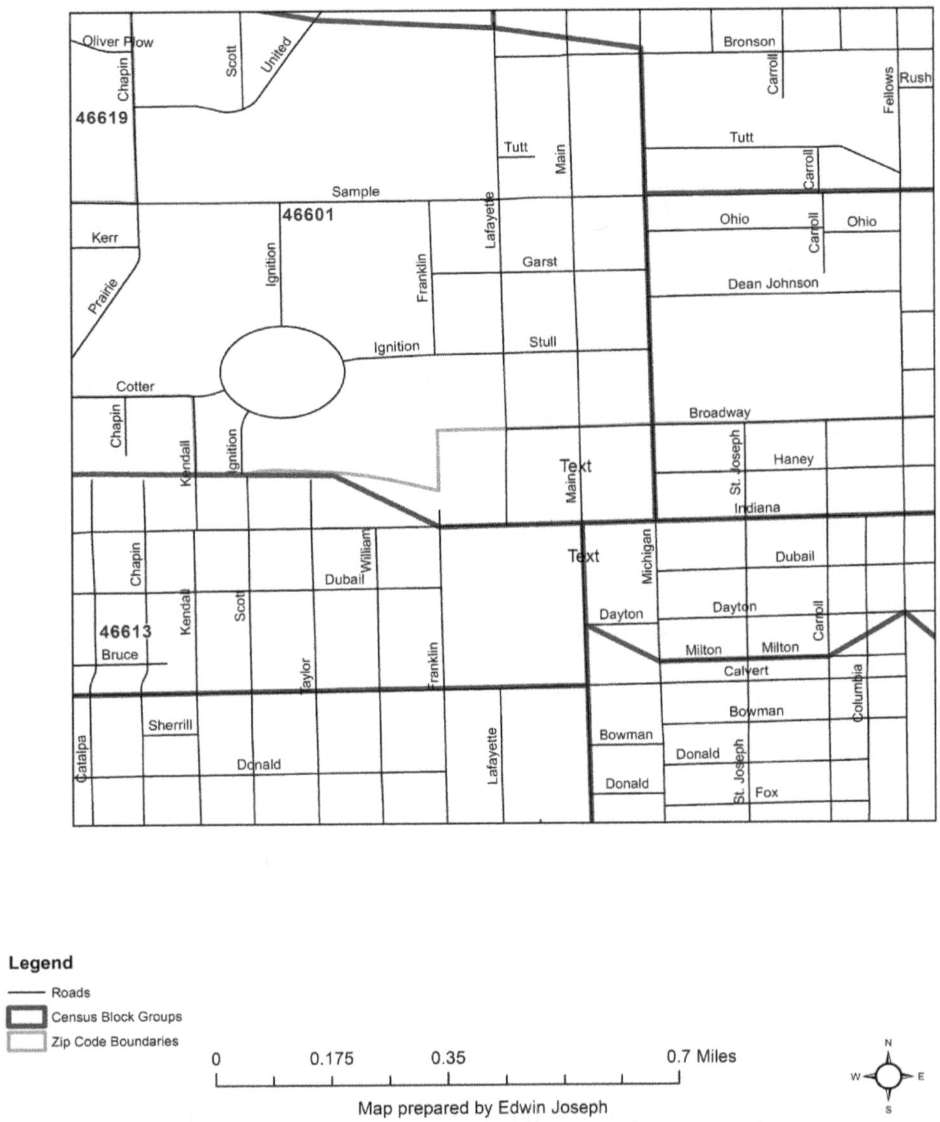

the use of Free and Open Source Software (FOSS) as valuable tools to help satisfy an overwhelming demand for spatial information.

In South Bend, spatial technologies will help maximize food production because of the ability to visualize spatial relationships and patterns, which will contribute to building a more sustainable environment (ArcUser, 2010). Through spatial analysis, we will understand the relationship among factors such as, soil type, pH, soil moisture, and nutrient deficiency. We can generate maps identifying where things are, how they move, and other important relationships.

Figure 14. Residents with income below the poverty level (64%): Located roughly along Lincoln Way West, and Ardmore Trail, zip code 46628 (62.7%-64.6% poverty)

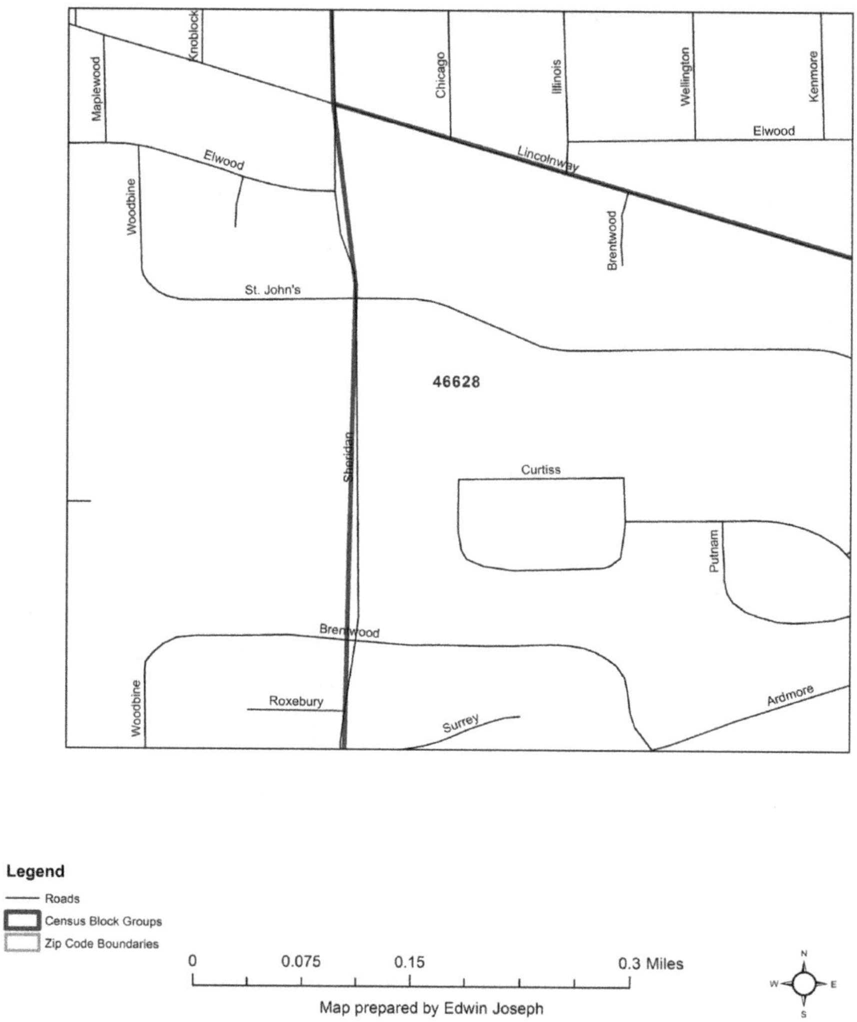

Spatial technologies will also connect community gardeners by introducing visual components to aid in site selection, planning, and modeling; and enable timely responses to issues that affect production. Integrating spatial technologies with urban gardening will provide an ideal platform for networking. As South Bend thrives to become a leader in sustainability, it has been shown how community participation will strengthen social relationships. South Bend needs to concentrate on growing more of its vegetables within the city. This will ensure healthier lifestyles and stronger communities. The aim is not just to improve health and decrease obesity, but create a more cohesive society with an equal distribution of incomes (Wilkinson & Pickett, 2009).

GIS and GPS technology has proven to be a quick and efficient approach for acquiring field data. GIS therefore, has the potential to become the hub or major planning tool for urban sustainability in

Figure 15. Residents with income below the poverty level (59.4%): Located roughly along West Washington St., & South William St., zip code 46601 (58.9%-60.8% poverty).

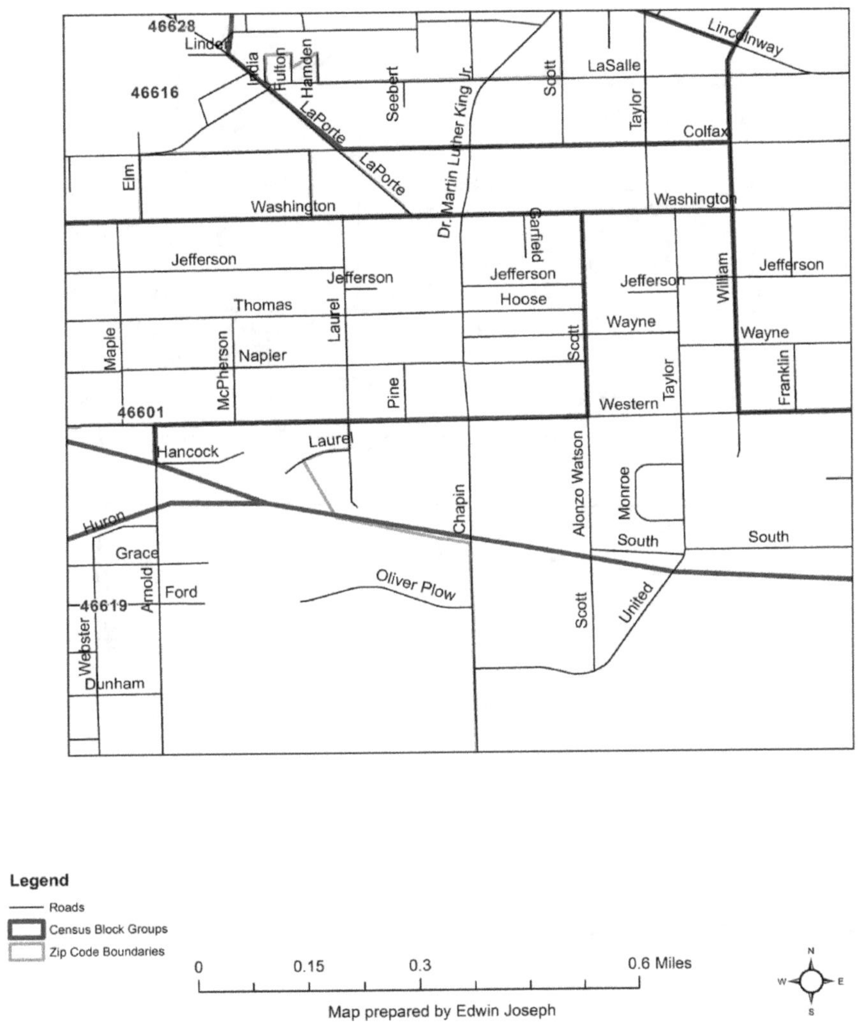

South Bend. When a GIS is combined with GPS, multimedia, and internet technologies, it provides a complete range of quantitative and qualitative research opportunities for future planning. Adopting these technologies would likely foster worthwhile sustainable investments, and would keep the momentum of GIS moving. However, the social, political, economic, and institutional climate must be supportive (North, 1990). A typical example of GIS in agriculture that is immediately applicable is pest and disease monitoring. Cell phones are now equipped with GPS, and a picture taken of a disease in the field can be crowdsourced and mapped in real time. Gardeners can seek information/solutions by uploading their images via community map servers where they can be viewed by experts.

There are many examples of communities transforming desolate landscapes into food oases. At Indiana University South Bend (IUSB), we recognize the importance of effective and sustainable urban land

Figure 16. Residents with income below the poverty level (49%): Located roughly within Lindsey St., Lincoln Way West, & North William St., zip code 46601 & 46616

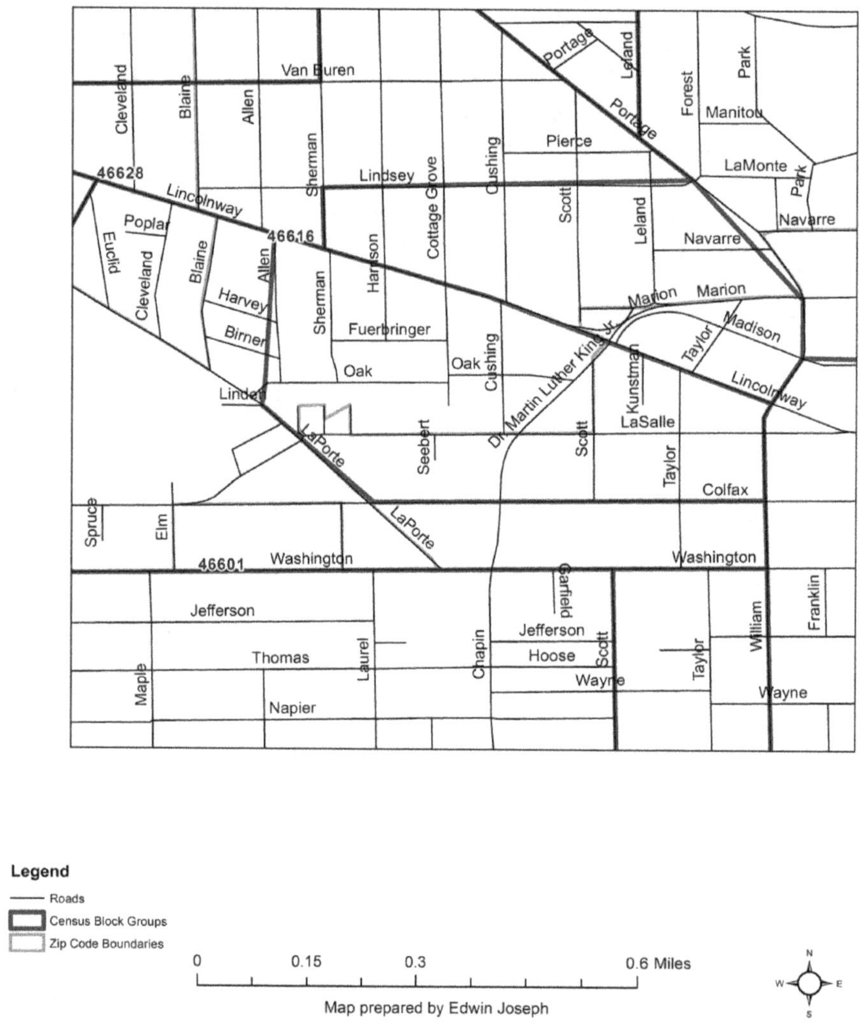

management practices, and have utilized geospatial technologies to map and monitor urban gardens in the city. As we look into the future we believe that the next wave of mapping technology would include AUVs (Aerial Unmanned Vehicles) for capturing landscape details that are otherwise unavailable to community gardeners. We believe that high resolution orthophotos (photographic maps) as overlays in Google Earth would help urban gardeners see the potential for online networking and collaboration. AUVs fitted with high-resolution visible-spectrum cameras and integrated GPS (Global Positioning System) are capable of acquiring overlapping aerial photos and elevations at garden sites. These would be invaluable for 3-D visualization, collaborative planning, and viewing temporal changes in the landscape.

It is hoped that in the near future, there would be more open access to city data so researchers would be able to share a platform that is built on collaboration. The current review of the spatial distribution of poverty in South Bend reveals a spatial pattern, but in-depth analysis was difficult because of incomplete and sometimes incorrect data entry from the data sources. Most of the analysis were limited to using on-line GIS resources that could not integrate new/external data, but simply reclassified in-house data, and generated outputs as static maps.

The extent of inner city agriculture can be easily mapped and monitored using crowdsourcing techniques. Crowdsourcing is an important data mining activity that taps into the collective intelligence of the public. Developing countries can take advantage of this platform to acquire data through user-generated media for example, blogs, Tweets, Wikipedia, Wikimapia, Facebook, and YouTube. GIS can be easily integrated onto these platforms. The integration of these services will enable the promotion of a network of food producers who can work together to alleviate poverty and malnutrition through community mapping and sustainable urban agriculture.

REFERENCES

AARP. (2016). *AARP Public policy Institute*. Retrieved from http://livabilityindex.aarp.org/search#South+Bend+IN+USA

ABC57 News. (2015, May 22). *Meet a descendant of one of South Bend's first African American families*. Retrieved from abc57 News: http://www.abc57.com/story/29135949/meet-a-decedent-of-one-of-south-bends-first-african-american-families

Andrews, W. L., & Adams, E. M. (2016). *North American Slave Narratives*. Retrieved from Documenting the American South: http://docsouth.unc.edu/neh/intro.html

ArcNews. (2014). *GIS Supports World Food Programme's Food Security Program*. Retrieved from http://www.esri.com/esri-news/arcnews/spring14articles/gis-supports-world-food-programmes-food-security-program

ArcUser. (2010). *The Next Generation of GIS. Dealing with design and space-time*. The Magazine for ESRI Software Users.

Areavibes. (2016). *South Bend IN, Livability*. Retrieved from http://www.areavibes.com/south+bend-in/livability/

Berman, E. (2011). *Local Food Systems: Selected Resources*. Retrieved from Science and Technology Resources on the Internet: http://www.istl.org/11-summer/internet.html

Buttigieg, P. (2015). *South Bend's New Economic Direction*. Retrieved from Pete Buttigieg, Mayor: http://www.peteforsouthbend.com/economicplan

Catanzarite, M. (2016, February 24). *Mayor Buttigieg delivers "State of the City" address*. Retrieved from http://www.wndu.com/content/news/Mayor-Buttigieg-delivers-State-of-the-City-addresss-369918351.html

CDW Services. (2015). *City-Data.com*. Retrieved from South Bend, Indiana (IN) Poverty Rate: http://city-data.com/poverty/poverty-South-Bend-Indiana.html

Center for the Homeless. (1988). Retrieved from Center for the Homeless: http://cfh.net/

Chappell, M. J., Wittman, H., Bacon, C. M., Ferguson, B. G., Barrios, L. G., Barrios, R. G., & Perfecto, I. (2013). Food sovereignty: An alternative paradigm for poverty reduction and biodiversity conservation in Latin America. *F1000 Research*, *2*, 235. doi:10.12688/f1000research.2-235.v1 PMID:24555109

Christensen, C. (1997). *The Innovator's Dilemma*. Harvard Business Review Press.

Citizen's Fair Housing Committee of South Bend-Mishawaka. (1963). *Transcript of public hearing concerning discrimination in the sale, rental and financing of private housing in South Bend and Mishawaka: held March 19, 1963 in Notre Dame Law School Auditorium. Discrimination in housing--Indiana--South Bend*. South Bend, IN: Citizen's Fair Housing Committee of South Bend-Mishawaka.

City of South Bend Common Council. (1965). *South Bend Fair Employment Practice Ordinance. City Ordinance No. 4091 prohibiting discrimination in employment, providing for the creation of the South Bend Fair Employment Practice Commission*. South Bend, IN: City of South Bend Common Council.

City of South Bend Indiana. (2016). *Interactive Mapping-GIS*. Retrieved January 2016, from City of South Bend: http://southbendin.gov/government/content/interactive-mapping-gis-0

Civil Rights Heritage Center. (2015). *Civil Rights Heritage Center*. Retrieved from https://www.iusb.edu/civil-rights/about-us/index.php

Department of Community & Economic Development. (2011). *2010 Census Report: South Bend-St. Joseph County Community Overview*. South Bend, IN: City of South Bend Department of Community & Economic Development Division of Community Development. Retrieved from https://www.southbendin.gov/sites/default/files/files/CI_Census2010Report.pdf

DESA. (2013). World Economic and Social Survey 2013: Sustainable Development Challenges. New York: United Nations Publication.

DUFI. (2015, August 15). Retrieved from Distributed Urban Farm Initiative: A community and economic growth project by Advent GX: http://dufi.adventgx.com/

ESRI. (2011, Spring). *South Bend, Indiana, Uses GIS for Brownfields Inventory*. Retrieved from ArcNews: http://www.esri.com/news/arcnews/spring11articles/south-bend-indiana-uses-gis-for-brownfields-inventory.html

Food Bank of Northern Indiana. (2016). *Foodbank of Northern Indiana*. Retrieved from http://www.feedindiana.org/mission

Foodbank of Northern Indiana. (2015). *About the Food Bank*. Retrieved from Foodbank of Northern Indiana: http://feedindiana.org/

FoodTank.com. (2013, October 12). Retrieved from Five Examples of Successful Urban Agriculture Done Differently Around the World: http://foodtank.com/news/2013/10/five-different-examples-of-urban-agriculture-from-around-the-world

GMDC. (2015, September 10). *Greenpoint Manufacturing and Design Center.* Retrieved from http://www.gmdconline.org/buildings/810-humboldt-street

Government Technology. (2016). *Disruptive Technology that Could Transform Government-Citizen Relationships*. Retrieved from Government Technology: http://www.govtech.com/budget-finance/The-Disruption-Pipeline-Government-Citizen-Relationships.html

Greensgrow. (2016). *What is Urban Farming*. Retrieved from http://www.greensgrow.org/: http://www.greensgrow.org/urban-farm/what-is-urban-farming/

Gundersen, C., Kreider, B., & Pepper, J. (2011). The Economics of Food Insecurity in the United States. *Applied Economic Perspectives and Policy*, *33*(3), 281–303. doi:10.1093/aepp/ppr022

Han, S. Y., & Kim, T. J. (1989). Can expert systems help with planning? *Journal of the American Planning Association*, *55*(3), 296–308. doi:10.1080/01944368908975417

Henderson, E., & Van En, R. (2007). *Sharing the harvest: a citizen's guide to Community Supported Agriculture*. Chelsea Green Publishing.

Hodgson, K., Campbell, M. C., & Bailkey, M. (2011). *Urban Agriculture: Growing Healthy, Sustainable Places*. APA Planning Advisory Service.

Indiana Demographics. (2016). *Indiana Cities by Population*. Retrieved from http://www.indiana-demographics.com/cities_by_population

Kumar, R. (2015, October 5). *How Urban Agriculture Is Revitalizing Local Economies*. Retrieved from The Blog: http://www.growingpower.org/

Lee, S. (2013, June 19). *The Chicago Council on Global Affairs*. Retrieved June 25th, 2015, from Global Food For Thought: http://globalfoodforthought.typepad.com/global-food-for-thought/2013/06/tackling-urban-food-security.html

Matthews, C. (2014, October 31). *Wealth inequality in America: It's worse than you think*. Retrieved from Fortune: http://fortune.com/2014/10/31/inequality-wealth-income-us/

Mayor's Biracial Committee Progress Report. (1963). *Progress report, July-October 1963, Mayor's Biracial Committee.* Retrieved from Michiana Memory: http://michianamemory.sjcpl.org/cdm/compoundobject/collection/p16827coll4/id/2286/rec/9

Montgomery, M. R., & Hewett, P. C. (2005). Urban poverty and health in developing countries: Household and neighborhood Effects. *Demography*, *42*(3), 397–425. doi:10.1353/dem.2005.0020 PMID:16235606

Newsweek. (2011, January 21). *America's Dying Cities*. Retrieved from http://www.newsweek.com/americas-dying-cities-66873

North, D. C. (1990). *Institutions, Institutional Change and Economic Performance*. Cambridge University Press. doi:10.1017/CBO9780511808678

Point2Homes. (2016). *South Bend*. Retrieved from pointhomes.com: http://www.point2homes.com/US/Neighborhood/IN/South-Bend-Demographics.html

Radican, E. A. (2010). *Growing Food Where it is Needed*. Indiana University, South Bend. South Bend Indiana: Center for a Sustainable Future (CSF). Retrieved November 10, 2015, from https://www.iusb.edu/csfuture/docs/WHITE%20PAPER%20NO.%202.pdf

Rivest, S. (2015, November 5). *ABC57 News*. Retrieved from www.abc57.com: http://www.abc57.com/story/30451531/local-businesses-work-to-combat-food-insecurity

Robinson, G. (2015). *Better Homes of South Bend: An American Story of Courage*. Charleston, SC: The History Press.

Schnell, S. M. (2007). *Food with a Farmer's Face: Community Supported Agriculture in the United States*. Retrieved from Researchgate.net: https://www.researchgate.net/publication/227701611_Food_with_a_Farmer's_Face_Community_Supported_Agriculture_in_the_United_States

Smart Growth America. (2016). *The Best Complete Streets Policies of 2015*. Retrieved from smartgrowthamerica.org: http://www.smartgrowthamerica.org/best-complete-streets-policies-of-2015

South Bend Region Economic Development. (2016). *Economic Profile*. Retrieved from southbendregion.com: http://southbendregion.com/index.php?src=gendocs&ref=EconomicProfile&category=Main

South Bend Tribune. (2014). *Can We Dream Big?* Retrieved from http://www.southbendtribune.com/app/economy/

South Bend Tribune. (2015, May 17). *South Bend's History Seen in Five Periods of Settlement*. Retrieved from http://www.southbendtribune.com/news/local/history/sb150/south-bend-s-history-seen-in-five-periods-of-settlement/article_79cb9dc6-f99e-11e4-b2c4-cbe41056da99.html

South Bend Tribune. (2016). *Did South Bend eserve the Harvard Ranking?* Retrieved from http://www.southbendtribune.com/news/local/did-south-bend-deserve-the-harvard-ranking/article_2000116e-5218-5175-8299-1d3a732165b2.html

STATSINDIANA. (2015, November). *STATSINDIANA Indiana's Public Data Utility*. Retrieved from Indiana Indepth Profile: http://www.stats.indiana.edu/profiles/profiles.asp?scope_choice=a&county_changer=18141

Suburban Stats. (2016). *Current South Bend, Indiana Population, Demographics and stats in 2016, 2015*. Retrieved from suburbanstats.org: https://suburbanstats.org/population/indiana/how-many-people-live-in-south-bend

The Chicago Council on Global Affairs. (2013). *Feeding an Urban World: A Call to Action*. Chicago: The Chicago Council on Global Affairs.

The History Museum. (2016). *Local African-American History*. Retrieved from historymuseusb.org: http://historymuseumsb.org/local-african-american-history/#

Links Incorporated. (2015). *The South Bend Area (IN) Chapter of The Links, Incorporated.* Retrieved from The Links Incorporated: http://southbendarealinksinc.org/About_Us.php

Human Relations Commission Praises County Housing Position. (1968, March 10). *The Reformer Newspaper.*

Trauger, A. (2009). Social agency and networked spatial relations in sustainable agriculture. *Area, 41*(2), 117–128. doi:10.1111/j.1475-4762.2008.00866.x

U.S. Bureau of Census. (1998, June 15). *Population of the 100 Largest Urban Places: 1920.* Retrieved from https://www.census.gov: https://www.census.gov/population/www/documentation/twps0027/tab15.txt

UBC News. (2014). *Study Reveals Unhappiest Cities in the US.* Retrieved from http://news.ubc.ca/2014/07/21/unhappy-cities/

Unity Gardens. (2009). *Unity Mission/Story.* Retrieved from Unity Gardens: http://theunitygardens.blogspot.com/p/unity-mission-story.html

USA.com. (2016). *South Bend, IN Income and Careers.* Retrieved from USA.com: http://www.usa.com/south-bend-in-income-and-careers.htm#Poverty-Level

USDA. (2015, September 8). *USDA ERS-Food Security in the US.* Retrieved from United States Department of Agriculture Economic Research Service: http://www.ers.usda.gov/topics/food-nutrition-assistance/food-security-in-the-us.aspx

Wikipedia. (2016, March 19). *Michiana.* Retrieved from https://en.wikipedia.org/wiki/Michiana

Wilkinson, R., & Pickett, K. (2009). *The Spirit Level: Why More Equal Societies Almost Always Do Better.* Allen Lane.

Wilson, W. J. (1987). *The Truly Disadvantaged.* Chicago, IL: Chicago Press.

WNDU. (2010, February 1). *Delve into South Bend's past on this Black History Month.* Retrieved from WNDU 16 South Bend Indiana: http://www.wndu.com/home/headlines/83231437.html

World Health Organization. (2015). *Trade, foreign policy, diplomacy and health.* Retrieved from WHO Food Security: who.int/trade/glossary/story028/en

zipcode.org. (2016). *City of South Bend, IN Zip Codes.* Retrieved from zipcode.org: zipcode.org/city/IN/SOUTHBEND

ENDNOTES

1 The data contained in this product are based on the Census 2000 sample. The data are estimates of the actual figures that would have been obtained from a complete count. Estimates derived from a sample are expected to be different from the 100-percent figures because they are subject to sampling and nonsampling errors.

2 The data are based on the Census 2000 sample. The data are estimates of the actual figures that would have been obtained from a complete count. Estimates derived from a sample are expected to be different from the 100-percent figures because they are subject to sampling and nonsampling errors.

3 The Livability Index is a signature initiative of the AARP Public Policy Institute to measure the quality of life in American communities across multiple dimensions: housing, transportation, neighborhood characteristics, environment, health, opportunity, and civic and social engagement (http://livabilityindex.aarp.org/?cmp=LVABLIDX_MAR25_015).

This research was previously published in Agricultural Development and Food Security in Developing Nations edited by Wayne G. Ganpat, Ronald Dyer, and Wendy-Ann P. Isaac, pages 263-299, copyright year 2017 by Information Science Reference (an imprint of IGI Global).

Chapter 13
The Economic Determinants of Food Security in the MENA Region

Ahmet Ali Koç
Akdeniz University, Turkey

Oznur Ozdamar
Adnan Menderes University, Turkey

Peyman Uysal
Akdeniz University, Turkey

ABSTRACT

This paper aims to analyse the determinants of the food insecurity examining the relationship of important economic,social and natural factors with an overall food insecurity index that derived by dividing the food imports value to the sum of total exports and the net remittance inflows. Overall analysis is made by employing panel-data methods using a dataset that covers 18MENA countries and Turkey for the period of 1990-2014.Emprical results support the evidence of the harmful effect of price increases on food security. Furthermore, significance of education related variables and fresh water withdrawals indicate that the region has had benefit from nearly 30 years investment on education and intensive water exploitation, although poor water management and great waste of irrigation will most likely have negative effects on food security in the near future.

INTRODUCTION

Middle East and North Africa (MENA) Region is a large geography that mainly consists of 18 countries[1]. Although these countries are different from each other regarding their economic development levels, they suffer from common environmental problems such as less amountof per capita arableland, improved water resources and their inappropriate usage which exarbatedfood insecurity. In addition to

DOI: 10.4018/978-1-5225-8063-8.ch013

these problems, the global food price shockoccurred in 2007-2008, was the main reason of worldwide attention to the food security phenomenon, especially for the MENA Countries where food insecurity has always been very serious concern.

Food security, as defined by FAO, is the condition that exists when all people, at all times have physical, social and economic access to sufficient, safe and nutritious food to meet their dietary needs and food preferences for an active and healthy life[1]. Food security can be disscussed in twofolds: a) Macro-Level Food Security and b) Micro-Level Food Security. The first one refers to a national phenomenon which exists when a nation has enough food supply to feed its population via domestic production, food imports or some combination of both. The second one is individual food security which can be defined as accession of all individuals to the adequate amount of food. These paper aims to analyse the determinants of macro-level food security taking into account various factors at country level for 19 countries (MENA plus Turkey). Evolved from the definition, food insecuritycan exist when one of those following conditions occurs; a) when access to food is not possible as food is not physically available where and when needed, b) when economic access to food is not available as people lack money to afford it even food is physically available, c) when people have physical and economical access to food that are not safe or dietary balanced, d) when none of the above conditions holds. In other words, food insecurity problem can be caused by both demand and supply side factors.Sen (1981) emphasizes on not only the role of supply side factors of food, but also demand side factors such as individuals incomes and assets that are also crucial for ensuring food security. In light of this view, FAO hadclarified four pillars of food security which are availibility, access, utilization and stability to guide addressing the problem with its supply and demand side and form appropriate policy solutions. Table 1 shows the detailed explanations on four pillars presenting the variables that are main components of these pillars.

According to the latest data, in terms of five components of availability dimension Israel is the most food secure country in the region. In terms of access dimension, oil exporting countries rank at the top as food secure while Yemen is in the most food insecure position. Oil exporter countries such as Israel, UAE, Oman, Saudi Arabia and Kuwait are the most ceral import dependent countries.

This study aims to investigate the relationship between some componentspresented in Table 1 and macro level food insecurity index that is derived by dividing the food imports value to the sum of total exports and the net remittance inflows. In IFPRI's definition, the index captures the ability of a country to finance food imports through exports of goods and services and the net remittances received (International Food Policy Research Institute IFPRI, 2010). The data on share of food imports in merchandise imports and the share of merchandise imports in GDP is obtained from the latest dataset of World Development Indicators Databank and used for finding the value of food imports value. The same procedure is followed to find the total exports value. Net remittance inflows are calculated as the difference between remittances received and remittances paid which is also available in the World Development Indicators Databank.

This macro-level food insecurity index will be the dependent variable, namely the outcome of the interest of our research. Some of the components of the aforementioned four pillars of food security, that are taken as independent variables, are chosen based on the data availability. The data limitation on thisissue, especially for the MENA Countries, is a common problem and not all data on the components of the pillars is available. Due to the limitation of the data availability, the variables included into analy-

Table 1. FAO food security indicators

Dimensions	Components
1. Availability	1.1 Average dietary energy supply adequacy
	1.2 Average value of food production
	1.3 Share of dietary energy supply derived from cereals, roots and tubers
	1.4 Average protein supply
	1.5 Average supply of protein of animal origin
2. Access	2.1 Percent of paved roads over total roads
	2.2 Road density
	2.3Rail lines density
	2.4 Gross domestic product per capita (in PPP)
	2.5 Domestic food price index
	2.6 Prevalence of undernourishment
	2.7 Share of food expenditure of the poor
	2.8 Depth of the food deficit*
	2.9 Prevalence of food inadequacy
3. Utilization	3.1 Cereal import dependency ratio
	3.2 Percent of arable land equipped for irrigation
	3.3 Value of food imports over total merchandise exports
	3.4 Political stability and absence of violence/terrorism
	3.5 Domestic food price volatility
	3.6 Per capita food production variability
	3.7 Per capita food supply variability
4. Stability	4.1 Access to improved water sources
	4.2 Access to improved sanitation facilities
	4.3 Percentage of children under 5 years of age affected by wasting
	4.4 Percentage of children under 5 years of age who are stunted
	4.5 Percentage of children under 5 years of age who are underweight
	4.6 Percentage of adults who are underweight
	4.7 Prevalence of anaemia among pregnant women
	4.8 Prevalence of anaemia among children under 5 years of age
	4.9 Prevalence of vitamin A deficiency in the population
	4.10 Prevalence of iodine deficiency

Source: FAO, 2013

*Depth of food deficit is measured by comparing the average amount of dietary energy that undernourished people get from the foods they eat with the minimum amount of dietary energy they need to maintain body weight and undertake light activity.

sis are; population growth, per capita GDP (PPP), per capita arable land, access to an improved water source, unemployment ratios, female literacy rate, improved sanitation access, government expenditures on education, inflation and exchange rates. The missing observation problem for those variables is solved using the "ice imputation technique" in STATA.

AN OVERVIEW: FOOD SECURITY IN MENA REGION AND RELEVANT DETERMINANTS

The report published by FAO in 2015 indicates that 805 million people in the world (about one in nine of the world's total population) are chronically undernourished with insufficient food for an active and healthy life. The sta7tistics for MENA region is about 33 million people. The report of IFPRI classifies the MENA countries and Turkey into five risk groups in which Bahrain, Kuwait, Qatar, Oman, Saudi Arabia and United Arab Emirates (UAE) are in low risk group while Turkey, Tunusia, Libya and Iran are in medium risk group. All the countries except the ones listed above are considered as suffering from seriousfood security challenges. According to an economic brief of African Development Bank (2012), once macro level indictors (such as dependence on food imports and fiscal position) are taken into consideration, all of the MENA countries seem to be harmed by food insecurity problems.

The countries that compose MENA region are extremely diversified both in terms of food supply and consumption. Several countries, including Turkey, Morocco and Israel, are major food exporters and self-sufficient in most agricultural products. On the other hand, the arid Gulf Arab states are heavily dependent on food imports, and their governments are therefore primarily focused on ringfencing supplies from abroad, while at the same time trying to expand food storage capacities domestically. Supporting especially the situation of Gulf States, Aşcı et al., (2014) found that food exportation has a decreasing trend with the increasing level of income. Income and wealth levels vary enormously across the countries in MENA, with GDP per capita at purchasing power parity (PPP) rates ranging from USD 3791 in Yemen to the USD 64,229 in UAE which is one of the richest Gulf States (World Bank, 2013).

As FAO (2013)indicates, one of the main indicators of vulnerability in terms of food security is cereal import dependency ratios, which are listed in Table 2 for the MENA countries. The lowest ratio belongs to Iran and Egypt (33-35%) and all other countries have high cereal dependency ratios which means that these countries tend to have big challenges in terms of supplying basic staples in the future.

The shares of food products in total exports and imports in MENA countries are presented in Table 3. According to the presented statistics, food accounts for the significant portion of total imports in MENA countries especially in Yemen, Egypt and Algeria.

Population and Population Growth Rates

MENA countries constitute about 5.6% of the world population with 404,2 million people according to World Bank data. The recent data shows that annual population growth rate of MENA countries is 2.25% and this growth rate is approximately twice as high as the world population growth rate of 1.01% (Table 4).The population structure of MENA countries include a majority of young people aged 15-64 which means potentially high amount oflabour supply but also a disadvantage as it brings the risk of high unemployment rates which are themain driving factor of food insecurity in the region.

Table 2. Cereal dependency ratio (%)

Countries	1990-92	1995-97	2000-02	2005-07	2007-09
Egypt	37,9	37,2	36,3	35,6	35,5
Tunisia	35,0	58,2	79,5	60,9	60,2
Syrian Arab republic	30,3	11,2	19,0	33,7	49,1
Lebanon	89,4	90,4	87,1	84,6	88,5
Jordan	93,7	93,9	96,0	98,1	100,1
Morocco	27,2	37,0	57,5	48,6	53,6
Algeria	62,4	66,9	80,5	67,5	70,7
Yemen	72,1	73,8	80,0	80,7	82,5
Kuwait	101,8	110,2	102,4	101,0	102,4
Libya	89,9	91,7	92,0	91,5	91,8
UAE	153,4	134,2	141,9	145,4	140,0
Saudi A.	48,4	75,1	71,5	77,1	82,9
Iraq	44,1	38,5	65,6	56,7	61,5
Israil	87,7	92,5	94,1	92,7	94,3
Iran	27,1	31,5	36,6	17,2	32,6

Source:FAO, 2013.

Regarding the urban and rural population shares in total population, rural population has a decreasing trend starting from early 1990s till today indicating an ongoing migration from rural areas to cities. Fertility rate is another potential indicator of food insecurity which generates population growth. Saudi Arabia has the highest fertility rate among MENA countries with a level of 80.3% in 2013. Although the population growth in the region is still a threating factor for the food insecurity, it is important to note that all MENA countries achieved a decrease in fertility rates with the help of national family planning programmes in 2000s compared to 1990s.

Economic Determinants

Although the average GDP per capita (PPP) in the region is about USD 52,892, in terms of income levels, absolute differences is observed among MENA countries. Considering per capita income, the first four countries in the relevant ranking are Israel, Kuwait, Qatar, and the United Arab Emirates which enjoy an average per capita GDP around 77,230 USD compared with 3,791 USD for Yemen, which is the poorest country in the region. Table 5 presents supporting statistics that MENA is an economically diverse region that includes countries with a common heritage but various stages of economic development and growth levels.

Despite undertaking economic reforms in many countries, the region's economic performance in the past 30 years has been below its potential. The purpose of this situation can be linked to the global economy and employment generation. A detailed exploration of the reasons for this outcome is needed to propose an agenda for urgent reforms.

Table 3. Share of food in imports and exports in MENA countries (Billion USD) (%)

Countries	2001				2005				2010				2014			
	Total Exports	Food Exports (%)	Total Imports	Food Imports (%)	Total Exports	Food Exports (%)	Total Imports	Food Imports (%)	Total Exports	Food Exports (%)	Total Imports	Food Imports (%)	Total Exports	Food Exports (%)	Total Imports	Food Imports (%)
Egypt	4.1	9.9	12.7	25.9	10.6	8.7	19.8	20.1	26.3	17.2	53.0	19.1	29.4	14.3	69.9	22.4
Tunusia	6.6	7.9	9.5	8.3	10.5	10.4	13.2	8.5	16.4	7.7	22.2	9.3	17.0	n.a	24.5	n.a
Syria	5.0	6.2	4.6	12.4	6.4	13.4	7.9	17.8	11.4	21.0	17.6	21.0	2.1	n.a	7.5	n.a
Lebanon	0.9	19.3	7.3	17.5	1.9	16.2	9.3	15.4	4.3	14.9	18.0	16.3	4.4	19.9	21.1	16.2
Jordan	2.3	14.9	4.9	17.7	0.0	15.5	0.0	13.6	7.0	16.6	15.3	16.2	7.9	18.5	20.7	17.4
Morocco	7.1	21.0	11.0	14.4	11.2	21.2	20.8	10.6	17.8	19.0	35.4	11.4	21.4	17.2	44.8	12.4
Algeria	19.1	0.1	9.9	26.1	26.1	0.2	20.4	19.3	57.1	0.6	41.0	16.3	71.9	0.4	50.4	19.9
Yemen	0.0	1.8	0.0	20.0	20.0	4.4	5.4	28.1	6.4	6.6	9.3	30.8	7.1	6.3	11.3	32.5
Bahrain	5.6	0.6	4.3	11.7	11.7	0.4	9.3	6.8	16.1	1.9	16.0	7.6	11.5	n.a	9.1	n.a
Kuwait	16.2	0.2	7.9	12.7	12.7	n.a	0.0	n.a	67.6	n.a	19.1	n.a	113.5	n.a	20.1	n.a
Qatar	10.7	0.0	3.8	11.1	11.1	0.1	10.1	6.6	75.1	0.0	23.3	8.2	133.4	n.a	26.2	n.a
Libya	0.0	n.a	0.0	n.a	n.a	n.a	0.0	n.a	36.4	0.0	17.7	12.1	58.4	n.a	19.6	n.a
UAE	0.0	0.8	0.0	10.7	10.7	1.1	80.8	7.1	161.2	n.a	161.5	n.a	213.3	n.a	211.6	n.a
Oman	68.0	0.6	31.2	15.6	15.6	0.6	59.5	14.8	251.1	1.2	106.9	15.7	388.4	0.9	155.6	13.9
Iraq	11.0	6.1	5.8	22.4	22.4	2.5	9.0	11.8	36.6	2.5	19.8	12.1	52.1	2.1	28.1	10.3
Israel	12.0	0.9	5.4	0.9	0.9	0.2	11.7	n.a	50.7	n.a	28.9	n.a	92.4	n.a	34.8	n.a
Iran	29.1	2.3	33.3	5.9	5.9	2.4	45.0	5.5	58.4	3.0	59.2	7.3	63.1	3.1	73.1	7.1
Turkey	31.3	12.7	41.4	3.7	73.5	10.5	116.8	2.9	114.0	10.6	185.5	4.0	152.5	10.8	236.5	4.5

Source: WDI, 2013; FAO 2013

In addition to the poor economic performance in most of the MENA countries, shocks on world agricultural commodity prices (especially on oil prices) and changes in exchange rates have beenalso important factors creating food security/vulnerability in MENA region. Recentpast global food price spikes have prompted countries and international institutions to think more about the new national food security strategies against price shocks.

The rise in the food prices in MENA countries is higher when compared to other countries in the world. Although all the countries were exposed to the price shock in 2007-2008, avarage world price returned to the price level of 1990's except in a few countries, but food prices in MENA region continued to rise even after 2010. Larson et.al (2013) state that the volatility of food prices will go on and the prices will never return to the level of 1990's in the region and these price shocks will create harmful risks in the future.

Another important economic determinant to achieve food security might be improving employment issues. Investment in employment and labor market developmentis needed in the region in order to reverse ongoing trend in unemployment rates. Especially increasing the number of employees in rural areas, the self-production of food can be achieved by agricultural workers as a contribution to their own food consumption. Second, the amount of income generated from any typesof employment determine the

Table 4. Population and population growth rates of MENA countries

Countries	Population (Million)		Rural Share		Growth Rate (%)	Population Structure		FertilityRate	
	1990	2014	1990	2014	1990-2014	Ages 15-65	Ages >65	1990	2013
UAE	1.8	9.4	20.9	14.7	422.2	83.8	0.5	54.6	27.3
Bahrain	0.5	1.3	11.9	11.3	160.0	76.4	2.3	26.4	13.6
Algeria	26.2	39.9	47.9	29.9	52.3	67.4	4.6	28.0	9.6
Egypt	56.3	83.4	56.5	56.9	48.1	63.2	5.9	77.5	42.0
Iran	56.4	78.5	43.7	27.1	39.2	70.7	5.4	89.9	31.5
Iraq	17.5	34.3	30.3	30.6	96.0	57.2	3.2	70.5	67.7
Israel	4.7	8.2	9.6	7.9	74.5	61.3	10.8	20.4	6.6
Jordan	3.2	6.6	26.7	16.6	106.3	62.7	3.6	52.1	26.0
Kuwait	2.1	3.5	2.0	1.7	66.7	72.9	2.4	23.5	14.4
Lebanon	2.7	4.5	16.9	12.3	66.7	71.1	8.7	43.8	11.9
Libya	4.3	6.3	24.3	21.6	46.5	65.7	4.9	9.5	2.4
Morocco	24.7	33.5	51.6	40.3	35.6	67.1	5.0	81.3	10.3
Oman	1.8	3.9	33.9	22.8	116.7	74.4	2.9	49.5	9.0
Qatar	0.5	2.3	7.2	0.8	360.0	85.4	1.0	69.1	10.0
Saudi A.	16.2	29.4	23.4	17.1	81.5	68.4	2.9	105.4	80.3
Syria	12.5	23.3	51.1	42.7	86.4	61.0	4.2	73.6	40.9
Tunisia	8.2	11.0	42.1	33.4	34.1	69.5	7.3	20.4	4.4
Turkey	54.0	75.8	40.8	27.1	40.4	67.2	7.5	64.9	29.4
Yemen. Rep.	11.8	25.0	79.1	66.0	111.9	57.4	2.9	150.0	45.6

Source: WDI, 2015

amount and quality of food that people can purchase. Especially for poor people who spend almost 70% of their income on food, being employed is really crucial[2].However, region faces high unemployment rates; particularly for the youth and womenacoording to World Bank data (see Table 6).

At the same time, providing job opportunities for the youth is a challenge due to the difficulties in employment generation which is typically slower than population growth of young people. Female employment rates are low as well. Globally, women benefit less than men from employment, mirroring a broader context of gender inequalities. Although female unemployment rates decreased between 2011 and 2012 in most of the countires in the region, they still remain high. According to the FAO (2016) study, women play a key role in food security. The report states that "…women produce between 60 and 80% of the food in most developing countries and are responsible for half of the world's food production, yet their key role as food producers and providers and their critical contribution to household food security is only recently becoming recognized…"

Environmental Conditions

Environmental problems could be one of the main reasons of food insecurity in MENA countries. For instance, fresh water resources are shared between countries and the rate of dependency to fresh water

Table 5. Gross domestic product values (current $) of MENA countries and GDP per capita growth

Countries	1990		2014		1990 to 2014
	GDP (Bilion $)	GDP per Capita($)	GDP (Billion $)	GDP per Capita($)	GDP per Capita Growth (%)
UAE	50.7	28,066	399.5	42,522	152
Bahrain	4.2	8,529	33.9	25,198	295
Algeria	62.0	2,365	213.5	5,361	227
Egypt	43.1	766	301.5	3,436	449
Iran	124.8	2,059	425.3	5,293	257
Iraq	179.9	10,270	223.5	6,433	63
Israel	52.5	11,264	305.7	37,032	329
Jordan	4.2	1,312	35.8	5,423	413
Kuwait	18.4	8,947	163.6		
Lebanon	2.8	1,050	45.7	10,139	966
Libya	28.9	6,785	41.1	6,575	97
Morocco	28.8	1,159	110.0	3,140	271
Oman	11.7	6,455	81.8	20,832	323
Qatar	7.4	15,446	210.1	93,397	605
Saudi A.	116.8	7,206	753.8	25,409	353
Syria	12.3	989			
Tunisia	12.3	1,507	48.6		
Turkey	150.7	2,791	798.4	10,543	378
Yemen	5.6	479			

Source: WDI, 2015.

is very high in all MENA countries. Moreover, inefficient usage of water in agricultural production and high population growth rates decrease the level of per capita fresh water to scarcity levels slowly (see Table 7 for more information). Although fresh water withdrawals have beenwell exploited since 30 years to produce food and reduce the food gap, MENA countries face great waste in irrigation water, poor management of water in agriculture sector to risk the future food productions in the region (AFESD, 2012). According to the reports of UN-ESCWA, the estimated water demand in the region will increase about 50% between 2000-2025. On the other hand,compared to today, the water sources are estimated to decrease about 50% by 2050 (World Bank, 2007). Besides the fresh water problem, there are other environmental problems in the region like droughts and desertification. While unconscious usage of meadows (for example excessive pasturage), devastation of forests and wrong application of agricultural techniques can be defined as human sourced problems, climate changes alsoworses off the food security in the region as a non human sourced problem.

Although the extent of droughts and desertification differ between countries, two thirds of the total land of the region is faced with the danger of desertification. The most desertificating area in the region

Table 6. Unemployment rates, (% of total labor force)

Countries	1991	1995	2000	2005	2010	2014
United Arab Emirates	3.0	1.8	2.3	3.1	4.2	3.6
Bahrain	4.2	4.2	4.5	4.4	3.6	3.9
Algeria	20.6	27.9	29.8	15.3	10.0	9.5
Egypt, Arab Rep.	9.6	11.3	9.0	11.2	9.0	13.2
Iran, Islamic Rep.	11.1	11.9	12.0	12.1	13.5	12.8
Iraq	19.8	19.4	18.2	18.0	15.2	16.4
Israel	10.6	6.9	8.8	9.0	6.6	6.1
Jordan	15.4	14.6	13.7	14.9	12.5	11.1
Lebanon	1.2	0.7	0.8	2.0	1.8	3.0
Libya	8.4	8.1	8.9	8.2	6.2	6.4
Morocco	18.8	19.0	19.7	19.5	18.5	19.2
Oman	12.5	14.7	13.6	11.0	9.1	10.2
Qatar	7.0	6.5	8.0	7.8	7.6	7.2
Saudi Arabia	0.5	0.4	1.0	1.1	0.4	0.3
Syrian Arab Republic	6.6	5.4	4.6	5.9	5.4	5.6
Tunisia	9.4	7.2	9.5	9.2	8.4	10.8
Turkey	16.8	14.5	15.7	14.2	13.0	13.3
Yemen, Rep.	8.2	7.6	6.5	10.6	11.9	9.2

is Arab Peninsula andthe North African countries are following it (Drakenberg & Wingqvist, 2010). The share of arable land is decreasing year by year in almost all MENA countries except Egypt (see Table 8 for more information).A similar efficiency loss caused by desertification and droughts is seen in aquaculture products in the coastal areas of the region. Almost all countries with coasts to Mediterranean Sea are facing problems because of the industrialization sourced chemical waste which causes pollution and loss of biodiversity in the region. For example Egypt with population of 84 millions, is the first in the ranking of MENA countries causing pollution. Evacuation of non refined industrial wastes and the chemicals used in agricultural production to Mediterranean Sea is one of the major problems in the region. Also Gulf countries are in the high risk group because of the activities of oil/energy plants and firms (AFED, 2008).

Besides these human caused problems, climate changes also have more impact on MENA countries than many countries in the world. MENA, is the most affected region from global warming in the world that is expected from the region to exceed the limit of water usage by 2025. As a result, ongoing and increasing water scarcity and droughts are striking a blow to food production in MENA region. As a contribution to thenon productiveness of food sector in the region, unemployment, poverty and vulnerability of food security are increasing with negative effects as well. In addition to those problems, lower health status of individuals caused by water and air pollution are also threatening the human capital potential of the region.

Table 7. Renewable internal freshwater resources per capita (m^3) and % of decrease between 1992-2014

Countries	1992	2014	% of Decrease
United Arab Emirates	74,3	16,5	77,8
Bahrain	7,7	2,9	62,3
Algeria	413,9	288,9	30,2
Egypt, Arab Rep.	30,5	20,1	34,1
Iran, Islamic Rep.	2210,6	1644,4	25,6
Iraq	1905,9	1011,1	46,9
Israel	146,4	91,3	37,6
Jordan	182,7	103,2	43,5
Kuwait		0,0	
Lebanon	1701,0	1055,7	37,9
Libya	152,3	111,8	26,6
Morocco	1121,1	854,9	23,7
Oman	707,1	330,5	53,3
Qatar	114,4	25,8	77,4
Saudi Arabia	137,9	77,7	43,7
Syrian Arab Republic	540,9	321,9	40,5
Tunisia	494,1	381,5	22,8
Turkey	4067,3	1519	26,5
Yemen, Rep.	158,6	80,2	49,4

Source: WDI, 2015

Data and Methodology

Measuring food security and determining its indicators as well as their impacts on food insecurity has been an interest for researchers especially since recent food price spike in 2007/2008. Some studies examine food insecurity level as a national case while others focus on the issue as a global problem. As an example to a national study, Applanaiduet. al., (2013) have analysed the dynamic relationship between selected macroeconomic variables (biodiesel production, exchange rate, government expenditure on rural development, GDP, food price index and population) and food security in Malaysia using VAR approach. The results of the study indicate that biodiesel production, exchange rate and government expenditure on rural development would give the highest shock to food security in ten years whereas exchange rate and population in five years and GDP in six years.

Garduno-Diazand Garduno-Diaz (2015) used Global Food Security Index (GFSI) which is a dynamic quantative and qualitative benchmarking model constructed from 28 indicators of 109 countries all over the World, to reveal the food security status of MENA countries in terms of three groups of food security indicators: affordability, availability and quality of food. According to the results of the study, MENA countries rank from 17 (Israel) to 91 (Yemen) out of 109 countries with household expenditure spent on food ranging from 15.5% (Israel) to 60% (Egypt).Study also reveals that lower spending on food as

Table 8. Land area, agricultural GDP, Arable land in MENA region

Countries	Land Area (km^2)	Agr. GDP (% of GDP)		Arable Land (% of Total Land)		Arable Land per Capita (ha)	
		1990	2013	1990	2012	1990	2012
United Arab Emirates	83,600	0,9	0,7	0,4	0,6	0,019	0,005
Bahrain	690	11,4		2,9	2,1	0,004	0,001
Algeria	2,381,740	19,4	10,5	3,0	3,2	0,270	0,196
Egypt, Arab Rep.	995,450	19,1	14,5	2,3	2,8	0,041	0,035
Iran, Islamic Rep.	1,628,550			9,3	10,9	0,270	0,232
Iraq	437,370			11,4	7,9	0,285	0,105
Israel	21,640	7,7		15,9	13,6	0,074	0,037
Jordan	88,240		3,4	2,0	2,4	0,057	0,034
Kuwait	17,820		0,3	0,2	0,6	0,002	0,003
Lebanon	10,230		7,2	17,9	20,2	0,068	0,047
Libya	1,759,540	19,3		1,0	1,0	0,424	0,279
Morocco	446,300	2,6	16,6	19,5	18,0	0,353	0,247
Oman	309,500		1,3	0,1	0,1	0,019	0,010
Qatar	11,610	5,7		0,9	1,1	0,021	0,006
Saudi Arabia	2,149,690	29,8	1,9	1,6	1,5	0,209	0,112
Syrian Arab Republic	183,780	17,7		26,6	25,4	0,392	0,208
Tunisia	155,360	18,1	8,6	18,7	18,3	0,357	0,263
Turkey	814,568	24,4	8,3	32,0	26,7	0,456	0,278
Yemen, Rep.	527,970	0,9		2,9	2,4	0,129	0,052

Source: WDI, 2013

a share of household consumption in most countries and better food safety net programs in the region have contributed a notable increase in food affordability.

In this study, the determinants of macro-level food securityis estimated by an econometric model using panel data for 18 MENA countries and Turkey over the period 1990-2014. This paper aims to analyse the links between macro level food insecurity index and macroeconomic variables such as population growth, per capita GDP, per capita arable land, access to an improved water source, unemployment ratios, female literacy rate, improved sanitation access, government expenditures on education, inflation and exchange ratesusing panel data estimation methods (e.g. GMM and FE). Since not all data of all countries for this period is available, the missing observation problem for those variables is solved using the "ice imputation technique" in STATA.

As it gives more reliable estimates and indicates special features for each country that aggregated or pooled data can not reach, panel regression methods are selected for empirical analysis. Panel data are also called longitudinal data or cross-sectional time-series data. These longitudinal data have "observations on the same units in several different time periods" (Kennedy, 2008). Panel data may have individual (group) effect, time effect, or both, but in our case, we employed fixed effect and GMMmodelsdue to data limitation. Panel data sets for economic research possess several major advantages over conventional cross-sectional or time-series data sets (Hsiao, 2003). Panel data allows to control omitted (unobserved or mismeasured) variables. For randomly missing observations, unbalanced panels have been dealt with in Fuller and Battese (1974), Baltagi (1985), Wansbeek and Kapteyn (1989) and Baltagi and Chang (1994).Other methods of dealing with missing data include: (1) imputing the missing values and analyzing the filled-in data by complete panel data methods, (2) discarding the nonrespondents and weighting the respondents to compensate for the loss of cases. In this study, the missing observation problem for relevant variables is solved using the "ice imputation technique" in STATA.

The panel data model has the following semi-logarithmic framework where the dependent variable FI_{it} denotes the food insecurity index reflecting countries' food insecurity status for each country i in period t:

$$FI_{it} = x_{it}\beta + \gamma_i + \mu_t + \upsilon_{it} \tag{1}$$

As an indicator of food security in MENA countries, IFPRI defined a macro-level which is derived by dividing the food imports value to the sum of total exports and the net remittance inflows (food imports / [total exports + net remittance inflows]). The macro-level indicator captures the ability of a country to finance food imports through exports of goods and services and the net remittances received. Share of food imports in merchandise imports data, share of merchandise imports in GDP and GDP data from the World Development Indicators databank is used to find the value of food imports value. The same procedure is followed to find total exports value. Net remittance inflows are calculated as the difference between remittances received and remittances paid, available from the World Development Indicators databank.

All potential determinants which can effect the food insecurity status of countries are included in vector x_{it}. Moreover, γ_i denotes a full set of country dummies and μ_t denotes a full set of year dummies. υ_{it} is an error term, capturing all other omitted factors, with $\mathrm{E}(\upsilon_{it}) = 0$ for all i and t.

Model is initially estimated using Pooled-OLS estimation method that excludes country dummies. As it is well known, strict exogeneity assumption is one of the crucial necessities for the unbiased and consistent estimates under Pooled-OLS specification. Strict exogeneity assumes that idiosyncratic error term (υit) is uncorrelated with the individual specific effects. Since the Pooled-OLS model neglects the heterogeneity across individuals and assumes that all individuals have a unique effect, Pooled-OLS estimator will be biased and inconsistent. Thus, results from Pooled-OLS estimatons are represented in the paper. However, the fixed-effect estimator will be consistent since it allows for the heterogeneity among countries. As an alternative to Pooled-OLS framework, we therefore use the fixed-effect estimation technique to control for the country specific time invariant characteristics.To check whether results from fixed effect estimation are robustto performing different estimation techniques,Arellano Bond GMM (General methods of Moment) is also employed. Furthermore, in order to test for stationary of the time series, a battery of panel unit root tests are applied for each variable in the models. Accordingly, we have first tested whether the time series are cross-section independent. Test statistics (Frees (1995), Friedman (1937) and Pesaran (2004)) however have shown mixed findings on whether cross-section dependence exists or not. Therefore, we apply both panel unit root testsIm-Pesaran-Shin (Im et al., 2003) and Fisher tests (referring to Maddala and Wu, 1999) and (Choi, 2001) that do not take into account cross-section dependence and Pesaran's second generation panel unit root test that takes into account cross-section dependence on the relevant series of all variables. Test results reject the null of all panel contains unit-roots and accept the alternative hypothesis of stationarity. For robustness checks, regressions are also run taking the first differences of all variables for the stationary concern.The variables, as potential determinants, used in the regression equation are population growth, per capita GDP, per capita arable land, access to an improved water source, unemployment ratios, female literacy rate, improved sanitation access, government expenditures on education, inflation and exchange rates. All the variables in the model are obtained from 2013 World Development Indicators Database.

RESULTS AND DISCUSSION

Scatter plots are not particularly enough to make predictions for a significant relationship between dependent and explanatory variables due to disregarded country specific factors and other determinants of dependent variable. However, before investigatingrelavant issues econometrically, it is always important to give some prior evidence on the relationship between some regressors and responding variable. Figure 1, presents evidence on the relationship between some variables with food insecurity where variables such as public expenditures on education, fresh water withdrawals, improved sanitation access indicate negative relationship and inflation shows positive relationship with food insecurity index.

Table 9 presents the fixed-effect estimation results on the determinants of food security.Turning to related variables, female literacy rate is negatively significant in the regression. The negativeimpact of female literacy rate on food insecurity status might indicate that, as FAO (2005) report also emphasizes,"…lack of education undermines productivity, employability and earning capacity, leading directly to poverty and hunger…" This reflects the human capital approach in which education is highly related with increasing amount of earnings and productivity, and therefore economic growth (Schultz 1961; Schultz, 1971; Becker, 1962). More specifically, in rural areas, education improves agricultural productivity, ensuring the food security (Jamison, Lau, & Lockheed, 1990; Pudasaini, 1983). Considering gender aspects, especially women's education is highly important for food security. Related literature suggests that girls who attend school and obtain at least the basic skills can transmit health and hygienic practices

Figure 1. Relationship between variables of food insecurity

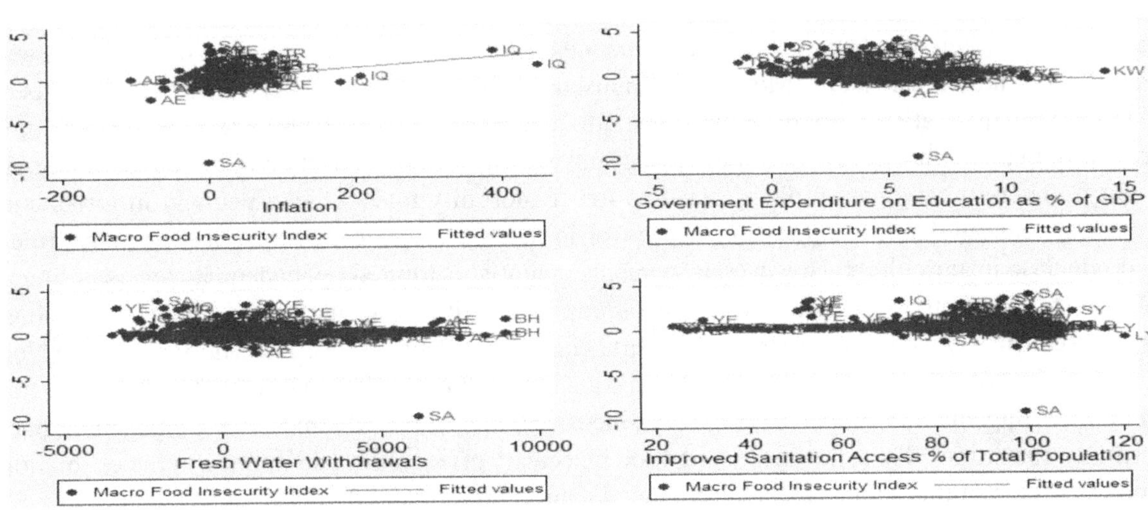

Table 9. Determinants of food insecurity (fixed-effect estimation)

Variables	Fixed Effect Estimates
Arable Land (Hectar Per Capita)	-0.941153
	(1.001584)
Total Unemployment Rate	0.028711
	(0.019459)
Public Expenditures on Education	-0.057806*
	(0.031720)
Female Literacy Rate	-0.011239**
	(0.004533)
Improved Sanitation Access	0.008843
	(0.006394)
Fresh Water Withdrawals	-0.000161*
	(0.000079)
Exchange Rates	0.000023*
	(0.000012)
Inflation Rate	0.005367***
	(0.001795)
Annual Population Growth	-0.006540
	(0.018185)
Log(Gross Domestic Product)	0.061933
	(0.135945)
Observations	468
Number of Countries	19
R-squared	0.214420

Note: All standard errors are robust for the arbitrary heteroscedasticity andrepresented in parentheses. One, two and theasterisk signs (*) (**) and (***) indicate significance at the 10, 5 and 1% level respectively.

to their children when they become mothers. This means that female education may have an additional direct effect on nutritional status. Moreover in households, women give higher value to the well-being of their children, allocating more resources to health, and nutrition (Sen, 1999). He also emphasized on the role of female literacy that is found to have a statistically significant reducing impact on under-five mortality". Therefore, women in family leadsto lower mortality rates whichespecially in low-income countries are caused due to malnutrition. Moreover in MENA countries, women play an active role in food production in agriculture. However, if women's potential is limited by inadequate levels of literacy which affects the way they access and utilize resources for sustainable agricultural production and household food security (Quisumbing, Brown, Feldstein, Haddad, & Peña, 1995). The Middle East and North Africa Region (MENA) has taken great strides in education. Thus, the region has quadrupled the average level of schooling since 1960, halved illiteracy since 1980 and achieved almost complete gender parity for primary education[3]. Thus, the decreasing food insecurity effect of female literacy in the estimations, can be the reason of this achievement in female education more than 30 years.

Furthermore the negative coefficient of fresh water on food insecurity might show the fact that the past three decades witnessed intensive exploitation of groundwater, including, non-renewable water, and construction of large number of dams in MENA. Irrigated areas are increased from about 9.5 million ha in 1980 to approximately 14.2 million ha in 2008. Thus, water has been the reason for the reduction in food gap, and has helped achieve self-sufficiency in food. However, nowadays, water is not carefully used and agriculture in most of the MENA countries faces a threat of great waste in irrigation water. More specifically, bad management and poor coordination between water and agriculture sectors and poor organization of major irrigation facilities are increasing in recent years. Even if it was not a serious threat on food insecurity in last decades, these factors may negatively affect the performance of water and agriculture management in the future since water has been continuously intensively exploited.

Moreover, the region in general has serious economic problems caused by inflation. In particular, high inflation deters investment and impairs household disposable income. In Egypt for instance, higher inflation has been experienced particularly. This can be the reason of the positive coefficient of inflation variable.Increasing food prices also pushed inflation up, affecting mainly the poor who spend a major proportion of their income on food. Especially food prices rose substantially during the second half of the 2000s, and even though prices decreased during the global economic and financial crisis, they recovered rapidly in 2010 before coming down again in 2011. Moreover, as international food prices increased significantly in recent years, almost all countries in the region experienced an increase in their domestic food prices as well (Ianchovichina, Loening, & Wood, 2014).

Similar to inflation rates, exchange rates in the region have also positive relationship with food insecurity during the years examined.More specifically exchange rates trend in the last years, make worse the food insecurity situation in the region. Exchange rate in the dataset is described astheratio of local currency units relative to the U.S. dollar. Once the local currency isundervalued, the food importation will be expensive which reduce the country spending on imported food.The outcome variable of the regressions here is the food security index that has a numerator which consists of food import.Positive coefficient of the exchange rate, therefore, indicates the decreasing trend of food import with undervalued local currency.

Moreover the coefficient of the government expenditures on education is significantly positive for a decreasing food insecurity based on fixed-effect estimations. This can depend on the recent improvements in education and increasing investment in public spending on education in MENA region that is comparable even with more developed countries in the rest of the world. MENA governments have shown strong commitment to funding public education. The average public investment in education across the region as a percentage of GDP is above 5.3% [4].

GMM estimates shown in Table 10, are also similar to fixed effect estimates with the exception of female literacy rate and improved sanitation access. In the contrary of fixed effect estimates, the coefficient of female literacy rate is insignificant. For stationary concern, the first-differenced fixed effect estimations are also shown at Table 11. First differenced regression results have similarity with fixed-effect estimates.

Moreover, improved sanitation access coefficient is negativeand significant based on GMM estimations meaning that food security increases with improved sanitation access. People need access to sufficient food but they also need to be able to absorb the nutrients from the food. Furthermore, being able to absorb enough nutrients depends on also water, hygiene and sanitation. Undernutrition causes over three million child deaths each year especially because of soil transmitted helminth infections or parasitic intestinal worms, due to lack of proper sanitation (Black et al., 2013).

Table 10. Determinants of food insecurity (GMM estimation)

Variables	GMM
Arable Land (Hectar Per Capita)	-0.3451
	(0.2870)
Total Unemployment Rate	-0.0010
	(0.0070)
Female Literacy Rate	-0.0039
	(0.0024)
Public Expenditures on Education	-0.1008***
	(0.0165)
Improved Sanitation Access	-0.0078***
	(0.0027)
Fresh Water Withdrawals	-0.0001***
	(0.0000)
Exchange Rates	-0.0000
	(0.0000)
Inflation Rate	0.0049***
	(0.0009)
Annual Population Growth	-0.0094
	(0.0144)
Log(Gross Domestic Product)	0.0006
	(0.0378)
Observations	468
Number of Countries	**19**

Note: All standard errors are robust for the arbitrary heteroscedasticityandrepresented in parentheses. One, two and three * indicate significance at the10, 5 and 1% level respectively.

CONCLUSION

Most countries in MENA region face food security challenges and this paper presents a macro level view to food security levels of MENA countries by constructing a food insecurity index which was developed firstly by IFPRI in 2012 in order to find the determinants of food insecurity and their impact levels on food security status. According to the results obtained from fixed effect estimation and GMM analysis, increasing female literacy rate and public expenditures on education will have positive impact on the food security level of the region. Also efficient usage of agricultural land and water sources is directly linked to food security issue. In addition, the results point that rising inflation and exchange rates are also important determinants of food security because of causing food affordability problems especially for countries which are dependent to food imports.

According to the overall results of both fixed effecet and GMM estimation, at macro level, countries in MENA region should be take the employment issues (especially youth and women employment) as

Table 11. Determinants of food insecurity (first difference fixed-effect estimates)

Variables	First Difference Fixed-Effect
Δ.Arable Land (Hectar Per Capita)	0.158264
	(0.634977)
Δ.Total Unemployment Rate	0.009255
	(0.013663)
Δ.Public Expenditures on Education	-0.004175
	(0.013843)
Δ.Female Literacy Rate	-0.005763**
	(0.002190)
Δ.Improved Sanitation Access	0.007566
	(0.006607)
Δ.Fresh Water Withdrawals	-0.000111*
	(0.000057)
Δ.Exchange Rates	0.000113***
	(0.000017)
Δ.Inflation Rate	0.001693*
	(0.000893)
Δ.Annual Population Growth	-0.023425
	(0.025319)
Δ.Log(Gross Domestic Product)	-0.003632
	(0.152781)
Observations	444
Number of cn	19
R-squared	0.139810

Note: All standard errors are robust for the arbitrary heteroscedasticityandrepresented in parentheses. One, two and three * indicate significance at the10, 5 and 1% level respectively.

a primary concern to reach a food secure status. As rural shares of the region still remain high, people living at rural areas should join labor force especially in agricultural sector in order to increase the self-production of food as a contribution to their own food consumption. In terms of increasing human capital, governments should increase expenditures on education. High population growth rates are another problem for ensuring food security in the region. The family planning programs should continue and governments should take actions in order to increase the awareness of people for a planned population growth. Limited water sources and their appropriate use is another important issue for ensuring food security in the region. Efficient management and coordination is needed between water and agriculture sector and irrigation facilities should be well organized. As food security is also related with the economic situation of countries, high inflation rates can be another important determinant of food insecurity in the region. High inflation rates and high food prices are in interaction and often trigger one another. High and volatile exchange rates can become a threat for ensuring food security for countries which are mainly dependent on food imports, therefore macroeconomic stability is critual for sustain food security.

REFERENCES

AFDB (African Development Bank). (2012). *The Political Economy of Food Security in North Africa.* AFDB Economic Brief.

AFED. (2008). Arab Environment: Future Challenges. In M.K. Tolba & N.W. Saab (Eds.), Report of the Arab Forum for Environment and Development.

AFESD. (2012). Meeting Water and Food Security Challenges In The Arab Countries: Insights From The Tunisian Experience. In A. Kubursi (Ed.), *Food and Water Security in the Arab World: Proceedings of the First Arab Development Symposium.* The International Bank for Reconstruction and Development/ World Bank and the Arab Fund for Economic and Social Development, Washington D.C.

Applanaidu, S. D., Abu Bakar, N., & Baharudin, H. A. (2013). An Econometric Analysis of Food Security and Related Macroeconomic Variables in Malaysia: A Vector Autoregressive Approach (VAR). In *Proceedings of the International Agribusiness Marketing Conference (IAMC '13)* (Vol. 1, pp. 93-102). Elsevier.

Aşcı, S., Koç, A. A., & Erdem, Ş. (2014). The Impact of Non-tariff Measures on Agro-food Export between MENA Countries and the EU. In Proceedings of the Euro-Mediterranean Service Congress, University of Alcalá, Madrid.

Baltagi, B. H. (1985). Pooling cross-sections with unequal time-series lengths. *Economics Letters, 18*(2-3), 133–136. doi:10.1016/0165-1765(85)90167-3

Baltagi, B. H., & Chang, Y. J. (1994). Incomplete panels: A comparative study of alternative estimators for the unbalanced one-way error component regression model. *Journal of Econometrics, 62*(2), 67–89. doi:10.1016/0304-4076(94)90017-5

Becker, G. S. (1962). Investment in Human Capital: A Theoretical Analysis. *Journal of Political Economy, 70*(October), 9–49. doi:10.1086/258724

Black, R. E., Victora, C. G., Walker, S. P., Bhutta, Z. A., Christian, P., de Onis, M., ... Uauy, R. (2013). Maternal and child undernutrition and overweight in low-income and middle-income countries.,The Maternal and Child Nutrition Study Group. *Lancet, 382*(9890), 427–451. doi:10.1016/S0140-6736(13)60937-X PMID:23746772

Choi, I. (2001). Unit Root Tests for Panel Data. *Journal of International Money and Finance, 20*(2), 249–272. doi:10.1016/S0261-5606(00)00048-6

Drakenberg, O., & Wingqvist, G. Ö. (2010). *Environmental and ClimateChangePolicyBrief- MENA'', Environmental Economics Unit (EEU).* Department of Economics, University of Gothenburg.

FAO. (2005). *The State of Food Insecurity in the World 2005: Eradicating World Hunger-Key to Achieving the Millennium Development Goals.*

FAO. (2013). Food Security Indicators.

FAO. (2016). Women: The key to Food Security. Retrieved from http://www.fao.org/docrep/x0171e/x0171e02.htm

Frees, E. W. (1995). Assessing Cross-Sectional Correlation In Panel Data. *Journal of Econometrics, 69*(2), 393–414. doi:10.1016/0304-4076(94)01658-M

Friedman, M. (1937). The Use of Ranks To Avoid The Assumption of Normality Implicit In the Analysis Of Variance. *Journal of the American Statistical Association, 32*(200), 675–701. doi:10.1080/01621459.1937.10503522

Fuller, W. A., & Battese, G. E. (1974). Estimation of linear models with cross-error structure. *Journal of Econometrics, 2*(1), 67–78. doi:10.1016/0304-4076(74)90030-X

Garduno-Diaz, S. D., & Garduno-Diaz, P. Y. (2015). Food Security in the Middle East and North Africa", World Academy of Science, Engineering and Technology, International Journal of Biological, Biomolecular, Agricultural. *Food and Biotechnological Engineering, 9*(1), 47–50.

Hsiao, C. (2003). *Analysis of Panel Data* (2nd ed.). Cambridge Universirty Press. doi:10.1017/CBO9780511754203

Ianchovichina, E. I., Loening, J. L., & Wood, C. A. (2014). How Vulnerable are Arab Countries to Global Food Price Shocks? *The Journal of Development Studies, 50*(9), 1302–1319. doi:10.1080/00220388.2014.928698

Im, K. S., Pesaran, M. H., & Shin, Y. (2003). Testing For Unit Roots In Heterogeneous Panels. *Journal of Econometrics, 115*(1), 53–74. doi:10.1016/S0304-4076(03)00092-7

International Food Policy Research Institute IFPRI. (2010). Food Security and Economic Development in the Middle East and North Africa (Discussion Paper No: 00985).

Jamison, D. T., Lau, L. J., & Lockheed, M. E. (1980). Farmer Education and Farm Efficiency: A Survey. *Economic Development and Cultural Change, 29*(October), 37–76.

Kennedy, P. (2008). *A Guide to Econometrics* (6th ed.). Malden, MA: Blackwell Publishing.

Larson, D.F., Lampietti, J., Gouel, C., Cafiero, C., & Roberts, J. (2014). Food Security and Storage in the Middle East and North Africa. *The World Bank Economic Review, 28*(1), 48-73.

Maddala, G. S., & Wu, S. (1999). A Comparative Study Of Unit Root Tests With Panel Data And A New Simple Test. *Oxford Bulletin of Economics and Statistics, 61*(s1Suppl. 1), 631–652. doi:10.1111/1468-0084.61.s1.13

Pesaran, M.H. (2004). General Diagnostic Tests For Cross Section Dependence In Panels (CESifo Working Papers No. 1229).

Pudasaini, S. P. (1983). The Effects of Education on Agriculture: Evidence from Nepal. *American Journal of Agricultural Economics, 65*(3), 509–515. doi:10.2307/1240499

Quisumbing, A. R., Brown, L. R., Feldstein, H. S., Haddad, L. J., & Peña, C. (1995). *Women Key to Food Security'' (No. 5). International Food Policy Research Institute.* IFPRI.

Schultz, T. W. (1961). Investment in Human Capital. *The American Economic Review, 1*(March), 1–17.

Schultz, T. W. (1971). *Investment in Human Capital*. New York: The Free Press.

Sen, A. (1981). *Poverty and Famines: An Essay on Entitlement and Deprivation*. Oxford University Press.

Sen, A. (1999). *Development as Freedom*. New York: Knopf.

Wansbeek, T. J., & Kapteyn, A. (1989). Estimation of the error components model withincomplete panels. *Journal of Econometrics*, *41*(3), 341–361. doi:10.1016/0304-4076(89)90066-3

World Bank. (2007). *Making the Most of Scarcity - Accountability for Better Water Management Results in the Middle East and North Africa*. Washington, DC: The World Bank.

World Bank. (2013). *World Development Indicators*.

ENDNOTES

1. MENA countries include Bahrain, Iran, Iraq, Israel, Jordan, Kuwait, Lebanon, Oman, Qatar, Saudi Arabia, Syria, the United Arab Emirates, Yemen, Algeria, Egypt, Libya, Morocco, and Tunisia.
2. http://www.un-documents.net/ocf-05
3. http://www.worldbank.org/en/region/mena/brief/education-in-mena
4. http://www.worldbank.org/en/region/mena/brief/education-in-mena

This research was previously published in the International Journal of Food and Beverage Manufacturing and Business Models (IJFBMBM), 2(1); edited by Constantin Zopounidis and George Baourakis, pages 1-19, copyright year 2017 by IGI Publishing (an imprint of IGI Global).

Chapter 14
A System Dynamics Model for Subsistence Farmers' Food Security Resilience in Sub-Saharan Africa

Benedict Oyo
Tshwane University of Technology, South Africa

Billy Mathias Kalema
Tshwane University of Technology, South Africa

ABSTRACT

Food security at subsistence farmers' level in sub-Saharan Africa has become an issue of concern due to increasing vulnerability caused by a number of factors such as: changing climate, resource scarcity (e.g. land and inputs), environmental degradation (e.g. declining soil fertility, deforestation, and surface water eutrophication), market failures and weak public/donor support initiatives. In light of these challenges, farmers must be prepared to survive by self-provisioning. To pursue the fastest and most practical route to improved food security, focus should be on resilience based initiatives at household and community levels. In this paper, the authors investigate the factors that have enabled subsistence farmers to succeed despite the previous shocks and stresses, and develop a system dynamics model for sustainable food security based on initiatives exclusive to the farmers. The model is used to examine the question: how can innovative subsistence farmers engage in better livelihood and market orientated production irrespective of external public or donor support?

1. INTRODUCTION

Worldwide, over 850 million people do not have regular access to the minimum calories they require on a regular basis, and most of those are in Asia and sub-Saharan Africa (Hammond & Dub, 2012). At the same time, the subsistence farmers are most vulnerable to any shocks to their agricultural system which is the sole source of their livelihoods. These farmers are not only isolated and deprived of access

DOI: 10.4018/978-1-5225-8063-8.ch014

to safety nets but are frequently exposed to pest and disease outbreaks and extreme weather events, which cause significant crop and income losses and aggravate food insecurity (Chauvin et al., 2012; Oyo, 2013). While agricultural growth has been the precursor to the acceleration of industrial growth in a number of emerging economies such as China, Brazil, and India, for sub-Saharan Africa, current agricultural productivity is low and there have been numerous failures in getting agriculture moving (Chauvin et al., 2012).

In contrast, the literature is rich with case studies where productivity led agricultural transformation is playing a significant role in economic transformation in general and making agriculture an important driver of growth in Africa (Thompson et al., 2007; Baiphethi & Jacobs, 2009; Haggblade, 2010; Africa Progress Panel Policy Brief 2010; Settle & Garba, 2011; Chauvin et al., 2012; Oyo, 2013). Studies have found that agriculture has been the engine of growth in most developing countries with causality running from agricultural growth to economy-wide growth in most cases, and that even small variations in agricultural productivity have had strong implications for the rate and pattern of economy wide growth (Chauvin et al., 2012). Furthermore, agricultural growth can provide the economy with much needed stimuli such as capital, labor, and foreign exchange to finance and fuel growth in non-agricultural sectors (Fan et al., 2013; Stave & Kopainsky, 2014).

In order to articulate the key issues in this paper, we provide definitions of key terms as outlined in Table 1.

1.1. Problem

Agricultural growth rate and food security in Africa has been volatile since the 1980s. This largely was caused by negligence by most governments and donors in the 1980s and 1990s (Chauvin et al., 2012).

Table 1. Definitions of key terms

Terminology	Definition
Resilience	The Resilience Africa Network (RAN) defines resilience as the capacity of people and systems to mitigate, adapt to, recover and learn from shocks and stresses in a manner that reduces vulnerability and increases well-being (RAN Report 2014). From a more systemic perspective, Walker et al. (2006) define resilience as the capacity of a system to experience shocks while retaining essentially the same function, structure, feedbacks, and therefore identity. The key issue of interest in these definitions that ties with food security at subsistence farmer level is ability to recover from shocks and hence being less vulnerable to stressors.
Food security	Food security has generally been defined as the access by all people at all times to enough food for an active healthy life (Anderson, 1990). According to the World Food Summit organized in Rome in 1996, food security exists when all people, at all times, have physical and economic access to sufficient, safe, nutritious food to meet their dietary needs and food preferences for an active life (FAO, 2011). At household level, food security refers to the availability of food in one's home which one has access to. In this case, a household is regarded as food secure when the members of the family do not live in hunger or fear of starvation.
Subsistence farming	Subsistence farming has been defined by Morton (2007) as self-sufficiency farming and a livelihood strategy where the main output is consumed directly, with a minor proportion of output marketed. It is also frequently and more generally used to describe rural producers, predominantly in developing countries, who farm using mainly family labor and for whom the farm provides the principal source of income (Fan et al., 2013).
System dynamics model	A model of a system is a tool used to study the behavior of a system without having to experiment on the real system. As such, a system dynamics model is a tool used to study the behavior of a system by describing the system and understanding how information feedback govern its behavior, and designing robust information feedback structures and control policies through simulation and optimization (Coyle, 1996). Since simulation and optimization are underpinned by mathematical equations, a model can also be defined as a collection of mathematical equations which represent behavior/characteristic of real system (Azar, 2012).

Renewed commitment to agriculture over that last decade, e.g., through the Comprehensive Africa Agricultural Development Programme (CAADP) and the Maputo declaration of 2003 in which African heads of state pledged to devote at least 10% of their national budgets to agriculture in an effort to raise agricultural growth to 6% a year, have had negligible impact. For instance, the benchmark for Maputo declaration with regard to 10% national budgets expenditure on agriculture has not been realized by majority of the signatory countries (Chauvin et al., 2012; Harvey et al., 2014). During the previous periods of declining public/donor intervention and unmet pledges, subsistence farmers developed resilience to shocks and stresses thus ushering in new trends of self-reliance/self-provisioning. What seems viable at this point is to pursue other alternatives to food security based resilience strategies developed over the years, with less expectation from government and donor intervention.

1.2. Objective

The dynamics of food production and hence food security and livelihoods are interlinked, and need to be studied as a whole. The interactional effects for food production at subsistence farmers' level needs to be studied in order to investigate appropriate strategies that could improve their production and hence livelihoods. Furthermore, food security and livelihood issues are complex and a thorough understanding of the underlying relationships is necessary. To help generate this understanding, conceptual and simulation models are necessary. In this paper, we investigate the factors that have enabled subsistence farmers to succeed despite the previous shocks and stresses, and develop a system dynamics model for sustainable food security based on initiatives exclusive to the subsistence farmers. Most importantly, the model examines possible successful interventions tailored to different types of subsistence farms and specific contexts in which they operate. Ultimately, the paper provides a plausible answer to the question: how can innovative subsistence farmers engage in better livelihood and market orientated production irrespective of external public or donor support?

2. EXISTING FOOD SECURITY INTERVENTIONS

A number of successful food security interventions in sub-Saharan Africa have been reported in the literature. For the purpose of the discussion in this paper, three approaches are emphasized, namely: sustainable production intensification, organic green revolution for Africa, and seed banking.

2.1. Sustainable Production Intensification

Sustainable production intensification (SPI) goal is to produce more from the same area of land while conserving resources, e.g., through the application of conservation agriculture practices, the use of good seeds of high yielding varieties, integrated pest management, plant nutrition based on healthy soils, efficient water management, and the integration of crops, pastures, trees and livestock (FAO, 2011; Settle & Garba, 2011). One of the key success cases of SPI is in diversifying smallholder farms which can provide farmers opportunities for rebuilding soil fertility by enhancing nutrient flows and efficiencies while introducing new sources of food and nutrition for local populations (Settle & Garba, 2011). This in return gives farmers a greater choice in what they can sell, trade, eat or feed to animals, which in the long run would lead to greater economic and ecological resilience. The SPI intervention is knowledge intensive and hence thrives in an environment of functional extension services.

Settle and Garba (2011) report of successful implementation of SPI through farmer field schools in which 116,000 rice, vegetable, cotton and other farmers were involved in four West African countries. According to Settle and Garba (2011), the programme resulted in improved yields and incomes and at the same time, substantially reduced the use of chemical pesticides while improving the use of fertilizers and organic amendments. As a long-term sustainability measure, the programme was successfully integrated into local, provincial and national structures.

2.2. Organic Green Revolution for Africa

The organic green revolution for Africa was proposed by Otsuka and Yamano (2005) as an extension of the very successful and well reported Asia's green revolution. The organic green revolution for Africa is a seed-livestock-agroforestry farming system meant for Africa whereby the use manure from cattle and leaves from agroforesty trees with nitrogen fixation capacity would increase soil fertility and hence yields. Initial adoption of this farming system in the Kenyan highlands area was successful. In practice, a typical farmer who employs this farming system cultivates feed crops and grows fodder trees with the capacity to fix nitrogen (i.e., agroforestry trees) on the farm, feeds the cattle in stalls, obtains manure from the stalls, and applies the manure or compost on crops such as maize or banana. When combined with improved maize varieties, the potential for doubling maize yield is high (Otsuka & Yamano, 2005). Another advantage of using manures on crops is that it facilitates retaining moisture on soil longer, which to a certain extent is an appropriate technology for the areas with less rainfall and irrigation facilities.

2.3. Seed Banking Initiative

Seed banking is a holistic approach for empowerment of smallholder farmers through preserving seeds for future planting, diversifying into higher value crops and saving money from sale of surplus crops concurrently (Oyo, 2013). Accordingly, seed banking ensures that smallholder farmers do not lack the quantity and quality of seeds they need during planting seasons as well as access their savings to invest in necessary farm inputs for higher productivity (Oyo et al., 2012; Oyo, 2013). In effect, seed banking services are provided by a seed bank which is a community based organization that also offers agro-processing, agricultural produce marketing and extension services. As such, the seed bank buys harvested crops from farmers at competitive prices, processes them and sells the processed crops for profit, while maintaining a threshold amount of money that is converted into improved seeds and other farm inputs which farmers access during the planting season.

According to Oyo (2013), results from initial implementation of seed banking confirmed that supporting smallholder farmers in production, processing, produce marketing and preservation of seeds for future planning through seed banking initiatives strengthens subsistence farming, making it more profitable. Using available data from initial roll-out of seed banking and simulating over a period of five years, Oyo (2013) found that seed banking can transform the subsistence farmers' population into a prosperous people with a two hundred per cent (200%) rise in earnings. Moreover, the improved income is possible in parallel with abundance of food for consumption and excess seeds preserved for future planting.

Evidence from the three cases discussed here suggest that food security is sustainable when the interventions are pursued by the farmers themselves with basic external support especially on infrastructural development and input subsidies. We examine this insight further in the next section.

3. THE CASE FOR RESILIENCE IN FOOD SECURITY INTERVENTIONS

Most food security programmes have been designed to increase productivity through technology driven intensification, while productivity is important, other related objectives must also be considered, especially by increasing resilience of the farmers themselves. Subsistence farming has always been a vulnerable enterprise due to a number of known factors that will not be reiterated here. As these factors escalate, farmers tend to endure because of their resilience through a variety of risk-coping strategies. Most importantly, successful subsistence farming occurs when external support from government, donors and NGOs complement farmers' self-provisioning initiatives. As such, the kind of external support varies from one farming community to another.

A deeper reflection on the effect of resilience and external support on food security reveals several unique conceptual behaviors as show in Figure 1.

The nature of external support implied in Figure 1 include the following:

- Subsidizing expensive farm inputs;
- Providing access to affordable fertilizers;
- Extension services intensification such as farmer field schools on better farming methods, soil conservation methods and post-harvest handling techniques;
- Infrastructure development through improved road network, small-scale irrigation schemes, and provision of small and medium size food processors;

Figure 1. The effects of resilience and external intervention on food security

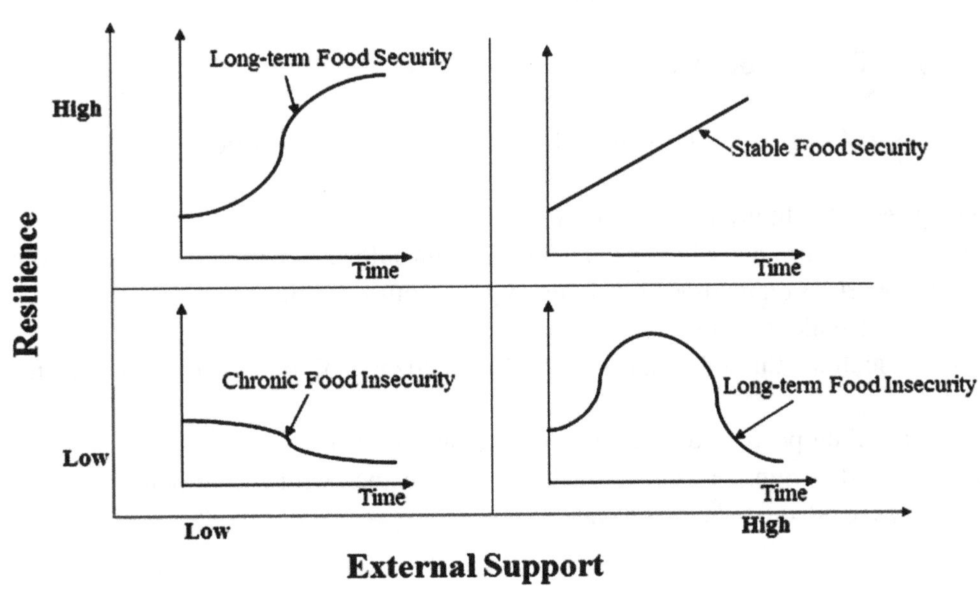

- Providing access to farm improvement finances/loans;
- Structuring of food/produce markets to ensure that farmers sell processed food and not raw food.

As depicted by Figure 1, four unique food security behaviors are possible when the nature of external support and resilience are considered simultaneously. The issues of concern for each behavior are as follows:

3.1. Chronic Food Insecurity

Chronic food insecurity in the context of Figure 1 would occur when both resilience and external support are low. In this case, the farmers are unable to meet their minimum food production requirements over a sustained period of time. As such, the farmers resort to copying strategies that further escalate their vulnerability such as: eating less food, reducing the number of meals per day, purchasing food, changing diet, selling assets to buy food, borrowing money for food, receiving food from relatives or the community, increasing consumption of wild plants and animals, leasing land to other farmers, taking children out of school, and receiving food aid from government/charity organizations. From our experience on livelihood projects involving farmers with the aforementioned copying strategies, increasing external support may alleviate the problem only for a short period of time, but cannot provide a long term solution unless their resilience is improved simultaneously.

3.2. Long-Term Food Insecurity

Over dependence on external intervention without self-supporting initiatives leads to short lived food security but long term food insecurity as shown in Figure 1. This category of farmers believe in constant flow of seeds for planting and other basic inputs without which their production capacity falls below minimum levels required for provision of household food needs. Again due to low resilience towards self-provisioning, external interventions cannot yield significant impact, resulting into food insecurity in the long-term or rather a better-before-worse behavior.

3.3. Long-Term Food Security

This is the category of self-reliant farmers with the following characteristics:

- Preserving seeds for future planting seasons;
- Storing food for future household needs in case of turbulences in projected production;
- Using soil conservation methods like mulching, contour farming;
- Applying small-scale irrigation initiatives;
- Invest in agriculture through improved seeds, fertilizers, pesticides, ox plough, and extension services;
- Focusing on selling processed produce whenever applicable;
- Using diversified farming practices by integrating livestock with crop production;
- Having access to adequate land and water.

The above characteristics apply to people who practice farming both as a business and for their livelihoods and therefore are able to thrive with minimal external intervention.

3.4. Stable Food Security

Stable food security is an ideal scenario whereby farmers' initiatives are complimented by external intervention in such a way that the net effect is a food secure community. In effect the farmers are self-reliant as in previous case and the external services provided, strengthens environmental resilience leading to stable food security.

In real-life, the achievable scenario given previous volatile external support to agriculture and food security in sub-Saharan Africa is the long-term food security that is based on farmers self-provisioning. We pursue this further in the next section.

4. SELF-PROVISIONING FOR LONG-TERM FOOD SECURITY

Most investments in subsistence agriculture and food security are realized by subsistence farmers themselves. This occurs through different modalities but mostly through labor investments to enlarge and improve the resource base, and to a lesser extent through personal savings and remittances from family members which are used for the acquisition of new/additional resources. However, these investments are limited since domestic needs receive priority when food, health or education expenditures are at risk. Innovative initiatives are thus needed to empower households into self-provisioning and long-term food security. We demonstrate these initiatives in Table 2.

The insights from Table 2 and Figure 1 further inform the system dynamics model discussed in the next section.

5. THE SYSTEM DYNAMICS MODEL PERSPECTIVE

In system dynamics terms, we might characterize the practices leading to the subsistence farmers' food security resilience in terms of identifiable stocks. As pointed out by Hassanien et al. (2015), any given stock might be sustainable if the flows in are greater or equal to the flows out of the stock, the agricultural production cycle from planting, harvesting, post harvesting, consumption and sale of surplus food, consists of several stocks that can be maintained by managing the inflows and outflows. Resilience in this cycle is achievable by considering adaptation strategies that would minimize any disturbances arising from sudden large changes to the outflows. For example, extended rains beyond the harvesting period would limit sunshine that is needed for drying the harvested crops and the farmers' response would be to process the raw food itself.

5.1. Related System Dynamics Models

Oyo (2013) analyses system dynamics models for food security over the period 2001-2011 and finds that these models are more qualitative than quantitative based. Interestingly, none of the models reviewed focused on Africa's situations. Between 2011 and 2015, however, system dynamics models for agricul-

Table 2. Strategies for self-provisioning and long-term food security

Strategies	Description	Implications to Subsistence Farming
Investment in inputs	All agricultural activities are investments and farmers need to cost their investment in farming and determine profit/loss	Every cost ranging from land, labor and inputs (seeds, fertilizers, pesticides, etc) to post harvest handling must be computed. In cases where the farmer uses own land, labor and seeds, then local market rates for these should be used. In doing so, a new perspective of farming as a business and means of livelihood becomes the foundation of subsistence practice.
Seeking appropriate extension services	Extension programmes can be unique to a community and in some cases applicable to several communities	For profitable farming, farmers should stick to better crop varieties that offer higher yields, but use less water, fertilizer or other inputs and are more resistant to drought, heat, submersion, and pests and diseases. Extension services should address specific and general needs of a farming community.
Ensuring access to better markets	The need for direct links between producers and consumers is necessary in ensuring that the producers sell higher value processed goods	Smallholder organizations and collective action are crucial for penetration into the domestic, national, and regional markets for higher returns. Higher returns becomes the goal seeking behavior for higher production.
Re-designing community self-empowerment programmes	Focus on community programmes initiated by the community for self-empowerment	Farmer groups need to pull resources for self-empowerment through improved seeds, small scale field machinery, and small scale food processing.
Preserving food and seeds for future consumption and future planting	Households are food secure when they can preserve food for future consumption and seeds for future seasons	It is increasingly becoming difficult for subsistence households to preserve food and seeds for future needs, except for resilient households. These exceptional households could be benchmarked by the rest in achieving holistic community culture of food and seeds preservation.
Diversifying livelihoods	Integration of livestock in the farming system	Livestock and crop farming complement each other. It has been found that most progressive subsistence farmers are engaged in crop and livestock production (Fan et al., 2013; Otsuka & Yamano, 2005).
Increasing saving from agricultural income	Formation of community saving schemes, e.g., saving clubs or saving societies to minimize shocks due to market failures	The community saving schemes are a necessary source for faster loans for investment in farming to increase production and income. Existing saving schemes focus on end of year celebrations only and these could be diversified.

tural development and food security in sub-Saharan Africa have been studied at an impressive rate. The reminder of this section focuses on these recent studies. Kopainsky et al. (2012) examine sustainable food security policies in sub-Saharan African countries from social dynamics perspective, maintaining that social dynamics override adoption of effective food security policies. In the same study area scope, Saldarriaga et al. (2013) explore stakeholders' views on dynamic complexity of climate change, agriculture and food security. Other studies have focused on individual sub-Saharan African countries such as: Oyo (2013) on system dynamics model for food security and livelihood at subsistence farmer level in the Uganda; and, Ayenew and Kopainsky (2014) on food insecurity issues using system dynamics simulation model for the effects of population, food production and markets in Ethiopia.

From the previous chapter, it is clear that research on food security issues in sub-Saharan Africa using system dynamics modeling is still at its infancy with the first published work dating 2012. This study therefore extends previous work by providing a generic system dynamics model using stocks and flows for sub-Saharan Africa's subsistence farmers' situations, leaving room for adaptation to specific environments with actual data.

5.2. Model Scope

The issues discussed in relation to Figure 1 formed the basis for the model scope. The focus of the model therefore is to experiment scenarios for the four unique food security behaviors categorized as: stable food security, long-term food security, long-term food insecurity and chronic food insecurity. These behaviors take place at household level and hence the model should be tested using household data. Since the model is meant for sub-Saharan Africa's subsistence farmers' situations, we present the generic stocks and flows in this first version, leaving room for adaptation to specific environments and subsequent simulation and analysis. These stocks and flows are based on underlying feedback for food security and food insecurity scenarios at subsistence farmers' levels as shown in Figure 2.

Figure 2 is the basic feedback loops depicting food security scenarios (green labeled loops, R2, R3 and R4) and food insecurity scenarios (red labeled loops, B1 and B2). The black labeled loop R1 is cross-cutting. In the context of this research, Figure 2 covers all the four unique food security behaviors, i.e, chronic food insecurity is depicted by loop B1 and loop B2; long-term food insecurity differs from chronic food insecurity by additional loop, R1. Similarly, long-term food security is represented by loops R1, R2 and R3 while stable food security is depicted by loop R4 on addition to long-term food security representative loops.

From the basic feedbacks in Figure 2, six important scenarios at subsistence farmers' level can be studied. These include:

Figure 2. Underlying feedbacks for food security/insecurity scenarios

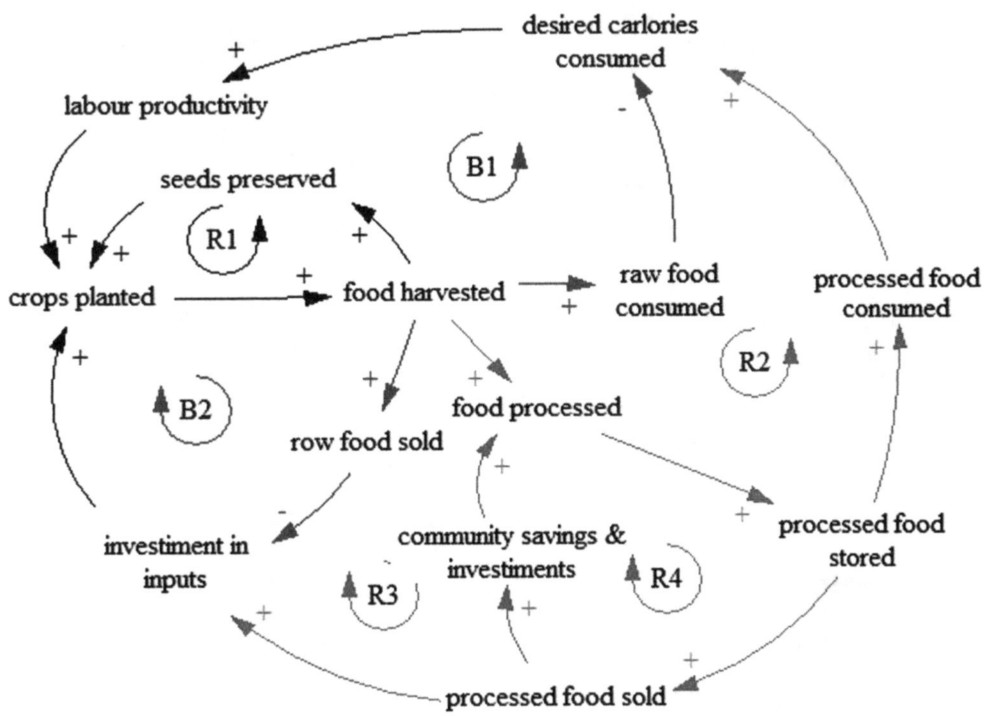

- Producing different types of food for domestic and commercial purposes;
- Resilience to shocks and stresses caused by climate change, soil fertility change, food production land change, and changes in food prices;
- Minimizing food losses in the garden, at harvest, and during post-harvest handling and storage;
- Maximizing income from food sales;
- Avoiding gaps in seeds for planting and basic inputs;
- Collective investment in extension services.

We explore these scenarios using stocks and flows in the next section.

5.3. Model Stocks and Flows

Following from Figure 2 and the six important subsistence farmers' scenarios discussed in the previous section, the conceptualized stocks and flows diagram is given in Figure 3.

Figure 3. Stocks and flows diagram exploring livelihood and commercial production

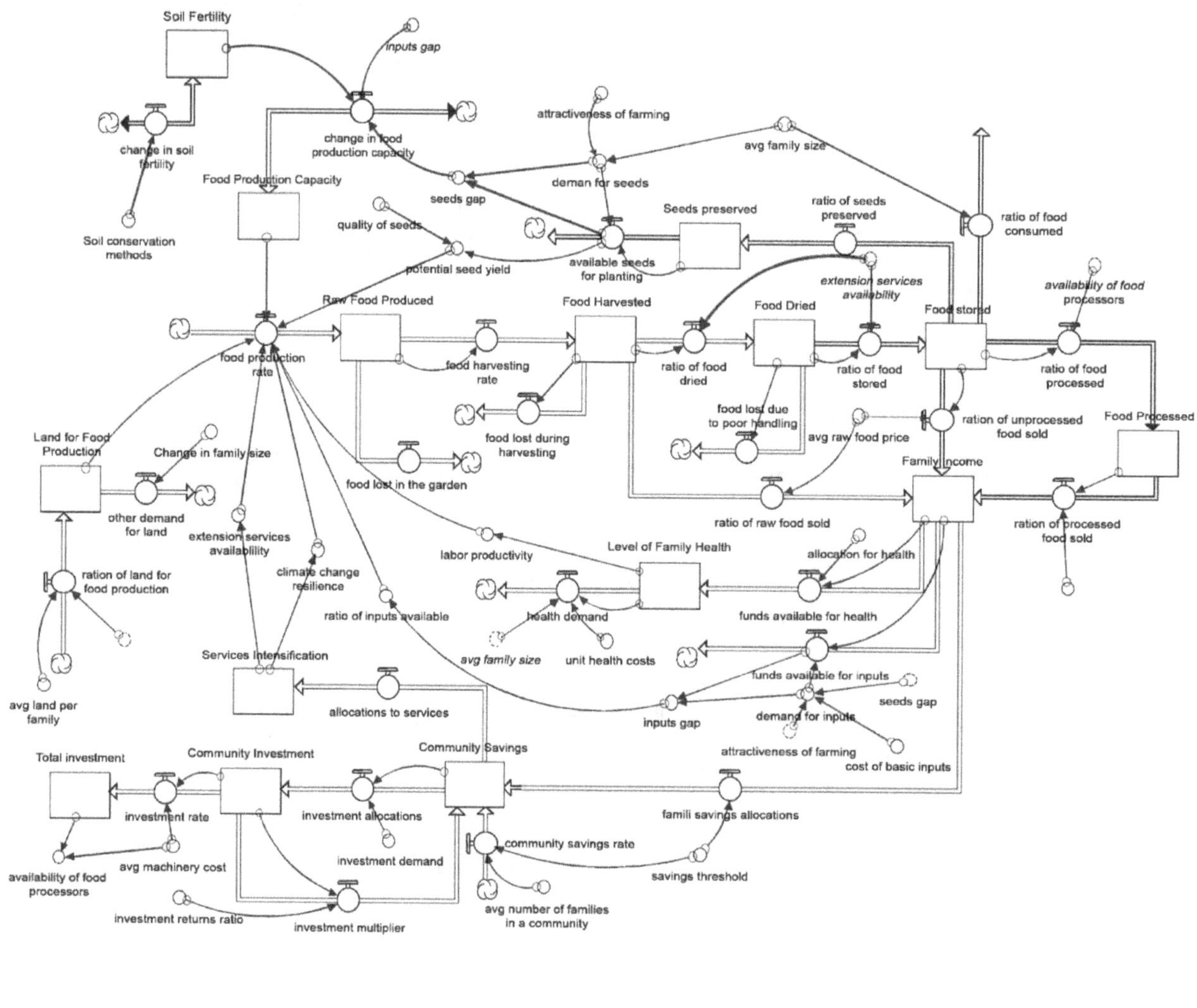

In discussing the model in Figure 3, we extract a part of Figure 3 in the context of each of the six areas as already outlined, i.e., food production management, promotion of resilience, minimizing food losses, maximizing income from food sales, minimizing gaps in inputs, and collective investment in extension services, and re-construct the underlying feedbacks structure whenever necessary.

5.3.1. Goal of Food Production

The default reason for subsistence farming is food production for domestic consumption. However, due to overlapping basic needs, subsistence farmers find themselves selling their food and leading lives of severe food shortages. The model in Figure 3 shows that livelihood and commercial production can be achieved concurrently when factors centered on collective action by individual farmers as well as farmers' groups, are emphasized. In addition, the importance of households preserving seeds for future planting while investing in inputs and saving income from food sales, culminate into the general goal for food production. This goal seeks to achieve adequate food for consumption, income from food sales and strengthening of food production capacity. The feedback structure corresponding with this general goal is given in Figure 4.

5.3.2. Nature of Resilience

In real-life, shocks and stresses are caused by climate change, soil fertility change, food production land change, vulnerability of food allocation for food and for sale. Building resilience to these shocks and stresses stems from advance preparations in anticipation of their occurrence as well as strong public mitigation and recovery mechanisms. Since the focus of the model is on household initiatives for self-provisioning through resilience, the indicators of this goal include: control factors for the *bi-flows* of food production capacity; level of service intensification and community investment stocks; ensuring

Figure 4. Feedbacks structure for basic food production goal

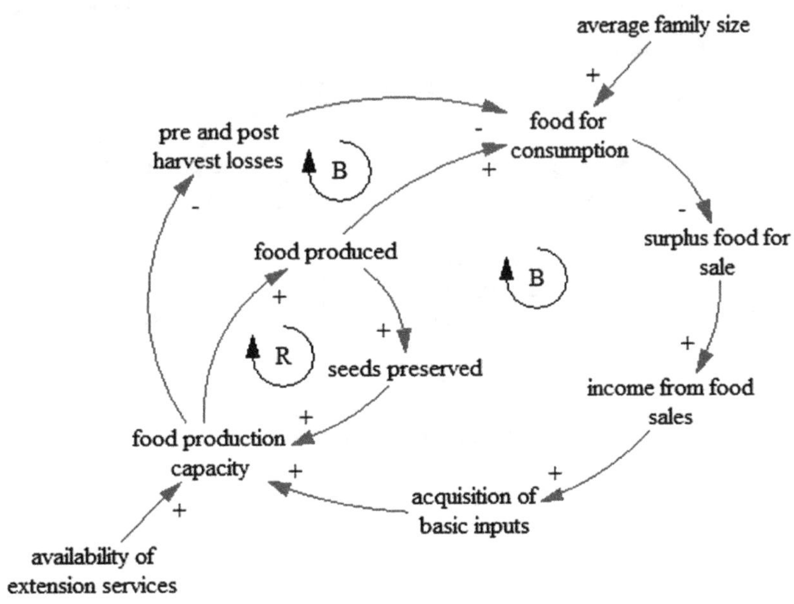

that the ratio of seeds preserved corresponds with household needs; and ensuring that the ratio of food processed is higher than the combined ratio of raw food sold and unprocessed food sold. The corresponding feedback structure for the resilience strategies as just highlighted is given in Figure 5.

One of the key strategies in attaining resilience is to separate production of food for cash and food for consumption, in order to understand farmers' priorities and design effective support mechanisms.

5.3.3. Minimizing Food Losses

The various levels of food losses as shown in Figure 3 are: in the garden, at harvest, and during post-harvest handling and storage. These losses should be at least minimized or completely eliminated. Following from Figure 4 and Figure 5, these loses could be mitigated by building food production capacity, i.e., farmers having the basic knowledge and resources to maximize their food production.

5.3.4. Maximizing Income From Food Sales

It is common knowledge that higher income is achievable when processed foods are sold as opposed to raw or unprocessed food. Getting farmers to achieve this milestone requires increasing access to basic food processors and eliminating buyers of unprocessed food. More investment into small-scale food processors by the farmers themselves through collective action is important. The model in Figure 3 depicts that farmers' collective savings could boost both food production and processing.

Figure 5. Feedbacks structure for resilience strategies

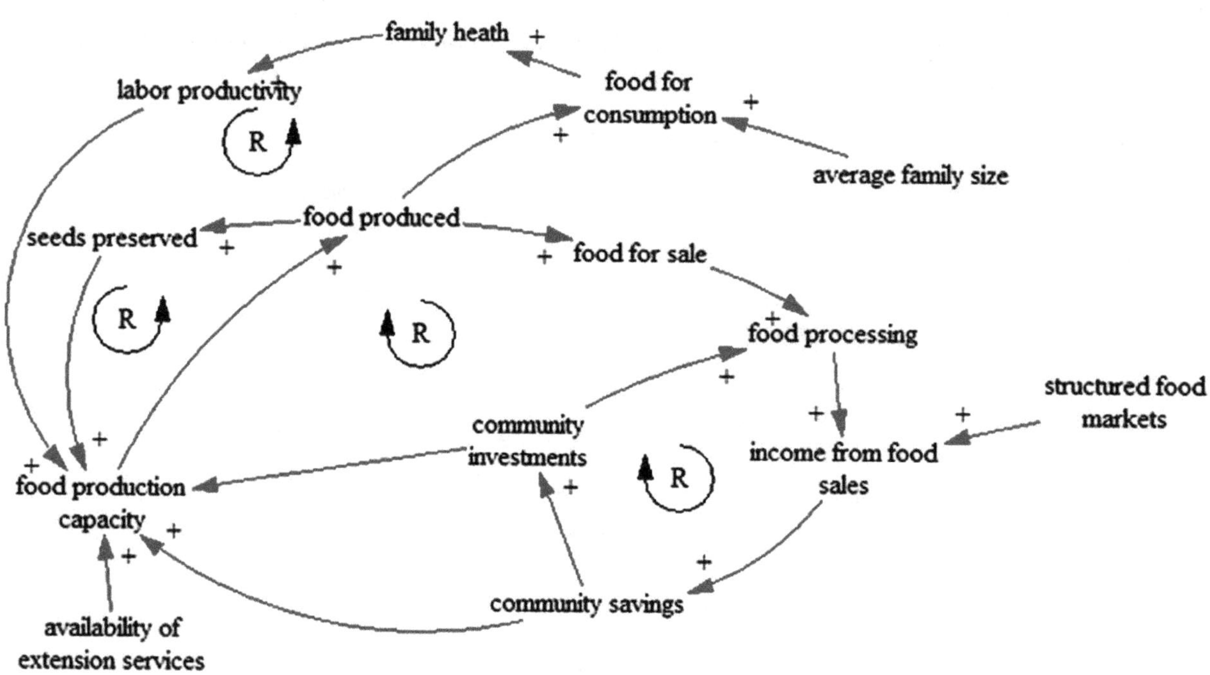

5.3.5. Avoiding Gaps in Seeds for Planting and Basic Inputs

Low production is in most cases caused by constraints in access to seeds for planting and other basic inputs. The model in Figure 2 shows how gaps in seeds and inputs arise, and therefore avoidance measures can be explored from the interacting factors.

5.3.6. Collective Investment in Extension Services

Subsistence farmers are known to have low/no savings and hence unable to access fixed and working capital to boost productivity. On the other hand, recent trends show that farmers can save collectively for end of year celebrations (Oyo, 2014). What is thus required is to form trusted farmers' groups that can persuade other farmers to extend the latter and related savings styles for procurement of extension services that would increase production and diversify their earnings. From the model in Figure 3, this insight is demonstrated by the section connecting *farmers' income* stock to *food production* flow through *services intensification* stock.

5.4. Policy Analysis

By convention, simulation experiments inform policy analysis. In the context of the generic model given in Figure 3, we propose some policy directions that can be investigated when the model is adapted to specific environments. These include:

- Separating production of food for cash and food for consumption as these are different goals and in some cases having unique requirements. For cash crops, production and processing need to be addressed concurrently, while for food crops, focus should be placed on production, post-harvest handling and storage;
- Building capacity for crop production at both family level and community level. The basic production requirements such as: access to land, preservation of seeds for future seasons and provision of human labor, should be addressed at family level. Other expensive production inputs such as acquiring high-yielding seed varieties, ensuring integrated pest management, and supporting small-scale irrigation, should be addressed at community level;
- Reducing waste and losses right from crop planting to food consumption. For instance, knowledge of seeds spacing is critical in avoiding planting excess seeds per unit land area. Similarly, families that are prone to food shortages are known to waste food for consumption when their harvests are good;
- Ensuring functional markets for processed food and not raw food.

The policy directions presented here can be examined further and new policies explored when the model is adapted to specific environments.

6. CONCLUSION

In this paper we have made and supported the claim that sustainable food security at subsistence farmers' level stems from building resilience, short of which renders the farmers susceptible to chronic food insecurity or at least long-term food insecurity. A conceptual analysis of the effect of resilience and external intervention revealed four unique behaviors categorized as: stable food security, long-term food security, long-term food insecurity and chronic food insecurity. The answer to the question examined in this paper on seeking strategies for which innovative subsistence farmers can engage in better livelihoods and market orientated production irrespective of external public or donor support, was tied to self-provisioning interventions for long-term food security. Eight key strategic issues were discussed to this effect, revealing that when subsistence farming is practiced both for business and livelihood, then, long-term food security is achievable even with minimal external support.

From a more practical perspective, a system dynamics stocks and flows model was developed to further examine how subsistence productions can be more commercial and market oriented. Subsequently, six scenarios for which the subsistence farmers can engage in livelihood and profit productions simultaneously were analyzed. These include: producing different types of food for domestic and commercial purposes; resilience to shocks and stresses caused by climate change, soil fertility change, food production land change, and changes in food prices; minimizing food losses in the garden, at harvest, and during post-harvest handling and storage; maximizing income from food sales; avoiding gaps in seeds for planting and basic inputs; and collective investment in extension services.

What remains to be seen is how the generic model given in Figure 3 can be extended and adapted to specific environments in sub-Saharan Africa using actual data. These case studies would then confirm or disqualify the propositions for achieving long-term food security at household level.

REFERENCES

Africa Progress Panel Policy Brief. (2010). Raising agricultural productivity in Africa Options for action, and the role of subsidies. Retrieved from http://www.cpahq.org/cpahq/cpadocs/Agricult%20 Productivity.pdf

Ayenew, M. M., & Kopainsky, B. (2014, July). Food Insecurity in Ethiopia: Population, Food Production and Market. *Paper presented at the 32nd International Conference of the System Dynamics Society*, Delft, The Netherlands.

Azar, A. T. (2012). System Dynamics as A Useful Technique for Complex Systems. *International Journal of Industrial and Systems Engineering*, *10*(4), 377–410. doi:10.1504/IJISE.2012.046298

Baiphethi, M. N., & Jacobs, P. T. (2009). The contribution of subsistence farming to food security in South Africa. *Agrekon*, *48*(4), 459–482. doi:10.1080/03031853.2009.9523836

Chauvin, N. D., Mulangu, F., & Porto, G. (2012). *Food Production and Consumption Trends in Sub-Saharan Africa: Prospects for the Transformation of the Agricultural Sector*. UNDP working paper, WP 2012-011. Retrieved from: http://news.heartland.org/sites/default/files/african_crop_trends.pdf

Coyle, D. F. (1996). *System Dynamics Modeling: A Practical Approach.* Florida: RCR Press. doi:10.1007/978-1-4899-2935-8

Fan, S., Brzeska, J., Keyzer, M., & Halsema, A. (2013). *From Subsistence to Profit: Transforming Smallholder Farms.* IFPRI Food Policy Report. doi:. doi:10.2499/9780896295582

FAO. (2011). *Save and Grow. A Policymakers' Guide to the Sustainable Intensification of Smallholder Crop Production.* Rome: Food and Agriculture Organization of the United Nations.

Haggblade, S., & Hazell, P. B. R. (2010). *Successes in African Agriculture: Lessons for the future.* IFPRI Issue Brief 63. Retrieved from http://www.ifpri.org/sites/default/files/publications/ib63.pdf

Hammond, R. A., & Dubé, L. (2012). A systems science perspective and transdisciplinary models for food and nutrition security. *Proceedings of the National Academy of Sciences of the United States of America, 109*(31), 1235–12363. doi:10.1073/pnas.0913003109 PMID:22826247

Harvey, C. A., Rakotobe, Z. L., Rao, N. S., Dave, R., Razafimahatratra, H., Rabarijohn, R. H., ... MacKinnon, J. L. (2014). Extreme vulnerability of smallholder farmers to agricultural risks and climate change in Madagascar. *Phil. Trans. R. Soc. B, 369.* doi:10.1098/rstb.2013.0089 PMID:24535397

Hassanien, A. E., Azar, A. T., Snasel, V., Kacprzyk, J., & Abawajy, J. H. (2015). Big Data in Complex Systems: Challenges and Opportunities. Berlin/Heidelberg, GmbH: Springer-Verlag. doi:10.1007/978-3-319-11056-1

Kopainsky, B., Tröger, K., Derwisch, S., & Ulli-Beer, S. (2012). Designing Sustainable Food Security Policies in Sub-Saharan African Countries: How Social Dynamics Over-Ride Utility Evaluations for Good and Bad. *Systems Research and Behavioral Science, 29*(6), 575–589. doi:10.1002res.2140

Morton, J. F. (2007). The impact of climate change on smallholder and subsistence agriculture. *Proceedings of the National Academy of Sciences of the United States of America, 104*(50), 19680–19685. doi:10.1073/pnas.0701855104 PMID:18077400

Otsuka, K., & Kalirajan, K. P. (2005). An Overview and Prospects for a Green Revolution in Sub-Sahara Africa. *Electronic Journal of Agricultural and Development Economics, 2*(1), 1–6.

Otsuka, K., & Kalirajan, K. P. (2006). Rice green revolution in Asia and its transferability to Africa: An introduction. *The Developing Economies, 44*(2), 107–122. doi:10.1111/j.1746-1049.2006.00009.x

Oyo, B. (2013, May). A System Dynamics Analysis of seed banking effectiveness for Empowerment of Small Holder Farmers. *Paper presented at the IST-Africa Conference. Nairobi, Kenya.*

Oyo, B., Kaye, M., & Nkalubo, N. (2014). Re-Conceptualisation of Agriculture Information System for Uganda. *TSEST Transaction on Electrical and Electronic Circuits and Systems, 4*(2), 39–44.

Oyo, B., Magara, S., & Andogah, G. (2012, May). *An Agriculture Information System for Uganda.* Paper presented at IST-Africa Conference on Digital Inclusion, Dar es Salaam, Tanzania.

Resilience Africa Network (RAN) Report. (2014). Retrieved from http://www.ranlab.org/resilience

Salami, A., Kamara, A. B., & Brixiova, Z. (2010). *Smallholder Agriculture in East Africa: Trends, Constraints and Opportunities*. ADB Working Paper No. Retrieved from http://www.afdb.org/fileadmin/uploads/afdb/Documents/Publications/WORKING201052020PDF20d.pdf

Saldarriaga, M., Kopainsky, B., & Alessi, S. M. (2013, July). *Knowledge analysis in coupled social-ecological systems. What do stakeholders in sub Saharan Africa know about the dynamic complexity of climate change, agriculture and food security?* Paper presented at the 31st International Conference of the System Dynamics Society, Cambridge, MA.

Settle, W., & Garba, M. H. (2011). Sustainable crop production intensification in the Senegal and Niger River basins of francophone West Africa. *International Journal of Agricultural Sustainability*, *9*(1), 171–185. doi:10.3763/ijas.2010.0559

Stave, K. A., & Kopainsky, B. (2014, July). *Dynamic thinking about food system vulnerabilities in highly developed countries: Issues and initial analytic structure for building resilience*. Paper presented at the 32nd International Conference of the System Dynamics Society, Delft, The Netherlands.

Thompson, J., Millstone, E., Scoones, I., Ely, A., Marshall, F., Shah, E., & Stagl, S. (2007) *Agri-food System Dynamics: pathways to sustainability in an era of uncertainty*. STEPS Working Paper 4, STEPS Centre, Brighton.

Thornton, P. K., Jones, P. G., Ericksen, P. J., & Challinor, A. J. (2011). Agriculture and food systems in sub-Saharan Africa in a 40C+ world. *Phil. Trans. R. Soc. A*, *369*(1934), 117–136. doi:10.1098/rsta.2010.0246 PMID:21115516

Walker, B. H., Anderies, J. M., Kinzig, A. P., & Ryan, P. (2006). Exploring resilience in social-ecological systems through comparative studies and theory development: Introduction to the special issue. *Ecology and Society*, *11*(1), 12.

This research was previously published in the International Journal of System Dynamics Applications (IJSDA), 5(1); edited by Ahmad Taher Azar, pages 17-30, copyright year 2016 by IGI Publishing (an imprint of IGI Global).

Chapter 15
Rights-Based Approach to Food and Nutrition Security in Nigeria

Clementina Oluwafunke Ajayi
Federal University of Technology – Akure, Nigeria

Kemisola O. Adenegan
University of Ibadan, Nigeria

ABSTRACT

This chapter focuses on the need for a rights-based approach to food and nutrition security in Nigeria. The topic is introduced with the definition of basic terms used throughout the chapter. The objectives of this chapter are to create awareness of the need to adopt a rights-based approach to food and nutrition security and help define the context of the right to food in Nigeria. The rest of the chapter sets out the roles of right holders, duty bearers, and accountable agents in food and nutrition security. It highlights their rights, obligations, and responsibilities, as well as voluntary guidelines and implications for a rights-based approach to food. The chapter also reviews policies that have been developed toward ensuring a right to food in Nigeria.

BACKGROUND

Food and nutrition insecurity is increasing in the world where 925 million people are undernourished. Today, over 900 million people are food insecure across the world despite the fact that the world food production has doubled during the past three decades (FAO, 2010). Helen (2002) noted, food security maintains political stability, and ensures peaceful coexistence among people while food insecurity results in poor health and reduced performance of both children and adult. Poor nutrition weakens children's immune systems and renders them more vulnerable to deadly diseases, which further deteriorates their nutritional status. Hence children get stuck in a cycle of undernutrition and infection, which adversely affects their health, development, and increases the risk of mortality.

The first and most important component of social and economic justice is adequate food production (Otaha, 2013). This can therefore improve food and nutrition security if viewed from rights-based

DOI: 10.4018/978-1-5225-8063-8.ch015

perspective with the ultimate goal of development to guarantee all human rights to everyone. A human rights-based approach is a conceptual framework for the process of human development that is normatively based on international human rights standards and operationally directed to encouraging and defending human rights (United Nations, UN, 2017). It seeks to analyze inequalities which lie at the core of development complications and amends discriminatory practices and unfair distributions of power that hinder development progress.

Under a human rights-based approach, the plans, policies and processes of development are anchored in a system of rights and conforming obligations established by international law. This helps to promote the sustainability of development work, empowering people themselves— especially the most marginalized—to participate in policy formulation and hold accountable those who have a duty to act. A rights-based approach to development is both a vision and a set of tools: human rights can be the means, the ends, the mechanism of evaluation and the central focus of sustainable human development (UN, 2002).

According to World Food Programme (WFP, 2017), here are ten things to know about Nigeria food and nutrition security status:

1. Nigeria is a food deficit country and is Africa's largest importer of rice.
2. One third of children under 5 are stunted. That's twice the rate of Thailand and three times that of Tunisia.
3. A child in the remote northwestern region of Nigeria, where stunting rates are around 55 percent, is four times more likely to experience malnutrition than a child in the south.
4. At the same time, child obesity is increasing and 33 percent of adults are obese or overweight. That is roughly on par with Singapore.
5. Nearly half of women of reproductive age (48.5 percent) are anaemic. This ranks Nigeria 172th best out of 185 countries.
6. Conflict with Boko Haram in northeastern Nigeria has left a large part of the population without access to enough food, water and health services.
7. Displacement, lack of access to many locations, high inflation and reduced purchasing power of communities are worsening the food security situation in northeastern Nigeria.
8. Over 3 million people are in urgent need of humanitarian assistance due to conflict.
9. In March 2016, WFP began providing cash transfers via mobile phones to displaced persons and host communities in critical areas. This gives them the opportunity to buy the food they need.
10. In partnership with the Government and other agencies such as UNICEF, WFP is scaling up its assistance in northeastern Nigeria to reach a total of 431,000 people in desperate need, including malnourished children and pregnant and nursing women.

The right to food and nutrition is a holistic concept which empowers people. The right to food and nutrition relates to diverse dimensions of human life, from the most basic access to sun light, seeds, land, water, which allows food production, up to the moment in which food reaches human beings. Food is not just a commodity, it is part of our life, as a social, cultural and economic right and it impacts our lives as communities: security, democracy, health, among others. It is in the light of these that this chapter aim at creating awareness of the need to adopt rights-based approach to food and nutrition security. It will also help define the context of the right to food in Nigeria so as to ensure attainment of food and nutrition security.

DEFINITIONS OF TERMS

Right to Adequate Food

The International Covenant on Economic, Social and Cultural Rights (ICESCR, 1999) defined the right to adequate food as what "every man, woman and child, alone or in community with others, have physical and economic access at all times to adequate food or means for its procurement". This definition is synonymous to that of food security by the World Food Summit (WFS, 1996) which states that 'a situation that exists when all people, at all times, have physical, social and economic access to sufficient, safe and nutritious food that meets their dietary needs and food preferences for an active and healthy life.' Also, the Constitution of the Federal Republic of Nigeria 1999, article 16 (2d) states that 'suitable and adequate shelter, suitable and adequate food, reasonable national living wage, old age care and pensions, and unemployment, sick benefits and welfare of the disabled are provided for all citizens.'

Food Security

Food security concept is not new in research studies. It has been defined in a variety of ways by different authors and organizations. It is a condition related to the supply of food, and individual's access to it. In 2009, the World Summit on Food Security stated that the "four pillars of food security are availability, accessibility, utilization, and stability" (Food and Agriculture Organization, FAO, 2009).

Food Availability

Food availability relates to the supply of food through production, distribution, and exchange (Gregory *et al.*, 2010). According to FAO (1997), food production is determined by a variety of factors including land ownership and use; soil management; crop selection, breeding, and management; livestock breeding and management; and harvesting.

Economic and Physical Access to Food

Food access refers to the affordability and allocation of food, as well as the preferences of individuals and households (Gregory *et al.*, 2010). An adequate supply of food at the national or international level does not in itself guarantee household level food security. Concerns about insufficient food access have resulted in a greater policy focus on incomes, expenditure, markets and prices in achieving food security objectives. The UN Committee on Economic, Social, and Cultural Rights noted that the causes of hunger and malnutrition are often not a scarcity of food but an inability to access available food, usually due to poverty (UN, 1999). Access depends on whether the household has enough income to purchase food at prevailing prices or has sufficient land and other resources to grow its own food (Garret and Ruel, 1999).

Utilization

Utilization refers to the metabolism of food by individuals (Tweeten, 1999). It is commonly understood as the way the body makes the most of various nutrients in the food. Once food is obtained by a household, a variety of factors affect the quantity and quality of food that reaches members of the household. In order to achieve food security, the food ingested must be safe and must be enough to meet the physiological requirements of each individual. Food utilization is also a function of how food is processed and stored, knowledge of good nutrition and care, and adequate health and sanitation services in the environment in which food is consumed *(Ecker and Breisinger 2012).*

Stability

Food stability refers to the ability to obtain food over time (FAO, 1997). Even if your food intake is adequate today, you are still considered to be food insecure if you have inadequate access to food on a periodic basis, risking a deterioration of your nutritional status. Adverse weather conditions, political instability, or economic factors (unemployment, rising food prices) may have an impact on your food security status.

For food security objectives to be realized, all four dimensions must be fulfilled simultaneously (WFS, 2009).

Nutrition

Nutrition is underpinned by intake, absorption and use of nutrients. There are many, often concurring, socio-political, economic and environmental factors that directly and indirectly impact on nutrition (World Health Organisation, WHO, 2013). Nutrition is the science that explains the role of food and nutrients in the human body during growth, development and maintenance of life. In its broader context, nutrition has also been defined as being concerned with "... how food is produced, processed, handled, sold, prepared, shared, and eaten and what happens to food in the body - how it is digested, absorbed, and used" (King and Burgess, 1993). Nutritional status refers to the nutritional state of the body, as expressed according to anthropometric measure such as weight, height, age or combinations of these. Only focusing on food availability or food accessibility is not sufficient and will not result in a well-nourished, healthy and productive population. Additional interventions are needed such as (environmental) health services, care practices, infrastructural facilities, adequate security and focus on nutritious food.

Malnutrition

Malnutrition arises from deficiencies of specific nutrients or from diets based on inappropriate combinations or proportions of foods; for example, goitre, scurvy, anaemia and xerophthalmia are forms of malnutrition caused by inadequate iodine, vitamin C, iron and vitamin A, respectively (Shetty, 2003). Malnutrition can also result from excess nutrient losses or utilization. It relates to imbalances in energy, and specific macro and micro nutrients intake in relation to the requirements. Malnutrition occurs when the intake of essential macro and micro nutrients does not meet or exceeds the metabolic demands for those nutrients.

Undernutrition

Undernutrition stems from the inadequate quantity and /or quality being consumed and/or repeated infection of disease resulting in improper absorption of vital nutrients. It manifests itself through wasting, stunting and micronutrients deficiencies (WHO, 2013). It is caused primarily by an inadequate intake of dietary energy, regardless of whether any other specific nutrient is a limiting factor (Shetty, 2003). It represents a state or condition of people that have to survive on less than 2100 kcal per day, which is estimated from food balance sheets (food production minus export plus import) – based on the FAO Food Balance Sheets, presented at national level and aggregated at global level. Chronic undernutrition is an inadequate nutrition coupled with recurrent or chronic illness over long period of time leading to failure of linear growth. The terms 'malnutrition' and 'undernutrition' are often used loosely and interchangeably. Malnutrition refers to all deviations from adequate and optimal nutritional status, including energy undernutrition and over-nutrition (obesity is a form of malnutrition). The term 'undernutrition' is used to refer to generally poor nutritional status, but also implies underfeeding.

Hunger

Hunger is a condition, in which people lack the basic food intake to provide them with the energy and nutrients for fully productive lives (Hunger Task Force, 2003)

Food and Nutrition Security

Food and Nutrition Security exists when all people at all times have physical, social and economic access to food, which is safe and consumed in sufficient quantity and quality to meet their dietary needs and food preferences, and is supported by an environment of adequate sanitation, health services and care, allowing for a healthy and active life (World Food Population, 2013). Improving food and nutrition security requires knowing where the most vulnerable are located and understanding what makes them vulnerable. Targeting is a key mechanism for reaching vulnerable populations and ensuring efficient and effective use of limited resources.

Human Rights

Human Rights are moral principles or norms which describe certain standards of human behavior, and are regularly protected as legal rights in municipal and international law. They are commonly understood as inalienable fundamental rights to which a person is inherently entitled simply because she or he is a human being and which are inherent in all human beings regardless of their nation, location, language, religion, ethnic origin or any other status. They are rights that every human being enjoys, agreement between states, norm about state-citizen relationships and can be protective devices or emancipatory tools but also political chips/tools or empty promises.

The Right to Food

The right to food is a human right protecting the right for people to feed themselves in dignity. This implies that sufficient food is available, that people have the means to access it, and that it adequately meets the individual dietary needs. The right to food protects the right of all human beings to be free from hunger, food insecurity and malnutrition. The right to food does not imply that governments have an obligation to hand out free food to everyone who wants it, or that everybody has a right to be fed. However, if people are deprived of access to food beyond their control such as in times of war, disaster etc., the right to food requires the government to provide food directly. The right to food originated from the Universal Declaration of Human Rights (1948) and underwent evolution of International Covenant on Civil and Political Rights (1966-1976), International Covenant on Economic, Social and Cultural Rights (1966-1976), General Comment no 12 of United Nation Commission on Economic, Social and Cultural Rights (1999) before metamorphosing into FAO guidelines on the Right to Adequate Food (RtAF), 2004 and FAO Panther principles (2006).

Right Holder

Every human being is inherently a right holder who should enjoy universal human rights that must be guaranteed (Ljungman, 2004). Every citizen of a country is a right holder for their basic food needs to be meet.'

Duty Bearer

According to Ljungman 2004, by endorsing the diverse United Nations human rights treaties, states (Government) automatically take up the principal roles of guaranteeing these rights. They are termed "principal duty bearers". The duty bearer has three obligations which are to respect, protect, and fulfill.

- To *respect* requires the State and all its organs and agents to abstain from carrying out, sponsoring or tolerating any practice, policy or legal measure violating the integrity of individuals or impinging on their freedom to access resources to satisfy their needs. It also requires that legislative and administrative codes take account of guaranteed rights. This means the duty bearer must respect existing access to adequate food and health and must not take any measures preventing or destroying such access.
- To *protect* the right to food is to take necessary measures to ensure that a third party does not deprive people of their access to adequate food. Where violations do occur the State must guarantee access to legal remedies.
- To *fulfill* involves issues of advocacy, public expenditure, governmental regulation of the economy, the provision of basic services and related infrastructure and redistributive measures. The duty of fulfillment comprises those active measures necessary for guaranteeing opportunities to access entitlements. To fulfill the right to food has two components: to facilitate or create social and economic environments that foster human development, and to provide food to people in an emergency or in circumstances when self-provisioning is beyond their control.

Even though states play the role of the principal duty bearer, there are other non-state entities (moral duty bearer) that have obligations to respect, protect and fulfill the rights of people. They are primary duty-bearers e.g. parents for children, teachers for students, police for crime suspects, doctors/nurses for patients, employers for employees; secondary duty-bearers e.g. institutions and organizations with immediate jurisdiction over the primary duty-bearers e.g. school principals, community organizations, hospital administrations, etc.; tertiary duty-bearers e.g. institutions and organizations at a higher level / more remote jurisdiction (Non-Governmental Organizations, aid agencies, private sector organizations); and external duty-bearers *e.g.* countries, institutions, organizations with no direct involvement e.g. United Nations, Security Council, African Union.

The constitutional recognition of the right to food in Nigeria makes it a normal standard that every human being is a right holder and enjoys obligations from the duty bearer. Also, "each state party to the present Covenant undertakes to take steps, individually and through international assistance and co-operation to the maximum of its available resources, with a view to achieving progressively the full realization of the rights recognized in the present Covenant" (Art. 2, ICESCR). To this end, there is need for every citizen to wake up to the call of ensuring their right to food is established.

Human Rights Principles: PANTHER

The PANTHER framework, developed by FAO in 2006 is a human rights–based approach. The framework stipulates the right to food and food security. As shown in Figure 1, it establishes the seven principles that should govern decision making and implementation processes of the right to food. The first letter of each principle formed the acronym PANTHER.

- *P*articipation requires that everyone have the right to subscribe to decisions that affect them.
- *A*ccountability requires that politicians and government officials be held accountable for their actions through elections, judicial procedures and other mechanisms.
- *N*on-Discrimination prohibits arbitrary differences of treatment in decision making.
- *T*ransparency requires that people be able to know processes, decisions and outcomes.
- *H*uman dignity requires that people be treated in a noble way.
- *E*mpowerment requires that they are in a position to exert control over decisions affecting their lives.
- *R*ule of law requires that every member of society, including decision makers, must comply with the law.

Figure 1. Human Rights PANTHER Principles
Source: FAO, (2006)

REVIEW OF FOOD PRODUCTION EFFORTS IN NIGERIA

Nigeria is challenged with rapid population growth which affects her ability to assure stable supply of, and access to food. Its population projected to rise from 182.3 million in 2015 to 262.3 million by 2030 to be the world's fifth most populous and the third most populous by 2050 (World Population Prospects, 2016). Also, research findings in Nigeria show that the returns for scaling up nutrition specific intervention has a compound rate of returns of >13% and benefit to cost ratio of 16:1 (Global Nutrition Report, GNR (2015)). The growing evidence food and nutrition insecurity in and around Nigeria calls for a critical analysis of potential challenges concerning citizens and how to cope with these challenges.

Nigeria has made several attempts to increase food production in both quantity and quality. Some of these attempts have cumulated into several programmes and projects such as Operation Feed the Nation, Agricultural Development Project, Fadama I, II and III, Vision 2020, Agricultural Transformation Agenda, The Agriculture Promotion Policy among others. These policies and programmes were aimed at boosting agricultural production thereby solving the problem of food insecurity and poverty.

The 1960s featured strong public intervention in agriculture, with development guidelines and plans established at the federal level and implemented in the states. The government's main concern at the time was to increase domestic production, particularly of cash crops. This period pushed Nigeria to the position of the world's top producer of groundnuts, rubber and palm oil, and the world's second-largest cocoa producer.

The 1970-1986 period, which coincided with intensive petroleum exploitation, was marked by a decrease in policies supporting agriculture. The strong drop in domestic agricultural production reduced the country to growing dependency on imported foodstuffs. In the wake of the major food crisis in the country in 1976, programmes such as "Feed the Nation" (1976-1979) and "Green Revolution" (1979-1983) were set up. These programmes focused on strengthening agricultural production, providing subsidized inputs, community development, and access to credit. However, they were implemented without a transparent framework to structure action, and the successive governments did not ensure continuity. The enactment of the Land Use Act in 1978 marked an historic turning point for land use management in Nigeria. The movement was reversed in 1987 with the structural adjustment programmes (SAPs) that sought to reduce the national economy's dependency on oil and promote the private sector as the engine driving growth. In 1998, the Nigerian government once again turned its attention to the agricultural sector. It adopted an agricultural policy that had the objective, among others, of ensuring food security for the population by developing local production.

Since the reference document "Agriculture in Nigeria: The New Policy Thrust" was issued in 2001, the government has assigned the agricultural sector an ambitious role in its strategic planning frameworks. The strategic document for reducing poverty in Nigeria, "National Economic Empowerment and Development Strategy" (NEEDS II 2008-2011) emphasized economic development driven by the private sector, and the "7-point Agenda", a framework guiding economic reform in the country that was adopted in May 2007. These two documents were the medium-term policy documents intended to help the country achieve the Millennium Development Goals of 2015 and its own "2020 Vision" plan. The latter aims to make Nigeria one of the top twenty economies in the world by 2020. For agriculture, this means increasing current domestic production by six folds. *The National Food Security Programme (NFSP)* issued in August 2008 by the Federal Ministry of Agriculture and Water Resources as it was called then was designed to help Nigeria attain food security by ensuring that all Nigerians have access to good-quality food while making Nigeria a major exporter of foodstuffs. The programme designated priority crops (cassava, rice, millet, wheat) for achieving food security and outlined objectives for all stages of these supply chains. The aim is to create more value in production, particularly downstream in the value chain by improving storage, processing, and access to agricultural markets. The programme also had plans for the creation of irrigation schemes (450,000 ha). The strategic frameworks in NEEDS II and the 7-point Agenda were been translated into short-to-medium-term programmes. The Ministry drew up a "5-point Agenda" for agriculture, a detailed roadmap of steps to be implemented to attain the objectives listed for agriculture in the 7-point Agenda.

Olusegun Obasanjo's administration also launched Presidential Initiatives in 1999 for seven agricultural products (cassava, rice, vegetable oil, sugar, livestock, cultivated trees and dry grains). The aim of these initiatives was not only to boost Nigeria's agricultural exports by taking advantage of preferential agreements in the framework of the World Trade Organization (WTO) and the Economic Partnership Agreements between the European Union and the Africa-Caribbean-Pacific countries but also to make the most of the potential regional market made up of neighboring countries. Although these measures have shown that investment in the agricultural sector can have concrete results in terms of increasing domestic production, their overall outcomes have been mixed in that only the "intensification of production" segment has been taken into account, ignoring the downstream segments of the value chain (such as product processing).

Support for agricultural inputs has been a central element of Nigerian agricultural policy since the 1950s. This support consists primarily of providing public subsidies so that farmers can more easily acquire inputs (fertilizers, improved seeds, phytosanitary products). The level of federal subsidies has followed a spiky path, with highs and lows, and methods of implementation have varied. In addition to federal subsidies, each state allocates its own subsidies for fertilizer. These vary greatly from one state to another in both amounts (50 to 150 kg per farmer) and subsidization rates (from 10% to 50%). Even so, many farmers still find it difficult to obtain good-quality inputs at an affordable price and at the time they are needed. The government has not yet managed to set up an effective regulatory and monitoring system to address quality issues and the diversion of subsidized inputs smuggled outside the country. Some states have been testing the distribution of input subsidy vouchers since 2008.

The Agricultural Transformation Agenda was an initiative of the Federal Ministry of Agricultural and Rural Development (previously named Federal Ministry of Agriculture and Water Resources) to support former President Goodluck Jonathan's Transformation Agenda. The goal of the ATA was to build commodity value chains and the institutions required to unlock the country's huge agricultural potentials with the targeted outcomes such as, add 20million tonnes of food to the domestic food supply by 2015, create 3.4 million jobs and ensuring import substitution through the acceleration of production of local staples; reducing dependence on food imports and turning Nigeria into a net food exporter. A major success story of the ATA in 2014 was the news that Nigerian food import bill had dropped by N466 billion thereby adding N780 billion to the economy during the period. Building on the successes and lessons from the ATA, the current Administration under the leadership of President Buhari, the Agricultural Promotion Policy (APP) has the vision to work with stakeholders to build an agribusiness economy capable of delivering sustained prosperity by meeting domestic food security goals, generating exports and supporting sustainable income and job growth between 2016 and 2020. Among the 11 guiding principles of APP are food as a human right focusing the policy instruments for enterprise development on the social responsibility of government with respect to food security, social security and equity in the Nigerian society; and compelling the government to recognize, protect and fulfil the irreducible minimum degree of freedom of the people from hunger and malnutrition.

Despite successive strategies and programmes implemented, food and nutrition insecurity are still rampant. The 2015 Global Food Security Index, which comprises 109 countries appraised by the level of food sector development (food affordability, availability, quality and safety); put *Nigeria* at 91st place with the average of 37.1 indexes as against 80th among 105 countries in 2011. The food security situation of the country thus call for rights based approach interventions. The APP has recognized food as human right, therefore, there is need to create the awareness among the right holders in order to make the duty bearers accountable.

ISSUES TO BE ADDRESSED WHEN ADOPTING THE RIGHTS-BASED APPROACH TO FOOD SECURITY IN NIGERIA

Adopting the human rights framework for addressing food security seems a natural extension of the progress already made in the health and human development movement. This can also be applicable to food security by paying attention to several key elements that are but not limited to equal and public participation, government accountability, an analytical framework that accounts for vulnerability and discrimination and stronger connections between policies and outcomes. These issues include:

Equal and Public Participation

According to ICESCR (Art. 2), each state party to the present covenant undertakes to take steps, individually as well as through international assistance and co-operation to the maximum of its available resources, with a view to achieving progressively the full realization of the rights recognized in the present covenant." The principle of public participation seeks and facilitates the involvement of those potentially affected by or interested in rights based approach to food security in its decision making process. This is in relation to the facilitator of collective intelligence and inclusiveness, shaped by the desire for the participation of the whole country. Participation in both programs and policy making means and goal should be embraced by all the right holders bearing in mind that ability to claim right to food is their right.

Promote Government Accountability

The human rights framework is premised on the concept of accountability of state authorities through social control to all stakeholders. Since evidence of food insecurity is on the increase in the country, there is no apparent linkage of the report's findings to any action plan to reduce rates. Measurement is a key component of tracking the magnitude of food insecurity. This includes ensuring that there are governmental actors charged with establishing these reference goals with clear timeframes for implementation of action plans to achieve them. Strategies that empower the people they are meant to serve should be embraced.

Address Conflict, Vulnerability and Discrimination

Conflict may be defined as a struggle or contest between people with opposing needs, ideas, beliefs, values, or goals. Conflict on teams is inevitable; however, the results of conflict are not predetermined. Conflict is compromising the food security and nutrition of millions of Nigerians. Vulnerability can be defined as the probability of an acute decline in access to food, or consumption, often in reference to some critical value that defines minimal values of human well-being (WFP, 2002). Vulnerability influences a household's behavior and means of coping with its current wealth and livelihood circumstances. It also has an impact on the state of food security at the individual and household levels. Certain groups by nature of socioeconomic conditions or previous discrimination, are more vulnerable to food insecurity than others. A human rights approach entails focusing on those who are most vulnerable, understanding what causes this vulnerability or susceptibility to adverse outcomes, and changing conditions to improve their situation.

According to the United Nation Food and Agriculture Organization (FAO), women are responsible for 50 percent of world food production, mainly for family consumption. The majority of rural women are "invisible" field workers in family plots, and as a result, they have no recognized independent status as farmers and their work is considered as secondary both in the family and in society. In sub-Saharan Africa, only 15 percent of landholders are women and they account for less than 10 percent of credit and 7 percent of extension services. According to estimates, policies that address gender inequalities could, conservatively, increase yields on women's farms by 2.5 to 4 percent as investing in rural women has been shown to increase productivity significantly and reduce hunger and malnutrition. The vulnerability of women to hunger and food insecurity has long been recognized in the human rights documents of the United Nations. One of the greatest concerns is the intergenerational transmission of malnutrition that is, pregnant women that are malnourished are more likely to have low-birth weight babies. As a result, their children are more susceptible to under nutrition and poor cognitive development, which in turn affects the children's ability to earn enough money to supportthemselves and their families when they become adults. Since women and children are especially vulnerable to food insecurity and to socioeconomic processes that cause it, ensuring women's rights is an important corollary to achieving the right to food.

In Nigeria the appalling food insecurity situation has degenerated to a level that it is listed among the 42 countries tagged "low-income food deficit countries." (Okunmadewa, 2003). Food insecurity disproportionately affects rural people particularly rural women, minorities and children, World Bank (2001) revealed that rural people face a high risk of food insecurity due to poverty, income inadequacies, limited access to resources, underemployment, and unemployment, and many barriers to self-sufficiency, which create family frailty and crisis.

Measurable Goals and Targets That Link Policies to Outcomes

A human rights framework applied to the right to food can interpret how food-related policies affect one's ability to purchase food and how such policies affect health and wellbeing. Although, many Nigerians have applauded the development in the nation's agricultural sector, many are also of the view that the claims to a drop in the nation's food import bill does not correspond with market development as the prices of food and agricultural commodities have remained on the rise despite claims to huge gains in the sector.

CONCLUSION AND RECOMMENDATIONS

The cost of inadequate diets to families and nations are considerably high. This includes increased vulnerability to diseases and parasites, reduced strength for tasks requiring physical effort, reduction of the benefit from schooling and training programmes and general lack of vigour, alertness and vitality.

These results in a reduction in the productivity of people in the short and long terms, sacrifice in output and incomes, and increasing difficulty for families and nations to escape the cycle of food and nutrition insecurity.

A right-to-food approach requires that states fulfil their obligation to ensure that safe, nutritionally adequate and culturally acceptable food is available. They must also respect and protect consumers as well as promote good nutrition for the deprived. In order to ensure the progressive realization of the right to food at the domestic level, it is imperative that constitutional principles and framework laws are established as a means of providing an appropriate institutional structure. The refreshed strategy- APP under the present administration can use the recognition of the right to food to benefit all Nigerians and promote gender equity in their policy thrust and objectives which are to solve the challenges of food security, import substitution, job creation and economic diversification. Achieving these through rights based approach can thus reinforce the nation's fundamental goal of leveraging the right holders' capabilities to ensure food and nutrition security as well as income security.

Rights-based approach to food security should ensure programs that focus on vulnerable, marginalized, disadvantaged and excluded groups, and that aim to reduce disparities between these groups and others. It is important, however, to note that advocacy works on two levels highlighted below:

- With the right holders to help them claim their right in legitimate ways while working on protecting the rights of others. Working especially with the disadvantaged and marginalized right holders is crucially important as part of helping them regain confidence in their ability to equally and fully participate in the decision making process.
- With the Duty bearers – be them state, or non-state – to be aware of, and determined by, the rights of people in their determinations to respect, protect and fulfill these rights; and do their best to not only fulfill these rights, but also, avail themselves accountable and quick to respond to the people in this regard.

REFERENCES

Ecker & Breisinger. (2012). *The Food Security System*. Washington, DC: International Food Policy Research Institute. Retrieved from http://www.ifpri.org/sites/default/files/publications/ifpridp01166.pdf

FAO. (2006). *The right to Food: The Human Right to Adequate Food*. Retrieved from http://www.fao.org/righttofood/about-right-to-food/human-right-principles-panther/en/

FAO. (2009). *Declaration of the World Food Summit on Food Security*. Rome: Food and Agriculture Organization of the United Nations. Retrieved from http://www.fao.org/fileadmin/templates/wsfs/Summit/Docs/Final_Declaration/WSFS09_Declaration.pdf

FAO. (2010). *Food insecurity in the World: addressing food insecurity in protracted crises*. Rome: Food and Agriculture Organization of the United Nations.

FMARD. (2016). *The Agriculture Promotion Policy (2016-2020): Building on the successes of the ATA, Closing Key Gaps*. Policy and Strategy Document of the Federal Ministry of Agriculture and Rural Development.

Food and Agriculture Organization (FAO). (1997). The food system and factors affecting household food security and nutrition. In *Agriculture, food and nutrition for Africa: A resource book for teachers of agriculture*. Rome: Agriculture and Consumer Protection Department. Retrieved 15 October 2013 from http://www.fao.org/docrep/W0078E/W0078E00.htm

FRN. (1999). Constitution of the Federal Republic of Nigeria.

Garrett, J., & Ruel, M. (1999). *Are Determinants of Rural and Urban Food Security and Nutritional Status Different? Some Insights from Mozambique*. Washington, DC: International Food Policy Research Institute. Retrieved 15 October 2013 from http://www.ifpri.org/sites/default/files/pubs/divs/fcnd/dp/papers/dp65.pdf

Global Nutrition Report. (2015). *Basic Global Nutrition Report*. Retrieved from https://www.globalnutritionreport.org/files/2015/10/Basic-GNR-2015-presentation_102815.pdf

Godfray, H. C. J., Beddington, J. R., Crute, I. R., Haddad, L., Lawrence, D., Muir, J. F., ... Toulmin, C. (2010). Food Security: The Challenge of Feeding 9 Billion People. *Science*, *327*(5967), 812–818. doi:10.1126cience.1185383 PMID:20110467

Gregory, P. J., Ingram, J. S. I., & Brklacich, M. (2005). Climate change and food security. *Philosophical Transactions of the Royal Society of London. Series B, Biological Sciences*, *360*(1463), 2139–2148. doi:10.1098/rstb.2005.1745 PMID:16433099

Helen, H. J. (2002). Food Insecurity and the Food Stamp Programme. *American Journal of Agricultural Economics*, *84*(5), 1215–1218. doi:10.1111/1467-8276.00382

Hunger Task Force. (2003). *Halving Hunger by 2015: A Framework for Action*. Interim Report, Millennium Project, UNDP.

ICESCR. (1999). *The International Covenant on Economic, Social and Cultural Rights. General comments 12. Twentieth session, 12th May, 1999*. Retrieved from http://www.fao.org/fileadmin/templates/righttofood/documents/RTF_publications/EN/General_Comment_12_EN.pdf

King, F. S., & Burgess, A. (1993). Nutrition for Developing Countries (2nd ed.). Oxford, UK: Oxford University Press. doi:10.1093/acprof:oso/9780192622334.001.0001

Ljungman, C. M. (2004). Applying a Rights-Based Approach to Development: Concepts and Principles. *Conference Paper: The Winners and Losers from Rights-Based Approaches to Development*, 6.

Okunmadewa, F. (2003). Risk and Vulnerability in Agriculture: concept and Context. *Staff seminar paper in the Department of Agricultural Economics*. University of Ibadan.

Otaha I. J. (2013). Food Insecurity in Nigeria: Way Forward. *Journal of African Research Review*, *7*(4), 26-35.

Shetty P. (2003). Malnutrition and Undernutrition. Science. *Medicine, 31*(4), 18-22. doi:10.1383/medc.31.4.18.27958

Tweeten, L. (1999). The Economics of Global Food Security. *Review of Agricultural Economics*, *21*(2), 473–488.

UN. (2002). *Rights-Based Approach to Development Programming: Training Manual*. Retrieved from http://www.un.org.ph/publications/RBAManual.pdf

UN. (2017). *What is a human rights-based approach? Human Rights-Based Approach Portal*. Retrieved from http://hrbaportal.org/faq/what-is-a-human-rights-based-approach

United Nations Committee on Economic, Social, and Cultural Rights. (1999). *The right to adequate food*. Geneva: United Nations.

WFP. (2002). Vulnerability Analysis and Mapping: Standard Analytical Framework. Rome: Author.

WFP. (2013). *Food and Nutrition Security in Democratic People's Republic of Korea. Vam Food security Analysis*. Retrieved from https://www.wfp.org/sites/.../files/DPRK_draft_FINAL%201%20October%20 2013.pdf

WFP. (2017). *Global Nutrition Report*. Retrieved from https://www.wfp.org/stories/10-fact-about-food-and-nutrition-nigeria

WHO. (2013). *Undernutrition: The Basics. Generation Nutrition 6*. Retrieved from http://www.who.int/nutgrowthdb/summary_jme_2013.pdf

World Bank. (2001). *Poverty and Hunger: Issues and Options for Food Security in Developing Countries*. Washington, DC: World Bank.

World Food Summit. (1996). *Rome Declaration on World Food Security*. Author.

World Population Prospects, (2016). *World Population Prospects: The 2015 Revision – Key Findings and Advance Tables*. United Nations Department of Economic and Social Affairs, Population Division.

This research was previously published in Food Systems Sustainability and Environmental Policies in Modern Economies edited by Abiodun Elijah Obayelu, pages 217-234, copyright year 2018 by Engineering Science Reference (an imprint of IGI Global).

Chapter 16
Indigenous Food Preservation and Management of Postharvest Losses Among the Akan of Ghana

Kofi Quan-Baffour
University of South Africa, South Africa

ABSTRACT

The rapid population increase has consequences on food security in Africa. The policies of the colonialists protected European markets and discouraged the growth of indigenous agro- industries in Africa. In Ghana much food is produced during the harvest seasons but greater part of it gets rotten due to lack of preservation or storage facilities. Despite the negative attitude of the colonialists towards local products indigenous food preservation continued unabated although limited to the aging population in the rural areas. The purpose of this chapter is to share the Akan heritage of indigenous food preservation as a strategy to manage postharvest losses and ensure food security and sustainable livelihoods. The chapter which emanated from an ethnographic study used interviews and observations for data collection. The study found that the Akan communities without agro-industries use their indigenous knowledge and skills to preserve food and create jobs. The chapter concluded that in this era of Africa's rebirth its people should utilize indigenous food processing skills to reduce postharvest losses and ensure food security. It was recommended that the government of Ghana should provide financial support to make indigenous agro-industries sustainable.

INTRODUCTION

Food security is a family, community, national and international concern. With the global rapid population increase food security has become a great challenge to humankind because of the inability of many developing countries to feed every mouth. In a developing country like Ghana farmers produce food in abundance but due to lack of modern preservation and storage facilities they go rotten and create scarcity

DOI: 10.4018/978-1-5225-8063-8.ch016

of food during the lean season. Thus many people in the country experience what can be referred to as poverty in the midst of plenty situation. The Akan ethnic group which is predominantly peasant farmers use their ingenuity and heritage to preserve some of their major foodstuffs to ensure that their families have access to food throughout the year.

This chapter gives the reader some insight into indigenous food preservation, storage and management techniques to ensure food security among the Akan of Ghana. The objective of the chapter is to:

- Outline the major food production activities among the Akan;
- Describe the various indigenous methods of preserving, storing and managing specific food items to ensure food security;
- Discuss some of the major challenges to indigenous food preservation and storage;
- Discuss some lessons that can be learnt from the Akan heritage by other developing communities; and
- Provide some recommendations for sustainable utilization of indigenous food security measures.

BACKGROUND

Food security is a concern for every family, community, government and nation all over the globe. Access to food is a right that should not be denied any human being because it borders on existence or survival and enhances development. Without food there can be no life and without life there would not be development. The rapid population increase in the developing world has consequences for food security in Ghana and Africa at large. Getting enough food to feed every mouth makes food production, preservation and its security crucial. Reutlinger (1986) for instance defines food security as 'access by all people at all times to enough food for an entire healthy life'. This definition of food security might shift the attention of the world from food production and importing capabilities at the micro or local-level to focus on individuals and their ability to avoid hunger and under nutrition. It also indicates the importance individuals and organisations, the world over, attach to food as the basic human right. The challenge in a developing country like Ghana is the ever increasing number of people to be fed. The country's population now stands at over 25 million and this has serious consequences for food security. The situation requires that people have both physical and economic access to basic food by either growing or buying it (Scanlan, 2001).

Agriculture (i.e. subsistence farming) is the mainstay of Ghana (Quan-Baffour, 2011) and in the rural areas food becomes abundant during harvest season but gets rotten due to lack of processing plants, preservation and storage facilities. Thus making postharvest loss a problem to farmers. Harvesting losses of grains alone range between 4.7 and 12.05% at farmer's fields (Appiah, Guisse & Dartey, 2011). About 55% of Ghana's population is directly engaged in agriculture and the sector contributes to 40% of the country's GDP (Bill & Melinda Gates, 2010). The colonial policy of discouraging local indigenous agro-industries in order to protect European market created a legacy of a lack of agro-based industries; a problem which successive governments have not been able to adequately address since independence in 1957. In the absence of formal food processing industries the indigenous mode of preserving food to ensure food security should be revived and popularised to fill in the vacuum. Unfortunately the knowledge and skill for indigenous food processing and preservation, a true Akan heritage, is limited to the aging population who might die with it.

In this era of African renaissance, Ghana a country with limited agro-based industries, should revisit (i.e. make a backward gaze), learn, popularise and intensify its indigenous food preservation methods (Quan-Baffour, 2012) as a strategy to curb postharvest loses and to ensure food security. Commenting on the value of indigenous food preservation Anderson (1999) attests that such enterprises are usually community based and have distinctive features both in their processes and objectives.

Although Ghana produces food on large scale much of it is lost or goes rotten within days due to lack of silos for storage and processing plants (Appiah etal, 2011). This appalling situation of poverty or hunger in the midst of plenty could be partly blamed on colonization and its concomitant slavery and slave trade. The manufacturing of indigenous goods (including food processing) were discouraged or outlawed by the colonialists under the pretext that they were unhygienic or heathen. In deed the Western *'incursion'* was not only a strategy to create and protect European market for manufactured goods (including processed or canned foods) but it also deliberately hindered the development of indigenous industries (Rodney, 2009) including food processing and preservation. The shipments of slaves were done by Europeans to markets controlled by them and for the interest of European capitalism and nothing else. African economic activity was affected both directly and indirectly by population loss emanating from slavery and slave trade (Rodney, 2009). The most economically active men and women were captured, lured or bought into slavery and shipped away to serve the interest of Western businesses and markets. In effect as noted by Rodney (2009) colonialism and its concomitant slave trade made Ghana and some African countries lose their battle to tame and harness nature- a battle which is at the basis of development. As Wiredu (2005) appropriately intimates, colonialism was not only a political imposition, but also a cultural one which gravely affected, or even perhaps infected African religious, [economic] and education systems.

This chapter is a case study that discusses the indigenous food preservation strategies among the Akan and how such a heritage could be learnt as a strategy to curb postharvest loses and to ensure food security. A case study was deemed appropriate for this chapter because the emphasis was on gathering of information to inform a specific practice or context (Ebersohn, Eloff & Ferreira, 2012). The purpose of the chapter is to share the experiences of Akan indigenous food preservation strategies to curb postharvest losses to ensure food security and to emphasise the need to utilise Akan heritage- indigenous knowledge systems- for the socio-economic advancement of the country.

THE STUDY SITE AND HISTORICAL REVIEW OF INDIGENOUS FOOD PRESERVATION

The Akan mostly occupy the central part of Ghana which is the most fertile region of the country because of much rainfall and forest. The study focuses on the Takyiman Municipality, a typical Akan area with ethnic diversity. The area lies in the forest- savannah zone at the geographical heart of Ghana's agricultural land. Takyiman is an ancient settlement which served as the main route and market for the trans-Saharan trade. Adu- Boahen (1975) attests that the first Akan state to emerge in the present day Ghana around the 13th century was the Bono, one of the 11 Twi speaking people. The area's long history of farming and trading attracted immigrants from the entire West African region.

The municipality has a population of 147,788 according to the 2010 census (Ghana Statistical Service, 2010). The people are predominantly peasant farmers with food farming employing over 75% of the residents in the Municipality. The area lacks formal agro-based industries; and although food is produced in abundance during harvest they get rotten within days or weeks. It is assumed that with proper training to equip more rural farmers with the skills in indigenous food technology the security of food can be assured in the area and throughout the country. Thus the utilisation of indigenous food preservation techniques can reduce shortage of food during the lean season. The production of food and the utilisation of indigenous methods to preserve it albeit on smaller scale has been an Akan cultural practice since time immemorial. Being predominantly rural and traditional area the people use indigenous knowledge side by side with modernity for survival. Among the Akan indigenous knowledge is an important resource hence they process their food for preservation through cultural practices.

Theoretical Framework

The study is underpinned by the Malthusian and the Techno-ecological theories because they are important determinants of food security in the technologically less developed countries in Africa. Malthus, an English clergy man and economist, propounded a theory on population dynamics and its relationship to the available resources (Gilbert, 1789). He posited that when population is unchecked it could lead to ecological disaster and famine. The central argument of the Malthusian theory is that human development can be limited by population pressure on the availability of food because population increases faster than the amount of food produced. Population, according Malthus as cited by Gilbert (1798), increases in a geometrical ratio while food production increases in arithmetic ratio. It is contended here that this paradox is a recipe for ecological disaster because of its potential negative effect on human ecology and possible food shortage. Malthus as cited by Gilbert (1798) argued that in order to avoid disaster population growth should be controlled. He however received some condemnation for his radical prediction that the factors that can control population growth are war, famine and the exercise of sexual restraint.

The neo-Malthusian theory which grew out of the Malthus theory shares his concerns but disagrees with the assertion that war and famine are the factors that can control population growth. Leading neo-Malthusian theorists which include Becker, Murphy & Tamura (1990) and Wrigley (1987) posit that instead of war and famine population growth can be controlled through family planning. They advocate family planning as a measure to control population because of their concern about the number of people the earth can hold and feed (Scanlan, 2001). The rapid population increase is a crucial and pertinent issue because it is associated with food production, consumption, the earth's carrying capacity and the threats to the global ecosystem.

Much of the efforts to reduce the impact of human activities on the environment focus on how technology can be used to bring about harmony in the ecosystem hence the relevance of techno-ecological theory. Techno-ecology is described here as the art and skill of employing technology to bring harmony to the ecosystem. Urban and Bakshi (2013), intimate that the ecosystem services (e.g. carbon and nitrogen cycles, freshwater and biodiversity) that support any activity are adequate and not deteriorated. For example, trees or vegetation can be grown and used as source of food. Thus the key to tackling potential threats to the earth's carrying capacity is human ingenuity for food security.

Human ingenuity here relates to how people can adapt to the changing environment by adopting the most appropriate technology to manage and control it for sustainability and improvement of life by producing and preserving enough food to feed every mouth. Indeed, it is through adaptation and human ingenuity that food security can be achieved in the less industrialised regions of the world. Scanlan (2001) affirms that the neo-Malthusian and the techno-ecological theories form the foundation for the critical examination of food security in the contemporary world. This chapter therefore contends that the ability of humans to adapt to the appropriate technology to the demands of population is not only crucial in bringing harmony to the global ecosystem but can also achieve food security.

In the developing world the adoption of indigenous technologies and methods of producing, processing and preserving food should be encouraged, learnt and intensified if humans are to change the seemingly paradox of 'poverty in the midst of plenty'. That is, harvesting plenty of food but experiencing postharvest losses because of lack of preservation skills, knowledge and facilities. The neo-Malthusian and the techno-ecological theories seem to complement each other because they are both concerned with population growth within limited resources such as food and space. Truly the central challenge to human development- *hunger*- cannot be addressed sufficiently without first addressing food security.

Research Questions

This chapter presents a case study of Akan indigenous food preservation techniques for the realisation of food security. For a period of two years the author conducted studies to explore the indigenous ways of food preservation among the Akan within the municipality of Takyiman in Ghana. The case study presented in this chapter provides answers to the following questions:

- What is the history behind food farming activities in the Takyiman area?
- What kind of foodstuffs are produced in the area and the motive for processing them for preservation?
- How are foodstuffs preserved or stored for the lean season?
- What are some of the major challenges to indigenous food preservation?
- What lessons can the Akan heritage offer other developing communities?

RESEARCH METHODOLOGY

This study was undertaken to explore the indigenous ways of food preservation and management of postharvest losses among the Akan of Ghana to ensure food security. By its nature the investigation was an ethnographic study which took the form of observation of the processes of food preservation at the selected sites and informal interviews involving 10 people who were engaged in indigenous food processing. Observation of the processes tied to informal interviews made the author get the first hand information about the indigenous ways of preserving food. The choice of this qualitative research approach was informed by the fact that the study involved direct interaction between the researcher and the participants. Such personal interaction can assist in acquiring the data needed to meet the objectives of the study. Again as a study that focused on the cultural practices of a particular ethnic group the ethnographic investigation was deemed suitable because it uncovers and explicates the ways a group of people live, understand their settings and take action to improve it (Quan-Baffour, 2013).

In order to understand the indigenous ways of processing and preserving food the researcher made a number of visits to the food processing sites within the municipality of Takyiman. He engaged in observation at the various sites where foods were processed for preservation and mingled the observation with informal interviews (something the author referred to as purposeful conversations). The conversations focused on the research questions outlined above.

DISCUSSION

The discussion presented in this chapter is based on the observation of the processes and the interviews which took place between the author and the entrepreneurs on the indigenous ways of preserving food, the challenges they face and the lessons that can be learnt by other African communities facing food insecurity as a result of postharvest losses. The chapter stresses that in the era of Africa's rebirth amidst inadequate food processing plants; an important heritage such as indigenous food preservation must be popularised in Ghana. The indigenous food processing and preservation skills which have been part of the Akan since time immemorial must be learnt from the aged before they die away. The Akan must '*gaze back*' to their cultural heritage to learn from the past; formalise and popularise the indigenous ways of preserving food to ensure its security. The transfer of the indigenous knowledge and skills to the young generation may reduce postharvest losses and ensure food security in the rural areas and the country at large. The processing and preservation of food may end the irony of poverty in the midst of plenty where lots of foodstuffs are produced during harvest but they get rotten and discarded within days or weeks.

Motive for Indigenous Food Processing

Being a predominantly rural and agricultural region the Akan area in general produces, in abundance, various local foodstuffs such as yam, cocoyam, cassava, maize, beans, oranges, banana, plantain, watermelon, cabbage and palm nuts. The motive for indigenous food processing which used to be social (cultural) has in recent times become economic as well. The Akan are culturally conservative but gradually open to innovation hence they accept the best practices from the West and integrate them into their traditional ways of life. Thus despite the Western incursion and deliberate attempt by the missionaries and the colonial administrators to annihilate African cultural practices the Akan fought to keep their cultural heritage-i.e. their way of life including the indigenous food processing and preservation methods. As part of Akan cultural practices food is preserved in order to make it easily available to the family at all times, particularly during the lean season.

In the recent years the increase in population, urbanisation, demand for food and lack of employment opportunities have motivated many people in the food producing areas to learn and use the indigenous methods to process and conserve food for economic purposes. The indigenous food processing and conservation has therefore become more of a business or entrepreneurship in recent times. Much of the *gari* processed in the Akan areas, particularly within the Takyiman Municipality is not only sold on the Ghanaian market but also exported to countries such as the Ivory Coast, Togo, Bourkina Faso, Benin, Guinea, Liberia, Sierra Leone, Niger, Mali, Chad and as far as South Africa, Europe and America. This indigenous entrepreneurship has created direct and indirect employment for thousands of people (retailers, wholesalers, drivers, farmers etc) and this has improved their socio-economic conditions.

The Process Involved in Processing Food for Preservation

Although a variety of foodstuffs are produced in the area this chapter covers the major ones that form the staple food of most Ghanaians which are *garri, kokonte, dokono and ngo*. The processing of raw cassava into *garri* (fried cassava dough) and cassava flour (*kokonte*), dry maize into kenkey (*dokono*), extraction of two types of oil from palm nuts – palm oil (*ngo*) and palm kernel *(adwe ngo*) are the main focus of this chapter. The processing of the above foodstuffs is done to ensure food security especially during the lean season and also to preserve the Akan unique heritage.

Garri and Kokonte Processing

The first section on the food processing, as presented in this chapter, gives details of how cassava is processed into two different kinds of food for preservation. The participant observation and the interviews that ensued between the author and the entrepreneurs made it clearer that the indigenous food preservation goes through laborious processes. It also revealed that once cassava is processed into *garri* or *kokonte* the products can last for over two years.

The process starts with the removal of the outer layer of the cassava tuber. After removing the thick outer layer, the inner juicy, whitish and fleshy part of the tuber is grated (in large quantities) into dough. The dough is put in large baskets covered with leaves or plastics and heavy stones or wooden planks are put on top of the baskets for 2-3 days. The weight of the stones or the wooden planks forces the starch in the dough to drain out naturally. Cassava contains too much starch and the draining process is an attempt to reduce the starch content of the food to make it better diet. After the starch is drained out the dough is fried in large quantities in big pans using hand ladles to stir it until the dough is completely fried into hard sand-like substance. The finished product- *garri*– is then packaged into sacks for market. There are wholesalers who buy *garri* from the producers in bulk and do their own distribution throughout the country and beyond.

Another food item made from cassava is *kokonte* (cassava flour). In making the *kokonte* a large amount of cassava tubers are peeled and cut into smaller pieces. The pieces of cassava are spread in open sun for one to two weeks for it to dry. The drying of the *kokonte* in the sun does not affect its nutritional value. It is done to quicken and enhance the process of preservation. Once dried, the large quantities of the cassava pieces are crashed, pounded, ground or milled into flour. The flour is then put in plastic or paper covers either for storage or for sale.

Processing Corn Into Kenkey

The second food of which the processing for preservation is discussed in this chapter is corn (maize). Farmers harvest dry corn and store it in barns for 3-5 months. The barns for storing the maize are built with sticks and palm branches and roofed with grass. The barns are either built in the farm or behind the farmers' houses. After harvesting the corn, the farmers do not remove the dry brown covers; they are left on the corn cobs and stored till the day the corn is removed from the barn. During the lean season the corn is taken out of the barn and shed. The corn is then shredded from the cobs and put in sacks for sale.

One common food made from the corn is moulded corn dough popularly referred to as *kenkey* (*dokono*) among Ghanaians. *Dokono*, which is made by women, goes through some lengthy processes. First the corn is soaked in water for 3-4 days for it to absorb water and become a little bit soft and easy to grind. After taking it out from the water the moist corn is either pounded or milled into corn flour. In preparing *kenkey or dokono* a measure of corn flour is soaked in water for 2-3 days to ferment. Part of the fermented dough is cooked and stirred with a hand ladle to make it a bit sticky.

The semi-cooked dough is added to an equal amount of raw dough by stirring it with a wooden ladle till the two types of dough are completely mixed. Once completely mixed a small cup-shaped calabash is used to measure and divide the soft dough into round shapes. Each of the ball–shaped dough is neatly wrapped in banana, plantain or corn leaves. About 20-30 (depending on the size of the cooking pot) of the wrapped dough are packed in a pot with water and cooked for 2-3 hours. Once properly cooked, *kenkey* or *dokono* can remain unspoiled for 3-6 weeks. *Dokono* or *Kenkey* can be eaten with gravy or soup made of meat or fried fish, onion, pepper and tomatoes or *shito* (smashed smoked fish). *Kenkey* can also be mashed into soft porridge (*Koko)* with sugar and milk and eaten as a snack. *Kenkey* which used to be made for domestic use has now become a commercial venture; made by women and sold in every food market, town and village in most parts of Ghana.

Processing Meat and Preservation of Fish

The third case presented in this chapter discusses indigenous ways of preserving meat and fish. In the rural areas some men engage in hunting and trapping expeditions where they kill grass cutters, deer, antelopes and zebras. The game is sold to market women who are meat sellers. In the same way local fishermen on the Black Volta around Babato villages sell fresh fish to fish mongers. Once the meat and fish have been bought the market women prepare them for preservation. They burn the fur and skin of the animals and cut them into sizable pieces. In the case of the fish the fins are trimmed before they are smoked. The women sprinkle salted water on the meat and the fish which they then smoke for hours in special ovens. The women turn the meat or fish at certain intervals to ensure that they are not burnt. The smoked meat and fish are packaged in big baskets and sold to wholesalers who distribute them to their clients. The smoked meat and fish which can last for over 4 months are exported to neighbouring countries and even overseas. The Ghanaians in diaspora buy smoked meat and fish in most cities in the countries where they live.

An important delicacy made from smoked fish is *shito* (gravy from smashed smoked fish). In preparing *shito* all the bones are removed from the smoked fish. Different types of smoked fish are smashed into powder-like substance. Tomatoes and onion are peeled and ground (in modern times grinder is used). The quantity of the ground tomato, pepper, salt and onion is mixed with the smashed smoked fish and boiled in oil for about two hours and bottled. The bottled *shito* is found on the markets in Ghana and abroad. It can last for a year and because it is a very chilli sauce only small amount can be served at any meal.

Processes of Making Edible Oils From Palm Nut Fruit

The processing of palm nut into edible oils is another important indigenous food preservation method presented in this chapter. The two types of edible oils made from the palm nut fruit and kernel are palm oil (*ngo*) and palm kernel oil (*adwe ngo*). The elaborate processes of making oils from the palm nut require specific skills and human ingenuity.

In the case of the palm oil quantities of palm nuts are boiled and pounded in a mortar to remove the succulent and juicy part from the nut. After pounding the palm nuts the cover is removed (manually) and sieved in water which turns into orange-like liquid. Large quantity of the liquid is put in large pots and cooked for about 2-3 hours. As the liquid boils the oil appears on its surface and from time to time the red-like oil is carefully scooped out of the cooked palm liquid basin and put in another container to cool down. The process continues until no more oil appears on the surface of the boiling palm. The oil is then re-cooked packaged into bottles or drums for sale.

In making the palm kernel oil stones are used in breaking or crashing the shell of the nut to take out the kernel. A large amount of the kernel is fried till almost burnt for the oil to ooze out. When the frying comes to an end the kernel which has now become brownish is taken out of the oil. The pot is taken out of the fire for the oil to cool down before it is put in drums and bottles for the market. Again the cooled fried kernels are taken to be milled into a paste. The paste is boiled on slow heat to exude more oil. The oil is then scooped out into another container to cool before bottling it for sale.

CHALLENGES FACING INDIGENOUS FOOD PROCESSING

The most important factor in production is capital. Manufacturing, whether indigenous or exotic, needs capital to start and expand. Unfortunately, the indigenous entrepreneurs engaged in food processing for preservation in Ghana do not have access to financial support or subsidy from either the government or financial institutions. Although motivated by self-employment most of the entrepreneurs engaged in indigenous food processing operate as individuals or as family business. A very large number of them are either illiterate or semi-literate and perhaps this might be one of the reasons why they are unable to engage in cooperative ventures. By operating alone as small business enterprises they do not only lack bargaining power to engage with the relevant government departments for support but also may not be able to access loans from financial institutions. Financial institutions require collateral security for loans and since most of them do not have it they are denied access to loans. Without access to finance production will always remain limited for many of the entrepreneurs.

Under the previous government the ministry of rural development organised *garri* manufacturers into cooperatives and granted them some financial support in the form of medium term loans but this has been discontinued when a new administration took power six years ago. Although the indigenous entrepreneurs engaged in food processing contribute immensely to the socio-economic development of the country lack of support will continue to limit production as they cannot modernise and increase production. As it is now they employ manual production methods with obsolete traditional gadgets which can highly increase production.

LESSONS LEARNED FROM THE AKAN HERITAGE OF FOOD PRESERVATION

In this era of African renaissance Africans should make a backward gaze to reclaim the useful past i.e. relevant skills and knowledge that can move the continent forward in industrialization. African communities are endowed with indigenous knowledge systems but because of Western influence people seem to down play their value. Africans therefore need to learn from each other. As Quan-Baffour (2008) warns, if Africans do not reclaim their useful past they will not only perpetuate poverty, disease and instability but may also be reduced to mental slaves and intellectual zombies in their own land.

The Akan heritage- indigenous industrialization- in the form food processing to ensure food security offers lessons for Africa at large. Governments and other developing communities in Africa need to learn from the Akan indigenous food processing project. Individuals and communities should do some introspection, use their ingenuity and rely on their indigenous knowledge systems to achieve food security.

In a continent where youth unemployment is a serious socio-economic and political concern in the midst of lots of agricultural products much can be learnt from the example of the Akan indigenous food preservation techniques. Kawagley and Barnhardt (1998) aptly assert that the depth of indigenous knowledge rooted in the long inhabitation of a particular place offers lessons that can benefit everyone, from educator to scientist, as we search for more sustainable way to live on this planet. Thus when learnt and adopted indigenous food processing industries may not only ensure food security but can also stem the tide of unemployment, reduce agitation and unguided radicalism among the unemployed youth and increase socio-economic development of communities and countries.

The learning of this lesson backed by government support can transform communities which lack western industrialization and set them on the pedestal of an African industrial revolution. Wiredu (2004) articulates this eloquently in the following words "this for us in Africa, must of necessity involve a complex of capabilities, drawing our heritage of indigenous knowledge on the one hand, and on the other, cognitive resources available in other cultures, East or West, provided they do us good''.

FUTURE RESEARCH DIRECTIONS

The future of Ghana's socio-economic development lies in the studying, researching and application of its indigenous knowledge systems. The country therefore must make a backward gaze to adopt and popularise its indigenous knowledge and skills and utilize them as the spring board for agro-based industrialization. To achieve this feat future research on food security in Ghana and Africa at large should focus on strategies to make indigenous food processing and preservation more sustainable. Future research projects on indigenous food processing and preservation that focus on raising funds for startups, managing, marketing, formation of co-operatives and the provision of basic education for indigenous entrepreneurs may go a long way to establish indigenous food processing firmly on the country's path to agro-based industrialization.

CONCLUSION

The four cases discussed in this chapter have emphasised some of the indigenous ways of processing food for preservation among the Akan of Ghana. The findings of the exploratory investigation described in this chapter underscore the importance of indigenous knowledge in Ghana and Africa at large. The colonial administrators discouraged the growth of indigenous industries in Ghana as a strategy to protect Western markets. In this era of Africa's rebirth and with too many mouths to feed, Africans need to gaze back to the past to reclaim their indigenous knowledge system in order to ensure food security. Through the application of indigenous food processing skills Ghanaians and other African communities can reduce postharvest losses and thus ensure food security.

This chapter offers lessons for African communities grappling with postharvest losses. It encourages communities without (Western) agro-based industries to make use of indigenous knowledge systems to ensure food security and also create jobs for the unemployed for the achievement of social cohesion and economic development. The socio-economic benefits of indigenous food preservation cannot be over emphasised. In spite of some challenges to the development of indigenous food preservation it is gaining currency among the rural folks in Ghana where there is paucity of modern agro-based industries. Indigenous food processing must therefore be encouraged to reduce food insecurity in developing countries.

RECOMMENDATIONS

In view of the importance of indigenous knowledge system in ensuring food security as discussed in this chapter the following recommendations were made:

- Indigenous food processing initiatives in Ghana must be supported by government and financial institutions to enable the entrepreneurs to expand their production for economic growth.
- The Ministries of education, trade and industries and rural development should offer the indigenous food processors with basic training in accounting and management to ensure sustainability of their businesses.
- The indigenous food processors should be encouraged to form cooperatives as a strategy to formalise their operations and also to approach the government as a common front for support.

REFERENCES

Adu-Boahen, A. (1975). *Ghana: Evolution and change in the nineteenth and twentieth centuries*. London: Longman Group Ltd.

Anderson, R. W. (1999). *Economic development among the Aboriginal Peoples of Canada: Hope for the future*. Toronto: Captus University Press.

Appiah, F., Guisse, R., & Dartey, P. (2011). Postharvest losses of rice from harvesting to milling in Ghana. *Journal of Stored Products and Postharvest Research*, *2*(4), 64–71.

Becker, G. S., Murphy, K., & Tamura, R. (1990). Human capital, fertility and economic growth. *Journal of Political Economy*, *98*(5), 12–37. doi:10.1086/261723

Bill and Melinda Gates Foundation Report. (2010). *Profiles and progress*. Retrieved from www.gatesfoundation.org/agriculture

Ebersohn, L., Eloff, I., & Ferreira, R. (2012). First steps in action research. In M. Kobus (Ed.), *First steps in research* (pp. 124–143). Pretoria: Van Schaik Publishers Ltd.

Ghana Population Census. (2010). *Statistical Services, Ghana*. Retrieved from www.ghanweb

Gilbert, G. (1789). Introduction. In *Malthus Thomas. An essay on the principle of population*. Oxford: Oxford World's Classics Reprint.

Kawagley, A. O., & Barnhardt, R. (1998). Education indigenous to place: Western science meets native reality. *Alaska Native Knowledge Network*, *15*(43), 117.

Quan-Baffour, K. P. (2008). The wisdom of our forefathers: Sankofaism and its educational lessons for today. *Journal of Educational Studies*, *7*(2), 22–31.

Quan-Baffour, K. P. (2011). Singing to farmers: Non-formal adult education through folksongs for food production in rural Ghana. *MUZIKI. Journal of Music Research in Africa*, *8*(1), 34–46.

Quan-Baffour, K. P. (2012). *Sankofa*: 'gazing back' to indigenous knowledge and skills for the socio-economic development of Ghana. *Studies in Tribes & Tribals*, *10*(1), 1–5.

Quan-Baffour, K. P. (2013). Demystifying the technological myth: Experiences of 'Techno-Phobians' in teaching adults through myUnisa. *Journal of Communication*, *4*(1), 1–8.

Reutlinger, S. (1986). *Poverty and hunger: issues and options for food security in developing countries. A World Bank Policy Study*. Washington, D.C: The World Bank.

Rodney, W. (2009). *How Europe underdeveloped Africa*. Abuja: Panaf Publishing, Inc.

Scanlan, S. J. (2001). Food availability and access in lesser- industrialised societies: A test and interpretation of Neo-Malthusian and techno- ecological theories. *Sociological Forum*, *16*(2), 231–262. doi:10.1023/A:1011000717740

Urban, R. A., & Bakshi, B. R. (2013). Techno-ecological synergy as a path toward sustainability of a North American residential system. *Environmental Science & Technology*, 47, 1–9. PMID:23294016

Wiredu, K. (2004). Prolegomena to an African philosophy of education. *South African Journal of Education*, *18*(3), 17–26.

Wrigley, E. A. (1987). *People, cities and wealth*. Oxford: Basil Blackwell Ltd.

KEY TERMS AND DEFINITIONS

***Adwe Ngo*:** Edible oil extracted from palm kernel.
Culture: The way of life of a group of people manifested in their customs, beliefs and practices.
***Dokono*:** Popular Ghanaian food from fermented corn dough.
Employment: Engagement in worthwhile socio-economic activity to earn a living.
Entrepreneur: One who initiates and engages in enterprises for profit.
Food Security: A state of food availability throughout the year.
***Garri*:** The kind of food made from fried cassava dough in Ghana.
Heritage: Something inherited from fore bearers.
***Kokonte*:** Ghanaian food made from cassava flour.
Livelihood: An activity people engage in for a living.
***Ngo*:** Kind of edible oil made from palm nut.
Preservation: To prepare something to be kept for a period of time without it spoiling or going bad.

This research was previously published in the Handbook of Research on Social, Cultural, and Educational Considerations of Indigenous Knowledge in Developing Countries edited by Patrick Ngulube, pages 334-345, copyright year 2017 by Information Science Reference (an imprint of IGI Global).

Chapter 17
Determinant of Food Security on Upland Agriculture Households in Paletwa Township, Chin State of Myanmar

San Lwin
Khon Kaen University, Thailand

Supaporn Poungchompu
Khon Kaen University, Thailand

ABSTRACT

This study aims to determine food security on upland agriculture households in rural area. Food security is concerned with the first two main goals of Sustainable Development Goals, No Poverty and Zero Hunger. Myanmar is Food Insecure State that showing 14.2% that is 7.7 million of 51 million population. Chin state is least developing and Paletwa is poorest out of 324 townships. Research is applied by both qualitative and quantitative approaches. 3 Villages and 1 Quarter are chosen for qualitative method and 141upland agriculture households are selected for field survey using random sampling method. The data are analyzed by logistic regression in SPSS 17 to determine food security. Age, education, schooling years of household head, size, second occupation and no. of working people in households are socio-economic determinant and own food production and fruit access are food security determinant.

INTRODUCTION

"Food Security exists when all people, at all times, have physical and economic access to sufficient, safe and nutritious food to meet their dietary needs and food preferences for an active and healthy life" (Food and Agriculture Organization[FAO], 1996). "It is now estimated that in 2009, there are 1.2 billion food insecure individual- the one sixth of world population". (Canadian International Development Agency

DOI: 10.4018/978-1-5225-8063-8.ch017

[CIDA], 2015). However, "during the past year, Food Security has improved almost in every region of the world". (Global Food Security Index [GFSI], 2015). The researcher from Asia and Pacific Policy Studies agrees with Global Food Security Index. "Both Southeast Asia and developing East Asia have made enormous progress in lowering food insecurity in the past two decades, with prevalence dropping by 18.7 percentage points and 9.3 points, respectively". (Timmer, 2014). In 2015, Myanmar's Food Security ranking is 78 with the overall score of 44.00. Compared with 2014, Myanmar increase food security +7.7 overall score that is the most increasing country after Egypt (Global Food Security Index [GFSI], 2015). Although Myanmar is middle food insecure country, Chinese scholars mentions Myanmar is less production country. "There are also countries that have favourable climates but produce less than average amount food per person: Bosnia and Herzegovina, El Salvador, Gabon, Georgia, Indonesia, Jamaica, and Myanmar" (Yu & You, 2013).

The Heads of State and Government and High Representatives, meeting at United Nations Headquarters in New York from 25 to 27 September 2015 as the Organization celebrates its seventieth anniversary, have decided Sustainable Development Goals with 169 associated targets which are integrated and indivisible "We are determined to end poverty and hunger in all their forms and dimensions, and to ensure that all human being can fulfil their potential in dignity and equality and in a healthy environment" (United Nations [UN], 2016). Food Security is sorely related with Sustainable Development Goals [SDG]. In their second SDG, Zero hunger extends to end hunger, achieve food security promote sustainable agriculture. This study will fulfil a tiny piece of this goal in study country, Myanmar.

In Myanmar, there are 5.7 million ha cultivated land that is 12.89% of the total land area of 676,578 sq.km (Myanmar Agriculture Census, 2003), but the World Bank reveal that cultivated land is 19.3% of the total land area (World Bank Group, 2013). 66% of the total population 51 million still live in rural area (FAO, 2015). Agriculture contribute 60% to Gross Domestic Product [GDP] and employ 65% of the population. (Encyclopedia of the Nations, 2016) Therefore almost all rural population engage in agriculture. Out of 14 States and Regions, Chin is least developing region. It has 0.5 million population and mostly settlement in rural area (Myanmar Institute of Integrated Development [MIID], 2014). The poverty rate is 73% and food insecurity rate is up to 52%. "Food insecurity is therefore a cyclical and chronic problem, which affects all townships of the State. The result of the 2013 World Food Programme [WFP] food security monitoring in Chin State showed that the food insecurity levels in the State ranged from moderate to severe" (World Food Program, 2013) "Chin has the smallest average and median rural land holding, 2.1 and 2.0 acre, respectively" (United Nations Development Program, 2015).

RESEARCH METHODOLOGY

Study Area

Myanmar is one of Association of Southeast Asian Nations [ASEAN] Country. It totally bases on agriculture. It has 51 million population on 676,578 sq.km which is the largest land area after Indonesia in Southeast Asia. She has tropical monsoon climate. The average temperature is from 22°C to 27°C year-round. The seasons can be categorized as: The cold and dry season, the hot-dry season and the wet season. Agriculture shares the economy constituting 44% of the GDP and 35% of export earnings. Total cultivated land is 18 million hectares. Rice, beans, pulses and sugar cane are the principal crops. Rice production takes into accounts for 25% of the GDP. The per capita GDP is 1416 United States Dollar

[USD] and Myanmar is lower middle income country. (World Economic Outlook, 2015) But the country has substantial human resources and economic potential as she has important geo-politic. Myanmar changes recently to open and democratic evolution.

Chin State is most western part of Myanmar. Total population is 478,690 that is 1% of Myanmar population (Department of Population, 2015). The area is 36,019 sq. km that is 5.23% of Myanmar area. Climate is mild hot and wet. April and May are the hottest months of the year. Monthly average temperature ranges from 15.5°C to 21°C. Average temperature in the cold season is 4.4°C. Chin state is divided by 3 districts into 9 townships which are made up of 505 village tract and quarters. Village has minimum 7 households to maximum 800 households. Most population in northern part of Chin State is Christian and in southern part is Buddhism. In Chin State, there are 6 major ethic groups namely, Asho, Zo, Khumi, Laimi, Mizo and Zoomi. Rice and Corn are main staple food and there are food security states in annually 8 months per year. Diversification is very wide from one place to another ([MIID], 2014)

Paletwa is most western part of Chin State, 18 km leads to Bangladesh border. Population is 96,889. (Myanmar Parliament, 2014). In Paletwa, there are 27,555 ha of cultivated lands. The mainly practice of agriculture is upland or hillside cultivation showing 21,957 ha (80% of total cultivated land). Most of upland crops are Upland Paddy (6,613 ha), Sesame (829 ha), Galangal or Gamoan (550 ha) (*Kaempferia galangal*), Mustard Seed (444 ha.) and Turmeric or Na-nwing (306 ha) (*Curcuma Longa*). The other upland crops are Corn, Food Legume, Chili, Thetkae for roofing the house, Tobacco, Ginger and other vegetable crops. Orchard or perennial crops are Mango, Lime, Orange and Banana (Settlement and Land Record Department, Paletwa, Chin State, 2016).

"In Chin State, most of the farmers carry out slash-and-burn or shifting cultivation or (*taungya*) farming method. This method destroys fauna, flora, and soil resulting in denudation of cultivable slope and its environment. This makes many flowing streams dry up in summer" (Than, 2010). Some credit slash-and-burn farming methods with "destroying the forests of the country, causing soil erosion and depletion of fertility," (Encyclopedia of the Nations, 2016). Therefore, this study will focus on the main practices of farming and will determine the factor analysis which affecting food security on upland agriculture households.

Study Design

Target population is 220 households in Paletwa and sample size (n) is 141 households based on Taro Yamane for quantitative research approach. Among 100 quarters and villages of Paletwa, 1 quarter and 3 villages are selected. Random sampling is used for data collection.

In order to examine each household has food security or not, their food expenditure and the months the household worry about enough food is compared with food poverty line at the household level. "Food poverty is worse diet, worse access, worse health, higher percentage of income on food and less choice from a restricted range of foods. Above all food poverty is about less or almost no consumption of fruit & vegetables" (Food Access Network, 2016). Among the national food poverty line of previous research, Karosaki *et al.* (2004) informs the value of 200kg rice per year, (Kyaw, 2009) regard 215 Myanmar Kyat (MMK) /person/day and the finding of UNDP and Ministry of National Planning and Economic Development of Myanmar (MNPED) gives 324 MMK/person/day that is 0.34 USD. In 2016, there is 10.3 USD/person monthly and 124 USD/person yearly (Kyaw, 2009).

Logistic regression was developed by statistician David Cox in 1958. Logistic regression measures the relationship between the categorical dependent variable and one or more independent variables by estimating probabilities using a logistic function which is the cumulative logistic distribution (Walker, 1967). The logistic regression is here employed to observe the determinant of socio economic and demographic characteristics of the households on their food security in studied area. Food security status was assumed as a big value of one (1) and zero (0), whereas one (1) means food secure and (0) means food insecure. This model explains the probability or the possible determinant variable of household on food security condition. The applied variables are used according to previous researchers from different places: they are (Sekhampu, 2013), (Mwema, Mutai, Lagat, Kibet & Maina, 2012), (Thanh, & Yapwattanaphun, 2015) and (Oo, Iwai & Saenjan, 2013). The logistic regression equation is shown as follow:

$$Y = \beta 0 + \beta 1X1 + \beta 2X2 + \beta 3X3 + \beta 4X4 + \beta 5X5 + \beta 6X6 + \beta 7X7 + \ell t$$

Whereas Y is food secure status (if food secure, it is 1 and if food insecure, it is 0), χt is the vector of variables and β_0 is the dependent variables and ℓ_t is the error term. The Scio-demographic information were used as following: β_1 the age,(number of years) β_2 education of household head (1 = no education, 2 = primary school, 3 = primary school, 4 = middle school, 5= high school, 6= higher education) β_3 household size (number of family members in household) and β_4 number of working people per household (number of people) and β_5 non-farm income (amount of earning in USD), β_6 their own food production (percentage of own food production per their requirement, 1 = up to 20%, 2 = 21% to 40%, 3 = 41% to 60%, 4 = 61% to 80% and 5= more than 80%) and β_7 fresh fruit access (1 = always, 2 = often, 3 = sometime, 4 = rarely and 5 = never) .

Former researcher applied the variables age of household head, gender of household head and education attained by household head in his study of "Determination of the Factors affecting Food Security Status of Households in Bophelong, South Africa" (Sekhampu, 2013). With above researcher, indicated that education of household head and household size are resulting significantly in the paper of "Contribution of Selected Indigenous Fruit on Households Income and Food Security in Mwingi, Kenya" (Mwema *et al.*, 2012). Age of household head, formal education of farmer, the main income and second income or off-farm income, the property size are employed in the study of (Thanh, & Yapwattanaphun, 2015). To improve food security of their family in Northeast of Thailand, the variables: age, education, average household size, Average labour and average cultivated area are investigated by researchers. (Oo et al., 2013) In sum up, the age, schooling years of household head or education of household head, household size, non-farm income, no. of working peoples, own food production and fresh fruit access are chosen for the variables in this study.

Among above 7 variables are expected to correlate with food security. If the household head have older age, he/she have more experience on farming and other income generation activities than the younger one that will result the positive determinant on food security. If the households head attained more education, he/she might know agriculture and technical knowledge as well as critical thinking and problem solving, therefore that family will be under food secure condition. Unlike previous two variables, the bigger household size might face food insecure as they have probably more children and elder. These variables are expected to determine negatively. The household size and no. of working people are different option here. Although more household's size is negative correlation, non-farm income may be positive. Almost all households depend on faming that is enough for food secure, the accessibility and amount of non-farm income may affect positively. The more they have working people, the more they secure food. It is predictable own food production and fresh fruit access are directly interrelated to food security in this study.

RESULT

The Demographic Information of Respondents

The data are analysed based on their response of the household. As shown in Table 1, the age of household head is 22 the youngest and 80 the oldest. The average age of respondents are 45 years old. Female household heads are 22 and the male 119 out of all household heads (n=141). The lowest education of household heads has not attained any education (no education and the highest is higher education. Generally, most of them are primary school education level. 72 households have got average primary school that is 51.1% of the whole population. Smallest household size is 2 and the largest one is 12. The mean of household size is 6. Every household has at least 1 working people and 6 is the most no. of working peoples. In every household, there are average 3 working peoples. Some households have no opportunity to earn non-farm income while some earn 2958 USD/year the most. The average non-farm income per household is 1263 USD/year. For their own food production, the lower group can produce up to 20% of their food requirement and the highest group 61% to 80%. The average production of household produce 21 to 40% of their requirement. Some of them never access fresh fruit and some always. In general, Fresh fruit are accessible sometime. Food expenditure is 210 USD/year the lowest and the highest 2880 USD/year. Average food expenditure per household is recorded 1164 USD/year.

Food Security Status of Households

Food Security status is analysed by using comparison of food expenditure and national food poverty line and check the probability of food security with the length of time that the households' concern on enough food. National Food Security Line of Myanmar is 124 USD/person/year. Analysing food expenditure, 46 households (32.6%) are below the line and other 95 households (67.4%) are more than national poverty line. The food insecure condition mostly can be seen in 5- member households. Out of 30 households that holding 5 family members, 11 households are below food poverty line.

Table 1. Minimum, Maximum, and Mean of demographic information of the households (2015-16) (n = 141)

Variables	Minimum	Maximum	Mean	Std. Deviation
Age of HHH	22	80	45.33	10.808
Gender of HHH	1	2	1.16	.364
Education of HHH	1	6	3.13	1.152
HH Size	2	12	5.91	2.077
Number of working people	1	6	2.77	1.193
No-farm Income	0	2950	1262	693.88
Own Food Production	1	4	1.65	765
Fresh Fruit Access	1	5	2.45	865
HH Food Expenditure	210	2880	1164	569.96

That 11 households are 24% of total food insecure 46 households and 37% of 5-memebr households. Examining their concern on enough food, 47 households (33.3%) worried about food more than four months and other 94 households (66.7%) worried less than four months. 4 households concern enough food severely the whole years as shown in Figure 1.

Therefore, copping with food expenditure below national food poverty line and their concern on enough food, 46 household are food insecure and 95 households are food secure that resulting 32.6% has not gain food security and 67.4% has gained as shown in Table 2.

Determinant of Food Security at Household Level

To determine what factors are influencing food security on upland agriculture households, 7 variables are analysed. Those are 4 Scio-demographic information and 3 food security indicators as shown in Table 3. The ages of household heads are one of the determinant of food security. They are correlated positively at the 5% significant level (p-value = 0.046). The household head become older, the food secure level of household will increase. Such as the household heads' age, the education they attained in their lives is related to food security status. The education of household heads shows the significant level 10% (p-value = 0.053) in positive way. The more the household head attained the education, the more the household food secure. Unlike the age and education of the household heads, the increasing family members in the household effect the food security condition reversely. With significant level 1% (p-value = 0.004) the household size and food security depend each other mutually. The capacity of

Figure 1. Food Security Status copping with the length of time (months) the households' concern on enough food

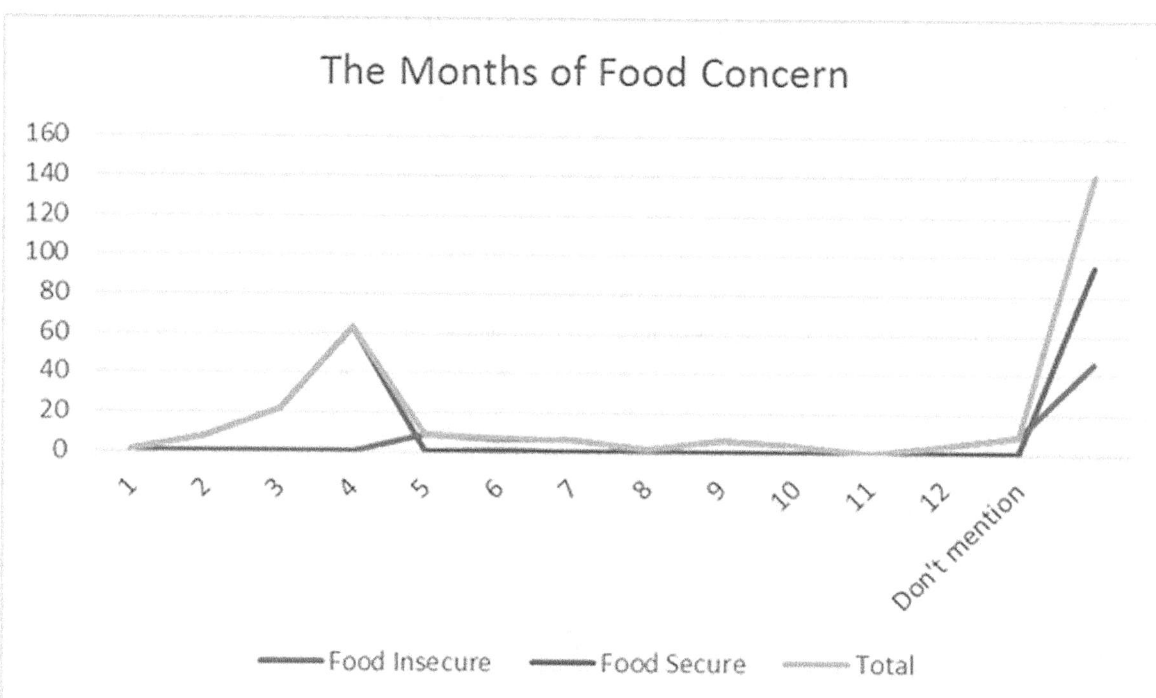

Table 2. Food security status of sampling households according their food expenditure on income under national food poverty line during last year (2015-16) (n = 141)

Comparison	Food Secure	Food Insecure	Households (n = 141)
Food Expenditure versus food poverty line	95	46	141
Percentage Food Secure and Insecure	67.4%	32.6%	100%

working force is also main factor to secure food. No. of working person in one household is interconnected to food security situation. This variable show the signification 5% level with p-value 0.049. Among 4 Scio-demographic variables, 3 variables: age, education and no. working people in household, can determine the household level food security directly. Otherwise, household size gives reversal linkage to food security determination.

When other 3 food security indicators are observed, non-farm income and their own production are significant at 1% and 5% respectively. The 2 variables are scanned as positive factors affecting food security. These 2 indicators are tightly contingent each other. Food can be attained by the mean of production such as farming, raising the livestock, fishing and hunting as well as earning some non-farm income like labouring, migration work, serving as employee and some seasonal jobs. Non-farm income of 1% significant level (p-value = 0.002) point out their food requirement is not enough themselves although they absolutely rely on agriculture. Regardless of their purchasing power, the household ensure its food necessity if own food production has been reached to some extent. The significant confirm 5% with p-value 0.045. Fresh fruit access is not weighty here in this study. In the study area, the stable food is rice and some forest products like bamboo shoot and vegetables. The nutritious and commonly eatable fruit like mango, papaya, banana and coconut, etc. are not available too much in township. The handy fruits they are able to pick up or, and purchase are local fruit. Therefore, the fresh fruit are not determinant of food security in this limited study.

Table 3. Determinants of food security on upland agriculture households in Paletwa Township, Chin State of Myanmar (2015-16) (N = 141)

Variables	B	S.E	Wald	Sig
Age of Household Head	.051	.026	3.998	.046**
Education of Household Head	.451	.233	3.747	.053*
Household Size	-.701	.242	8.410	.004***
No. of working people	1.154	.587	3.867	.049**
Non-Farm Income	.002	.001	9.769	.002***
Own Food Production	.782	.390	4.018	.045**
Fresh Fruit Access	-.384	.305	1.588	.208
Constant	1.195	1.742	2.876	.090

Source: Field Survey, Paletwa, Chin State, July 2016

Note: *, ** and *** means coefficients significant at 10%, 5% and 1% respectively.

DISCUSSION

Household size and non-farm income are most significant factors affecting food security. As there are more human resource in one households, earning income and employment opportunity is uncertain due to their lack of opportunities. The increase number of household members results in the decrease in food security. It is very difficult to maintain their food security status when the household size becomes bigger and bigger. This study contracted with other researcher (Sekhampu, 2013), (Babatunde, Omotesho & Sholotan, 2007) (Babatunde, Omotesho & Sholotan, 2007b), (Omonona, & Agoi, 2007), (Oluyole & Taiwo, 2015), (Bashir, Schilizzi & Pandit, 2012), (Zakari, Ying & Song, 2014), (Abu, & Soom, 2016) and (Arene & Anyaeji, 2010).

Result of analysis with logistic regression is seen that non-farm income is linked with food security positively at 1% level significance. Previous studies are some different that "off-farm activity is negatively related to food security" (Abu & Soom, 2016) and "food security is negatively affected by off-farm income" by Oluyole & Taiwo (2015) and Babatunde et al., (2007). The household cannot solely depend on farming, the access of other income source support the increasing status on food security. Although all household in this region are agro-based, their main food production and income from farm work is not enough for their families. Therefore, non-farm income like daily-labour, collecting forest product, serving the working in nearby Indian Harbour Project in Plaetwa is very important for them.

The age of household heads can determine food security positively with 5% significant level. While the household heads become older and older, he or she have gained a lot of experience to find food and money for his or her beloved food. Sekhampu, 2013 approves this finding and mention in the paper "the age of household head was positively associated with food security the coefficient for age (B = -.678) was positive and significant at 1%". This study agreed the former researcher, (Arene & Anyaeji, 2010), (Zakari & Song, 2014) and (Babatunde et al., 2007b) express their conclusion that the increase age of household head had a positive effect on food security status. However, other previous research including (Babatunde et al., 2007) and (Babatunde et al., 2007b) show also the age of household head and food security is negatively correlated in study of (Bashir, Schilizzi & Pandit, 2012) in Pakistan, and (Omonona & Agoi, 2007) in Nigeria, (Oluyole & Taiwo, 2015), (Kuwornu, Suleyman, & PK, 2013), (Abu & Soom, 2016) and (Bashir, Schilizzi & Pandit, 2012).

The increasing number of working people in household lift up of food security. The more they have the people who can work and possible to earn income, the more they have more secure for food. Positive relation of number of working people go with Bashir et al. (2012) that use as the total earning household members (TEHH).

Own food production drive the food security situation in positive way. The result of this study is consistent with (Kuwornu et al., 2013), (Babatunde et al., 2007) and (Ojogho, 2010). The food security is determined by education of household in so many different ways positively. Bashir, et al. (2012) proved the information of significance of education of household head while (Sekhampu, 2013) study is not significant.

CONCLUSION

This study cover only 3 villages and 1 quarter in Paletwa Townships out of 100 village's tracts, and it is very limited to study most of food security indicators, therefore it is impossible to represent all population. Further afford to employ research able to scope all population should be tried. Extension of this limited study may be the few step and guiding star for the ambitious researcher and scholar who have desire to end poverty with zero huger by the name of sustainable development goals.

The age and education of household heads, household size, no. of working people, non-farm income and own food production are observed as the determinant of food security with the analysis of logistic regression in this study.

When the age and education of household heads increase, the food security level increase. The older households have become more experience on farming and to sustain his or her household food security. It is clear that the household heads learn everything through time. Every household heads should be given the opportunity to gain experience and knowledge like as Farmer Field School (FFS) or agricultural extension services. Everybody in that area should be encouraged to study more and to extend their formal education as well as the households heads should be extended their informal and non-formal education to attain easier secure food in their whole life. The increasing number of household is driving their food insecurity condition. This means when their family become bigger and bigger they cannot control their food security status. They are lacking income generation activities or ideas. This factor is directly related with number of working peoples and non-farm income. As greater number of above two factors can save food security condition, government and some humanitarian group should promote the livelihood activities and income generation activities. Their food production is relatively low and it effects directly on food security and also it is cause of lower level of education and their lesser chance to upgrade their faming system. Government ought to give service of agriculture extension and some remarkable loan based on their need and ought to provide provision of inputs like good quality seed, fertilizers, pest and insect management as well as some agricultures implements and materials.

ACKNOWLEDGMENT

Here I would like to express my gratitude to International Program of Master of Rural Development Management (MRDM), Khon Kaen University. I deeply want to thank to the Scholarship organization Southeast Asia Regional Centre for Agriculture and Research. (SEARCA).

REFERENCES

Abu, G. A., & Soom, A. (2016). Analysis of Factors Affecting Food Security in Rural and Urban Farming Households of Benue State, Nigeria. *International Journal of Food and Agricultural Economics*, *4*(1), 55.

Arene, C. J., & Anyaeji, C. (2010). Determinants of food security among households in Nsukka Metropolis of Enugu State, Nigeria. *Pakistan Journal of Social Sciences*, *30*(1), 9–16.

Babatunde, R. O., Omotesho, O. A., & Sholotan, O. S. (2007). Factors influencing food security status of rural farming households in North Central Nigeria. *Agricultural Journal*, *2*(3), 351–357.

Babatunde, R. O., Omotesho, O. A., & Sholotan, O. S. (2007b). Socio-economic characteristics and food security status of farming households in Kwara State, North-Central Nigeria. *Pakistan Journal of Nutrition*, *6*(1), 49–58. doi:10.3923/pjn.2007.49.58

Bashir, M. K., Schilizzi, S., & Pandit, R. (2012). The determinants of rural household food security: The Case of Landless Households of the Punjab (Working Paper 1208). Pakistan, School of Agricultural and Resource Economics.

Canadian International Development Agency [CIDA]. (2015). Increasing Food Security.

Central Statistical Organization. (2003). *Myanmar Agriculture Census*. Yangon, Myanmar: Settlement and Land Records Department.

Department of Population. (2015). 2014 Myanmar Population and Housing Census - The Union Report (Vol. 2). Ministry of Immigration and Population of Myanmar, Nay Pyi Taw, Myanmar. Retrieved from http://www.dop.gov.mm/moip/index.php?route=product/product&path=54_52&produc

Encyclopedia of the Nations. 7. (2016). Retrieved from http://www.nationsencyclopedia.com

Food and Agriculture Organization [FAO]. (1996). World Food Summit.

Global Food Security Index [GFSI]. (2015).

Group, W. B. (2013). *Doing Business 2014: understanding regulations for small and medium-size enterprises* (Vol. 11). World Bank Publications.

Kuwornu, K. M. J., Suleyman, D. M., & Amegashie, D. P. K. (2013). Analysis of food security status of farming households in the forest belt of the Central Region of Ghana. *Russian Journal of Agricultural and Socio-Economic Sciences*, *13*(1), 26–42. doi:10.18551/rjoas.2013-01.04

Kyaw, D. (2009). Rural Households' Food Security and Coping Strategies to Food Insecurity in Myanmar.

Maxwell, S. (n. d.). Food policy old and new. *Development Policy Review*.

Mwema, C. M., Mutai, B. K., Lagat, J. K., Kibet, L. K., & Maina, M. C. (2012). Contribution of selected indigenous fruits on household income and food security in Mwingi, Kenya. *Current Research Journal of Social Sciences*, *4*(6), 425–430.

Myanmar Institute of Integrated Development [MIID]. (2014). Chin State Local Social Development Plan Vol I and II. Yangon, Myanmar: UNICEF.

Myanmar Parliament. (2014). Nay Pyi Taw, Myanmar Food and Agriculture Organization [FAO]. (2015). FAO Statistical Pocketbook 2015. Food and Agriculture Organization of the United Nations.

Ojogho, O. (2010). Determinants of Food Insecurity among Arable Farmers in Edo State. Nigeria. *Agricultural Journal*, *5*(3), 151–156. doi:10.3923/aj.2010.151.156

Oluyole, K. A., & Taiwo, O. (2015). Socio-economic Variables and Food Security Status of Cocoa Farming Households in Ondo State, Nigeria. Asian Journal of Agricultural Extension. *Economia e Sociologia (Evora, Portugal)*, *9*(1), 1–7. doi:10.9734/AJAEES/2016/10077

Omonona, B. T., & Agoi, G. A. (2007). An analysis of food security situation among Nigerian urban households: Evidence from Lagos State, Nigeria. *Journal of Central European Agriculture*, *8*(3), 397–406.

Oo, A. N., Iwai, C. B., & Saenjan, P. (2013). Food Security and Socio-economic Impacts of Soil Salinization in Northeast Thailand. *Environmental and Rural Development*, *4*(2), 76–81.

Sekhampu, T. J. (2013). Determination of the Factors Affecting the Food Security Status of Households in Bophelong, South Africa. *The International Business & Economics Research Journal*, *12*(5), 543. Retrieved from http://search.proquest.com/docview/1418721319/abstract/14189E8F6CC132D5D3/17?accountid=13598

Than, T. S. (2010). Development of Chin State.

Timmer, C. P. (2014). *Food security in Asia and the Pacific: the rapidly changing role of rice*. Asia & the Pacific Policy Studies.

United Nations. (2016). Follow-up to the outcome of the Millennium Summit. In *Integrated and coordinated implementation of and follow-up to the outcomes of the major United Nations conferences and summits in the economic, social and related fields* (p. 4). United Nation. [UN]

United Nations Development Program. (2015). Human Development Report 2015 Work for human development.

Van Thanh, N., & Yapwattanaphun, C. (2015). Banana Farmers' Adoption of Sustainable Agriculture Practices in the Vietnam Uplands : The Case of Quang Tri Province. *Italian Oral Surgery*, *5*, 67–74. doi:10.1016/j.aaspro.2015.08.010

Walker, S. H., & Duncan, D. B. (1967). Estimation of the probability of an event as a function of several independent variables. *Biometrika*, *54*(1-2), 167–179. doi:10.1093/biomet/54.1-2.167 PMID:6049533

World Food Program. M. (2013). Chin Operational Brief (Pamphlet). Yangon, Myanmar.

Yu, B., & You, L. (2013). A typology of food security in developing countries. *China Agricultural Economic Review*, *5*(1), 118–153. doi:10.1108/17561371311294810

Zakari, S., Ying, L., & Song, B. (2014). Factors Influencing Household Food Security in West Africa: The Case of Southern Niger. doi:10.3390u6031191

This research was previously published in the International Journal of Knowledge Society Research (IJKSR), 8(2); edited by Miltiadis D. Lytras and Linda Daniela, pages 44-53, copyright year 2017 by IGI Publishing (an imprint of IGI Global).

Section 5

Organic Agriculture and Consumption

Chapter 18
The Role of Private Labels in the Organic Food Market

Miguel Llorens
Universidad Complutense de Madrid, Spain

Sonia Carcelén
Universidad Complutense de Madrid, Spain

ABSTRACT

The organic food market has become one of the most rapidly growing sectors in developed economies around the world in the last decade, but it has not grown at the same pace in all the countries. The review of literature clearly indicates that the main motivations of the Spanish consumer to buy organic food are Health, Taste and Quality; also reveals that the main barriers are related to Price and Availability. The organic Store Brand appears as an opportunity for retailers to overcome those barriers, the price gap and the lack of availability. This study investigates the role of Store Brands in the development of the organic food product category. The authors provide a comprehensive picture of the current status in the Spanish distribution channels and review some PL strategies for the retailers to develop the category in different marketing areas such as branding, labelling, pricing, merchandising and promotion.

INTRODUCTION

In order to understand organic products and the current status of the food industry, we must look back over recent history. In 1900, the world's population was blighted by poverty, and infectious and nutrient-deficiency diseases were common. The 20th century witnessed unprecedented political, social, and economic changes. At the same time, scientific and technological advances took place at a faster pace than ever before. A combination of the above factors impacted the way food is grown, processed and marketed today.

In a comprehensive review of 20th-century milestones in the food processing evolution, Welch and Mitchel (2000) distinguished two very distinctive periods:

DOI: 10.4018/978-1-5225-8063-8.ch018

- The first half (1900–1950s) was characterized by profound political, social, and economic upheavals, when the major concerns were ensuring an ongoing supply of food to sustain the population, preventing or alleviating deficiency diseases, and reducing the incidence of food-borne diseases.
- The second half (1951–1999) was a time of political and economic stability, paired with increasing prosperity, which changed the focus. At the same time, improvements in medicine, in particular the mass production of antibiotics and the peacetime baby boom, led to an increase in population growth.

The changes took place across the different stages of the food production chain, from agriculture to processing and marketing. The modern agricultural period began after the Second World War, when industrialized countries concentrated on increasing productivity of their crops in order to be more self-sufficient and thus avoid the supply problems experienced during the war. From 1945, developing countries achieved stunning increases in agricultural production in what today is commonly referred to as the "green revolution", principally by using new varieties of crops (crossed plant varieties) such as wheat, rice, and corn, and increased use of pesticides and oil-based fertilizers combined with mechanization.

In the last decade, concerns regarding health issues have intensified, with the articulation of civil movements around consumer associations and non-profit organizations to raise awareness of the health issues associated with the new farming and processing methods. The modern production techniques not only affect agriculture, but also livestock, and imply selection of the most productive seeds and animal species, leaving the rest to disappear gradually due to a lack of use and interest.

The mass production of food also brought with it the concentration of trade in the 1980s and the development of large retail chains with cross-national interests. These big market players in the current scenario place large orders of products that can only be delivered by large-scale production, creating a virtuous circle of big buyers and farmers concentration to meet large-scale orders.

Finally, the availability of mass-produced products has generated changes in consumer behavior and brought about a standardization of lifestyles. Food security campaigners are now more concerned than ever that farmers are becoming dependent on large multinational corporations (MNCs) that use genetically modified organisms (GMOs) to the same end of increasing productivity by making crops either withstand direct application of herbicide, and/or produce their own insecticide.

Ironically, while large retail groups are in part to blame in the development of modern agricultural methods and GMOs, which arose as part of the quest for more productive crops and longer-lasting groceries to meet demand and endure longer distribution lead times, organic Private Labels from large retailers could be also a good way of helping the development of the organic food market, although this may seem contradictory.

This chapter focuses on the role that can be played by Private Labels (PLs) in the development of the organic category in Spain. To that end, the authors review the structural barriers, as well as consumer attitudes towards organic products, in order to obtain a comprehensive picture of the current status in the distribution channels and review some PL strategies for the retailers to develop the category in different marketing areas such as branding, labelling, pricing, merchandising and promotion. Finally there are some recommendations which develop the findings of the study.

BACKGROUND TO THE ORGANIC MARKET IN SPAIN AND EUROPE

It is a fact that the consumption of organic products is growing worldwide and the organic food market has become one of the most rapidly growing sectors in developed economies around the world (Organic Monitor, 2012). The global market for organic products increased more than threefold between 2000 and 2012. With a growth of nearly 10% in 2012, the US accounts for 44% of global sales, followed by the EU (41%) and Canada (4%). In fact, the production of organic products in the US grew by 240% between 2002 and 2011, while the market for non-organic products grew only 3% (Organic Monitor, 2012).

Sales of organic food and non-food products in the US broke another record in 2014, totaling USD 39.1 billion, up 11.3% from the previous year (OTA, 2014). Organic food sales posted an 11% rise in 2014, at USD 35.9 billion, while organic non-food sales, at USD 3.2 billion, jumped almost 14% for the largest annual increase in six years.

Organic fruits and vegetables continued to be the biggest-selling organic category in 2014, with USD 13 billion in sales, up 12% from the previous year, and making up more than 36% of all organic food sales. Of all the produce now sold in the US, 12% of it is organic – a market share that has more than doubled in the past 10 years when organic-product sales accounted for only 5% of the fruit and vegetable market. The organic-dairy sector posted an almost 11% jump in sales in 2014 to USD 5.46 billion, the biggest percentage increase for that category in six years (OTA, 2014).

In 2014, sales of organic non-food products, which represent 8% of the total organic market, posted the biggest percentage gain in six years, with sales of organic fiber and organic personal care products the stand-out categories (OTA, 2014) (Figure 1).

Europe is both a significant consumer and a large-scale producer of organic foods. Organic production accounts for 2.3% of European farmland, and 5.6% of farmland within the EU. In 2012, the area of organic farmland increased by 6% to 11.2 million ha. Sales of organic products in the EU also increased

Figure 1. Total US organic sales 2004–2014 (USD billions)
Source: OTA, 2014

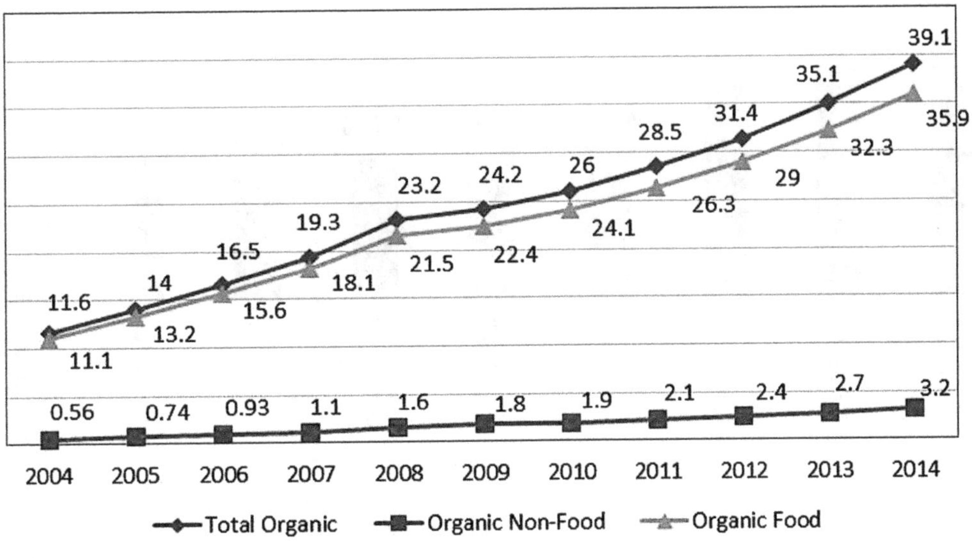

by 6% to GBP 19.7 billion. Germany (31%) and France (18%) account for nearly half of all sales, followed by the UK (8%), Italy (8%), and Switzerland (7%). The three biggest European markets outside the UK all achieved healthy growth in 2012.

In 2013 sales in Italy were up 9%, while the German and French markets both grew by 7%. The highest per-capita consumption is in Scandinavian and Alpine countries, with Switzerland and Denmark out in front. The highest average consumption per person was in Switzerland (GBP 163), Denmark (GBP 137), and Luxembourg (GBP 123). Finally, the highest share of the food and drink market for organic products was in Denmark, followed by Switzerland and Austria (Organic Market Report, 2014).

Worldwide, about 94% of organic food consumption is concentrated in 15 markets, led by the US, Germany, and France (MAGRAMA, 2012). If we focus on Spain, organic food consumption is residual in comparison to in the EU and US. Despite strong demand growth and the fact that the country is the largest producer of organic products in Europe, only 1% of food consumed in Spain is organic; this is very low when compared with other European countries such as Denmark and Germany, where organic market represent 7% and 4%, respectively (MAGRAMA, 2014) (Figure 2).

Per-capita organic food consumption in Spain has consolidated at around 20 or 21 euros per year (MAGRAMA, 2014), which leaves the country at the bottom of the ranking in relation to the main consumer countries (Figure 3).

However, despite the low penetration of these products, Spain is the European leader in market growth, with a rate of 25%, exceeding the average for the whole of the EU, which stands at 12% above countries with strong roots of ecological culture, such as Germany and Denmark. This growth rate is over a much lower base than those of the aforementioned countries, and therefore also represents high potential. Indeed, it is estimated that the market for organic products in Spain will grow by around 12% annually through 2020, rising from 905 million euros in turnover in 2008 to 12,182 million euros expected in 2020 (EVERIS, 2012).

Figure 2. Organic food expenditure as a percentage of total food expenditure
Source: MAGRAMA, 2014, 2012. All data is for 2012 except for Sweden and Denmark (2010)

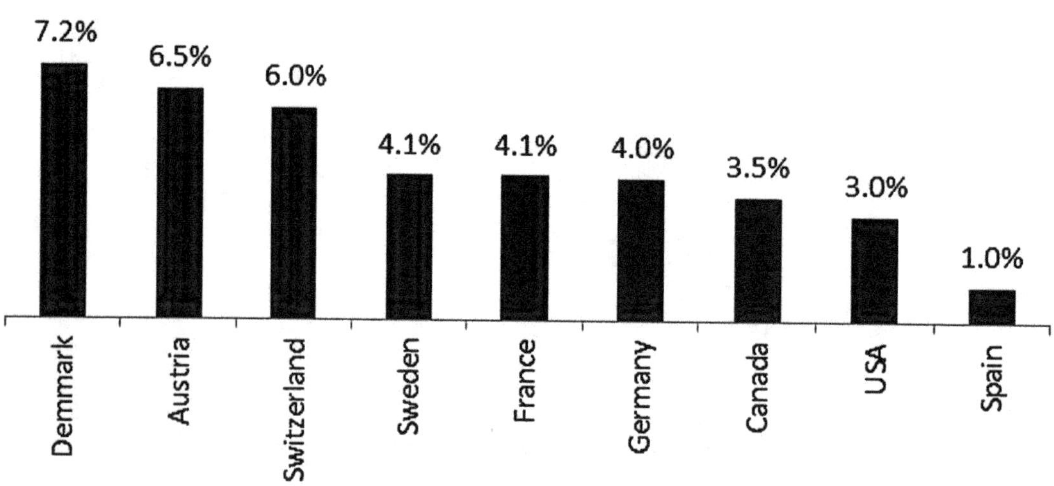

Figure 3. Per-capita organic food consumption (euros/individual/year)
Source: MAGRAMA, 2014; Willer & Lernoud, 2014. All data from 2012.

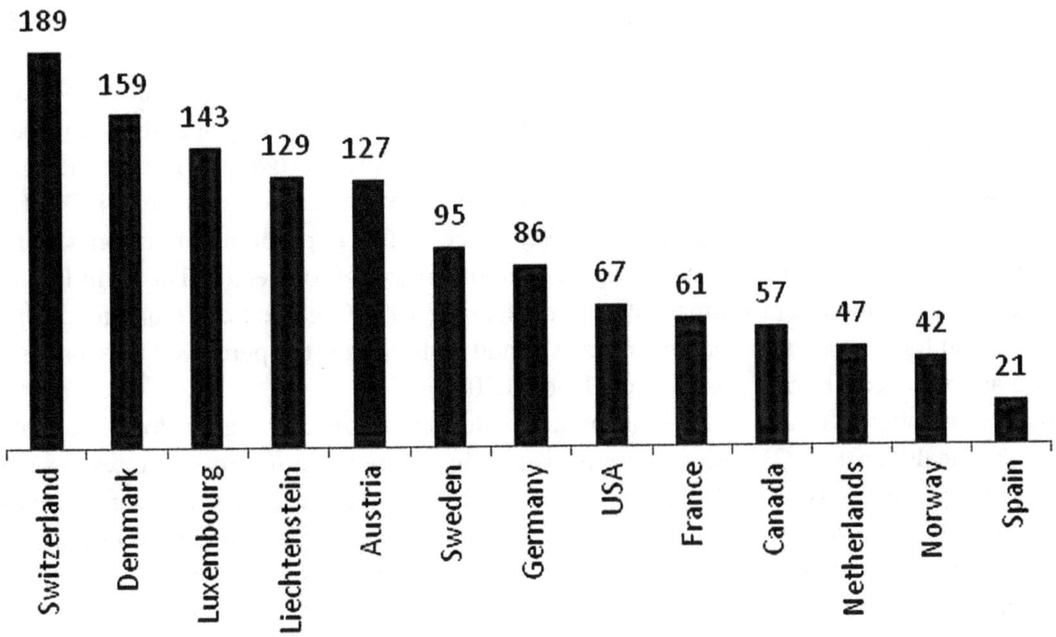

Nearly three-quarters (73%) of organic shoppers among UK consumers surveyed in a YouGov poll for the Soil Association in February 2014 said that they expected to buy more organic products that year compared to the previous year, with fruit and vegetables (45% of respondents), eggs (32%) and chicken (29%) most likely to attract increased spending. Furthermore, evidence from recent years reaffirms that organic production has the potential to be a profitable way forward for many farmers (Soil Association, 2014).

Public policies are fundamental to empower citizens for sustainable food consumption because they can reduce important structural constraints, such as the distribution system and relative prices that prevent consumers from changing their behavior toward more sustainable food choices (Thøgersen, 2005). A study by Jeff (2014) highlighted the lack of support given by national and European policies to increase organic food consumption in Spain. The policy support for the organic sector is essentially defined by the Common Agricultural Policy (CAP), which has been key to further its development in Spain, as well as in other EU member states (Sanders, Stolze, & Padel, 2011).

According to Armesto-López (2008), almost all Spanish farmers received European financial support for conversion from a conventional to an organic method of production in the first six years of the 21st century. Organic farming also provided an opportunity for farmers to charge higher prices for their produce. Moreover, Spanish producers receive a premium when selling in foreign markets because they can sell organic foodstuffs at a higher price compared to in the domestic market (Gil et al., 2001; Sanjuan et al., 2003).

MAIN FOCUS OF THE CHAPTER

Consumer Attitudes Towards Organic Foods

On the consumer side, the current growth of consumer adoption of higher-priced organic products comes from the realization that mass-produced food could be a threat to our health due to the use of pesticides on vegetables and antibiotics in animals as a way of increasing productivity. Schroeder (2014) carried out an extensive review of market research published over the past 25 years by organic- and natural-product marketers, and revealed that perceived safety concerns tied to pesticides, hormones, antibiotics, and GMOs are the critical component driving sales in the organic food sector. This is in line with the point made by Kay Hamilton, of Promar International, at the 1999 Organic Food Conference in the US, that the potential for growth in the organic market would be limited if the perceived threats to safe food production are removed. (Forrer, Avery & Carlisle, 2000).

There is an ongoing argument on the issue of how to determine whether organic food is actually more nutritious and healthier than GMOs, as there is no conclusive research to prove one way or the other. The organic food industry is allegedly funding anti-GMO/pesticide advocacy organizations to disseminate concerns on mass production food safety. The truth is that researchers cannot be certain either way when it comes to the safety of today's first-generation genetically engineered crops, and probably will not know their impact for another generation (Hirshberg, 2013).

Although it has not yet been proven that organic food is healthier for consumers, the relevant issue is consumer perception, which in general considers organic food to be healthier, safer, better tasting, and higher quality than conventional foods, and is connected with the environment and the support of organic farmers (Haas et al., 2014).

Consumer behavior is influenced by different sociodemographic factors, such as age, nationality, education, household size, and income. Aged people with more education and income are willing to pay a price premium for organic products because these products are believed to provide benefits to health (Muhammad, 2015; Jimenez et al., 2014). Moreover, organic food consumption is influenced by an overall holistic healthy lifestyle, including a healthy diet and exercise (Goetzke et al., 2014).

In most of the studies and publications consulted there is some consensus that for most consumers of organic products, the main motivations for purchase are health and nutrition issues (Baker, Thompson & Engelken, 2004; Fotopoulos, Krystallis & Ness, 2003; Fuentes & López de Coca, 2005; Gershoff & Frels, 2015; Lockie et al., 2002; Nath et al., 2014; Van Doorn & Verhoef, 2015; Zanoli & Naspetti, 2002). Other studies also highlight taste and concern for the environment as key drivers (Baker et al. 2004; Fraj & Martinez, 2007; Gracia & De Magistris, 2007).

Díaz & Bernabeu (2012) study the Spanish consumer and state that attitudes toward the purchase of organic foods are directly related to consumer lifestyle and are influenced by consumers' attitude towards the environment as well. In this sense, lifestyle and environmental attitudes have a positive influence on the purchase of organic food.

Some studies also mention organic product knowledge as an important factor determining organic food purchases. Consumers who report having higher levels of organic product knowledge will be more willing to buy organic food products (Gracia & De Magistris, 2007). Organic knowledge is achieved through consumer´s education, lifestyle related to healthy food habits and information provided in the market on organic food products.

In Spain a study by Soler, Gil and Sánchez (2002) discover that as more accurate information is provided, consumers´ acceptability of labeled organic food products increases and that WTP (willingness to pay) is highly correlated with consumption habits variables.

However, other research indicates that environmental motives are not important for consumers in the Spanish market and that the term "organic" plays an important role as a heuristic due to superiority, irrespective of the consumer's knowledge about the real features of organic food (Hughner et al. 2007; Vega Zamora et al., 2014).

Thus, the principal motivations for this type of consumption can be summarized according to three aspects: altruistic (environment, animal welfare, and rural and local environment), egoistic (individual health and food safety), and hedonistic (quality or flavor) (Aertsens, 2009; Kareklas, 2014; Vega Zamora et al., 2013).

In the US, organic consumers are more likely to focus on food quality to improve their health, including: choosing healthful options (57%), eating lean meats and/or fish (56%), choosing natural foods (54%) and choosing organic foods (47%) (OTA, 2014).

In Spain organic consumers are still "early adopters" and are considered to represent a niche market with high growth potential. This contrasts to the situation in Europe, where consumers of organic product have already reached a critical mass, or represent a large section of the population. Some of the most important differences between other European and Spanish consumers of organic products are found in their buying motivations and consumer profile. While for the Spanish consumers health, taste and quality are the main reasons for buying (these products are perceived as more natural and higher quality) for other Europeans environmental awareness is also very relevant (Everis 2012; MAGRAMA, 2014).

Gil, Gracia and Sanchez (2000) and Díaz and Bernabeu (2012) point out that consumers concerned about healthy diet and environmental degradation are the most likely to buy organic food in Spain, and are willing to pay a higher premium. Organic attributes are easily identified in perishable products as the price premium consumers would pay for organic meat, fruits and vegetables is higher.

Regarding the main barriers to consumption of organic products, the literature points at the price (Buder et al., 2014; Dupupet et al., 2010; Roddy, Cowan & Hutchinson, 1996; Rödiger & Hamm, 2015, Tregear, Dent & McGregor, 1994; Van Doorn & Verhoef, 2015) and its influence on repeat purchase (Marian et al., 2014); poor accessibility to such products (Jolly, 1991; Roddy et al., 1996; Soler, Gil & Sánchez, 2002; Tregear, 1994; Wandel & Bugge, 1997) and the obstacles that influence consumer intention to purchase organic food at the point of purchase (Henryks et al., 2014). However, if organic products are perceived as difficult to find and expensive, most consumers judge them positively (Zanoli & Naspetti, 2002).

Though 65% of the Spanish population is willing to consume these products, the barriers, in addition to high price and difficulty in finding these products in retail outlets outside of shops, include ignorance and lack of information about them (33% of the Spanish population are still unaware of the existence of organic products). Consumers ask for more information and want to choose with more freedom and knowledge (they are also interested in certification and think labeling is a good starting point). Most of all, they wish to understand and to be aware of how organic production and processing differ from conventional methods, and how organic products can be distinguished. (Zanoli & Naspetti, 2002).

Although one third of Spaniards are unaware of the existence of this type of product, it is important to note that current consumers of organic products are usually very well informed, read the labels of products they buy, and know how to identify real organic products (MAGRAMA, 2014).

In order to understand the lower development of the organic market in Spain, it is important to point out that the price and accessibility barriers mentioned above disappear in other European countries. This is because, on one hand, organic products are more accessible, as they are distributed by large groups that also promote consumption as a means of preserving the environment and reinforce its position by incorporating organic category strategy into their sustainability reports. On the other hand, the price differential of these products is reduced, often making them available at less than half the price at which they are sold in Spain. The price differential between organic and non-organic products in Spain is 74%, while in other countries, such as Denmark, Germany, the UK, and France, the difference is 10%, 9%, 21%, and 51%, respectively (Everis 2012).

According to some studies, Spain has relatively cheap food prices compared to the EU average, and has the lowest food price level index (PLI) relative to the largest consumers of organic food in Europe (Table 1). The price gap between conventional and organic products arises because the producers' price strategy is to obtain more or less the same price level in both domestic and foreign markets in order to earn the same premium in each. As a result, the price that Spanish consumers have to pay for organic food products is relatively higher in comparison to other European countries, and consumers are not willing to pay these prices (Gil et al., 2001; Sanjuán et al., 2003).

PLIs are shown in purchasing power parities, and provide a comparison of countries' price levels relative to the EU average: if the PLI is higher than 100, the country concerned is relatively expensive compared to the EU average; if the PLI is lower than 100, the country is relatively cheap compared to the EU average. Spain has the lowest PLI for food, which explains the larger gap between conventional and organic food prices relative to other European countries. Therefore, organic food is too expensive for Spanish consumers, and organic producers prefer to export their produce to countries in which food prices are higher.

There is a business opportunity in the fact that Spanish organic food distribution is still mainly in the hands of specialist organic stores (see Figure 6). The percentage of organic food sold in conventional stores is very low compared to other European countries.

Table 1. Price level index for food

Country	Price Level Index
Norway	184
Switzerland	159
Denmark	139
Luxembourg	119
Sweden	124
Austria	121
France	110
Germany	106
Netherlands	97
Spain	94

Source: Data source: Eurostat (2013). All data from 2012.

Consumer Profile and Segmentation

In the US, the Natural Marketing Institute (2008) has classified consumers according to their level of interest in products that promote health and environmental safety. These categories include: "lifestyles of health and sustainability", who are the heaviest purchasers of organic products and are environmental stewards dedicated to personal and planetary health; "naturalites", who make most purchase decisions based on benefits to their personal health; "drifters", who are motivated by the latest trends, and for whom commitment to sustainability is constantly shifting; "conventionals", who are well educated, waste conscious, practical, rational, and motivated by frugality rather than environmental goals; and "the unconcerned", who do not feel much environmental responsibility unless it is seen to imminently affect their livelihood.

Van Doorn and Verhoef (2015) find that there is a greater interest in organic products in a specific segment composed by consumers with higher education, women and small households. Also Ngobo (2011) clarified the relationships between demographics and the demand for organic products and found that the probability of purchasing an organic product is greater among the high income, college educated, and older families as well as among consumers holding high-level occupations. Hence larger families are less likely to buy organic products. Overall, financial resources play an important role in the purchase of organic products.

If we focus in Spain, consumers of organic products are usually aged 45–55 years and are upper or upper-middle class. Their education level and household structure are very similar to those of the general population (in 2011, their education level was in fact slightly higher than the average of the general Spanish population), and they are committed to the environment (MAGRAMA, 2014)

In Spain, four consumer segments of organic products have been identified according to their attitudes: "Ecologist", who show greatest concern for the environment and are developing related life habits and more responsible consumption; "the unconcerned", for whom organic consumption is determined by trends or fashions; "the convinced", whose defense of organic food is based on both its healthy nature and the guarantee of sustainable development that is respectful of the environment; and "the worried about health ", whose focus is on food quality and health (Figure 4).

Figure 4. Typology of consumers of organic products in Spain
Source: MAGRAMA, 2014

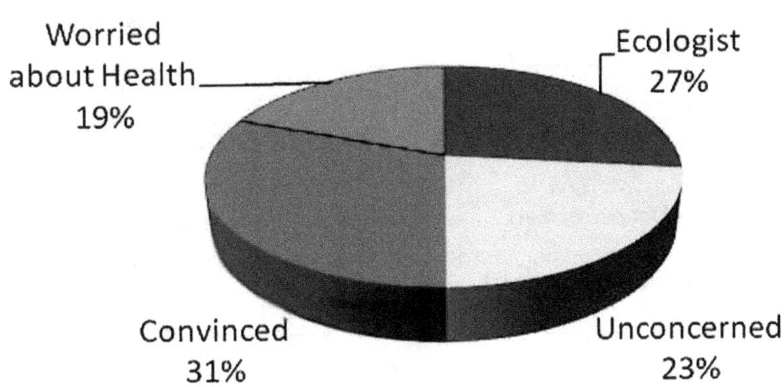

ORGANIC PRODUCT DISTRIBUTION

The distribution of organic products in Spain can be divided into two main channels: modern or conventional retail distribution, and specialized retailing. Regardless of the channel used, there is a great number and variety of formats in which to offer the product to the final consumer, as can be seen in Figure 5.

The introduction of organic products into mainstream retailing has drastically changed patters of distribution. Responding to the shift in consumer demand, mainstream retailers are becoming increasingly involved in the marketing of organic products (Jonas & Roosen, 2004). The maturing organic market has gradually shifted from short, channels that distribute directly to specialized organic stores, to large retail chains in Europe and in the US (Haas et al., 2014). In the US the modern channel has a greater presence as there is a large variety and number of major conventional store chains (e.g. Walmart or Target) and specialists (e.g. Trader Joe's or Whole Foods) offering a wide range of organic products, and which have a long tradition and experience in this market.

If we focus on Spain the picture changes significantly, since the largest weight is on specialized distribution (particularly to dedicated supermarkets and traditional organic food stores) versus conventional distribution, which currently still has a very limited range and is far removed from what happens in other European countries, such as Switzerland, Denmark, the UK, Germany, or France (Figure 6).

In Spain the share of organic products sold via the conventional retail distribution channel has grown from 20% in 2008 (see Figure 6) to 35–40% in 2014, and this is estimated to reach 58% by 2020. On the other hand the specialized channel currently accounts for 50–60% of sales (MAGRAMA, 2014). The increase of Modern distribution share may be due, first, to the fact that large retail chains (e.g. Car-

Figure 5. Distribution channels of organic products
Source: Adapted from MAGRAMA, 2012

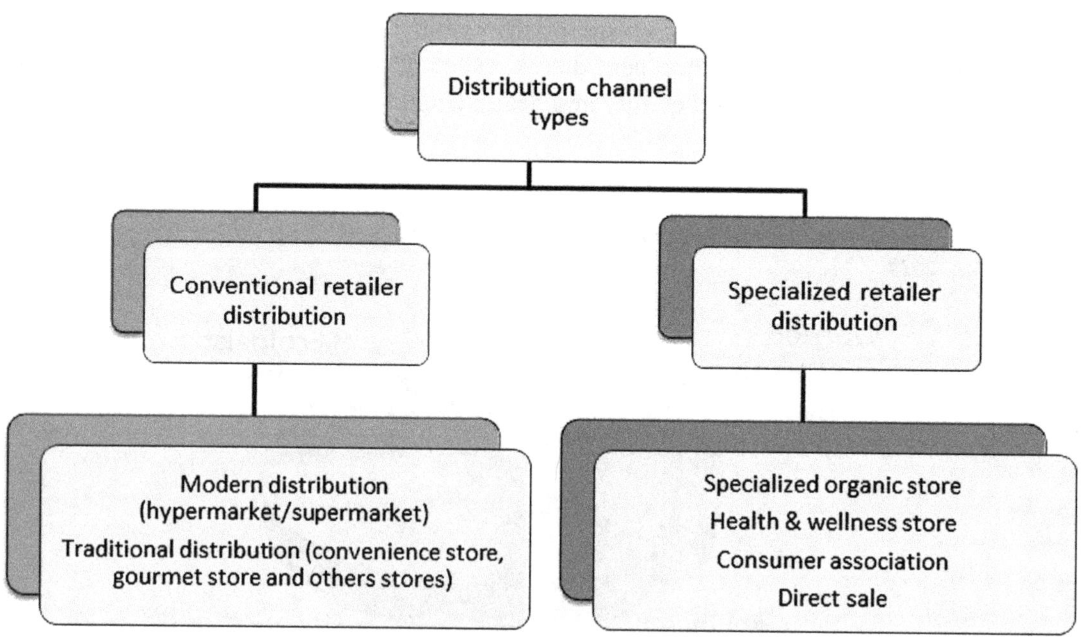

Figure 6. Market share of organic food and beverage sales of different distribution channels (2008)
Source: OTA 2008, cited in Haas et al, 2014
**Note: Conventional USA retail comprises conventional mass-market retailers and national natural food chains.*

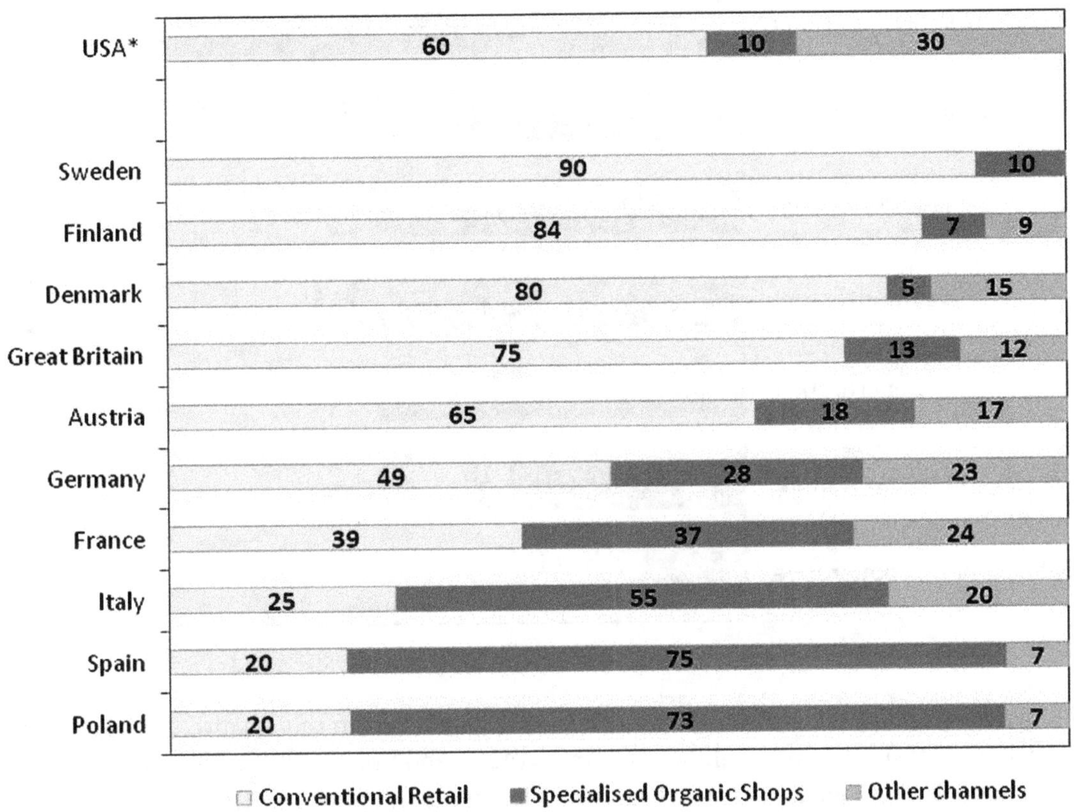

refour, Auchan, and El Corte Ingles) are increasing their ranges of new organic products in response to growing consumer demand. Second, these same retailers are betting on the development and marketing of organic PL products. Hence they increase their organic range and at the same time they complement it with their PL.

Figure 7 shows that although in Spain the conventional distribution channel (hyper- and supermarkets) and specialty stores carry great weight, it is also important to note how in recent years the direct channel (direct purchase from farmers) has grown.

With the increased distribution of organic products by conventional retailers, two of the most significant barriers noted above are being reduced. On the one hand, the price level of organic products is decreasing, which encourages consumers to buy more of these items, and on the other, improved accessibility towards these products is allowing consumers to find organic products more easily and comfortably.

PRIVATE LABELS AND ORGANIC PRODUCTS

Following the American Marketing Association's definition of a brand (www.ama.org), a PL or store brand is considered to identify the goods and services of a retailer and differentiate it from competitors.

Figure 7. Channels used by Spanish consumers to buy organic products
Source: MAGRAMA, 2014

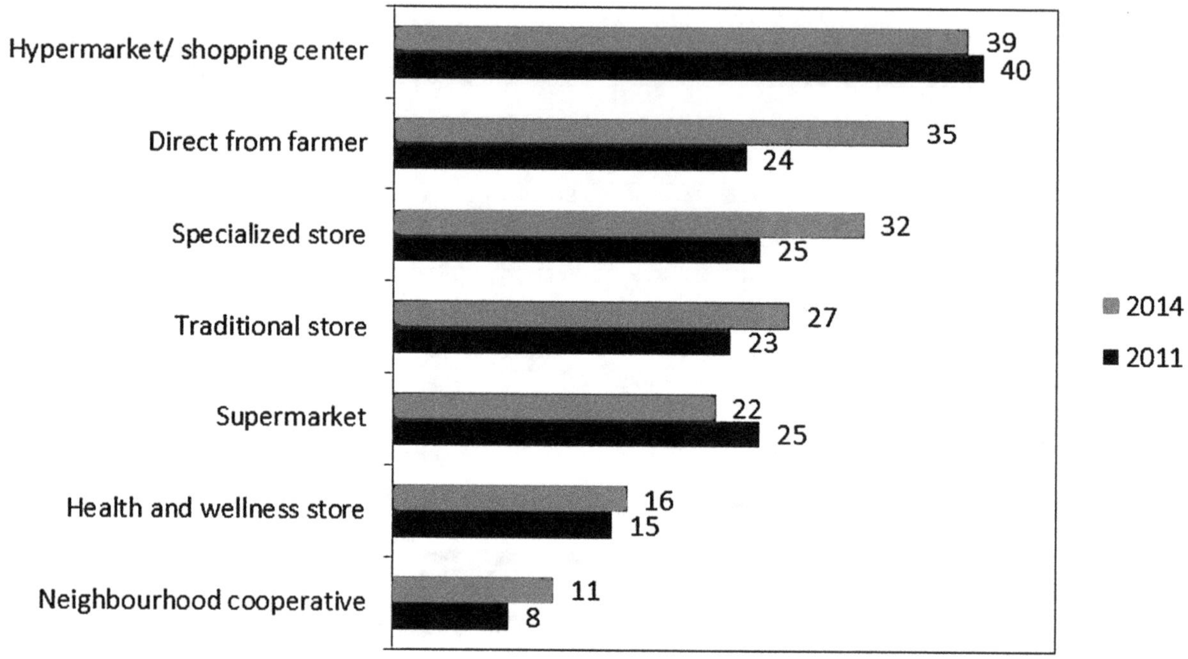

PLs have been gaining market share in Europe in recent years. In this regard, the UK occupied the highest position (51%) as of 2014, followed by Spain (42%) and France (29%). However, in 2014 manufacturer brands (MFs) started to regain ground from PLs for the first year since the start of the 2008 crisis; this has been reflected throughout Europe, with the share of PLs being reduced in value and volume by 0.1 percentage points compared to 2013, reaching 38.7% and 48.9%, respectively (IRI, 2015).

The lines between MFs and retailer PLs are becoming increasingly blurred in the global organic food industry. Retailers are taking advantage of organic PLs because they are winning consumers by marketing products at competitive prices. Organic MF brands are having to reinvent themselves, with many focusing on new values in order to broaden consumer demand. With increasing commoditization of organic products, there is growing pressure for leading brands to differentiate themselves.

Overall, the strategies followed by European retailers are moving away from large discounts and promotions (except in Spain, being the country with the highest percentage, since 30.3% of promotions are provided by PLs), and are increasing prices to meet the demand of those consumers seeking premium and higher quality products, including organic products. In this sense, organic PLs could be a great business opportunity for large retail chains, which can offer products at lower prices while they also improve accessibility – two of the main barriers limiting organic consumption (Dimitri & Greene, 2000; Puelles et al., 2008; Vicente et al., 2007). Moreover, as in some European countries, distributors could use this strategy to improve their image and positioning (Jonas & Roosen, 2004, 2005). Only some large Spanish retailers, such as Carrefour, Eroski, El Corte Ingles and Auchan, have introduced organic products in their outlets, but with a low level of penetration.

In most product categories, PL products are outselling branded products of organic food. Food retailers are becoming increasingly involved in the organic market and have launched PLs consisting solely of organic products (Jonas & Roosen, 2004). In fact, many of the most successful organic brands in Europe are PLs (Haas et al., 2012).

Retailers are very successful with PLs as they enable organic products to be marketed at competitive prices. PLs for organic food are most successful in Germany, where they have been introduced by discounters. In some cases, organic products are even cheaper than conventional ones (e.g. in Lidl some products are priced lower compared to similar conventional products of a leading supermarket). So, in most categories, PLs products are outselling branded products of organic food.

The advantages of PLs are that they improve overall store performance by providing a point of differentiation in the market competition, building store loyalty and acting as a form of advertisement for the store (Bhasin et al., 1995; Corstjens & Lal, 2000; Mejri & Bhatli, 2014).

As noted above, high retail price is frequently cited as a key obstacle to greater demand for organic foods, as organic and non-organic foods are generally viewed as substitutable (Tan & Cadeaux, 2006). The demand for organic food is suggested to be highly contingent on its pricing relative to conventional products (Halpin & Parkinson, 2004). The problem usually arises when the price premiums for organic products are above what consumers are willing to pay. Such price premiums tend to be lower when PLs are present. Tan & Cadeaux (2006) found some evidence in the Australian market that conventional supermarkets are pursuing a "lower-price" strategy (with PLs), and that the emergence of organic PL products might potentially change the competitive environment within a product category.

Minnens (2013) studies the Belgian organic market and the role of organic Private Labels, concluding that consumers' price perceptions of organic products are negatively correlated with their purchase intention of organic food. This author suggests that those who perceive the prices of organic food as being too high, or beyond their budget, are more likely to have a positive attitude toward organic food from PLs. This implies that for consumers who refrain from buying organic food because of the price premium compared to conventional food, the organic PL alternative may be of interest.

A greater supply of organic products within PLs from large Spanish distribution chains would be a clear response to the growing demand for such products by consumers looking for premium products with higher prices, higher quality, and above all more health benefits, but without having to pay the high price differentials found in specialty stores that sell these products.

If the concerns of Spanish retailers are considered more closely, what the current issues are and why they are difficult to overcome can be clearly seen. The introduction of organic products within conventional distribution channels has faced several barriers derived from the nature and characteristics of the Spanish market, which has slowed down the development of the organic market compared to other similar countries.

According to a report by MARM (2009), the main reasons for Spanish retailers to sell organic products (45%) included:

- Increasing consumer demand.
- Desire to improve company image.
- Desire to beat competitors.

In the same report, the main reasons for Spanish retailers not selling organic products (55%) were due to:

- Lack of demand.
- Lack of reliable sourcing.
- Higher cost price.

From the above, it is clear that there is a dilemma for retailers in terms of what comes first: consumer demand or availability in stores. Retailers seem to agree on the existence of the demand, but at the same time feel that this is not strong enough to hold shelf space. Ultimately, retailers need to sell products at high enough volumes to produce profit in order to optimize store room, so they are reluctant to allocate more room to products that sell in lower volumes unless their price premiums compensate their lower rotation on the shelves.

In addition to this low rotation, and perhaps because of it, as discussed earlier, the fact that overall average food prices are lower in Spain than in other European countries makes the premium price of organic food more significant for the Spanish consumer. Spanish producers would rather sell their produce abroad where they can charge higher prices, or sell them in Spain at the same price they get abroad (see Table 1); this represents an inflated cost to retailers in relation to conventional food, which in turn is passed on to the final retail price paid by the consumer. These inflated prices as a consequence of foreign demand and willingness to pay more has created a structural barrier that is preventing the Spanish organic market from developing further.

Some interviews with retailers have indicated that the food industry is not meeting consumer demand with respect to social concerns, namely fair trade, organic production, the environment, and animal welfare. Therefore, retailers can develop these new product categories because of their proximity to consumers, and their large product-development and marketing departments (European Commission, 2011).

Lavorata and Mejri (2008) highlight the potential of customer loyalty to retailers by improving the perceived quality of their PLs. Hence, PLs are no longer "me-too" brands, but are beginning to acquire a higher status. The marketing of organic and fair trade PLs by retailers reveals their willingness to assign the status of "ambassador of environment and social values" to their own brands.

The use of organic seals has encouraged development of the organic segment in different countries, including the US and Spain, where organizations are taking on the role of certifying and awarding organic seals at either international (EU), national, regional, or local levels.

In the EU, the use of the organic farming logo (Figure 8) and correct labeling is obligatory for all organic pre-packaged food produced within EU borders. The organic symbol is protected from being used on non-organic products throughout the EU. This enhances fair competition in the market, and of course, consumer protection.

In Spain, the different administrative regions have created regulatory bodies to trace and certify organic agriculture. Their seals are shown in Figure 9.

Alternatively, the role of certifying can be taken over by large retailers, who already have consumer trust in terms of guarantying the quality and safety of their product offerings. This is another area in which PLs have a role to play in developing organic product categories.

If the UK is considered as a developed organic market, according to an Organic Market Report (2014), 71.3% of organic products are sold through multiple grocers, and only 14.5% by independent retailers such as specialist stores. In terms of the UK's largest organic retailers, PL share of organic sales is very high at 74% in Morrisons, 69% in Sainsbury's, 61% in Waitrose, 60% in Tesco, and 45% in the online grocer Ocado.

Figure 8. EU organic farming logo
Source: European Commission. Agriculture and Rural Development (2015).

Figure 9. Organic agricultural logos in different Spanish regions
Source: http://www.ecomprometete.com

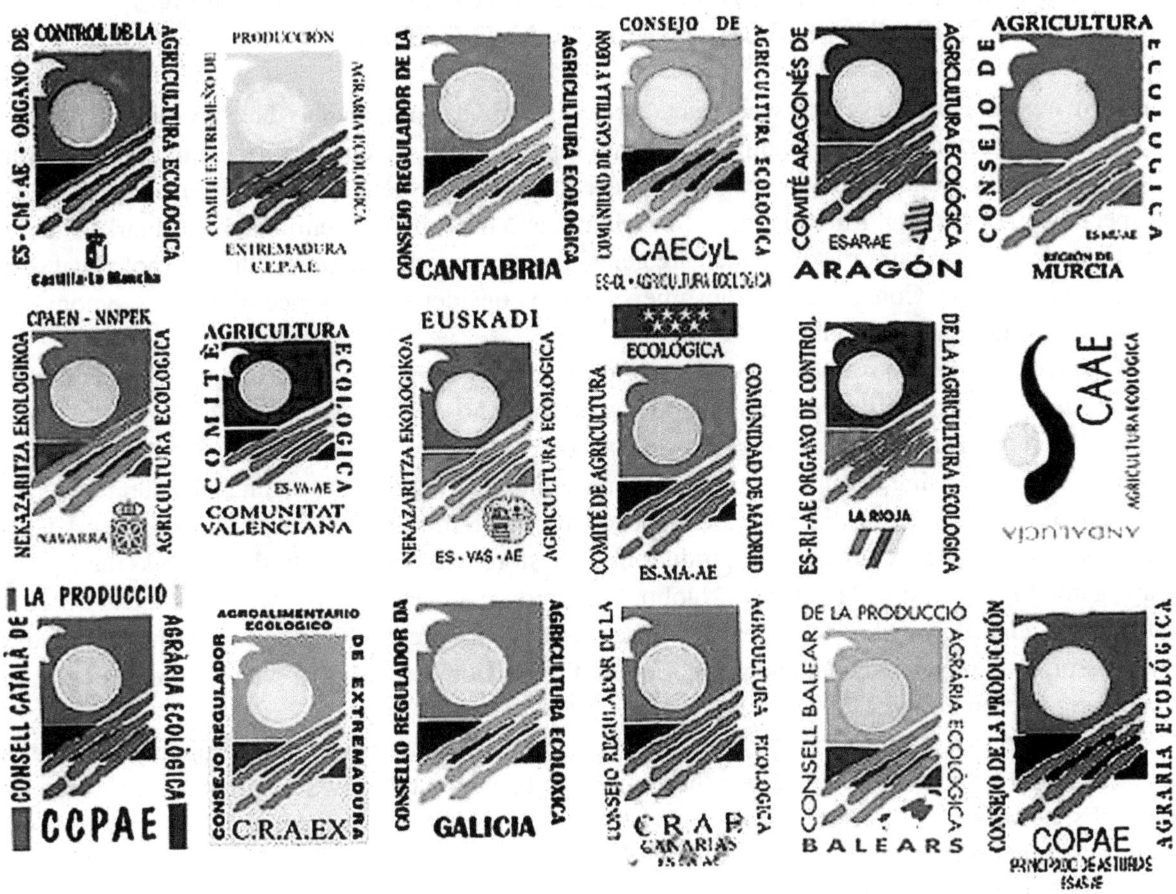

As noted above, in Spain the organic market is more skewed towards specialist organic stores, with higher prices and less availability, and organic PL are less developed within the Spanish retailers.

Puelles, Llorens and Talledo (2014) used the Theory of Planned Behaviour (TPB) to evaluate the differences in consumer attitudes towards organic products for MFs versus PLs. They found that perception of control is crucial in determining the buying behavior related to organic products, and highlighted that a greater variety of products on the shelves, affordable price, and additional information about the health benefits at the point of sale can positively influence buying behavior towards organic products.

Ajzen (2002) introduced a new component into TPB, known as "perception of control", based on a study by Bandura (1977), in order to study the influence of environmental factors (e.g. marketing stimuli) as moderators of intention arising from beliefs and norms. The basis for introducing this new factor was a realization that the execution of a particular behavior depends jointly on motivation (intention) and ability (behavioral control).

Puelles, Llorens and Talledo (2014) highlighted the important role to be played by organic PLs in developing the organic food segment in Spain, as the value of the perceived control factors in PLs is not as high as for MFs, which means that there is a different perception of the price barrier.

ORGANIC PRIVATE LABEL STRATEGIES: INITIATIVES IN BRANDING, LABELLING, PRICING, MERCHANDISING, AND PROMOTION

Marketing mix variables have different effects for organic goods compared to the conventional products. So, it´s very important for both manufacturers and retailers to carry out specific strategies to promote their organic PLs. Ngobo (2011) indicates that most consumers associate organic products with specific and specialized stores (e.g. Mundo Orgánico in Spain) and not with conventional hypermarkets and supermarkets (e.g. Carrefour or Lidl), which are generally associated with good value but not always with high-quality products. Consequently, consumers tend to consider the presence of a PL as synonymous to a poor (organic) quality product.

To change negative consumer perceptions and to improve their corporate image, retailers can distinguish themselves by selling premium products, such as organic food, that are considered being of superior quality and are offered at a higher price. Also, retailers can restore consumer trust and preserve customer loyalty by using organic PL products thus demonstrating that the chain cares about the safety and health of their customers, and cares about the environment. In particular retailers need to work harder to persuade consumers to buy organic product in their store and additionally to persuade them to buy their organic PLs instead of MF Brand (Ngobo, 2011).

Retailers can follow different strategies for their organic private label products in terms of branding, labelling, pricing, merchandising and promotion. In this section the authors review some strategies that retailers are implementing.

One specific aspect of the retailer's brand strategy has become particularly important in the last decade as a tool for retailers to influence their image and develop their own brand name. Most retailers carry MFs; however, they are also increasingly offering PL products. Motivations for offering PLs is the higher profit margins they provide to retailers; the negotiating leverage they provide over manufacturers; and the implicit assumption that providing PL brands will engender loyalty to the retailer (Ailawadi & Keller, 2004). Store image and corporate reputation enhance the image and perceived quality of PL brands (Calvo & Lang, 2015).

Ngobo and Jean (2012) find that the type of images consumers develop about a store influences the demand for organic products from that store. In addition, the influence of store image on the demand for PL organic products depends on the PL branding strategy. They also suggest that although own brands are accepted in stores with quality produce and with quality PL, they are less likely to be adopted in stores with varied selections. Furthermore, the own-brand strategy (the use of the retailer's own name) is not always an effective branding strategy for organic products, except in some stores.

There is a range of organic PL branding strategies: many retailers give their own name to their PL, whereas others use different names for these products. For instance, in 2005 Safeway launched its "O Organics" brand (see Figure 10), which is becoming the leading organic food brand in the US (the private label now houses over 300 certified organic products and exceeds USD 400 million in sales) (www.safeway.com), and Walmart introduced its Wild Oats PL at prices that undercut brand-name organic competitors by at least 25% (www.walmart.com).

Other retailers, such as Whole Foods, a US supermarket chain specializing in organic food, have used their own name for their organic PLs (Figure 11).

Finally, retailers such as Carrefour and Auchan have used their own names for their organic PLs, but have added the word "Bio" and "Eco", respectively (Figure 12).

Figure 10. Products bearing the Safeway Organics logo
Source: www.safeway.com

Figure 11. Products bearing the logo Whole Foods
Source: www.wholefoods.com

Figure 12. Product bearing the Carrefour Bio logo
Source: www.carrefour.es

There are both advantages and disadvantages to each of the above brand name PL strategies. On one hand, using the same name and perhaps even the same package design for products in a wide array of categories across the store strengthens awareness and recall of the retail brand, and may facilitate consumers' decision making. On the other, consumer may doubt whether the retailer can provide a good value, strong product in so many different product categories (Ailawadi & Keller, 2004).

Alongside the brand strategies, it is very important for retailers the use of a specific "signal" in an organic food package, such as using the word "organic" or adding the logo of an independent certification agency in the package. This is one strategy that both manufactures and retailers resort to send messages to consumers that their organic products are safer and healthier (Chinnici et al., 2002; Hutchins & Greenhalgh, 1995; Padel & Foster, 2005; Squires et al., 2001). An organic label can influence much more than health views; in fact, perceptions of taste, calories, and value can be significantly altered when a food is labeled "organic" (Lee et al., 2013). This means that use of package labels to promote organic product claims are the most important influence for consumers when they purchase.

Vega et al (2014) study the Spanish consumer and find that the term "organic" performs an important evocative role for consumers because with a single word they attribute a set of positive characteristics to these products that endow them with superiority over conventional ones: they think that organic food are better, even if they are not quite sure why.

However, Janssen & Hamm (2012) find that very few consumers trusted the generic labelling with the prefix 'organic' without a certification logo. It is important that the product labelling has an organic certification logo because is a tool for signaling consumers that a product is a certified organic product. So, an instrument to gain consumer trust in credence goods is third-party certification of the supply-side (Roe & Sheldon, 2007). Consumers prefer logos that are well-known and trusted with perceived strict organic standards and a strict control system and the WTP is significantly higher than for the generic labelling (Janssen & Hamm, 2012).

Also, Janssen and Hamm (2012) find that consumer perceptions of organic labelling schemes turned out to be of subjective nature and in many cases not based on objective knowledge. They conclude that it is advisable to label organic products with well-known organic certification logos that consumers trust.

Along those lines, Bauer, Heinrich and Schäfer (2014) find that the positive effects of organic labeling are also supported for consumers' purchase intentions and their WTP a price premium. Notably, they suggest that PLs are more adept at profiteering from the use of organic labels than global and local brands. This may stand for the fact that consumer assign less trust to PLs and therefore their proposal needs to be reinforced by a certified organic label.

There are other strategies related to price as discussed previously in the chapter. Those consumers who perceive the prices of organic food as being too high, or beyond their budget, are more likely to have a positive attitude toward organic food PLs. (Minnens, 2013). There are also studies (Ngobo, 2011; Buzby & Skees 1994; Hay 1989) relative to how high prices influence purchase quantity, and find that prices provide indications about the quality of the organic brands inputs, although there is a limit to the use of price differences in the selling of organic products. This indicates that the value of price as a quality signal should not exceed a certain price point.

Ngobo (2011) posits an important recommendation to retail managers on how to sell organic products to consumers. Managers should view price level as an instrument to increase the organic brand sales, and also that it is important to maintain a noticeable difference with conventional brands

In Spain the price gap with conventional food is much larger than in the rest of EU as seen in the consumer behavior section. Hence the sacrifice requested to the Spanish consumer is relatively higher being a barrier for switching from conventional to organic food purchase. This is an opportunity for retailers to offer organic food at competitive prices, as consumers justify the price difference between PL and MF products by the retailer bargaining power and savings in communication costs.

The authors already mentioned the lack of availability and difficulty to find in the store as main barriers for the development of the organic market in Spain. It is also clear that organic food is not usually top of mind, which underline the importance of facilitating the access to these products in the store.

There are two main strategies relative to the merchandising of the organic food category. One is to allocate the different products in a dedicated section in the store which helps the current organic consumer to identify the products easier saving shopping time; the alternative is to place each organic product next to the conventional products of each different category. This may trigger impulse buying among not current organic consumers. In Spain, the main conventional store chains like El Corte Ingles, Auchan, Eroski and Carrefour choose the first strategy and have a dedicated section in their stores.

Finally, there are some important strategies related to communication variable. There are differences in consumer reaction to communication according to the category where the organic food falls. In the reviewed literature, virtue and vice products are typically conceptualized in relation to each other as relative virtues and relative vices. Relative vices (also known as "wants") are products that provide immediate pleasurable experience (such as the good taste of a chocolate cake), but contribute to negative long-term outcomes (such as future weight gain and related health problems). Relative virtues (also known as "shoulds") are less gratifying and appealing in the short term but have less negative long-term consequences than vices and therefore are a more prudent choice (Milkman et al., 2008; Okada, 2005; Wertenbroch, 1998).

The vice/virtue nature of a product category is likely to influence responses to an organic claim. For example, Bezawada and Pauwels (2010) find that consumers are more responsive to promotions of organic virtue food.

Van Doorn and Verhoej (2015, 2011) find lower shares of organic products in vice categories and in categories with high promotional and higher shares in fresh versus processed categories. They suggest that virtue categories with a low promotional intensity are the best candidates for new organic product introductions. In addition, consumers are less likely to buy organic products in concentrated categories because consumers are loyal to big players and the smaller organic brands may face a double jeopardy phenomenon. Also, consumers are less disposed to buy organic products in categories where products are often promoted in store flyers (feature ads) and displays (Ngobo, 2011).

The main factors explaining organic food purchase are consumers´ attitude towards health, taste and environmental benefits provided by organic foods. Some authors claim that the main motivations are self-oriented (health and taste), while others posit that buying organic food is only motivated by other-oriented attitudes and values (Thøgersen, 2011). According to this a retailer promotional strategy to foster organic food consumption should design communication campaigns to make consumers more aware of the environmental and health benefits provided by organic food.

SOLUTIONS AND RECOMMENDATIONS

The development of an organic category by retailers should be part of their marketing strategy, since organic PLs are opportunities for retailers to improve their corporate image and appear as socially responsible organizations that are committed to sustainable food production. In addition, PLs bring retailers closer to consumers and allow retailers to increase their educational function by building consumer awareness of organic food benefits and their potential link to locally grown products as a way of reducing the food carbon footprint by shortening distribution lead times.

Consumer education has a role to play in the development of the organic market, as some studies of Spanish consumers (MAGRAMA, 2014) reveal a lack of understanding and awareness of what "organic" stands for, and a lack of recognition of the official seals promoted by the EU and regional governments. The lack of promotion that organic food has received from the public administration is a key issue here; thus, it is clear that government policies should promote awareness and education jointly with retailers and manufacturers in order to foster growth of the organic market.

Retailers should plan their PL branding strategy carefully, as some retailers can leverage their own name to give credibility to their organic PL, while others may not have the right positioning to sustain an organic brand using their own name. In the latter case, they may be better off choosing a different brand name for their organic food category.

Organizations owning an organic labeling strategy should put effort into measures for increasing consumer awareness of the logo and forming consumer perceptions and attitudes regarding the underlying scheme in terms of standards and control regime. We also recommend that retailers owning a labelling scheme for consumer goods invest in marketing communication and public relations for increasing consumer awareness of the logo and shaping consumer attitudes towards the certification scheme behind it.

The price strategy is paramount, as the reduction of the price premium for organic products is key to developing the organic category in Spain to attract consumers who refrain from buying organic food because of the price gap with conventional food, which is much larger than in the rest of EU. The strategy should be to narrow the price gap, and can be done if some economies of scale are achieved and also by promotions that give the opportunity to new consumers to enter the category and sample the products.

The promotional strategy of the retailer for organic food can either focus on self-oriented consumer motivations (health and taste) or on altruistic motivations (environmental awareness). In Spain most organic food consumers have self-oriented motivations (Puelles et al, 2011), hence communication message should aim at reinforcing the health and taste attributes of organic products.

In-store execution is also very important for the organic category, as there is a need for Spanish consumers to easily access organic food, not being organic food top of mind on a shopping trip. This can be addressed by retailers identifying organic products within the store by using specific symbols and merchandising features. On this point collaboration of manufacturers with retailers is very advisable and can be articulated around a Category Management project to help to identify the products in the store and increase the impulse buying.

The authors consider there is an opportunity to develop the online channel to connect with organic consumers, who have a similar profile in terms of age and lifestyle to that of online shoppers. There are several successful examples in the UK of specialized organic retailers that sell exclusively online, like Ocado.

FUTURE RESEARCH DIRECTIONS

There are several related areas that were not covered in this chapter but are very relevant in terms of consumer trends for the near future. One of these is the adoption of conscious decision making by the consumer, with preference given to local products as a way of developing the local economy and helping the sustainability of food production.

Although concern about the sustainability of food production is still relatively lacking in Spain, it is growing. Such concern relates to the "carbon footprint" of products until they reach supermarket shelves, meaning the amount of greenhouse gas emissions caused by a product or organization in moving that product from the area of production to the consumption point. The argument is that although a product may be grown according to organic standards, if it has to travel very far (to another continent, for instance) before it can reach the consumer, then its carbon footprint will offset the expected gain to the environment of growing the product organically. This issue is fostering interest on buying locally grown products in order to both reduce the carbon footprint and stimulate local economy.

In this context, retailers have an opportunity to improve their corporate image by offering products that help the environment and the local economy while following sustainability standards, not only in the food category but also in other growing organic categories, such as personal care and beauty products, as well as other non-food categories, such as household cleaners and pet food.

CONCLUSION

The authors consider that local and organic food is a consumer answer to an ever increasingly industrialized global food supply. Technological and organizational innovations over the last 70 years have significantly changed the way food is produced, marketed and consumed. The system has created an abundant food supply at affordable prices, but there have been consumer reactions to the negative consequences of this industrialization, such as health related issues (food safety scandals, increasing obesity rates), and quality issues (loss of natural flavors, questionable freshness).

One common belief is that Spanish consumers are not very environmentally friendly or particularly concerned with health; indeed, according to the above mentioned literature, environmental concern is not a major motivation for organic product consumption in Spain, as it is for other European consumers. However, the authors believe this is not the main obstacle for the development of the organic food market in Spain. In addition to consumer barriers, there are structural issues that hinder the sector's development, and one important point is that although Spain is the largest producer of organic food, most of it is sold abroad where higher prices can be charged. Spanish producers can get more or less the same income by selling in the domestic market; the relatively low price index of conventional food in Spain compared to the high prices of organic food mean that the sacrifice for Spanish consumers is bigger than for consumers in other countries, and Spanish consumers are not willing to pay such prices.

Another structural issue is the larger share in the organic food market of specialty shops, which charge higher prices and reduce availability compared to other countries in which modern distribution has a much larger share of the organic market. Finally, there is also an issue pertaining to lack of consumer awareness and lack of support from public authorities to educate consumers on what organic stands for, and the potential benefits of organic produce.

The study reveals that PLs currently play a key role in Spanish consumers' food choices, and that organic PLs will impact the extent to which retailers focus on the organic category as a premium earner and a way to leverage their corporate image. This strategy will be successful as long as consumers perceive PLs as reducing the main barriers to access the organic category.

There is an opportunity for retailers to use organic PLs to differentiate themselves from competitors and improve their positioning, projecting a corporate image of an organization committed to sustainability. The aim behind this positioning is to subscribe to the idea of responsible environmental behavior and try to minimize the social costs associated with conventional farming and mass-production methods.

Of the Spanish top retailers, some like Carrefour, El Corte Ingles, Eroski and Auchan, are using an umbrella brand strategy with their organic products, adopting their own name in their organic PL but adding words like "bio" or "eco" to differentiate their organic PL category from the rest of their PLs, in order to help consumers to correctly identify these products. Others like Mercadona and Dia do not have an organic category within their assortment. This may be a sign undevelopment of the Spanish market. In more developed organic markets such as USA, UK and Germany, some retailers also use specific brand names for their organic PLs.

REFERENCES

Abb, J. (2014). *Increasing organic food consumption in Spain. An analysis of policy obstacles in national and European policy*. (Unpublished doctoral dissertation). University of Lund, Sweden.

Aertsens, J., Verbeke, W., Mondelaers, K., & Van Huylenbroeck, G. (2009). Personal determinants of organic food consumption: A review. *British Food Journal, 111*(10), 1140–1167. doi:10.1108/00070700910992961

Ailawadi, K. L., & Keller, K. L. (2004). Understanding retail branding: Conceptual insights and research priorities. *Journal of Retailing, 80*(4), 331–342. doi:10.1016/j.jretai.2004.10.008

Ajzen, I. (2002). Perceived Behavioral Control, Self-Efficacy, Locus of Control, and the Theory of Planned Behavior. *Journal of Applied Social Psychology, 32*(4), 665–683. doi:10.1111/j.1559-1816.2002.tb00236.x

Alonso, L. E., Rodríguez, C. J. F., & Rojo, R. I. (2014). Crisis y nuevos patrones de consumo: discursos sociales acerca del consumo ecológico en el ámbito de las grandes ciudades españolas. *Empiria. Revista de metodología de ciencias sociales, 29*, 13-38.

Armesto-López, X. A. (2008). Organic farming in Spain. Two case studies: Catalonia and Galicia. *Journal of Sustainable Agriculture, 31*(4), 29–55. doi:10.1300/J064v31n04_04

Baker, S., Thompson, K., Engelken, J., & Huntley, K. (2004). Mapping the values driving organic food choice: Germany vs the UK and the UK vs Germany. *European Journal of Marketing, 38*(8), 995–1012. doi:10.1108/03090560410539131

Bandura, A. (1977). Self-efficacy: Toward a Unifying Theory of Behavioral Change. *Psychological Review, 84*(2), 191–215. doi:10.1037/0033-295X.84.2.191 PMID:847061

Bauer, H., Heinrich, D., & Schäfer, D. B. (2013). The effects of organic labels on global, local, and private brands. More hype than substance? *Journal of Business Research, 66*(8), 1035–1043. doi:10.1016/j.jbusres.2011.12.028

Bezawada, R., & Pauwels, K. H. (2010). *Are organics that special? Managing organic and conventional products across categories*. Working paper, Ozyegin University, Istanbul.

Bhasin, A., Dickinson, R., & Nandan, S. (1995). Retailer brands: a channel perspective: the United states. *Journal of Marketing Channels*, *4*(4), 17–36. doi:10.1300/J049v04n04_02

Buder, F., Feldmann, C., & Hamm, U. (2014). Why regular buyers of organic food still buy many conventional products: Product-specific purchase barriers for organic food consumers. *British Food Journal*, *116*(3), 390–404. doi:10.1108/BFJ-04-2012-0087

Calvo Porral, C., & Lang, M. F. (2015). Private labels: The role of manufacturer identification, brand loyalty and image on purchase intention. *British Food Journal*, *117*(2), 506–522. doi:10.1108/BFJ-06-2014-0216

Chinnici, G., D´Amico, M., & Pecorino, B. (2002). A multivariate statistical analysis on the consumers of organic products. *British Food Journal, 104*(3/4/5), 187-199.

Corstjens, M., & Lal, R. (2000). Building store loyalty through store brands. *JMR, Journal of Marketing Research*, *37*(3), 281–291. doi:10.1509/jmkr.37.3.281.18781

Council Regulation. (2007). *Organic production and labelling of organic products and repealing Regulation*. Retrieved May 15, 2015, from http://ec.europa.eu/agriculture/organic

Diaz, M., & Bernabéu, R. (2012). Consumer atttitudes to organic foods. A spanish case study. *Estudios de Economía Aplicada*, *30*(2), 755–775.

Dimitri, C., & Greene, C. (2002). *Recent growth patters in the US. Organic foods market. Market Economic Research Service*. USDA.

Dupupet, A., Valor, C., & Labajo, V. (2010). Comercialización minorista de productos biológicos ¿Está el marketing respondiendo a las necesidades de los consumidores? *Distribucion y Consumo*, *112*, 1–10.

European Commission. (2011). *Sustainable food consumption and production in a resource constrained world*. Retrieved from April 12, 2015, from http://ec.europa.eu/research/agriculture/scar/

Everis. (2012). *Estudio sobre la situación actual y el potencial del mercado ecológico en España*. Retrieved April 12, 2015, from http://www.everis.com

Fotopoulos, C., Krystallis, A., & Ness, M. (2003). Wine produced by organic grapes in Greece: Using means-end chains analysis to reveal organic buyers purchasing motives in comparison with the non-buyers. *Food Quality and Preference*, *14*(7), 549–566. doi:10.1016/S0950-3293(02)00130-1

Fraj, E., & Martinez, E. (2007). Ecological consumer behaviour: An empirical analysis. *International Journal of Consumer Studies*, *31*(1), 26–33. doi:10.1111/j.1470-6431.2006.00565.x

Fuentes, C., & López de Coca, E. (2005). El consumo de los productos ecológicos. *Distribución y Consumo*, *99*, 5–24.

Gershoff, A. D., & Frels, J. K. (2015). What Makes It Green? The Role of Centrality of Green Attributes in Evaluations of the Greenness of Products. *Journal of Marketing*, *79*(1), 97–110. doi:10.1509/jm.13.0303

Gil, J. M., Gracia, A., & Sanchez, M. (2001). Market segmentation and willingness to pay for organic products in Spain. *International Food and Agribusiness Management Review*, *3*(2), 207–226. doi:10.1016/S1096-7508(01)00040-4

Goetzke, B., Nitzko, S., & Spiller, A. (2014). Consumption of organic and functional food. A matter of well-being and health? *Appetite*, *77*, 96–105. doi:10.1016/j.appet.2014.02.012 PMID:24630940

Gracia, A., & De Magistris, T. (2007). Organic food product purchase behaviour: A pilot study for urban consumers in the south of Italy. *Spanish Journal of Agricultural Research*, *5*(4), 439–451. doi:10.5424jar/2007054-5356

Grunert, K. G., Hieke, S., & Wills, J. (2014). Sustainability labels on food products: Consumer motivation, understanding and use. *Food Policy*, *44*, 177–189. doi:10.1016/j.foodpol.2013.12.001

Haas, R., Sterns, J., Meixner, O., Nyob, D. I., & Traar, V. (2014). Do US Consumers' Perceive Local and Organic Food Differently? An Analysis Based on Means-End Chain Analysis and Word Association. *International Journal on Food System Dynamics*, *4*(3), 214–226.

Haas, R., Weaver, R., & Pöchtrager, S. (2012). Private Labels: A Strategic Tool for Retailers? In Food Value Chain Networks in the 21st Century (pp. 225–252). Madrid: Editoral Agricola Española.

Halpin, D., & Parkinson, J. (2004). Beyond the farm gate: the perspectives of processors, wholesalers and distributors in the organic industry. *The Australian Organic Industry A Profile, 47*.

Henryks, J., Cooksey, R., & Wright, V. (2014). Organic food at the point of purchase: Understanding inconsistency in consumer choice patterns. *Journal of Food Products Marketing*, *20*(5), 452–475. doi :10.1080/10454446.2013.838529

Hughner, R. S., McDonagh, P., Prothero, A., Shultz, C. J., & Stanton, J. (2007). Who are organic food consumers? A compilation and review of why people purchase organic food. *Journal of Consumer Behaviour*, *6*(2-3), 94–110. doi:10.1002/cb.210

Hutchins, R. K., & Greenhalgh, L. A. (1995). Organic confusion: Sustaining competitive advantage. *Nutrition & Food Science*, *95*(6), 11–14. doi:10.1108/00346659510103584

Janssen, M., & Hamm, U. (2012). Product labelling in the market for organic food: Consumer preferences and willingness-to-pay for different organic certification logos. *Food Quality and Preference*, *25*(1), 9–22. doi:10.1016/j.foodqual.2011.12.004

Jiménez, A. M., Escalera, M. E., & Campos, M. Á. V. (2014). La edad como factor del comportamiento del consumidor de productos orgánicos. *European Scientific Journal, 10*(7).

Jolly, D. (1991). Differences between buyers and non buyers of organic produce and willingness to pay organic price premiums. *Journal of Agribusiness*, *9*, 97–111.

Jonas, A., & Roosen, J. (2004). *Private labels for premium products. The example of organic food.* (Unpublished doctoral dissertation). University of Kiel, Germany.

Kareklas, I., Carlson, J. R., & Muehling, D. D. (2014). I eat organic for my benefit and yours: Egoistic and altruistic considerations for purchasing organic food and their implications for advertising strategists. *Journal of Advertising*, *43*(1), 18–32. doi:10.1080/00913367.2013.799450

Lavorata, L., & Mejri, C. A. (2008). *Retailers sustainable development practices: a French consumers perspective*. Paper presented at the meeting EGOS. Amsterdam.

Lee, W. C. J., Shimizu, M., Kniffin, K. M., & Wansink, B. (2013). You taste what you see: Do organic labels bias taste perceptions? *Food Quality and Preference*, *29*(1), 33–39. doi:10.1016/j.foodqual.2013.01.010

Lockie, S., Lyons, K., Lawrence, G., & Mummery, K. (2002). Eating 'green': Motivations behind organic food consumption in Australia. *Sociologia Ruralis*, *42*(1), 23–40. doi:10.1111/1467-9523.00200

Marian, L., Chrysochou, P., Krystallis, A., & Thøgersen, J. (2014). The role of price as a product attribute in the organic food context: An exploration based on actual purchase data. *Food Quality and Preference*, *37*, 52–60. doi:10.1016/j.foodqual.2014.05.001

MARM. (2009). *Marketing y alimentos ecológicos. Manual de aplicación a la venta detallista*. Retrieved January 10, 2015, from http://www.mapa.es

Mejri, C. A., & Bhatli, D. (2014). CSR: Consumer responses to the social quality of private labels. *Journal of Retailing and Consumer Services*, *21*(3), 357–363. doi:10.1016/j.jretconser.2013.08.001

Milkman, K. L., Rogers, T., & Bazerman, M. H. (2008). Harnessing our inner angels and demons: What we have learned about want/should conflicts and how that knowledge can help us reduce short-sighted decision making. *Perspectives on Psychological Science*, *3*(4), 324–338. doi:10.1111/j.1745-6924.2008.00083.x PMID:26158952

Ministerio de Medio, A., & Alimentacion y Medio, A. MAGRAMA (2012). *Caracterización del mercado de productos ecológicos en los canales especialistas de venta*. Retrieved February 20, 2015, from http://www.magrama.gob.es

Ministerio de Medio, A., & Alimentacion y Medio, A. MAGRAMA (2014). *Evolución de la caracterización y perfil sociodemográfico del consumidor de alimentos ecológicos en España*. Retrieved February 20, 2015, from http://www.magrama.gob.es

Minnens, F. (2013). *Consumers attitude on private label organic food products: a study of Flemish consumers*. (Unpublished doctoral dissertation). University of Ghent, Belgium.

Muhammad, S., Fathelrahman, E., & Ullah, R. U. (2015). Factors Affecting Consumers' Willingness to Pay for Certififed Organic Food Products in United Arab Emirates. *Journal of Food Distribution Research*, *46*(1).

Nath, V., Kumar, R., Agrawal, R., Gautam, A., & Sharma, V. (2014). Impediments to Adoption of Green Products: An ISM Analysis. *Journal of Promotion Management*, *20*(5), 501–520. doi:10.1080/10496491.2014.946200

Natural Marketing Institute (2008, Spring). Connecting values with consumers. *LOHAS Journal*, 19-22.

Ngobo, P. V. (2011). What drives household choice of organic products in grocery stores? *Journal of Retailing*, *87*(1), 90–100. doi:10.1016/j.jretai.2010.08.001

Ngobo, P. V., & Jean, S. (2012). Does store image influence demand for organic storebrands? *Journal of Retailing and Consumer Services*, *19*(6), 621–628. doi:10.1016/j.jretconser.2012.08.003

Okada, E. M. (2005). Justification effects on consumer choice of hedonic and utilitarian goods. *JMR, Journal of Marketing Research*, *42*(1), 43–53. doi:10.1509/jmkr.42.1.43.56889

Organic Trade Association. (2014a). *Organic Industry Survey*. Retrieved February 20, 2015, from http://www.ota.com

Organic Trade Association. (2014b). *U.S. Families Organic Attitudes and Beliefs Study*. Retrieved February 20, 2015, from http://www.ota.com

Padel, S., & Foster, C. (2005). Exploring the gap between attitudes and behaviour: Understanding why consumers buy or do not buy organic food. *British Food Journal*, *107*(8), 606–625. doi:10.1108/00070700510611002

Puelles, J. A., Gómez, M., & Puelles, M. (2011). *Marcas de Distribuidor: concepto, evolución, protagonistas y adaptación de los ciclos económicos*. Madrid: Ediciones Pirámide.

Puelles, M., Briz, T., & Labajo, V. (2008, October). *Marcas del distribuidor ecológicas en productos alimentarios. Iniciativas empresariales mas novedosas en mercados europeos y EEUU y repercusiones sobe los sectores implicados*. Paper presented at the meeting of the IV Congreso Internacional de Ingenieros Agrónomos y Profesionales de la Agronomía, Madrid, España.

Puelles, M., Llorens, M., & Talledo, H. (2014). El factor de la percepción de control como determinante en la intención de compra de productos ecológicos. *Innovar*, *24*(54), 139–152. doi:10.15446/innovar.v24n54.46655

Roddy, G., Cowan, C. A., & Hutchinson, G. (1996). Consumer attitudes and behaviour to organic foods in Ireland. *Journal of International Consumer Marketing*, *9*(2), 41–63. doi:10.1300/J046v09n02_03

Rödiger, M., & Hamm, U. (2015). How are organic food prices affecting consumer behaviour? A review. *Food Quality and Preference*, *43*, 10–20. doi:10.1016/j.foodqual.2015.02.002

Sanders, J., Stolze, M., & Padel, S. (2011). *Use and efficiency of public support measures addressing organic farming*. Academic Press.

Sanjuan, A. I., Sanchez, M., Gil, J. M., Gracia, A., & Soler, F. (2003). Brakes to organic market enlargement in Spain: Consumers' and retailers' attitudes and willingness to pay. *International Journal of Consumer Studies*, *27*(2), 134–144. doi:10.1046/j.1470-6431.2003.00295.x

Schroeder, J. (2014). *Organic Marketing Report*. Retrieved February 20, 2015, from http://www.academicsreview.org

Soil Association. (2013). *Organic market report 2013*. Retrieved March 25, 2015, from http://www.soilassociation.org

Soil Association. (2014). *Organic market report 2014*. Retrieved March 25, 2015, from http://www.soilassociation.org

Soler, F., Gil, J. M., & Sánchez, M. (2002). Consumers acceptability of organic food in Spain: Results from an experimental auction market. *British Food Journal*, *104*(8), 670–687. doi:10.1108/00070700210425921

Squires, L., Juric, B., & Cornwell, T. B. (2001). Level of market development and intensity of organic food consumption: Cross-cultural study of Danish and New Zealand consumers. *Journal of Consumer Marketing*, *18*(5), 392–409. doi:10.1108/07363760110398754

Tan, L., & Cadeaux, J. (n.d.). *Private Labels Beyond Convenience Consumer Goods: The Case of Organic Foods Retailing in Australia*. Retrieved May 15, 2015, from http://www.anzmac.org/

Tarkiainen, A., & Sundqvist, S. (2005). Subjective norms, attitudes and intentions of Finnish consumers in buying organic food. *British Food Journal*, *107*(11), 808–822. doi:10.1108/00070700510629760

Thøgersen, J. (2005). How may consumer policy empower consumers for sustainable lifestyles? *Journal of Consumer Policy*, *28*(2), 143–177. doi:10.100710603-005-2982-8

Thøgersen, J. (2011). Green Shopping: For Selfish Reasons or the Common Good? *The American Behavioral Scientist*, *55*(8), 1052–1076. doi:10.1177/0002764211407903

Tregear, A., Dent, J. B., & McGregor, M. J. (1994). The demand for organically grown produce. *British Food Journal*, *96*(4), 21–25. doi:10.1108/00070709410061032

Vaclavik, T. (2007). *Latest trends in the european organic retail market*. Retrieved May 15, 2015, from http://www.orgprints.org

Van Doorn, J., & Verhoef, P. C. (2011). Willingness to pay for organic products: Differences between virtue and vice foods. *International Journal of Research in Marketing*, *28*(3), 167–180. doi:10.1016/j.ijresmar.2011.02.005

Van Doorn, J., & Verhoef, P. C. (2015). Drivers of and Barriers to Organic Purchase Behavior. *Journal of Retailing*, *91*(3), 436–450. doi:10.1016/j.jretai.2015.02.003

Vega Zamora, M., Parras-Rosa, M., Murgado-Armenteros, E., & Torres-Ruiz, F. J. (2013). The Influence of the Term 'Organic' on Organic Food Purchasing Behavior. *Procedia: Social and Behavioral Sciences*, *81*, 660–671. doi:10.1016/j.sbspro.2013.06.493

Vega-Zamora, M., Torres-Ruiz, F. J., Murgado-Armenteros, E. M., & Parras-Rosa, M. (2014). Organic as a heuristic cue: What Spanish consumers mean by organic foods. *Psychology and Marketing*, *31*(5), 349–359. doi:10.1002/mar.20699

Vicente, A., Izaguirre, J. & Tamayo, U. (2007). Análisis de precios de alimentos ecológicos en distintos formatos comerciales: el caso de Vizcaya. *Conocimiento, innovación y emprendedores: camino al futuro*, 2427-2443.

Wandel, M., & Bugge, A. (1997). Environmental concern in consumer evaluation of food quality. *Food Quality and Preference*, *8*(1), 19–26. doi:10.1016/S0950-3293(96)00004-3

Welch, R. W., & Mitchell, P. C. (2000). Food processing: A century of change. *British Medical Bulletin*, *56*(1), 1–17. doi:10.1258/0007142001902923 PMID:10885101

Wertenbroch, K. (1998). Consumption self-control by rationing purchase quantities of virtue and vice. *Marketing Science*, *17*(4), 317–337. doi:10.1287/mksc.17.4.317

Willer & Lernoud. (2014). *The world of organic agriculture. Statistics & Emerging trends 2014*. Retrieved March 25, 2015, from https://www.fibl.org/

Zanoli, R., & Naspetti, S. (2002). Consumer motivations in the purchase of organic food. A means-end approach. *British Food Journal*, *104*(8), 643–653. doi:10.1108/00070700210425930

KEY TERMS AND DEFINITIONS

Carbon Footprint: The total amount of greenhouse gases produced to directly and indirectly support human activities, usually expressed in equivalent tons of carbon dioxide (CO_2).

Eco-Labeling: A voluntary method of environmental performance certification and labeling that is practiced around the world. Eco-labels identify overall, proven environmentally friendly origin of a product or service within a specific product/service category.

Ecological Consumer Behavior: Behaviour that takes into account the need to make more responsible purchases and looks to institutions to behave in this way too.

Environmental Sustainability: This refers to harvesting renewable resources that can be continued indefinitely, minimizing pollution creation, and avoiding non-renewable resource depletion.

Genetically Modified Organism (GMO): An organism or microorganism whose genetic material has been altered by means of genetic engineering.

Organic Agriculture: A holistic production management system that promotes and enhances agro-ecosystem health, including biodiversity, biological cycles, and soil biological activity.

Organic Farming: A type of farming that recognizes the direct connection between health and how food is produced. Artificial fertilizes are banned from organic farms and farmers develop fertile soil by rotating crops and using compost, manure, and clover.

Organic Food: Food produced using environmentally and animal-friendly farming methods on organic farms. These methods are legally defined and any food sold as organic must be strictly regulated. Animals used in organic food production must enjoy the highest welfare standards: they must be truly free range and have plenty of space and access to fields.

Organic Trade Association (OTA): A membership-based business association for organic agriculture and products in North America. Its members include growers, shippers, processors, certifiers, farmers' associations, distributors, importers, exporters, consultants, retailers, and others. OTA's member businesses work together through networking, advocacy, and other initiatives to encourage and protect organic farming practices, and to share messages about the positive environmental and nutritional attributes of organic products with consumers, the media, and policymakers.

Soil Association: The UK´s leading membership charity campaigning for healthy, humane, and sustainable food, farming, and land use. It was founded in 1946 by a group of farmers, scientists, and nutritionists who were concerned about the health implications of increasingly intensive agricultural systems following the Second World War.

This research was previously published in the Handbook of Research on Strategic Retailing of Private Label Products in a Recovering Economy edited by Mónica Gómez-Suárez and María Pilar Martínez-Ruiz, pages 359-387, copyright year 2016 by Business Science Reference (an imprint of IGI Global).

Chapter 19
Green Consumption: A Study to Understand Consumers' Organic Food Consumption

Ozan Kaya
Mugla Sitki Kocman University, Turkey

Feridun Duman
Mugla Sitki Kocman University, Turkey

ABSTRACT

The main purpose of this chapter is to reveal the profile of organic food consumers and the factors that affect consumers' motivation for organic food consumption as being one of the most important aspects of green marketing and green consumption. Therefore, this study first dealt with green consumers and green consumption and then, organic food consumption and the state of organic food market were evaluated and finally, an empirical research was conducted with 393 respondents in order to better determine the consumers of organic food consumption in Turkey. According to the findings, four motivations were found in influencing the behavior of organic food consumption. These are: health, ecological and social welfare, sensory appeal and natural contents. More specifically, this research reveals that those that have children participated in this research frequently consume organic food.

INTRODUCTION

Today's world witnesses the two sides of the technologic and global advances. Industrialization process stimulated the ultimate aim of increasing the well-being of the societies and improving existing conditions in 18th century, hence countries indifferently focused on integrating resources to production and consumption operations, without considering the potential social outcomes. Particularly after 1970s, the industrial revolution had resulted in an exponential increase in the human consumption of resources and an increase in health, wealth and population. That is to say, baby-boom period of post war time lead to excessive need for energy resources, and the higher need for more goods and services in many societies. In parallel, the attention kept focusing on the scarcity of the resources under the pressure of

DOI: 10.4018/978-1-5225-8063-8.ch019

rapid consumption. In other words, research and development focused progress of countries and societies engender the inevitable dilemma of benefiting from the environment or exploiting it. However, benefiting from the nature, had been taken as a one-way approach, and humanity ignored to renew the resources, that had been consumed. The primary principle of economics scarcity of resources had been neglected, and by time nature called for aid. According to Bahro (1986), ecological problems signify the dead end of capitalism. Hence, extended consumption, without attempt to renew the natural resources is unfortunately threatening the well-being of next generations. In a reactive manner, societies tried to identify the environmental problems, and offer solutions to ensure the ongoing relationship between humanity and the nature.

The negative issues emerged in environmental socio-cultural and economic fields caused by the traditional business philosophy, which was first time recognized that there was a need for sustainable development at the Stockholm conference in 1972 (Mebratu, 1998). Later in the 1980s, sustainable development focused on the relationship between saving the Earth and meeting human needs. Brundtland at his report in 1987 used similar statements by expressing meeting today's human needs without compromising future generations' ability to meet their own needs at their time (Santori, et al., 2014). In parallel to this, Rio conference in 1992 and (Rio+10) Johannesburg conference in 2002 led to a sustainable development understanding, which was built on the three pillars of economic, environmental and socio-culture developments in order to meet human needs now and in the future. This sustainable understanding was later adopted by the whole world (Islam et al., 2003).

Growing awareness on environmental disruption in past decades expose an attention in this issue. Kirkpatrick (1990) stated this environmental attention as "the biggest business issue of the 1990s". According to another view, the concept of sustainability and green marketing has been known for a long time, but not necessarily by their current names. For example, in the early 1970s, Kotler & Zaltman (1971) put forward the idea of societal marketing which encompassed the idea of the environment. While this idea was originally in response to a natural disaster, it is not news that 'green' marketing has gained ground in recent times. However, closely linked to green marketing concepts are environmentalism (Chamorro et al., 2009; Peattie, 2001) and the green consumer (McDonald & Oates, 2006; McEachern & McClean, 2002; Cope & Winward, 1991). When the ultimate aim of satisfying human needs and wants matches with technological advances, the result is two-fold. In other words, humanity enjoyed technologic advances by improving living conditions. On the other hand, attempts of production and trade over hundreds of years had negative impact on natural resources resulted in pollution and depletion. Especially, in the last 30 years, both the industry and the society are worried about the potential outcomes of the underestimated importance of the scarcity of nature (Straughan & Roberts, 1999).

Similarly, food production and consumption exert a growing pressure on the environment. Although, concerns about product safety and health induced many people to change their consumption patterns towards green products including organic food (Millock & Nauges, 2014). For instance, appearance of food-borne illnesses on the media, consumers become more sensitive about what they eat, and the belief that organic food is safer than the conventional food. According to Foodborne Diseases Active Surveillance Network of Centers for Disease Control and Prevention, in 2006 food supply was challenged by pathogens like Listeria (25% case fatality rate) and Shigella (causes Shigellosis which develops fever and stomach cramps). Since infections caused by Salmonella and E.coli are attributed to consumption of egg, poultry, meat and chicken, concern about food safety reflects itself mostly in the demand of organic eggs, poultry and meat (Akbas, 2009).

In this context, the main purpose of this study is to reveal the profile of organic food consumers and to determine consumers' organic food consumption motivations, which is a part of green marketing studies and also one of the most important aspects of green consumption. Therefore, this study primarily dealt with green consumers and green consumption, then the organic food consumption and the organic food market were evaluated and an empirical study to determine the end-consumer of organic food consumption was presented

GREEN CONSUMERS AND GREEN CONSUMPTION

Due to the developments of mass media and technology in general, economic and cultural elements have become free to move around the world without any borders. Increased economic activity, started with the industrial revolution previously, has recently intensified with the globalization process. Firms, in order to survive within this intense competition, have had to produce products with a better quality at a lower cost. This has brought a threatening situation for the world in which we live (Yurtseven & Kaya, 2010). In parallel to these developments, the firms that follow the concept of sustainability have left the aims of gaining pure profit and market share growth aside and they have more focused on a sustainable environment in which people could live in a more economic way. Especially, after the industrial revolution, natural (the destruction of the natural environment, global warming, etc.) and socio-economic (gap in income distribution, poverty, etc.) problems have called into questioning the existing understanding of doing business due to the expansion of global businesses around the world.

Within this context, the concerns have increased about its effects on wild life and the world since the 1970s. The question that whether there are enough natural resources for future generations has widely spread in the minds of consumers. As a result of environmental damage caused by production processes and environmental disasters, the current environmentalism has become a major phenomenon over the last 30-year period. For consumers, the 1960s was the 'awareness' period, the 1970s was the 'take action' period, the 1980s was the 'responsibility' while the 1990s was the 'power of consumers' in the market. Especially, the number of consumers described themselves as 'environmentalist' increased in demanding environmentally friendly products by expressing their concerns upon environment and the scarcity of natural resources towards the end of 1980s (Kalafatis et al., 1999). In this sense, consumers that thought of the scarcity of natural resources in the world turned into green consumers. Green consumers, when compared with the other consumers, are the ones that reflect more environmental concerns over their consumption. Green consumers, moreover, are interested with not only the scarcity of natural resources but also post product consumption processes (Zinkhan & Carlson, 1995).

Green consumers, compared with non-green consumers, mean that they prefer purchasing green products beyond the standard alternatives available or purchasing environmentally friendly products. Green consumers believe that they are intuitive for their internal control and open minded so that they would not leave their responsibilities to protect environment to governments, business communities, environmentalists and scientists (Shamdasani et al., 1993:491). Green consumers are the ordinary people thinking that business operations have a vital role in fighting to protect environment in general. Moreover, these consumers believe that individuals' consumption makes a difference on environmental issues. When green consumers demand products and services, firms combine environmental and social costs of their production and consumption. They also provide green management and green marketing practices by creating job opportunities in green consumption (Straughan & Roberts, 1999). Therefore, green consumers could be said that they are the individuals targeting sustainability by using their purchasing power.

The rise of the green consumption phenomenon in its historical development has pushed firms to take steps on this direction by pressurizing the consumer market. In this context, firms more focused upon the environmental issues in order to better market and to make better management decisions to compete with their rivals. For example, the companies such as 3M, DuPont and McDonald's managed to reduce their waste constantly (Menon et al., 1999). In the future, the areas will be focused like how marketing processes to be turned into green marketing, how to combine marketing strategies with the environmental issues, how to target and segment the green consumers, how to improve the production and distribution channels of ecologically oriented products and services, internal and external logistics (recycling, re-use and waste management) (Crane, 2000).

Looking at the previous academic studies in order to better understand green consumption, Bei & Simpson (1995) emphasized the importance of environmental issues and consumers being motivated by the environment. According to this, firms are required to explain to the consumers about the effectiveness of protecting the environment by purchasing green products. In another study, green consumers' purchasing characteristics and their impact of advertising strategies were examined and the findings showed that the green consumers are more informed and careful buyers (Shrum, et al., 1995). Peattie (2001) stated that green consumers are willing to pay more to purchase green products. Diamantopoulos et al., (2003) found in their study that demographic variables are weak as to determine the profile of green consumers. More specifically, female consumers participate and they are more concerned with the environment. The married couples are more likely to engage with the environment. In this study, there is a positive correlation between education, information, attitude and behavior. However, there is a negative correlation between age and attitude towards the environment. Moreover, the hypothesis looking at the relationship of participation with the green activities, environmental quality, being a member of the higher social class and having environmental information is not supported.

Overall, green consumption means avoiding from purchasing the products listed below (Hailes & Makower, 1990):

- The products that threaten individuals and community health,
- The products that harm the environment during their manufacturing, consumption or post-consumption,
- The products that require excessive resource for their production, consumption or post-consumption,
- The products that are produced with excessive packaging, unnecessary product features or polluting materials and waste due to their ready-made consumption nature,
- The products that expose abuse or mistreating animal during their production,
- The products that adversely affect other countries, especially the developing countries.

Consumers using green products are said not to be a homogenous group. Within this context, Bergin-Seers and Judith (2009) divided green consumers into three main groups as translators, exceptors and selectors. As seen in Table 1, although green consumers are divided into different groups, they all have green consumer awareness and knowledge about green products. However, selectors -as a most common group- are green or ethical in one aspect of sustainability only, exceptors have a complex understanding of a wide range of interdependent sustainability ideas, and translators' awareness usually translates into action and they gradually include more and more activities.

Table 1. Typologies of green consumers

Translators	Exceptors	Selectors
In this consumer group, awareness usually translates into action: • They feel very guilty about not doing it before • They gradually include more and more activities • Their concern is often at the level of products rather than companies or industries • They are prepared to make some sacrifices and are open to change • if they can see the impact of their actions • Their information seeking is largely passive • Word of mouth and opinion leaders are important to this group • They are uncritical of information sources	Sustainability is a priority. This group have a complex understanding of a wide range of interdependent sustainability ideas: • They are change seeking • Their information seeking is active, company level and very • critical • They are comfortable with non-mainstream outlets, products and • information sources • BUT: There is one exception to their green lifestyle: • This is usually a conscious exception • It is usually a 'small' exception • During the purchase process for this item they will completely ignore their usual green and/or ethical criteria • They will have a specific justification for this purchase which allows them to be happy with their decision	This group is green or ethical in one aspect of sustainability only - Greenpeace OR recycling OR green energy OR organic: • Probably the most common group • This would explain why green marketing fails and green marketing research can give conflicting results • Support for Peattie's notion of a context dependent portfolio of (possibly inconsistent) purchases • Could be a starting point for the other groups • Information seeking is selective, ad hoc and can be active, depending on the issue.

Source:. Bergin-Seers, Suzanne & Mair, Judith, 2009

ORGANIC FOOD CONSUMPTION AND ORGANIC FOOD MARKET

The provision of food, its preparation and consumption has constantly played an important role in human life. However, it is seen that the developments of the living conditions have changed the relationship between food and people. Food, once seen as a physical necessity and the basic element of survival, is nowadays encountered as art, hobbies, cultural items, identity, profession, social activities etc. (Hall & Mitchell, 2002).

The changing relationship between food and people finds its way in today's economic system. But, what we see is that the sustainability of present global food systems faces with a big question mark. Especially, industrial food causes environmental, socio-cultural and economical problems and these problems, suggested by the number of writers, are tried to solve by organic food and using food sources produced locally (Hall & Wilson, http://canterbury-nz.academia.edu, 2014; Sims, 2009; Li et al., 2014). Moreover, today's new consumer profile questions how food is produced from the beginning to dinner table e.g. where it comes from and how it is processed. In other words, today's individuals are more interested in the food and beverage information and locations from which they come (Kanık, 2014).

As a physiological need, everybody without any exception must eat to survive. When it is looked in this way, the eating aspect of life is very crucial in everyday's life. This brings an issue that conventional farming practices with a global identity raise many controversies related to the environment. Especially, the extraordinary changes beginning with the transition from the traditional framing practices to the conventional farming practices have caused new geographic and economic boundaries in distributing raw materials and products across the world. This practice in turn results in eliminating the seasonal product identification with availability of any product in all seasons of the year (Kanık, 2014). However, the production and distribution on a global scale have brought out many environmental problems e.g.

global warming, scarcity symptoms of food, increase in environmental pollution etc. In this situation, consumer behavior proves once more its importance.

During and post conventional production of food reveals some problems that bring out the concept of organic food and organic food production. Organic production is defined as the ecological production management system targeting biodiversity, biological life cycle and development of biological activities in the soil. Organic production is also based upon the management practices aiming at protecting ecological harmony, its implementation and developments with use of off-farm inputs at the lowest possible level (Winter & Davis, 2006). Gottschalk & Leistner (2013) stated that organic products do not contain contaminants of organic products; therefore, they have a positive impact on health and the environment.

The first stage of organic food production involved the small farms producing locally and distributing these fresh farm products daily. As for today, organic food production has become a large and multi-dimensional system comprising of small and big food producers with local and global distribution networks producing fruit and vegetables, meat and animal food products as well as processed food (Winter & Davis, 2006). The reasons behind the rapid rise of the organic food consumers are that first, consumers feel more confident on organic food and second, concerns of conventional product production gradually increase as they put an impact on the environment and human health. In recent years, the food crisis such as BSE and food and mouth disease has changed the attitude towards food products in general. Especially, conventionally produced products containing pesticides, chemicals and antibiotics and other chemicals used during the production have shaken the consumers' confidence (Winter & Davis, 2006).

The resulted health problems have contributed to the growth of organic agricultural production. The health problems experienced in the world have also created some variations on consumer preferences and they started preferring more natural and organic food products. In particular, the consumers' preferences are created by the fact that the food products are chemical free agricultural products and the way they are produced is healthy and natural for consumers. Due to the health problems, the demand for organic products increased and the rise of production for organic food, in turn, was observed. However, the availability of agricultural land for organic farming is limited and this determines agricultural policies to be applied. This even shapes production techniques for organic production accordingly. Especially, the producers use chemicals in order to achieve maximum product from agriculture land available. To manage this, they use fertilizers and pesticides onto food products. These used chemicals and pesticides cause serious human health problems as well as irreversible environmental problems.

The growing importance of human health and the environment causes consumers to prefer more organic products. The organic product market is divided into two main groups regarding the consumer characteristics. The first market group is the health market as being the largest one. The consumers in this market purchase organic products as they find these products healthier than the conventional products. The consumers in the second market group prefer organic products due to the reasons of social effects to their lives. For example, a cleaner environment, less harmed agricultural land and locally produced products are important for these consumers (Knudson, 2007). Indeed, McIver (2004), in his study, found that the consumers purchased organic products due to the reasons that organic products are good for themselves as well as for the environment, and some specific health problems and limited use of agricultural fertilizers and pesticides contributed for consumers' preferences on organic products.

Another important factor influencing the choice of consumers' organic products is their norms. A norm could be explained as 'a person taking on board some responsibilities in order to change the present situation and predicting harmful results to be caused by their actions in his/her life'. The studies indicate that individuals' attitudes are based upon a moral reasoning in general (Honkanen et al., 2006). Consumers' choice for organic products includes the models of moral and emotional factors, individual's belief and values, life styles, value-driven consumer behavior and purchasing behavior. Vegetarian and vegan food preferences are the good examples for the values, beliefs and principles of consumers (Kihlberg & Risvik, 2007).

Williams & Hammitt (2000) compared the organic product consumers with the conventional product consumers and they found that those who purchased more organic products than convential products displayed more environmentally friendly behaviors. In addition, the purchasers of organic products showed willingness to pay more for organic products in order to reduce the perceived food safety risk. However, while consumption is an individual act, it leads to a direction affecting the social environment of that individual. The studies show that there is a strong and direct link between social identity and organic food consumption. This is an important connection between social identity and product selection. This is because those individuals saying that they would not harm the environment and that they would use healthy products (Bartels & Onwezen, 2014).

Lee and Worsley (2005) in their research on consumers in Australia found that the organic products are healthier and more delicious than the conventional products and also friendly to the environment. However, the organic products are more expensive and not being available in everywhere is considered as the reasons of the organic products are not consumed frequently. In addition, women in the study were found to display more positive attitudes about organic products than men.

Another study was conducted to research the consumers' decision making process to purchase organic products in Italy in 2008 (Magistris & Gracia, 2008). It was concluded that the consumers' attitudes towards organic products are one of the determining factors in purchasing organic product process. It was also found that the consumers should be better informed about organic products in order to manage them to display a positive attitude towards organic products. The consumers that prefer a healthy diet or balanced nutrition programme exhibit positive attitudes towards organic products when compared to the other consumers, and this finding should be noted as one of the important findings of the research (Magistris & Gracia, 2008). The demographic characteristics of consumers also lead to differences in the attitudes and behavior towards organic food consumption. In a related work, it was found that women exhibit a more positive approach to organic food than men and women are also seen that they purchase and consume more organic products than men (Storstad & Bjorkhaug, 2003). When the consumers' income increase, they purchase more organic products (Turjusen et al., 2001), the consumers that buy organic food are better educated than those who do not buy frequently (Storstad & Bjorkhaug, 2003). Within this context, first, the factors that affect organic food consumption in a country, then, its strategies are required to be adjusted accordingly.

Organic food consumption is a part of the green consumer behavior. The studies since 1970s have showed that consumers' growing awareness about the environment affects the attitude towards environmentally friendly products and this in turn increases the demand for organic products (Kanchanapibul et al., 2014). Organic agriculture and organic food are becoming more important against the potential damages caused by conventional agriculture and conventional products. In recent years, new rules and new regulations were introduced about the environment, then, the increase in the number of organic consumers was considered as the outcome of these new rules and new regulations (Gottschalk & Leistner, 2013; Giannakas, 2002; Padel & Foster, 2005). The academic studies showed that the organic agriculture sector was more developed and the incentives increased towards the investments made by individuals in this area (Costa et al., 2014; Giannakas, 2002; Wier & Calverley, 2002; Fotopoulos & Krystallis, 2002; Batte et al., 2007; Jang et al., 2011). Moreover, it was noticed that the organic food market is one of the fastest growing sector in most developed and developing countries. The organic food consumption behavior, which was pioneered in North America and Western Europe, remains the consumers' popularity throughout the world. In addition, organic products have recently taken their places in the shelves of supermarkets and their market share has in turn increased and they are now directly in competition with the conventional products (Gottschalk & Leistner, 2013).

In the last 20 years, organic food market experienced an enormous growth worldwide because of the rising consumer awareness and demand for organic food. The global market for organic food and beverages was valued as 27.8 bn US dollar in 2004. But in 2012 only United States organic food sales generated nearly 31.32 bn US dollars with %11.5 annual sales growth (www.statista.com, 2016). According to the organic market report prepared by Soil Association, (2016), the organic product market grew at 4.9% in the UK and the revenue from the organic product market has nearly reached up to 2 billion Sterling Pounds annually (Soil Association, 2016). Moreover, IFOAM (2008) stated that the demand for organic products across Europe has increased beyond the existing production capacity. Therefore, producers in the developing countries would have significant opportunities to level this increased demand for organic products in Europe. European Union organic food sector was calculated as the retail value of around 25-35 billion Euros in 2005. This indicates that the organic product market is no longer a niche market segment and it looks like gaining a main stream market status overall. In 2002, 130 countries in the world were active in organic agricultural production. 90 out of 130 countries having the extremely favorable environmental conditions for developing organic farming and production were the developing countries. This rapid growth in organic production market requires the issues like training employees in the agricultural sector, consultancy and information services to farmers, auditing and certification procedures, the reforms in the organic supply chain etc. (Fotopoulos & Krystallis, 2002). Lockie et al., (2002: 23) stated that 'success of strategies to create and stabilize organic food market will be dependent on the ability to mobilize people as 'organic consumers' by providing foods that materially and symbolically satisfy and/or influence those peoples' needs, desires and pleasures'. This overall emphasizes the necessity of understanding organic food consumers well.

METHODOLOGY

Research Instrument

The main objective of the study is to reveal the profile of the organic food consumers and to determine motivations for organic food consumption. As a sub-objective, the study is to look at the factors affecting organic food consumption. In the data collection, the questionnaire technique was used. The convenience sampling method was used and the survey questionnaires were sent to the respondents through internet. The respondents were included in the study sample on condition that they have consumed organic food before. The questionnaire consisted of 3 parts. The demographic questions were in the first part. The questions of organic food consumption behavior were in the second part. The last part of the questionnaire consisted of the statements assessing the respondents' organic food consumption motivations. For the last part of the questionnaire, the study variables were measured with a 5-point agree/disagree Likert scale.

The questionnaire was formed through examining the previous studies and questionnaires in the literature (Lee & Yun, 2015; Lindeman & Vaananen, 2000; Yadav, & Pathak, 2016; Pomsanam, et al., 2014; Mohammad, et al., 2014). An 18-item scale was developed. The data were collected on internet from March 20016 to May 2016. The questionnaire survey targeted the respondents living in Turkey and 393 fully answered questionnaires were obtained from the respondents. The sample size is generally accepted and sufficient with 5% error and 5% confidence level (Yazicioglu & Erdogan, 2004). The scale of importance for the study was found as being, 807 through Cronbach's Alpha values. This suggests a high level of confidence for the questionnaire (Tabachnick & Fidel 2007).

As seen in Table 2, the measurement instrument was presented. According to this, the most important factors in the respondents' organic food consumption motivations were numbered as 1, 3, 7 and the lowest level motivational factors were itemized as 2, 4 and 15. In addition, the average value of all the expressions was calculated as 3.92.

RESULTS

Sample Profile

According to Table 3, of the respondents, most respondents were male (57.5%); two third were married (65.1%); one third have bachelor's degree (35.6%). Although there is not a huge difference between the two groups of the respondents proportionally, 60.1% of the respondents' monthly personal income is less than 1500 USD. A similar situation is seen at the profession of the respondents. By looking at the

Table 2. Measurement instrument

No.	Items	Mean*	SD
1	Organic food keeps me healthy	4,14	,824
2	Organic food has been produced in a way that animals' rights have been respected	3,40	1,242
3	Organic food product does not have any side effects when eaten	4,25	,730
4	Organic food has been produced in a way that animals have not experienced pain	3,64	1,195
5	Organic food contains a lot of vitamins and minerals	4,07	,835
6	Organic food is packaged in an environmentally friendly way	3,83	1,017
7	Organic food has a pleasant texture	4,21	,618
8	Organic food tastes good	4,12	,732
9	Organic food looks nice	3,89	,889
10	Organic food is high in protein	4,06	,773
11	Organic food is nutritious	4,09	,874
12	Organic food contains no additives	3,88	,947
13	Organic food supports local communities	3,96	,952
14	Organic food contains no artificial ingredients	3,98	,898
15	Organic food supports social justice	2,99	1,393
16	Organic food contains natural ingredients	4,13	,861
17	Organic food supports traditional farming practices	4,02	,843
18	Organic food has been produced in a way which has not shaken the balance of nature	3,82	1,079

*Mean measured with a 5-point agree/disagree Likert scale.

Table 3. Demographic profile of respondents (N = 393)

Category		n	%
Gender	Female Male	167 226	42.5 57.5
Marital Status	Married Single	256 137	65.1 34.9
Age	15-24 25-34 35-44 45-54 55-64 65 +	92 83 79 55 38 46	23.4 21.1 20.1 14.0 9.7 11.7
Education	Secondary High School Vocational School Bachelor Graduate Post Graduate	20 122 94 140 16 1	5.1 31.0 23.9 35.6 4.1 .3
Occupation	Student Retired Student Officer Housewife Worker Professional Other	59 97 29 54 49 34 37 34	15.0 24.6 7.4 13.7 12.5 8.7 9.4 8.7
Monthly Personal Income (USD)	0-500 501-1000 1001-1500 1501-2500 2501-3000 3000 +	74 92 73 57 50 47	18.8 23.4 18.6 14.5 12.7 11.9
Children	Yes No	126 267	32.1 67.9

distribution of the professions, the study found as retired (24.6%), students (15%), officers (civil servants) (13.7%) and housewives (12.5%). In addition, 32.1% of the respondents have children

Organic Food Consumption Behavior

In the second part of the questionnaire, three variables were used to determine the level of organic product consumption (Table 4). Here, the frequency of organic food consumption, purchased organic food types and from where they were bought were questioned. As seen in Table 4, the majority of the consumers (28.2%) stated that they consume organic food at least once a week. 24.4% of the respondents consume organic food more than once a week and 21.1% of the respondents only consume organic food once a month. The respondents had the choices to mark more than one choice in the organic food categories and the places that organic food are sold. In this context, the organic food that the respondents were most likely to purchase is organic eggs (71.7%). The second most preferred organic food is fruit (62.3%) and vegetables (59%). The consumers prefer purchasing organic food from supermarkets (67.4%).

Table 4. Organic food consuming behavior

Items	n	%
Frequency of Organic Food Purchase		
Everyday	33	8.4
Several times a week	96	24.4
Once a week	111	28.2
Once in several weeks	65	16.5
Once a month	83	21.1
Once in several months	5	1.3
Purchased Organic Food Categories		
Eggs	282	71.7
Fruit	245	62.3
Vegetables	232	59
Milk and dairy products,	180	45.8
Meat and poultry	163	36.4
Bread and cereal products	125	31.9
Place of Purchase		
Supermarkets	265	67.4
Public Bazaars	214	54.5
Organic markets	183	46.6
Health shops	142	36.1
Organic Farms	113	28.8

The frequency of the demographic data of the respondents, which consisted of marital status, gender and having children, and organic food consumption were tested by chi-square. The relationship between organic food consumption frequency and consumers' having children is shown in Table 5. According to the chi-square test, this relationship is significant (X^2 =35,245, p<0.001). As seen in Table 5, 17.5% of the respondents that have children consume organic food every day, whereas 4.9% of the respondents with no children only consume organic food every day. Similarly, 34.1% of the respondents with children consume organic food several times a week while 19.9% of the respondents with no children consume organic food several times a week. As the time frequency becomes longer, then, the category of the respondents with no children is observed to increase organic food consumption.

When we look at the relationship between marital status and consumers' organic food consumption Frequency (Table 6). It could be said that the married respondents consume organic food more often than the single respondents (X2 = 114.932, p <0.001). When the groups are compared among themselves, 35.9% of the married group several times a week and 33.2% of the same married group once a week consume organic food, while 42.3% of the single group once a month and 29.2% of the same single group once in two weeks consume organic food. According the gender variable, there is no significant difference between women and men in the frequency consumption (X^2 =4,478, p>0.05).

Motivation factors leading to consume organic food: The principal component analysis with varimax rotation method was used for 19 items listed in the study questionnaire. Before the factor analysis, the

Table 5. Comparison of consumption frequency and having children

			Children		Total
			Yes	No	
Organic Food Consumption Frequency	Everyday	Count	22	11	33
		% within Children	17.5%	4.1%	8,4%
	Several times a week	Count	43	53	96
		% within Children	34.1%	19.9%	24,4%
	Once a week	Count	24	87	111
		% within Children	19.0%	32.6%	28,2%
	Once in two weeks	Count	16	49	65
		% within Children	12.7%	18.4%	16,5%
	Once a month	Count	20	63	83
		% within Children	15.9%	23.6%	21,1%
	Once in several months	Count	1	4	5
		% within Children	.8%	1.5%	1,3%
Total		Count	126	267	393
		% within Children	100.0%	100.0%	100.0%

X^2 =35,245; df: 5; p: .000

Table 6. Comparison of consumption frequency and marital status

			Marital Status		Total
			Married	Single	
Organic Food Consumption Frequency	Everyday	Count	24	9	33
		% within Marital Status	9.4%	6.6%	8.4%
	Several times a week	Count	92	4	96
		% within Marital Status	35.9%	2.9%	24.4%
	Once a week	Count	85	26	111
		% within Marital Status	33.2%	19.0%	28.2%
	Once in two weeks	Count	25	40	65
		% within Marital Status	9.8%	29.2%	16.5%
	Once a month	Count	25	58	83
		% within Marital Status	9.8%	42.3%	21.1%
	Once in several months	Count	5	0	5
		% within Marital Status	2.0%	.0%	1.3%
Total		Count	256	137	393
		% within Marital Status	100.0%	100.0%	100.0%

X^2 =114,932; df: 5; p: .000.

suitability of the data for the analysis was evaluated by using the Kaiser Meyer Olin (KMO) and Barlett's test of sphericity. The KMO measure of sampling adequacy is .796 that exceeds the suggested threshold value of 0.60 (Tabachnick & Fidel 2007). Barlett's test of sphericity was significant (X^2 =2233,915, p<0.001), indicating that the between-item correlations were sufficiently large for principal component analysis. Four factors affecting organic food consumption were found. Factor loadings range from 502 to 866. All the factors are above Cronbach's alpha 0.70. This shows that the limit values over 0.70 for reliability (Hair et al., 2006). As a result of Factor analysis, four motivation factors obtained from the study show 57.79% of the total variance. According to the factors containing the items, the groups are named as Health, Ecological and Social welfare, Sensory Appeal and Natural Content. As seen in Table 7, the factors are explained by the variances of 20,090%, 18,436%, 10,207% and 8,532%.

By examining the descriptive data, the total scale mean is revealed as 3,933. When we look at the descriptive statistics of the factors, the Health factor has the highest arithmetic mean as 4.12. This is

Table 7. Factor analysis

	Factors			
	Health	**Ecological and Social Welfare**	**Sensory Appeal**	**Natural Content**
Organic food contains a lot of vitamins and minerals	.866			
Organic food keeps me healthy	.793			
Organic food is nutritious	.782			
Organic food is high in protein	.754			
Organic food product does not have any side effects when eaten	.749			
Organic food has been produced in a way which has not shaken the balance of nature		.833		
Organic food is packaged in an environmentally friendly way		.712		
Organic food has been produced in a way that animals have not experienced pain		.664		
Organic food has been produced in a way that animals' rights have been respected		.661		
Organic food supports traditional farming practices		.659		
Organic food supports local communities		.590		
Organic food supports social justice		.571		
Organic food looks nice			.845	
Organic food has a pleasant texture			.809	
Organic food tastes good			.648	
Organic food contains no additives				.711
Organic food contains natural ingredients				.638
Organic food contains no artificial ingredients				.502
Cronbach Alpha	.865	.798	.752	.747
% of Variance Explained	20.090	18.436	10.207	8.532
Mean	4.1208	3.6972	4.0681	3.9947

% Total Variance Explained: 57.779 KMO: .766
Bartlett's Test of Sphericity: 2233,915 p: .000

followed by the Sensory Appeal factor with 4.06. The Natural Content factor with 3.99 is placed in the third rank. Finally, the Ecological and Social Welfare factor comes with the value of 3.69.

The analysis made above by the authors of this study shows that there is a frequency difference between having children and organic food consumption ($p<0.001$). With the application of t-test, it was looked at whether there was similar differences in factor groups and it was found that there is no significant differences between Ecological, Social Welfare, Sensory Appeal and Natural Content factors at all ($p>0.05$). According to the analysis, the Health factor only has a significant difference ($p<0.05$). Table 8 shows the comparison between Health and Having Children factors. It was found that there is no significant difference between having children and having no children according to the Health factor ($t=2,68$, $p<0.05$). It was also found that the respondents with children ($\overline{x}=4,26$) give higher importance to the health factor than the respondents with no children ($\overline{x}=4,06$) when they consume organic food. The similar analyses were made for marital status and it was found that there was no statistically significant difference.

CONCLUSION

The environmental, socio-cultural and economic problems caused by the global economic system and doing business in a traditional way have pushed consumers to look for different sources during their purchasing processes. In the 1960s, the search named as awareness move has turned the traditional consumers into the green consumers (Kalafatis, et al., 1999; Zinkhan & Carlson, 1995). Therefore, today, the business philosophy benefiting from the environment at a maximum level is not welcome by the existing consumers (Yurtseven & Kaya, 2010).

Organic production processes comprise of avoiding the use of synthetic chemicals and rejecting any management practices endangering human health and environment. Moreover, organic food and production processes should not only support the local farmers and regional development but they should also have the social responsibilities of the green marketing. With this aspect of organic food production, it should further contribute to human health as well as socio-economic development in the area. In this study, therefore, organic food consumption behavior as one of the most important applications of green marketing has been researched.

Various academic studies about organic food defined as one of green marketing elements have been researched (Gottschalk & Leistner, 2013; Storstad & Bjorkhaug, 2003; Winter & Davis 2006; Mohammad, et al., 2014; Yadav & Pathak, 2016). By comparing the previous studies with this study's findings, it was found that the motivation factor is consistent with these earlier studies. The Factor analysis has the highest mean (4.12) and therefore, the Health factor was found as the basic consumer motivation in this study. In order to support this with the literature, several studies have found the Health factor as the most important motivational tool (Hjelmar, 2011; Chryssohoidis and Krystallis, 2005).

Table 8. Comparison of having children and health factors

Children		N	Mean	Std. Deviation	df	t	p
Health	Yes	126	4.2600	.49404	369	2.683	0.008
	No	267	4.0621	.70382			

Consumers find that organic food tastes better than conventional food and they also have a better appearance and texture because they grow within their own seasons (Kihlberg & Risvik, 2007). In this study, the factor named as Sensory Appeal shows the perceived food quality dimensions and it was found as the second most important factor for organic food consumers.

In general, Organic food refers to food that does not include the use of chemicals and artificial contents with conventional methods. With this aspect, organic food's basic feature is its natural content. This feature is widely accepted by consumers and it is one of the purchasing factors (Lee & Yun, 2015). Fotopoulos et al., (2003) state that natural ingredients with less use of pesticides and chemicals affect the organic food purchasing behavior and this also is a function of organic food consumption by making organic ingredients available. According to the results of the study, the naturalness possessed by organic food affects the organic food consuming behavior.

Most of the studies in the literature show that there is a direct relationship between organic food consumption and environmental and sociological welfare (Padel & Foster, 2005; Verhoef, 2005; Gracia & Magistris, 2007; Chen, 2009). However, Chryssohoidis & Krystallis (2005) found that there is no connection between the Greek consumers purchasing organic food and the environment. Similarly, Baker et al. (2004) also found that the British consumers do not have any connection between organic food and the environment. However, the German consumers see a connection between organic food and the environment. In this study, the environment and social welfare is the least important factor in organic food consumption behavior.

The studies of green consumption and purchasing behavior show that demographic factors like age, gender and education etc. do not have any relationship with consumption or they might have a limited relationship with consumption (Gracia & de Magistris, 2007; Lea & Worsley, 2005). Within this aspect, this study looks at the frequency of organic food consumption between having children and marital status. It was found that the married respondents consume more frequently organic food than the single respondents (X^2 =114,932, p<0.001). A similar relationship was found in the having children variable in this study, and the having children group consume more organic food than the group with no children (X^2 =35,245, p<0.001). This is consistent with the results of the earlier studies (McEachern & Willock 2004; Freyer & Haberkorn, 2008).

The identification of key elements related to the organic food consumption and any efforts towards marketing and production will contribute towards the sustainability approach. In addition, in order to meet the growing demand for organic food markets, marketers and policy makers should find out consumers' profiles, motivations and preferences to create a successful strategy (Lee & Yun, 2015). Therefore, organic food retailers should adapt the results of this study in their marketing practices in order to operate successfully in the consumers' markets.

REFERENCES

Akbas, S. M. (2009). *Should Old Technologies Be Banned: The Case Of The Organic Food Market* (Unpublished M.A. Thesis). Sabancı University, Istanbul.

Bahro, R. (1986). *Building the Green Movement*. London: Heretic Books.

Baker, S., Thompson, K. E., Engelken, J., & Huntley, K. (2004). Mapping the values drivingorganic food choice: Germany vs the UK. *European Journal of Marketing*, *38*(8), 995–1012.

Bartels, J., & Onwezen, M. C. (2014). Consumers willingness to buy products with environmental and ethical claims: The roles of social representations and social identity. *International Journal of Consumer Studies*, *38*(1), 82–89. doi:10.1111/ijcs.12067

Batte, M. T., Hooker, N. H., Haab, T. C., & Beaverson, J. (2007). Putting their money where their mouths are: Consumer willingness to pay for multi-ingredient, processed organic food products. *Food Policy*, *32*(2), 145–159. doi:10.1016/j.foodpol.2006.05.003

Bei, L., & Simpson, E. M. (1995). The determinants of consumers' purchase decisions for recycled products: An application of acquisition-transaction utility theory. *Advances in Consumer Research. Association for Consumer Research (U. S.)*, *22*(1), 257–261.

Bergin-Seers, S., & Judith, M. (2009). Emerging green tourist in Australia: Their behaviours and attitudes. *Tourism and Hospitality Research*, *9*(2), 109–119. doi:10.1057/thr.2009.5

Chamorro, A., Rubio, S., & Miranda, F. J. (2009). Characteristics of research on green marketing. *Business Strategy and the Environment*, *18*(4), 223–239. doi:10.1002/bse.571

Chen, M. F. (2009). Attitude toward organic foods among Taiwanese as related to health consciousness, environmental attitudes, and the mediating effects of a healthy lifestyle. *British Food Journal*, *111*(2), 165–178. doi:10.1108/00070700910931986

Chryssohoidis, G. M., & Krystallis, A. (2005). Organic consumers personal values research: Testing and validating the list of values (LOV) scale and implementing a value-based segmentation task. *Food Quality and Preference*, *16*(7), 585–599. doi:10.1016/j.foodqual.2005.01.003

Cope, D., & Winward, J. (1991). Information failures in green consumerism. *Consumer Policy Review*, *1*(2), 83–86.

Costa, C., García-Lestón, J., Costa, S., Coelho, P., Silva, S., Pingarilho, M., ... Teixeira, J. (2014). Is organic farming safer to farmers health? A comparison between organic and traditional farming. *Toxicology Letters*, *230*(2), 166–176. doi:10.1016/j.toxlet.2014.02.011 PMID:24576785

Crane, A. (2000). Facing the Blacklash: Green marketing and strategic reorientation in the 1990s. *Journal of Strategic Marketing*, *8*(3), 277–296. doi:10.1080/09652540050110011

Diamantopoulos, A., Bodo, B. S., Rudoldf, R. S., & Greg, M. B. (2003). Can socio-demographics still play a role in profiling green consumers? A Review of the evidence and an emprical investigation. *Journal of Business Research*, *56*(6), 465–480. doi:10.1016/S0148-2963(01)00241-7

Elkington, J., Hailes, J., & Makower, J. (1990). *The Green Consumer*. London: Penguin Books.

Fotopoulos, C., & Krystallis, A. (2002). Organic product avoidance. Reasons for rejection and potential buyers identification in a countrywide survey. *British Food Journal*, *104*(3-5), 233–260. doi:10.1108/00070700210425697

Fotopoulos, C., Krystallis, A., & Ness, M. (2003). Wine produced by organic grapes in Greece: Using means-end chains analysis to reveal organic buyers purchasing motives in comparison with the non-buyers. *Food Quality and Preference*, *14*(7), 549–566. doi:10.1016/S0950-3293(02)00130-1

Freyer, B., & Haberkorn, A. (2008). Influence of young children (3-6 years) on organic food consumption in their families. *16th IFOAM Organic World Congress; Cultivating the Future Based on Science*.

Giannakas, K. (2002). Information asymmetries and consumption decisions in organic food product markets. *Canadian Journal of Agricultural Economics*, *50*(1), 35–50. doi:10.1111/j.1744-7976.2002.tb00380.x

Gottschalk, I., & Leistner, T. (2013). Consumer reactions to the availability of organic food in discount supermarkets. *International Journal of Consumer Studies*, *37*(2), 136–142. doi:10.1111/j.1470-6431.2012.01101.x

Gracia, A., & de Magistris, T. (2007). Organic food product purchase behaviour: A Pilot study for urban consumers in the South of Italy. *Spanish Journal of Agricultural Research*, *5*(4), 439–451. doi:10.5424jar/2007054-5356

Gracia, A., & de Magistris, T. (2007). Organic food product purchase behaviour: A pilot study for urban consumers in the South of Italy. *Spanish Journal of Agricultural Research*, *5*(4), 439–451. doi:10.5424jar/2007054-5356

Hall, C. M., & Mitchell, R. (2002). The changing nature of the relationship between cuisine and tourism in Australia and New Zealand: From fusion cuisine to food Networks. In A. Hjalager & G. Richards (Eds.), *Tourism and Gastronomy* (pp. 186–207). London: Routledge.

Hall, C. M., & Wilson, S. (2014). *Scoping paper: Local food, tourism and sustainability*. Retrieved 01.11.2014, from http://canterbury-nz.academia.edu/CMichaelHall

Hjelmar, U. (2011). Consumers purchase of organic food products: A matter of convenience and reflexive practices. *Appetite*, *56*(2), 336–344. doi:10.1016/j.appet.2010.12.019 PMID:21192997

Honkanen, P., Verplanken, B., & Olsen, S. O. (2006). Ethical values and motives driving organic food choice. *Journal of Consumer Behaviour*, *5*(5), 420–431. doi:10.1002/cb.190

IFOAM. (2008). *Organic directory online*. Bonn, Germany: International Federation of Organic Agriculture Movements.

Islam, S., Munasġnghe, M. N., & Clarke, M. (2003). Making longterm economic growth more sustainable: Evaluating the costs and benefits. *Ecological Economics*, *47*(2-3), 1–19. doi:10.1016/S0921-8009(03)00162-9

Jang, Y. J., Kim, W. G., & Bonn, M. A. (2011). Generation Y consumers selection attributes and behavioral intentions concerning green restaurants. *International Journal of Hospitality Management*, *30*(4), 803–811. doi:10.1016/j.ijhm.2010.12.012

Kalafatis, S. P., Pollard, M., East, R., & Tsogas, M. H. (1999). Green marketing and Ajzens theory of planned behaviour: A Cross-market examination. *Journal of Consumer Marketing*, *16*(5), 441–460. doi:10.1108/07363769910289550

Kanchanapibul, M., Lacka, E., Wang, X., & Chan, H. K. (2014). An empirical investigation of green purchase behaviour among the young generation. *Journal of Cleaner Production*, *66*(1), 528–536. doi:10.1016/j.jclepro.2013.10.062

Kanik, İ. (2014). Yerel beslenme hareketi ve bir kitle iletişim anlatisi dizayni: Amerika Birleşik Devletleri (ABD) örneği. *Journal of Tourism and Gastronomy Studies*, *2*(4), 4–22.

Kihlberg, I., & Risvik, E. (2007). Consumers of organic foods. Value segments and liking of bread. *Food Quality and Preference*, *18*(3), 471–481. doi:10.1016/j.foodqual.2006.03.023

Kirkpatrick, D. (1990). Environmentalism: The new crusade. *Fortune*, *121*(February), 44–51.

Knudson, W. A. (2007). *The organic food market* (Working Paper 01-0407). Michigan State University, Product Center: Business Innovation in Agriculture, Food and Natural Resources, The Strategic Marketing Institute.

Kotler, P., & Zaltman, G. (1971). Social marketing: An approach to planned social change. *Journal of Marketing*, *35*(3), 3–12. doi:10.2307/1249783 PMID:12276120

Lea, E., & Worsley, T. (2005). Australians organic food beliefs, demographics and values. *British Food Journal*, *107*(11), 855–869. doi:10.1108/00070700510629797

Lee, H., & Yun, Z. (2015). Consumers perceptions of organic food attributes and cognitive and affective attitudes as determinants of their purchase intentions toward organic food. *Food Quality and Preference*, *39*, 259–267. doi:10.1016/j.foodqual.2014.06.002

Li, P. A. E., Bruwer, J., & Lyons, G. (2014). Exploring consumer motivations towards buying local fresh food products. *British Food Journal*, *116*(10), 1533–1549. doi:10.1108/BFJ-04-2013-0083

Lindeman, M., & Vaananen, M. (2000). Measurement of ethical food choice motives. *Appetite*, *34*(1), 55–59. doi:10.1006/appe.1999.0293 PMID:10744892

Lockie, S., Lyons, K., Lawrence, G., & Mummery, K. (2002). Eating green: Motivations behind organic food consumption in Australia. *European Society for Rural Sociology*, *41*(1), 23–40. doi:10.1111/1467-9523.00200

McDonald, S., & Oates, C. J. (2006). Sustainability: Consumer perceptions and marketing strategies. *Business Strategy and the Environment*, *15*(3), 157–170. doi:10.1002/bse.524

McEachern, M., & McClean, P. (2002). Organic purchasing motivations and attitudes: Are they ethical? *International Journal of Consumer Studies*, *26*(2), 85–92. doi:10.1046/j.1470-6431.2002.00199.x

McEachern, M. G., & Willock, J. (2004). Producers and consumers of organic meat: A focus on attitudes and motivations. *British Food Journal*, *106*(7), 534–552. doi:10.1108/00070700410545737

Mclver, H., (2004, February). Organic hip: popular picks at health food stores. *Better Nutrition,* 58-61.

Mebratu, D. (1998). Sustainability and sustainable development: Historical and conceptual review. *Environmental Impact Assessment Review*, *18*(6), 493–520. doi:10.1016/S0195-9255(98)00019-5

Menon, A., Menon, A., Chowdhury, J., & Janovich, J. (1999). Envolving paradigm for environmental sensitivity in marketing programs: A Synthesis of theory and practice. *Journal of Marketing Theory and Practice*, *7*(2), 1–12. doi:10.1080/10696679.1999.11501825

Millock, K., & Nauges, C. (2014). Household behaviour and food consumption. In *OECD, Greening Household Behaviour: Overview from the 2011 Survey* (Revised edition). Paris: OECD Publishing. doi:10.1787/9789264214651-10-en

Mohamada, S. S., Rusdi, S. D., & Hashim, N. H. (2014). Organic food consumption among urban consumers: Preliminary results. *Procedia: Social and Behavioral Sciences, 130*, 509–514. doi:10.1016/j.sbspro.2014.04.059

Padel, S., & Foster, C. (2005). Exploring the gap between attitudes and behaviour understanding why consumers buy or do not buy organic food. *British Food Journal, 107*(8), 606–625. doi:10.1108/00070700510611002

Peattie, K. (2001). Golden gooseor wild goose? The hunt for the green consumer. *Business Strategy and the Environment, 10*(4), 187–199. doi:10.1002/bse.292

Pomsanam, P., Napompech, K., & Suwanmaneepon, S. (2014). Factors driving thai consumers intention to purchase organic foods. *Asian Journal of Scientific Research, 7*(4), 434–446. doi:10.3923/ajsr.2014.434.446

Sartori, S., Latrônĝco, F., & Campos, L. M. S. (2014). Sustainability and sustainable development: A taxonomy in the field of literature. *Ambiente & Sociedade, 17*(1), 1–22. doi:10.1590/1809-44220003490

Shamdasani, P., Chon-Lin, G. O., & Richmond, D. (1993). Exploring green consumers in an oriental culture: Role of personal and marketing mix factors. *Advances in Consumer Research. Association for Consumer Research (U. S.), 20*, 488–493.

Shrum, L. J., McCarty, J. A., & Lowery, T. M. (1995). Buyer characteristics of the green consumer and their implication for advertising strategy. *Journal of Advertising, 24*(2), 71–90. doi:10.1080/00913367.1995.10673477

Sims, R. (2009). Food, place and authenticity: Local food and the sustainable tourism experience. *Journal of Sustainable Tourism, 17*(3), 321–336. doi:10.1080/09669580802359293

Soil Association. (2016). *Organic Market Report*. Edinburgh, UK: Author.

StatistaR. (2016). Retrieved from http://www.statista.com/statistics/196962/organic-food-sales-growth-in-the-us-since-2000/

Storstad, O., & Bjorkhaug, H. (2003). Foundations of production and consumption of organic food in Norway: Common attitutes among farmers and consumers. *Agriculture and Human Values, 20*(2), 151–163. doi:10.1023/A:1024069627349

Straughan, R. D., & Roberts, J. A. (1999). Environmental segmentation alternatives: A look at green consumer behavior in the new millennium. *Journal of Consumer Marketing, 16*(6), 558–575. doi:10.1108/07363769910297506

Tabachnick, B. G., & Fidell, L. S. (2007). *Using Multivariate Statistics* (5th ed.). New York: HarperCollins.

Turjusen, H., Lieblein, G., Wandel, M., & Francis, A. C. (2001). Food system orientation and quality perception among consumers and producer of organic food in Hedmark country, Norway. *Food Quality and Preference, 12*(3), 207–216. doi:10.1016/S0950-3293(00)00047-1

Verhoef, P. C. (2005). Explaining purchases of organic meat by Dutch consumers. *European Review of Agriculture Economics*, *32*(2), 245–267. doi:10.1093/eurrag/jbi008

Wier, M., & Calverley, C. (2002). Market potential for organic foods in Europe. *British Food Journal*, *104*(1), 45–62. doi:10.1108/00070700210418749

Williams, P., & Hammitt, J. (2000). A Comparison of organic and conventional fresh produce buyers in Boston area. *Risk Analysis*, *20*(5), 735–746. doi:10.1111/0272-4332.205066 PMID:11110219

Winter, C. L., & Davis, S. F. (2006). Organic foods. *Journal of Food Science*, *71*(9), 117–124. doi:10.1111/j.1750-3841.2006.00196.x

Yadav, R., & Pathak, G. S. (2016). Intention to purchase organic food among young consumers: Evidences from a developing nation. *Appetite*, *96*, 122–128. doi:10.1016/j.appet.2015.09.017 PMID:26386300

Yazicioğlu, Y., & Erdoğan, S. (2004). *SPSS Uygulamalı Bilimsel Araştırma Yöntemleri*. Ankara: Detay Yayıncılık.

Yurtseven, H. R., & Kaya, O. (2010). Could slow business be an alternative to obtain sustainability? *6th International Symposium on Business Administration: Global Economic Crisis and Changes-Resstructuring Business System: Strategic Perspectives for Local, National and Global Actors*. University in Opava.

Zinkhan, G. M., & Carlson, L. (1995). Green advertising and the reluctant consumer. *Journal of Advertising*, *24*(2), 1–6. doi:10.1080/00913367.1995.10673471

KEY TERMS AND DEFINITIONS

Green Consumer: The types of consumers preferring products do not harm the eco-system, human and animal health and socio-economic system.

Green Marketing: A holistic management process is responsible for meeting consumers' needs with profit and sustainability.

Green Product: Green product is a product, which is non-polluting, consuming less natural resources and recyclable.

Organic Food: Food with no genetically modified and no pesticides, chemicals and additives during their growing process.

Organic Production: An ecological production system aiming at developing biodiversity, biological life cycle and biological activities in the soil.

Sustainability: A process transferring economic, socio-cultural and environmental resources to the next generation.

Traditional Business: A way of doing business without any sustainability concern.

This research was previously published in Green Marketing and Environmental Responsibility in Modern Corporations edited by Thangasamy Esakki, pages 42-67, copyright year 2017 by Business Science Reference (an imprint of IGI Global).

Chapter 20
Moderating Role of Demographics on Attitude Towards Organic Food Purchase Behavior:
A Study on Indian Consumers

Arpita Khare
Indian Institute of Management, India

ABSTRACT

The purpose of the current research is to examine moderating role of demographics on attitude towards organic food purchase behavior. Environmental attitude components were classified as actual, verbal, and affect commitment. Data was collected through survey technique in six cities across India. The findings revealed that consumers' attitude towards organic food purchase was influenced by attitude components of Actual and Verbal commitment and moderated by demographic factors of income, gender and age. The findings can be of use to firms marketing organic food brands in India. Environmental attitude factors and demographic factors like income, age, and gender can be used for profiling consumers. With increased growth of organic food market in the country, green marketing and organic food products are upcoming research areas. There is limited research on Indian consumers' attitude towards organic food products. The findings can provide valuable insights to companies marketing these products in the country.

INTRODUCTION

The concept of organic farming is not a novel concept as in the past many regions in India practiced organic farming. However, after independence there was food scarcity in the country. To tackle the food shortage, government invested on chemical fertilizers in order to boost agricultural productivity. In recent years, ancient concept of organic cultivation has again gained momentum due to awareness among consumers about deteriorating effects of pesticide residues on environment and human health

DOI: 10.4018/978-1-5225-8063-8.ch020

(Chandrashekhar, 2010). Organic food is defined as food item prepared or grown without using harmful pesticides, preservatives and other chemicals. As organic food consumption is becoming popular among health and environment conscious consumers (Technopak, 2012), it was assumed that understanding factors affecting Indian consumers' attitude towards organic food would be useful for companies marketing organic food. In recent years, there has been increased demand for organic food products among metropolitan consumers. The monthly expenditure on organic products was reported to be highest in Mumbai followed by Delhi, Bangalore, and Chennai. Recent press releases suggest that sixty two percent of metropolitan consumers buy organic food. Retailers state that health and environmental concern are major reasons for consumers preferring organic food (Sally, 2013).

Organic vegetable is most popular food product among Indian consumers and accounts for sixty eight percent followed by fruits, pulses, juices, grains and milk (Technopak, 2012). Organic packaged food and beverages are other product items popular with Indian consumers. Most Indian consumers felt that organic food is healthier than conventionally grown food. Scientific and health related factors influenced consumers' attitude towards organic food. Indian consumers' attitude towards organic food items was affected by availability, price, region where it was grown, processing, packaging and total time taken from farm to market (Sally, 2013).

Organic food market is expected to grow at Compound Annual Growth Rate of around nineteen percent during year 2012-2017 (PRWEB, 2013). The global organic food and beverages market is projected to grow at USD 104.5 billion by 2015. India ranks tenth in the world based on total cultivable land under organic certification and area under organic food certification is currently 5.21 million hectare (The Economic Times, 2012). India produced around 1.34 million certified organic products in last year. Organic food items like sugarcane, cotton, basmati rice, pulses, tea, spices, coffee, oil seeds, fruits, and other value added products are being cultivated across different states in the country. The increased production of organic food is driven by increased demand for organic food in domestic market. Province/state of Madhya Pradesh ranked first in organic food cultivation followed by Rajasthan and Uttar Pradesh (APEDA, 2013). The other Indian states involved in organic farming are Gujarat, Kerala, Karnataka, Uttaranchal, Sikkim, Maharashtra, Tamil Nadu, and Himachal Pradesh (Chandrashekar, 2010; The Economic Times, 2012). These reports clearly indicate that the organic food market in India has high untapped potential and thus provides a huge opportunity for researchers as well as practitioners.

RESEARCH OBJECTIVES

Research in western countries on organic products has discussed the influence of attitudes and past environmental buying behavior. Factors like environmental attitude, health concern, lifestyle, and peer influence affect pro-environment behaviour (Bigne´, 1997; Fraj & Martı´nez, 2002; Junaedi, 2007; Fraj & Martı´nez, 2007; Pickett-Baker and Ozaki, 2008; Smith and Paladino, 2010; Akehurst et al. 2012). This research primarily focuses on the moderating role of demographic factors on environmental attitude towards organic food buying behavior among Indian consumers. Given that organic food is a nascent market and likely to grow substantially in the coming years (PRWEB, 2013), it was assumed that understanding consumers' attitude towards organic food would help companies marketing these products to understand consumers better. Researchers in western countries have examined multiple dimensions like socio-demographics, food buying behaviour, and nutritional composition that influence awareness level and purchase of organic foods (Hill & Lynchehaun, 2002; Brugarolas Molla`-Bauza` et al., 2005; Krystallis & Chryssohoidis, 2005; First & Brozina, 2009). However, there is limited research in the Indian context on organic food buying behavior. Chakrabarti (2010) in his research on Indian consumers' purchase of organic food found health motivation as important factor affecting organic food. Attitudes like conviction about utility of organic food, consumer innovativeness, word-of-mouth activity for organic food, and commitment towards stores selling organic food were important predictors to consumers' preference for organic food. He had used expert panel to understand consumers' preference for organic food. Expanding on existing work on organic food attitudes, the current research extends Chakrabarti's work and attempts to understand consumer organic food attitude and its relationship with environmental attitude. There was limited discussion on role of demographics on organic food purchase. Drawing from earlier research on organic food, it was assumed that influence of demographics on organic food purchase cannot be ignored. Thus, focus of the current research is to investigate the role of demographics on environmental attitude dimensions on organic food purchase behavior. While there are several psychological and behavioural factors that influence buying patterns, attitude dimensions play a significant role in determining purchase behavior. Attitudinal dimensions like affect, verbal and actual commitment were considered for discussion. These dimensions were drawn from earlier researches on organic food products in Western countries.

LITERATURE REVIEW

Environmental Friendly Buying Behavior

Researchers have discussed influence of several factors like ecological concern, pro-environmental values, past green buying, lifestyles, and demographics on consumers environmental friendly purchase behaviour. Fraj and Martinez (2006) suggest that self-fulfillment values play an important role in predicting Spanish consumers' ecological behaviour. Thogersen and Olander (2003) examined relationship between Danish consumers' environment-friendly behaviour and its impact on other behavioural domains. They indicate that 'spillover' of environment-friendly behaviour occurs in case of consumers who place importance to universalistic values of environment consciousness. Jansson et al. (2010) found that values, beliefs, norms, and habits were positively related to Swedish consumers' curtailment behaviour and willingness to adopt green innovations. Zabkar and Hosta (2013) suggest influence of pro-social environment perceptions of Slovenian consumers' on willingness to buy green products. Environmental related information, attitudes, and beliefs affected consumers' self-perception and commitment towards green product purchase. Similarly, Harland et al. (2007) states that social norms, awareness about environment, conviction that environment is suffering generates sense of responsibility to exhibit pro-environmental attitudes. Personal and social norms were vital in creating consumers' commitment towards environment. These researchers discuss individual and social factors influencing green or environmental friendly buying. The current research focuses only on consumers' attitude towards organic food products. The following section examines factors affecting consumers' preference for organic food.

Attitude Towards Organic Food

With increased interest of consumers for organic products, firms are interested in understanding factors influencing consumers' attitudes towards organic food. Organic production combines best environmental practices, biodiversity, preservation of natural resources, and animal-welfare practices (de Magistris & Gracia, 2008). Organic food consumption decisions are driven by consumers' concern for environment, health, improved taste, and nutritional value (Misra et al., 1991; Davies et al., 1995; Kristensen, 1997; Brugarolas Molla`-Bauza` et al., 2005; Krystallis & Chryssohoidis, 2005).

Aertsens et al. (2009) applied Schwartz's Value theory and Theory of Planned Behaviour (TPB) to understand organic food consumption. Consumers associated organic food with values like security, hedonism, universalism, benevolence, stimulation, self-direction and conformity. Subjective, personal norms and perceived behavioural control affected organic food consumption. In case of TPB, other researchers have discussed role of subjective norms, personal values, and emotions in predicting individual's organic food consumption (Thøgersen & Olander, 2006; Dean et al., 2008; Park & Sohn, 2012; Kim & Chung, 2011). Johnston (2008) reported that egocentric values were stronger motivators than altruistic values. It enabled individuals to show their commitment towards environmental issues.

First and Brozina (2009) examined influence of cultural values on organic food consumption in Western European countries. The findings revealed "health protection (including health improvement and prevention), care for other society members (including child health care, local development contribution, care for farmers, retailers and local agriculture), hedonic pleasure (including taste), and environmental protection (including environmental protection and responsibility to the world" as important concerns for organic food consumption. Countries having lower scores on individualism and assertiveness were likely to consider environment protection as important motive for organic food purchase. de Magistris and Gracia (2008) found that health concern and attitude towards environment were important factors influencing for organic food purchase. Consumers having positive attitudes towards environment protection, pollution, and environmental damages are likely to be involved in pro-environmental activities. They are involved in environmental protection, recycling, and willing to buy organic food. Further, attitude towards organic food was important in predicting consumers' willingness to buy organic food (Padel & Foster, 2005; Durham & Andrade, 2005). Hill & Lynchehaun (2002) state that personal factors like personality, values and lifestyles, and attitudes, extrinsic and intrinsic factors related with organic food packaging, taste, quality, price, and safety, cultural values, and knowledge about organic food affects consumption decisions. Bartels and van den Berg (2011) categorized organic food consumers as 'non, light, and heavy'. The differences among consumers were on factors like innovativeness, social identification, and attitudes towards antioxidants in fresh fruits and vegetables. Heavy users were conscious about health issues about organic food. Whereas, non-users did not seek information about sustainable food products, antioxidants, and health issues related to consuming un-organic food items. A recent study highlighted that collectivistic values and environmental attitudes are strong determinants of Chinese consumers' attitudes towards green foods (Perrea et al., 2014). Another study pointed out that organic food consumption and an active lifestyle has a high correlation (Goetzke & Spiller, 2014). Tung et al. (2015) demonstrated that individual vegetarianism influence organic food attitude. Also individuals living with young children demonstrate more positive attitudes toward organic food.

Squires et al. (2001) examined the relationship between health, diet concern, environmental concern, confidence in conventional food industry, demographics, and nature of organic food industry in New Zealand and Denmark. Green self-perception and personal eco-identity were more important in predicting consumers' preference for organic food rather than environmental concern. New Zealanders concerned about health and diet placed importance to organic food. In more mature organic markets, environmental concerns were important in predicting consumers' preference for organic food. Makatouni (2002) posit that life values centered on human beings, animal well-being, and environment were important in predicting British consumers' decision to purchase organic food. Values related to human beings were health, responsibility for health and well-being for self and family, relaxed, satisfaction, long life, and happiness. These values focused on individual and family. The protection of animal and environment on health were other two important values.

H_{1a}: Consumers' environment friendly attitude like affect commitment would positively influence their organic food purchase behavior.

H_{1b}: Consumers' environment friendly attitude like verbal commitment would positively influence their organic food purchase behavior.

H_{1c}: Consumers' environment friendly attitude like actual commitment would positively influence their organic food purchase behavior.

Influence of Demographic Factors on Organic Food Purchase

Tsakiridou et al. (2008) found that Greek consumers were informed about organic food products. Demographic factors like education and income affected attitudes and consumption of organic food. This was primarily driven by desire to protect the environment. Older consumers were more concerned about nutritional value of organic food than young consumers. Environmental and health related factors were motivating factors influencing consumers' perceptions, attitudes, and consumption of organic food. Padel and Foster (2005) state that consumers' associated organic food with fruits and vegetables. Motivations for purchasing organic food were primarily due to health and environmental concerns. Households with kids used complex criteria like health to motivate kids to eat organic food. Young working women and middle aged women were informed about organic vegetables. They were more likely to buy organic fruits and vegetables. Barriers to buying organic food were price, availability, visual presentation of product, eating habits, quality, and mistrust about organic food. Fotopoulos and Krystallis (2002) posit that education affected consumers' awareness about organic food. They clustered consumers' preference for organic food according to region, income levels, and diet habits. Lea and Worsley (2005) examined relationship between Australian consumers' beliefs about organic food, personal values, and socio-demographic factors. They posit consumers perceived organic food items to be healthier, tastier, and beneficial for environment. Women had more positive attitude towards organic food than men. Personal values like concern for environment and equality were important in predicting consumers' preference for organic food. Drawing from these studies, it was assumed that demographic factors would moderate the attitude components in influencing Indian consumers' preference for organic food products.

H_2: Demographic factors like age, income, gender, education and marital status would moderate consumers' attitude towards organic food purchase.

Organic Food Purchase Behavior

Organic food purchase behavior was considered as the dependent variable. Several dimensions of organic food purchase behavior have been discussed by researchers. They can be classified under non-chemical ingredients, recyclable and health related benefits. Aertsens et al. (2011) found that Belgian consumers' motivation for consuming organic vegetables was because it did not contain synthetic pesticides, was better for the environment, healthy, of superior quality and tasted better. Objective and subjective knowledge about organic food was positively related to a positive attitude towards organic food and greater experience with products. Consumers purchasing organic food felt that they were members of "ecological organization". Attitude towards organic food was influenced by subjective knowledge, ecological organization membership, norms, motivations and female gender. The consumption of organic vegetables was positively affected by ecological organization membership, subjective knowledge, attitudes, and presence of kids in the household.

Roddy et al. (1994) in their research on Irish consumers state that organic food was associated with superior taste, flavor, quality, and free from pesticides. Negative perceptions about organic food were related to unavailability, high price, and promotional issues. Krystallis and Chryssohoidis (2005) attempted to understand whether factors affecting Greek consumers' willingness to purchase conventional and organic food were similar and did their preference for organic food differ according to product type. Consumers' willingness to buy organic food was affected by food security, quality, trust on certificate, and brand name. Consumers were willing to pay more for organic food products because they believe that these products are of higher quality. However, consumers having concerns organic food were unwilling to pay for them. In another research, Zakowska-Biemans (2011) examined consumers' preference for organic food among Polish consumers. Consumers conceptualized organic food as healthy, safe, and having no chemical ingredients. Chen (2009) posits positive relationship between health consciousness and environmental attitudes on Taiwanese consumers' attitude towards organic foods. In another research on British consumers, Sirieix et al. (2013) compared consumers' perceptions and attitude towards sustainable and other label food brands. They were skeptical about unfamiliar brands and terminologies like 'climate friendly'. Trust and familiarity with brands were important criteria for buying organic food labels. An examination of literature on factors suggests influence of environmental friendly attitudes on consumers' organic food purchase behavior (Padel and Foster, 2005; Durham and Andrade, 2005; Tsakiridou et al. 2008; Johnston, 2008; Sirieix et al. 2013). An examination of literature provided interesting insights about organic food purchase.

The hypothesized model is depicted in Figure 1.

Summary of the major research conducted on consumers' attitude towards organic food is provided in Table 1.

Figure 1.

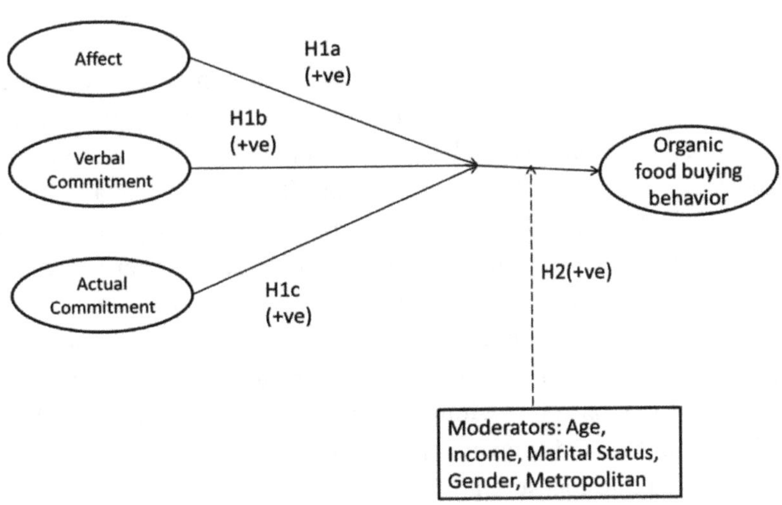

Table 1. Major studies in organic food purchase behavior

Author	Country	Contribution
Roddy et al. (1994)	Irish consumers	Organic food was associated with superior taste, flavour, quality, and free from pesticides. Negative perceptions about organic food were related to unavailability, high price, and promotional issues
Squires et al. (2001)	Denmark and New Zealand consumers	Green self-perception and personal eco-identity were more important in predicting consumers' preference for organic food rather than environmental concern
Makatouni (2002)	British consumers	Life values centred around human beings, animal well-being, and environment were important in predicting consumers' decision to purchase organic food
Krystallis and Chryssohoidis (2005)	Greek consumers	Consumers' willingness to buy organic food was affected by food security, quality, trust on certificate, and brand name
Lea and Worsley (2005)	Australian consumers	Women had more positive attitude towards organic food than men.
Tsakiridou et al. (2008)	Greek consumers	Demographic factors like education and income affected attitudes and consumption of organic food
First and Brozina (2009)	Western European consumers	Cultural values influence organic food consumption
Chen (2009)	Taiwanese consumers	Positive relationship between health consciousness and environmental attitudes on consumers' attitude towards organic foods
Aertsens et al. (2011)	Belgian consumers	Objective and subjective knowledge about organic food has a positive relationship to attitude towards organic food and greater experience with products
Zakowska-Biemans (2011)	Polish consumers	Consumers conceptualized organic food as healthy, safe, and having no chemical ingredients
Sirieix et al. (2013)	British consumers	Trust and familiarity with brands were important criteria for buying organic food labels
Perrea et al. (2014)	Chinese consumers	Collectivistic values and environmental attitudes are strong determinants of Chinese consumers' attitudes towards green foods
Goetzke & Spiller (2014)	German consumers	Organic food and an active lifestyle has a high correlation
Tung et al. (2015)	Taiwanese consumers	Individual vegetarianism influence organic food attitude. Individuals living with young children demonstrate more positive attitudes toward organic food

RESEARCH METHODOLOGY

Measures

Drawing from research on environmental attitudes and their influence on environmental friendly buying, Fraj and Martinez (2007) had conceptualized attitude under three components: affect commitment, verbal commitment, and actual commitment. The scales were adapted from their research. To measure environmental commitment, attitude model comprising of affect, cognitive, and behaviour components has been used by researchers (Chan, 2000; Fraj and Martinez, 2007). Fraj and Martinez (2007) in their research on Spanish consumers, had examined influence of affect and verbal commitment factors on actual commitment to purchase green products. Actual commitment factor had been used as dependent factor. However, in current research, these three factors were taken as three constructs measuring consumers' environment friendly attitude. The influence of these three factors on organic food purchase was examined. The organic food purchase behaviour was measured using scale developed by Lee (2009). The items were modified to understand consumers' attitude towards organic food and purchase intention. Five point Likert scale was used to measure all factors.

Sample

Prior to selecting the sample, different organized retailers across five major cities were identified. The selection of the stores was based upon their readiness to stock and sell organic food items. Discussion with the store managers revealed that organic food was a popular category sold in their stores. Most consumers were interested in purchasing organic vegetables and fruits on a regular basis. However, the store managers could not provide the monthly sales data of organic food items due to confidential reasons. The stores selected were Subhiksha, Reliance Fresh, Easy Day, and Food Bazaar. These stores have chains across the country and stock similar organic food product categories. This ensured similarity across sample distribution and would reduce biasness.

Convenience sampling method was used. Data collection was done during months of November 2013-January 2014. Consumers were randomly approached in six major cities in India (Bangalore, Chennai, Delhi, Lucknow, Jaipur, & Kolkata). Selection of cities was based on convenience sampling with an objective to get representation from eastern, western, northern and southern regions of India. The store managers were approached and the objective of the research was communicated to them. Their help was sought for data collection. Respondents were contacted randomly at the stores identified (Subhiksha, Reliance Fresh, Easy Day, and Food Bazzar) while they were shopping for food items. Data was collected at different days of the week and both in morning and evening in order to reduce biasness. Sample inclusion was based on consumers' awareness about 'organic food'. Following the procedure of earlier studies (Chan, 2000; Fraj & Martinez, 2007), consumers were presented with three simplified definitions of organic food and asked to indicate which according to them explained organic food. The respondents providing the correct answer were presented with questionnaire on organic food (The socio-demographic details of respondents are presented in Table 2). Total respondents contacted were 1,500 people and 666 people qualified for sample inclusion.

FINDINGS AND DISCUSSIONS

Bhalla and Lin's (1987) methodology was followed for pretesting the scales. In the attitude scale, two items from each Affect and Actual commitment sub-scale, and three items from Organic purchase behavior scale were removed because they failed to meet the desired 0.5 factor loading and failed to fit. All items from Verbal commitment sub-scale were retained as they met Nunnally's (1978) recommended level of internal consistency for scale development. The scale validation methodology followed in other studies is used (Chan, 1999; Fraj & Martinez, 2006). Cronbach alpha values for the factors is above acceptable limit of 0.60 (Malhotra, 2008) and ranged between .674-.806.

For each of the constructs, the measurement model has been checked. The value for $\chi2$, degrees of freedom and fit indices for Actual Commitment sub-scale are $\chi2$ =50.460, df =14; CFI =0.926, GFI=0.978; AGFI=0.956; RMSEA = 0.063, for Verbal commitment sub-scale the values are $\chi2$ =29.305, df =14; CFI = 0.971, GFI=0.987; AGFI=0.975; RMSEA = 0.041, for organic food purchase behavior the values are $\chi2$ =21.164, df =9; CFI =0.976, GFI=0.990; AGFI=0.976; RMSEA = 0.045 and for Affect sub-scale the values are $\chi2$ =14.272, df =5; CFI = 0.959, GFI=0.991; AGFI=0.974; RMSEA = 0.053. The fit indices clearly indicate that the models are having good fit for the data (Hair et al. 1995).

CFA indicated that all factor loadings ranged from 0.628 to 0.859 and corresponding t-values are statistically significant ($p < 0.001$) providing support for convergent validity. In relation to goodness of

Table 2. Demographic Description of Respondents

Variable		Frequency	Percentage
Gender	Male	376	75.0
	Female	125	25.0
Age (years)	18-21	74	14.8
	22-25	124	24.8
	26-30	130	25.9
	31-40	128	25.5
	41-50	40	8.0
	51 and above	5	1.0
Marital Status	Married	251	50.1
	Single	250	49.9
Education	Higher Secondary	103	20.6
	Graduation	290	57.9
	Post Graduation	108	21.6
Household Income (monthly)	Below INR 10,000 (below $153)	45	9.0
	INR 10,000- 20,000 ($153-307)	104	20.8
	INR 21,000- 30,000 ($307-461)	160	31.9
	INR 31,000- 40,000 ($461-615)	86	17.2
	INR 41,000 – 50,000 ($615-769)	43	8.6
	Above 50,000 ($ 769)	63	12.6
Total		501	

fit scales, the residuals are inferior to .05 for all components, which suggests that the adjustment of the factor analysis was acceptable (Grande, 2000; Fraj & Martinez, 2006). Chi-square significance level (p) for all factors is .000. Goodness-of-fit indices were within acceptable range (Hair et al. 1995).

Organic food purchase behavior was taken as dependent factor. Step wise regression analysis was run to understand the moderating role of demographic factors on affect, actual and verbal commitment factors on organic food purchase behavior. The results are shown in Table 3.

Stepwise regression model revealed that demographic factors moderated consumers' attitude towards organic food purchase. In the first model, 'actual commitment' emerged as the predictor variable for organic food purchase behavior (R^2=.415, $p < 0.00$). The first model suggests that actual commitment towards environment accounts for 41.5 percent of Indian consumers' attitude towards organic food purchase behavior (see Table 3).

The results support earlier researches (Roddy et al. 1994; Aertsens et al. 2011; Zakowska-Biemans, 2011). In the second model, 'actual commitment and gender' factors emerge as predictors (R^2 =.431, $p<0.00$), and account for 43.1 percent of attitude towards organic food purchase behavior. In the third model, 'verbal commitment' was introduced. Actual commitment, verbal commitment and gender emerged as predictors and account for 44.6 percent of (R^2=.446, $p<0.00$) attitude towards organic food purchase behavior. In the fourth model, 'income' was introduced. Actual commitment, verbal commitment, gender and income emerged as predictors and account for 45.9 percent of (R^2=.459, $p<0.00$) attitude towards

Table 3. Step-wise Regression

Model	Variable	B	R^2	Adjusted R^2	Significance
1.	**First regression (Dependent Variable: Organic food buying behaviour)**		.415	.414	
	Actual Commitment	.644			.000*
N=665; d.f=1; F=471.171; significant F=.000					
2.	**Second regression (Dependent Variable: Organic food buying behaviour)**		.431	.429	
	Actual Commitment	.626			.000*
	Gender	.128			.000*
N=665; d.f=2; F=251.239; significant F=.000					
3.	**Third regression (Dependent Variable: Organic food buying behaviour)**		.446	.443	
	Actual Commitment	.518			.000*
	Gender	.128			.000*
	Verbal commitment	.163			.000*
N=665; d.f=3; F=177.624; significant F=.000					
4.	Forth regression **(Dependent Variable: Organic food buying behaviour)**		.459	.456	
	Actual Commitment	.489			.000*
	Gender	.166			.000*
	Verbal commitment	.164			.000*
	Income	.123			.000*
N=665; d.f=4; F=140.327; significant F=.000 ***significant at 0.05 level**					
5.	Fifth regression **(Dependent Variable: Organic food buying behaviour)**		.464	.460	
	Actual Commitment	.483			.000*
	Gender	.149			.000*
	Verbal commitment	.159			.000*
	Income	.116			.000*
	Age	-.074			.014*
N=665; d.f=5; F=114.358; significant F=.000 ***significant at 0.05 level**					

organic food purchase behavior. In the fifth model, Actual commitment, verbal commitment, gender, income and age emerged as predictors (R^2=.464, p<0.00), and account for 46.4 percent of attitude towards organic food purchase behavior. Therefore H_{1b}, H_{1c} and H_2 are accepted. The β value for age variable is negative which suggest that there is a negative relationship between age of consumers and their attitude towards organic food. The younger consumers are more likely to purchase organic food. The findings are in tandem with findings of Padel and Foster (2005). They posit that younger consumers are likely to exhibit favourable attitude towards organic food. Similar conclusions can be drawn from the current research. The findings support other researches which suggest that women and income affects attitude towards organic food (Squires et al. 2001; Lea & Worsley, 2005; Tsakiridou et al. 2008).

The research findings suggest that attitude components like actual commitment, verbal commitment towards environment are moderated by demographic factors like income, age and gender. Individuals involved in environmental issues and concerned about environment protection are likely to have favorable attitude towards organic food products. The findings support earlier researches which suggest influence of pro-environmental values and attitudes on purchase decision (Makatouni, 2002; Harland et al. 2007; Fraj & Martinez, 2007; First & Brozina, 2009; Zabkar & Hosta, 2013). The results indicate that demographic variables play an important role in influencing consumers' choice towards organic food purchase.

The findings posit that companies marketing organic food products must use attitudinal and demographic profiling for targeting consumers. Marketing of organic food should focus on providing detailed information about ingredients and health related implications. Since consumers' interested in organic food belong to high income and younger generation segments, targeting them would involve focusing on product quality and its nutritional composition. They would be therefore interested in reading the labels in order to understand organic food product composition. Women are conscious about health of their family and are likely show more interest in organic food. The findings support earlier research where demographic and attitudinal factors are important in predicting organic food purchase behavior (Padel & Foster, 2005; Durham & Andrade, 2005; de Magistris & Gracia, 2008; Perrea et al., 2014; Tung et al. 2015).

MARKETING IMPLICATIONS

Researches on Indian consumers' attitude towards organic food have discussed importance of health and environmental concerns (Technopak, 2012; The Economic Times, 2012). Chakrabarti (2010) in his research on Indian consumers had suggested that choice for organic food products was influenced by utility, reputation of store, and certification. The findings of current research add to his work. This research attempts to understand the moderating role of demographic factors in influencing organic food purchase decision. Earlier research on Indian consumers has not examined the role of demographic and attitudinal factors on organic food purchase. The findings can help in understanding influence of both psychographic and demographic factors on purchase decisions.

The research findings posit that income levels determine consumers' perception about organic food items. They are aware about impact of consuming organic food items on health and environment and are therefore cautious about its content. Their preference for organic fruits, vegetables, and other food items is based on ingredients, quality, and authenticity. Companies marketing organic food should place information about composition and ingredients on packages. This would enable consumers to check ingredients, composition, and quality aspects of the product before purchasing them. Since organic food consumption has health related implications, consumers prefer to analyze information before purchasing. Information about ingredients on packages helps them in understanding the health benefits of those items. This enables consumers to take an informed decision.

The implications of current research for companies are two-fold. Organic food consumption is increasing (The Economic Times, 2012; Sally, 2013) and therefore marketing of organic food becomes important. Companies should pay special attention to labeling and packaging of the products as it enforces quality, authenticity, and freshness of food items. Packaging would enhance utilitarian and hedonic value of products. It would help companies build brand awareness for organic products. The socio-demographic factors like income, age and gender were found to be important factors influencing consumers' attitude towards organic food. Firms engaged in manufacturing or selling organic food products need to maintain and enhance health benefits, freshness and taste factors. Lifestyle and income related consumer profiling can be done in order to target consumers.

Family health and well-being can be used as popular themes to build awareness about organic food. In India, currently organic food marketing, promotions, and advertisements are limited. Companies are not using national or state level advertisements to promote organic food. Using promotions and advertisements to communicate health and environmental benefits of adopting organic food can generate greater awareness among non-users. The companies can offer more value to consumers by launching products in different packaging sizes, flavors, and providing information about where the product was grown, processing, and health related benefits associated with its consumption. Detailed information about ingredients and composition can enable consumers compare different product features and purchase items relevant for their family needs. Visual images on packages about freshness, quality, certification, and naturalness of organic food can help in improving consumers' perception about organic food brands.

The increased market share of organic food shows that consumers are aware of the ill-effects of non-organic food (The Economic Times, 2012; Sally, 2013). They are willing to pay a high price premium for organic food. Higher income groups are not willing to compromise on health and environment related issues. They are willing to spend on quality food items that do not have any side-effects and preservatives. Organic food costs about forty to sixty percent more than conventional food items. Gurtoo (2013) found that sixty two percent of high income households prefer organic food items. This suggests a clear need for companies to focus enforcing on quality and non-chemical ingredients while marketing organic food. Consumers associate organic food with good health and "health-halo effect" (Mielach, 2013). They believe that organic food are healthy and do not contain any harmful ingredients.

In India, many firms are trying to market their organic food categories. The major organized retail chains like Nilgiris, Godrej Nature's Basket, More, Spencers, Food Bazaar, and Fab India stock organic foods (Chatterjee, 2012). While designing marketing strategy, firms should understand that though organic food products are gaining popularity, it is still a niche market. The promotional campaigns should spread awareness about product attributes, its benefits, and non-chemical ingredients. There is a need to educate people about organic food because they often think natural and organic food is the same. They purchase fresh food from farms and equate them to organic. They lack awareness that fresh food items like vegetables and fruits can contain chemical/fertilizer residues if they are not cultivated with the help of organic farming techniques. Firms should provide information about organic food products on the labels. This will enable consumers to understand the actual composition of organic food items and its impact on health and environment.

LIMITATIONS AND FUTURE RESEARCH DIRECTIONS

Research in India on organic food product category is in nascent stages. Though the organic food market is predicted to grow in coming years, research on understanding consumers' attitude, lifestyle and values in predicting consumption of organic food is limited. Academic researches taken up in this area can help in providing valuable insights about consumers' predisposition towards organic food. The current research focused only on influence of three attitude components (affect, behaviour, and cognition) and demographics on Indian consumers' attitude towards organic food. It presents limited set of factors for understanding consumer attitudes. Factors like past environmental friendly behaviour, ecological concern, social influence, and psychographics can be taken up to provide a comprehensive understanding of consumer attitude. Study on products of organic food retailers can be conducted to understand product related dimensions affecting consumers' purchase decision. Future research can be conducted to understand difference between consumers' attitude towards organic and non-organic food. Consumers' perception and attitude towards specific organic food brands sold at the retail stores can be taken up for further study. This would help in understanding the brand related attributes and consumers' attitude towards them.

REFERENCES

Aertsens, J., Mondelaers, K., Verbeke, W., Buysse, J., & Van Huylenbroeck, G. (2011). The influence of subjective and objective knowledge on attitude, motivations and consumption of organic food. *British Food Journal, 113*(11), 1353–1378. doi:10.1108/00070701111179988

Aertsens, J., Verbeke, W., Mondelaers, K., & Van Huylenbroeck, G. (2009). Personal determinants of organic food consumption: A review. *British Food Journal, 111*(10), 1140–1167. doi:10.1108/00070700910992961

Akerhurst, G., Afonso, C., & Gonclaves, H. M. (2012). Re-examining green purchase behaviour and green consumer profiles: New evidences. *Management Decision, 50*(5), 972–988. doi:10.1108/00251741211227726

APEDA. (2013). *National Programme for organic food production.* Retrieved on 22nd February 2014 from http://www.apeda.gov.in/apedawebsite/organic/Organic_Products.htm

Bartels, J., & van den Berg, I. (2011). Fresh fruit and vegetables and the added value of antioxidants: Attitudes of non-, light, and heavy organic food users. *British Food Journal, 113*(11), 1339–1352. doi:10.1108/00070701111179979

Bigne´, J. E. (1997). *El consumidor verde: bases de un modelo de comportamiento.* Esic Market.

Brugarolas Molla-Bauza, M., Martinez-Carrasco Martinez, L., Martinez Poveda, A., & Rico Perez, M. (2005). Determination of the surplus that consumers are willing to pay for an organic wine. *Spanish Journal of Agricultural Research, 3*(1), 43–51. doi:10.5424jar/2005031-123

Chakrabarti, S. (2010). Factors influencing organic food purchase in India- expert survey insights. *British Food Journal, 112*(8), 902–915. doi:10.1108/00070701011067497

Chan, K. (2000). Market segmentation of green consumers in Hong Kong. *Journal of International Consumer Marketing, 12*(2), 7–24. doi:10.1300/J046v12n02_02

Chandrashekar, H. M. (2010). Changing scenario of organic farming in India: an overview. *International NGO Journal, 5*(1), 34-39. Retrieved on 22nd February 2014 from. uni-mysore.ac.in/14660/1/Changing Scenario of Organic Farming in India.pdf

Chatterjee, B. (2012). *Organic Foods enter India to a low-key welcome*. Retrieved on 19th March 2014 from http://www.businesseconomics.in/?q=node/1267

Chen, M.-F. (2009). Attitude toward organic foods among Taiwanese as related to health consciousness, environmental attitudes, and the mediating effects of a healthy lifestyle. *British Food Journal, 111*(2), 165–178. doi:10.1108/00070700910931986

Churchill, G. A., Iacobucci, D., & Israel, D. (2010). *Marketing research: A south Asian perspective*. New Delhi, India: Cengage Learning.

Davies, A., Titterington, A., & Cochrane, C. (1995). Who buys organic food? A profile of the purchasers of organic food in Northern Ireland. *British Food Journal, 97*(10), 17–23. doi:10.1108/00070709510104303

de Magistris, T., & Gracia, A. (2008). The decision to buy organic food products in Southern Italy. *British Food Journal, 110*(9), 929–947. doi:10.1108/00070700810900620

Dean, M., Raats, M. M., & Shepherd, R. (2008). Moral concerns and consumer choice of fresh and processed organic foods. *Journal of Applied Social Psychology, 38*(8), 2088–2107. doi:10.1111/j.1559-1816.2008.00382.x

Durham, C. A., & Andrade, D. (2005). *Health vs. environmental motivation in organic preferences and purchases*. Paper presented at the American Agricultural Economics Association Annual Meeting, Providence, RI.

First, I., & Brozina, S. (2009). Cultural influences on motives for organic food consumption. *EuroMed Journal of Business, 4*(2), 185–19. doi:10.1108/14502190910976538

Fotopoulos, C., & Krystallis, A. (2002). Organic product avoidance: reasons for rejection and potential buyers' identification in a countrywide survey. *British Food Journal, 104*(3-5), 233-260.

Fraj, E., & Martı'nez, E. (2002). *Comportamiento del Consumidor Ecolo'gico*. Madrid: Esic Editorial.

Fraj, E., & Martinez, E. (2006). Environmental values and lifestyles as determining factors of ecological consumer behaviour: An empirical analysis. *Journal of Consumer Marketing, 23*(3), 133–144. doi:10.1108/07363760610663295

Fraj, E., & Martinez, E. (2007). Ecological consumer behaviour: An empirical analysis. *International Journal of Consumer Studies, 31*(1), 26–33. doi:10.1111/j.1470-6431.2006.00565.x

Goetzke, B. I., & Spiller, A. (2014). Health-improving lifestyles of organic and functional food consumers. *British Food Journal, 116*(3), 510–526. doi:10.1108/BFJ-03-2012-0073

Gurtoo, H. C. (2013). *Taste for organic foods*. Retrieved on 19th March 2014 from http://www.hindustantimes.com/lifestyle/food/taste-for-organic-foods/article1-1109620.aspx

Harland, P., Staats, H., & Wilke, H. A. M. (2007). Situational and Personality Factors as Direct or Personal Norm Mediated Predictors of Pro-environmental Behavior: Questions Derived from Norm-activation Theory. *Basic and Applied Social Psychology, 29*(4), 323–334. doi:10.1080/01973530701665058

Hill, H., & Lynchehaun, F. (2002). Organic milk: Attitudes and consumption patterns. *British Food Journal, 104*(7), 526–542. doi:10.1108/00070700210434570

Jansson, J., Marell, A., & Nordlund, A. (2010). Green consumer behavior: Determinants of curtailment and eco-innovation adoption. *Journal of Consumer Marketing, 27*, 358–370.

Johnston, J. (2008). The citizen-consumer hybrid: Ideological tensions and the case of whole foods market. *Theory and Society, 37*(3), 229–270. doi:10.100711186-007-9058-5

Junaedi, M. F. S. (2007). The roles of Consumer's knowledge and emotion in ecological issues: An Empirical Study on Green Consumer Behavior. *Gadjah Mada International Journal of Business, 9*(1), 81–99.

Kim, H. Y., & Chung, J.-E. (2011). Consumer purchase intention for organic personal care products. *Journal of Consumer Marketing, 28*(1), 40–47. doi:10.1108/07363761111101930

Kristensen, L. (1997). *Implementation of organic food in the public sector in New Zealand* (MCom thesis). School of Horticulture, Christchurch Polytechnic, New Zealand.

Krystallis, A., & Chryssohoidis, G. (2005). Consumers willingness to pay for organic food: Factors that affect it and variation per organic product type. *British Food Journal, 107*(5), 320–343. doi:10.1108/00070700510596901

Lea, E., & Worsley, T. (2005). Australians organic food beliefs, demographics and values. *British Food Journal, 107*(11), 855–869. doi:10.1108/00070700510629797

Lee, K. (2009). Gender differences in Hong Kong adolescent consumers green purchasing behaviour. *Journal of Consumer Marketing, 26*(2), 87–96. doi:10.1108/07363760910940456

Makatouni, A. (2002). What motivates consumers to buy organic food in UK? British Food Journal, 104(3-5), 345-352.

Malhotra, N. K. (2008). *Marketing research: An applied orientation*. Pearson Education India. doi:10.1108/S1548-6435(2008)4

Mielach, D. (2013). *Consumers Overestimate benefits of organic food*. Retrieved on 19th March, 2014 from http://www.businessnewsdaily.com/4257-organic-food-labels.html

Misra, S. K., Huang, C. L., & Ott, S. L. (1991). Consumer willing to pay for pesticide-free fresh produce. *Western Journal of Agricultural Economics*, (16): 218–227.

Nunnally, J. C. (1967). *Psychometric Theory* (2nd ed.). New York, NY: McGraw-Hill.

Padel, S., & Foster, C. (2005). Exploring the gap between attitudes and behaviour. *British Food Journal, 107*(8), 606–625. doi:10.1108/00070700510611002

Park, S.-Y., & Sohn, S. H. (2012). Exploring the normative influences of social norms on individual environmental behavior. *Journal of Global Scholars of Marketing Science: Bridging Asia and the World, 22*(2), 183–194. doi:10.1080/12297119.2012.655142

Perrea, T. G., Grunert, K., Krystallis, A., Zhou, Y., Huang, G., & Hue, Y. (2014). Testing and validation of a hierarchical values-attitudes model in the context of green food in China. *Asia Pacific Journal of Marketing and Logistics*, *26*(2), 296–314. doi:10.1108/APJML-09-2013-0106

Pickett-Baker, J., & Ozaki, R. (2008). Pro-environmental products: Marketing influence on consumer purchase decision. *Journal of Consumer Marketing*, *25*(5), 281–293. doi:10.1108/07363760810890516

PRWEB. (2013). *Organic Food market in India set for incessant growth finds TechSci Research report.* Retrieved on 22nd February, 2014 from http://www.prweb.com/releases/2013/9/prweb11174562.htm

Roddy, G., Cowan, C., & Hutchinson, G. (1994). Organic food: A description of the Irish market. *British Food Journal*, *96*(4), 3–10. doi:10.1108/00070709410060998

Sally, M. (2013). *Increase in consumption of organic food products: ASSOCHAM Survey.* Retrieved on 22nd February 2014 from http://articles.economictimes.indiatimes.com/2013-05-23/news/39475623_1_organic-food-organic-products-organic-sector

Sirieix, L., Delanchy, M., Remaud, H., Zepeda, L., & Gurviez, P. (2013). Consumers perceptions of individual and combined sustainable food labels: A UK pilot investigation. *International Journal of Consumer Studies*, *37*(2), 143–151. doi:10.1111/j.1470-6431.2012.01109.x

Smith, S., & Paladino, A. (2010). Eating clean and green? Investigating consumer motivations towards the purchase of organic food. *Australasian Marketing Journal*, *18*(2), 93–104. doi:10.1016/j.ausmj.2010.01.001

Squires, L., Juric, B., & Cornwell, T. B. (2001). Level of market development and intensity of organic food consumption: Cross cultural study of Danish and New Zealand Consumers. *Journal of Consumer Marketing*, *18*(5), 392–409. doi:10.1108/07363760110398754

Technopak. (2012). *Organic Food Market in India.* Retrieved on 22nd February 2014 fromhttp://www.globalorganictrade.com/files/market_reports/OTA_India_Market_Report_2012.pdf

The Economic Times. (2012). *India's organic food market growing at over 20%.* Retrieved on 22nd February 2014 from http://articles.economictimes.indiatimes.com/2012-08-23/news/33342399_1_organic-food-organic-cotton-chemical-free-food

Thogersen, J., & Olander, F. (2003). Spillover of environment-friendly consumer behaviour. *Journal of Environmental Psychology*, *23*(3), 225–236. doi:10.1016/S0272-4944(03)00018-5

Thøgersen, J., & Olander, F. (2006). The dynamic interaction of personal norms and environment-friendly buying behavior: A panel study. *Journal of Applied Social Psychology*, *36*(7), 1758–1780. doi:10.1111/j.0021-9029.2006.00080.x

Tsakiridou, E., Boutsouki, C., Zotos, Y., & Mattas, K. (2008). Attitudes and behaviour towards organic products: An exploratory study. *International Journal of Retail & Distribution Management*, *36*(2), 158–175. doi:10.1108/09590550810853093

Tung, S., Tsay, J. C., & Lin, M. (2015). Life course, diet-related identity and consumer choice of organic food in taiwan. *British Food Journal, 117*(2), 688–704. doi:10.1108/BFJ-11-2013-0334

Zabkar, V., & Hosta, M. (2013). Willingness to act and environmentally conscious consumer behaviour: Can pro-social status perceptions help overcome the gap? *International Journal of Consumer Studies, 37*(3), 257–264. doi:10.1111/j.1470-6431.2012.01134.x

Zakowska-Biemans, S. (2011). Polish consumer food choices and beliefs about organic food. *British Food Journal, 113*(1), 122–137. doi:10.1108/00070701111097385

This research was previously published in Business Analytics and Cyber Security Management in Organizations edited by Rajagopal and Ramesh Behl, pages 279-295, copyright year 2017 by Business Science Reference (an imprint of IGI Global).

APPENDIX

Table 4.

Affect: Fraj and Martinez (2007)	
It frightens me to think that much of the food I eat is contaminated with pesticides.(retained)	Retained
It genuinely infuriates me to think that the government doesn't do more to help control pollution of the environment.	Retained
I become incensed when I think about the harm being done to plant and animal life by pollution.	Retained
I get depressed on smoggy days.	Retained
I rarely ever worry about the effects of smog on myself and family	Retained
When I think of the ways industries are polluting, I get frustrated and angry.	Deleted
The whole pollution issue has never upset me too much since I feel it's somewhat overrated.	Deleted
Verbal commitment: Fraj and Martinez (2007)	
I'd be willing to ride a bicycle or take the bus to work in order to reduce air pollution.	Retained
I would be willing to use a rapid transit system to help to reduce air pollution	Retained
I would donate a day's pay to a foundation to help improve the environment.	Retained
I would be willing to stop buying products from companies guilty of polluting the environment, even though it might be inconvenient.	Retained
I'd be willing to write local authorities weekly concerning ecological problems.	Retained
I wouldn't go house to house to distribute literature on the environment.	Retained
I would not be willing to pay a pollution tax even if it would considerably decrease the smog	Retained
Actual Commitment: Fraj and Martinez (2007)	
I guess I've never actually bought a product because it had a lower polluting effect.	Retained
I keep track of my congressman and senator's voting records on environment issues.	Retained
I have contacted a community agency to find out what I can do about pollution.	Retained
I make a special effort to buy products in recyclable containers.	Retained
I have switched products for ecological reasons.	Retained
I have never joined a cleanup drive.	Retained
I subscribe to ecological publications.	Retained
I have attended a meeting of an organization specifically concerned with bettering the environment.	Deleted
I have never attended a meeting related to ecology.	Deleted
Organic food buying behavior: Adapted from Lee (2009)	
I often buy food products that are against animal-testing	Retained
I often buy food products that contain no or fewer chemical ingredients	Retained
When I consider buying a product, I will look for a certified environmentally-safe or organic stamp	Retained
I often buy food products that support fair community trades	Retained
I understand that chemicals are being used in food products that are harmful	Retained
I prefer to purchase food items that do not contain preservatives	Retained
I often buy food products that are labeled as environmentally safe	Deleted
I often buy products that use recycled/ recyclable packaging	Deleted
I am interested in buying food products that do not contain artificial chemicals	Deleted

Section 6
Sustainable Agriculture

Chapter 21
New Food Industries Toward a New Level of Sustainable Supply:
Success Stories, Business Models, and Strategies

Vittorio D'Aleo
University of Messina, Italy

Francesco D'Aleo
University of Messina, Italy

Roberta Bonanno
University of Messina, Italy

ABSTRACT

"Novel foods" indicates all those foods that do not fall within the traditional western diet. The authors underline that there is an important economic potential linked to this new sector. After having introduced the topic by identifying what is meant by new food, food sustainability, and entrepreneurship in this sector, the analysis focuses on the qualitative aspects of novel foods. The authors emphasize the advantages of these sources, their qualities, and the impact on the environment. In this chapter, the authors demonstrate that novel food sources can be a substitute for other food that require production processes, and thus make a big impact on the environment. After figuring out the advantages of novel food sources, the authors analyze some success stories from the companies that already operate in this area. They also investigate existing business models on the novel food market and the way they operate and create wealth. Summarizing the results of their research, the authors underline the economic potential of the new foods exploitation in the Mediterranean area.

DOI: 10.4018/978-1-5225-8063-8.ch021

INTRODUCTION

Feeding is needful but the relationship between people and food is different from country to country, depending on the culture of the population and evolving changes in the different historical periods. The term "new foods" or "novel foods" concerns the use of non-conventional Western and Mediterranean sources of food (Burlingame, Charrondiere, & Mouille, 2009), which are already used since ancient times in other countries of Southeast-Asia area, for example, India, China, and Thailand. It is statistically estimated that in 2050 the world population could grow to about 9 billion people (Brookes, 2007), consequently, the food request could rise enormously. Therefore, there is a need to spread information about the existence of the alternative food sources, which will enrich those which are currently in use. In particular, the economic experts think (World Economic Forum and Food and Agriculture Organization of the United Nations (FAO)) that the food request could increase so high, that a new farmland larger than the territory of Brazil would be needed. The paucity of food resources could be a big problem for whole humanity in the imminent future, and novel foods could provide a possible solution. An example: the oceans cover about the 70% of the Earth even if actually they supply only a smaller quantity of food in comparison to the land surface. Therefore, from Southeast Asian Sea we could import Jellyfishes as a new potential future worldwide food. In fact, actually, the economists estimated (FAO and World Economic Forum) that the potential of this food is enormous. Some countries could import insects or breed them like a new source of food proteins, while African countries could use algae like a supplement of proteins. Spirulina microalgae is an example of food used like a nutrient supplement against the food shortages in many African countries. Hopefully, new farms concentrated in South America and other countries like China and Thailand could be established in other countries as soon as possible

BACKGROUND

Food Market Trends

Before evaluating the aspects of the new supply in detail, it is important to underline the food market trends to better understand the perspective of the global market. Therefore, the authors emphasize the following trends:

- Healthier food. More responsibility regards to animal welfare and land sustainability and more attention to the environment (environmentally friendly).
- Consumer needs access to detailed and clear information.
- Necessity to adapt the alimentation to the lifestyle of everyone.
- More and more consumers would like to feel connected to nature using some natural products.
- Homemade products named "0-kilometer foods" which are provided from their homeland.
- New consumption experience and opportunity to test new flavors like artisanal treats.
- Increase in home-cooking and home entertaining.

World economic traders could try to approach to these new trends and to expand existing market across new consumption occasions. New start-up companies related to "novel foods" should adapt their products to new market requirements.

Novel Foods Market

According to the ISPRA (Brookes, 2007), the global market for novel foods is actually about €35 billion (sales volume), most of which derived from food ingredients with the new primary molecular structure or genetically modified. The European Union (EU) represents about 40% of global novel foods market, an equivalent of €14 billion in sales. Enterprises related to the novel foods sector employ about 90,000 people worldwide. Research and development (R&D) are the most important costs, and novel foods enterprises usually spend an average of 3-5% of their turnover on R&D (range between 2% and 8%). Even if an enterprise wants to increase its profit it has to wait for a longer period of time (4-10 years) compared to the companies in traditional foods sector (1-3 years), because novel foods business is related to innovations. Life expectancy is better for new products (5-15 years) than those which are on the market already (1-3 years) (Hepburn et al., 2008).

European Regulation

The European Parliament has ratified that novel foods might be marketed in all countries of the EU. Receiving an authorization from the individual states will not take long, but the European Food Safety Authority will have to authorize the definitive marketing. What exactly are novel foods? Larvae, algae, scorpions, and spiders are considered novel. These products are already marketed in the Netherlands, Denmark, Belgium, and Great Britain. Within this group of foods, insects and algae are not the only potential resources of food. There are other supplies as well that could be used in purpose to eradicate malnutrition in the world. The FAO researchers all agreed that this could be a new possibility. To support this, people have to remember that there are many hundreds of tons of jellyfishes in the Mediterranean Sea that represents an important source of water and proteins. Actually, Professor Stefano Piraino of the University of Salento (Italy) coordinates the European project named "Med-Jellyrisk", which scope is exploitation of jellyfishes as a source of food. In Europe, jellyfishes and some other products were not considered safe food before 1997 (European Parliament, 1997) and therefore were not subject to any analysis by the European Community which usually advised by the EFSA before to be approved for marketing and consumption. Novel foods and all related products consumed inside the EU are governed by the Regulation 258/97 approved in 1997 (European Parliament, 1997).

This Regulation is a safety tool allowing the entry of novel food only after detailed controls and inspections and the subsequent final authorization in order to drive away unsafe products and foodstuffs.

According to the Regulation 258/97, a product is recognized as novel when it relates to one of the four categories:

- Foods and food ingredients with a new primary molecular structure or genetically engineered.
- Foods and food ingredients which are constituted or isolated from microorganisms (bacteria, fungi, or algae).
- Foods and food ingredients which are composed of plants, animals, or parts of them.
- Products and food ingredients which could perform a change in the nutritional value of the current diet and produced with an innovative or not currently used manufacturing process.

Procedures for requesting the authorizations and for marketing of novel foods within the EU as well as in certain member states are rather complex (Hermann, 2009). Companies willing to deal with novel foods have to submit an application to the related body of the member state, which scientifically reviews it, completes an initial assessment report, and then sends it to the EU commission. The commission members draw up another report in which provide assessments and objections. Applying companies must satisfy all requests of the commission before obtaining authorization for their new products. The final act is the decision for authorization (or denial) published in the Community Official Gazette. Actually, the procedure does not distinguish among innovative genuine products and those used outside the EU which could be imported under different names from the countries of their origin. Whenever a company advances a product which has already been authorized, it is also required to make another submission by the commission of a member state. However, in some cases, the procedure is simplified because of the acknowledgment of "substantial equivalence" of the product compared to the one already authorized. The same procedure may also be followed to certify an equivalence on the market of novel food compared to traditional food.

European Start-Ups' Market

Actually, 28 markets altogether represent the European start-ups' market (D'Aleo & Morabito, 2015). Europe offers funding for entrepreneurs, supports them, and grants network opportunities, but the clutches between 28 submarkets of the EU member states cause a loss of homogeneity unlike other markets (the United States are the most obvious example) (Darkow, Foerster, & von der Gracht, 2015). However, the market of the EU is more complex and diversified compared to the USA, because every member state has its recipes and characteristic variations, which could result in a loss of a market. Even if Europe is in the fifth place among the top 10 most innovative countries worldwide, multiple data indicate that the conditions of European economic could not be ideal for entrepreneurs and for fast-growing companies. This appears due to the fragmentation of the European market, denied access to the capital sources, and a supportive shortage of the initiatives. Furthermore, the absence of the European law on data protection complicates the expansion and displacement of companies that would like to enter into commercial negotiations with European regions. There are European countries wherein venture capital and bank loans are easily accessible: Ireland, Sweden, Luxembourg, the Netherlands, Germany, France, Belgium, Austria, the United Kingdom, Slovak Republic, and Bulgaria. Meanwhile, there are other European countries wherein the capital is more difficult to obtain: Portugal, Romania, Poland, Denmark, Lithuania, Italy, Greece, Slovenia, Spain, Croatia, and Hungary. During 2003-2013, Europe produced 30 "unicorns" (slang term which refers to the companies that have been evaluated billion dollars), compared with 39 in the USA. There are some sectors in Europe that encourage the development of new business; examples are music, financial technologies, and games. The most recent successes have been reported recently: Tradeshift, Soundcloud, Supercell. Start-ups operating in the food sector represent only a small percentage, but the EXPO could symbolize a new stimulus.

Special Focus: Italian Environment for Start-Ups

In Italy, there is a special department for start-up companies that identifies new innovative start-ups. In March 2015, the segment of "innovative start-ups" (Law decree 179/2012) (Government of Italy, 2012) counted about 3,711 companies with an increase of 530 units compared to 2014. Innovative start-ups represent 0.25% of the total number of companies in Italy. The total value of these new companies is estimated to be €192 million. However, when the authors analysed that segment of business, they have observed that most of the innovative start-ups provide business services or consultancy (73%), while others operate in the spheres of industrial manufacturing or new products creation (18.2%), and only a few start-ups work in distribution of new products (Table 1).

New start-ups are dominantly managed by men, while women head only 477 companies, which is equivalent to 12.9% of the total number of start-ups. Furthermore, young people are the founders of only 879 new companies (23.7% of the total). There are only 88 start-ups founded by foreigners in Italy (2.4% of the total). Start-ups are mainly located in the north of Italy. Lombardia is the Italian region that holds the most of the new companies (808, or 21.8% of the total) followed by Emilia Romagna with 451 new companies (12.2% of the national total). Lazio is the most important region in Central Italy with 347 new companies (9.4% of the total). Veneto and Piemonte have 274 and 266 new companies, equal to 7.4% and 7.2%, respectively. Italian regions with the smallest number of new companies are Basilicata with 20 start-ups, Molise with 16, and Valle d'Aosta with 10. Among the southern regions, Campania is the first with 214 new companies (5.8%) followed by Sicily and Puglia each of them hosts 156 new companies (4.20%). Calabria has only 91 new companies or 2.5% of the total number of a start-up in Italy (Table 2).

Big cities are more attractive for new companies. Milan is the main city in Italy chosen to start a new company. 533 companies are now located in Milan, a which is an equivalent to 14.4% of the total number of companies in the entire region. Meantime, Roma, Torino, Bologna, and Naples are also important in terms of presence of start-ups.

New companies have four partner founders and employ 2.50 people in average; return on investment (ROI) and return on equity (ROE) indicators of profit for innovative start-ups record negative values.

Typically, the evolution of start-ups in food sector occurs in two phases:

- Innovations of "incremental", an occurrence that introduces a new tool or channel for the marketplace.
- Creation of unusual service which is substantially discontinuous with the recent past.

Table 1. Breakdown of start-ups by sectors in Italy

Sector	Number of companies
Consulting and business services	2,708
New products production	677
Sells and commerce	153
Building	42
Agricultural and connected activity	13
Tourism	9
Logistics	10
Insurance and finance	6
Other sectors	93
Total	3,711

Source: Camere di Commercio d'Italia, 2015

Table 2. Geographic dispersion of start-ups in Italy

Region	Number of start-ups
Lombardia	808
Emilia Romagna	451
Lazio	347
Veneto	274
Piemonte	266
Toscana	238
Campania	214
Sicilia	156
Puglia	155
Marche	152
Trentino Alto Adige	128
Friuli Venezia Giulia	110
Sardegna	105
Calabria	91
Abbruzzo	70
Liguria	57
Umbria	43
Basilicata	20
Molise	16
Valle D'Aosta	10

Source: Camere di Commercio d'Italia, 2015

When a company starts to change existing product, it tries to create new products and services based on new unpublished experience.

In order to extend the marketplace, new companies should interconnect the real and the virtual market and produce new devices or apply innovations to traditional tools.

Nowadays, the market of "new products" is a subject to an increase of availability of new foods or "niche foods". Therefore, new companies could provide a wide variety of new products and incorporate them into the traditional distribution channel.

MAIN FOCUS OF THE CHAPTER

Insects

There are about sixty thousands vertebrate species on the Earth, and even if the humans (Homo sapiens) apparently dominate the world in term of their impact on the environment and use of resources, insects are the true rulers of the world with about one million of species. Biologically, the great diversity of insects is due to their co-evolution with plants and environment. The insects class is further divided into thirty orders such as Coleoptera, Diptera, or Hemiptera, so at last, it is estimated that about 75% of all animals that live on the Earth are insects. These organisms belong to a bigger systematic group; they are the most different species of animals living on the Earth, except open oceans, and occur inside all ecological niches. As pollinators, they play an important role in the reproduction of plants, biodegradation of organics substrates, waste, and humans' food. Various species of insects serve as carriers for pollens and concur to reproduction system of plants. Ingram, Nabhan, and Buchmann (1996) identified one hundred thousand pollinator species, and almost all of these (98%) are insects. Over 90% of 250.000 species of flowering plants depend on pollinators. An important role is also played by insects in biodegradation of waste (Simon, 2008). Beetles (4,000 species play a significant role in decomposition of dung), flies, ants, and termites break down organic matter until it is fit to be consumed by fungi and bacteria (Van Huis, 2003). Honey and silk are the most commonly known insect products. Insects provide to humans a wide variety of valuable products, for example, about 1.2 million tons of commercial honey is made by bees and 90.000 tons of silk is produced by silkworms every year (Van Huis, 2003; Yong-Woo, 1999). Insects are important for the production of colour red carmine and other colorants used to colour foods, textiles, and pharmaceuticals and for medical applications including maggot therapy (Sherman & Pechter, 1988; Wollina, Karte, Herold, & Looks, 2000), where they are used to treat infected wounds and burns (Van Huis, 2003). However, some of the insects are pests which inflict damage to humans, farm animals, and crops. Some insects pose a serious threat to humans' health, for example, mosquito, a vector of the malaria parasite, kills one million people every year. In agriculture, insects inflict severe damages and cause failure of one-fifth of the world's total crop production every year (Shelton, Zhao, & Roush, 2002; World Health Organization, 2014). The introduction of alien pests into new habitats due to the global exchange causes another severe damage (Snowden, 2010). Bringing a pest to a new geographical area, in most cases, may lead to critical complications (Snowden, 2010). Very particular example is Red palm weevil (Rhynchophorus ferrugineus Olivier, 1790). In Europe and the USA, this pest destroys all ornamental plants belonging to Phoenix species' palms, and it seems to be also involved in the destruction of other palms species. However, in some countries, people consume larvae of these species, and Red palm weevil is the most popular edible beetle in the tropics. Insects belong to Arthropoda's phylum along with crabs; lobsters and shrimps are common foods worldwide. In 2013, Jongema published a list (updated in June 2015) of edible insects of the world, where he presented about 2,000 of species of edible insects that are known to be consumed from the native populations in many countries (Jongema, 2015). In 2013, Van Huis et al. (2013) in FAO/INFOOD conference in Rome published a database of the biological food and included insects. This database is a collection of analytical data on biodiversity food from the published and unpublished sources. It reports 471 entries on about 5,700 values. Even if this database represents the largest compilation of data for this animal group, it seems to be incomplete so providing a definitive idea on the number of edible insect species seems to be difficult. The largest number of edible insects seems to come from North America, where Ramos-Elorduy, Moreno, and Camacho (2009), reported 549 species of edible insects from Mexico (Vantomme, 2015). The true problem about edible insects is that some species cannot be identified especially at the larval stage. Among the thirty orders of insects, only eighteen are considered edible, the most of these belonging to Coleoptera (31%) order followed by Lepidoptera (18%) and Hymenoptera (15%) (Vantomme, 2015).

Entomophagy, the word that indicates consumption of insects as food, is common in the world, mostly in Africa and Asia (Katayama et al., 2008; Aiking, 2011). It is also accepted in North America, but not well accepted among the Western European populations. In 2050, the global population is estimated to be about 9 billion people, and populations from developed countries will have to adapt to alternative sources of animal proteins because traditional breeding will become unsustainable (Aiking & De Boer, 2004; Nowak, Persijn, Rittenschober, & Charrondiere, 2014). In 2013, Caparros Megido et al. (2013) published the first study and results on a possible acceptance of entomophagy among Belgian people. The results were that even if 46.6% of respondents had a negative attitude to eating insects, about 77.7% of them were willing to try. Surprisingly, respondents in the age group from 19 to 45 years after trying some insects declared their readiness to cook them at home. After Caparros Megido et al. (2013) published the paper, a new brand smart-farm for insects breeding (Insect Europe) was established in the Netherlands. The company breeds, sells, and promotes edible insects, mainly crickets and grasshoppers, as human and animal food. In France, three people opened a shop (Europe-entomophagy) selling edible bugs. In January 2014, Jarrod Goldin with his brothers launched a new start-up company (Next Millennium Farms) located in Ontario. He was inspired by a landmark United Nation report, which recommended insects for human consumption. Next Millennium Farms (now Entomo Farms) began in 2014 with 5,000 square feet for mealworms and finished with $65,000 in sales. Insects is a future source of animal proteins. Establishing of new brand and start-ups for breeding and selling insects in Europe is the first step to increasing the number of potential consumers (Rumpold & Schlüter, 2013; De-Magistris, Pascucci, & Mitsopoulos, 2015).

Some business success stories for insect breeding are provided below.

1. Insect Europe BV. The company breeds and elaborates edible insects like crickets, mealworms, and grasshoppers for the food industry. Insect Europe BV undertook the edible insects business some years ago. Nowadays, products of the company are distributed and marketed worldwide. The target of Insect Europe BV is the output of "premium" insects for the food industries, with a focus on high quality and safety standards. Insect Europe BV could get the leadership among the edible insect's trade and could get the top of edible insects culture. Insect Europe BV created two different brands to sell its products: "Bugs for Pets" and "Delibugs". The two brands have two different markets: "Bugs for Pets" is for animal consumption (pet animals); "Delibugs" are products for human consumption. "Delibugs" spread and sell products derived by worms, crickets, grasshoppers, ants, scorpions, and butterflies.
2. Entomo Farms. The company elaborates, harvests, and distributes insects and their derivatives of the highest quality and provide alternatives to farmers producing healthier and certifiable organic fish and chicken. Entomo Farms has been the first business that used the "Entomophagy" term. They follow a way for the dissemination of insect farming methods and marketing of insect protein. Entomo Farms aims to be the world leader in providing protein alternatives in the form of the finest organic and excellent insect proteins with great tasting products and recipes.
3. Entonote was is founded in April 2014 by Giulia Maffei (biologist) and Giulia Tacchini (freelance) with a precise target to offer services and introduce entomophagy in Italy. This is the first experience in Italy. Even if it is not a start-up, it represents the only entity that actually organizes events and provides consultancy in the sphere of entomophagy.

Jellyfishes

Jellyfishes are aquatic, pluricellular organisms (salt and sweet waters) that belong to Cnidaria or Ctenophora phylum. They appeared in the oceans about six hundred million years ago (before the appearance of dinosaurs) and still populate the oceans today (Uye, 2008). Scientists classified about two hundred species of jellyfishes (distributed in 19 families and 50 genera) (Hsieh, Leong, & Rudloe, 2001). The edible jellyfishes are consumed in many countries of Southeast Asia but mostly in China, where domestic demand is increasing. China is the first country to process jellyfish for human consumption. Chinese people have been catching jellyfishes for 1,700 years, and consider them as a delicacy. Jellyfishes are fished in different methods (Iversen & Hale, 1992). Despite the fact that jellyfishes are consumed since ancient time, only recently this business has become commercially active. After catching, jellyfishes are sent to the local farm and then exported as soon as possible. Kitamura and Omori (2010) quote a short list of the six major edible jellyfishes harvested in Southeast Asia. They provide an identification key for a correct taxonomic identification. Fresh jellyfish readily spoil at ambient temperature. Therefore, processing of jellyfish is carried out preferably within a few hours of capture while the animals are still alive. The body of jellyfish consists of a hemispherical transparent umbrella. The mouth is on the under surface of the umbrella and is protected by fused oral arms, commonly known as legs. The umbrella and oral arms of jellyfish are separated immediately after catching. Jellyfishes are cleaned with seawater, scraped to remove mucus membranes and gonadal material. Both umbrella and oral arms are used in processing. Traditional methods of processing involve a step-wise reduction of the water content using salt and alum. A salt mix containing about 10% alum is used for initial salting of jellyfish using about 1 kg salt-alum mix for 8–10 kg of jellyfish. Salted jellyfishes are then left in the brine for 3–4 days, followed by several transfers to another container salted with a fresh mix containing a smaller amount of alum. Salted jellyfish can then be heaped and left dry on a draining rack at room temperature for 2 days, and the heap is turned upside down several times during that period to allow excess water to drain out through compression by its own weight. The entire process requires 20–40 days to produce a salted final product with 60–70% moisture and 16–25% salt (Subasinghe, 1992). The processed jellyfish has a yield of about 7–10% of the raw weight depending on the species and processing formula. Preservation of jellyfish in a mixture of salt and alum is necessary to obtain products of desirable structure and texture. Alum reduces pH, acts as a disinfectant and hardening agent, giving and maintaining a firm texture by precipitating protein. Salt aids in reducing water content and keeping the product microbial stable. Salt or alum used singularly in the processing of jellyfish does not produce a product of satisfactory properties. Extensive liquidation of tissue occurs in the absence of salt, while disagreeable odors develop in the absence of alum. Lowered pH greatly reduces the possibility of microbial growth and enhances the shelf life of the product. Colour of freshly processed jellyfish is creamy white but gradually turns yellowish as the product ages. In the Asian jellyfish market, a whiter colored product has higher retail value. Yellow, but not brown color, is acceptable. The longer the sample remains around, the darker it becomes. If it remains too long, it turns brown and the product is unacceptable. The salted jellyfish has a stable shelf life up to one year at room temperature. The shelf life can be increased to more than two years if the product is kept cool. Chinese wedding or formal banquet is rarely completed without a jellyfish salad. The Chinese have various methods for preparing jellyfish, cooked or uncooked. Fresh unprocessed jellyfishes are rich in minerals such as Na, Ca, K, and Mg, but processed product is depleted of salts after desalting in fresh water. Lipid content of the whole jellyfish ranges from 0.0046% to 0.2% on a wet weight basis (Subasinghe, 1992). Chinese products contain an average of 5.5 and 6.8 g of protein per 100 g of salted product, respectively (Morikawa, 1984).

Consumption in Italy: unofficial sources suggest that Asian restaurants located in Italy serve local jellyfish. Animals, caught by small boats, are treated with a substance similar to flour and carried to Chinese restaurants. Jellyfishes are also dried and exported.

Market characteristics: analysis of the industries (few) allowed detecting the strengths and weaknesses of the most important international companies. AS.MP Jellyfish Processing & Trading Com is a company for processing and trading salted jellyfish. The company is on the market since 1970, however, it sells its products in local markets in the Philippines. Actually, the company tries to sell its products on the international market with over 20 tons of jellyfishes. There is no another company of such caliber existing in Europe.

Algae

Algae were the first "plants-like" to appear on the Earth. Billions of years ago, they transformed the atmosphere allowing to other forms of life to evolve. The term "algae" is an imprecise word. In biology, with the term "algae" one can define a heterogeneous group of generally simple unicellular or multicellular eukaryotic or prokaryotic organisms, some smaller than a cell, others bigger than a football field. Algae are complex organisms fixing carbon dioxide and releasing oxygen. They are also among the biggest producers of organic substance. Even if all algae contain chlorophylls, pigments distinctly color them. The variety of their coloration is so wide that pigmentation may play an important role in algae classification (Buschmann, Troell, & Kautsky, 2001). Algae belong to one of three groups of organisms that biologists divided in red-algae, brown-algae, and green-algae (Kawashima, 1993; Lüning, Yarish, & Kirkman, 1990). The estimated total value of cultivated algae was \$7.8 billion in 2008, which was equivalent to approximately 15.8 million tons of fresh weight, 99.6% of which is dominated by seaweeds. The total of farmed algae in 2010 was estimated at 19 billion tons with a value of \$5.7 billion, in comparison to wild harvested algae, which was equivalent to approximately 800 thousand tons (Neori et al., 2004). Red algae are very delicate organisms and often appear graceful, with a branched structure and with a color range from pink to red made by a phycoerythrin pigment that hides chlorophylls. Even if some organisms inside this group are microscopic, many of them grow up to several centimeters. Some of them grow as thin filaments or leaflike structures. Red Algae includes several edible organisms as Porphyra (known as Nori) genera. There are approximately two thousand and five hundred species of red algae. Brown Algae, approximately about 1,800 species, are the smallest group of algae. This group includes macroscopic members only, with the one big seaweeds known (Kelp's greenwood). Their color ranges from brown to yellow because of the unique pigment, fucoxanthin. There are different brown algae, but only a few are consumed by humans, the main is Saccharina japonica known as Kombo. Green Algae is the most known kind of algae in the human diet. They grow as filaments or branching fronds in the oceans and seas worldwide. The most commonly eaten seaweeds in this algae group are marine lettuce (Ulva lactuca) and Enteromorpha intestinalis (Aonori). Most of the green algae dwell in freshwater, but the most consumed ones come from the sea. There are approximately ten thousand species of green algae, five thousand of which of marine origin (Lüning et al., 1990).

Cyanobacteria are sometimes called blue-green algae, but they are prokaryotic organisms and are not true algae. However, Arthrospira platensis, a cyanobacterium, is the main microalgae provider in terms of commercial production. Its cultivation is particularly attractive for several reasons. Humans eat some varieties of algae for ages. Edible algae, that the ancients used as food for centuries raise from coastal regions or internal lakes. Only a small part of these species of freshwater macro and micro-algae are used for diet because many of them are poisonous. Seaweeds are traditionally consumed as food in North and Southeast Asia. In the USA, some companies began cultivating seaweeds inside tanks for human consumption, and their markets grew so much that they exported these products to Asian countries.

However, in the last few years, there was a small spread of seaweed and freshwater microalgae into the European and occidental cuisine, where the interest for tropical and non-conventional food is now growing.

Red Algae for Food Industry

Among the Red Algae group, there are some species interesting for economic development. Chondrus crispus is a common organism which can be found on rock surfaces. Red seaweed was found on both sides of the Atlantic Ocean, along the Canadian line and from Iceland to Spain and Portugal. Chondrus crispus was also known as Irish Moss is a red macro-alga with a dark red-brown color, but bleaching may occur. Irish Moss extract has been used in Ireland since the beginning of the 19th century as a folk remedy for some illnesses. The name "carrageen" seems to be introduced around 1829; it probably came from Carrigan Head in Co. The use of Irish Moss or Carrageen Moss spread from Ireland, and its use was exported to the USA by emigrates. Small amounts are still collected in Ireland for cooking and as a health drink. Historically, carrageenan was extracted from wild seaweeds Chondrus crispus. However, supply from wild seaweed was insufficient. This has prompted a cultivation of carrageenan-containing seaweeds in tropical oceans. In 2008, about thirty countries launched carrageenan biomass production. Nowadays, according to FAO statistics, red seaweed production is 21% of the total production of all cultivated seaweeds, an equivalent of 9 million tons. The main red seaweed species under cultivation are Kappaphycus and Eucheuma (primarily raw materials for carrageenan), Gracilaria (used to extract agar), and Porphyra, or Nori (mainly for direct human consumption) (Buschmann, Correa, Westermeier, del Carmen Hernandez-Gonzalez, & Norambuena, 2001).

Agar and carrageenan are called hydrocolloids and are used as food additives and in medical or research laboratory. In the past decade, demand for hydrocolloids has grown with an increased consumption of processed food. Nevertheless, there was an insufficient supply of wild seaweed. For this reason, there was a motivated experimentation of cultivating carrageenan seaweeds in tropical waters. This resulted in a rapid expansion of Kappaphycus and Eucheuma cultivation from 944.000 wet tons in 2000 (48% of total red seaweed cultivation) to 5.6 million wet tons in 2010 (63%). Among various carrageenan-containing seaweeds, only Eucheuma has been mainly cultivated. Carrageenan has been used as a natural food additive for over 600 years, and today it is recognized as a harmless food additive without nutritional value. A vast amount of carrageenans is applied for dairy and water products. Nowadays, it is used for stabilization, thickening, and gelation of processed foods. It was used by the food industry in the USA and Europe and for new food products. It was used worldwide to enhance production of milk, toothpaste, flans, puddings, custards, jellies, confections, ice creams, chocolate, and nutritional beverage mixes. Carrageenan is also an approved food additive in many countries for humans and animals. With the development of the agar industry, interest for Gracilaria increased further (Armisen, 1995). The annual yield of agar is about 7,000 tons in the world, the most part constituted by Gracilaria species as a raw material. In many species, the content of agar is about 20-30%, or even more, and the quality is good. Experiments on artificial cultivation have been carried out in many countries (Kapraun, 1999). China, India, and Taiwan, the foremost countries in Gracilaria culture. It is carried out in China since 1984 (Guiry & Blunden, 1991).

Artificial cultivation of Gracilaria is carried out by two methods according to its biological characteristics. The first method is used for the species that can be propagated by the vegetative form and can be harvested to time during the whole growing period (examples are Gracilaria parvaspora and Gracilaria blodgettii). The second method is used for the species that may propagate through spores only (examples are Gracilaria asiatica and Gracilaria tenuistipitata). Agar was the first phycocolloid discovered and prepared as a purified extract. It has an ability to form gels upon cooling of a hot solution to 30–40°C and to melt to sols upon heating to 90–95°C. Agar has a wide variety of uses, other than microbiological use. It is engaged in the food industry, dairy industries, meat packing, pharmaceuticals, and in biomedicine. Agar is used in food industry predominantly for its stabilizing, thickening, and gelling characteristics (Armisen, 1995; Dawes, 1995; Friedlander & Levy, 1995).

Nori or Porphyra spp. is a red algae that grows as a very thin, flat, reddish blade. It is found in the most temperate zones worldwide, where it is consumed by the local populations. It is among the most nutritious seaweeds, with a protein content of about 50% (with the large amounts of three amino acids: alanine, glutamic acid, and glycine) and low content of sugars (below 1%). In addition, vitamin content is very high. During processing, most salt is washed away, so the sodium content is low. The manufactured products are Nori sheets each of about 20×20 cm size and 3.5–4.0 g weight. In ancient times, Porphyra could be collected by hand from natural sources, but now the most are derived from artificial cultivation. Good quality Nori is in demand in the Republic of Korea, where production methods differ between the northern and southern parts of the country. In the Republic of Korea, annual production volume is 270,000 wet tons (Armisen, 1995; Oliveira, Alveal, & Anderson, 2000). There are no data for North Korea. In the Republic of Korea, Nori lightly fried with a little oil is consumed as a popular snack with beer. The culture of Porphyra in non-Asian countries, in particular in the western and eastern coasts of the USA, used to be successful but became unviable commercially when residents of the shore areas objected to the presence of seaweed farms and access to sufficient space to expand the pilot farm was refused. On the east coast of Maine, cultivation of Porphyra had many problems with indigenous species that slowed the progress. Therefore, in the USA the biggest company in the sphere of food biotechnology was reorganized and started developing other businesses. Cultivation of Porphyra (Nori) in Japan yields about 500,000 wet ton per year, or $1.5 billion. Dried nori is oversupplied in Japan, and producers and dealers encourage nori consumption outside national borders, primarily in the USA. Production and markets in China are expanding, although the quality of the product is not always as good as that coming from the Republic of Korea and Japan. In China, Nori industry produces 210,000 wet tons per year (Oliveira et al., 2000). In many countries, processing of wet Porphyra into dried sheets of Nori are highly mechanized, rather like an adaptation of the paper-making process. Wet Porphyra is rinsed, chopped, and stirred in a paste. It is then poured onto a cloth and run through a dryer. Adjusting of the temperature carefully controls the rate of drying. The sheets are removed from the clothes and packed in about ten pieces to be distributed. This product is called "hoshi-nori" which distinguishes it from "yaki-nori", which is toasted and distributed in sealed packages. Nori is used as a luxury food mainly. In Japan, it is consumed with rice balls and sushi, a typical Japanese consisting food, or cut into small pieces and eaten with noodles. It can be incorporated into soy sauce to make a luxury sauce. In China, it is mostly used in soups and for seasoning fried foods.

Brown Algae for Food Industry

There are about 1,800 species of brown algae and their brown color results from the dominance of the xanthophyll pigment fucoxanthin, which masks other pigments, like Chlorophylls. There are no known unicellular organisms inside this group of algae. However, the biggest algae are known, for example, some members of the Macrocystis genus, may grow up to 70 meters and form prominent submerged kelp forests. Brown algae are adapted to a wide variety of marine ecological niches including rocky marine pools and intertidal zones. Only four species are restricted to dwell in freshwater. Among the Brown Algae group, there are some species perspective for economic development. Brown Giant Kelp algae Macrocystis pyrifera, Ascophyllum nodosum, and Laminaria hyperborean are harvested on the coasts of Atlantic coast, and they are used for alginate extraction (Lüning et al., 1990). Alginates are used mainly for commercial use as toothpaste, soap, and other applications. Its primary function is as a binder, stabilizer, and emulsifier agent. Saccharina japonica (Laminaria) is produced in Japan, China, and Southeast Asia for food and alginate production. "Kombu" is the Japanese name for the dried seaweed that is derived from a mixture of diverse Laminaria/Saccharina species. "Haidai" is the Chinese name for Saccharina japonica, the most cultivated species. There are three techniques for cultivation of kelp algae: vertical, horizontal, and mix method. These are similar in each of the countries that produce kelp products.

Vertical culture method is a technique which assumes exploitation of water space. It is a simple, easily managed procedure. Among the kelp line, there is a space that allows the light pass through them, thus stimulating sporophyte growth. However, this method has a drawback, indeed, the light decreases with depth, and the plants at the lower level may not get enough light to meet their requirements for good growth.

Horizontal culture method can be used in lower sea areas. Horizontally positioned culture line give greater and more distributed light to all of the attached plants compared to the vertical method. The disadvantage of this model is interwoven of algae line. This usually happens during storms (Oliveira et al., 2000).

Mixed culture method combines vertical and horizontal systems. Harvesting is an intensive procedure and depends on the seawater temperature. Kelp has to be harvested only when the temperature reaches 21°C. Naturally growing plants are biennial and are ready for harvesting after 20 months. Harvesting is made by manual operations: unfastening the algae line from the floating barges, loading the kelp into boats, and transporting them to the seashore for sun-drying or inside farm to process them. There are two main processing methods: immediate sun-drying or salting. Saccharina japonica species contain about 10% proteins and 2% percent fat, as well as an amount of minerals and vitamins. The cost of seedlings is currently about $0.50-0.60 per seedling-curtain (each with about 40,000–50,000 seedlings). No data on total production costs are available. According to Oliveira et al. (2000), global production of Saccharina japonica was about 5.9 million tons in 2013. Despite Saccharina japonica naturally grows and cultivated in the Republic of Korea, the demand in the country is lower because Koreans prefer consuming of Undaria pinnatifida. This species inhabits the intertidal zone to a depth of about 20 meters. It is introduced in Australia, by not consumed there. Undaria is highly invasive species; it grows rapidly and has the potential threat to native algal species. Harvesting is about 800,000 wet tons of three different species, and about 50% of this is for wakame. Produced from different brown seaweed, Undaria pinnatifida, it is grown in a similar fashion to Saccharina in China where this species was not as popular as Undaria (Yamanaka & Akiyama, 1993). Japan has developed different hybrids with higher growth and nutritional characteristics.

Green Algae for Food Industry

Green algae are the biggest group of algae and they may be unicellular (one cell), multicellular (many cells), and colonial (existing like an aggregation of cells). Most green algae are aquatic and are found commonly in freshwater and marine habitats, even if some are terrestrial. Some species live inside other organisms, like a symbiotic cell, an example is lichens. There are about 4,500 species, including about

2,500 Chlorophyceae (mostly freshwater organisms) and 250 marine Ulvophyceae, or seaweeds. Three important macroscopic edible Green Algae genera are marketed: Monostroma spp., Enteromorpha spp., and Ulva spp. First two genera are now cultivated in Japan and other countries of Southeast Asia, where they are consumed as Aonori (dried seaweed). They are cultivated in calm waters like bays and estuaries, but they may also be grown in deeper waters using floating lines (Monostroma spp.), like some species of Red Algae. They contain about 20% proteins and little fat and have a high content of vitamins and minerals (Blunden, 1991).

It is collected from the wild because it grows naturally, usually attached to rocks, but it may be cultivated using different methods. Ulva spp. has an important role for many coastal communities, and are currently used for numerous purposes. Various coastal communities utilized this resource since ancient time. Ulva is consumed worldwide mainly in soups and salads. Ulva spp. has higher protein content (proportion varies up to 40%) and lower lipid content than other two marketed Green macroscopic algae. Cultivation of Ulva spp. on a commercial scale has a low economic value, due to the high natural abundance (Guiry & Blunden, 1991). However, some cultivation method is selected. Open water cultivation is the oldest, but this genera may be cultivated with a suspended technique which is attained through the attachment of the seaweed to ropes, lines, floating rafts, and nets, or enclosed in cages. Semi-suspended cultivation, basket culture, raft culture, and long line cultivation are other techniques of cultivation. Among microscopic Green Algae, Chlorella genus belonging to the phylum Chlorophyta is the most important for food supply (Iversen & Hale, 1992). Cells of this microscopic species are spherical, subspherical or ellipsoid, single or forming colonies. This genus of freshwater green algae is composed of organisms which are the incredibly sustainable food source offering many important nutrients coupled with the high and complete value of proteins. Chlorella is another alga with high protein content (60%); it provides large quantities of folate, vitamin B-12, iron, and other minerals. However, the cell wall of Chlorella is rigid and indigestible. Therefore, harvested chlorella cells need to be processed in order to break the cell wall before using it as a dietary supplement for its nutrient content.

Green-Blue Algae (Cyanophyta)

Spirulina originated about 3.5 billion years ago. It is a primitive organism which established an ability to use carbon dioxide dissolved in seawater as a nutrient source for its reproduction. Spirulina is a Cyanophyta, or blue-green algae and is not an algae but bacterium. It is found in soil, marshes, freshwater, brackish water, and thermal springs. Alkaline, high temperatures (Spirulina shows an optimum growth between 35°C and 39°C), saline water (>30 g/l) with high pH values (11.0) and high level of solar radiation allowing it to grow rapidly, reaching a maximum density. Spirulina's main photosynthetic pigment is phycocyanin, but it also contains chlorophyll and carotenoids. Characteristic of the genus is a helical shape of the filaments (or trichomes). Spirulina naturally grew in the lakes of Texcoco, Chad, and Niger and under normal water conditions. This organism is one of many algal species, but it thrives in the alkaline lakes where it is difficult or impossible for other microorganisms to survive. Most commercial production systems are based on shallow raceways (open or closed), in which Spirulina cultures are mixed by a paddle wheel. However, there are still some examples of Spirulina which are harvested commercially from natural sites and local populations. Aztecs living in the Valley of Mexico collected Spirulina from a lake and made a like-paste "techuitlatl". Since 1967, Sosa-Texcoco Ltd has been harvesting Spirulina maxima from an area of Texcoco lake. The first pilot plant that produced 150 tons of dry spirulina biomass per year started production in 1973. Sosa-Texcoco Ltd has invested about $5 million in the development of production since 1977. In 1995, Sosa-Texcoco ceased production of Spirulina. Four volcanic lakes with natural Spirulina blooms in Myanmar have been reported to be used as a production site. Production began in 1988, and since 1999 increased to 100 tons/year. About 70% is harvested from boats on the surface of the lake, and about 30% is grown in outdoor ponds alongside the lake. Habib, Parvin, Huntington, and Hasan (2008) suggested that the mass culture of algae (labeled as one of the "future foods") would help to overcome global protein shortages. However, two of the main barriers to the small-scale culture of Spirulina are the costs and availability of inorganic nutrients. Gami, Naik, and Patel (2011) investigated the use of a medium formulated for mass production of Spirulina that is different from the laboratory medium. This newly formulated medium contains single super phosphate (1.25 g/litre), sodium nitrate (2.50 g/litre), sodium chloride (0.50 g/litre), magnesium sulfate (0.15 g/litre), calcium chloride (0.04 g/litre), and sodium bicarbonate (8 g/litre). There are some alternatives for Spirulina culturing, especially from waste effluents, which may be available in rural locations. Mass cultivation of Spirulina is usually carried out in shallow raceway ponds, equipped with paddle wheels to mix the culture. Two types of open raceway ponds are typically used: the first is lined by concrete and is, therefore, expensive, the second is a shallow earthen tunnel lined with polyvinyl-chloride (PVC) or some other durable plastic material. The first type even being expensive for manufacturers could bring bacterial or other contaminations. The surface of commercial raceways varies from 0.1 to 0.8 hectares, and culture depth is usually maintained at 10–25 cm (D'Aleo & D'Aleo, 2015a, 2015b). The paddle wheel, large or small with a speed from 10 rpm is the most common stirring device. If the paddle wheel is not sufficient for shifting the culture or the culture is over-growing, self-shading governs the light availability to the single cell of the culture. Thus a certain part of the culture always fails to receive enough light to fulfill photosynthesis needs, the light will be limited. Torzillo, Carlozzi, Pushparaj, Montaini, and Materassi (1993) showed the Spirulina growing in polyethylene tubes in Italy. Spirulina is produced in at least 22 countries: Benin, Brazil, Burkina Faso, Chad, Chile, China, Costa Rica, Côte d'Ivoire, Cuba, Ecuador, France, India, Madagascar, Mexico, Myanmar, Peru, Israel, Spain, Thailand, Togo, the USA, and Vietnam. According to Van Huis et al. (2013), total industrial production of Spirulina in China was 19.1 million tons in 2003 and 41.60 million tons in 2004 ($7.6 million and $16.6 million, respectively). China started industrial production of Spirulina in 1990, and now the country has over 200 factories. Other countries with significant Spirulina production facilities the USA and France. The USA have the largest intensive farms in the world, mainly located in Hawaii and California. Earthrise has a plot of about 50 hectares located in the Sonoran Desert that produced about 450 tons in 2012. It uses water from the mineral-rich Colorado river. Cyanotech located in Hawaii produces 350 tons of Spirulina annually and use a combination of fresh water and deep ocean water (Guiry & Blunden, 1991; Iversen & Hale, 1992).

Some business success stories for Spirulina production are provided below:

1. Micro-Life. In Italy, Spirulina is produced by Micro-Life farm only which sells different algae products and has approximately €1 million in revenues. Micro-Life's products do not contain any kind of animal ingredients and hence are 100% vegan. Micro-Life is the first Italian company that may produce organic Spirulina. In 2014, it obtained organic certification. High revenues are derived from cosmetics made with Spirulina.

2. There are two new start-ups in southern Italy: Spirulina-Microfarm in Calabria and Apulia Mundi in Puglia. Spirulina Microfarm is specialized in consulting and in the creation of innovative and sustainable systems for Spirulina production. The company is equipped with a laboratory producing small amounts of high-quality product.

Business Model for Novel Food Start-Ups

The aim is to identify the general business model for the entire novel foods, even if the data are scarce. The companies which operate in this field should own some common characteristics. Most of these companies are founded by researchers, biologists, and life cycle experts. It is not possible to imagine a new start-up that operates without a scientist (Aceleanu, Serban, & Burghelea, 2015). These companies, at least initially, have a close relationship with the university or research center. There are no suppliers of such novel food as jellyfishes in Europe, hence there are foreign companies that supply the market (Alvarez, Mackalski, Loeb, & Mazzanti, 2015). However, there are some European suppliers for algae. These new start-ups (Belz & Binder, 2015) are able to connect profits and environmental resources to preserve the nature and contain some of the most important food industry trends: healthier foods, environmentally friendly, natural products, new consumption experience and the opportunity. Initially, the only diffusion channels were the Internet (the potential market is infinite) or small shops dedicated to the sustainable foods. The companies that operate in the algae production may sell their products in pharmacies or health food stores which are very broad distribution channels (D'Aleo, 2016). The Mediterranean products are considered high-quality commodity (Frewer et al., 2001). The researches are important sources for these companies; they provide counsels, trainings, and information. Usually, these companies offer packages, meetings, or guides for all consumers who would like to become small producers. The typical consumer of these products is a man/woman aged between 20 and 40 years with higher education and high disposable income (Hallström, Carlsson-Kanyama, & Börjesson, 2015). These consumers know that novel foods contain higher amounts of nutritional compounds like vitamins, antioxidants, and polyunsaturated fatty acids and smaller amounts of unwanted nutritional compounds such as heavy metals, mycotoxins, and pesticide residues. Therefore, novel foods are better for health and environment than conventional ones in terms of the essential elements and active environmental protection (Frewer, Risvik, & Schifferstein, 2001). Initial costs to start a new company in this area are usually high. Such high costs are because of required permissions (such costs present in all food sectors), high skilled human resources, and machinery. These high initial costs allow getting in active about after 3/4 years or collapsing about at the first months of the second year of activity.

We are sure that in common with all food startups in this sector there are the essential and fundamental elements to give more chance to a business project:

1. Solve a logistical limit.
2. Involve more suppliers.
3. Offer a flexible service to customers.
4. Give guarantees on quality.
5. Propose different methods of taste and try.
6. Reach agreements with local entities.

7. Reduce purchasing choice stress.
8. Draw aggregation tools Win/Win.
9. Strong marketing strategy to create awareness.

SOLUTIONS AND RECOMMENDATIONS

It is proved, therefore, that food tastes evolve from genetic factors such as the preference for sweet or salty and the refusal to bitter or sour, and it is also known that there is an innate and automatic regulation of appetite. The neuroendocrine mechanism behind these individual variations expressed through the interrelationship between hormones and neurotransmitters so totally unique and genetically predetermined for each of us. However, from birth (in truth already in utero), genetics in its psychophysical manifestations (such as body size, perceptions, behavior) is directed and shaped by life experiences. Experiences will determine whether caloric intake gradually matches the energy expenditure. It is essential, in order to perceive the new foods as "normal", since childhood, even in nations with strong culinary traditions and food, that these new foods are gradually introduced into the daily diets. Therefore, the families must be the first promoters of these new sources, along with schools with proper food education. Certainly, it is appropriate that the state invests in specific information campaigns so that the adult population, skeptical about these new foods, can enjoy all the necessary information. The potential economic impact of these new sources must be conveyed by the state that must invest in measures able to stimulate investment in this sector (Marsden & Smith, 2005). Large industries that operate in the food sector have no interest in this type of novel food that may create distortions in supply chain and demand side. Thus, only small startups supported by government subsidies and an adequate information campaign that starts on the level of families and school altogether represent the driving force for this still embryonic sector.

FUTURE RESEARCH DIRECTIONS

In order to establish a complete vision of the issue, it is necessary to understand the nature of evolution in food requirements for target populations. Indeed, the researchers must be able to tell if it is better to focus on the marketing and commercial strategies, to specific segments of the population and consumers (Kim & Mauborgne, 2015), who may be more sensitive to the trial, or if a general marketing strategy is most appropriate. Researchers should focus future studies to understand how to make these new foods more acceptable. In fact, the way of presenting, serving, preparing, and packaging of novel food become crucial variables.

CONCLUSION

Novel foods could be subjects to boycott by new consumers who could get wrong information from mass media or Internet. These new products are priced higher in comparison to the conventional ones. They possess greater organic value but because consumers get little information through the conventional channels, novel foods have a small market. Professionals and experts in this field promote the reputation of novel foods. Therefore, this knowledge is deeply niched and restricted to regular consumers of novel food only. Food market in Europe and Mediterranean countries is saturated by traditional food products; it is difficult to enter this mature market with novel foods. Therefore, legislation in the sphere of promotion of novel foods is needed. It should help developing logistic and distribution, establishing of sector agreements and ensuring growth in the novel food sphere along with adequate quantitative controls and awareness of potential consumers. Furthermore, related national institutions should provide special funds for start-ups which would like to enter to the novel foods market, a marketplace of high economic potential.

REFERENCES

Aceleanu, M. I., Serban, A. C., & Burghelea, C. (2015). Greening the Youth Employment – A Chance for Sustainable Development. *Sustainability, 7*(3), 2623–2643. doi:10.3390u7032623

Aiking, H. (2011). Future Protein Supply. *Trends in Food Science & Technology, 22*(2), 112–120. doi:10.1016/j.tifs.2010.04.005

Aiking, H., & De Boer, J. (2004). Food Sustainability: Diverging Interpretations. *British Food Journal, 106*(5), 359–365. doi:10.1108/00070700410531589

Alvarez, J.B., Mackalski, R., Loeb, A., & Mazzanti, L. (2015). Leading Sustainable Change: An Organizational Perspective. *Strategic Management Journal, 34*(13), 1209–1231.

Armisen, R. (1995). Worldwide Use and Importance of Gracilaria. *Journal of Applied Phycology, 7*(3), 231–243. doi:10.1007/BF00003998

Belz, F. M., & Binder, J. K. (2015). Sustainable Entrepreneurship: A Convergent Process Model. *Business Strategy and the Environment.* Retrieved November 26, 2016, from http://onlinelibrary.wiley.com/doi/10.1002/bse.1887/abstract

Blunden, G. (1991). Agricultural Uses of Seaweeds and Seaweed Extracts. In M. D. Guiry & G. Blunden (Eds.), *Seaweed Resources in Europe: Uses and Potential* (pp. 65–81). Chichester, UK: John Wiley & Sons Ltd.

Brookes, G. (2007). *Economic Impact Assessment of the Way in which the EU Novel Foods Regulatory Approval Procedures Affect the EU Food Sector*. Retrieved November 26, 2016, from http://www.pgeconomics.co.uk/archive.php

Burlingame, B., Charrondiere, R., & Mouille, B. (2009). Food Composition is Fundamental to the Cross-Cutting Initiative on Biodiversity for Food and Nutrition. *Journal of Food Composition and Analysis, 22*(5), 361–365. doi:10.1016/j.jfca.2009.05.003

Buschmann, A. H., Correa, J. A., Westermeier, R., del Carmen Hernandez-Gonzalez, M., & Norambuena, R. (2001). Red Algal Farming in Chile: A Review. *Aquaculture (Amsterdam, Netherlands), 194*(3), 203–220. doi:10.1016/S0044-8486(00)00518-4

Buschmann, A. H., Troell, M., & Kautsky, N. (2001). Integrated Algal Farming: A Review. *Cahiers de Biologie Marine, 42*(1–2), 83–90.

Caparros Megido, R., Sablon, L., Geuens, M., Brostaux, Y., Alabi, T., Blecker, C., ... Francis, F. (2013). Edible Insects Acceptance by Belgian Consumers: Promising Attitude for Entomophagy Development. *Journal of Sensory Studies*, *29*(1), 14–20. doi:10.1111/joss.12077

D'Aleo, F., & D'Aleo, V. (2015a). *Preliminary Study in Open Raceway for Production of Spirulina Biomass in Southern Italy. First Experience in Calabria.* Poster presented at the 110° Congresso della Società Botanica Italiana, Pavia.

D'Aleo, F., & D'Aleo, V. (2015b). *Start-up Company for Production of Spirulina Biomass in Southern Italy. First Experience in Calabria.* Poster presented at the 43° Congresso Società Italiana di Microbiologia, Napoli.

D'Aleo, V., & Morabito, G. (2015). Verso il prototipo di start-up mediterranea. Confronto fra modelli di politiche pubbliche, incubatori e acceleratori. *Proceedings of the BIC-Italia Net Congress & EBN AGM*. Milan: BIC; EBN AGM.

DAleo, V. (2016). Supply Chain Management: Overview, Competition and Competences, How to Exploit the Hidden Capabilities. *International Journal of Management and Network Economics*, *3*(4), 336–346. doi:10.1504/IJMNE.2016.079878

Darkow, I. L., Foerster, B., & von der Gracht, H. A. (2015). Sustainability in Food Service Supply Chains: Future Expectations from European Industry Experts Toward the Environmental Perspective. *Supply Chain Management: An International Journal*, *20*(2), 163–178. doi:10.1108/SCM-03-2014-0087

Dawes, C. P. (1995). Suspended Cultivation of Gracilaria in the Sea. *Journal of Applied Phycology*, *7*(3), 303–313. doi:10.1007/BF00004004

De-Magistris, T., Pascucci, S., & Mitsopoulos, D. (2015). Paying to See a Bug on my Food: How Regulations and Information Can Hamper Radical Innovations in the European Union. *British Food Journal*, *117*(6), 1777–1792. doi:10.1108/BFJ-06-2014-0222

di Commercio d'Italia, C. (2015). *Infocamere.* Retrieved December 12, 2016, from http://www.camcom.gov.it/

European Parliament. (1997). *Regulation No 258/97 on Novel Foods and Novel Food Ingredients.* Retrieved December 12, 2016, from http://eur-lex.europa.eu/legal-content/EN/TXT/PDF/?uri=CELEX:31997R0258&from=EN

Frewer, L. J., Bergmann, K., Brennan, M., Lion, R., Meertens, R., Rowe, G., ... Vereijken, C. (2011). Consumer Response to Novel Agri-Food Technologies: Implications for Predicting Consumer Acceptance of Emerging Food Technologies. *Trends in Food Science & Technology*, *22*(8), 442–456. doi:10.1016/j.tifs.2011.05.005

Frewer, L. J., Risvik, E., & Schifferstein, H. (2001). Food, People and Society: A European Perspective of Consumers' Food Choices. Dordrecht: Springer Science+Business Media B.V. doi:10.1007/978-3-662-04601-2

Friedlander, M., & Levy, I. (1995). Cultivation of Gracilaria in Outdoor Tanks and Ponds. *Journal of Applied Phycology*, *7*(3), 315–324. doi:10.1007/BF00004005

Gami, B., Naik, A., & Patel, B. (2011). Cultivation of Spirulina Species in Different Liquid Media. *Journal of Algal Biomass Utilization*, *2*(3), 15–26.

Government of Italy. (2012). *Testo del decreto-legge 18 ottobre 2012. Pubblicato nel supplemento ordinario n. 194/L alla 19 ottobre 2012, No. 245. Coordinato con la legge di conversione 17 dicembre 2012, No. 221. Recante: "Ulteriori misure urgenti per la crescita del Paese" (Publication No. 179).* Rome: Gazzetta Ufficiale.

Guiry, M. D., & Blunden, G. (Eds.). (1991). *Seaweed Resources in Europe: Uses and Potential*. Chichester, UK: John Wiley & Sons Ltd.

Habib, M. A. B., Parvin, M., Huntington, T. C., & Hasan, M. R. (2008). *A Review on Culture, Production and Use of Spirulina as Food for Humans and Feeds for Domestic Animals*. Rome: Food and Agriculutre Organization of the United Nations.

Hallström, E., Carlsson-Kanyama, A., & Börjesson, P. (2015). Environmental Impact of Dietary Change: A Systematic Review. *Journal of Cleaner Production*, *91*, 1–11. doi:10.1016/j.jclepro.2014.12.008

Hepburn, P., Howlett, J., Boeing, H., Cockburn, A., Constable, A., Davi, A., ... Wal, J. M. (2008). The Application of Post-Market Monitoring to Novel Foods. *Food and Chemical Toxicology*, *46*(1), 9–33. doi:10.1016/j.fct.2007.09.008 PMID:17950974

Hermann, M. (2009). The Impact of the European Novel Food Regulation on Trade and Food Innovation Based on Traditional Plant Foods from Developing Countries. *Food Policy*, *34*(6), 499–507. doi:10.1016/j.foodpol.2009.08.005

Hsieh, Y. P., Leong, F. M., & Rudloe, J. (2001). Jellyfish as Food. *Hydrobiologia*, *451*(1–3), 11–17. doi:10.1023/A:1011875720415

Ingram, M., Nabhan, G., & Buchmann, S. L. (1996). Our Forgotten Pollinators: Protecting the Birds and Bees. *Global Pesticide Campaigner*, *6*(4), 1–8.

Iversen, E. S., & Hale, K. K. (1992). Species for Human Food. In E. S. Iversen & K. K. Hale (Eds.), *Aquaculture Sourcebook* (pp. 35–204). New York, NY: Springer US. doi:10.1007/978-1-4684-1428-8_3

Jongema, Y. (2015). *List of Edible Insects of the World*. Retrieved November 26, 2016, from http://www.wur.nl/en/Expertise-Services/Chair-groups/Plant-Sciences/Laboratory-of-Entomology/Edible-insects/Worldwide-species-list.htm

Kapraun, D. F. (1999). Red Algal Polysaccharide Industry: Economics and Research Status at the Turn of the Century. In *Proceedings of the Sixteenth International Seaweed Symposium*. Cebu City: Springer-Science+Business Media, B.V. 10.1007/978-94-011-4449-0_2

Katayama, N., Ishikawa, Y., Takaoki, M., Yamashita, M., Nakayama, S., Kiguchi, K., ... Force, S. A. T. (2008). Entomophagy: A Key to Space Agriculture. *Advances in Space Research*, *41*(5), 701–705. doi:10.1016/j.asr.2007.01.027

Kawashima, S. (1993). Cultivation of the Brown Alga, Laminaria "Kombu". In M. Ohno & A. T. Critchley (Eds.), *Seaweed Cultivation and Marine Ranching* (pp. 25–40). Yokosuka: Japan International Cooperation Agency.

Kim, W. C., & Mauborgne, R. (2015). *Blue Ocean Strategy, Expanded Edition: How to Create Uncontested Market Space and Make the Competition Irrelevant*. Brighton, MA: Harvard Business Review Press.

Kitamura, M., & Omori, M. (2010). Synopsis of Edible Jellyfishes Collected from Southeast Asia, with Notes on Jellyfish Fisheries. *Plankton and Benthos Research*, *5*(3), 106–118. doi:10.3800/pbr.5.106

Lüning, K., Yarish, C., & Kirkman, H. (1990). *Seaweeds: Their Environment, Biogeography, and Ecophysiology*. Hoboken, NJ: John Wiley & Sons.

Marsden, T., & Smith, E. (2005). Ecological Entrepreneurship: Sustainable Development in Local Communities through Quality Food Production and Local Branding. *Geoforum*, *36*(4), 440–451. doi:10.1016/j.geoforum.2004.07.008

Morikawa, T. (1984). Jellyfish. *FAO Infofish Marketing Digest*, *1*, 37–39.

Neori, A., Chopin, T., Troell, M., Buschmann, A. H., Kraemer, G. P., Halling, C., ... Yarish, C. (2004). Integrated Aquaculture: Rationale, Evolution, and State of the Art Emphasizing Seaweed Biofiltration in Modern Mariculture. *Aquaculture (Amsterdam, Netherlands)*, *231*(1), 361–391. doi:10.1016/j.aquaculture.2003.11.015

Nowak, V., Persijn, D., Rittenschober, D., & Charrondiere, U. R. (2014). Review of Food Composition Data for Edible Insects. *Food Chemistry*, *193*, 36–46. PMID:26433285

Oliveira, E. C., Alveal, K., & Anderson, R. J. (2000). Mariculture of the Agar-Producing Gracilarioid Red Algae. *Reviews in Fisheries Science*, *8*(4), 345–377. doi:10.1080/10408340308951116

Ramos-Elorduy, J., Moreno, J. M. P., & Camacho, V. H. M. (2009). Edible Aquatic Coleoptera of the World with an Emphasis on Mexico. *Journal of Ethnobiology and Ethnomedicine*, *5*(11). Retrieved from https://ethnobiomed.biomedcentral.com/articles/10.1186/1746-4269-5-11 PMID:19379486

Rumpold, B. A., & Schlüter, O. K. (2013). Potential and Challenges of Insects as an Innovative Source for Food and Feed Production. *Innovative Food Science & Emerging Technologies*, *17*, 1–11. doi:10.1016/j.ifset.2012.11.005

Shelton, A. M., Zhao, J. Z., & Roush, R. T. (2002). Economic, Ecological, Food Safety, and Social Consequences of the Deployment of Bt Transgenic Plants. *Annual Review of Entomology*, *47*(1), 845–881. doi:10.1146/annurev.ento.47.091201.145309 PMID:11729093

Sherman, R. A., & Pechter, E. A. (1988). Maggot Therapy: A Review of the Therapeutic Applications of Fly Larvae in Human Medicine, Especially for Treating Osteomyelitis. *Medical and Veterinary Entomology*, *2*(3), 225–230. doi:10.1111/j.1365-2915.1988.tb00188.x PMID:2980178

Simon, J. Y. (2008). Detoxification Mechanisms in Insects. In J.L. Capinera (Ed.), Encyclopedia of Entomology (pp. 1187–1201). Dordrecht: Springer Science+Business Media B.V.

Snowden, A. L. (2010). Post-Harvest Diseases and Disorders of Fruits and Vegetables: Vol. 1. *General Introduction and Fruits*. Boca Raton, FL: CRC Press.

Subasinghe, S. (1992). Jellyfish Processing. *INFOFISH International*, *4*, 63–65.

Torzillo, G., Carlozzi, P., Pushparaj, B., Montaini, E., & Materassi, R. (1993). A Two-Plane Tubular Photobioreactor for Outdoor Culture of Spirulina. *Biotechnology and Bioengineering, 42*(7), 891–898. doi:10.1002/bit.260420714 PMID:18613137

Uye, S. I. (2008). Blooms of the Giant Jellyfish Nemopilema Nomurai: A Threat to the Fisheries Sustainability of the East Asian Marginal Seas. *Plankton and Benthos Research, 3*(Supplement), 125–131. doi:10.3800/pbr.3.125

Van Huis, A. (2003). Medical and Stimulating Properties Ascribed to Arthropods and Their Products in Sub-Saharan Africa. In E. Motte-Florac & J. M. C. Thomas (Eds.), *Insects in Oral Literature and Traditions* (pp. 367–382). Paris: Peeters.

Van Huis, A., van Itterbeeck, J., Klunder, H., Mertens, E., Halloran, A., Muir, G., & Vantomme, P. (2013). *Edible Insects: Future Prospects for Food and Feed Security*. Rome: Food and Agriculture Organization of the United Nations.

Vantomme, P. (2015). Way forward to Bring Insects in the Human Food Chain. *Journal of Insects as Food and Feed, 1*(2), 121–129. doi:10.3920/JIFF2014.0014

Wollina, U., Karte, K., Herold, C., & Looks, A. (2000). Biosurgery in Wound Healing – The Renaissance of Maggot Therapy. *Journal of the European Academy of Dermatology and Venereology, 14*(4), 285–289. doi:10.1046/j.1468-3083.2000.00105.x PMID:11204517

World Health Organization. (2014). *World Malaria Report 2013*. Geneva: World Health Organization.

Yamanaka, R., & Akiyama, K. (1993). Cultivation and Utilization of Undaria Pinnatifida (Wakame) as Food. *Journal of Applied Phycology, 5*(2), 249–253. doi:10.1007/BF00004026

Yong-Woo, L. (1999). *Silk Reeling and Testing Manual*. Rome: Food and Agriculture Organization of the United Nations.

ADDITIONAL READING

Bixler, H. J. (1996). Recent Developments in Manufacturing and Marketing Carrageenan. *Hydrobiologia, 326*(1), 35–57. doi:10.1007/BF00047785

Crouch, I. J., & Van Staden, J. (1993). Evidence for the Presence of Plant Growth Regulators in Commercial Seaweed Products. *Plant Growth Regulation, 13*(1), 21–29. doi:10.1007/BF00207588

Jiang, L., Hoegg, J., Dahl, D. W., King, J., Metzger, M., Owen, S., & Kahle, L. R. (2015). Balanced Innovations in New Product Development: Sunny Faces, Sunny Technology. In *Proceedings of the 2010 Academy of Marketing Science (AMS) Annual Conference*. New York, NY: Springer International Publishing. 10.1007/978-3-319-11797-3_165

Jonker, J. (2015). The Future of Corporate Sustainability: Towards an Ecology of Organisations Focused on Sustainability. In L. O'Riordan, P. Zmuda, & S. Heinemann (Eds.), *New Perspectives on Corporate Social Responsibility* (pp. 23–47). Wiesbaden: Springer Fachmedien Wiesbaden. doi:10.1007/978-3-658-06794-6_2

Kenis, A., & Lievens, M. (2015). Greening the Economy or Economizing the Green Project? When Environmental Concerns are Turned into a Means to Save the Market. *Review of Radical Political Economics*. Retrieved November 26, 2016, from http://rrp.sagepub.com/content/early/2015/07/09/0486613415591803.full.pdf+html

Larsen, H. B., Lindkvist, K. B., Sánchez-Hernández, J. L., & Trondsen, T. (2015). Analysing Food Chain Development: A Theoretical Framework. In K. B. Lindkvist & T. Trondsen (Eds.), *Nordic-Iberian Cod Value Chains* (pp. 7–22). Basel: Springer International Publishing. doi:10.1007/978-3-319-16405-2_2

Leonidou, L.C., Christodoulides, P., Kyrgidou, L.P., & Palihawadana, D. (2015). Internal Drivers and Performance Consequences of Small Firm Green Business Strategy: The Moderating Role of External Forces. *Journal of Business Ethics*, 1–22. Retrieved November 26, 2016, from file:///C:/Users/Vasily/Downloads/10.1007%252Fs10551-015-2670-9.pdf.

Mepham, B. (2000). A Framework for the Ethical Analysis of Novel Foods: The Ethical Matrix. *Journal of Agricultural & Environmental Ethics*, *12*(2), 165–176. doi:10.1023/A:1009542714497

Rollin, F., Kennedy, J., & Wills, J. (2011). Consumers and New Food Technologies. *Trends in Food Science & Technology*, *22*(2), 99–111. doi:10.1016/j.tifs.2010.09.001

Salafsky, N., Salzer, D., Stattersfield, A. J., Hilton-Taylor, C., Neugarten, R., Butchart, S. H., ... Wilkie, D. (2008). A Standard Lexicon for Biodiversity Conservation: Unified Classifications of Threats and Actions. *Conservation Biology*, *22*(4), 897–911. doi:10.1111/j.1523-1739.2008.00937.x PMID:18544093

Schneider, U. A., Havlík, P., Schmid, E., Valin, H., Mosnier, A., Obersteiner, M., ... Fritz, S. (2011). Impacts of Population Growth, Economic Development, and Technical Change on Global Food Production and Consumption. *Agricultural Systems*, *104*(2), 204–215. doi:10.1016/j.agsy.2010.11.003

Shahidullah, A. K. M., & Haque, C. E. (2015). Green Microfinance Strategy for Entrepreneurial Transformation: Validating a Pattern towards Sustainability. *Enterprise Development and Microfinance*, *26*(4), 325–342. doi:10.3362/1755-1986.2015.027

Sili, C., Torzillo, G., & Vonshak, A. (2012). Arthrospira (Spirulina). In B. A. Whitton (Ed.), *Ecology of Cyanobacteria II: Their Diversity in Space and Time* (pp. 677–705). Heidelberg: Springer Verlag. doi:10.1007/978-94-007-3855-3_25

Taylor, J., Garat, J. P., Simreen, S., & Sarieddine, G. (2015). An Industry Perspective: A New Model of Food Safety Culture Excellence and the Impact of Audit on Food Safety Standards. *Worldwide Hospitality and Tourism Themes*, *7*(1), 78–89. doi:10.1108/WHATT-12-2014-0041

Verbeke, W. (2005). Consumer Acceptance of Functional Foods: Socio-Demographic, Cognitive and Attitudinal Determinants. *Food Quality and Preference, 16*(1), 45–57. doi:10.1016/j.foodqual.2004.01.001

Wilcock, A., Pun, M., Khanona, J., & Aung, M. (2004). Consumer Attitudes, Knowledge and Behaviour: A Review of Food Safety Issues. *Trends in Food Science & Technology, 15*(2), 56–66. doi:10.1016/j.tifs.2003.08.004

KEY TERMS AND DEFINITIONS

Alternative Sources: A food that is regarded as a healthier alternative to food that is prepared or produced by conventional methods. Alternative sources of food which mean that substitute food can replace the principal food can help to solve food problems. Many kinds of alternative food sources can be helpful to solve food problems such as starvation, lack of nutrition, and economy and business crisis.

Economic Impact: An effect that an event, policy change, or market trend will have on economic factors such as consumer confidence, stock market activity, and others. Events such as regulatory changes, supply shortages or introduction of new products can have a significant economic impact due to the way that they affect business activities.

Emerging Trend: A trend in which people are becoming more interested over a period of time.

Food: Anything serving for consumption or use.

Green Start-Ups: Used to describe someone who is not experienced, either in a particular area or in an overall sense. A business functioning in a capacity where no negative impact is made on the local or global environment, the community, or the economy.

Novel Foods: Newly developed, innovative food or food produced using new technologies and production processes as well as food traditionally eaten.

Safety Foods: Refers to the conditions and practices that preserve the quality of food. These practices prevent contamination and foodborne illnesses.

This research was previously published in Establishing Food Security and Alternatives to International Trade in Emerging Economies edited by Vasily Erokhin, pages 74-97, copyright year 2018 by Business Science Reference (an imprint of IGI Global).

Chapter 22
Human Overpopulation and Food Security:
Challenges for the Agriculture Sustainability

Rishikesh Singh
Banaras Hindu University, India

Pratap Srivastava
Banaras Hindu University, India

Pardeep Singh
University of Delhi, India

Shweta Upadhyay
Banaras Hindu University, India

Akhilesh Singh Raghubanshi
Banaras Hindu University, India

ABSTRACT

World population is rapidly growing and expected to reach in between 8.5 to 12 billion by 2100. More than 75% of the population is expected to inhabit in the African and Asian countries having most of the developing nations. The overpopulation leads to a state of food insecurity that induced the evolution of resource-exhaustive agriculture causing irreparable environmental damages. Now the challenge is to feed more with less environmental damages. Adoption of technologically-sound, traditional knowledge inclusive, socio-economically sensible recommended agricultural practices can be the basis for achieving future dietary demands. However, before wider recommendation, their environmental impact assessment at various sustainability issues is necessitated for a holistic understanding of the future agriculture. The challenges of overpopulation and food security can only be managed by identifying the core areas of research and development under different agricultural sectors. The present chapter will provide a brief dimension on some of these key issues.

DOI: 10.4018/978-1-5225-8063-8.ch022

INTRODUCTION

Agriculture is the oldest way where humans interact with natural systems for crop and livestock production by alteration of natural resources (land, water, nutrients, biomass and energy) for their well-being (Dale et al., 2013). It is the world's largest industry occupying about 38% (including cropland and pasture land) of the Earth's terrestrial surface (Robertson & Swinton, 2005; Ramankutty et al., 2008; FAOSTAT, 2011; Foley et al., 2011). In addition, it utilizes about 70% of global water withdrawals from freshwater bodies (FAO, 2011). It involves various interrelated activities like land management, settlement patterns, crop selection and livestock production throughout its span (Dale et al., 2013). It supports basic human needs and is governed by human activities (Figure 1) (Robertson & Swinton, 2005). For example, the interrelationship of human and agriculture can be visualized by the present growth of human population (during Industrial Revolution, 1750) followed by increased agricultural production (during Green Revolution, 1950) (Eikelboom, 2013). The present human population of more than seven billion is impending pressure on the capacity of agriculture to fulfil the food requirements without compromising the natural resources for their ability to nourish future generations (Robertson & Swinton, 2005). Moreover, the relative increase in urban population (about 55% of total population) is of more concern in this regard (Figure 2) (Eikelboom, 2013; Milder et al., 2014). A 2/3rd increase in urban population is expected by the year 2025 (Hamlett, 2011; Eikelboom, 2013). It would cause a pressure on agriculture due to mass transition from producer to consumer category, thus loss of agricultural workforce and production. It may have severe consequences as the present overpopulated world is already facing twin challenges of food security and environmental degradation.

The environmental degradation has been regarded as the prime cause of collapses of several past civilizations (Ehrlich & Ehrlich, 2013). Therefore, to satisfy the basic demands (food, fibre and fuel) of

Figure 1. General overview of the population growth, agriculture and its imperative effects

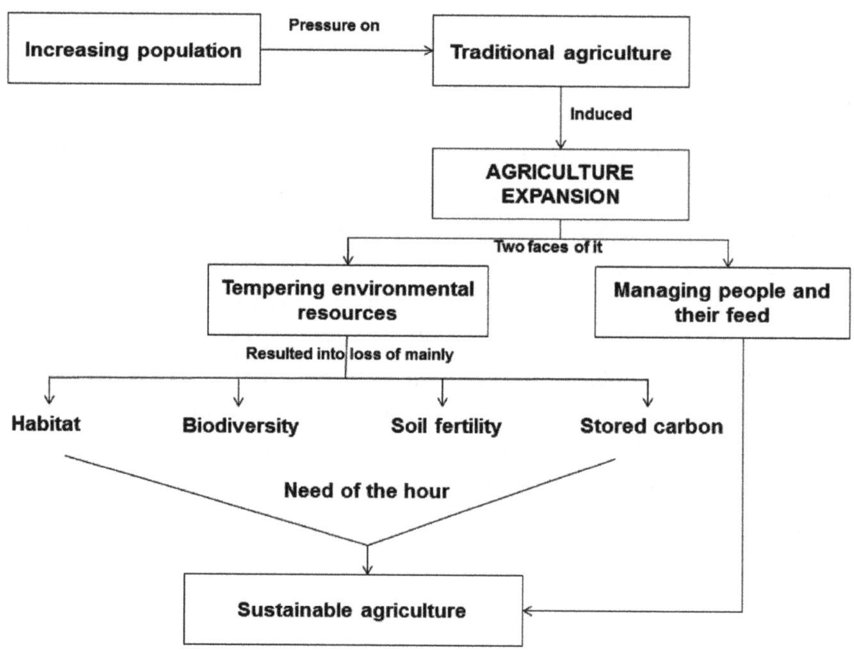

Figure 2. An illustrative representation of world population growth (as bars) and subsequent growth in urban population (as dots) from 1950-2016 (in billions) (Worldometer, 2016)

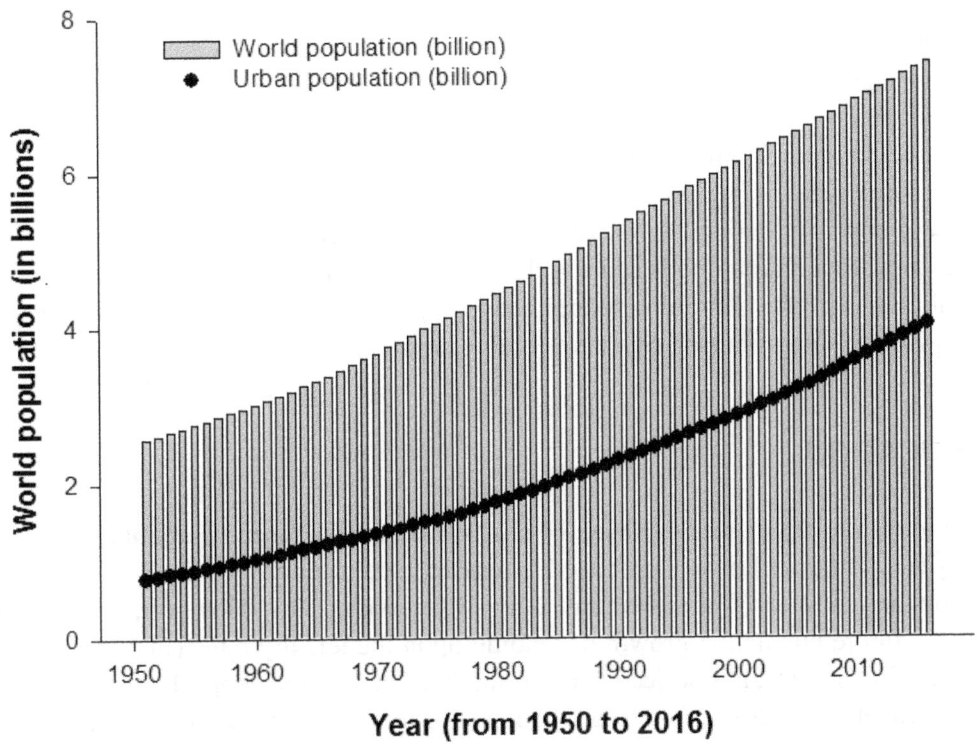

the ever-growing population and its collapse, an intensive agriculture production is urgently required (Barea, 2015). However, the expansion of intensive agriculture for meeting these demands at the cost of compromised ecosystem services may further lead to severe environmental degradations (Figure 3) (Laurance et al., 2014; Milder et al., 2014). For example, the present agricultural production attained a pace with the present growing population due to various external inputs in the form of synthetic chemicals (especially N-fertilizers and pesticides) (Srivastava et al., 2016a). Such externalized control of agriculture is deleterious to the soil productivity and causes environmental pollutions (Erisman et al., 2013; Seneviratne & Kulasooriya, 2013; Alqarawi et al., 2014; Lal, 2015; Singh, 2015). Therefore, there is a great pressure on agriculture to meet various socio-economic and climatic challenges with minimal

Figure 3. Major visible impacts of intensive agriculture practices on natural resources

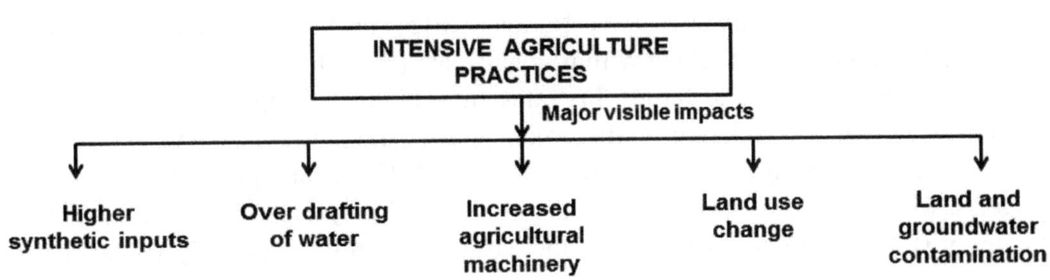

environmental impacts (Foley et al., 2005; Bommarco et al., 2013). Moreover, to attain the future food security, the future agriculture production should grow faster than the population growth by adopting resource-intensive agriculture practices (FAO, 2011). Therefore, there is an urgent need for sustainable thinking in agriculture (Srivastava et al., 2016a).

The present chapter would highlight a set of contemporary issues, such as:

1. The overpopulation, with their projected increase;
2. Food insecurity, with the measure to attain food security;
3. The present impact of agriculture on various natural resources;
4. A nexus of overpopulation-food security-climate change-agriculture; and
5. Sustainable agriculture and a few recommended agricultural practices for meeting future food demand without compromising environmental resources.

OVERPOPULATION

World population is rapidly increasing since the past few decades. Presently, doubling time for population is 40-50 years. The population is projected to remain in between 8.5 to 12 billion (depending upon fertility rate) by the year 2100 (United Nations, 2011; Eikelboom, 2013). However, different projections are available for future population growth depending upon the fertility rates (Figure 2). For example, with the medium fertility rate, it is projected to get stabilized at 8 billion people by the year 2300 (United Nations, 2011; Eikelboom, 2013). These projections are important in the context of Earths carrying capacity, which is the maximum number of individuals which can be sustained indefinitely by the natural resources (Turner II et al., 1990; Eikelboom, 2013). Recent studies on Earth's carrying capacity (biocapacity or the ecological footprint) revealed that we already overshoot the carrying capacity which actually has capacity to sustain approximately 4 billion people at an average living standard (Pimentel et al., 2010; Eikelboom, 2013; World Population Balance, 2013). The present human population of 7.3 billion (as on July-2015) is projected to reach to 8.5 billion and 9.7 billion by 2030 and 2050, respectively (Figure 4) (United Nations, 2015), which would lead to a condition of overpopulation. Overpopulation is a condition when the inhabiting number of individuals surpasses the carrying capacity of a region or planet. The population of the last decade increases with a rate of 1.24% adding about one billion people, however, now a slight decrease (1.18%) is observed in population growth (United Nations, 2015).

The half of the projected increase in world population in between 2015-2050 is expected to be accounted by only nine countries, mostly belonging to Asian and African continents (Figures 4 and 5) (United Nations, 2015). These regions already shared more than three forth of world population, of which only two countries, viz., China (1.4 billion) and India (1.3 billion) accounted for 19 and 18% of total world population, respectively (United Nations, 2015). Further, the population of India is expected to exceed the population of China within seven years (Ehrlich & Harte, 2015; United Nations, 2015). However, in spite of having lower fertility rates, most of the developed nations are considered as overpopulated because their population is still remain above carrying capacity (United Nations, 2011; Eikelboom, 2013). Interestingly, at present more than half (62%) of total population stands in the age group of 15-59, whereas remaining population is either below 15 years (26%) or older than 59 years (12%) (Figure 5) (United Nations, 2015). Thus, increasing population, especially young population, is

Figure 4. An illustrative representation of present and projected world population under different continents: A. in 2015; B. in 2030; C. in 2050; and D. in 2100. (United Nations, 2015)

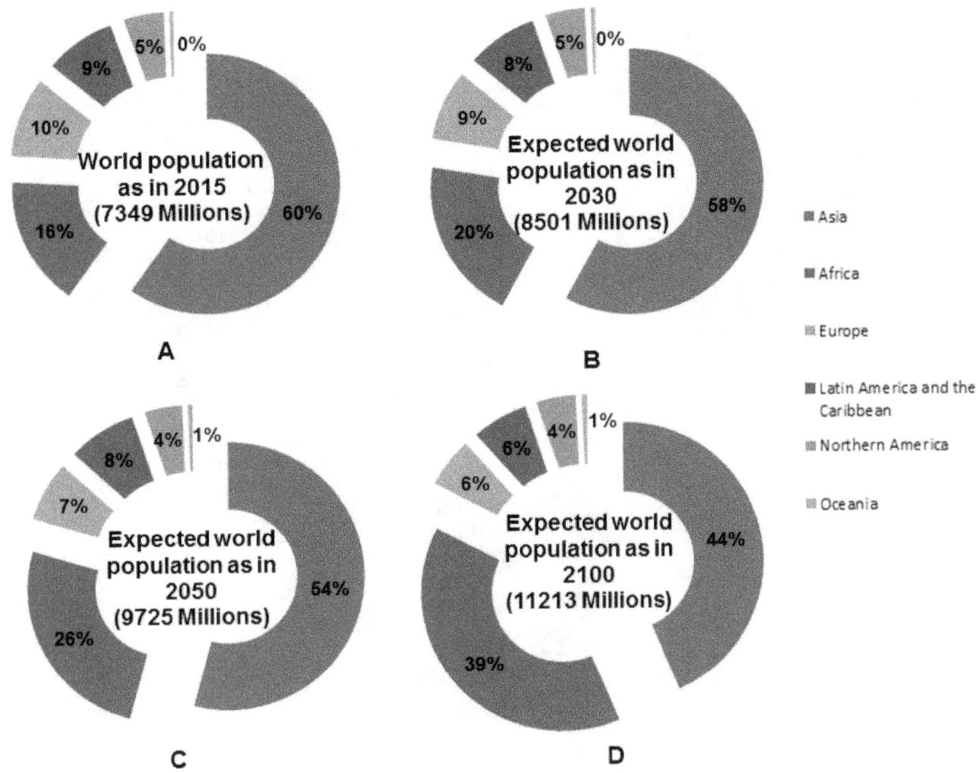

mounting unprecedented pressure on agriculture and natural resources for food security, resulting into overconsumption of resources (Foley et al., 2011). Moreover, unequal distribution and consumption of resources in between developed and developing countries had further worsened the natural resources in the past (Weld & Wright, 1996; Eikelboom, 2013). Therefore, a wise approach is the urgent need of the hour for managing food security for the growing population in future. In the following sections, food security will be discussed followed by a discussion on overpopulation-food security-climate change-agriculture productivity nexus and the future sustainable agriculture goals.

FOOD INSECURITY

Ecological footprint is often used for quantifying the amount of resources (WWF, 2012; Eikelboom, 2013). It is the amount of productive land that provides resources for an individual consumption and for assimilation of resultant metabolic by-products (Eikelboom, 2013). It provides a base for devising carrying capacity of the Earth (Wackernagel & Rees, 1996; Global Footprint Network, 2012; Eikelboom, 2013), which revealed that the present global living standard of human calls for additional half of this planet, whereas for those of US level standard would require four to five such planets (Ehrlich & Ehrlich, 2013). There are a daunting challenges of sufficient production and equitable distribution of agricultural

Figure 5. An illustrative representation of distribution of world population: A. under developed and developing nations; B. under different continents; and C. with age-group. (United Nations, 2015)

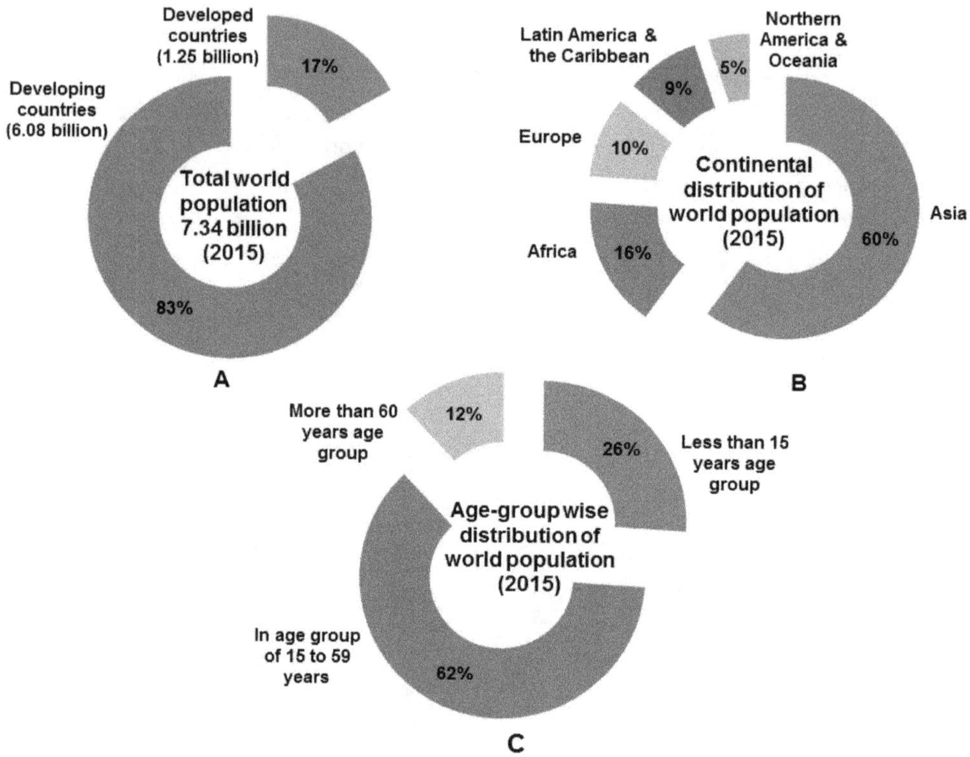

and natural products to meet the basic demand of projected world population increase of about 30% by 2050 (Ehrlich & Harte, 2015). World population have already surpassed the carrying capacity of the Earth and this projected increase in population may further worsen the effects on the Earth's resources (Eikelboom, 2013). As per the World Bank report (2012), there are more than one billion people deprived of their basic needs i.e., chronically undernourished, thus living under absolute poverty (World Bank, 2012; World Food Programme, 2012; Eikelboom, 2013; Sadik, 2016). In addition, greater number of people, though physically not hungry, but are suffering from health consequences of malnutrition, especially in the developing (and under-developed) countries due to lack of essential nutrients in their feed (Godfray & Garnett, 2014). Of the total undernourished worlds' population, about two-thirds of the population is accounted by only seven countries:

- Bangladesh,
- China,
- the Democratic Republic of Congo,
- Ethiopia,
- India,
- Indonesia and
- Pakistan) (PAI, 2011).

This may increase further with the increase in future population. Thus, there is a great challenge of overcoming food insecurity, which literally means the situation of proneness to threats due to lack of protection (Ehrlich & Harte, 2015), in the coming decades.

The Food and Agriculture Organization (FAO, 2002, 2010) defines food security as: "the situation that exists when all people, at all times, have physical, social and economic access to sufficient, safe and nutritious food that meets their dietary needs and food preferences for an active and healthy life". It, thus, depends upon food availability, stability, accessibility and its utilization (Schmidhuber & Tubiello, 2007). The average per capita world food requirement is 2780 kcal day^{-1} (as in 2003), whereas it remain still less than 2200 kcal day^{-1} in the poor countries (Pretty et al., 2006). A significant growth in food production is observed with an increase by 145% in the aggregate world agricultural production since the mid of 20th century (Pretty, 2008). Moreover, by the 2050, the world food production is expected to grow further by 70% (by 100% in developing nations) as of 2009 (FAO, 2011). This growth would include additional annual production of about 1000 and 200 million tons of cereals and livestock products, respectively (Bruinsma, 2009; FAO, 2011). However, it further needs to be increased for attaining the future food security. According to a recent projections of International Food Policy Research Institute's (IFPRI's) International Model for Policy Analysis of Agricultural Commodities and Trade (IMPACT) (Rosegrant et al., 2002), the estimated cereal and livestock production by 2050 needs to be increased by 56 and 90%, respectively, of which about 93 and 85% would be accounted by developing nations (Rosegrant & Cline, 2003).

According to a recent Millennium Development Goals report (2015), out of 795 million undernourished people of the world, 780 million people belong to developing nations. However, despite of their population growth rate, the rate of poverty and undernourishment is continuously declining (from 23.3% to 12.9% in last two decades) in developing nations. For example, the possible reasons for this significant decline in the share of undernourished population from 24% in 1990-92 to 15% in 2014-16 in India are the management of the agricultural productivity, food wastage and economic development. Moreover, there is a concern of healthy and nutrient-rich food production; and its equitable distribution is needed to achieve the precise food security (Godfray et al., 2010; Godfray & Garnett, 2014). Several studies raised major concern of yield gaps, which is the difference between highest possible yield with best management practices and current yield obtained (Foley et al., 2011; Grinsven et al., 2015), in contemporary agriculture (Lobell et al,. 2009; Foley et al., 2011; Mueller et al., 2012; Dale et al., 2013). For example, Ray et al. (2012) reviewed the available literature and found a yield stagnation of wheat and rice in most of the Asia and Australia due to interrelated factors such as:

- Location-specificity,
- Climate change,
- Temperature variations,
- Diminishing soil fertility and salinization, and
- Depleting water resources.

It needs to be undertaken considerably to meet the future food production. However, the inclusion of livestock food products along with cereals would help to achieve the global food requirements by 2050 (Grinsven et al., 2015), in addition to manage the yield gaps (Foley et al., 2011; Mueller et al., 2012; Grinsven et al., 2015). Further, the ability to feed the population is majorly determined by the population pressures as well as arable land and water availability (FAO, 2011; Sadik, 2016).

Globally, the average per capita cultivated land availability is about 0.23 ha, however, it is 0.37 ha for high-income countries and only 0.17 ha for low-income countries (FAO, 2011). The reduced per capita availability of the poor countries might be due to limited accessibility of natural resources as well as high populations (Sadik, 2016). Thus, there is a great challenge for agriculture to produce more food with less resource utilization and related environmental impacts. In the following section, a brief highlight has been given on the major challenges of agriculture in the coming decades.

CHALLENGES FOR AGRICULTURE AND ITS ENVIRONMENTAL IMPACTS

There are several contemporary challenges for agriculture to address the growing food demand with minimal environmental damages (Robertson & Swinton, 2005; Foley et al., 2011; Sadik, 2016; Barea, 2015). The most important challenges (Figure 6) for agriculture (especially in developing nations) are:

- To meet the dietary demand of growing population (especially increasing urban population),
- Changing diet preferences,
- Adapting modern technologies,
- Scarcity of resources,
- Reducing greenhouse gas emissions and
- Mitigating the projected climate change (Pingali, 2012; Godfray & Garnett, 2014).

The present agriculture production could not be attainable without the use of synthetic fertilizers, irrigation water, agricultural machinery and land use changes (Tilman et al., 2002; Pretty, 2008). It leads to significant addition of nitrogen and phosphorus input to the terrestrial ecosystems (Vitousek et al., 1997; Carpenter et al., 1998; Tilman et al., 2002), which is expected to further increase if the similar agricultural practices continue to grow (Cassman & Pingali, 1995; Tilman et al., 2001, 2002). Thus, it resulted into various pollutant input to the land and groundwater resources from synthetic fertilization, excessive water use, soil degradation, biodiversity loss, and greenhouse gaseous emissions related climate change (Eikelboom, 2013; Foley et al., 2011; Pingali, 2012; Wezel et al., 2014). Moreover, the conversion of natural vegetation for agriculture land has the major environmental impact. Agriculture expansion engulfed several natural ecosystems such as:

- The grassland (70%),
- The savannah (50%),
- The temperate deciduous forest (45%), and
- The tropical forest biome (27%) (Ramankutty & Foley, 1999; Ramankutty et al., 2008; Foley et al., 2011).

As per a FAO projection, to meet the future agriculture demand, the cultivated land is expected to increase by about 5% in the developing countries (FAO, 2010; Dale et al., 2013). Future agriculture expansion, especially in tropics, is expected from the conversion of forests (Gibbs et al., 2010; Foley et al., 2011). It may have further resulted into a loss of habitats, biodiversity, soil fertility and carbon storage (Tilman et al., 2002; Foley et al., 2005; Millenium Ecosystem Assessment, 2005; Steinfeld et al., 2006). Thus, approaching food security without further land expansion can only be possible by producing

Figure 6. Major challenges to the modern agriculture

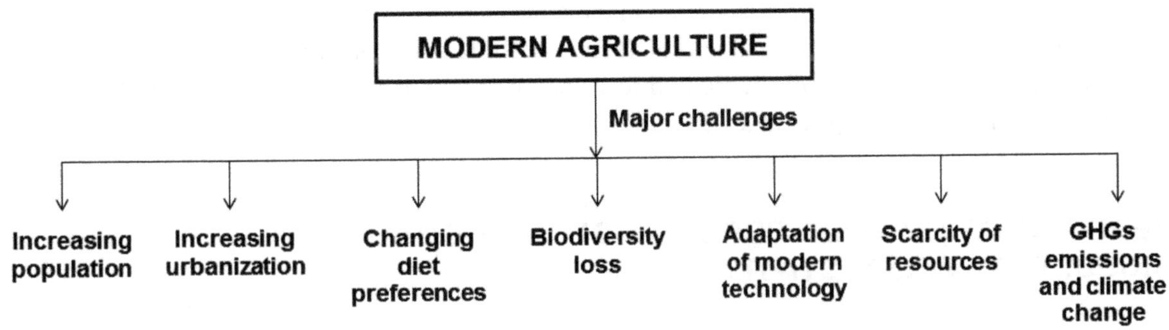

more grains from less cropland; which on one hand would improve food production and on other hand would sequester more carbon to mitigate climate change (Robertson et al., 2000; Vermeulen et al., 2012).

The recent research results have shown that conventional agronomic practices are water and nutrient intensive and ultimately leading towards unsustainability in agriculture. Globally, agriculture is the biggest user of freshwater resources (Calzadilla et al., 2013). For example, about 70% global freshwater withdrawal is diverted to irrigation (FAO, 2011; Foley et al., 2011). About 30% of total freshwater is used alone for rice cultivation worldwide. Further, flood-irrigated rice consume more than 45% of total freshwater used in Asian countries, which is 2-3 times higher than for maize and wheat like cereal crops (Singh, 2013). In the last five decades, the cropland area under irrigation is approximately doubled (FAO, 2011; FAOSTAT, 2011; Rosegrant et al., 2002; Gleick, 2003; Foley et al., 2011). Such resource-intensive agriculture resulted in 2.5 to 3 fold increase in the crop yield (FAO, 2011). However, it has directly and indirectly aggravated the problems in agriculture due to its effect on natural resources (air, soil and water) and climate. For examples, rain-fed agriculture consumes significant amount of water resources (Gordon et al., 2005; Rockstrom et al., 2009; Foley et al., 2011). However, the shortage of water (which is highly vulnerable in the present climate change scenario), has threatened the present agricultural production and it may further affect it (Calzadilla et al., 2013). For example, the depleting water availability is increasingly getting one of the most important constraints due to fast population growth and climate change problem (FAO, 2011). Thus, the present agriculture is facing a challenge of shortage of water, degradation of existing cropland areas, in addition to climate change (Millennium Ecosystem Assessment, 2005; Pretty et al., 2010; FAO, 2011; Alroe et al., 2016). Moreover, water scarcity has more impact over crop production than land degradation in areas having inadequate rainfall (FAO, 2011). Due to rapid depletion of water resources under changing climate conditions, the sustenance of conventional irrigation practices is limited in most of the countries of the world (Du et al., 2015). The studies suggest that the global food security can only be achieved by sustainable utilization of water resource (Biswas, 2008; García, 2008; Du et al., 2015). This can only be possible by improving water-use-efficiency of the crops (Tilman et al., 2002).

In addition, declining soil fertility and quality is another major global constraint for achieving increased agricultural production, though more severely in tropical and sub-tropical regions of the world (Lal, 2015). Further, the conversion of productive cropland to urbanization, desertification, salinization and unsustainable land management may further affect the agricultural productivity (Fargione et al., 2008; Godfray et al., 2010). The present average agricultural productivity is enough or slightly higher to feed the present population, however, it is low under some of the African countries. Moreover, it is a great challenge to achieve the projected 70% increase in food production most possibly by crop intensification, arable land expansion and land use reallocation (FAO, 2010; Koohafkan et al., 2012; Dale et al., 2013). In addition, identification of crop-specific and location-specific adaptation for agricultural production is the urgent need of the hour (Ray et al., 2012).

OVERPOPULATION-FOOD SECURITY-CLIMATE CHANGE-AGRICULTURE PRODUCTIVITY NEXUS

Now-a-days the world is facing complex but interrelated challenges of overpopulation, food insecurity, impoverished agriculture productivity, environmental degradation and climate change (Figure 7) (Ehrlich & Ehrlich, 2013). In nutshell, there is a need to produce more food from less utilization of natural resources for feeding more people with less environmental impacts (Garnett, 2013). Thus, the focus of future research should include these complex relations in integrity to make the way forward for future generations (Sadik, 2016). In the present section, we would give a brief insight on these interrelated challenges which ultimately calls for a more integrated system of agriculture systems for attaining sustainability (Foley et al., 2011). It is now established that the population pressures and climate change would continue to regulate the development of agriculture, thus, the balance of proper land and water management in most of the developing countries (Gregory et al., 2005; Ehrlich & Harte, 2015; Sadik, 2016). On the other hand, population growth and agricultural intensification is also considered as the most important contributors of greenhouse gaseous emissions and on-going climate change (Easterbrook, 2009; IPCC, 2014). Though, green revolution improved crop productivity in between 1961 to 2005 which indirectly led to reduction of 161 Gt C emissions, mainly by reducing the conversion of marginal and vegetated land upto some extent (Burney et al., 2010; Vermeulen et al., 2012), however, its dependency

Figure 7. An illustrative representation of the agriculture and overpopulation-food security-environmental degradation-climate change nexus

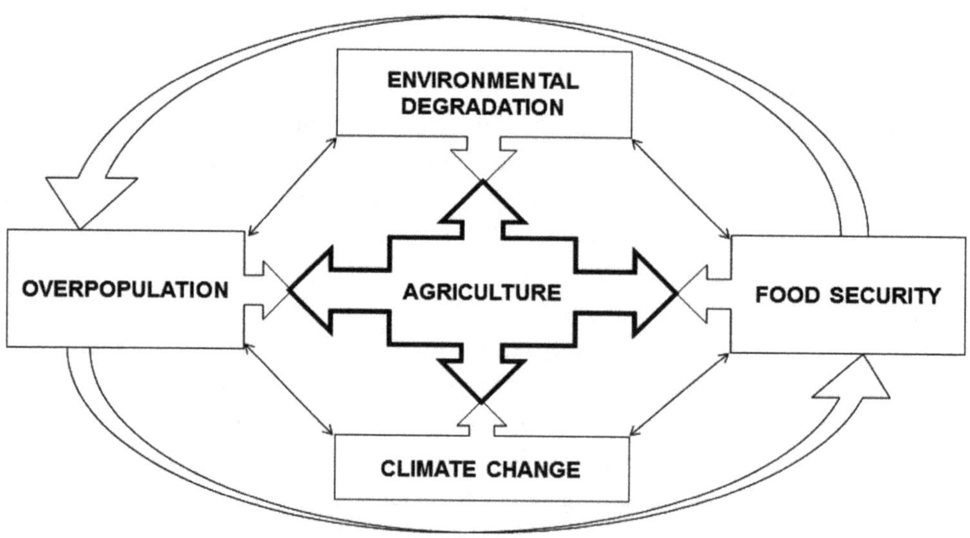

on various externalized inputs resulted into several environmental concerns (Singh et al., 2016; Srivastava et al., 2016a). Further, climate change is expected to regulate all the major dimensions of food security, thus, the most important constraints for the future agriculture (FAO, 2003; Schmidhuber & Tubiello, 2007; World Bank, 2008; Godfray et al., 2010). For example, it governs the agriculture production either directly by affecting the important resource availability (e.g., changing precipitation, temperature, water availability, CO_2 fertilization and climatic variability) or indirectly by affecting the socio-economic status of the people (Schmidhuber & Tubiello, 2007; World Bank, 2007; Calzadilla et al., 2013).

Human and environmental health is interlinked with diet (Tilman & Clark, 2014). They show a complex interrelationship (Koohafkan et al., 2012). Recently, replacement of traditional diets with refined diets is a very common tradition which is regarded as the major contributor (about 80%) of total greenhouse gaseous emissions from agriculture sector by 2050. Therefore, the diet preferences will also be a major challenge in relation to food productivity and environmental health, for the coming decades (Tilman & Clark, 2014). Moreover, the proposed agricultural productivity is thought to be achieved by crop intensification by adopting resource-use-efficient, precise and sustainable land management agronomic practices (FAO, 2011). The adoption of modern agronomic practices based on integrated approaches (inclusive of biological, ecological, techno-skilled, efficient, and socio-economic) can help in reduction of greenhouse gas emissions from agriculture sectors (Godfray et al., 2010) and in common agricultural and environmental problems (Pretty 2008; Srivastava et al., 2016a). Therefore, a need for sustainable agriculture satisfying considerably with a set of agro-ecological, socio-economic and environmental standards is proposed (Koohafkan et al., 2012; Singh et al., 2016; Srivastava et al., 2016a).

AGRICULTURE SUSTAINABILITY

The green revolution practices based on externalized inputs has doubled the global cereal production in the past four decades (WHO, 1990; FAO, 2001; Tilman et al., 2001, 2002), however, at the increased environmental cost. Now there is a global awareness for the adoption of sustainable agriculture practices for meeting the future world food demand (Altieri, 2004; Barea, 2015). Sustainability in agriculture (or sustainable intensification or ecological intensification) means smart (accessible and efficient) use of natural resources for improved food productivity that have negligible effects on environmental resources and suitable for local adaptation (Tilman et al., 2002; Pretty et al., 2006; Pretty, 2008; Godfray et al., 2010; Pretty et al., 2011; Tittonell, 2014). Tilman et al. (2002) define sustainable agriculture as "the practices that meet current and future societal needs for food and fibre, for ecosystem services, and for healthy lives, and that do so by maximizing the net benefit to society when all costs and benefits of the practices are considered". Pretty (2008) suggests some key principles for agricultural sustainability which include integrated biological and ecological processes, minimal use of natural resources, local understanding by farmers based on their traditional knowledge and skills, as well as collective resource utilization and collaborative work by different stake-holders (Figure 8). It integrates the traditionally established practices with positive environmental impacts along with the modern technologies for improved production and reduced environmental impacts (Ponisio et al., 2015). The benefits from sustainable agriculture can only be possible by enhanced crop yields, improved resource-use-efficiency, judicious use of environmental resources and integrated livestock production (Tilman et al., 2002). Further, it involves such an ecological understanding in which wise substitution of externalities in agriculture with the self-adjusted intrinsic mechanisms for improving yield and ecosystem services are evolved (Bommarco et al., 2013).

Figure 8. Basics of the sustainable agriculture (Based on Pretty, 2008)

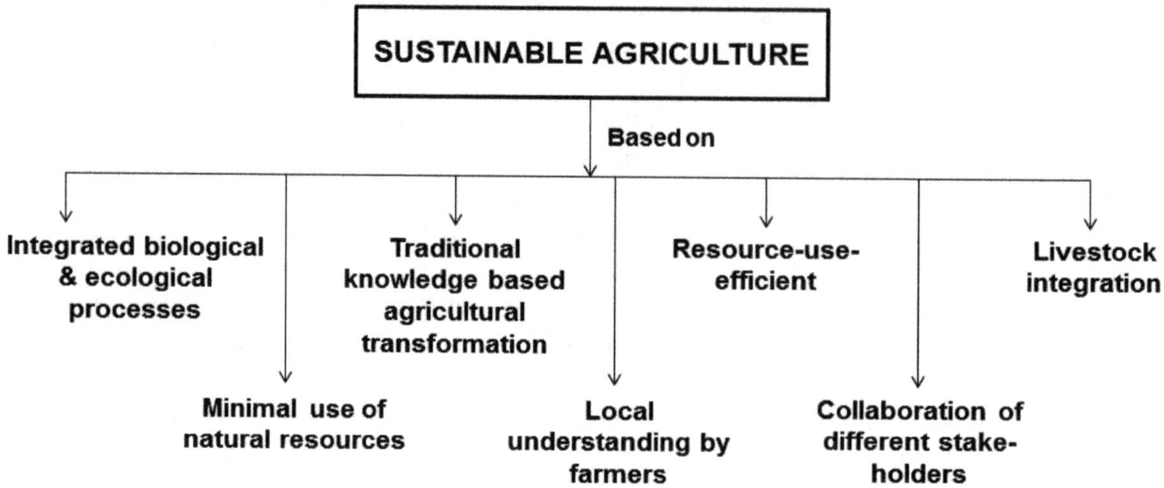

In a nutshell, sustainable agriculture means a package of traditionally evolved innovative technologies which ensures agricultural productivity and food safety on one hand and environmental resource conservation on the other, for a sustainable future (Araujo et al., 2014; Singh, 2015). Based on these principles and recommendations, a few agronomic practices have evolved which are discussed in the later section. Moreover, their environmental impacts have also been discussed for better understanding about these practices before their recommendations to a particular ecological region.

RECOMMENDED PRACTICES FOR AGRICULTURE SUSTAINABILITY

The current agriculture practices affect the food production and environmental health status (Tilman et al., 2002). There are several approaches proposed for achieving sustainability in agriculture under the present climate change scenario (FAO, 2011; Koohafkan et al., 2012). For example:

- Organic farming,
- Biofertilizers,
- Integrated nutrient and pest management,
- Direct seeding, conservation agriculture (e.g., crop rotation, intercropping and mixed cropping),
- Deficit irrigation,
- Precision farming and better nutrient use (nutrient efficiency),
- Improved agricultural waste utilization,
- Agroforestry and aquaculture systems, and
- Crop-livestock integration are considered for sustainable resource management, carbon sequestration and agriculture (Pretty, 2008; Phalan et al., 2011; Koohafkan et al., 2012; Garnett, 2013; Smith et al., 2013; Wezel et al., 2014; Singh, 2015).

The basic principle underlying for recommended agronomic practices (Figure 9) include the improvement of soil fertility and quality by enhancing soil organic C and N pools, biological diversity, better soil structural development, whereas reducing soil erosion and degradation (Lal, 2015). In addition to improved crop yield, soil fertility and biodiversity, these practices may increase socio-economic status of the farmers (Mtengeti et al., 2015). Moreover, identification of various sustainability indicators has been proposed by Singh et al. (2016) which need to be assessed before recommending various agronomic practices at wider scale for a particular geographic location. In the following sub-sections, a brief highlight on a few recommended agronomic practices with their possible environmental impacts has been described.

Agronomic Practices Used for Cereals and Water Productivity

Rice, wheat, maize, cotton and sugarcane are among the most important 17 major crops grown worldwide. Of these crops, rice and wheat covered 63% of global irrigated cropland area with 59% consumption of irrigation water; whereas, cotton, sugarcane and maize consume about 30% of irrigation water (West et al., 2014). In the following sub-sections, a few commonly used agriculture practices for major cereal crops used for conserving natural resources and improving yield have been addressed:

Aerobic Rice

A study was conducted by Sarwar et al. (2013) and the results of the study pointed out that flooded condition is not necessarily required for whole rice crop growth period. Therefore, a shift from continuously flooded rice cultivation to aerobic rice cultivation with proper water scheduling can be a beneficial alternative to conserve water without yield penalty. Aerobic rice is used for growing rice in upland areas for rice cultivation under near saturated soil conditions (Yang et al., 2005). Now-a-days it is also applied for lowland areas where water shortage is seen (Bouman & Tuong, 2001; Tabbal et al., 2002). In comparison to conventional flooded system, aerobic rice system is more suitable for those soils having high seepage and percolation rates (Nguyen et al., 2009). In Brazil, aerobic rice system is used for upland rice cultivation using sprinkler irrigation. This practice is economically advantageous over conventional rice cultivating practices in terms of water-use-efficiency; however, a marginal decrease in yield is reported in a few studies of tropical countries (Bouman et al., 2002; Singh, 2013).

Figure 9. Basic Principles for the recommended agronomic practices (Based on Lal, 2015)

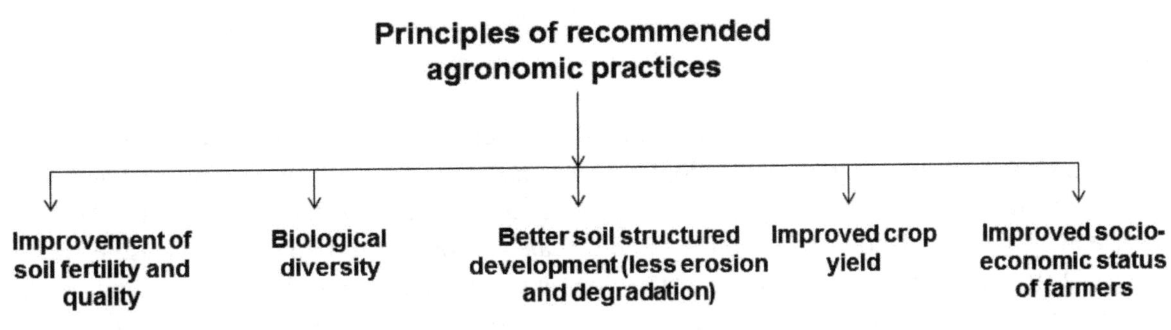

Dry Direct Seeded Rice System

It is a system of growing rice crop directly (without transplantation) in semi-saturated or non-saturated soils for conserving water (Tuong & Bouman, 2003). In this system of rice cultivation, less intensive tillage is required than conventional plots resulting into favourable environment for crop establishment as well as significant reduction in CO_2 from fuel burning. In a two-year field study, Pathak et al. (2013) found that the grain yield in direct-seeded rice system plots is comparable to conventional plots in addition to 45-52% labour saving, 32% water saving and 46-58% reduction in tractor use. In addition, maintenance of aerobic environment in this system leads to subdued or significantly negative effect on methane emission (Pathak et al., 2013). It is more suitable for flat fields, which need less water, and drought prone areas. However, studies on the effects of this cropping system on SOC and aggregate development, soil biological activities and nutrient dynamics needs further attention (Singh et al., 2016).

System of Rice Intensification

Systems of rice intensification (SRI) is reported to get increasing interest in various countries for the efficient resource utilization by managing the soil carbon and reducing carbon footprint without compromising the agricultural production (Mishra et al., 2013). In SRI systems, the reported increase in yield is due to its effects on changing the management of plants, soil, water, and nutrients. Such effects are even observed using new, higher-yielding seeds or agrochemical inputs which are less vulnerable to pests and disease losses as well as to climatic stresses (http://sri.ciifad.cornell.edu; Uphoff, 2012). Increase in yield by 20-40% and decrease in water use up to 50% have been reported in SRI system than the conventional cropping system (Uphoff et al., 2011). The increased yield and water productivity in this system is due to improved physiological efficiency of plants (Stoop et al., 2002; Mishra et al., 2013). As reviewed by Uphoff et al. (2011) in Indo-Gangetic plains, efforts have been made to integrate SRI with conservation agriculture practice and the results were found significant. In a review on SRI in Cambodia, Ly et al. (2012) found that the Cambodian farmers have adapted to a SRI like practice for maximization of rice yield and resource conservation and they named those practices as "SRI-influenced practices-SRII". In SRI, single plants/hill seedlings plantation and intermittent irrigation for maintaining aerobic environment which favours root growth and development and enhances tillering (Yang et al., 2004; Stoop, 2005) and, thus, minimizes the water shock to the seedlings (San-oh et al., 2006; Mishra & Salokhe, 2008). Further, aerobic environment in SRI leads to avoid reducing environment, thus, formation of ferrous iron and reduced organic molecules, which have detrimental effect over root growth and development (Yoshida, 1981).

Deficit Irrigation

The cereal crops are the basis for future food production, therefore, several measures have been proposed for managing the irrigation water use by adopting water-use-efficient and water productive agronomic practices. Several irrigation measures have been proposed for managing water requirement and efficient water use in agriculture. Deficit irrigation is a micro-scale approach of improving water-use-efficiency (Du et al., 2015). This practice save water by developing potential plant physiological adaptation (such as stomatal conductance), and is widely used for crop and fruit production worldwide (Du et al., 2015).

Agronomic Practices Used for Maintaining Soil Fertility and Health

Soil fertility and health management is one of the imperative aspects for managing the long term sustainability of agriculture. The soil organic carbon (SOC) pool is the most important indicator of soil quality and health (Manlay et al., 2007; Rajan et al., 2010; Krupenikov et al., 2011; Lal, 2015). Several practices have been proposed for managing soil health such as organic farming, conservation agriculture, addition of soil ameliorants like biofertilizers, vermicomposts, biochar, etc. A few of them are discussed below:

Conservation Agriculture

The conservation agriculture is one of the leading agronomic practices which is followed by a range of cropping systems worldwide for land management in the era of climate change (Scopel et al., 2013). It has been reported to provide several benefits (such as reduced production cost, soil erosion, SOC improvement, energy saving, improved soil biological activity, C sequestration, etc.), if applied properly (Lal, 2015; Scopel et al., 2013). Conservation agriculture is mainly based on three basic principles,

1. Less or no mechanical disturbances to soil (i.e., reduced or no-tillage),
2. Soil protection by permanent soil cover by crop residue (i.e., mulching or residue retention), and
3. Crop diversification (i.e., crop rotation and cover cropping) (Scopel et al., 2013; Lal, 2015).

Integrated nutrient management is also considered as one of the intrinsic components of conservation agriculture (Lal, 2015). Under no-tillage system, seeds or seedlings are shown into a mulched system without or less disturbing the soil profile, which reduces soil erosion and, soil compaction, whereas, increases soil biota and SOC (Wezel et al., 2014). Further, crop residue retention along with no-tillage system preserves fresh organic matter at the upper soil surface; thus, improve soil health (Wezel et al., 2014). Cover cropping and crop rotation (especially legumes or green manure) along with the no-tillage systems are regarded as the best approach to minimize the yield gaps along with better soil fertility (Derpsch & Benites, 2003; Derpsch et al., 2014). Proper adoption of all the principle of conservation agriculture (in combination) also helps in maintaining appropriate soil moisture (and water use efficiency), which is thought to be most affected by the present climate change (Pittelkow et al., 2014). However, effective benefits of conservation agriculture can only be achieved by following all the principles properly, but not in isolation (Scopel et al., 2013).

Organic Agriculture

Organic agriculture, an on-farm inputs (organic and natural products) based agriculture improving the local soil biological conditions, has been widely promoted as an alternative of off-form inputs (synthetic agro-chemicals) based conventional agriculture (Knight & Newman, 2013; Ponisio et al., 2015). It is now practiced on about 0.9% of the total agricultural land (Willer & Kilcher, 2011; Ponisio et al., 2015). World Health Organization and Food and Agriculture Organization (WHO & FAO, 2007) defined organic agriculture as “a holistic production management system which promotes and enhances agroecosystem health, including biodiversity, biological cycles, and soil biological activity”. It is considered as one of the best means of agriculture alternatives to achieve the sustainability in agriculture (Ponisio et al., 2015). The benefits of this cropping system includes better soil biological diversity and activity, improved soil fertility and efficiency in use of nutrient-, water- and energy (Mader et al., 2002; Nissen & Wander, 2003; Lynch et al., 2011; Ponisio et al., 2015). More specifically, organic farming systems promote the resiliency of soil system in addition to boost the production for maintaining the sustainability than the ever-input dependent conventional agriculture (Mäder et al., 2002; Grinsven et al., 2015). Previous studies find variable effects of organic agriculture on yield in different geographic locations and economic status of the country. For examples, 8-9% yield gaps from developed, whereas about 180% yield gains from developing countries are reported under organic farming systems as compared to conventional farming systems (Stanhill, 1990; Badgley et al., 2007; Ponisio et al., 2015). However, yield gap or reduced yield in organic agriculture than conventional agriculture is one of the major constraints for this system. In a meta-analysis of yields from the global organic and conventional based cropping systems, Ponisio et al. (2015) find 19.2% (±3.7%) lower yields in organic systems than the conventional cropping systems, which was smaller than the previously reported yield gaps (20-25%) (de Ponti et al., 2012; Seufert et al., 2012; Ponisio et al., 2015). These yield gaps are the results of differences in soil types, crop types and field management approaches (Ponisio et al., 2015). Further, lack of nitrogen fertilizers is considered as a basic cause of yield penalty in organic agriculture (Smil, 2002; Grinsven et al., 2015). Therefore, rather than a sole organic approach, combined organic agriculture with moderate inorganic fertilization can be considered as a better alternative to manage the yield and soil productivity at a time (Srivastava et al., 2016b). Further, Ponisio et al. (2015) suggested that the yield gaps can be managed significantly by adopting crop diversification (or multi-cropping) and crop rotation systems (Ponisio et al., 2015) in addition to intensive crop-livestock agriculture. Thus, a combination of conservation agriculture and organic farming could be a better alternative for closing the yield gaps and managing the food security with sustainability in agriculture.

Integrated Nutrient Management (INM)

Integrated nutrient management (INM) is reported to improve the SOC pools, thus, helps in maintaining soil productivity and health in addition to soil C sequestration (Lal, 2015). It has an additional benefit of nutrient management than the sole organic fertilization that sometimes has been constrains due to slow and/or late release of nutrients at the peak of crop demands, resulting into yield penalty (Tilman et al., 2002).

Soil Amendments

To improve the soil fertility and crop productivity to feed the world population, the role of soil microbial community is imperative (Barea, 2015). Addition of soil microbial inoculants (such as rhizospheric bacteria and fungi) to improve the soil nutrient conditions and their use efficiency by enhancing efficient microbial diversity of agro-ecosystems is reported (Barea, 2015; Singh, 2015). Further, addition of different consortium of microbial inoculants under different problem soils or degraded ecosystems have been considered as an approach to manage the future agricultural productivity and sustainability (Singh, 2015). Similarly, vermicompost is a semi-digested biological-organic mixture which has been used as an alternative to the chemical fertilizers. It has ability to improve soil health and also help in waste management (Singh et al., 2011). There are many more biological approaches used for managing soil health, however, they are beyond the scope of this chapter.

Biochar is one of the soil ameliorating agent recently being promoted for improving soil health, waste management and soil C sequestration (Singh et al., 2015a). It is a C rich material having tendency to nourish basic soil properties and remain in soil for decades to centennials. After its application to soil, the plant material gets concentrated into biochar providing potential to improve soil mineral matter (Bruun et al., 2011). It increases basic soil nutrients and ultimately increased crop yield (Li et al., 2012; Zhang et al., 2013; Singh et al., 2015a). It has a highly porous structure which increases soil porosity, soil aeration, water holding capacity, whereas decreases soil aggregation, soil strength and bulk density after its application to soil. Various studies on biochar amended soil have reported the increase in bacterial, fungal and micorrhizal communities and activity (Warnock et al., 2007; Li et al., 2011). Moreover, integration of biochar with conservation agriculture is reported to improve the soil fertility, SOC level, and mitigate climate change by potential soil C sequestration (Singh et al., 2015b). However, its application and benefits depends on soil basic properties. Depending upon soil properties, it may act as a boon for agronomic benefits, however, the risks related with biochar application also needs to be taken equally for its wider adaptation (Kuppusamy et al., 2016).

POSSIBLE SOCIO-ECONOMIC AND ENVIRONMENTAL IMPACTS OF RECOMMENDED SUSTAINABLE AGRICULTURE PRACTICES

The above-mentioned agronomic systems, having high potential for enhancing agriculture sustainability, still need wider adoptability by the farmers in the coming decades (Wezel et al., 2014). Several factors are important in wider adaptability of these cropping systems (Derpsch et al., 2014; Pittelkow et al., 2014). The adoption of recommended agricultural practices, technological incorporation in addition to traditional knowledge in agriculture resulting into an inclusive growth, might form the basis for such positive change. However, environmental impact assessment of recommended agriculture practices and its socio-economic interventions is required for its holistic understanding and wider acceptability, and thus, in sustainability of the agriculture. On the basis of available literature on different agronomic practices, a few major limitations have been figured out. These can be:

1. Physiological and biochemical activities are reported to constrain the plant growth resulting into decline in yield under aerobic rice system than the conventional systems. The water stress at the time of transplantation, in various less water used agronomic practices, lead to transplanting shock to the seedlings resulting into reduced grain yield (Pasuquin et al., 2008). It can be managed by intermittent irrigation at critical growth stages in aerobic rice (Ghosh et al., 2012).
2. Broad understanding of different principles and their applications of conservation agriculture practices is lacking (Scopel et al., 2013; Pittelkow et al., 2014).
3. The yield gaps reports for different major crops under recommended agricultural practices are the major cause of concern for future sustainability in food production and agricultural demands (Ray et al., 2012). For example, a significant yield penalty (about 25-30%) is reported under conservation agriculture systems when applied in isolation (Pittelkow et al., 2014). Further, it is suggested to apply no-tillage systems in the field already adapted to residue retention and crop rotation principles (Pittelkow et al., 2014).
4. Conservation agriculture is facing adaptation problems from small land-holder farmers due to their more mechanised approach. Also, the severe chances of crop failure in earlier years of cropping

lead to less interest of poor farmers in spite of its several environmental benefits. However, it can be managed by providing appropriate financial, technical and technological assistance to the small holder farmers (Scopel et al., 2013). In addition, it can be a better approach for dryland areas, which are thought to be more affected under climate change scenario, when all the principles are applied in synergy (Pittelkow et al., 2014).

5. The organic agriculture systems are not independent; therefore, it always depends on other agronomic perspectives (Alroe et al., 2016). Moreover, the effect of organic agriculture on soil microbial properties are still limited which calls for more researches on mechanistic exploration of soil organisms under this system (Tuck et al., 2014).
6. Reduction of future agricultural land, increase in urban population and the modified consumption patterns are alarming the food demands in developing countries and agricultural productivity from the sustainable agronomic systems (Pretty, 2008).
7. Biochar application has been reported to have several risks which limit its wider application (Singh et al., 2015a, 2015b; Kuppusamy et al., 2016).

Overall, the future agriculture in the realm of present climate change should be an internally regulated system encompassing microscale ecological interactions (Figure 10) (Srivastava et al., 2016a). It can be achieved by a proper blending of traditionally successful agricultural practices with modern technological interventions. Moreover, a deep socio-economic evaluation of different agronomic practices in addition to the coordinated involvement of all the stakeholders of the agriculture is needed for achieving sustainable production for growing population.

Figure 10. A conclusive remark on the overpopulation, food security, and agriculture sustainability in the era of climate change

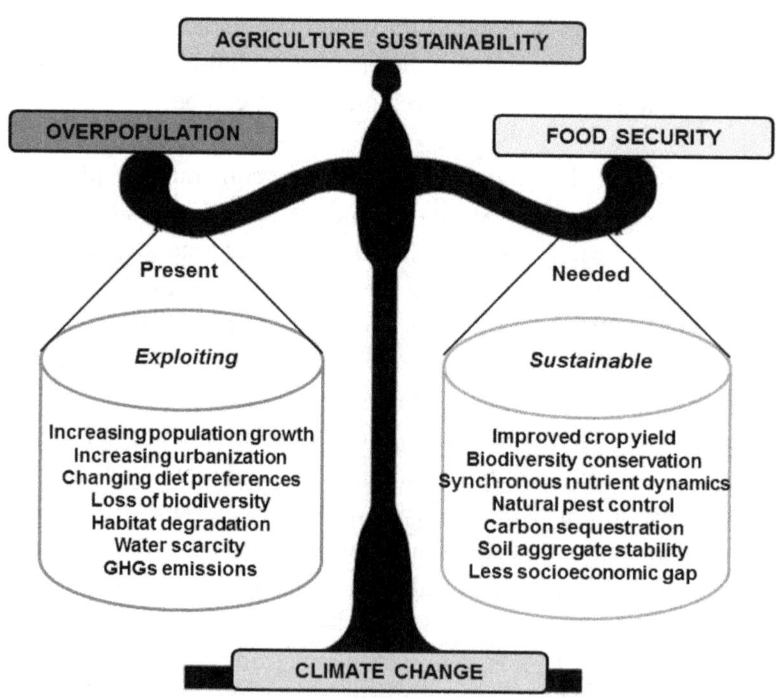

CONCLUSION

Overpopulation, food security, climate change, environmental degradation and inherent economic disruptions are the synergistic problems the entire world is facing today. The agriculture is facing a quadruplicated pressure from these problems. The challenges of overpopulation and food security can only be managed by identifying the core areas of research and development under different agricultural sectors. Approaching food security by conserving natural resources is of immense importance to achieve sustainable development. More specifically, to achieve the future food demand of the continuously growing population at low or no environmental cost, there is an urgent need of sustainable agricultural systems. Such systems should be inclusive of sustained production based on soil biological functioning, using resources efficiently to feed the world population. The recommended agronomic practices have potential to feed the world, but have limited adaptability so far due to yield loss. There is a great need on more research focus on devising the policy measures for wider applicability of recommended agriculture practices. The integration of local and traditional knowledge along with the recommended agronomic practices would be of immense importance for attaining sustainability in agriculture. Therefore, for long term development and food security management, all the stakeholders of agriculture need to be work in a coordinated and complementary sharing and response basis under various geographical locations.

REFERENCES

Alqarawi, A. A., Abd Allah, E. F., & Hashem, A. (2014). Alleviation of salt-induced adverse impact viamycorrhizal fungi in Ephedra aphylla Forssk. *Journal of Plant Interactions*, *9*(1), 802–810. doi:10.1080/17429145.2014.949886

Alrøe, H. F., Moller, H., Læssøe, J., & Noe, E. (2016). Opportunities and challenges for multicriteria assessment of food system sustainability. *Ecology and Society,* 21-38. Doi:10.5751/ES-08394-210138

Altieri, M. A. (2004). Linking ecologists and traditional farmers in the search for sustainable agriculture. *Frontiers in Ecology and the Environment*, *2*(1), 35–42. doi:10.1890/1540-9295(2004)002[0035:LEATFI]2.0.CO;2

Araujo, A. S. F., Borges, C. D., Tsai, S. M., Cesarz, S., & Eisenhauer, N. (2014). Soil bacterial diversity in degraded and restored lands of Northeast Brazil. *Antonie van Leeuwenhoek*, *106*(5), 891–899. doi:10.100710482-014-0258-5

Badgley, C., Moghtader, J., Quintero, E., Zakem, E., Chappell, M. J., & Aciles-Vasquez, K. (2007). Organic agriculture and the global food supply. *Renewable Agriculture and Food Systems*, *22*(02), 86–108. doi:10.1017/S1742170507001640

Barea, J. M. (2015). Future challenges and perspectives for applying microbial biotechnology in sustainable agriculture based on a better understanding of plant-microbiome interactions. *Journal of Soil Science and Plant Nutrition*, *15*, 261–282.

Biswas, A. K. (2008). Integrated water resources management: Is it working? *Water Resources Development*, *24*(1), 5–22. doi:10.1080/07900620701871718

Bommarco, R., Kleijn, D., & Potts, S. G. (2013). Ecological intensification: Harnessing ecosystem services for food security. *Trends in Ecology & Evolution*, *28*(4), 230–238. doi:10.1016/j.tree.2012.10.012

Bouman, B. A. M., & Tuong, T. P. (2001). Field water management to save water and increase its productivity in irrigated rice. *Agricultural Water Management*, *49*(1), 11–30. doi:10.1016/S0378-3774(00)00128-1

Bouman, B. A. M., Wang, H., Yang, X., Zhao, J., & Wang, C. (2002). Aerobic rice (Han Dao): A new way of growing rice in water-short areas. In *Proceedings of the 12th International Soil Conservation Organization Conference*. Beijing, China. Tsinghua University Press.

Bruinsma, J. (2009). *The resource outlook to 2050: By how much do land, water use and crop yields need to increase by 2050?* Expert Meeting on How to Feed the World in 2050. Retrieved from ftp://ftp.fao.org/docrep/fao/012/ak542e/ak542e06.pdf

Bruun, E. W., Müller-Stöver, D., Ambus, P., & Hauggaard-Nielsen, H. (2011). Application of biochar to soil and N2O emissions: Potential effects of blending fast-pyrolysis biochar with anaerobically digested slurry. *European Journal of Soil Science*, *62*(4), 581–589. doi:10.1111/j.1365-2389.2011.01377.x

Burney, J. A., Davis, S. J., & Lobell, D. B. (2010). Greenhouse gas mitigation by agricultural intensification. *Proceedings of the National Academy of Sciences of the United States of America*, *107*(26), 12052–12057. doi:10.1073/pnas.0914216107

Calzadilla, A., Rehdanz, K., Betts, R., Falloon, P., Wiltshire, A., & Tol, R. S. J. (2013). Climate change impacts on global agriculture. *Climatic Change*, *120*(1-2), 357–374. doi:10.100710584-013-0822-4

Carpenter, S. R., Caraco, N. F., Correll, D. L., Howarth, R. W., Sharpley, A. N., & Smith, V. H. (1998). Nonpoint pollution of surface waters with phosphorus and nitrogen. *Ecological Applications*, *8*(3), 559–568. doi:10.1890/1051-0761(1998)008[0559:NPOSWW]2.0.CO;2

Dale, V. H., Kline, K. L., Kaffka, S. R., & Langeveld, J. W. A. (2013). A landscape perspective on sustainability of agricultural systems. *Landscape Ecology*, *28*(6), 1111–1123. doi:10.100710980-012-9814-4

De, P. T., Rijk, B., & Van, I. M. K. (2012). The crop yield gap between organic and conventional agriculture. *Agricultural Systems*, *108*, 1–9. doi:10.1016/j.agsy.2011.12.004

Derpsch, R., & Benites, J. R. (2003). Situation of conservation agriculture in the world. In *Proceedings of II World Congress on Conservation Agriculture.*

Derpsch, R., Franzluebbers, A. J., Duiker, S. W., Reicosky, D. C., Koeller, K., Friedrich, T., ... Weiss, K. (2014). Why do we need to standardize no-tillage research? *Soil & Tillage Research*, *137*, 16–22. doi:10.1016/j.still.2013.10.002

Du, T., Kang, S., Zhang, J., & Davies, W. J. (2015). Deficit irrigation and sustainable water-resource strategies in agriculture for China's food security. *Journal of Experimental Botany*, *66*(8), 2253–2269. doi:10.1093/jxb/erv034

Easterbrook, S. M. (2009). *Population growth vs. emissions growth.* Retrieved August 5, 2013 from http://www.easterbrook.ca/steve/2009/06/population-growth-vs-emissions-growth/

Ehrlich, P. R., & Ehrlich, A. H. (2013). Can a collapse of global civilization be avoided? *Proceedings of Royal Society of London Biology Sciences*, *280*(1754), 20122845. doi:10.1098/rspb.2012.2845

Ehrlich, P. R., & Harte, J. (2015). Food security requires a new revolution. *The International Journal of Environmental Studies*, *72*(6), 908–920. doi:10.1080/00207233.2015.1067468

Eikelboom, J. A. J. (2013). *An essay on the ecological and socio-economic effects of the current and future global human population size* (MSc Thesis). Utrecht University.

FAO. (2003). World agriculture towards 2030/2050. Rome, Italy: FAO.

Fargione, J., Hill, J., Tilman, D., Polasky, J., & Hawthorne, P. (2008). Land Clearing and the Biofuel Carbon Debt. *Science*, *319*(5867), 1235–1238. doi:10.1126cience.1152747

Foley, J. A., DeFries, R., Asner, G. P., Barford, C., Bonan, G., Carpenter, S. R., & Chapin, F. S. (2005). Global consequences of land use. *Science*, *309*(5734), 570–574. doi:10.1126cience.1111772

Foley, J. A., Ramankutty, N., Brauman, K. A., Cassidy, E. S., Gerber, J. S., Johnston, M., ... Zaks, D. P. M. (2011). Solutions for a cultivated planet. *Nature*, *478*(7369), 337–342. doi:10.1038/nature10452

García, L. E. (2008). Integrated water resources management: A 'small' step for conceptualists, a giant step for practitioners. *Water Resources Development*, *24*(1), 23–36. doi:10.1080/07900620701723141

Garnett, T. (2013). Food sustainability: Problems, perspectives and solutions. *The Proceedings of the Nutrition Society*, *72*(01), 29–39. doi:10.1017/S0029665112002947

Ghosh, A., Dey, R., & Singh, N. (2012). Improved management alleviating impact of water stress on yield decline of tropical aerobic rice. *Agronomy Journal*, *104*(3), 584–588. doi:10.2134/agronj2011.0278

Gibbs, H. K., Ruesch, A. S., Achard, F., Clayton, M. K., Holmgren, P., Ramankutty, N., & Foley, J. A. (2010). Tropical forests were the primary sources of new agricultural land in the 1980s and 1990s. *Proceedings of the National Academy of Sciences of the United States of America*, *107*(38), 16732–16737. doi:10.1073/pnas.0910275107

Gleick, P. H. (2003). Global freshwater resources: Soft-path solutions for the 21st century. *Science*, *302*(5650), 1524–1528. doi:10.1126cience.1089967

Godfray, H. C. J., Beddington, J. R., Crute, I. R., Haddad, L., Lawrence, D., Muir, J. F., ... Toulmin, C. (2010). Food Security: The challenge of feeding 9 billion people. *Science*, *327*(5967), 812–818. doi:10.1126cience.1185383

Godfray, H. C. J., & Garnett, T. (2014). Food security and sustainable intensification. *Philosophical Transactions of the Royal Society of London. Series B, Biological Sciences*, *369*(1639), 20120273. doi:10.1098/rstb.2012.0273

Gordon, L. J., & Steffen, W., & Jönsson. (2005). Human modification of global water vapor flows from the land surface. *Proceedings of the National Academy of Sciences USA*, 102, 7612-7617. 10.1073/pnas.0500208102

Gregory, P. J., Ingram, J. S. I., & Brklacich, M. (2005). Climate change and food security. *Philosophical Transactions of the Royal Society of London. Series B, Biological Sciences*, *360*(1463), 2139–2148. doi:10.1098/rstb.2005.1745

Grinsven, H. J. M., Erisman, J. W., Vries, W., & Westhoek, H. (2015). Potential of extensification of European agriculture for a more sustainable food system, focusing on nitrogen. *Environmental Research Letters*, *10*(2), 025002. doi:10.1088/1748-9326/10/2/025002

Hamlett, E. (2011). *Overpopulation: Disease prevalence*. Academic Press.

IPCC. (2014). *Climate change 2014 synthesis report*. Retrieved June 26, 2016, from http://ar5-syr.ipcc.ch/

Knight, K. W., & Newman, S. (2013). Organic agriculture as environmental reform: A cross-national investigation. *Society & Natural Resources*, *26*(4), 369–385. doi:10.1080/08941920.2012.687070

Koohafkan, P., Altieri, M. A., & Gimenez, E. H. (2012). Green agriculture: Foundations for biodiverse, resilient and productive agricultural systems. *International Journal of Agricultural Sustainability*, *10*(1), 61–75. doi:10.1080/14735903.2011.610206

Krupenikov, I., Boincean, B., Dent, D., Krupenikov, I., Boincean, B., & Dent, D. (2011). Humus- fertility and global carbon sink. In *The black earth: Ecological principles for sustainable agriculture on guardian of Chernozem Soils. International year of planet earth* (pp. 39–50). Dordrecht, The Netherlands: Springer. doi:10.1007/978-94-007-0159-5_7

Kuppusamy, S., Thavamani, P., Megharaj, M., Venkateswarlu, K., & Naidu, R. (2016). Agronomic and remedial benefits and risks of applying biochar to soil: Current knowledge and future research directions. *Environment International*, *87*, 1–12. doi:10.1016/j.envint.2015.10.018

Lal, R. (2015). Restoring soil quality to mitigate soil degradation. *Sustainability*, *7*(5), 5875–5895. doi:10.3390u7055875

Laurance, W. F., Sayer, J., & Cassman, K. G. (2014). Agricultural expansion and its impacts on tropical nature. *Trends in Ecology & Evolution*, *29*(2), 107–116. doi:10.1016/j.tree.2013.12.001

Li, D., Hockaday, W. C., Masiello, C. A., & Alvarez, P. J. J. (2011). Earthworm avoidance of biochar can be mitigated by wetting. *Soil Biology & Biochemistry*, *43*(8), 1732–1737. doi:10.1016/j.soilbio.2011.04.019

Lobell, D. B., Cassman, K. G., & Field, C. B. (2009). Crop yield gaps: Their importance, magnitudes, and causes. *Annual Review of Environment and Resources*, *34*(1), 179–204. doi:10.1146/annurev.environ.041008.093740

Ly, P., Jensen, L. S., Bruun, T. B., Rutz, D., & de Neergaard, A. (2012). The system of Rice Intensification: Adapted practices, reported outcomes and their relevance in Cambodia. *Agricultural Systems*, *113*, 16–27. doi:10.1016/j.agsy.2012.07.005

Lynch, D. H., MacRae, R., & Martin, R. C. (2011). The carbon and global warming potential impacts of organic farming: Does it have a significant role in an energy constrained world? *Sustainability*, *3*(12), 322–362. doi:10.3390u3020322

Mader, P., Fliessbach, A., Dubois, D., Gunst, L., Fried, P., & Niggli, U. (2002). Soil fertility and biodiversity in Organic farming. Science, 296, 1694-1697. doi:10. 1126/science.1071148

Manlay, R., Feller, C., & Swift, M. (2007). Historical evolution of soil organic matter concepts and their relationships with the fertility and sustainability of cropping systems. *Agriculture, Ecosystems & Environment*, *119*(3-4), 217–233. doi:10.1016/j.agee.2006.07.011

Milder, J. C., Arbuthnot, M., Blackman, A., Brooks, S. E., Giovannucci, D., Gross, L., ... Zrust, M. (2014). An Agenda for assessing and improving conservation impacts of sustainability standards in tropical agriculture. *Conservation Biology*, *29*(2), 309–320. doi:10.1111/cobi.12411

Millennium Development Goals Report. (2015). *The millennium development goals report 2015.* United Nations. Retrieved from http://www.un.org/millenniumgoals/2015_MDG_Report/pdf/MDG%202015%20rev%20(July%201).pdf

Millennium Ecosystem Assessment. (2005). Ecosystems and Human Well-Being: *Scenarios: Findings of the scenarios working group (vol. 2).* Island Press.

Mishra, A., Kumar, P., & Noble, A. (2013). Assessing the potential of SRI management principles and the FFS approach in Northeast Thailand for sustainable rice intensification in the context of climate change. *International Journal of Agricultural Sustainability*, *11*(1), 4–22. doi:10.1080/14735903.2012.658648

Mishra, A., & Salokhe, V. M. (2008). Seedling characteristics and the early growth of transplanted rice under different water regimes. *Experimental Agriculture*, *44*(03), 1–19. doi:10.1017/S0014479708006388

Mtengeti, E. J., Brentrup, F., Mtengeti, E., Eik, L. O., & Chambuya, R. (2015). Sustainable intensification of Maize and Rice in smallholder farming systems under climate change in Tanzania. In R. Lal & ... (Eds.), *Sustainable Intensification to Advance Food Security and Enhance Climate Resilience in Africa* (pp. 441–465). doi:10.1007/978-3-319-09360-4_24

Mueller, N. D., Gerber, J. S., Johnston, M., Deepak, K. R., Ramankutty, N., & Foley, J. A. (2012). Closing yield gaps through nutrient and water management. *Nature*, *490*(7419), 254–257. doi:10.1038/nature11420

Nguyen, H. T., Fischer, K. S., & Fukai, S. (2009). Physiological responses to various water saving systems in rice. *Field Crops Research*, *112*(2-3), 189–198. doi:10.1016/j.fcr.2009.03.004

Nissen, T. M., & Wander, M. M. (2003). Management and soil-quality effects on fertilizer-use efficiency and leaching. *Soil Science Society of America Journal*, *67*(5), 1524–1532. doi:10.2136ssaj2003.1524

Pasuquin, E., Lafarge, T., & Tubana, B. (2008). Transplanting young seedlings in irrigated rice fields: Early and high tiller production enhanced grain yield. *Field Crops Research*, *105*(1-2), 141–155. doi:10.1016/j.fcr.2007.09.001

Pathak, H., Sankhyan, S., Dubey, D. S., Bhatia, A., & Jain, N. (2013). Dry direct seeding of rice for mitigating greenhouse gas emission: Field experimentation and simulation. *Paddy and Water Environment*, *11*(1-4), 593–601. doi:10.100710333-012-0352-0

Phalan, P., Onial, M., Balmford, A., & Green, R. E. (2011). Reconciling food production and biodiversity conservation: Land sharing and land sparing compared. *Science*, *333*(6047), 1289–1291. doi:10.1126cience.1208742

Pimentel, D., Whitecraft, M., Scott, Z. R., Zhao, L., Satkiewicz, P., Scott, T., ... Moe, T. L. (2010). Will limited land, water, and energy control human population numbers in the future? *Human Ecology*, *38*(5), 599–611. doi:10.100710745-010-9346-y

Pingali, P. L. (2012). Green Revolution: Impacts, limits, and the path ahead. *Proceedings of the National Academy of Sciences of the United States of America*, *109*(31), 12302–12308. doi:10.1073/pnas.0912953109

Pittelkow, C. M., Liang, X., Lindquist, B. A., Van Groenigen, K. J., Lee, J., Lundy, M. E., ... van Kessel, C. (2015). Productivity limits and potentials of the principles of conservation agriculture. *Nature*, *517*(7534), 365–368. doi:10.1038/nature13809

Ponisio, L. C., M' Gonigle, L. K., Mace, K. C., Palomino, J., de Valpine, P., & Kremen, C. (2015). Diversification practices reduce organic to conventional yield gap. *Proceedings of the Royal Society B: Biological Sciences*. 10.1098/rspb.2014.1396

Population Action International. (2011). *Why population matters to food security*. Washington, DC: Author.

Pretty, J. (2008). Agricultural sustainability: Concepts, principles and evidence. *Philosophical Transactions of the Royal Society of London. Series B, Biological Sciences*, *363*(1491), 447–466. doi:10.1098/rstb.2007.2163

Pretty, J., Sutherland, W. J., Ashby, J., Auburn, J., Baulcombe, D., Bell, M., ... Pilgrim, S. (2010). The top 100 questions of importance to the future of global agriculture. *International Journal of Agricultural Sustainability*, *8*(4), 219–236. doi:10.3763/ijas.2010.0534

Pretty, J., Toulmin, C., & Williams, S. (2011). Sustainable intensification in African agriculture. *International Journal of Agricultural Sustainability*, *9*(1), 5–24. doi:10.3763/ijas.2010.0583

Pretty, J. N., Noble, A. D., Bossi, D., Dixon, J., Hine, R. E., & Penning De Vries, F. W. (2006). Resource-conserving agriculture increases yields in developing countries. *Environmental Science & Technology*, *40*(4), 1114–1119. doi:10.1021/es051670d

Rajan, K., Natarajan, A., Kumar, K., Badrinath, M., & Gowda, R. (2010). Soil organic carbon—The most reliable indicator for monitoring land degradation by soil erosion. *Current Science*, *99*, 823–827.

Ramankutty, N., Evan, A. T., Monfreda, C., & Foley, J. A. (2008). Farming the planet: 1. Geographic distribution of global agricultural lands in the year 2000. *Global Biogeochemical Cycles*, *22*(1), GB1003. doi:10.1029/2007GB002952

Ramankutty, N., & Foley, J. A. (1999). Estimating historical changes in global land cover: Croplands from 1700 to 1992. *Global Biogeochemical Cycles*, *13*(4), 997–1027. doi:10.1029/1999GB900046

Ray, D. K., Ramankutty, N., Mueller, N. D., West, P. C., & Foley, J. A. (2012). Recent patterns of crop yield growth and stagnation. *Nature Communications*, *3*, 1293. doi:10.1038/ncomms2296

Robertson, G. P., Paul, E. A., & Harwood, R. R. (2000). Greenhouse gases in intensive agriculture: Contributions of individual gases to the radiative forcing of the atmosphere. *Science*, *289*(5486), 1922–1925. doi:10.1126cience.289.5486.1922

Robertson, G. P., & Swinton, S. M. (2005). Reconciling agricultural productivity and environmental integrity: A grand challenge for agriculture. *Frontiers in Ecology and the Environment*, *3*(1), 38–46. doi:10.1890/1540-9295(2005)003[0038:RAPAEI]2.0.CO;2

Rockstrom, J., Steffen, W., Noone, K., Persson, A., Chapin, F. S. III, Lambin, E. F., ... Foley, J. A. (2009). A safe operating space for humanity. *Nature*, *461*(7263), 472–475. doi:10.1038/461472a

Rosegrant, M. W., & Cline, S. A. (2003). Global food security: Challenges and policies. *Science*, *302*(5652), 1917–1919. doi:10.1126cience.1092958

Rosegrant, M. W., Meijer, S., & Cline, S. A. (2002). *International Model for Policy Analysis of Agricultural Commodities and Trade (IMPACT): Model description.* IFPRI. Retrieved from www.ifpri.org/themes/impact/impactmodel.pdf

Sadik, N. (2016). *Population growth and the food crisis*. FAO. Retrieved June 26, 2016 from http://www.fao.org/docrep/U3550T/u3550t02.htm

San-Oh, Y., Sugiyama, T., Yoshida, D., Ookawa, T., & Hirasawa, T. (2006). The effect of planting pattern on the rate of photosynthesis and related processes during ripening in rice plants. *Field Crops Research*, *96*(1), 113–124. doi:10.1016/j.fcr.2005.06.002

Sarwar, N., Ali, H., Ahmad, A., Ullah, E., Ahmad, S., Mubeen, K., & Hill, J. E. (2013). Water wise rice cultivation on calcareous soil with the addition of essential micronutrients. *Journal of Animal and Plant Sciences*, *23*, 244–250.

Schmidhuber, J., & Tubiello, F. N. (2007). Global food security under climate change. *Proceedings of the National Academy of Sciences of the United States of America*, *104*(50), 19703–19708. doi:10.1073/pnas.0701976104

Scopel, E., Triomphe, B., Affholder, F., Macena, F., Corbeels, M., & Valadares, J. H. (2013). Conservation agriculture cropping systems in temperate and tropical conditions, performances and impacts. A review. *Agronomy for Sustainable Development*, *33*(1), 113–130. doi:10.100713593-012-0106-9

Seneviratne, G., & Kulasooriya, S. A. (2013). Reinstating soil microbial diversity in agroecosystems: The need of the hour for sustainability and health. *Agriculture, Ecosystems & Environment*, *164*, 181–182. doi:10.1016/j.agee.2012.10.002

Seufert, V., Ramankutty, N., & Foley, J. A. (2012). Comparing the yields of organic and conventional agriculture. *Nature*, *485*(7397), 229–232. doi:10.1038/nature11069

Singh, J. S. (2015). Microbes: The chief ecological engineers in reinstating equilibrium in degraded ecosystems. *Agriculture, Ecosystems & Environment*, *203*, 80–82. doi:10.1016/j.agee.2015.01.026

Singh, R., Babu, J. N., Kumar, R., Srivastava, P., Singh, P., & Raghubanshi, A. S. (2015a). Multifaceted application of crop residue biochar as a tool for sustainable agriculture: An ecological perspective. *Ecological Engineering*, *77*, 324–347. doi:10.1016/j.ecoleng.2015.01.011

Singh, R., Srivastava, P., Singh, P., & Raghubanshi, A. S. (2016). Sustainability indicators of the emergent agronomic practices: An Indian perspective. In B. R. Gurjar & L. T. Molina (Eds.), *Environmental Science & Engineering: Climate Change and Sustainable Technology* (vol. 12, pp. 142–166). Studium Press LLC.

Singh, R., Srivastava, P., Upadhyay, S., Singh, P., Raghubanshi, A. S., & Babu, J. N. (2015b). Integrating biochar as conservation agriculture tool under climate change mitigation scenario. In *Proceedings of National Conference on Climate Change: Impacts, Adaptation, Mitigation Scenario and Future Challenges in Indian Perspective.*

Singh, R. P., Embrandiri, A., Ibrahim, M. H., & Esa, N. (2011). Management of biomass residues generated from palm oil mill: Vermicomposting a sustainable option. *Resources, Conservation and Recycling*, *55*(4), 423–434. doi:10.1016/j.resconrec.2010.11.005

Singh, Y. V. (2013). Crop and water productivity as influenced by rice cultivation methods under organic and inorganic sources of nutrient supply. *Paddy and Water Environment*, *11*(1-4), 531–542. doi:10.100710333-012-0346-y

Smil, V. (2002). Nitrogen and food production: Proteins for human diets. *Ambio*, *31*(2), 126–131. doi:10.1579/0044-7447-31.2.126

Smith, P., Haberl, H., Popp, A., Erb, K.-H., Lauk, C., Harper, R., ... Rose, S. (2013). How much land-based greenhouse gas mitigation can be achieved without compromising food security and environmental goals? *Global Change Biology*, *19*(8), 2285–2302. doi:10.1111/gcb.12160

Srivastava, P., Singh, R., Tripathi, S., & Raghubanshi, A. S. (2016a). An urgent need of sustainable thinking in agriculture - An Indian scenario. *Ecological Indicators*, *67*, 611–622. doi:10.1016/j.ecolind.2016.03.015

Srivastava, P., Singh, R., Bhadouria, R., Tripathi, S., Singh, P., Singh, H., & Raghubanshi, A. S. (2016b). Organic amendment impact on SOC dynamics in dry tropics: A possible role of relative availability of inorganic N-pools. *Agriculture, Ecosystems & Environment*, *235*, 38–50. doi:10.1016/j.agee.2016.09.036

Stanhill, G. (1990). The comparative productivity of organic agriculture. *Agriculture, Ecosystems & Environment*, *30*(1-2), 1–26. doi:10.1016/0167-8809(90)90179-H

Steinfeld, H., Gerber, P., Wassenaar, T. D., Castel, V., & de Haan, C. (2006). *Livestock's long shadow: Environmental issues and options*. Rome, Italy: Food & Agriculture Organization.

Stoop, W. A. (2005). *The 'System of Rice Intensification' (SRI); Results from exploratory field research in Ivory Coast: Research needs and prospects for adaptation to diverse production systems of resource-poor farmers*. Retrieved from http://ciifad.cornell.edu/sri/

Tabbal, D. F., Bouman, B. A. M., Bhuiyan, S. I., Sibayan, E. B., & Sattar, M. A. (2002). On-farm strategies for reducing water input in irrigated rice: Case studies in the Philippines. *Agricultural Water Management*, *56*(2), 93–112. doi:10.1016/S0378-3774(02)00007-0

Tilman, D., Cassman, K. G., Matson, P. A., Naylor, R., & Polasky, S. (2002). Agricultural sustainability and intensive production practices. *Nature, 418*(6898), 671–677. doi:10.1038/nature01014

Tilman, D., & Clark, M. (2014). Global diets link environmental sustainability and human health. *Nature, 515*(7528), 518–522. doi:10.1038/nature13959

Tilman, D., Fargione, J., Wolff, B., D'Antonio, C., Dobson, A., Howarth, R., ... Swackhamer, D. (2001). Forecasting agriculturally driven global environmental change. *Science, 292*(5515), 281–284. doi:10.1126cience.1057544

Tittonell, P. (2014). Ecological intensification of agriculture - sustainable by nature. *Current Opinion in Environmental Sustainability, 8*, 53–61. doi:10.1016/j.cosust.2014.08.006

Tuck, S. L., Winqvist, C., Mota, F., Ahnstrom, J., Turnbull, L. A., & Bengtsson, J. (2014). Land-use intensity and the effects of organic farming on biodiversity: A hierarchical meta-analysis. *Journal of Applied Ecology, 51*(3), 746–755. doi:10.1111/1365-2664.12219

Tuong, T. P., & Bouman, B. A. M. (2003). Rice production in water-scarce environments. In J. W. Kijne, R. Barker, & D. Molden (Eds.), *Water productivity in agriculture: Limits and opportunities for improvement* (pp. 53–67). CABI Publishing. doi:10.1079/9780851996691.0053

Turner, B. L. II, Clark, W. C., Kates, R. W., Richards, J. F., Mathews, J. T., & Meyer, W. B. (1990). *The earth as transformed by human action: Global and regional changes in the biosphere over the past 300 years*. Cambridge University Press.

United Nations. (2011). *World demographic trends.* Report of the Secretary-General.

United Nations. (2015). *World Population Prospects: The 2015 Revision, Key Findings and Advance Tables*. Working Paper No. ESA/P/WP.241. United Nations.

Uphoff, N. (2012). Supporting food security in the 21st century through resource-conserving increases in agricultural production. *Agriculture and Food Security, 1,* 18. Retrieved from http://www.agricultureandfoodsecurity.com/content/1/1/18

Uphoff, N., Kassam, A., & Harwood, R. (2011). SRI as a methodology for raising crop and water productivity: Productive Adaptations in rice agronomy and irrigation water management. Paddy and Water Environment, 9, 3-11. doi: s10333-010-0224–4

Vermeulen, S. J., Aggarwal, P. K., Ainslie, A., Angelone, C., Campbell, B. M., Challinor, A. J., ... Wollenberg, E. (2012). Options for support to agriculture and food security under climate change. *Environmental Science & Policy, 15*(1), 136–144. doi:10.1016/j.envsci.2011.09.003

Vitousek, P. M., Mooney, H. A., Lubchenco, J., & Melillo, J. M. (1997). Human domination of Earth's ecosystems. *Science, 277*(5325), 494–499. doi:10.1126cience.277.5325.494

Wackernagel, M., & Rees, W. (1996). *Our ecological footprint: Reducing human impact on the Earth.* Gabriola Island, BC: New Society Publishers.

Warnock, D. D., Lehmann, J., Kuyper, T. W., & Rillig, M. C. (2007). Mycorrhizal responses to biochar in soil – Concepts and mechanisms. *Plant and Soil, 200*(1-2), 9–20. doi:10.100711104-007-9391-5

Weld, M., & Wright, W. (1996). *Confronting the population crisis: Responses to the twenty-one most commonly used arguments to confound the issue.* Retrieved June 30, 2013 from http://www.aldridgeshs.eq.edu.au/sose/modrespg/populate/pop/index.htm

West, P. C., Gerber, J. S., Engstrom, P. M., Mueller, N. D., Brauman, K. A., Carlson, K. M., ... Siebert, S. (2014). Leverage points for improving global food security and the environment. *Science, 345*(6194), 325–328. doi:10.1126cience.1246067

Wezel, A., Casagrande, M., Celette, F., Vian, J.-F., Ferrer, A., & Peigné, J. (2014). Agroecological practices for sustainable agriculture. A review. *Agronomy for Sustainable Development, 34*(1), 1–20. doi:10.100713593-013-0180-7

WHO. (2008). Public health impacts of pesticides used in agriculture. WHO in collaboration with the United Nations Environment Programme.

WHO & FAO. (2007). Codex alimentarius: Organically produced foods (3rd ed.). Rome, Italy: Authors.

Willer, H., & Kilcher, L. (2011). The world of organic agriculture, statistics and emerging trends. Bonn, Germany: International Federation of Organic Agriculture Movements (IFOAM).

World Bank. (2007). *World development report 2008: Agriculture for development.* Washington, DC: World Bank.

World Bank. (2008). *World Development Report 2008: Agriculture for development.* Washington, DC: World Bank.

World Bank. (2012). *World Bank sees progress against extreme poverty, but flags vulnerabilities.* Retrieved July 5, 2013, from http://www.worldbank.org/en/news/press-release/2012/02/29/world-bank-sees-progress-against-extreme-poverty-but-flags-vulnerabilities.

World Food Programme. (2012). *The year in review 2011.* Rome, Italy: World Food Programme.

World Population Balance. (2013). *Current population is three times the sustainable level.* Retrieved July 31, 2013 from http://www.worldpopulationbalance.org/3_times_sustainable

Worldometers. (2016). *World population.* Retrieved June 26, 2016 from http://www.worldometers.info/world-population/

WWF. (2012). *Living planet report.* Retrieved July 31, 2013 from http://wwf.panda.org/about_our_earth/all_publications/living_planet_report/2012_lpr/

Yang, C., Yang, L., Yang, Y., & Ouyang, Z. (2004). Rice root growth and nutrient uptake as influenced by organic manure in continuously and alternately flooded paddy soils. *Agricultural Water Management, 70*(1), 67–81. doi:10.1016/j.agwat.2004.05.003

Yang, X., Bouman, B. A. M., Wang, H., Wang, Z., Zhao, J., & Chen, B. (2005). Performance of temperate aerobic rice under different water regimes in north China. *Agricultural Water Management, 74*(2), 107–122. doi:10.1016/j.agwat.2004.11.008

Yoshida, S. (1981). *Fundamentals of Rice Crop Science*. Manila, Philippines: IRRI.

Zhang, A., Bian, R., Hussain, Q. B., Li, L., Pan, G., Zheng, J., ... Zheng, J. (2013). Change in net global warming potential of a rice–wheat cropping system with biochar soil amendment in a rice paddy from China. *Agriculture, Ecosystems & Environment*, *173*, 37–45. doi:10.1016/j.agee.2013.04.001

This research was previously published in Environmental Issues Surrounding Human Overpopulation edited by Rajeev Pratap Singh, Anita Singh, and Vaibhav Srivastava, pages 12-39, copyright year 2017 by Information Science Reference (an imprint of IGI Global).

Chapter 23
Integrating Environment, Food Systems, and Sustainability in Feeding the Growing Population in Developing Countries

Abiodun Elijah Obayelu
Federal University of Agriculture – Abeokuta (FUNAAB), Nigeria

ABSTRACT

Food is indispensable to life. It plays an important role in the economy but what is not well known is the impact of production and consumption that food has on the environment. The nexus of food systems and the environment are complex and driven by many economic, socio-cultural, and environmental factors considered to be important in the contemporary global arena. As the world population grows, there is an increased demand on the already stretched food system and fragile environment. Processes along the food chain from agricultural production to food consumption produce outputs other than consumable food that are returned to the natural environment such as pollution or waste. This chapter sheds light on the links in food systems and environment in developing countries. A major finding is that the existing food systems that were supposed to produce adequate food for all are placing major stress on environmental assets including soil, water, fisheries, and biodiversity. For food systems to be sustainable, all hands must be on deck.

INTRODUCTION

One of the main challenges facing most developing countries is how to simultaneously provide enough food while conserving the natural resources to produce food for the present and future generations. Fact have emerged that despite production of enough food to feed the population by some countries, almost 800 million people (representing about 12.9 percent of the population in developing regions) are hungry (FAO, IFAD and WFP, 2015); 161 million under-five year olds have been estimated to be stunted (WHO, 2013); and over 2 billion suffer from micronutrient deficiencies, in particular vitamin A, iodine, iron and zinc (Ng, *et al.*, 2014; FAO, 2013). At the same time, the number of overweight/obese people has

DOI: 10.4018/978-1-5225-8063-8.ch023

reached more than 1.4 billion adults globally (representing about 30 percent of the total adult population). Obesity-related health conditions are rising rapidly both in developing and developed countries (WHO, 2015); and around 30 percent of the food produced worldwide (about 1.3 billion tons) is lost or wasted every year (FAO, 2011). This shows that a lot of people are probably not getting enough food and most eat diets poor in quality because the type of food systems currently are not sustainable. Food waste alone represents around 3-5% of global warming impacts, more than 20% of biodiversity pressure, and 30% of all of the world's agricultural land (EU, 2014). A food system governs what we eat; and there has been increasing concern at all levels of governance and in different policy sectors, civil society, academia and business that the food system today is not sustainable and endangers both health and the future of the planet (Kickbusch, 2010). Food system is said to be sustainable if it ensures food security and nutrition for all in such a way that the economic, social and environmental bases to generate food security and nutrition of future generations are not compromised (HLPE, 2014). Food system therefore is at the heart of the 2030 Agenda for Sustainable Development. Food production places enormous demand upon the environment. Food systems vary significantly from country to country in terms of actors, technology and type of resources used (UNEP, 2016). Food systems comprise all aspects of food production (the way the food is grown or raised; harvested or slaughtered; processed, packaged, or otherwise prepared for consumer purchase), food distribution (where and how the food is sold to consumers and how the food is transported) and food consumption (Ericksen, 2008; Farmar-Bowers, 2013; Pearson *et al.,* 2014). The various outcomes of the food system contribute to food security and health on the one hand and environmental degradation on the other (Ericksen, 2008).

Most of the food we consume is no longer produced in self-sufficient, but travels (and often a long way) from producer to consumer. Most of the food consumed are processed elsewhere and arrives in packaged forms through trade or the exchange of technologies or resources.

This chapter provides an overview of the current status of food systems, their interaction with the environment, and an assessment of the sustainability of current agricultural practices and effects on food production, underlying causes of unsustainable production and consumption patterns.

SITUATION ANALYSIS OF CURRENT FOOD SYSTEMS IN DEVELOPING COUNTRIES

The form of food system is a major issue in the politics of sustainable consumption and production (SCP) because of its impact on the environment, individual and public health, social cohesion, and the economy. Current food systems are not delivering food security and healthy food for everyone, nor are they sustainably using the limited resources. Emphasis is placed on 'modern' food systems in industrialized and emerging regions while 'traditional' food systems are still more prominent in rural areas of developing countries. Modern food systems are observed to have resulted in polluted soil, air, and water; eroded soil; dependence on imported oil; and loss of biodiversity. This variety in food systems, in combination with the social and natural environment in which they operate, has important implications on the possible pathways towards sustainable food systems and on the logic of intervention. In developing regions, there is a rapidly evolving replacement of traditional food systems by modern food systems. This trend is driven by macro-trends such as urbanization, increased wealth and other socio-economic and demographic developments. Countries with the biggest nutritional challenges are India, Nigeria and South Africa with the Food Sustainability Index (FSI) of 40.51, 45.60 and 49.34 respectively (Economist

Intelligence Unit, 2015). India has a high rate of nutritional deficiency, and high prevalence of under- and malnourishment, while South Africa has a high adoption of fast foods, and limited purchasing power for buying healthy foods.

One of the findings by Liverpool-Tassie *et al.,* (2016) is that urban and rural households in Nigeria have rapidly transforming diets. Consumption has diversified greatly, shifting beyond self-sufficiency to heavy reliance on food purchases and with a heavy shift into consumption of processed foods'. Food system transformation is characterized with supermarket growth in cities and imports still dominate the processed foods sector. Supermarkets can lead to higher consumption of processed foods and may be associated with higher rates of obesity among adults (Rischke *et al.,* 2015). The extent of supermarket penetration in food markets is therefore relevant for food policy.

Minten *et al.* (2013); Hassen *et al.* (2016) documented Ethiopia's food systems and how changes in the systems were driven in part by public investments to develop modern varieties of teff which is the main staple cereal. Ghana's traditional food systems are based on varieties of staple crops including cassava, maize, millet, plantain, rice, and sorghum. Ghanaian diets are just shifting across staples as people move to cities. The urban consumers care about packaging as a sign of product quality (Hollinger and Staatz, 2015).

MAJOR FINDINGS

Food Systems-Environment-Sustainability Nexus

Drawing from the United Nations Secretary General's Zero Hunger Challenge, a food system is defined as a system that embraces all the elements (environment, people, inputs, processes, infrastructure, institutions, markets and trade) and activities that relate to the production, processing, distribution and marketing, preparation and consumption of food and the outputs of these activities, including socio-economic and environmental outcomes (HLPE, 2014). (See Figure 1). A productive food system is one that produces crops, livestock, and fisheries using the fewest resources possible. The food we grow, produce, consume, trade, transport, store and sell is the essential connecting thread between people, and the environment. The food we produce and the food choices we make are creating stresses on our environment through climate, wildlife and habitats, driving unethical trade agreements, challenging animal welfare and putting pressure on water usage. Ericksen (2008) Ingram (2011) and Tendall *et al.*(2015) considered food systems as social-ecological systems that consist of mutually interacting biophysical and social components. It encompasses the activities of food production, processing, distribution, retail, and consumption and involve multidimensional (e.g., social, economic, institutional, environmental political, cultural) processes.

Food systems are inextricably linked to the environment and sustainability. Food environment' being the physical, social and economic surroundings influence what people eat and plays a major role in determining food systems in terms of food consumption pattern. Food companies, restaurants, food vendors and retailers are actively influencing this food environment to tempt people to make certain choices. Sustainability in any given food system encompasses a broad range of factors such as farming practices, local and seasonal sourcing, supply chain policies, consumer decisions, waste and packaging (Ng, 2015). There is no one model of a sustainable food system, but a set of principles that constitute sustainability. Food systems (production, processing, packaging, distribution, retailing, consumption and waste management) are webs of people and the resources and behaviors they affect. It is interconnected with the environment and sustainability. Food system does not only overlap in the area of food production, but also comprise the diverse set of institutions, technologies and practices that govern the way food is marketed, processed, transported, accessed and consumed. The food system is linked to the environment through a range of inputs at each stage and outputs that are returned to the environment as greenhouse gases, waste water, packaging and food waste, as well as land degradation and loss of biodiversity. Sustainable food system according to Pothukuchi and Jufman, (1999) is secure, reliable and resilient to change (including climate change, rising energy prices, etc) and accessible and affordable to all members of society. It is energy efficient, an economic power generator for farmers, whole communities and regions', healthy and safe and environmentally beneficial. Sustainable food system uses creative water reclamation and conservation strategies for agricultural irrigation, balances food imports with local capacity, adopts regionally-appropriate agricultural practices and crop choices, works towards organic farming, contributes to both community and ecological health. It builds soil quality and farmland through the recycling of organic waste, supports multiple forms of urban as well as rural food production, and ensures that food processing facilities are available to farmers and processors, celebrated through community events, markets, restaurants. It also preserves biodiversity in agro-ecosystems as well as in the crop selection; has a strong educational focus to create awareness of food and agricultural issues, and is fairly traded by providing a fair wage to producers and processors locally and abroad.

Sustainable food system ensures that all residents have access to healthy and affordable food options, minimizes the environmental impact of food production and transport; facilitates and encourages local food production and processing; creates local jobs that provide fair working conditions and a living wage; benefits local economies by supporting local food producers, retailers and businesses and maximizes resources through collection and reuse of organics (compost) and other food related by products (i.e. fats, oils, grease). These activities lead to a number of food and nutrition security outcomes including access to, and utilisation of food, in addition to food availability; and to other socioeconomic outcomes (e.g. employment). Food systems crucially depend on the environment: land, soil, water, terrestrial and marine biodiversity, minerals (essential nutrients for crops and animals) and fossil fuels.

A food system therefore also encompasses the interdependent sets of enterprises, institutions, activities and relationships that collectively develop and deliver material inputs to the farming sector, produce primary commodities, and subsequently handle, process, transport, market and distribute food and other agro-based products to consumers. Food systems differ from one location or country to another in terms of actors involved and characteristics of their relationships and activities (UNEP, 2016). Examples of activities that support sustainable food systems: farmers' markets, community gardens, regional food hubs, healthy corner store initiatives and/or programs to increase access to supermarkets, farm-to-school programs, food policy and/or security councils, local food procurement policies, food waste collection programs.

These processes not only produce food but also other outputs that are returned to the natural environment, such as greenhouse gases, waste water, as well as packaging and food waste. The link that exists in food systems, the environment and sustainability are therefore complex. It encompasses scientific, economic, cultural and social factors. Consumers in both developed and developing countries are more interested in knowing what lies beyond their food market. They want to know where the foods come from, and how such food was grown because of the consciousness of their health which is a function of the food they eat. Sustainability of the environment and food systems which is multi-faceted issue has therefore become an important international issue as more foods are needed to meet up with the

Figure 1. Components of a food system. Source: (HLPE, 2014)

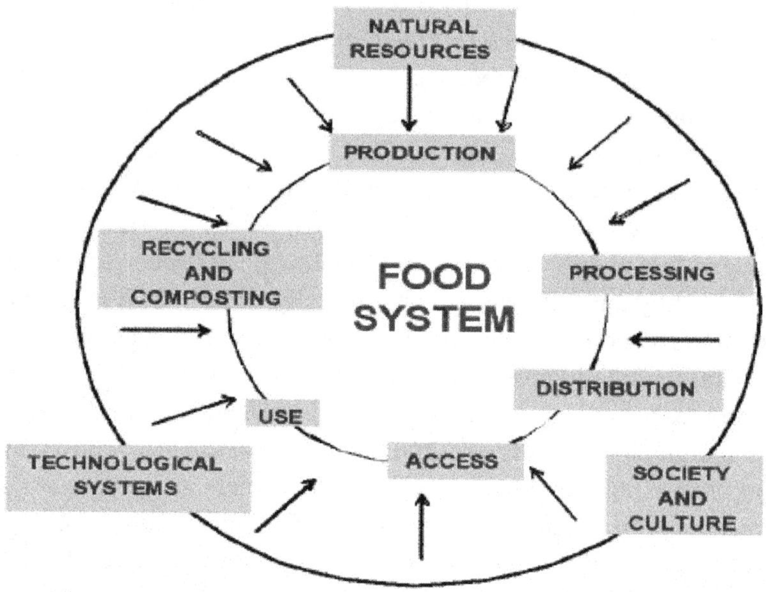

increasing population. The current systems of increasing food production through over-utilization of resources such as overfishing in many parts of the world has exhausted most of the world's fisheries. Overgrazing is leading to increasing desertification and over-cultivation has degraded soils. Felling of forests is disrupting rainfall and water supplies thereby enhancing climate change. All these activities are altering food system in most countries, with implications for yield, quality and affordability.

The failure to observe the simple principle of sustainability poses an ever-growing threat to food supplies that is needed to feed future generations. Sustainable food production uses systems and practices that can be continued indefinitely without depleting non-renewable resources (such as soil, energy, and biological diversity), causing ecological harm, or widening social inequities (within and across communities, countries, or generations). This by implication means that a sustainable food system does not utilize chemical pesticides, synthetic fertilizers, or genetically modified seeds or participate in practices that might contaminate the air, water, or soil. Sustainable farmers grow a variety of crops, use crop rotation, and manage waste which benefits not only the ecosystem but also the farm workers, consumers, and general public.

Food Systems and Value Chains

Food systems differ greatly across and within region and nation. They are increasingly interconnected through global and regional markets. They involve billions of actors at different levels along the food value chain. Food production activities comprises such factors as land use and tenure, soil management, crop breeding and selection, crop management, livestock breeding and management and harvesting (FAO, 1997). Food distribution involves a series of post-harvest activities including the processing, transportation, storage, packaging and marketing of food as well as activities related to household purchasing power, traditions of food use (including child feeding practices), food exchanges and gift giving and public food distribution. Activities related to food utilization and consumption include those involved in the preparation, processing and cooking of food at both the home and community levels, as well as household decision-making regarding food, household food distribution practices, cultural and individual food choices and access to health care, sanitation and knowledge. Diet shifts and growing demands for processed foods in developing countries is creating opportunities for adding value to primary agricultural products.

Determinants of Food System and Sustainability

The food system is highly complex and is driven by many economic, cultural and environmental factors. Better understanding of these drivers and how they interact could help to improve public policies. A food system operates within, and is influenced by, the social, political, economic and environmental context (Goodman, 1997; UNEP, 2016). Food systems gathers all the elements (environment, people, inputs, processes, infrastructures, institutions) and activities that relate to the production, processing, distribution, preparation and consumption of food and the outputs of these activities, including socio-economic and environmental outcomes (HLPE, 2014). All these factors therefore determine the nature of the food system. Achieving sustainable food systems depends on the ability to ensure zero food waste, 100 per cent access to food and increasing smallholder's income. Any policy measures for sustainable food systems must be able to link food production, distribution, consumption and the environment within which these activities occur (Figure 2).

A food system's sustainability is influenced by natural and human factors. These factors interact with each other within a food system. For example, the availability of water and land for food produc-

Figure 2. Determinants of the Food System. Source: Nugent et al., (2011)

tion is influenced by human actions such as use of fossil fuel, while human choices are influenced by environmental conditions. Food loss and waste have also been observed as determinants of sustainability of food systems in both economic and social terms.

Management of Food Systems in Developing Countries

Developing countries are going through significant changes in their food systems (Regmi and Meade, 2013; Reardon, 2015; Tschirley *et al.*, 2015a). The new food system model has implications for education and employment policies, as labor markets shift to respond to changing production systems both on- and off-farm (Tschirley et al., 2015b). For instance, the Nigerian traditional food system is characterized by low return on investments, crude and ineffective farm implements, low irrigation, and expensive inputs such as fertilizers, improved planting materials, low yielding plants and livestock. The food system is heavily reliant on fossil fuels. Large amounts of fuel are required to power farm equipment, to produce artificial fertilisers, and to transport food to supermarkets. Small producers often lack access to affordable inputs, roads and transport, electricity, irrigation, extension services, appropriate technology and market infrastructure. For example, the use of fossil fuels leads to carbon dioxide emissions (and to air pollution, depending on the burning process), while the use of minerals typically leads to nutrient emissions to ground and surface water. In Kenya, poor land preparation, inadequate soil management, and ineffective pest and disease control, leads to significant losses in food production. The inability to effectively monitor and enforce compliance and rules governing offshore territorial waters has also curtailed full exploitation of the offshore fishing potential (Alila and Atieno, 2006).

Sustainable Food Systems in the Existing Policy Frameworks and Development Agendas

Food systems are not only complex but are also characterized by many difference perceptions about the critical issues it involved, the challenges as well as way forward. Sustainable food system is complex systems that depend on a range of policy areas such as effect of climate change, water, ecosystems, the supply chain relationships, global economics, nutrition and obesity. The form of food systems carried out by people determine access to nutritious food and a way of achieving the United Nations Secretary-General Zero Hunger Challenge and Sustainable Development Goals 2, 3, 5, 6, 9, 12, 14, and 15 (Table 1)

Both developed and developing countries currently have a large number of laws, financial and other regulations that are influencing directly or indirectly on food systems.

Drawing from the transition theory (De Haan and Rotmans, 2011) three governance dynamics or ‘pathways’ can be distinguished that have the power to reshape current food systems, and hence their interactions with natural resources. These governance dynamics are (i) reforms by governments and international institutions; (ii) adaptations by food system actors; and (iii) alternative (niche) innovators.

CONCLUSION AND RECOMMENDATIONS

Food systems are characterized by interactions among system components and among factors, processes, and outcomes which are often non-linear. They are characterized by time and space lags and thresholds, which make it difficult to implement effective policies for promoting food system sustainability. A sus-

Table 1. SDGs and food system challenges

SDGs	Food System Challenges
Goal 2: End hunger, achieve food security and improved nutrition and promote sustainable agriculture	• Access to affordable and nutritious food • Food fortification and vitamin supplementation programmes to needy population
Goal 3: Ensure healthy lives and promote well-being for all at all ages.	• Earlier years' nutrition education for mothers • Encouraging exclusive breast feeding • Regulating marketing and sales obesogenic food • Public education campaigns on optimal diets
Goal 5: Achieve gender equality and empower all women and girls.	• Gender empowerment in food production, processing and distribution • Gender equality in access to agricultural inputs
Goal 6: Ensure availability and sustainable management of water and sanitation for all.	• Sustainable water usage • Ensuring good sanitation for all
Goal 9: Build resilient infrastructures, promote inclusive and sustainable industrialization, and foster innovation.	• Ensuring infrastructure for storage, transport, and logistics • Investment in innovations through R&D
Goal 10: Reduced inequalities	Nutritional deficits in the early years of life can cause lifelong deficits such as stunting and impaired cognitive development, deepening inequality cycles as malnourished children are unable to participate in labour force
Goal 12: Ensure sustainable consumption and production patterns.	• Sustainable use of arable land • Sustainable water management practices • Limiting agriculture- related pollution and emissions
Goal 13: Take urgent action to combat climate change and its impacts	Agriculture is both a cause of climate change through emissions it produces, and a victim as changes in temperature and rainfall impact crop growth and agricultural productivity
Goal 14: Conserve and sustainably use the oceans, seas and marine resources for sustainable development.	Sustainable use of the oceans, seas and marine resources
Goal 15: Protect, restore and promote sustainable use of terrestrial ecosystems, sustainably manage forests, combat deforestation, and halt and reserve land degradation and halt diversity loss.	• Managing deforestation related to food and non-food activities including livestock and biofuels • Lower use of harmful chemicals and related substances in agriculture

tainable food system therefore embraces the interconnectedness of all the food-related activities and the environment within which these activities occur. The production, distribution and consumption of food are dynamic process involving multiple inputs, outputs and stakeholders. The current global distribution of food is already unequal (about 800 million people worldwide do not have access to sufficient food, while 1.9 billion adults are overweight or obese), and this is projected to become even worse if we continue with "business as usual". There are obvious health and ethical implications to all of these. One of the major challenges facing many developing countries is how to ensure a food system and a food future that is more secure, sustainable, just and healthy. Lack of nutrition education and poor processing contribute significantly to nutrient losses and food systems in many developing countries. These types of systems are imperative to feed the growing population and minimize the effects of climate change. How food is grown and produced, the types of foods consumed, and quantity of food wasted have major impacts on the sustainability of the food system in developing countries. A given food system therefore offers multiple potentially competing and complementary points for intervention

Integration of food systems, the environment and sustainability require the support from a vast range of stakeholders from the farmers to the final consumer, from governments to private sector to civil society. Investments and innovations are needed. This includes fighting food waste and food loss, promoting knowledge and technology-sharing practices for agricultural producers, investing in farm-to-market infrastructures in developing countries, and advancing research and development (R&D) into new techniques and technologies to improve yield and lower environmental impact. Reducing food loss and waste also requires efforts by retailers and food processors.

Policy makers could help smallholders gain a better position in global food supply chains dominated by agro-food corporations; reduce food losses in line with FAO, (2012); invest in women's agriculture, reform agricultural input markets; improve resource efficiency, participation, and accountability in water and energy systems and invest in nutrition and health. Investment in management practices and research development will enable a more effective food production and process in developing countries. There is need for a regulation to prevent nutrient losses at all stages especially in the livestock sector, proper regulation on access to and use of water, biodiversity, and ecosystems services; removal of subsidies that encourage unsustainable production or practices. Under-nutrition can be reduced significantly when the traditional food systems in developing countries is improved using a combination of strategies including nutrition education.

REFERENCES

Alila, P. O., & Atieno, R. (2006). *Agricultural Policy in Kenya: Issues and Processes.* Paper presented for the Future Agricultures Consortium workshop, Nairobi, Kenya.

De Haan, J., & Rotmans, R. (2011). Patterns in transitions: Understanding complex chains of change. *Technological Forecasting and Social Change, 78*(1), 90–102. doi:10.1016/j.techfore.2010.10.008

Economist Intelligence Unit. (2015). *Fixing: Towards A More Sustainable Food System.* Available online at http://foodsustainability.eiu.com/whitepaper/

Ericksen, P. (2008). Conceptualizing food systems for global environmental change research. *Global Environmental Change, 18*(1), 234–245. doi:10.1016/j.gloenvcha.2007.09.002

European Commission. (2014). Impact assessment on measures addressing food waste. Review of EU Waste Management Target. EC.

FAO. (1997). The food system and factors affecting household food security and nutrition. In Agriculture food and nutrition for Africa - A resource book for teachers of agriculture. Rome: FAO.

FAO. (2011). *Global food losses and food waste – Extent, causes and prevention.* Rome: FAO. Available at: http://www.fao.org/docrep/014/mb060e/mb060e00.htm

FAO. (2012). *The state of food insecurity in the World.* Rome: Food and Agriculture Organisation.

FAO. (2013). *The State of Food and Agriculture. Food Systems for Better Nutrition.* Rome: FAO.

FAO, IFAD, & WFP. (2015). The State of Food Insecurity in the World 2015. In *Meeting the 2015 international hunger targets: Taking stock of uneven progress.* Rome: FAO.

FAO and Biodiversity International. (2010). *Biodiversity and Sustainable Diets United against Hunger.* International Scientific Symposium, FAO/Biodiversity International.

Farmar-Bowers, Q. (2013). Framing the research needs for food security in Australia. In Q. Farmar-Bowers, V. Higgins, & J. Millar (Eds.), *Food security in Australia: Fixing Business Media*. doi:10.1007/978-1-4614-4484-8_15

Goodman, D. (1997). World-scale processes and agro-food systems: Critique and research needs. *Review of International Political Economy*, *4*(4), 663–687. doi:10.1080/09672299708565787

Hassen, I. W., Dereje, M., Minten, B., & Hirvonen, K. (2016). *Diet transformation in Africa: the case of Ethiopia*. Ethiopia Strategy Support Program (ESSP). Working Paper no. 87, March. International Food Policy Research Institute.

HLPE. (2014). *Food losses and waste in the context of sustainable food systems. A report by the High Level Panel of Experts on Food Security and Nutrition of the Committee on World Food Security*. Rome: HLPE.

Hollinger, F., & Staatz, J. M. (2015). *Agricultural Growth in West Africa Market and Policy Drivers*. Rome: African Development Bank and Food and Agriculture Organisation.

Ingram, J. A. (2011). food systems approach to researching food security and its interactions with global environmental change. *Food Security*, *3*(4), 417–431. doi:10.100712571-011-0149-9

Kickbusch, I. (2010). *The food system: A prism of present and future challenges for health promotion and sustainable development*. Triggering the debate – white paper. Bern: Promotion Santé Suisse, 2010.

Liverpool-Tasie, L., Adjognon, A., & Reardon, T. (2016). *Transformation of the food system in Nigeria and female participation in the Non-Farm Economy (NFE)*. 2016 Annual Meeting, Boston, MA.

Minten, B., Tamru, S., Engida, E., & Tadesse, K. (2013). *Ethiopia's value chains on the move: the case of teff*. Ethiopia Strategy Support Program (ESSP). Working Paper no. 52, April. International Food Policy Research Institute.

Ng, M., Fleming, T., Robinson, M., Thomson, B., Graetz, N., Margono, C., … Abera, S.F. (2014). Global, regional, and national prevalence of overweight and obesity in children and adults during 1980–2013: a systematic analysis for the Global Burden of Disease Study 2013. *The Lancet*.

Ng, R. (2015). *Making Sustainable Food Choices: Influencing Consumer Behaviour for Food Sustainability in Oxford*. OxPolicy.

Nugent, R. (2011). *Bringing agriculture to the table: How agriculture can play a role in preventing chronic disease*. The Chicago Council on Global Affairs.

Pearson, D., Friel, S., & Lawrence, M. (2014). Building environmentally sustainable food systems on informed citizen choices: Evidence from Australia. *Biological Agriculture and Horticulture*, *30*(3), 183–197. doi:10.1080/01448765.2014.890542

Reardon, T. (2015). The hidden middle: The quiet revolution in the midstream of agrifood value chains in developing countries. *Oxford Review of Economic Policy*, *31*(1), 45–63. doi:10.1093/oxrep/grv011

Regmi, A., & Meade, B. (2013). Demand side drivers of global food security. *Global Food Security*, *2*(3), 166–171. doi:10.1016/j.gfs.2013.08.001

Reisch, L., Eberle U., & Lorek S. (2013). Sustainable food consumption: an overview of contemporary issues and policies. *Sustainability: Science, Practice, & Policy, 9*(2), 7–25.

Rischke, R., Kimenju, S. C., Klasen, S., & Qaim, M. (2015). Supermarkets and food consumption patterns: The case of small towns in Kenya. *Food Policy, 52*, 9–21. doi:10.1016/j.foodpol.2015.02.001

Tendall, D. M., Joerin, J., Kopainsky, B., Edwards, P., Shreck, A., Le, Q. B., ... Six, J. (2015). Food system resilience: Defining the concept. *Global Food Security, 6*, 17–23. doi:10.1016/j.gfs.2015.08.001

Tschirley, D., Reardon, T., Dolislager, M., & Snyder, J. (2015a). The rise of a middle class in east and southern Africa: Implications for food system transformation. *Journal of International Development, 27*(5), 628–646. doi:10.1002/jid.3107

Tschirley, D., Snyder, J., Dolislager, M., & Snyder, J. (2015b). Africa's unfolding diet transformation: Implications for agrifood system employment. *Journal of Agribusiness in Developing and Emerging Economies, 5*(2), 1–48. doi:10.1108/JADEE-01-2015-0003

UN-DESA. (2015). *World Population Prospects: The 2015 Revision, Key Findings and Advance Tables.* Available online at http://esa.un.org/unpd/wpp/publications/files/key_findings_wpp_2015.pdf

UNEP. (2016). *Food Systems and Natural Resources.* A Report of the Working Group on Food Systems of the International Resource Panel.

Westhoek, H., Ingram, J. S. I., Van Berkum, S., Özay, L., & Hajer, M. (2016). *Food Systems and Natural Resources. A Report of the Working Group on Food Systems of the International Resource Panel.* UNEP Nairobi.

WHO. (2013). *UNICEF-WHO-The World Bank Joint Child Malnutrition Estimates 2013.* Available at: http://www.who.int/nutgrowthdb/summary_jme_2013.pdf?ua=1

WHO. (2015). *Global Health Observatory (GHO) data.* Available at: http://www.who.int/gho/ncd/risk_factors/obesity_text/en/

This research was previously published in Food Systems Sustainability and Environmental Policies in Modern Economies edited by Abiodun Elijah Obayelu, pages 1-14, copyright year 2018 by Engineering Science Reference (an imprint of IGI Global).

Section 7
Urban Agriculture

Chapter 24
Urban Farming Movement

Pierluigi Nicolin
Politecnico di Milano, Italy

ABSTRACT

The expression "urban agriculture" refers to the emergence in many cities of areas cultivated by farmers who distribute the fruits of the land they work in the environs of the zone of production. The movement, born in response to a range of real needs, has become a global phenomenon, and has taken on an organized form in a large number of cities: from Mumbai to Beijing, London, New York, Detroit, São Paulo, Rosario, Vancouver, Tokyo, San Francisco, etc. The urban farming movement, with its production of food, its educational aims and the idea of creating sustainable situations, has been able to take root in many cities and metropolises as it is closely integrated with the urban ecosystem. For the most part it is the poor and women who, working on small farms located both inside and outside the city, are nurturing this politico-cultural movement. Their agricultural settings are creating new and interesting landscapes that need to be analyzed from an aesthetic perspective as well for the influence that they might have on contemporary landscape architecture. The phenomenon could have repercussions on the visual conventions of the urban and suburban environment and even affect the behavior and lifestyles of city dwellers should it develop on a larger scale.

The first effects of this phenomenon can already be seen in the way that some people living in cities are trying to recover the sense of the day/night cycle, of seasonal rhythms. Out of this may come the happy discovery that the urban environment can also participate in cycles of nature, such as snow-rain-sun, spring-summer-fall-winter and others. In short with urban agriculture people would go back to experiencing, but in a conscious, reflective and aesthetic manner, what has always been present in the rural world, but in a superficial, habit-bound way lacking any true aesthetic awareness. Since the world of peasant farming has almost totally vanished in our advanced countries, replaced by a pervasive agricultural industry that has left only a few residues of the old farming environments and the old customs, it seems to be up to the city, paradoxically, to attract something of that world. Thus at times in Milan flcks of birds can be seen taking possession of certain large trees that now only survive in urban parks. The horizon of landscape in which this essay moves stimulates some particular considerations on the role of urban agriculture in answering a need that seems to have been neglected by the programs of landscaping carried out in the

DOI: 10.4018/978-1-5225-8063-8.ch024

last few decades. For example the tendency to revive the approach of the 18th-century Accademia dei Georgofili could be understood as a backlash against the over aestheticizing currents of contemporary landscape architecture with their purely picturesque treatment of environments that ask only to be seen.

Called on to solve the particular problems that afflict the production of a location in a deterritorialized world, the art of the landscape architect, left to itself, and even if inspired by the images and patterns of the agricultural landscape, could do no more than stage simulacra through evocation of the various techniques for the production of location: the organization of paths and passageways, the creation or removal of fields and vegetable gardens, ceremonies of naming, the symbolic mapping of spaces of transhumance and hunting grounds, in a scenario that obviously lacks the social organisms of the actors corresponding to the functions evoked. In short the landscape architect now has to do without the neighborhood, i.e. an actually existing social form in which the location, as dimension or value, is realized. Neighborhood communities linked to intensive forms of urban agriculture have been present since the most ancient civilizations, more "environmentally friendly" than our own, and it is no surprise that some think of today's biointensive urban cultivations as a "medicine" capable of creating new neighborhood communities, of contributing to energy saving and of fostering environmental sustainability. If with the development of industrial capitalism in the 19th century the separation between town and country was identified as the prime cause of the division between manual labor and intellectual work, today's urban agriculture with the manual skills it requires can help to alleviate that historical opposition. And yet that opposition is evoked once again in the expression "urban agriculture," given that "town" and "country" with their contradictions still act as basic divisions of human settlement. So the phrase sounds like an oxymoron, in the manner, for example, of "hellish heaven, heavenly hell," used by Giovan Battista Marino to define love in his *Adone* (1623). Moreover the figure of the oxymoron, we learn from Jacques Derrida, is intrinsic to the meaning of the Greek term *pharmakon*, which is at once the remedy and the poison, what restores health and at the same time brings sickness and death, just as there are two faces into which love is split over time, both farsighted and blind, sacred and profane. In fact if *pharmakon* is translated as "remedy," "therapeutic medicine," the emphasis is placed on the rational, scientific value of the term, thereby disregarding the magical component, thepeculiar and uncontrollable occult power of enchantments and charms, whose consequences cannot be totally predicted as it has the effect of a spell capable of generating lethal torpor, narcosis and paralysis. For its part urban agriculture could have aesthetic as well as therapeutic consequences on the city, and produce a rebalancing with regard to the merely decorative aspects of urban green spaces. Certainly the shift that has led to the idea of a planetary garden in the manner of Gilles Clément, put forward in the celebrated exhibition "Le Jardin Planétaire" at the Grand Halle de la Villette (Paris 1999), to the idea of a "planetary vegetable garden" implicit in the revival of the "Georgophilic" tendency, initially announced publicly as the underlying philosophy for the Milan Expo in 2015, entails the choice of a humanistic option in favor of cultivation, rejecting, de facto, the more radical positions of deep ecology that as is well known regard farming as an unwarranted anthropocentric intrusion into the course of natural events. Thus the new "Georgophilic" enthusiasm has nothing to do with Gilles Clément's idea of the third landscape, however fascinating it may be. The expression "third landscape" indicates all the "places abandoned by man": parks and nature reserves, the large uninhabited areas of the planet, but also disused industrial sites where brambles and brushwood grow; the weeds in the center of a median strip … different spaces that have only the absence of any human activity in common, but that taken as a whole are fundamental for the conservation of biological diversity. The new passion for the vegetable garden, for urban or suburban agriculture, draws on old and new knowledge of nature, reviving old techniques of cultivation. It wants to establish a new pact with these living organisms, adapting them to an urban botany, selecting them, grouping them, having them on the doorstep, cultivating them, harvesting their fruits and, not least, reproducing a certain kind of community around these actions. While remaining in the city, or precisely because of this, the champions of urban agriculture would like to put into effect particular aesthetic and behavioral codes in such a way as to create a sense of neighborhood by making the most of certain material practices and tackling the production of locations in the by and large deterritorialized panorama of the metropolis.

What is the meaning of the words "we must cultivate our garden" in Voltaire's *Candide*? Is Voltaire saying that cultivating our garden is like "minding our own business?" Is there something in this declaration that concerns the recent spread of urban agriculture? By putting these words in the mouth of his protagonist, Voltaire is arguing that we should do what is within our capacity without yielding to impossible dreams or imagining a nonexistent reality. Amidst so many horrors humanity can also be the master of its own destiny on condition that it does not give into the illusion of false theories or utopian revolutions that clash with the reality of our human nature and of the evils of the various established societies. If this is how things stand, do we have to frame the phenomenon of the new urban agriculture in the narrative of the end of the great illusions? Is it the response to these years of crisis in Western countries? Deeply affected by sad events in his own life as well as by terrible disasters like the Lisbon earthquake of 1755 that had destroyed the city, causing thousands of deaths and undermining his faith in Providence, Voltaire wrote *Candide* under the influence of the pessimism of the moment. Only at the end of the story does the philosopher yield to the hope expressed by the celebrated words "il faut cultiver notre jardin" ("we must cultivate our garden"). Once again, in the current phenomenon of cultivation of one's own garden, there is an accentuation of the collective, community dimension, as well as a request not to remain confined to the outskirts of the city as in the tradition of the urban allotments of the last century, but to see with pride these activities as a contribution to the safeguarding of the planet. Cultivating one's garden would represent a political phenomenon of dissidence for those who seek to save the world by defending the supply of food, in a moment of dramatic oscillation between bulimia and anorexia, but once again the figure of the oxymoron raises its head in this discussion, as in the title of Serge Latouche's latest book, *Vers une société d'abondance frugale* ("Toward a Frugal Abundance"), "a fiery tirade against the catastrophe produced by bulimia of consumption." So some have come to think that in our time cooking with the simple products of nature, and growing some of them yourself, can be regarded as a subversive act. When "*Manifesto: The Mad Farmer Liberation Front*" by the poet, writer and farmer Wendell Berry was published in the *Whole Earth Catalog*, the mother of all alternative magazines, in 1969 it was immediately evident that it contained all the best that had come out of the American student revolt and the hypothesis of a return to nature and abandon of consumer society. Forty years have passed since then and his environmental message has become legendary, but in reality we do not know whether we have to respond to the excesses of consumerism or to the new shortages produced by the grave economic crisis underway by turning to alternative sources of energy. The new urban agriculture certainly has a symbolic and therapeutic value. It can help us to modify many points of view on the world, but we have to realize that we will not be able to rely much on its production capacity. And yet the new urban agriculture undoubtedly wants to maintain the character of a place of compensation typical of the garden, although with the addition of a particular pathos due to the discovery that in addition to having a form the garden yields fruits. Who in Italy does not admire the agricultural landscapes present in the great tradition of painting, painting that has laid the foundations of the idea of the Italian landscape? The cultivation of open urban space seems an exciting proposition and probably also an aesthetic culture medium in which to try out new forms of living together and developing fertile community relations, and this might have consequences. It will be interesting to see how many vegetable gardens will spring up in the neglected zones of our city and whether someone is going to start making its existing gardens cultivable. The aesthetic appreciation of the vegetable garden presupposes a shift in sensibility from the visual and mental sphere of recent landscape architecture to the physiological and empirical one more characteristic of the rural habitat, where getting your hands dirty, but also tasting, eating, toiling, etc. implies an opening to the contamination between different realms of the sensible world. Finally, why not? "Cultivating our garden" might be a *pharmakon*, a "remedy," an interesting therapeutic medicine from which we cannot exclude the emergence of something whose consequences cannot be wholly predicted.

REFERENCES

Brand, S. (1971). *The Last Whole Earth Catalog*. Portola Institute, Inc.

Haney, D. H. (2010). *When Modern Was Green, Life And Work Of Leberecht Migge*. London: Routledge.

Ingersoll, R. (2010). Civic Agriculture: Gartenbau zwischen Sinneslust und sozialer Statdterneuerung [Civic Agriculature: Urban Gardening between Sensual Pleasures and Social Renewal of the City]. In Grunstadt [Urban Green]. Basel: Birkhauser.

Lascelles, M. (2010). *The Union Street Urban Orchard Book. A Case Study Of Creative Interim Use*. London: The Architecture Foundation.

Poole, S. (2006). *The Allotment Chronicles, A Social History Of Allotment Gardening*. Northants, UK: Silver Link Publishers.

Sanderson, E. W. (2009). *Mannahatta. A Natural History Of New York City*. New York: Abrams.

This research was previously published in Design Innovations for Contemporary Interiors and Civic Art edited by Luciano Crespi, pages 186-189, copyright year 2017 by Information Science Reference (an imprint of IGI Global).

Chapter 25
Developing an Economic Estimation System for Vertical Farms

Yiming Shao
University of Nottingham, UK

Tim Heath
University of Nottingham, UK

Yan Zhu
University of Nottingham, UK

ABSTRACT

The concept of vertical farming is nearly twenty years old, however, there are only a few experimental prototypes despite its many advantages compared to conventional agriculture. Significantly, financial uncertainty has been identified as the largest barrier to the realization of a 'real' vertical farm. Some specialists have provided ways to calculate costs and return on investment, however, most of them are superficial with calculations based on particular contextual circumstances. To move the concept forwards a reliable and flexible estimating tool, specific to this new building typology, is clearly required. A computational system, software named VFer, has therefore been developed by the authors to provide such a solution. This paper examines this highly flexible, customised system and results from several typical vertical farm configurations in three mega-cities (Shanghai, London and Washington DC) are used to elucidate the potential economic return of vertical farms.

INTRODUCTION

The 'vertical farm', a relatively new concept in the realm of urban agriculture that, proposes high-rise buildings as vehicles to cultivate plants in sealed, artificial indoor conditions with advanced greenhouse technologies such as hydroponics and areoponics (Despommier & Ellingsen, 2008). A number of advantages of such a method can be realized compared with conventional outdoor earth-based agriculture. Significantly, these can include:

DOI: 10.4018/978-1-5225-8063-8.ch025

1. Saving land (stacking up floors and several levels on each floor);
2. All-year-round, high yield production;
3. Protection from severe weather events enabling secured production;
4. No use of pesticides or fertilizers;
5. Saving water (using 70%-95% less water);
6. Saving financial and environmental logistic costs (local fresh production minimizing transportation); (Despommier, 2011; Heath, Zhu, & Shao, 2012)

As a result, in the past few years, vertical farms have drawn unprecedented attention from academia to business communities. Despite the aforementioned advantages, proposals are generally still at the conceptual stage except for a few experimental small-scale research-based and pilot examples existing in developed countries. Significantly, Heath et al. (2012) identify that the largest barrier to the realization of vertical farming is not the capability or availability of technology, but the uncertainty of its economic feasibility. Since a vertical farm concept integrates multidisciplinary knowledge from architecture and building services to agriculture and other specialist areas, a detailed financial analysis of start-up costs, operation costs, and revenue is difficult to establish. Only a few designers, engineers, economists have made their own approximate calculations either on capital budget or operating costs. Scientist and anti-global warming activist, Monbiot (2010) calculated that the cost of providing enough supplementary light to grow the grain required for a single loaf of bread to be almost $10 in the US. In addition, according to Omafra (2010), the initial cost could easily be over $100 million, for a 60-hectare vertical farm. Also, Graff (2011) estimated the return on investment (ROI) for of a 10-storey vertical farm in Toronto to be approximately 8 per cent which is lower than the minimum acceptable ROI, at 10-12 cent, for most investors. These calculations are based on certain particular circumstances, which are difficult to transplant to other cases or locations. The lack of precedent or pioneering examples therefore creates challenges in justifying the ROI for potential developers or investors. Clearly, developers will require some financial evidence before financially committing to the development of detailed design drawings. As a result, without information on the potential ROI, developers and investors are likely to view vertical farm projects as a too risky investment, therefore, a reliable financial estimation tool is required that can overcome the lack of existing detailed design drawings or historical data from similar projects.

Focusing on the above issues, this paper presents an integrated knowledge-based software system developed by the authors, called 'VFer' (short for 'vertical farmer'), to enable better cost and revenue estimations of vertical farms. VFer has two basic functions. The first is for preliminary design estimating, whereby, the user inputs some of the initial decisions and settings of a potential vertical farm project, such as site location, construction information, system selection, plant selection and so forth. Then the VFer system will automatically select the rest of the information necessary for estimating from its knowledge-based library according to user's choices and restrictions. In this phase, the user's selections are integrated with expert knowledge, construction codes, current cost data, system performances, labor productivity, and so forth. Finally, a cost and revenue report can be generated. The total processes for this stage can be completed in approximately 3 minutes. The second function of VFer is for a vertical farm project that has a fixed budget. A cost-to-revenue sensitivity analysis will be completed by the system with design alternatives before the optimal solution is provided. VFer will then simulate and present the value of the maximum area of the vertical farm that can be built under constraints of the given budget.

SYSTEM DEVELOPMENT

Conceptual Cost Estimate

According to different planning and design phases, cost estimation can be classified into a number of types. The American Association of Cost Engineers (AACE) lists five types of cost estimates: 1) order-of-magnitude estimate; 2) study estimate; 3) preliminary estimate; 4) definitive estimate; and 5) detailed estimate (Coker, 2007). A conceptual estimation is in between the first and second definition of AACE, which functions as a crucial reference for the investor, contractor, designer or lending entities to evaluate the feasibility and create an initial budget for a project (Sonmez, 2004). VFer is positioned as an evaluation system at a conceptual estimation level and because it will be used at the beginning of project planning, it does not require a detailed design proposal. It can therefore provide flexible design alternatives and a manifestation of associated results as a reference for a user's decision-making process. However, the difficulties and challenges are also obvious during the development of VFer. The conceptual cost estimating models employed needs to comprehensively consider the strategic issues, the limited information, the calculation speed, the control of accuracy, and may also need to assist the estimator in making subjective but reasonable judgments.

Overview of Estimating Methods

A wide range of literature on estimating methods was studied before the development of VFer. Some authors classify estimating methods with estimate levels, such as March (2009), whilst many more focus on techniques, such as the case-based reasoning model (Kim & Kang, 2007), the regression analysis model (Draper & Smith, 1998), and artificial intelligence (AI) modeling methods including expert systems, fuzzy logic, artificial neural network (ANN) and genetic method (Jackson, 1998). Although there are plenty of techniques, they can be categorised into two groups. The first group is an analogous method, vividly called a 'top-down method', which is based on historical data. With a historical data library, the top-down method analyses similarities between the proposed project and projects in the library, and selects those that share the most characteristics, assuming they have similar cost and performance levels. This method can be relatively accurate based on the condition that the library is large enough, and the proposed project type can be found in that library. The other group, a "bottom-up method", uses elemental and parametric techniques to detail the components and then to combine them. This method is effective as long as all the components or attributes are identified and information for each element is available.

The proposal for VFer integrates these two types of methods by dividing the estimation into three key parts: construction; system; and others. Since the vertical farm is a new building typology both in terms of architectural design and construction estimation, its function or cost has never been articulately documented, due to the limited number of practical examples. Therefore, separately computing the main components with a proper method is a relatively accurate and more convincing approach to cost estimation. As such, an analogous method was applied to the construction part, supposing the building as a 'container', while an elemental parametric method was used for the system part, by synthesizing the necessary subsets of the systems applied in a vertical farm. A 'user-defined' method was applied to other costs (such as construction-related costs and capital costs), which are highly dependent upon a user's decisions and preferences.

In general, there were 3 steps in the development of VFer:

Step I: Literature Review and Data Collection

1. Reviewing the research methods associated with cost estimation methods, mathematical /statistical/ analytical methods and so on;
2. Collecting background information on construction, including building structure types, construction project processes in different countries, cost data sources and cost models (building elemental costs for different functional types) and so on;
3. Collecting background information in agricultural science including: unit cost rates of agricultural cultivation service systems; yield of various hydroponic-grown crops under standard conditions (with optimal temperature, humidity and photosynthetically active radiation and proper supply of water and nutrients); the wholesale and retail prices of plants in the market; labor costs; and so on.

Step II: 'VFer' Programming Process

1. Outlining all of the variables that may significantly influence a vertical farm's cost and performance;
2. Creating a built-in database for the input variables and format for output variables;
3. Programming the built-in mathematics and logic interrelationships among these variables;
4. Creating the interface and controlling AI of the program.

Step III: Validation and Analysis

1. Evaluating the accuracy of VFer by comparing the results simulated by the software with actual values from publications;
2. VFer was evaluated by specialists in different fields during and after the Vertical Farming and Urban Agriculture (VFUA) 2014 international conference. The feedback was then collected and analyzed in order to refine the system.

The system processing structures are shown in Figure 1 and Figure 2, which represent the two main functions of VFer. The interactive processes of cost and ROI estimation between the user and system are illustrated in Figure 1. In this function, the user is required to interact with the process by making general decisions, such as whether to apply certain kinds of technologies or design strategies, at different phases. The rest of the information needed will be supplied by the library, which includes detailed values and specifications. Therefore, the requirements of a user's background knowledge are relatively low in operating VFer. Also, help information and warnings are displayed to help the user proceed or prevent any unreasonable decision-making. Based on trials, using VFer takes around 3 minutes to finish the streamlining and achieve the final ROI report. Subsequently, the user can proceed further to sensitivity analysis, which indicates the importance of a group of selected factors that may effectively influence a vertical farm's performance. This sub-function was designed to give the user a better understanding of changes to ROI and budget based upon a changing market.

The second function is the simulation of the maximum buildable area (see Figure 2). Area simulation refers to the previously mentioned second function: whereby given a fixed budget, it can estimate the scale of vertical farm that can be built. To realise this function, firstly, other construction costs and monetary costs were subtracted from the budget to get the value that could be directly spent on construction and systems. After receiving the necessary initial settings (to determine the main strategies) from the user,

Figure 1. System processing structure of Function 1: cost and ROI estimate

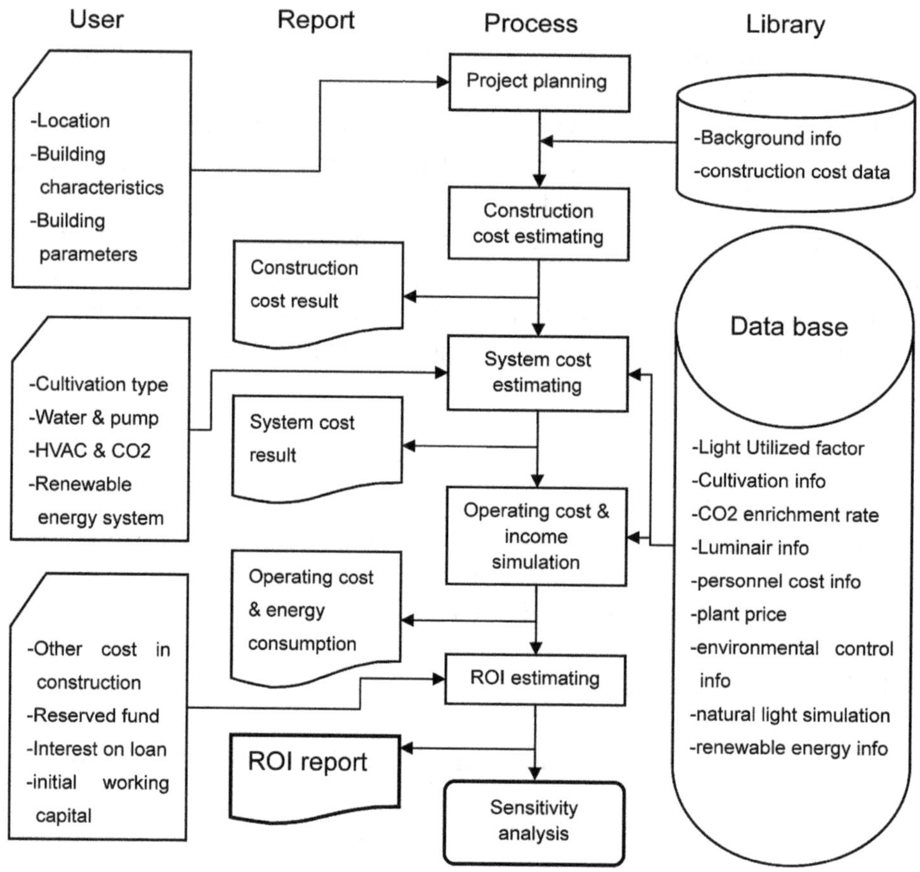

VFer can then be used to conduct a rough calculation to give a predicted value of area. Subsequently, a method of successive approximations is applied, with 1 per cent iterations. Finally, the maximum value of area (in m^2) is presented with a 'guarantee' that the total budget is not exceeded.

Attributes List

A full attributes list should be the first and most important step of a financial estimate for a vertical farm project. Various forms of estimating manifests or reports can be found from different countries or regions and therefore in order to develop VFer as a uniform system applicable to different locations around the world, whilst not compromising its accuracy, various estimating regulations were investigated and compared. After a study of the typical forms of estimation in the US, UK, and China it was found that the main attributes of construction estimates are basically the same, although differences or sequences can be seen in some subsets (APUC, 2011; GB 50500, 2008; Ohno & Harada, 2006). During the development of VFer, these common attributes were extracted, while differences in some subsets were left to the third part, 'other costs', to enable a user to customize the estimation tool according to their specific needs. The selected attributes and subsets of VFer are:

Figure 2. System processing structure of Function 2: simulation of maximum buildable area

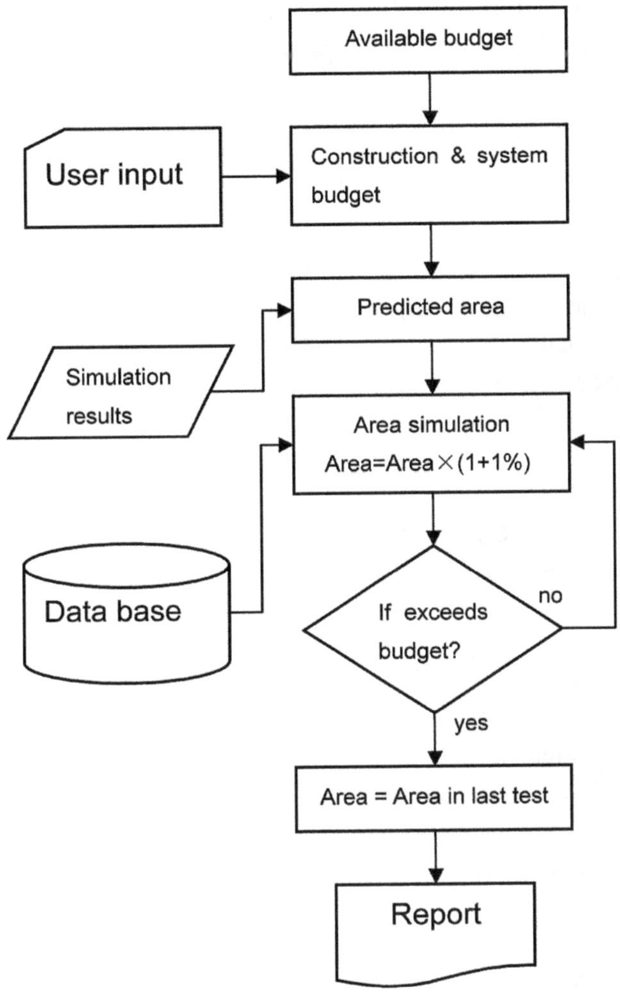

Part 1 Construction

- Structure:
    - Structural type;
    - Floor area;
    - Floor-to-floor height;
    - Number of floors;
    - Number of basement levels;
    - Width-to-depth ratio.
- Level of Finishes:
    - A combination of factors including: floor, ceiling and façade surface finishes and opening finishes.

- Appliances:
    - A combination of factors including: electrical, water and drainage, gas, fire protection, HVAC, smart control, lift and other appliances.

Part 2 System

- Growth light system:
    - Radiometric watts;
    - Electrical watts;
    - Photosynthetic photon flux (PPF);
    - Power supply efficiency;
    - Growth light price;
    - Life time.
- Growing area system:
    - Growing system;
    - Cooling system, vent door, air circulation;
    - Testing equipment and growing supplies;
    - CO2 enrichment;
    - Insect exclusion.
- Germination and clean system:
    - Sowing machine;
    - Tray cleaning machine;
    - Germination cabins;
    - Others.
- Water and nutrient system:
    - Smart controller;
    - Accumulator tank;
    - Water recovery system;
    - Piping system;
    - Others.
- Food and processing system:
    - Multi-produce washers;
    - Roll stock poly bagger;
    - Stretch wrapping & Produce wrapping;
    - Conveyors;
    - Computer system.
- Waste management system:
    - Biogas digesters;
    - Fermentation tanks;
    - Pipes;
    - Biogas generators;
    - Gas cleaning.

- Renewable energy system:
    - Solar system;
    - Wind system;
    - Hydro system;
    - Biomass system;
    - Geothermal system.

Part 3 Others

- Other Construction Costs:
    - Land associated (demolition, relocation compensation, infrastructure, etc.);
    - Management (tender, temporary structure, survey and design, research and testing, project management, insurance, environmental impact, labor safety and hygiene);
    - Other Fees (joint commission, production preparation, office furniture purchase, etc.).
- Reserved funds;
- Interest on loans;
- Initial working capital.

Creation of the Database

The database comprises of a series of datasets stored in Microsoft Excel files that are used as input information to enable the process of the VF's simulations. Basically, there are two types of data used for the library in the development of VFer: location-related data; and non-location-related data. The location-related data, such as climate, energy price, construction cost, labor cost and so on, varies for different regions in the world. Non-location-related data includes other factors that are not influenced or much less influenced by location, such as the PAR level that a particular species of plant needs for the photosynthesis process, the total illuminance emission by a module of LED growth light, the optimal spacing between two plants using indoor hydroponics, and so on. The location-related data needs to be collected for each potential location, which proves to be time-consuming considering there are a series of datasets and a number of variables in each dataset. Moreover, some of the data can only be acquired by computational simulation, such as building energy consumption and indoor solar radiation on certain surfaces. Therefore, in this pilot study, only three international mega cities, Shanghai, London and Washington DC, were selected as the sample cities to develop the initial library of VFer. The number of locations can be expanded and connected easily into the flexible platform in further research.

Unit Cost of Construction

In the construction cost estimation phase, the conceptual estimate is achieved by multiplying the unit cost rate by the building area. Since the building gross area can be calculated easily, the unit cost rate is the most important variable that may substantially determine the estimating accuracy and as such, published cost models were used in the creation of the datasets. Data collection work commenced with the dataset of Shanghai where the data source was mainly collected from Prices and Indices of Shanghai Construction Works (SCCMIAD, 2012). SCCMIAD (2012) is a government report published by Shanghai Construction & Construction Material Industry Administration Department, which encompasses the latest unit cost rates, such as labor, materials and equipment, and representative cost models of different types of buildings developed from surveys and statistics. The construction cost model composites of the three main parts: structure; finishes; and appliances and installation. The costs of these three parts were calculated separately and then combined to achieve the construction cost estimate. In order to develop the cost model in VFer, seven cost models of different use types, building footprints, and building height (all with relatively large-span spaces) were selected from SCCMIAD (2012) to mimic the similar sized vertical farms (see Table 1).

To create the dataset of the building structure cost, the average structure cost of a car garage and an industrial plant was used to represent a single storey vertical farm, office 1 and office 2 for a multi-storey vertical farm, office 3 and office 4 for a high-rise vertical farm, office 5 for a super high-rise vertical farm, respectively. Unit cost rates shown in Table 1 are based on reinforced concrete. Other material types of structures were predicted based on structure cost comparisons studies (Li & Zhang, 2000; Lau & Yam, 2007; RLB, 2013). The structure costs in Shanghai is shown in Table 2.

Table 1. Representative cost models of different types of buildings in Shanghai

	Car garage	Industrial plant	Office 1	Office 2	Office 3	Office 4	Office 5
Number of Storey	4	3	6	8+1 *	15+1	25+2	37+4
Structure type	Frame	Frame	Frame	Frame	Frame-tube	Frame-tube	Braced Frame- tube
Building Height (m)	20.2	14.5	22.8	31.6	64.4	99.7	167.3
Gross Floor Area (m^2)	15060	6708	7675	13067+8176#	22041+4410	37838+7132	71126+23706
Construction Cost (USD)	135.10	123.71	206.23	195.38	269.03	278.82	357.03
Finishing Cost (USD)	56.26	58.05	125.44	191.54	604.14	258.02	225.09
Appliances & Installation Cost (USD)	79.25	21.45	147.12	180.99	241.29	230.02	357.07
Total Cost (USD)	270.60	203.22	478.78	567.91	1114.46	766.86	939.19

Notes: * 8+2 means 8 floors and 2 levels of basement
#13067+8176 means 13067 m2 above the ground and 8176 m2 area of basement
Source: Produced by the authors from SCCMIAD data (2012)

Table 2. Dataset of structure cost in Shanghai

Main Material	Single Floor	Multi-floor	High-rise	Super High-rise
Concrete	128	192	272	/
Steel	176	232	314	354
Concrete and Steel	128	192	272	358
Brick	80	86	/	/
Timber	118	176	/	/
Other	160	192	/	/

Note: Values in USD/m^2; Multi-floor = total height below 24 m; High-rise= total height between 24m to 100m; Super High-rise = total height > 100m

Finishing costs fluctuate much more flexibly than the structure cost. These include the finishing work of ceilings and floors, walls and columns, extruded covers, doors and windows, painting or wallpapering, and others. In the creation of finishing cost dataset, the finishing cost values from Table 1 were sorted. The lowest finishing cost for a car garage and the highest value from office 3 were selected to represent the *Very Low* and *Excellent* quality levels respectively. The Other values were created from interpolation calculations to quantify the fuzzy logic method (see Table 3).

Appliances and installation costs cover a series of systems engineering work, including electrical engineering, water supply and drainage, gas supply, fire control, HVAC system, intelligent systems, lift engineering, and others. Much like the construction finishing work, the selection of appliances is highly determined by the client and the designer. Therefore, the cost can also vary widely. In order to measure the representative costs for different appliances and installation quality levels, the cost breakdown in this category of the selected seven cost models was investigated. For each element in the dataset, values were sorted and then the maximum and the minimum value were eliminated. Then the remaining five values, as shown in Table 4, represent the different quality levels. Finally, the values were added up to achieve the total cost.

The basic construction cost, excluding preliminaries and on-site costs, was calculated as:

$$\text{Basic Construction Cost} = \text{Cost}_{\text{structure}} + \text{Cost}_{\text{finishing}} + \text{Cost}_{\text{appliance}} \quad (1)$$

Construction cost estimates of other locations were calculated with reference to Shanghai using international construction comparison survey (Turner & Townsend, 2012). The construction cost of a certain country was calculated as:

Table 3. Dataset of finishing costs in Shanghai

Finishing Quality	Very Low	Low	Medium	High	Excellent
Cost per m^2	54.40	192.00	329.60	464.00	600.00

Note: Values in USD/m^2

Table 4. Dataset of appliances and installation cost in Shanghai

Appliances	Very Low	Low	Medium	High	Excellent
Electrical	16.48	67.36	70.56	73.76	102.40
Water	9.12	9.12	9.12	12.64	25.60
Gas	/	/	/	/	/
Fire Control	7.84	18.08	19.68	24.96	29.76
HVAC	/	48.16	50.24	61.92	65.76
Intelligent Systems	/	5.92	24.96	37.92	64.80
Lift	/	8.96	18.40	19.04	21.28
Total	33.44	157.60	192.96	230.24	309.60

Note: Values in USD/m^2

$$\text{Cost}_{\text{city } i} = \text{Cost}_S \times \frac{\text{Max Cost}_i + \text{Min Cost}_i}{\text{Max Cost}_s + \text{Min Cost}_s} \times \text{F}_{\text{region}} \tag{2}$$

Where $\text{Cost}_{\text{city } i}$ is the construction cost of the target city; Cost_S is the construction cost for Shanghai (either structure, or finishing or appliances and installation cost); Max Cost_i is the typical unit cost rate of an office in a business park and Min Cost_i is the typical unit cost rate of a warehouse in the target country; Max Cost_s and Min Cost_s are unit cost rates for an office and a warehouse in Shanghai, respectively. These values are given in USD per square meter for the year of 2012 in Turner & Townsend (2012); F_{region} is an adjustment factor which reflects the construction cost difference between the target city and the capital city of the country or mega cities in that country. If the target city is the capital city or a metropolitan economically alike, F_{region} equals to 1. Otherwise, F_{region} should be calculated with further information of cities' comparison surveys.

Information on Growing and Service Systems

The unit costs of the growing system, including the growing set, the HVAC system, the environmental controls, the electrical panel and technical support, the growing supplies, the CO2 enrichment, and the insect exclusion, were collected from CropKing (2013). The unit costs of the germination and cleaning system, the water and nutrient systems, and the food processing system were collected from Banerjee (2012). Information on grow lights was collected from selected manufactures' (Philips™ and Illumitex™) product brochures including datasets on the Photosynthetic Photon Flux Density (PPFD), the radiometric watts, the electrical watts, the life time, and the price of the luminaire. Information on renewable energy systems and energy production were collected from IRENA (2013).

Information on Plant Cultivation

In order to simulate the annual yield of different crops and thus to predict the revenue of a vertical farm, a dataset that contains the information on plant cultivation is necessary. This dataset provides the input values of a series of fundamental variables, such as photosynthetically active radiation (PAR), photoperiod (how many hours the plant needs to be exposed under the rated PAR), the germination and growth period, and so on. In order to estimate the operating cost, another group of variables are needed, such as water usage, plant spacing, the distance between the growth light and the plant canopy, and so on as shown in Table 5.

During the creation of this dataset, the method of information capture was diverse but applied in a sequence. The first step was searching for published papers in academic journals to examine whether there is any scientific research of hydroponics for the selected plant species. If there is any necessary, then the input/output data from that paper were extracted and applied in the dataset. For instance, there is research data on the growth of lettuce (A. Shah & S. Shah, 2009; Li & Kubota, 2009; Lina et al., 2013); data on the growth of tomatoes (Fan et al., 2013); and data on the growth of cucumbers (Grewal, Maheshwari, and Parks 2011). If there was not such data available, then online databases, research reports, handbooks and theses were searched. For instance, Both (2002) presents a regression model on the growth of lettuce, whilst Molinar, Yang, & Moura (2005) indicate the cost and return of producing cherry tomatoes,

Table 5. Information on plant cultivation

Plants	PPFD µmol /m2/s	PAR mol/ m2/ day	Photop-eriod hours	Days before Harvest	Dry Mass/ plant kg	DM:FM	Plants/ m²	Gross yield kg/m2/ year	Water L/DM kg plant	FM per plant kg	Water Use L/ M²/ year	Plant spacing m	Height of shoot m	Depth of root m	Light Use Efficiency kg/mol/m2	Harvest Index
Tomatoes	313	13.5	12	75	0.291	0.060	4.8	113.1	300	4.9	2036	0.5	1.2	0.2	0.0242	0.50
Lettuce (butterhead)	295	17.0	16	30	0.007	0.048	43.1	78.5	402	0.1	1515	0.2	0.2	0.2	0.0127	0.67
Spinach	295	17.0	16	30	0.002	0.080	43.1	10.5	402	0.0	337	0.2	0.1	0.2	0.0017	0.67
Cucumber	250	16.2	18	58	0.331	0.045	1.7	80.4	172	7.4	618	0.8	1.5	0.3	0.0136	0.20
Broccoli	250	10.8	12	64	0.085	0.090	4.8	25.7	198	0.9	458	0.5	0.5	0.4	0.0065	0.33
Celery	250	14.4	16	97	0.034	0.050	43.1	110.2	450	0.7	2480	0.2	0.4	0.2	0.0210	0.75
Eggplant	250	10.8	12	81	0.108	0.080	4.8	29.0	198	1.3	459	0.5	0.6	0.1	0.0073	0.33
Lettuce (Iceberg)	250	14.4	16	73	0.018	0.040	15.5	56.6	402	0.5	565	0.3	0.3	0.2	0.0108	0.67
Sweet Pepper	250	10.8	12	68	0.101	0.100	10.8	58.4	150	1.0	876	0.3	0.8	0.2	0.0148	0.25
Bok-Choy	295	17.0	16	49	0.030	0.080	24.2	67.6	402	0.4	2173	0.2	0.3	0.1	0.0109	0.67
Strawberry	255	11.0	12	85	0.042	0.080	10.8	24.1	198	0.5	382	0.3	0.2	0.2	0.0060	0.33
Watermelon	350	15.1	12	85	0.960	0.080	0.7	34.7	198	12.0	549	1.2	0.6	0.3	0.0063	0.33
Chinese Leaf	295	17.0	16	85	0.084	0.070	15.5	79.8	480	1.2	2681	0.3	0.3	0.1	0.0129	0.80

Note: DM = Dry Mass; FM= Fresh Mass; the Light Use Efficiency represents the fresh mass of fruits harvested per unit PAR every square meter; Harvest Index is the ratio that edible weight to the gross harvest weight

Source: Author produced according to various publications and databases and Alexander (2013) produced a growing guide for a number of different vegetables and crops. The third step in this search meant acquiring the remaining data from abroad study of other media, such as websites that contains hydroponic information, manufacturers' brochures, homegrown societies, and so on. The final data sources referenced for the creation of database are summarized in Table 6.

Simulation on Heating and Cooling Loads

The energy consumption of the environmental controls (heating and cooling loads) varies even if the same building configuration is applied to different climatic zones. In order to achieve relatively accurate values of energy consumption of the environmental controls, five variables were selected for parametric tests of benchmark vertical farms in different configurations and locations. Descriptions on variables and initial settings are shown in Table 7. Computational simulations were conducted over 200 times using Autodesk Ecotect® software in order to create a data library where the simulation results of the benchmark vertical farms (in kWh/m^2/year) can be representative to those of the target VF which has similar specifications.

Simulation of Solar PAR on Growing Trays

Unlike the artificial lighting type of vertical farm where the PAR on plants is controlled by electric power, PAR in a sun-fed vertical farm is mainly determined by the insolation of natural light, which fluctuates daily. Therefore, it is necessary to simulate how much PAR there will be on growing trays with different orientations and height levels to estimate the annual yield and revenue. In order to fulfill this

Table 6. Data sources on plant cultivation

Plants	Data Sources
Tomatoes	Fan et al. (2013); Banerjee (2012); Alexander (2013); Holliman (2006) Bastin & Henken (1997); Willis (1992)
Lettuce (butterhead)	Li & Kubota (2009); Lina et al. (2013); Brechner and Both (1998); Albert (2011)
Spinach	Banerjee (2012); Alexander (2013); Willis (1992); Albert (2011)
Cucumber	Grewal et al. (2011); Bastin & Henken (1997); Albert (2011)
Broccoli	Le Strange et al. (2010); Alexander (2013); Bastin & Henken (1997); Albert (2011)
Celery	Alexander (2013); Bastin & Henken (1997); Albert (2011)
Eggplant	Bastin & Henken (1997); Islama et al. (2010); Willis (1992)
Lettuce (Iceberg)	Banerjee (2012); Both (2002); A. Shah & S. Shah (2009) Bastin & Henken (1997); Willis (1992)
Sweet Pepper	Banerjee (2012); Willis (1992); Albert (2011)
Bok-Choy	Pant, Radovich, & Arancon (2012); Zhu, Gerendas, & Sattelmacher (1997); Cho & Son (2007); Willis (1992)
Strawberry	Banerjee (2012); Bastin & Henken (1997)
Watermelon	Bastin & Henken (1997); Willis (1992); Albert (2011)

Table 7. Descriptions on variables and settings

Variables	Description
Scale of floor area	Three scales: 10m×10m; 20m×20m; 40m×40m;
Building height	Three levels: ■ Single floor – less than 10m; ■ Multi-floor – 6 floors; ■ High-rise – for 10m×10m footprint, N/A*; for 20m×20m footprint, 20 floors; for 40m×40m footprint, 30 floors
Insulation level	Three levels: ■ Poor insulated – 220 mm brick (U-value 2.02 W/m^2.K) ■ ETFE (transparent) – double skin (U-value 1.6 W/m^2.K) ■ Well insulated – 150 mm foil-faced, glass-fiber, 75 mm air gap, 110 mm brick (U-value 0.12 W/m^2.K)
Air change rate per hour (ACH)	Three levels: ■ Air tight – 0.25 ■ Medium – 0.5 ■ Normal -1
Power of growth lights in W/m^2	Three levels: ■ Sun-fed type – 0 W/m^2 ■ Artificial lighting 4-6 layers of growing trays – 200 W/m^2 ■ Artificial lighting 6-8 layers of growing trays – 300 W/m^2
Other settings	Floor to floor height is 4.5 m Temperature range: 18-30 °C

* 10m×10m footprint may not have a high-rise configuration due to structural stiffness

objective, three configurations of sun-fed vertical farm, A-frame, façade system and columnar system, were simulated separately using the Solar Access Analysis tool in Autodesk Ecotect ®.

Figure 3 shows an example of the 3D model of an A-frame vertical farm used for simulation purposes. One of the middle frames (the second one from the east) was selected for the simulation because it is an ideal representative of the average situation where middle frames may be somewhat shaded by the frames on both sides. Finally, the simulation on each separate layer of the frame was conducted.

Calculation Process

In VFer, the most used of the models in calculating the costs are rooted on the principle of the quantity of work or material and the unit cost rate (Skitmore & Marston, 1999).

$$C = \sum_{i=1}^{n} q_i \cdot r_i \tag{3}$$

Where C is the total estimated cost; q_i is the quantity of work i (representing the floor area or growing area in units of square meters); r_i is the unit cost rate of work i (cost or energy consumption per square meter), which comes from the created database.

There are 5 steps to achieve the total capital cost.

Figure 3. 3D model of A-frame prototype in Ecotect ® Solar Access Analysis ®

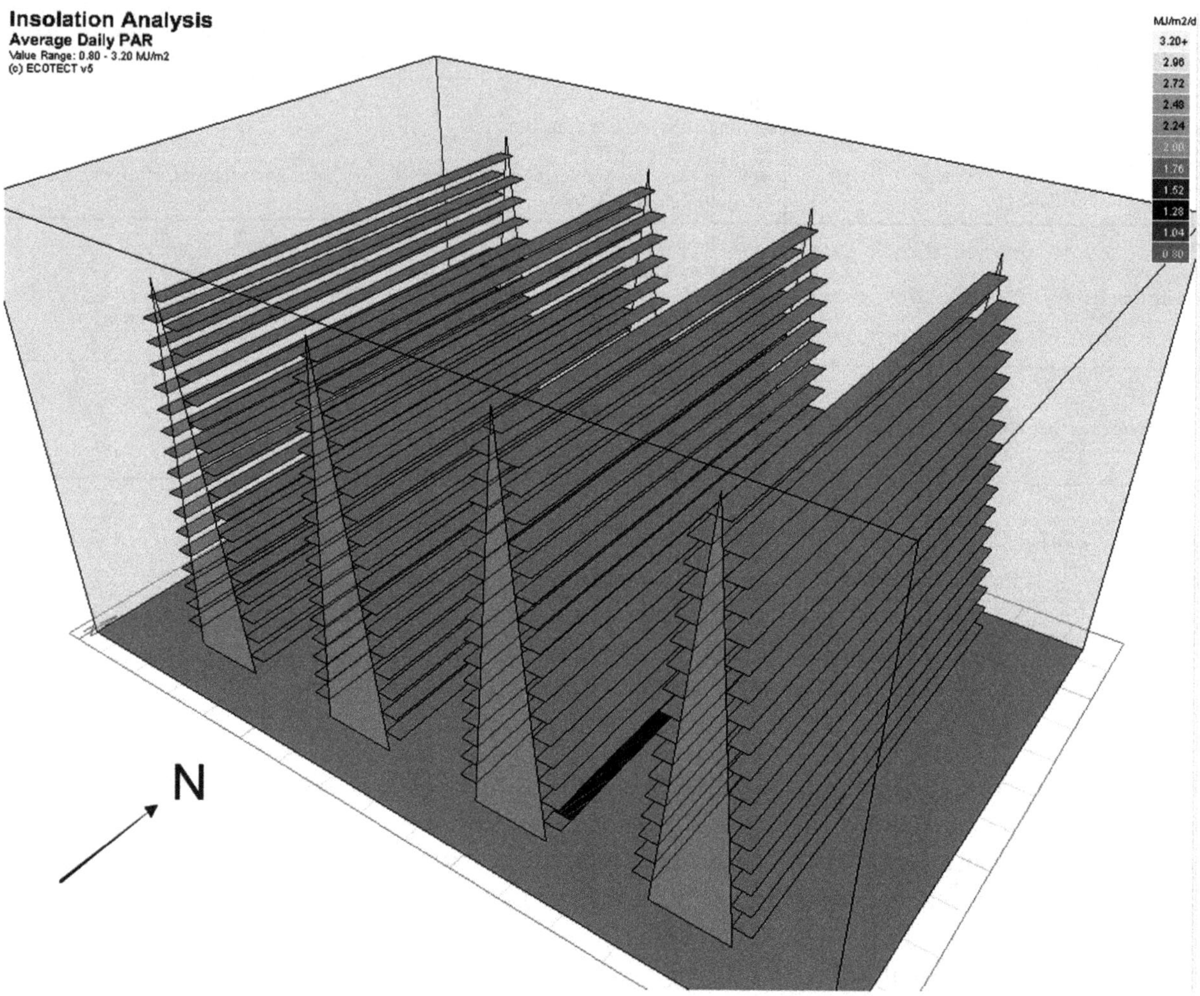

Step 1: The land price, representing the average price of business parks in the peri-urban area, was collected from commercial surveys in order to calculate the Land Acquisition Cost.
Step 2: Calculating the construction cost:

Construction Cost = Structure Cost + Finishing Cost + Appliance Cost (4)

Step 3: Calculating the system cost:

System Cost = Grow Light Cost + Growing Area Cost + Germination & Clean Cost + Water & Nutrient Cost + Food Processing Cost + Waste Management Cost + Renewable Energy Cost (5)

Step 4: Calculating other costs including other construction costs, reserve funds, interests on loans, and initial working capital. As aforementioned, these sub-items are highly user-defined, although some suggested values are provided by the system.

Step 5: As the entire sub-items of systems were calculated, the total cost can be achieved by adding them up as:

Total Capital Cost = Land Acquisition Cost + Construction Cost + System Cost + Other Construction Cost + Reserve Fund + Interests on Loan +Initial Working Capital (6)

Then, the annual operating cost can be calculated as:

Operating Cost = Bill_Growth_Lights + Bill_Environmental_Controls + Bill_Miscellaneous_Energy + Water Bill + Seed Cost + Nutrient Cost + Personnel Cost + Maintenance Cost+ CO_2 Cost - Reduction from Renewable Energy (7)

The annual yield of a particular plant is:

Adjusted Plant Annual Yield = Standard Annual Yield × Plant Area × PAR Factor × Increment by CO_2 Enrichment × (1- Failure Rate) × Temperature Factor (8)

Where the Standard Annual Yield is the validated value from literature which is recorded in Table 4; the PAR Factor is a ratio of the actual PAR delivered to the plants' canopy to the theoretical PAR requirements. For an artificial lighting type of vertical farm, the value was 1, which means the PAR can be always controlled at the optimal level. However, for sun-fed types, the value of actual PAR was acquired by Ecotect ® simulations. Temperature Factor is a factor that reflects the reduction of yield caused by overheating or freezing of the growing area if the indoor temperature is uncontrolled by HVAC or other systems. The value was set as 0.9 for the preliminary estimation; Failure Rate was set as 5%.

Therefore, annual income from plant production is:

Plant Annual Income = Plant Price × Price Index × Adjusted Plant Annual Yield × Price Share Rate (9)

The Price Index is the ratio that the price of products from a vertical farm to the average retail price from current market. This was set at 1 if not specified by the user. The Price Share Rate is the ratio that the revenue is shared between the farm and other marketing process. This was introduced to reflect the potential cost savings of transporting produce to market from rural farms and from the reduction in food supply chain that are significant cost savings but cannot be directly included at this stage due to the wide range of locations of the source of traditionally grown produce (Bloom & Hinrichs, 2011). If not specified by the user, the Price Share Rate was set at 0.6 (approximately three times as high as rural farms), assuming 60% of the revenue will be shared by the farm.

Operation of the System

VFer processes the following sequence:

1. The user selects the location of the proposed project. Initial settings are automatically imported by the system, although all the values can be reassigned by the user if necessary.
2. The user determines the main strategies of the project and specifies the parameters of the building.
3. The user determines the technology strategies which encompass the type of energy input (the artificial lighting type or the sun-fed type), the water recycling method, the HVAC system, the CO_2 enrichment, and the application of renewable energy systems.
4. The user determines the plant types to be grown and the percentage of each plant type. VFer will automatically filter plants that are not suitable for the selected cultivation framework.
5. The user determines the percentage of construction cost and monetary costs.
6. VFer displays the ROI result and cost breakdown charts for both fixed costs and operating costs (see Figure 4 as an example). Then a full report can be created automatically.
7. VFer conducts a sensitivity analysis on the ROI test and the Budget test. Two sensitivity charts are then created to show the effect of changing the 12 selected factors on the ROI or the total budget.

Validation

It is not yet possible to validate VFer against a completed 'real project' since there is no published data available. Even if available, considering the great flexibility of VFer, segments of news or pieces of information are far less than required for full validation needs, because:

- The simulation results depend on many parameters and different settings that may lead to totally different results;

Figure 4. ROI results displayed by VFer

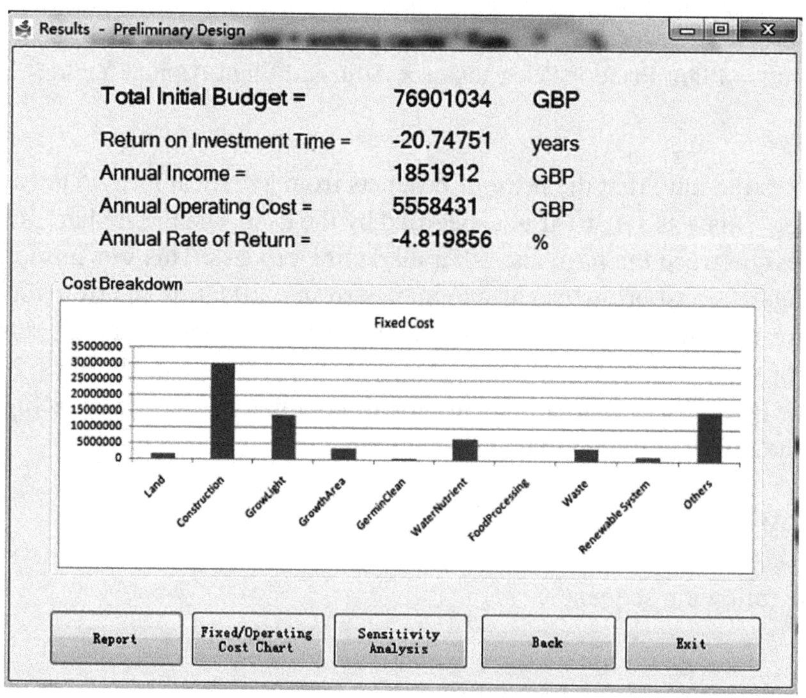

- As a new typology, the vertical farm is still 'under testing'. 'Real-time' data could therefore reveal information or even commercial secrets hence the unavailability at present of the limited data that may exist. Besides, if test results are not stable, they still cannot be used in the validation process.
- Some external factors, such as plant prices, are changing all the time and can be considerably different from place to place. The simulation may therefore have some 'lags' compared to the latest circumstances.

Instead of validation as a 'whole', VFer was validated 'part-by-part'. For instance, in the construction estimation part, data from Turner & Townsend (2012) were used for the cross validation (See Table 8).

Mean absolute percentage error (MAPE), the most important parameter for accuracy evaluation, was used to examine the model effectiveness by applied to different cases (An et al., 2007). It can be obtained using Equation 10:

$$\mathrm{MAPE} = \left(\sum_{i=1}^{n} \frac{x_i' - x_i}{x_i} \times 100\% \right) / n \tag{10}$$

where x_i is the actual cost; x_i' is the predicted cost; and n is the total number of cases.

It was assessed that the accuracy of construction cost estimation part could be controlled within an acceptable range (the best was +0.38%, and the worst was +32.84%). Mean absolute percentage errors (MAPE) was calculated to be 13.14% which is a good result for a conceptual estimation.

System information was taken directly from literature, government databases, or manufacturers' brochures, and this can be controlled within +/-10%. An accuracy of +30% and -20% is expected from VFer as a whole and this level of accuracy is satisfactory, considering a typical range of a conceptual estimation is between +50% and -30% (Phaobunjong, 2002).

Table 8. Construction cost cross validation with the cost models from Turner & Townsend (2012)

China (Shanghai)			
Cost model	**Average Cost (USD/m²)**	**Predicted Cost (USD/m²)**	**Error Rate**
Office - business park	795	798	+0.38%
Warehouse/factory unit -basic	437	478	+9.38%
High-tech facility/ laboratory	995	886	-10.95%
UK (London)			
Cost model	**Average Cost (USD/m²)**	**Predicted Cost (USD/m²)**	**Error Rate**
Office - business park	2490	2534	+1.77%
Warehouse/factory unit -basic	1328	1664	+25.30%
High-tech facility/ laboratory	2490	2878	+15.58%
US (Washington DC)			
Cost model	**Average Cost (USD/m²)**	**Predicted Cost (USD/m²)**	**Error Rate**
Office - business park	1936	1946	+0.52%
Warehouse/factory unit -basic	883	1173	+32.84%
High-tech facility/ laboratory	1787	2173	+21.60%

THEORETICAL CASE STUDY

Performance of Different Types on Different Locations

The ROI estimations of different types of vertical farm in three selected mega-cities (Shanghai, London and Washington DC) are shown in Figure 5. It is not suggested here that these three cities are necessarily ideal or proposed cities for vertical farm projects, rather they provide worked examples, with readily available data for this research, from three different continents. Applying current average plant prices, it seems that most types of vertical farms will not currently be commercially profitable in these locations. Exceptions can be found in the façade system group. However, since that group has a mixed function including offices then the cultivation area is quite limited, and it is the office rental return that is the contributor of a positive ROI, rather than income from plant production. An exception can also be seen in Washington DC with the artificial growth light types (but only 3.5% return for a bed system, and 3.46% for a drum system).

The budget comparisons for the selected sites are illustrated in Figure 6. Taking London's figures as a standard value '1', a vertical farm project in Shanghai costs less (with an index varies between 0.6 to 0.8). Washington DC has similar values to London, but shows some advantages in single-floor sun-fed types.

Which Plant Is More Profitable to Produce?

Income from plant production was simulated to discover the 'profit margins' with the same cultivation system (artificial bed system, in order to eliminate the influence of differences in solar radiation) and the same cultivation area. Comparison results are shown in Figure 7 (taking butter-head lettuce as standard value '1'). Shanghai has the lowest income according to China's low price index with celery, strawberries and butter-head lettuce being top three of the list. In London, the most profitable plant should be

Figure 5. ROI of different types of vertical farm growing butter-head lettuce

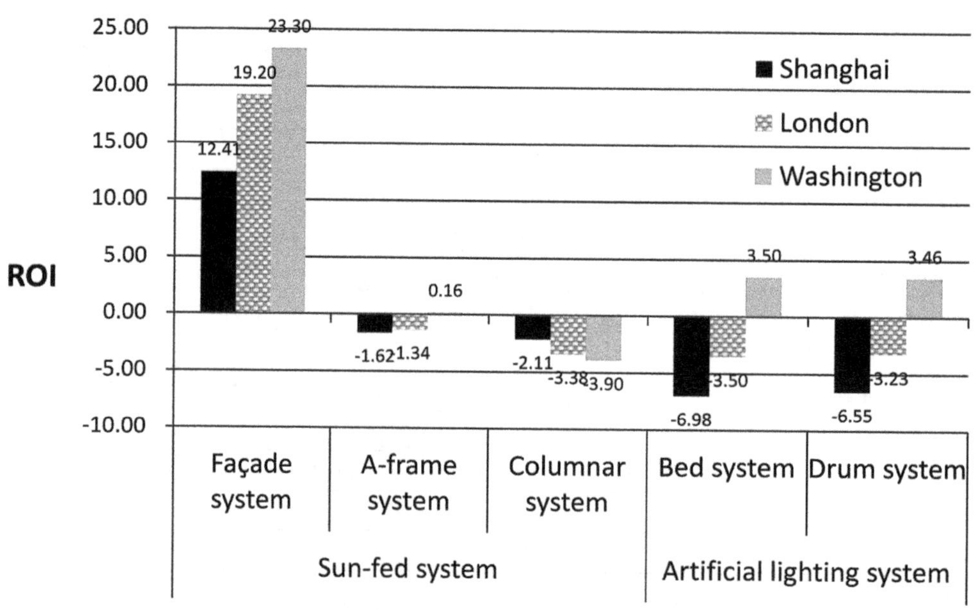

Figure 6. Budget comparisons of different types of vertical farm growing butter-head lettuce

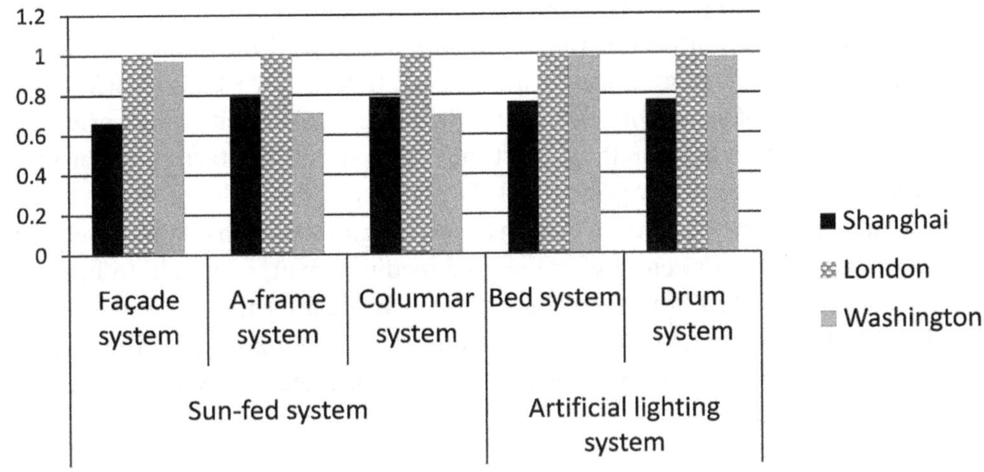

Figure 7. Plant income comparisons

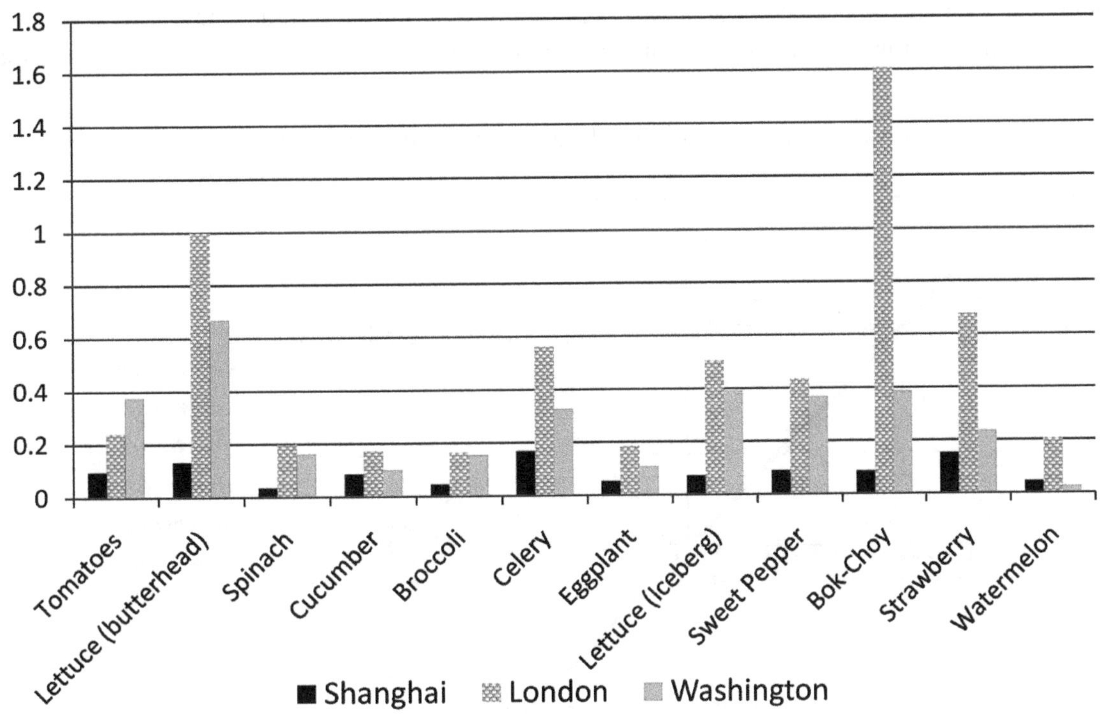

bok-choy (retail price of 5.6 GBP per kg), which is 1.6 times more profitable than lettuce. However, its supply at present is relatively low and a significant increase in supply could lead to a lower price per kilo. For London, butter-head lettuce and strawberries occupy the second and third places. In Washington DC, butter-head lettuce is top, followed by iceberg lettuce, bok-choy, sweet peppers and tomatoes having similar plant incomes.

Findings From Sensitivity Analysis

Sensitivity analysis indicates the effects of various factors on the ROI (see Figure 8) and budget (see Figure 9, the larger the slope, the more sensitive a factor is). In terms of ROI, in artificial growth light types, the three most important sensitivity factors are the price of electricity, plant price, and the cost of CO_2 enrichment. In sun-fed types, the three most important sensitivity factors are plant price, floor area, and CO_2 enrichment and electricity price equal. The quality of the building, is also worthy of mention, as this is also quite important in certain circumstances. Significantly, a quality level (including the equipment quality level) 10-15 per cent higher than the medium level, may help to improve the ROI. However, any higher quality is too 'luxurious' for a sun-fed vertical farm since it will reduce the ROI. In terms of budget, for both artificial types and sun-fed types, floor area and quality levels are crucial factors. In artificial types, growth light price is the second in the sensitivity factor list, and therefore needs to be given full attention.

How to Improve the Economic Performance?

The first and most direct way to improve the economic performance of a vertical farm is to set a higher product price. Considering some of the advantages of a vertical farm's produce (fresh, without any pesticides or fertilizers) it is reasonable to set and expect to achieve a higher price than the average market

Figure 8. ROI sensitivity analysis of an artificial lighting type of vertical farm

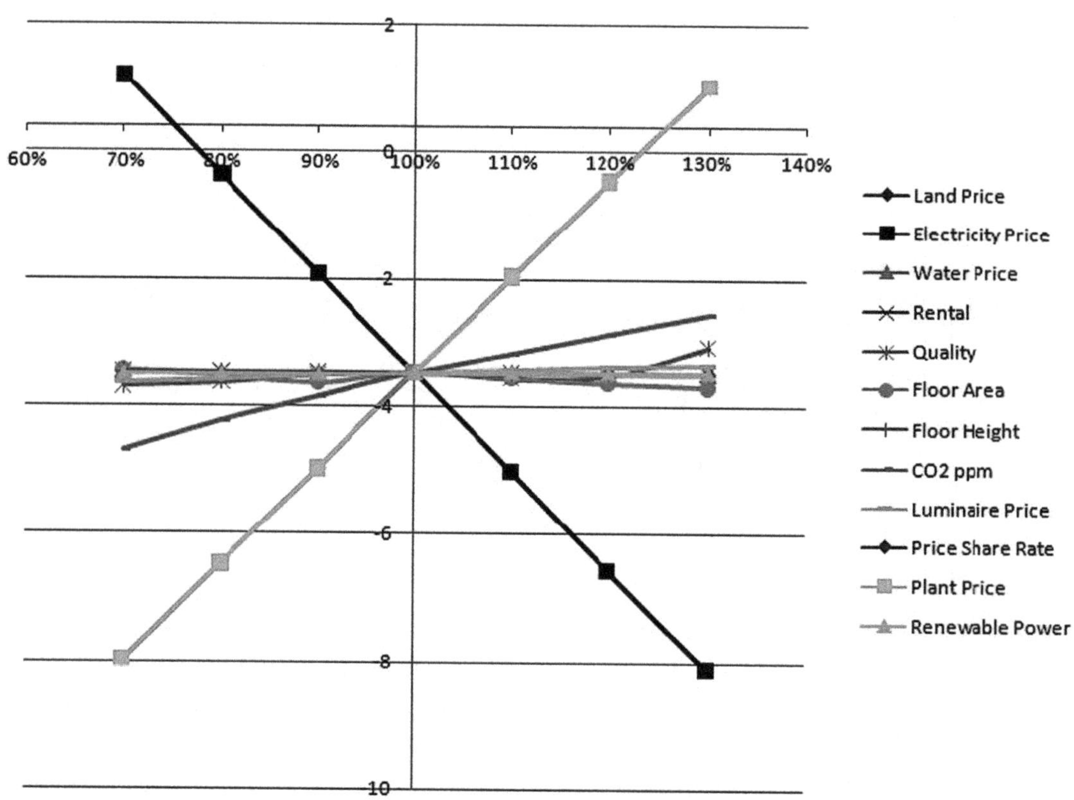

Figure 9. ROI sensitivity analysis of a sun-fed type of vertical farm

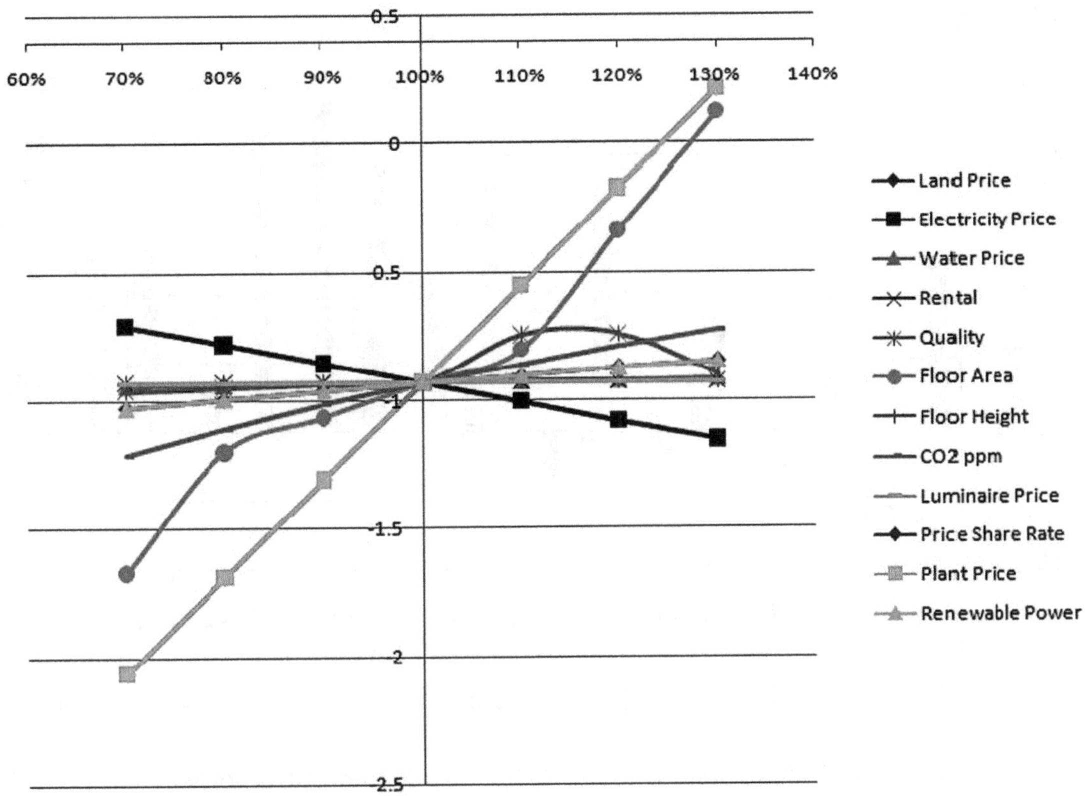

price, as tends to be the case with organically produced foods. Predictions of ROI on changing plant price are shown in Figure 10 (growing butter head lettuce). 'PI' stands for price index and a value of '2' in the PI means the price is doubled, '3' means tripled, and so on. It is clear that although the ROI is negative at current price index (PI=1), it becomes positive value when PI is equal to 2. In artificial lighting types, the ROI can then be 12-14 per cent (in London and Washington DC), which already becomes a profitable business that can justify commercial investment. The results show that artificial lighting types of vertical farms show greater potential for ROI growth compared with sun-fed ones. In addition, regional differences are obviously presented. Given the input values in this research, Shanghai does not currently appear to be a good location to develop a vertical farm in comparison to other selected cities.

The second method is to find the optimal strategy, including:

- Selecting the 'right plant' according to the market and cost-effective simulation result from VFer;
- Applying smart ventilation systems and strategies. It was predicted that applying an advanced ventilation system with water spray rather than simply using air-conditioning in an artificial lighting type of vertical farm may effectively reduce the energy consumption for environmental controls, as cooling energy accounts for more than 1/3 energy demand in total;
- 'Think twice' before adding a renewable energy system. Since renewable energy technologies are still costly, they may deteriorate the ROI, although they can reduce the project's carbon footprint.

Figure 10. ROI comparisons with different price indices (PI)

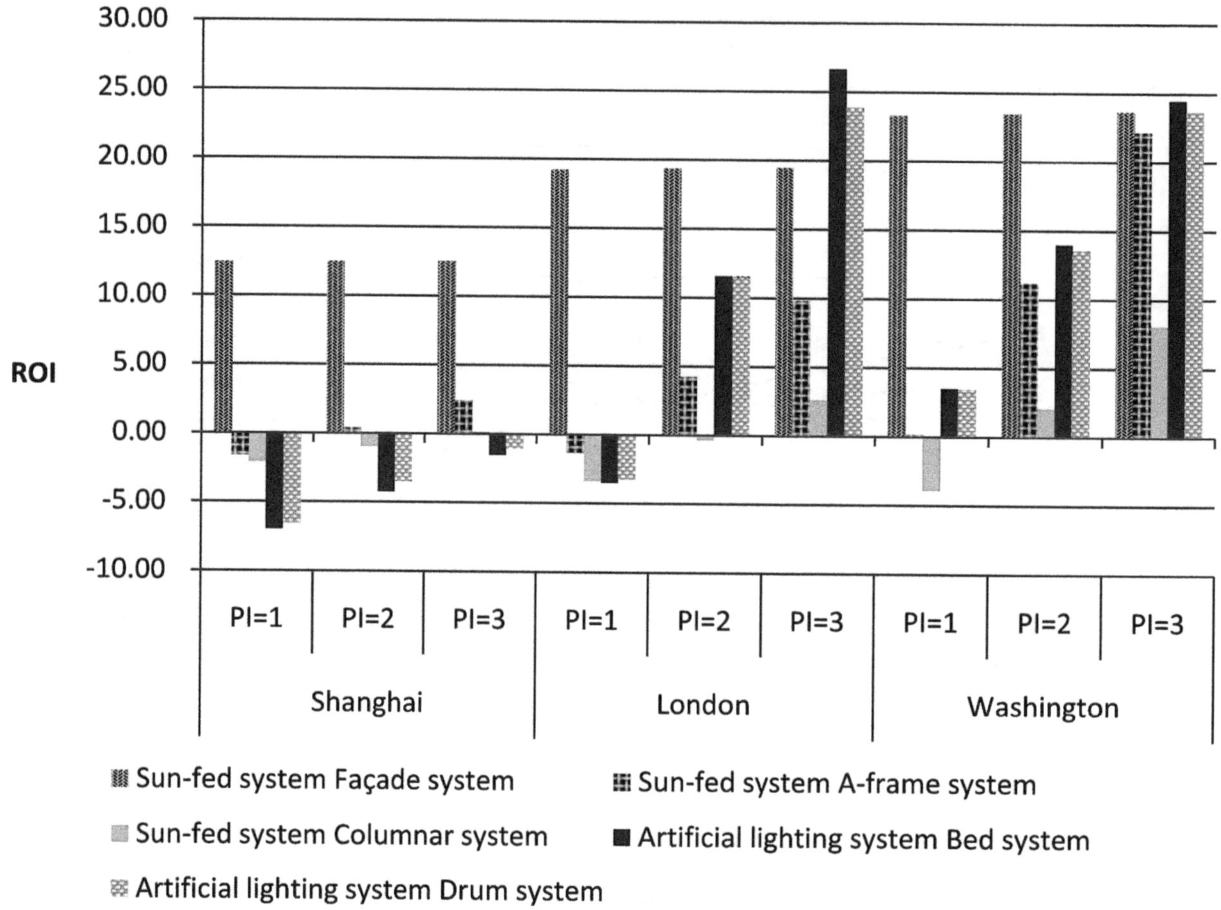

Nevertheless, certain kinds of renewable energy system may be more suitable for particular sites. However, it is recommended to apply them after a full evaluation on their cost-effective performance is made;

- Applying supplemental growth light to sun-fed vertical farms in order to stabilize the photosynthetically active radiation (PAR) on the plants' canopies and thus to improve the yields.

Thirdly, the ROI can also be improved through advances in scientific development or as a result of evolution in technology. For example, testing the hypothesis that smaller plant spacing in a bed tray may benefit the efficiency use of light, thus improving the yields, can be of great research interest in vertical farming. Besides, every new generation of growth light, either LED or other advanced sources, is continually changing the balance sheet. As a result, with time higher efficiencies are expected, and therefore a more attractive ROI is likely to be achieved.

Last but not least, financial support or subsidy from government, for example, could be a catalyst that enables more pioneering projects to be developed. In some countries or cities, direct support could become a priority as part of planning regulations, completed infrastructure, supplementary allowance and so forth. In addition, indirect support could include the development of industry standards to regulate the market, certificating products from a vertical farm to acknowledge the high quality, and propagating the vertical farming concept to build a more knowledgeable environment which in turn will give more confidence to developers and investors (Heath & Shao, 2014).

CONCLUSION

VFer provides an innovative and highly flexible, customised system to estimate a vertical farm's cost and performance. In this research, the results from several typical configurations were used to elucidate the economics of vertical farms. With respect to the sun-fed type of vertical farms, under current average plant price levels, they appear to barely break even in terms of financial return. Indeed, their economic performance largely relies on the availability of solar radiation and other external factors. Since they are actually variants of a single-floor vertical farm (assuming the façade system as a "vertically designed" single floor farm, which is quite shallow in depth), the scale and production are inevitably constrained. However, in some specific locations, where there is plenty of natural light available, the plant price is relatively high, and there are physical or fertile land resource constraints, such as Singapore or Middle East countries, this could be a viable system.

In terms of the artificial lighting type, estimates show that the ROI varies from case to case, but all are negative with current technology and current average food prices. However, this trend is likely to change within 3 to 5 years if the efficiency of LEDs or the price of products produced in a vertical farm can be doubled (similar to the trend for the price of organic produce). In such a scenario, the ROI can be 12 to14 per cent, and this does not include the extra profit of potentially selling the land saved from by adopting vertical agriculture. Therefore, although pure artificial-energy-driven farms may be not so optimistic at present, they could still be a viable alternative agricultural solution in the near future, in particular locations, especially as food security is becoming an increasingly global concern.

The next important step for research into vertical farming is to work with the existing small-scale pioneers of urban agriculture and vertical farm projects to develop and refine increasingly accurate cost models. This in turn will enable more informed decisions on the part of developers, investors and local governments and should in turn lead to the establishment of this new urban building and land-use typology.

REFERENCES

Alexander, S. (2013). *A Guide to Vegetable Growing*. Dublin: Agriculture and Food Development Authority.

An, S.-H., Kim, G.-H., & Kang, K.-I. (2007). A case-based reasoning cost estimating model using experience by analytic hierarchy process. *Building and Environment*, *42*(7), 2573–2579. doi:10.1016/j.buildenv.2006.06.007

APUC. (2011). Guide to Procuring Construction Projects. Retrieved from http://www.apuc-scot.ac.uk/uploads/Docs/pdf/Guide%20to%20Procuring%20Construction%20Projects%2029-6-11.pdf

Banerjee, C. (2012). *Market Analysis for Terrestrial Application of Advanced Bio-Regenerative Modules: Prospects for Vertical Farming*. Bonn, Germany: Rheinische Friederichs Wilhelms Universitat.

Bastin, S., & Henken, K. (1997). Water Content of Fruits and Vegetables. Cooperative Extension Service UK, ENRI-129.

Bloom, J. D., & Hinrichs, C. C. (2011). Moving local food through conventional food system infrastructure: Value chain framework comparisons and insights. *Renewable Agriculture and Food Systems*, *26*(1), 13–23. doi:10.1017/S1742170510000384

Both, A. J. (2002). Ten Years of Hydroponic Lettuce Research. Retrieved from http://ecaaser3.ecaa.ntu.edu.tw/weifang/lab551/vegetable/culturalpractice/ten%20years%20of%20hydroponic%20lettuce%20research.pdf

Brechner, M., & Both, A. J. (1998). *Cornell Controlled Environment Agriculture: Hydroponic Lettuce Handbook*. Ithaca: Cornell University.

Cho, Y. Y., & Son, J. E. (2007). Estimation of leaf number and leaf area of hydroponic pak-choi plants (Brassica campestns ssp,chinensis) using growing degree-days. *Journal of Plant Biology*, *50*(1), 8–11. doi:10.1007/BF03030593

Code of Valuation with Bill Quantity of Construction Works GB 50500. (2008). Ministry of Housing and Urban-Rural Development of People's Republic of China.

Coker, A. K. (2015). *Ludwig's Applied Process Design for Chemical And Petrochemical Plants* (4th ed.). London: Gulf Professional Publishing.

CropKing. (2013). CropKing Greenhouse Package: NFT System. Retrieved from http://www.cropking.com/sites/CropKing.com/files/pdfdocs/NFTPricing.pdf

Despommier, D. (2011). *The Vertical Farm: Feeding the World in the 21st Century*. London: Picador.

Despommier, D., & Ellingsen, E. (2008). The Vertical Farm: The sky-scraper as vehicle for a sustainable urban agriculture. *Paper presented at the CTBUH 8th World Congress*, Dubai.

Draper, N. R., & Smith, H. (1998). *Applied Regression Analysis*. New Jersey: John Wiley. doi:10.1002/9781118625590

Fan, X., Xua, Z., Liu, X., Tang, C., Wang, L., & Han, X. (2013). Effects of light intensity on the growth and leaf development of young tomato plants grown under a combination of red and blue light. *Scientia Horticulturae*, *153*, 50–55. doi:10.1016/j.scienta.2013.01.017

Graff, G. (2011). *Skyfarming* [Thesis]. University of Waterloo, Canada.

Grewal, H. S., Maheshwari, B., & Parks, S. E. (2011). Water and nutrient use efficiency of a low-cost hydroponic greenhouse for a cucumber crop: An Australian case study. *Agricultural Water Management*, *98*(5), 841–846. doi:10.1016/j.agwat.2010.12.010

Heath, T., & Shao, Y. (2014). Vertical farms offer a bright future for hungry cities. *The Conversation*. Retrieved from https://theconversation.com/vertical-farms-offer-a-bright-future-for-hungry-cities-26934

Heath, T., Zhu, Y., & Shao, Y. (2012). *Vertical Farm: A High Rise Solution to Feed the City?* Paper presented at the CTBUH 9th World Congress Asia Ascending: Age of the Sustainable Skyscraper City, Shanghai.

Holliman, J. B. (2006). *An Economic Analysis of Integrating Hydroponic Tomato Production into an Indoor Recirculating Aquacultural Production System*. Auburn, Alabama: Auburn University.

IRENA. (2013). Renewable Power Generation Costs in 2012: An Overview: International Renewable Energy Agency. Retrieved from http://costing.irena.org/media/2769/Overview_Renewable-Power-Generation-Costs-in-2012.pdf

Islama, S., Tomokoa, H., Ishikawaa, K., Takedaa, N., Azada, M. A. K., & Miyauchib, K. (2010). Growth and fruit quality responses of hydroponically cultivated eggplants to mineral controlled deep sea water. *Journal of Plant Nutrition*, *33*(13), 1970–1979. doi:10.1080/01904167.2010.512055

Jackson, P. (1998). *Introduction To Expert Systems*. Boston: Addison-Wesley.

Lau, E., & Yam, K. S. (2007). A Study of the Economic Value of High-rise Office Buildings. Retrieved from http://www.fig.net/pub/fig2007/papers/ts_4g/ts04g_02_lau_yam_1670.pdf

Le Strange, M., Cahn, M. D., Koike, S. T., Smith, R. F., Daugovish, O., Fennimore, S. A., . . . Takele, E. (2010). Broccoli production in California: University of California, Vegetable Research and Information Center. Retrieved from http://anrcatalog.ucdavis.edu/pdf/7211.pdf

Li, G., & Zhang, J. (2000). Cost Comparison Analysis of High-Rise Buildings with Steel and Reinforced Concrete in Shanghai. *Journal of Building Structures*, *21*(2), 75–79.

Li, G., & Zhang, J. (2000). Cost Comparison Analysis of High-Rise Buildings with Steel and Reinforced Concrete in Shanghai. *Journal Of Building Structures*, *21*(2), 75–79.

Li, Q., & Kubota, C. (2009). Effects of supplemental light quality on growth and phytochemicals of baby leaf lettuce. *Environmental and Experimental Botany*, *67*(1), 59–64. doi:10.1016/j.envexpbot.2009.06.011

Lin, K.-H., Huang, M.-Y., Huang, W.-D., Hsu, M.-H., Yang, Z.-W., & Yang, C.-M. (2013). The effects of red, blue, and white light-emitting diodes on the growth, development, and edible quality of hydroponically grown lettuce (Lactuca sativa L. var. capitata). *Scientia Horticulturae*, *150*, 86–91. doi:10.1016/j.scienta.2012.10.002

March, C. (2009). *Finance and Control for Construction*. New York: Taylor & Francis.

Molinar, R. H., Yang, M., & De Moura, R. L. (2005). Sample Costs to Produce Cherry Tomatoes. *San Joaquin Valley – South Small Farm*. University of California Cooperative Extension. Retrieved from http://coststudyfiles.ucdavis.edu/uploads/cs_public/67/e5/67e5f286-deeb-4081-965a-6c16714995d6/tomatochsjv2004.pdf

Monbiot, G. (2010). Towering Lunacy. Retrieved from http://www.monbiot.com/archives/2010/08/16/towering-lunacy/

Ohno, T., & Harada, Y. (2006). A Comparison of Tendering and Contracting Systems for Public Works between Japan, the United States and EU Countries. *Government Auditing Review*, *13*, 49–71.

Pant, A., Radovich, T. J. K., Hue, N. V., & Arancon, N. Q. (2012). Pak choi (Brassica rapa, Chinensis Group) yield, phytonutrient content, and soil biological properties as affected by vermicompost to water ratio used for extraction. *HortScience*, *47*, 395–402.

Phaobunjong, K. (2002). *Parametric cost estimating model for conceptual cost estimating of building construction projects* [Ph.D. thesis]. University of Texas, Austin.

RLB. (2013). Quarterly Construction Cost Update: Rider Levett Buchnall. Retrieved from http://rlb.com/wp-content/uploads/2014/01/rlb-hong-kong-cost-report-december-2013.pdf

SCCMIAD. (2012). Prices and Indices of Shanghai Construction Works, Shanghai Construction & Construction Material Industry Administration Department. Retrieved from http://www.fangce.cn/show-124-47943.html

Shah, A. H., & Shah, S. H. (2009). Cultivation of Lettuce in Different Strengths of The Two Nutrients Solution Recipes in A Non-Circulating Hydroponics System. *Journal of Agriculture*, *25*(3), 419–428.

Skitmore, R. M., & Marston, V. K. (1999). *Cost Modelling*. London: E & FN Spon.

Sonmez, R. (2004). Conceptual cost estimation of building projects with regression analysis and neural networks. *Canadian Journal of Civil Engineering*, *31*(4), 677–683. doi:10.1139/l04-029

Turner & Townsend. (2012). International Construction Cost Survey 2012. Retrieved from http://www.turnerandtownsend.com/TT_ICC_Report_Single_Pages_zKJM

Willis, C. R. (1992). Hydroponics - the solution to the food problem. Retrieved from http://www.androidworld.com/prod26.htm

Zhu, Z., Gerendas, J., & Sattelmacher, B. (1997). Effects of replacing of nitrate with urea or chloride on the growth and nitrate accumulation in pak-choi in the hydroponics. *Plant Nutrition for Sustainable Food Production and Environment*, *78*, 963–964. doi:10.1007/978-94-009-0047-9_313

This research was previously published in the International Journal of Agricultural and Environmental Information Systems (IJAEIS), 7(2); edited by Petraq Papajorgji and François Pinet, pages 26-51, copyright year 2016 by IGI Publishing (an imprint of IGI Global).

Chapter 26
Building an Urban Food System Through UDC Food Hubs

Sabine O'Hara
University of the District of Columbia, USA

Dwane Jones
University of the District of Columbia, USA

Harris B. Trobman
University of the District of Columbia, USA

ABSTRACT

The Landgrant College of the University of the District of Columbia embodies the university's unique mission as the only exclusively urban land-grant university in the United States. With most of the world's population now living in urban areas, this mission is relevant to cities worldwide. The UDC urban food hubs reimagine our food system as diversified, urban, and encompassing food production, food preparation, food distribution, and waste and water recovery. The hubs utilize bio-intensive hydroponic and aquaponic systems and green roofs to maximize productivity on small urban spaces; kitchens as business incubators and training facilities for food processing and nutrition education; waste and water reuse through composting, rain water capture, and green infrastructure. Each of these components offers opportunities for business startups and capacity building. The hubs also re-connect urban neighborhoods to nature. This chapter describes the urban food hubs, their locations, and the training, wellness, and leadership opportunities they offer to UDC students and DC residents.

INTRODUCTION

The Urban Food Hubs of the University of the District of Columbia re-envision the United States food system as urban, decentralized, and integrated. This stands in sharp contrast to the common association of food production as rural, highly concentrated, and as segregated into discrete entities of food production, food processing, food distribution, and food consumption. Their innovative approach won the Urban Food Hubs recognition by the National Extension Committee on Organization and Policy (ECOP)

DOI: 10.4018/978-1-5225-8063-8.ch026

of the Association of Public and Landgrant Universities (APLU) as one of the top four innovations in Cooperative Extension in the United States.

The UDC Urban Food Hubs consist of four integrated components - food production, food preparation, food distribution, and waste and water recovery (O'Hara 2015; 2017). These four components incorporate the entire food system value chain:

1. Food production takes place in highly efficient urban food production sites where small spaces are put to use employing more traditional production methods and new technologies.
2. Food preparation takes place in commercial kitchens where raw produce is turned into processed food products, and where educational programs on healthy food options and age appropriate diets are offered to improve nutritional health.
3. Food distribution takes place through a range of options from farmers markets, to contracts with restaurants and specialty grocery stores, to cooperative models like community supported agriculture (CSA) where participating households receive weekly deliveries of fresh produce.
4. Waste and water recovery includes green infrastructure that remediates urban soil and improves water use efficiency, that reuses plant and nutrient waste, and that recovers and reuses water run-off and turns it into a valuable resource.

Each Urban Food Hub is conceived as a business incubator that has the potential to create jobs, provides access to education and training, improves health outcomes, and ultimately improves the quality of life of urban populations, and especially of those who lack access to everything from fresh food, to jobs, and opportunity.

The Urban Food Hubs thus are an expression of the mission of the University of the District of Columbia (UDC) and its College of Agriculture, Urban Sustainability, and Environmental Sciences (CAUSES) as the only public university in Washington D.C., and the only landgrant university in the United States that serves an exclusively urban population. Contrary to the over 100 other landgrant universities in the United States (see Figure 1) whose home states incorporate urban as well as rural areas, Washington DC encompasses no rural areas. CAUSES has therefore a unique mission of re-imagining its own regional food system, and the global food system as urban. With more than 60% of the world's population now living in cities and metropolitan areas, this mission is critical not only to Washington DC, but to urban communities across the nation and the world.

The UDC Urban Food Hubs are both conceptual and physical, and offer multiple educational opportunities and quality of life benefits to UDC students, DC residents, and neighborhoods. Consistent with its unique mission, CAUSES offers academic programs in nutrition and dietetics, urban sustainability, urban agriculture, water resources management, health education, nursing and architecture. The latter is unusual for landgrant colleges, but urban design, green buildings, and low impact development are of great importance to a program focused on the urban food system and urban sustainability. In addition, the College offers a wide range of community outreach and cooperative extension services including healthy eating, wellness, and green infrastructure demonstrations, health fares, soil and water testing services, and presentations at schools, community centers, and farmers markets. UDC's Extension Programs also provide non-credit-bearing education and training opportunities in agriculture, gardening, nutrition, food safety, and natural resources management, but also in health and wellness, low impact development, urban sustainability, and resiliency.

Figure 1. United States Landgrant colleges and universities (USDA NIFA)

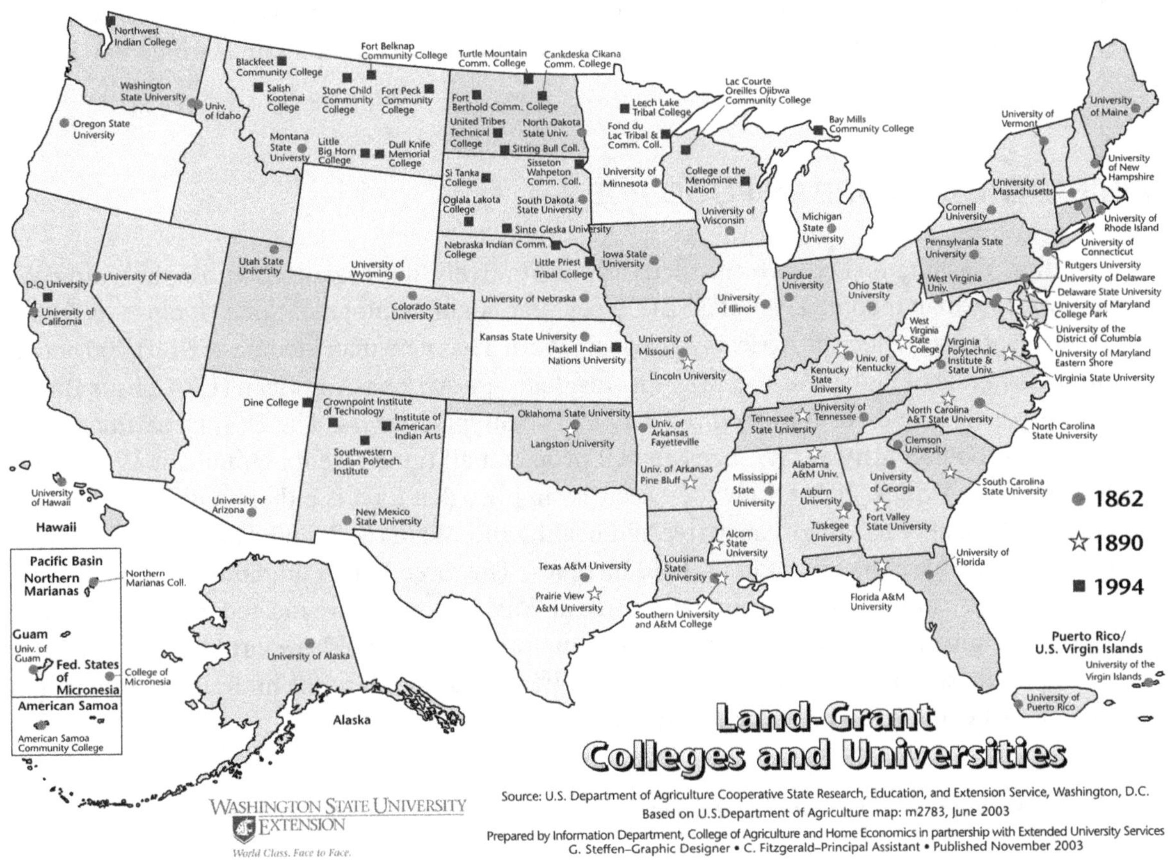

The cooperative extension programs also work closely with the applied research arm, or agricultural research services, of the landgrant university system. This research and community outreach partnership forms the basis for bringing practical knowledge to the people, communities, and organizations the landgrant system serves. The goal of the collaborative relationships between research and community outreach is to create tangible positive change as captured in the CAUSES mission to offer "…research based academic and community outreach programs that improve the quality of life and economic opportunity for people and organizations in the District of Columbia, the nation, and the world."

The CAUSES landgrant programs are offered through five centers that are well aligned with the unique urban focus of the college: the Center for Urban Agriculture and Gardening Education, the Center for Sustainable Development and Resilience, the Center for Nutrition, Diet and Health, the Center for Architectural Innovation and Building Science, and the Center for 4-H and Youth Development.

In 2013, CAUSES adopted its tagline *Healthy Cities – Healthy People* to affirm its capacity building mission as inextricably liked to improving the environmental/physical condition (= Healthy Cities) and the social/cultural condition (= Healthy People) of the urban community it serves. To give practical expression to this commitment CAUSES developed its Urban Food Hubs model.

This chapter provides a brief description of the four components of the UDC Urban Food Hubs, their locations across Washington DC, and their direct and indirect benefits. Beyond their implications for

Washington D.C., the Urban Food Hubs can serve as a model for sustainable development that focuses on the triple bottom line of economic, social, and environmental benefits; and as a model for experiential learning that uses physical locations as living learning laboratories that facilitate learning beyond what could be achieved in an on-campus setting.

DEFINING THE UDC URBAN FOOD HUBS

Like most cities, Washington D.C. is bifurcated. Administratively, it is organized into eight Wards. Ward 8 has the lowest median income level of $32,000 per year, an unemployment rate of close to 20% and 90% of its residents are African or African-American. Ward 3 has a median income of $110,000 per year, an unemployment rate of under 4%, and 5% of its residents are African-American (US Census Bureau).

Similar disparities are evident in food security and health parameters. The U.S. Department of Agriculture defines food security as "… access by all people at all times to enough nutritious food for an active, healthy life" (USDA, 2014). Low food security implies that food is either unavailable altogether or that it lacks the quality to support an active and healthy life. Wards 5, 7, and 8 are the three Wards in Washington DC that have the lowest household incomes. The three Wards are home to 34% of the DC population, yet less than 10% of all the grocery stores in Washington DC are located there (see Figure 2). It is therefore not surprising that 13 percent of Washington DC households report being food insecure. 19 percent reported experiencing food hardship; and 37% of households with children reported that they were unable to get enough food (D.C. Department of Health, 2013).

Figure 2. Full-service grocery stores and urban food hub Locations in Washington D.C.

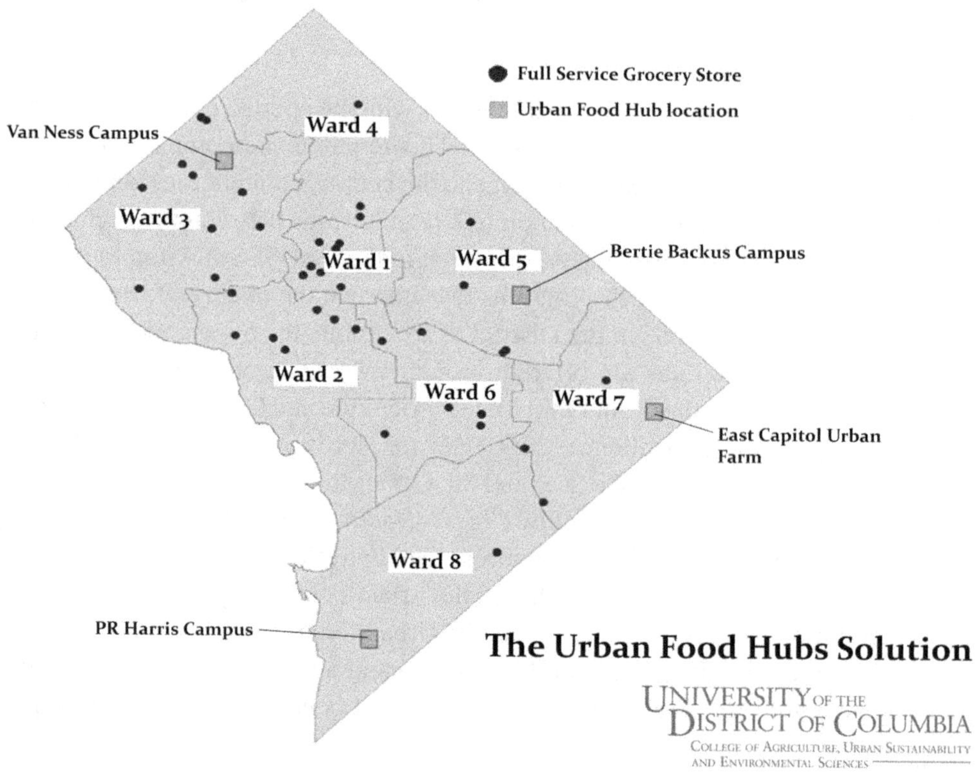

Cities are not places where we have historically grown much of our food (Gorgolewski, Komisar, & Nasr, 2011; UNDP, 1996). The UDC Urban Food Hubs seek to change that. They seek to provide a model for commercially viable food production that can grow meaningful amounts of food in small spaces where the majority of consumer live, namely in urban areas. Yet the Urban Food Hubs also expand the focus beyond food production to a comprehensive food systems model that incorporates the entire value chain of food, from food production, to food preparation, food distribution, and returning waste and water run-off to productive uses.

Similar to the Cooperative Extension offices that have shaped the capacity building mission of landgrant universities in rural counties across the United States, the Urban Food Hubs seek to build capacity in urban neighborhoods across Washington DC by focusing on the complexities of the urban food system including urban agriculture, health and wellness, urban design, low-impact development, and green infrastructure.

According to the Agricultural Marketing Service of the U.S. Department of Agriculture, a food hub is "a business or organization that actively manages the aggregation, distribution, and marketing of source-identified food products primarily from local and regional producers to strengthen their ability to satisfy wholesale, retail, and institutional demand" (USDA AMS, 2012). This definition emphasizes the distribution component of the food system. The UDC Urban Food Hubs broaden this definition to include uniquely urban considerations. Many urban dwellers have, for example, become distanced from their food source to no fault of their own. Since fresh unprocessed food is often unavailable, there is a growing disconnect between consuming food and knowing where it comes from and how it is grown. Often the interest in the quality and origin of food is first triggered by an interest in cultural associations of food, or by a concern for food-related illnesses like obesity, diabetes, and hypertension.

Urban food production also intersects uniquely with urban sustainability considerations like storm-water management and waste reduction. Urban gardens, for example, increase permeable surfaces and green roofs absorb storm water run-off. Both reduce pressure on urban storm water systems that are at capacity in many of our cities. (Agudelo-Vera et.al. 2011; Lovell, 2010; James et.al., 2015). When food is produced in densely populated urban communities rather than in sparsely populated rural areas, issues like integrative pest management and mitigating heat islands also come into focus. The four components of the UDC Urban Food Hub address these complex connections. They include:

- Food production through bio-intensive, hydroponic, aquaponic methods and green roofs;
- Food preparation through kitchens that add value and offer nutrition education;
- Food distribution through farmers markets, CSAs, restaurants and niche markets;
- Closing the loop through composting, water harvesting, rain gardens and other green infrastructure installations.

Each of these four components offers opportunities for business development, training, and improved resilience (see Figure 3). The Urban Food Hubs seek to form a decentralized network of local food centers that improve food security, nutritional health, job creation and resilience especially in urban neighborhoods that have deficits on all counts.

Each food hub then comprises a cluster of business opportunities for entrepreneurs who ideally come from the neighborhood where a hub is located. The local entrepreneurs receive training and technical support to implement their business plans. These can range from a health-focused business that maximizes nutrient yield and offers health assessment and nutrition counseling; to growing micro-greens and

Figure 3. The urban food hubs as comprehensive food systems model

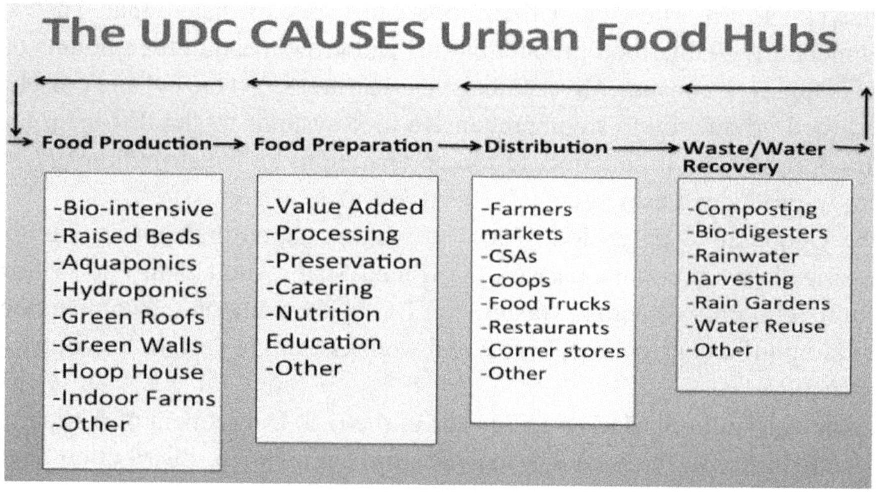

herbs for high-end restaurants; to ethnic crop production for local niche restaurants and grocery stores; green roofs that serve as food production and event space; and native plant seedlings grown for urban parks and rain gardens.

While each Urban Food Hub features the four components of food production, food preparation, food distribution, and waste and water recovery, their specific design varies based on the urban density, site-specific characteristics, and business objectives of each location. What follows is a brief description of the four components at the UDC Urban Food Hubs that are currently in various stages of implementation in Wards 3, 5, 7 and 8 in Washington D.C.

The Ward 3 Urban Food Hub

The University's Van Ness campus is located in Ward 3 at a high traffic intersection next to a metro station. Space is very limited in this high-density neighborhood. The food production component of the food hub is a 20,000-square-foot green roof that was installed on an existing roof structure (see Figure 4).

Planters placed around the perimeter of the roof support the production of a wide range of food plants, from tomatoes to peppers, beans, okra, eggplant, and even berry bushes. The interior of the roof is limited to three to four inches of soil depth that can support the production of leaf lettuce, microgreens, and herbs. In its start-up phase the shallow beds are used mostly to grow sedums, a plant that requires little maintenance. The sedums absorb water and contribute to insulating the building, lowering the need for heating and air conditioning, and reducing storm water runoff. The roof also features a greenhouse and a hydroponic system that is arranged vertically to fully utilize the limited space (see Figure 5).

Hydroponics refers to a method of growing plants in nutrient enriched water rather than soil. Nutrient levels are maintained by adding liquid fertilizer. Hydroponics systems produce substantial amounts of food in small spaces, including indoors, and can be used to grow a wide variety of vegetables including herbs, lettuce, spinach, kale, and other greens, but also tomatoes, peppers, cucumbers, beans, okra and squash. Since no soil is needed, hydroponic production can be used even in areas where soil contamination may be an issue.

Figure 4. Retrofitted food producing green roof at the UDC urban food hub in ward 3

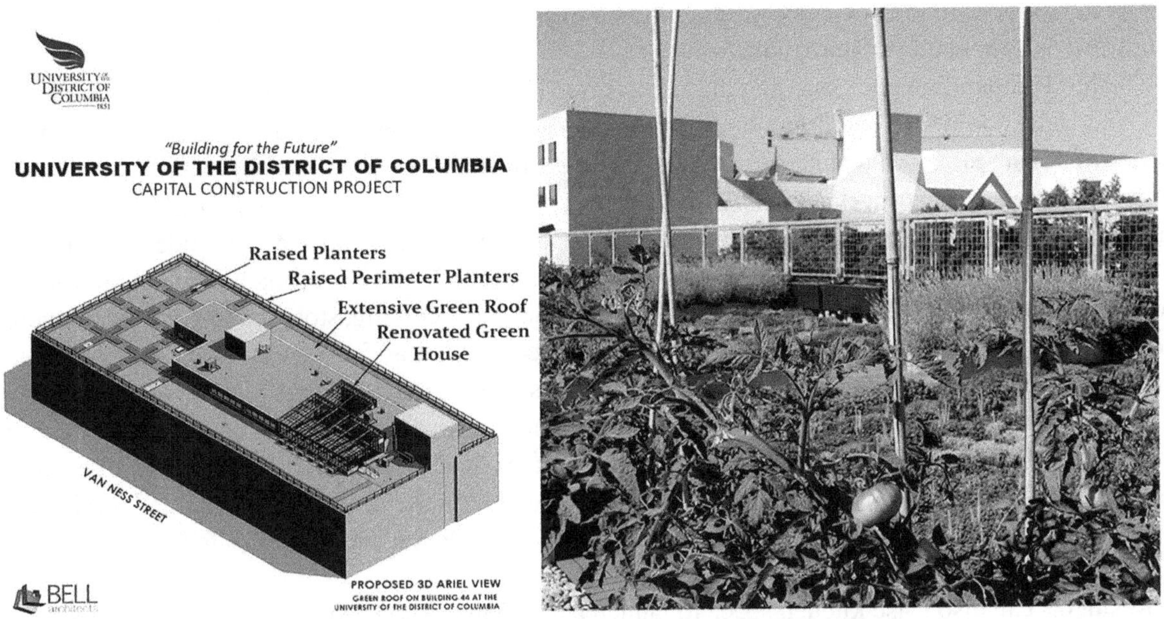

Figure 5. Hydroponics and aquaponics at the UDC Van Ness urban food hub

The second technology-driven food production method that CAUSES utilizes in its Urban Food Hubs is aquaponics. Aquaponics combines growing fish (aquaculture) and growing vegetables without soil (hydroponics). The systems, pioneered by CAUSES, are highly energy efficient and stable. The central component is an aeration device called the Flo-Vex®, which is a patented invention of Emeritus Professor Thomas Kakovitch, a physicist and environmental engineer who taught environmental science at UDC for over 40 years (Kakovitch Industries 2016). The Flo-Vex emulsifies atmospheric oxygen with the water that circulates through the aquaponic and hydroponic systems and eliminates the need for mechanical compression. By using the excrement from the fish as fertilizer for the plants, aquaponic systems eliminate the need to add fertilizer. As the plants absorb the nutrients from the water, it can be directed back to the fish tanks for reuse. The co-production of vegetables and fish creates significant water savings, high productivity levels, and reduced waste. The CAUSES aquaponic systems use only 10% of the water used in conventional agriculture and since the fish waste is used as plant fertilizer, very little waste is generated (O'Hara, under review).

The food preparation component of the Urban Food Hubs comprises two main areas: nutrition education, and value addition. To address them effectively each Urban Food Hub ideally has a well-equipped kitchen that serves as a teaching facility to improve information about healthy eating, healthy food preparation, and age-appropriate diets. The kitchens are also business incubators for those interested in starting a catering business or turning tomatoes and peppers into salsa or other value-added products. The kitchens are designed to be functional, energy efficient, and food safety compliant. Demonstration areas provide visible workspaces, and well-defined workstations for receiving, storage, preparation, recycling, and other functional areas to provide training for proper food handling, food safety and food management.

Figure 6 shows the kitchen facility at the Van Ness Food Hub. Activities at the kitchens include nutrition education workshops, cooking classes, food demonstrations, certifications in food handling, food safety, Hazard Analysis and Critical Control Point (HACCP), and food preservation like canning and pickling. The space can also be used to advance research on the determinants of safe food handling behaviors, risk perception, and cultural barriers that may impede the adoption of safe food handling and healthier eating habits.

Residents can also lease kitchen space to launch a catering or food processing and food preservation business. Nutrition education itself may also offer viable business opportunities. Recent legislative changes allow dieticians to prescribe a therapeutic diet. This is consistent with the increased focus on prevention, rather than treatment that offers new opportunities to qualified dieticians and nutrition educators.

Food distribution at the Urban Food Hubs takes a variety of forms. Local farmers markets and Community Supported Agriculture (CSA) programs that sell directly to the end consumer are popular options. CSAs deliver fresh produce, typically once a week, to a central pickup location or directly to CSA member homes. Rather than charging a price per pound, CSAs charge their customers a flat fee at the beginning of the growing season. In exchange, CSA members receive a weekly delivery of in-season produce. Some CSAs also offer a lower price to customers willing to assist with crop maintenance, harvesting, and delivery (O'Hara & Stagl 2002).

The Urban Food Hubs also support collaborative efforts like gleaning days and harvest collections. One such initiative, the City Orchard project, established a fruit orchard at the UDC Firebird Farm to

Figure 6. Van Ness food hub kitchen

grow a variety of berries, apples, and Asian pears for clients of a local food bank. The food bank in turn helps maintain the orchard through weekly volunteer days, which have the added benefit of introducing food bank clients to a farm often for the first time. Collaborations with local schools, culinary institutes, and other hospitality sector partners also expand the food distribution opportunities of the Urban Food Hubs. Like any small producer, urban food producers will want to diversity their revenue stream to be successful and stabilize their markets. Events like business lunches, wedding and other celebrations offer great opportunities for a diversified business model in densely populated urban areas.

A prime example is Up-Top-Acres (2017) an urban agriculture business that was launched by three graduates of Wilson high school, one of the DC public schools. Up-Top Acres produces food on green roofs in the DC Stadium neighborhood. The successful business was recently cited by Forbes in their top-thirty-under-thirty report, which recognizes top businesses in the United States that are run by entrepreneurs who are under the age of thirty. Up-Top-Acres produced locally grown lettuce, arugula, spinach, herbs, tomatoes and various other crops, but it also offer events like health and wellness classes and roof top lunches, and it captures storm water run-off to cash in on the storm-water credits offered by the District of Columbia government to individuals and businesses whose practices contribute to reducing the storm water load that is added to the DC storm water system.

The example of Up-Top-Acres already speaks to component four of the Urban Food Hubs, namely resource recovery through waste and water management. This component consists of various water quality and quantity control methods. Water quality measures ensure minimal damage to the District's watersheds. This includes planning, designing, constructing, maintaining and managing rain gardens, bio-retention areas, green roofs and other green infrastructure practices. The measures also offer opportunities for entrepreneurship and improved aesthetics. The social benefits of this food hubs component thus go beyond their soil and water management benefits.

As the urban food hubs continue to evolve each Hub may eventually be equipped with an alternative energy source that takes the Urban Food Hubs off the grid and turns them into resiliency hubs that can secure food, water, and energy during times of crisis.

Urban soils are often contaminated and may require soil amendments before they are suitable for food production. Composting is key and each of the urban food hubs has a small composting site where plant and food waste can be turned into healthy soil. A recent business start-up in the D.C. area, Compost Cab, picks up food waste from residential households for a monthly fee and turns it into compost. Since many of the members of Compost Cab are apartment dwellers, they have no use for the compost generated from their food waste and make it available to Compost Cab (2017), which sells the product to

Figure 7. Water capture at Van Ness and rain water management at east capitol urban farm

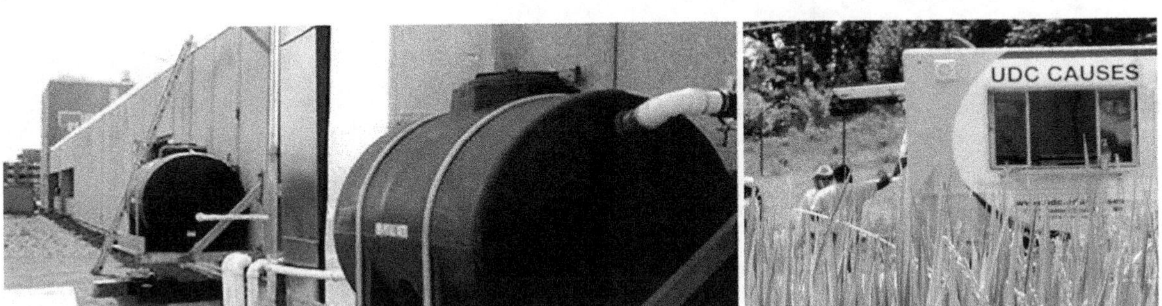

other home owners who want to contribute to sustainable practices in the District of Columbia. Compost Cab is a prime example of the business start up opportunities associated with the forth component of the UDC Urban Food Hubs – waste and water recovery and reuse.

The Ward 5 Urban Food Hub

The University's Bertie Backus campus is located in a neighborhood with lower urban density than its Van Ness campus counterpart. Food production at the Ward 5 location takes place in two 30 x 90 ft hoop houses, one with a hydroponic system and one with an aquaponic system (see Figure 8). Like the Van Ness location, the Bertie Backus location is easily accessible by Metro. It also features a third hoop house with a native plant nursery. This offers additional business opportunities not strictly related to the food system. Native plants are needed by to combat evasive species in public parks and recreation areas.

The Ward 7 Urban Food Hub

The East Capitol Urban Farm is the UDC Urban Food Hub location in Ward 7 (see Figure 9). It features a community garden with 60 raised beds available to residents at no cost, a children's garden and outdoor STEM education area, a nature trail, and an aquaponic system installed in a 30 x 90 foot hoop house. The D.C. housing authority made the three-acre site available to UDC for the purposes of developing a model for temporary urban farming that improves the condition of vacant land while future development decisions are being made.

The East Capitol Urban Farm is located in a food desert neighborhood. The one grocery store that was within reasonable distance of the site closed in the summer of 2016. While the farm is only in its second growing season, and the aquaponic system is not yet in full production, CAUSES launched a farm stand in 2017 at the request of local residents.

Kitchens are expensive to build and operate and given the temporary use model of the East Capitol Urban Farm, the construction of a commercial kitchen facility was not feasible. A more affordable option is a food truck that can serve as a mobile food demonstration and teaching facility in addition to serving as a mobile farmers market that brings fresh produce to food desert neighborhoods (see Figure 10). The UDC CAUSES food truck serves these multiple roles and it is a regular feature at the East Capitol Urban Farm.

Figure 8. The UDC urban food hub at Betrie Backus in ward 5

Figure 9. The east capitol urban farm site plan

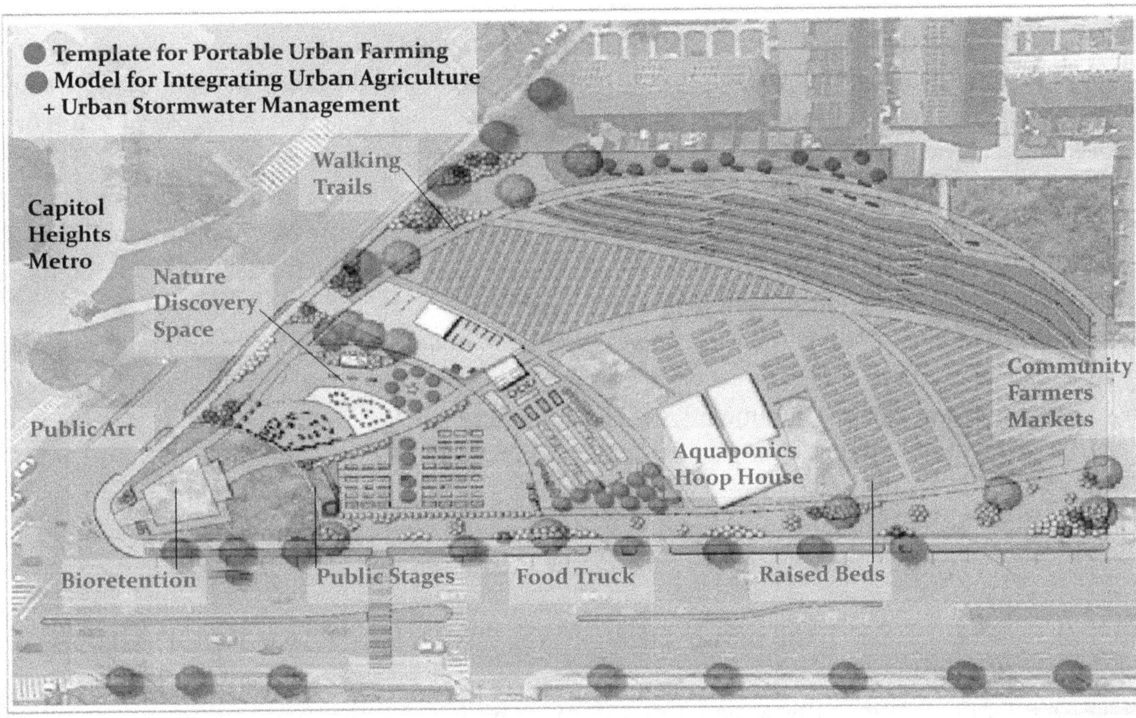

Figure 10. The east capitol urban farm food hub in ward 7

The Ward 8 Urban Food Hub

The Ward 8 Urban Food Hub is located at the PR Harris Education Campus. A former elementary school the campus features a large windowless building that was designed in the brutalist architecture tradition. Brutalist architecture was prominent in the 1960s and 70s and is associated with the use of beton or raw concrete, which earned it the French label brut or raw (Le Corbusier; Banham)

The land surrounding the building is park-like and within easy walking distance of a small park and picnic area that serves as a location for occasional neighborhood gatherings and outdoor celebrations. The neighborhood is a food desert area with the closest grocery store, which offers a very limited selection of fresh produce, being located in neighboring Maryland.

The PR Harris Urban Food Hub features two large hoop houses, one with a hydroponic system and one with an aquaponic system (Figure 11). While the location is not yet developed enough to support a weekly farmers market, some community events are being launched during the 2017 growing season including a farm stand, food demonstrations, educational activities for children and youth and other activities that focus on the food access, nutrition education, and community building aspects of the Urban Food Hubs model.

TANGIBLE AND LESS TANGIBLE BENEFITS OF THE URBAN FOOD HUBS

The UDC Urban Food Hubs offer a model to diversify and localize the food and green infrastructure sectors of metropolitan communities. The model thus contributes to the triple bottom line of economic, social, and environmental benefits for cities and metropolitan areas. Yet peri-urban and rural communities

Figure 11. The hydroponic- and aquaponic hoop houses at the PR Harris urban food hub

too have expressed interest in the model. For example, the Delaware Nation of Oklahoma has entered into an agreement with UDC and has expressed interest in building a Food Hub on their reservation, which is as much in need of an integrated local food system as food desert neighborhoods in Washington DC.

While the economic viability of the Urban Food Hubs will depend on the specific business model and site characteristics of each Hub, the social benefits of improved health outcomes of underserved urban populations are indisputable. Seeing how food is grown can be an effective catalyst for adopting more healthful eating habits and thus reducing public health expenditures associated with food related illness.

Michigan State University's Cooperative Extension Service identified seven benefits of locally grown food that are not captured in the food price. Locally grown food (1) is more flavorful; (2) reconnects us with the seasons; (3) is more nutrient dense; (4) supports the local economy; (5) benefits the environment; (6) promotes a safer food supply; (7) connects consumers and growers and improves awareness of where food comes from (Halweil, 2002; Klavinski, 2013;). A report published by the University of California Davis also attributes considerable positive social, health, and economic impacts to urban agriculture (2015), and an MIT study documents the economic impact of urban aquaponics (Goodman, 2012). The field of ecological economics offers especially useful examples for calculating the value of the social and environmental side benefits of economic activity (O'Hara 1995, 1997, 2001, 2014). What follows is a brief discussion of some of the key benefits of the Urban Food Hubs.

Food Security

Food security is a serious and growing concern for populations across the United States and the world. Food security speaks to the need for every person to have consistent and dependable access to highly nutritious food that supports a healthy and active life. Eight census tracks in Washington DC are considered food desert neighborhoods, meaning that there is no grocery store that offers fresh produce within a more than a mile. Households who rely on public transportation are particularly hard hit by food insecurity. They will have to get their groceries via bus or metro even as they balance the tight scheduling constraints of work, childcare, school, and livelihood needs. Affordability too is a concern, since high quality produce is often prohibitively expensive for low-income households.

Concerns about food security are not unique to urban areas either. Rural areas too, even when they are the center of grain or corn production, may have food insecure households who cannot get access to fresh produce since the local production may be quite homogeneous and may be focused almost entirely on the wholesale or large scale food processing market. The needs of local populations go in turn unmet.

Additional food security concerns are related to the quality of food. Health concerns, for example, are associated not only with a lack of access to nutritious food, but with access to the wrong food that is highly processed, high in fact and high in salt content (Dixon et al 2007). This food is often referred to as 'empty calories' and it is prevalent in urban food desert neighborhoods that feature fast food and take-out places rather than fresh produce. Health concerns also stem from the declining nutrient content of some of our food (USDA, 2013). This low nutrient density can at least in part be attributed to the increased time that elapses between food harvest and food consumption. To address the problem, more nutrient dense produce must be grown closer to where the majority of consumers live, namely in metropolitan areas like Washington DC. Growing perishable food closer to where the majority of consumers live also reduces the energy use associated with food transportation (Canning et al. 2010; Weber & Matthews, 2008).

Some of the benefits of locally grown food will depend on the specific production methods applied. Some growers, for example, may be committed to growing seasonal food plants while others will grow the same salad mix year around. Environmental benefits may vary with the energy used for food production. Almost regardless of the specific production methods, the UDC Urban Food Hubs improve freshness, and reduce water use and storm water run-off regardless of the specific food production method used. The urban food hubs can also improve the aesthetics of a neighborhood. Raised bed gardens, trellises, and vertical growing systems can bring tremendous enhancements to an urban neighborhood. Other benefits may result from the reuse of unused or underutilized buildings that can be successfully repurposed for indoor food production (Royte 2015).

Water Management and Green Infrastructure

Cities have changed the drainage system that handles rainfall and snow melt since city surfaces are mainly comprised of impermeable materials like roads, sidewalks, parking lots and roofs. As a result, the amount of storm water runoff that enters the storm water systems of cities and metro areas has increased substantially. For many cities the problem is exacerbated due to aging infrastructure that has not kept pace with increased urbanization (Niemczynowicz 1999). This is certainly true for Washington DC. The city's storm water system was built for about half the number of residents that the city has today.

The Urban Food Hubs help ameliorate the problem of excess storm water run-off by adding permeable surfaces that increase infiltration and reduce the storm water runoff problem (Barthel & Isendahl 2013). Urban green spaces like community gardens, raised beds, green roofs and other growing spaces add filtration capacity; composting and cover crops build up healthy soils and mitigate soil pollutions, which can be a problem in urban communities due to past land use practices; and green infrastructure installations like rain gardens and water harvesting cisterns reduce storm water runoff. The Urban Food Hubs also reduce water consumption through water harvesting, water reuse, and water saving irrigation practices like drip irrigation that reduce evaporation.

Green infrastructure (GI) strategies mimic the storm water management processes of nature. They use vegetation, soils, and rainwater harvesting to reduce or eliminate storm water runoff, help maintain water quality, and create healthier urban environments. Conventional 'gray' storm water infrastructure such as catch basins, drainage pipes, and ponds move urban storm water away from buildings, sidewalks and roads, and direct it to nearby water bodies or drainage systems. Water runoff also carries trash, bacteria, fertilizers, road salt, heavy metals, and other pollutants that degrade water quality. GI practices reduce the pollutants entering the storm water system by capturing them in the green infrastructure installations before they reach the storm water system. GI practices also contribute additional benefits including heat island mitigation, air quality management, and restoring ecosystem services. All of these contributions benefit urban environments and urban populations. The Urban Food Hubs thus can have a measurable positive impact on water quality and water management as well as of the quality and availability of other natural resources.

Education, Training, and Job Creation

Central to the capacity building mission of the UDC landgrant programs are the workforce development programs, training certificates, workshops and demonstrations offered through the five landgrant centers. Several of these non-credit bearing programs are offered at the Urban Food Hub locations and at the UDC research farm, the Firebird Farm, which is located in Beltsville, MD. Workshop and certificate topics include introduction to urban agriculture, bio-intensive growing methods, hydroponic and aquaponic production, soil maintenance, and integrative pest management; but also food safety, age appropriate nutrition, food preservation, composting, and rain garden design and maintenance.

For many of these programs, the Urban Food Hubs serve as hands-on learning centers and filed trip sites. The Food Hubs currently under development feature four hydroponic and four aquaponic facilities, a food producing green roof, two commercial kitchens, a food truck, two rain gardens, and a water capture and reuse system. Two additional food Hubs in Wards 7 and 8 are currently in the planning stage.

Funding for the Urban Food Hubs come from a sustainable development grant of the District of Columbia, from the Anacostia Economic Development Corporation, and from the University's landgrant funds. In the tradition of landgrant universities, CAUSES also provides technical assistance and training for the business development aspects of the Urban Food Hubs. Once the Urban Food Hubs have been fully implemented the food production and food preparation facilities will be available to aspiring businesses through licensure agreements between the University and private sector partners. Priority will be given to partner businesses that stem from and plan to contribute to the urban neighborhoods where the Food Hubs are located.

The UDC Urban Food Hubs thus offer a model for business development from the ground up that identifies local entrepreneurs who wish to utilize the business opportunities the Hubs offer. An on-line application process invites prospective entrepreneurs to apply for the use of the facilities, and a review panel interviews applicants to select the most promising entrepreneurs. Since the Urban Hood Hub facilities were financed through grants, the new business start-ups launched through the food hubs will have to carry only operating expenses. This should improve their financial viability and future expansion potential as well as their expected success rate.

A pilot project to test the University's entrepreneurship based model was launched at the UDC Firebird Farm where a group of students built and managed a hydroponic system installed in a 30 foot x 60 foot hoop house. The students successfully produced head lettuce and basil for sale at local farmers markets and the training model is now being transferred to the UDC Urban Food Hubs.

In addition to its business development mission, CAUSES also seeks to create jobs through partnerships with District of Columbia agencies. One such recent initiative is the green infrastructure program launched in collaboration with DC Water. The program trains underemployed and unemployed DC residents in the design, installation and maintenance of Green Infrastructure installations. After completing the training program, graduates have the option to sit for a national exam. Those who pass the exam will be offered employment opportunities by the local DC water authority or its contractors. Similar programs are under development in collaboration with other DC agencies.

Community Development

Metropolitan universities have long been at the forefront of investing in their surrounding communities. Their capacity building investments include renovation and reuse of residential properties, business incubators, neighborhood storefronts, and more. Frequently these initiatives put un- and under-utilized buildings back in use as resident halls, offices, classrooms, and businesses (Rucinski 2001).

The UDC Urban Food Hubs initiative can be viewed as a similar investment, whereby the specific focus of the initiative is the urban food system and the commitment to turn food desert neighborhoods into thriving local food systems. The vehicle through which the UDC Urban Food Hubs realize their capacity building mission are the university-owned food production, food processing, and green infrastructure facilities that form the physical backbone of the Hubs. Similar to a business incubator that makes space, equipment, and shared services available to business start-ups, the Urban Food Hubs make an upfront investment in the urban food and green infrastructure sector. This improves the likelihood of success for start-up businesses that would find it difficult to make the capital investment necessary to install a hydroponic or aquaponic system, a green roof, or a commercial kitchen.

Even less capital-intensive production methods that utilize raised bed gardens may require considerable up front investments. When land that was previously used as a building site, is converted to agricultural land, the up-front soil remediation and site preparation costs can be substantial. The Urban Agriculture toolkit published by the USDA calculates the projected investment at $100,000 to $250,000 for a half-acre parcel of land (USDA, 2016). The costs of the kitchen facilities associated with the food preparation component of the Urban Food Hubs can be considerably higher. Yet community development initiatives that focus on building rehabilitation can be even more costly, and few have as direct of an impact on the health and economic empowerment of underserved urban populations as the Urban Food Hubs initiative.

Resiliency

Urban resilience is the capacity of a city and its residents, neighborhoods and organizations to prevent, survive, and adapt to chronic stress and acute shocks. These can include such sudden and acute events as a hurricane, an earthquake, or the sudden outbreak of a highly contagious disease; and they can include such slow and eroding events like high unemployment, endemic violence, and food and water shortages (Burton 2009; Meerow, Newell & Stults 2016).

Most people are naturally resilient yet populations rarely know just how resilient they can be until they are tested through a volatile experience or unexpected event. Resiliency often happens instinctively rather than planned. Planned resilience, however, can be far more effective and can mean the difference between merely surviving a volatile or catastrophic events, and thriving in the face of these adversities. Such thriving includes proper planning, adequate resources, and a robust network of communication to rapidly address the emerging needs both individual and communal. The Rockefeller Resilient Cities initiative defines the seven characteristics of resilient cities as reflective, resourceful, robust, redundant, inclusive, integrated and flexible (Rockefeller 2017). These characteristics can be planned, designed, implemented, maintained and managing to ensure that a community is better prepared in response to a volatile experience. This also includes the capacity to forecasting the intensity of the experience, wherever possible.

Washington DC was recently inducted into the family of 100 resilient cities and the UDC Urban Food Hubs are quickly becoming a basis for Urban Resiliency Hubs. The food hubs are self-sustained systems within the city and by design they seek to provide social, economic and environmental resiliencies. Like food security, the vulnerability to chronic and acute shock too is not evenly distributed across the District of Columbia. Populations at risk have higher sensitivity and lower adaptive capacity to shock. The easternmost Wards, 5, 6, 7, and 8, are home to the largest number of vulnerable populations, and they are also most vulnerable to flooding. During extreme weather events or in emergencies, available services and infrastructure can become especially impacted in these vulnerable Wards.

Since the Urban Food Hubs are embedded in the community, they have the potential of reducing community vulnerability to extreme weather events and emergencies. During such shock events, each hub can become a community hub through which food, water, energy, communication, and protection can be extended. A decentralized, diversified, local food system is key to addressing resiliency needs at the local level by minimizing the need for transport of goods and services, maximizing the potential for food production in controlled environments, and engaging the community as part of the process. Introducing canning and food preservation into the Urban Food Hubs will further increase their ability to serve the community during times of emergencies. Additionally, solar power with battery storage can be used to power the sites off the grid, and to provide a community resource such as on site charging stations and net power contributors to the grid.

The use of green infrastructure is especially critical in extreme weather events. GI helps to reduce flooding on site and water harvesting and re-use provides water quality and water management assets that provide supplemental water during drought events for irrigation and other non-potable uses. Additional green infrastructure options are currently under consideration that would add larger filtration capacity to the Urban Food Hubs to provide potable water in emergency situations. Adding this water quality and water management capacity would offer further opportunities for workforce development and business start ups, while also addressing a critical need for the city. The Urban Food Hubs locations can also be effectively aligned with existing safety resources such as police/ security/ emergency responders, health centers, and shelters.

Biophilic Cities and Nature Deficit Disorder

Biophilia is the human tendency or need to interact with or be closely associated with nature. "Bio" refers to processes that occur in living things or beings; "philia" means a feeling of love, affection, or admiration; and "ic" is having the character or form of relating. Biophilic thus refers to the human need to interact with and love nature. Biophilic cities are urban communities committed to introducing or re-introducing urban populations to nature, and to facilitating their relating to nature (Beatley 2010).

Urban populations are commonly nature deprived and suffering from Nature Deficit Disorder, a phrase coined by Richard Louv in his book 'Last Child in the Woods'. Louv pointed out that all humans, but especially children, are spending less time outdoors resulting in a wide range of behavioral issues. Adding green space has many benefits including improved neighborhood safety, better walkability, more exercise, and added public space where neighbors can meet and socialize.

The UDC Urban Food Hubs include components that can reduce the negative effects of nature deprivation. Landscaping, green-scaping, and green infrastructure are all ways to introduce green spaces to the urban scape. And there is nothing that connects humans as effectively to nature as our food. Growing food and knowing where it comes from are essential reminders of the connectedness of humans and nature. The urban forestry and green infrastructure programs of the CAUSES Center for Sustainable Development and Resilience, and the urban agriculture and gardening education work of the CAUSES Center for Urban Agriculture offer programs and neighborhood demonstrations that reconnect urban residents to green spaces in general and to food production in particular. One of the programs of the Center for Urban Agriculture is the Master Gardener program, which trains volunteer educators to work within their local communities to encourage and promote environmentally sound horticulture practices through sustainable landscape management education and training. More recently CAUSES has added a food production component and an urban agriculture certificate to its training programs.

LEARNING AND LEADERSHIP

The urban food hubs enhance the learning opportunities of UDC students. They can serve as a common focal point for service learning, student and faculty research, leadership development, and professional networking. Students can also enroll in the certificate courses and workshops offered through the University's landgrant programs to prepare them for success as entrepreneurs in the local food economy or the green infrastructure sector.

Given the breadth and comprehensive nature of the Urban Food Hubs it is not surprising that students from a wide range of academic fields have been engaged in the work of CAUSES and of its Urban Food Hubs project. Nutrition and Dietetics students are an obvious fit with the Urban Food Hubs. Several undergraduate students participated in collaborative research projects at the UDC Firebird Farm and at the Van Ness Urban Food Hub. Some of these research projects also included participation from the USDA Beltville Agricultural Research Center (BARC). The CAUSES Center for Nutrition, Diet and Health (CNDH) involved nutrition students in piloting a new nutrition curriculum designed to address the food related public health issues of African-Americans and descendants of Africans across the diaspora. The program, entitled 'A Taste of African Heritage' (Oldways 2015) focuses on nutrition education, African heritage foods, and cultural history to reconnect African-Americans to the history, culture and foods associated with their ancestors.

Students in the Professional Science Masters degree program in Water Resources Management take advantage of the green infrastructure and soil and water testing opportunities of the Urban Food Hubs to hone not only their knowledge of hydrological science, but also their communication skills, problem solving skills, entrepreneurship talents, and innovation capacity. For example, PSM students worked with landgrant staff to test the garden soils of DC residential properties and community gardens across the city. They took over 300 samples and analyzed them under the supervision of laboratory technicians in the Environmental Testing Laboratory of CAUSES and reported the results back to their community clients.

Architecture students have been engaged in designing the site plans for the Urban Food Hubs and in developing a mobile, modular Urban Food Hubs model. Business majors have developed project management plans and potential business plans for the urban food hubs production sites. A psychology major worked at the Van Ness and East Capitol Farm Food Hubs as well as at the UDC Firebird Farm and is now involved in a research project on the impact of food and nutrition on mental health.

The opportunities the Urban Food Hubs and the landgrant programs in general offer to UDC students are endless and limited only by the constraints in time, resources and creativity the students themselves and their faculty and staff mentors face. More formal program collaborations include faculty, student and staff research collaborations, co-teaching opportunities in the cooperative extension programs of CAUSES, internships in the CAUSES landgrant centers, experiential learning opportunities, and service learning projects at the Urban Food Hubs, or at one of the other community projects sites of CAUSES. As the University's landgrant college, CAUSES is positioned to serve as the experiential learning hub for the University by utilizing the research, community outreach, and capacity building opportunities of its landgrant centers. The UDC Urban Food Hubs cut across all of the CAUSES landgrant centers and across a wide range of academic fields. They most clearly exemplify the practical, community based focus of the University and its commitment to its urban mission. Table 1 summarizes enrollments in the UDC landgrant programs, whereby individuals may be enrolled in more than one certificate program or workshop.

Table 1. Enrollment in UDC Landgrant programs in all eight wards in 2015

Landgrant Center	Enrollment[1]	%
Center for Urban Agriculture	1,717	5%
Center for Sustainable Development	535	2%
Center for Nutrition, Diet and, Health	4,690	13%
• SNAP-Ed[2]	14,750	42%
• Institute of Gerontology	5,152	15%
Center for 4-H and Youth Development	8,294	24%
Total	35,138	100%

CONCLUSION

The Urban Food Hubs of the University of the District of Columbia and its College of Agriculture, Urban Sustainability and Environmental Sciences are literally changing the Urban Landscape of Washington DC. The Urban Food Hubs re-envision the United States food system as urban, decentralized, comprehensive, and integrating the entire value chain of the food system from food production to food preparation, food distribution, and closing the loop through waste and water recovery and reuse.

The UDC Urban Food Hubs are not merely a conceptual framework for a re-envisioned food system, they are actually taking shape in four locations across the District of Columbia with two additional locations currently in the planning stages. The food production component of the Urban Food Hubs utilizes common sustainable agricultural practices like bio-intensive growing systems and raised bed gardens; but also more cutting-edge practices like food production on a retrofitted green roof, and in high-tech hydroponic and aquaponic systems. Food preparation takes place in commercial kitchens, food science laboratories and food trucks that serve as teaching facilities for nutrition education and culturally appropriate healthy eating, as well as business incubators for food processing and food preservation related ventures. The food distribution component focuses on organizational and marketing skills, but also on market analysis and on exploring the best ways to distribute different types of locally grown produces from farmers markets, to contracts with restaurants and specialty stores, to community supported agriculture models. And last but not least, each food hub serves as a demonstration site and training center for green infrastructure installations that remediate urban soil and improve water use efficiency by replicating nature and its remediating qualities in the midst of the urban scape.

Each Urban Food Hub is conceived as a business incubator that is committed to a socially and environmentally responsible triple-bottom-line business model that creates jobs, provides access to education and training, improves health outcomes, and ultimately improves the quality of life of urban populations, especially in food desert neighborhoods that lack access to everything from fresh food, to community health, jobs, and economic development opportunities. Each Urban Food Hub is also an expression of the unique urban landgrant mission of the University of the District of Columbia as the only public university in Washington D.C., and the only exclusively urban landgrant university in the United States.

The unique Urban Food Hubs project of the University and its College of Agriculture, Urban Sustainability and Environmental Sciences (CAUSES) is re-imagining not only the urban food system of Washington DC, but that of cities across the United States and around the world. More than 50% of the world's population now live in cities and metropolitan areas, and most of the additional two billion people that we expect to add to the global population in less than thirty years are expected to live in urban areas as well (The World Bank, 2016). The pressures of urbanization and population growth are further exacerbated by the decline of arable land (The Guardian 2015; Bruinsma 2014). This calls for a larger re-envisioning of the global food system as local and regional and as rural, peri-urban and urban.

The UDC Urban Food Hubs offer multiple benefits to Washington DC residents and UDC students. These include the opportunity to be on the cutting-edge of building an urban food system that creates such benefits for underserved urban neighborhoods as job creation, improved health outcomes, and better resources management; and benefits for students that range from collaborative applied research opportunities, to experiential learning, service and leadership development opportunities. In addition to these more tangible benefits, the Urban Food Hubs create critically important benefits including community engagement, and a renewed connection to our food, to nature, and to a more resilient and just future.

REFERENCES

Agudelo-Vera, C. M., Mels, A. R., Keesman, K. J., & Rijnaarts, H. H. (2011). Resource management as a key factor for sustainable urban planning. *Journal of Environmental Management*, *92*(10), 2295–2303. doi:10.1016/j.jenvman.2011.05.016 PMID:21641714

Banham, R. (2012, July 27). The new brutalism by Reyner Banham. *The Architectural Review*.

Barthel, S., & Isendahl, C. (2013). Urban gardens, agriculture, and water management: Sources of resilience for long-term food security in cities. *Ecological Economics*, *86*, 224–234. doi:10.1016/j.ecolecon.2012.06.018

Beatley, T. (2010). *Biophilic Cities: Integrating Nature into Urban Design and Planning*. Washington, DC: Island Press.

Bruinsma, J. (2014). The resources outlook: by how much do land, water and crop yields need to increase by 2050? In Looking ahead in world food and agriculture. FAO.

Burton, C. (2009). *Building resilient communities: Risk management and response to natural disasters through social funds and community-driven development operations. Research Report*. Washington, DC: The World Bank.

Canning, P., Charles, A., Huang, S., Polenske, K., & Waters, A. (2010). *Energy use in the US food system. United States Department of Agriculture*. Economic Research Service.

Compost Cab. (2017). *Community composting made easy*. Retrieved from https://compostcab.com./

District of Columbia Department of Health. (2013). *District of Columbia community health needs assessment*. Retrieved from http://doh.dc.gov/sites/default/files/dc/sites/doh/page_content/attachments/2nd%20Draft%20CH NA%20%28v4%202%29%2006%2004%202013%20-%20Vol%201.pdf

Dixon, J., Omwega, A. M., Friel, S., Burns, C., Donati, K., & Carlisle, R. (2007). The health equity dimensions of urban food systems. *Journal of Urban Health*, *84*(1), 118–129. doi:10.100711524-007-9176-4 PMID:17401697

Goodman, E. R. (2011). *Aquaponics: community and economic development* (Doctoral Dissertation). Massachusetts Institute of Technology. Retrieved from http://hdl.handle.net/1721.1/67227

Gorgolewski, M., Komisar, J., & Nasr, J. (2011). *Carrot City: Designing for urban agriculture*. Academic Press.

Halweil, B. (2002). *Home grown: The case for local food in a global market* (Vol. 163). Worldwatch Institute.

Horowitz, C. (2012). *Looking up: How green roofs and cool roofs can reduce energy use, address climate change, and protect water resources in southern California*. Academic Press.

James, P., Magee, L., Scerri, A., & Steger, M. (2015). *Urban sustainability in theory and practice: circles of sustainability*. Routledge.

Kakovitch Industries. (2016). *The Flo-Vex Company*. Retrieved from http://www.flo-vex.com

Klavinski, R. (2013). *The benefits of locally grown food*. Michigan State University Cooperative Extension Service. Retrieved from http://msue.anr.msu.edu/news/7_benefits_of_eating_local_foods

Le Corbusier, C. (1959). *L'urbanisme des trois établissements humains*. Paris: Editions de Minuit.

Louv, R. (2005). *Last child in the woods: Saving our children from nature-deficit disorder*. New York, NY: Workman Publishing.

Lovell, S. T. (2010). Multifunctional urban agriculture for sustainable land use planning in the United States. *Sustainability*, *2*(8), 2499–2522. doi:10.3390u2082499

Meerow, S., Newell, J. P., & Stults, M. (2016). Defining urban resilience: A review. *Landscape and Urban Planning*, *147*, 38–49. doi:10.1016/j.landurbplan.2015.11.011

Niemczynowicz, J. (1999). Urban hydrology and water management–present and future challenges. *Urban Water*, *1*(1), 1–14. doi:10.1016/S1462-0758(99)00009-6

Nord, M., Coleman-Jensen, A., Andrews, M., & Carlson, S. (2010). *Household food security in the United States, 2009. ERR-108*. U.S. Dept. of Agriculture, Econ. Res. Serv.

O'Hara, S. (1995). Sustainability: Social and ecological dimensions. *Review of Social Economy*, *53*(4), 529–551. doi:10.1080/00346769500000017

O'Hara, S. (1997). Toward a sustaining production theory. *Ecological Economics*, *20*(2), 141–154. doi:10.1016/S0921-8009(96)00024-9

O'Hara, S. (2001). Urban development revisited: The role of neighborhood needs and local participation in urban revitalization. *Review of Social Economy*, *59*(1), 23–43. doi:10.1080/00346760110036265

O'Hara, S. (2014). Everything needs care: Toward a context-based economy. In M. Bjørnholt & A. McKay (Eds.), *Counting on Marilyn Waring: New Advances in Feminist Economics* (pp. 37–55). Bradford, Canada: Demeter Press.

O'Hara, S. (2015). Food security: The urban food hub solution. *Solutions, 6*(1), 42-52. Retrieved from https://www.thesolutionsjournal.com/article/food-security-the-urban-food-hubs-solution/

O'Hara, S. (2017, Winter). The urban food hubs solution: Building capacity in urban communities. *Metropolitan Universities*, *28*(1). doi:10.18060/21477

O'Hara, S. (under review). *The urban aquaponics Tool Kit.* United States Department of Agriculture. National Institute of Food and Agriculture.

O'Hara, S. U., & Stagl, S. (2002). Endogenous preferences and sustainable development. *Journal of Socio-Economics*, *31*(5), 511–527. doi:10.1016/S1053-5357(02)00134-8

Oldways. (2015). *Inspiring Good Healthy Through Cultural Food Traditions*. Boston, MA: Author.

Rockefeller Foundation. (2017). *100 Resilient Cities*. Retrieved from http://www.100resilientcities.org/#/-_/

Royte, E. (2015, April 27). Urban Agriculture is booming but what does it really mean: The benefits of city-based agriculture go far beyond nutrition. *ENSIA Magazine*.

The Guardian. (2015). *Earth has lost a third of arable land in past 40 years, scientists say. Soil*. Retrieved from https://www.theguardian.com/environment/2015/dec/02/arable-land-soil-food-security-shortage

United Nations Development Program. (1996). *Urban Agriculture - Food, Jobs and Sustainable Cities*. New York: United Nations Development Programme Publication Series for Habitat II, Volume One.

United States Bureau of the Census. (2014, July 10). *Population*. Retrieved fromhttp://www.census.gov/topics/population.html

United States Department of Agriculture, Agricultural Research Service. (2014). *USDA Food and Nutrient Database for Dietary Studies 2011-2012*. Food Surveys Research Group Home Page. Retrieved from http://www.ars.usda.gov/ba/bhnrc/fsrg

University of California Davis. (2015). *Social, Health, and Economic Impacts of Local Food*. Retrieved from http://asi.ucdavis.edu/programs/sarep/publications/food-and-society/ualitreview-

Up-Top-Acres. (2017). Retrieved from http://uptopacres.com/about-us.html

Weber, C. L., & Matthews, H. S. (2008). Food-miles and the relative climate impacts of food choices in the United States. *Environmental Science & Technology*, *42*(10), 3508–3513. doi:10.1021/es702969f PMID:18546681

World Bank. (2016). The World's *Cities (London, England)* 2016. Author.

ENDNOTES

1 Enrollment indicates the number of participants in certificate programs, workshop, symposium etc.; figures may include individuals enrolled in multiple programs.

2 SNAP-Ed is a program of the Center for Nutrition, Diet, and Health.

This research was previously published in Changing Urban Landscapes Through Public Higher Education edited by Anika Spratley Burtin, Jeffery S. Fleming, and Pamela Hampton-Garland, pages 116-143, copyright year 2018 by Information Science Reference (an imprint of IGI Global).

Chapter 27
Geospatial Evaluation for Urban Agriculture Land Inventory:
Roanoke, Virginia USA

Tammy E. Parece
Colorado Mesa University, USA

James B. Campbell
Virginia Tech, USA

ABSTRACT

Urban agriculture is recently being recognized as a distinctive urban land use contributing to greenspaces and food security. The land inventory forms the critical first step in identifying sites for urban agriculture. The authors' analysis greatly expands on prior land inventory strategies, first analyzing land cover to identify all open areas available for siting urban agriculture. Then in GIS, the authors completed a land use suitability analysis, and finally a demographic analysis to assess potential sites for contribution to food security of lower income populations. Results show that Roanoke includes 2,312 hectares suitable for schoolyard gardens, urban farms, community gardens, orchards, and home gardens, of which 189.4 hectares are found in neighborhoods with extremely high rates of poverty. The authors' inventory strategy can be implemented elsewhere without special data or software. A detailed inventory offers opportunities for long-range planning, and broadening participation of stakeholders.

INTRODUCTION

Recently, research on urban agriculture has expanded exponentially because of widespread food insecurity and expanding urban areas across the world. Although urban expansion is attributed to increasing populations, rates of conversion to urban land uses exceed rates of urban population increases (Lincoln Land Institute, 2015). Expanding urbanization causes loss of vegetated lands, expansion of impervious surface cover, increasing demands on existing infrastructures, warmer air temperatures as compared to adjacent rural areas (the urban heat island effect), and increasing demands on municipal services such as waste management and provision of potable water (Pickett et al., 2001). Urban areas also must

DOI: 10.4018/978-1-5225-8063-8.ch027

reach well beyond their political boundaries to import food, energy, and clean water to meet basic needs of their populations and, as such, adverse effects of urbanization reach into regions outside the urban setting (Aitkenhead-Peterson, Steele, & Volder, 2010; Deelstra & Girardet, 2000; Pickett et al., 2001; McGranahan & Satterthwaite, 2003).

Urban agriculture, through its varied manifestations, can offer a multi-dimensional solution to these adverse effects. It improves access to healthy and nutritious fresh food, revitalizes inner cities, promotes social interactions, improves urban hydrology, and mitigates adverse climatic effects. The first step in establishing new urban agriculture sites requires identifying open, available, and suitable locations - a *land inventory* (Dongus & Drescher, 2006; Dubbeling & Merzthal, 2006; Dubbeling, van Veenhuizen, & de Zeeuw, 2010).

Although urban agriculture studies include land inventories to identify potential urban agriculture sites, most inventories are limited in scope, often emphasizing identification of vacant lands. In addition, most researchers conduct them to assist local municipalities wishing to incorporate urban agriculture into their planning and/or sustainability initiatives, so tailor their methods to that specific locality. Although some incorporate other stakeholders (i.e. local non-profits or local urban farmers) into their inventory process by asking them what characteristics require inclusion, these results, again, often focus upon specific local characteristics. Although a local perspective is valuable in recognizing unique dimensions of natural, social, and built landscapes, use of geospatial analysis allows us to develop a more generally applicable methodology. Such a framework can be relevant for any urban area, and can include a wider suite of data that nonetheless permits incorporation of those specific parameters necessary when examining an individual locale.

We expand on prior land inventory analyses by incorporating, first - land cover, second - land use, third – urban agriculture form, and finally – demographics, into our evaluation. In our literature review, we introduce and define urban agriculture in its several forms. Next, we identify existing land inventories and parameters researchers included in their geospatial analysis. We then describe our study site and highlight why it forms an appropriate locale for developing our inventory strategy. We step the reader through our analysis, and report results. These results identify locations within the city for siting different forms of urban agriculture (and, at different scales). Our results also include an evaluation of the extent these potential sites are proximally located to assist lower income residents with food security. Our conclusions discuss the significance of our procedures, and their relevance for other urban areas.

URBAN AGRICULTURE

Urban agriculture is the production of agricultural products (for food, fuel, and other uses) and rearing of livestock in urban and peri-urban areas (Mougeot, 2000). It is not a new phenomenon; people began producing agricultural products within urban areas when first establishing them thousands of years ago (van Leeuwen, Nijkamp, & de Noronha Vaz, 2010). In the United States, its history exceeds 100 years (Lawson, 2005), intensifying during periods of national crisis, such as both World Wars and the Great Depression (Deelstra & Girardet, 2000; Iaquinta & Drescher, 2010).

Urban agriculture is now increasing in a context driven by parallel trends of expanding urbanization and increasing food insecurity. Worldwide, one in nine people suffers from chronic malnutrition due to food insecurity (FAO, 2014). In the United States, more than one in ten households suffers from food insecurity (Brown et al., 2002; Coleman-Jensen, Nord, Andrews & Carlson, 2011), and this rate increases to almost one in five (19.5%) in households with children (USDA, 2015). Since, the majority of people now live in urban versus rural areas (in 2009, the percentage of people living in urban areas exceeded 50%, and estimates place this percentage at 66% by 2050 (United Nations, 2014)), many of the food insecure live in urban areas.

Although income factors, significantly, into food choices (Hofferth & Curtin, 2005; Rose, 1999), disparity exists over access to healthy and nutritious foods. This disparity manifests itself spatially as access appears related to differing socio-economic status and types of neighborhoods, i.e. segregated by race or ethnicity (Moore & Diez Roux, 2006; Morland, Diez Roux, & Wing, 2006; Seaman, Jones, & Ellaway, 2010). A lack of supermarkets and commercial food establishments offering healthy choices complicates accessibility to healthy and nutritious foods (Bader, Purciel, Yousefzadeh, & Neckerman, 2010; Eisenhauer, 2001; Galvez et al., 2008; Moore & Diez Roux, 2006; Morland et al., 2006; Rose & Richards, 2004). Urban agriculture is well-documented as contributing to alleviation of food insecurity in lower income urban populations (Eisenhauer, 2001; Mougeot, 2000; RUAF, 2010; Smit, Nasr, & Ratta, 2001).

In addition, urban agriculture forms a greenspace, and as a greenspace, it provides a wide range of benefits. For example, urban agriculture makes productive use of vacant lands (Balmer et al., 2005; Grewal & Grewal, 2012; McClintock, Cooper, & Khandeshi, 2013; New York State Energy Research and Development Authority (NYSERDA), 2013), or idle public lands (Smit & Nasr, 1992); increases economic stability for gardeners, from selling agricultural products or from releasing income which can be used elsewhere (Bellows, Brown, & Smit, 2003; Smit et al., 2001); reduces stormwater runoff, increases groundwater recharge, and improves water quality (Aitkenhead-Peterson, Steele, & Volder, 2010; Beer, 2010; Botkin & Beveridge, 1997; van Leeuwen at al., 2010); reduces urban heat island effects (Beer, 2010; van Leeuwen et al., 2010); generates health benefits from increased physical exercise and stress reduction (Barton, 2009; Botkin & Beveridge, 1997; Kuo, 2010); increases social interactions between urban residents (Beer, 2010; Kuo, 2010; van Leeuwen et al., 2010); promotes environmental knowledge of urban residents (Fuller, Irvine, Devine-Wright, Warren, & Gaston, 2007; Lehmann, 2012; Seaman et al., 2010; Viljoen, Bohn, & Howe, 2005); provides residents with a sense of place (Mok et al., 2014; White, 2014); and generates ecosystem services (Bolund & Hunhammar, 1999; van Leeuwen et al., 2010). The extent that urban agriculture produces these benefits depends on its scale and form.

Urban Agriculture Forms

Urban agriculture ranges in size from micro-gardening (i.e., containers on balconies and patios), meso-scale (i.e., shared garden plots) and macro-scale (i.e., urban farms) (Pearson, Pearson, & Pearson, 2010). Each of these scales possesses specific characteristics, but not exclusively and overlaps exist. For instance, urban farms normally operate for profit, but people gardening in containers on patios and balconies can sell their products and some urban farms exist as part of non-profit food banks. While differing scales are significant as they relate to production capability, start-up costs, and man-power, when conducting a land inventory for sites appropriate for urban agriculture, discussing form is important.

Home gardens are the most common form (Smit et al., 2001). Home gardens are usually referred to as backyard gardens, but they include side yards and front yards, and consist of micro-gardening in containers on windowsills, patios and balconies (Pearson et al., 2010). A home garden's major characteristic is an individual or family growing food for household consumption in an area adjacent to their residence (Drescher, Holmer, & Iaquinta, 2006).

Community gardens exist as a very popular form of urban agriculture in the Americas (Lawson, 2005), are found in some European cities (Hardman & Larkman, 2014), and are expanding in Asia and Africa (Parece & Campbell, *forthcoming*). A community garden consists of an area of land divided into plots shared by members of a community (Lawson, 2005). Community members share some type of relationship – e.g. church, neighborhood, school, or employer. Each member works their own plot but contributes to maintenance and care of common areas. While the land upon which the garden is located can be owned by a community garden association, normally, another entity (governments, churches, non-profit organizations) owns the land and grants gardening privileges, sometimes for only a limited period of time (Drescher et al., 2006; Iaquinta & Drescher, 2010; Smit & Nasr, 1992).

Informal gardening is closely related to community gardens although opportunistic in nature (in some locations, deemed *guerilla gardens*). Local residents engaged in informal gardening find a vacant lot, clean it up, cultivate it, revitalize the area with agriculture – all accomplished without permission of the land owner (Hardman & Larkman, 2014). This form of urban agriculture is prevalent in the rapidly expanding urban areas of the Global South. In many cases, the property owner or municipal authorities, later, reclaim the land (Hardman & Larkman, 2014). Figure 1 shows such an example for the City of Roanoke, Virginia USA.

Allotment gardens are similar to community gardens, practiced extensively in Europe (Andersson, Barthel, & Ahrné, 2007; Hardman & Larkman, 2014; Petts, 2001) and Japan (Matsuo, 2000). Allotment gardens, also a community project, differ from community gardens because the local government usually sponsors and manages the garden. Initially started as food production sites for the lower income working class, the demographics for those now using allotment gardens varies much more broadly (Petts, 2001).

School gardens are located on public and private school properties. The food produced usually supports school lunch programs, or supplements students' home food supplies. School gardens are also used for student education. In rare instances, community gardens are located on school properties (Moore, Wilson, Kelly-Richards, & Marston, 2015; Pudup, 2008; Smit et al., 2001).

Urban farms account for the greatest areal extent of urban agriculture (Smit et al., 2001). Urban farms include greenhouses, rooftop gardens, and community-supported agriculture, and in most instances, operate for-profit by families or commercial enterprises. Because of start-up and maintenance costs of green roofs and greenhouses, operators produce and sell higher-priced or specialized products to restaurants and florists. In the practice of community-supported agriculture, members purchase a share of produce grown by a farmer, rarely participating in the growing process (Doron, 2005; Smit et al., 2001).

Figure 1. Left: 2011 Virginia Base Mapping Program aerial photo shows a grassy area. Center: 2012 National Agriculture Imagery Program (NAIP) aerial photo shows a garden in this lot; Right: 2015 street-level photo shows the area reclaimed for grass (Photo – second author 2015)

Orchards, consisting of fruit or nut-bearing trees, are also a form of urban agriculture; in the Global South, orchards provide sources of fuelwood and building materials (Smit & Nasr, 1992; Smit et al., 2001). Orchards are rarely discussed within the literature as a form of urban agriculture. Nor are fruit- or nut-bearing trees discussed within the literature specific to expanding urban forests.

LAND INVENTORY ANALYSES

Incorporating urban agriculture into a specific urban area begins with a *land inventory* - identifying those areas within a city available for agricultural production (Dongus & Drescher, 2006; Dubbeling & Merzthal, 2006; Dubbeling, Veenhuizen, & Zeeuw, 2010). Such efforts have begun in many areas of the United States and Canada (Table 1). Most *land inventories* are accomplished specifically to identify sites for urban agriculture, but some inventories are accomplished as part of studies to quantify urban agriculture production capability within a specific locality (indicated in Table 1 by an *). All studies use geospatial technologies for their analysis, most using just GIS, but some also incorporating remote sensing (e.g. Colasanti, 2010; McClintock et al., 2013; Nipen, 2009; NYSERDA, 2013).

Table 1. Land inventory analyses for potential urban agriculture

Author (date)	Location	Variables included in analysis
Balmer et al. (2005)	Portland, Oregon	Land tenure, water access, slope, accessibility, proximity to existing agriculture
Kaethler (2006)	Vancouver, British Columbia	Land tenure, under-utilized lands, tree canopy cover, sunlight, impervious surfaces, soils, proximity to existing agriculture, accessibility, opportunity for community buildings, size (only those areas over 1000 square meters)
MacKenzie & Cohen (2007)	Portland, Oregon	Further evaluation of sites identified by Balmer et al., (2005). Excluded protected natural resource sites, areas of open spaces for recreation, locations of springs and steep slopes. Included portions of parks for community gardens.
Horst (2008)	Seattle, Washington	Lands classified as vacant, unused, or excess right of way, utility right of ways, public schools and parks, evaluated only those suitable for community gardens (> 2,000 feet^2)
Nipen (2009)	Halifax, Nova Scotia	Eliminated impervious surfaces and areas shaded from buildings, through use of lidar data. Identified those areas without structures.
Taggart, Chaney, & Meaney (2009)	Cleveland, Ohio	Vacant lands, proximity to existing agriculture, access to water and transportation, prime farmland soils, and eliminated forested and riparian areas.
Colasanti (2010)*	Detroit, Michigan	Vacant lands identified from city datasets and cross referenced against 2005 aerial photos. Excluded lands owned by City of Detroit Recreation Department.
MacRae et al., (2010)*	Toronto, Canada	Zoning regulations, then overlaid on aerial photos for size, shape, site coverage, accessibility, proximately to water courses, to roads, use of park space, and rooftops.
Grewal & Grewal (2012)*	Cleveland, Ohio	Vacant lots, commercial and industrial rooftops and residential areas. Calculated a portion of each of these identified areas in their production capability study.
Meenar & Hoover (2012)	Philadelphia, Pennsylvania	Number of vacant lands identified within each Census Tract.
Patel &MacRae (2012)	Toronto, Canada	Follow up to MacRae, et al., (2010), identified locations greater than 1 acre, large enough for Community-Supported Agriculture.
McClintock, et al. (2013)*	Oakland, California	Identification of vacant and under-utilized locations, included access to water, slopes, aspect, proximity to schools and transportation, and used aerial photos to visually inspect and eliminate areas that appeared in use or with dense vegetation.
NYSERDA (2013)*	New York, New York	Comprehensive evaluation – see below for more details.

One of the most notable inventories is the extremely complex New York State Energy Research and Development (NYSERDA) (2013) study. A production capability report, like some other studies, it presents the most comprehensive land inventory completed, including ground, rooftop, and indoor capabilities. Researchers, for this report, identified vacant public and private lands, roof tops, under-utilized open spaces, surface parking areas, open areas in public housing developments, and existing community gardens. They eliminated wetlands, heavily forested areas, areas adjacent to landfills, and vacant lands that appeared to be in "active use" via overlays on aerial photos and satellite imagery. They also noted other possible available locales - traffic islands and median strips for fruit trees, and yard space as home gardens, but did not quantify these areas. They identified areas of possible soil contamination, but did not eliminate these as potential sites. They discussed many other factors, e.g., slopes, hydrology, energy consumption or reduction, but these were not factors affecting their land inventory.

These studies are much more limited in scope than our study because they conduct a land use analysis, whereas we start with land cover and then proceed to a land use evaluation. Land cover represents the surface features of a parcel of land such as forest, water, or impervious surfaces. In contrast, land use refers to the economic utility that society makes of the same land area, for instance the land use of a forest cover might be for the production of fuel or for recreational purposes (Parece & Campbell, 2015). Our approach screens to eliminate parcels with land cover unsuitable for the practice of urban agriculture as the first step, then we consider land use.

We acknowledge and evaluate current and past land uses, (e. g., commercial locations in active use cannot provide sites for urban farms nor home gardens, but can be actively cultivated as community gardens for business owners and employees). We also include orchards and schoolyard gardens in our urban agriculture form evaluation. Through this inventory strategy, we deliver a comprehensive, multifaceted, perspective on the resources potentially available to a municipality for urban agriculture. In addition, we also identify potential sites in lower income areas to assist in alleviating food insecurity – schools in proximity to urban farms, orchards and community gardens, and locales in high poverty rate census block groups -- a factor that only one study considered (e.g. Meenar & Hoover, 2012).

STUDY SITE

Important qualities highlight the City of Roanoke, Virginia USA as a setting to illuminate and investigate potential sites for urban agriculture. Roanoke is the largest metropolitan region in southwestern Virginia (Figure 2). It is characterized by a variety of urban land uses, with a history as a transportation hub for rail and road traffic, and services and industries supporting the rail system, as well as finance, distribution, trade, manufacturing, and healthcare industries. The City of Roanoke's land cover includes 47.9% tree canopy (Pugh, 2010), and 31.9% impervious surfaces (as calculated by first author).

Roanoke's demographics favor examination of relationships between urban agriculture and its potential to address food insecurity; its poverty rate, K-12 student qualification for a free or reduced price school lunch, and food insecurity rates are all higher than the surrounding county, and both state and national percentages (Table 2). The USDA identifies most census tracts within the city as having limited access to food, i.e. a food desert (USDA, 2014). Furthermore, a study completed by the University of Wisconsin ranked the City of Roanoke 115th out of 133 counties and cities in Virginia for health factors such as obesity, diabetes, and physical inactivity (2014).

Figure 2. Reference Map, City of Roanoke, Virginia USA (Sources – Virginia Geographic Information Network, 2004; TigerLine® Shapefiles, 2014)

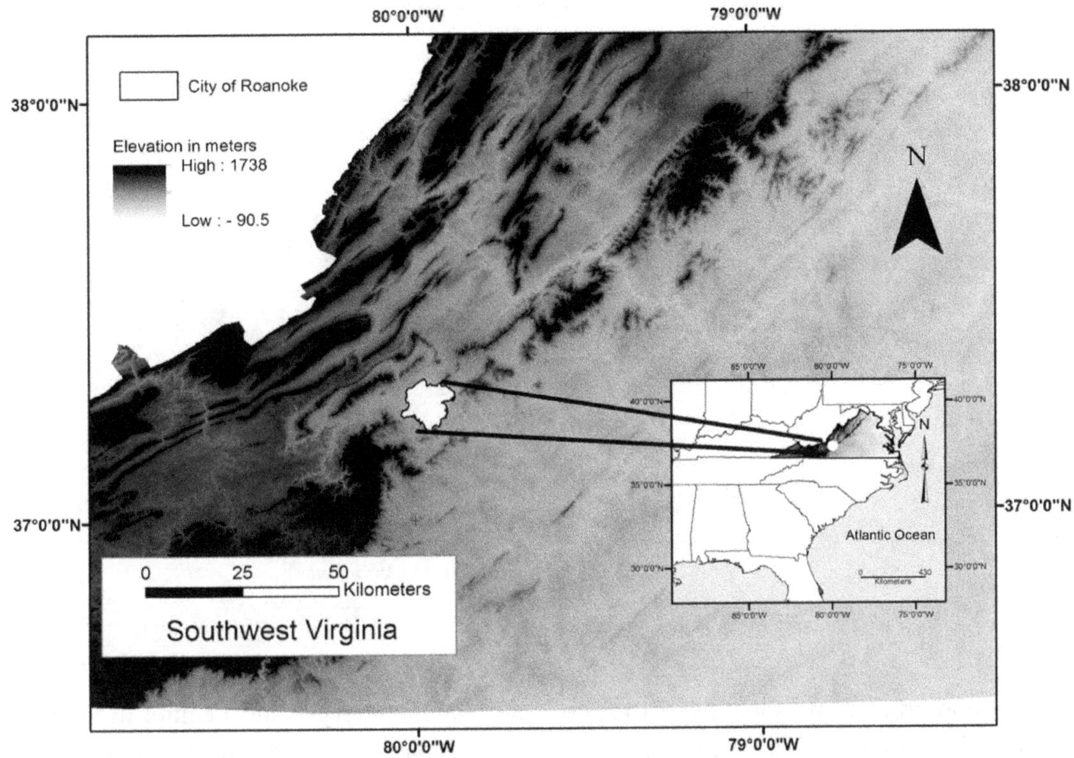

Table 2. Comparison of City of Roanoke to Roanoke County, Virginia, and United States

	Roanoke	Roanoke County	VA	United States
Households below Federal Poverty Level (U.S. Census, 2012)	20.9%	5.1%	10.3%	13.8%
Food Insecurity Rates (Feeding America, 2014)	16.9%	8.3%	12.1%	14.3% (USDA, 2015)
National School Lunch Program qualification (Virginia, 2015b)	74. 8%	26.5%	42.0%	42.9% (USDA, 2013)

Urban Agriculture in Roanoke, Virginia

Roanoke's urban agriculture scene embraces many forms – community gardens, urban farms, and includes two very active local-food organizations - Roanoke Community Garden Association and Roanoke Natural Foods Co-op. However, as Figure 3 demonstrates, these two urban agricultural forms are not well-distributed across the city.

The Roanoke Natural Foods Co-op operates one urban farm at Heritage Point in northeast Roanoke. The land, purchased by the Co-op in 2012, is approximately 7.2 hectares and located near an industrial park. In addition, the Co-op has leased another 1.4 hectares from the city, adjacent to the farm (Ress,

Figure 3. Sites of two urban farms and several community gardens within the City of Roanoke (Please note the size of the polygon does not represent the size of the garden or farm -- symbology has been enhanced)

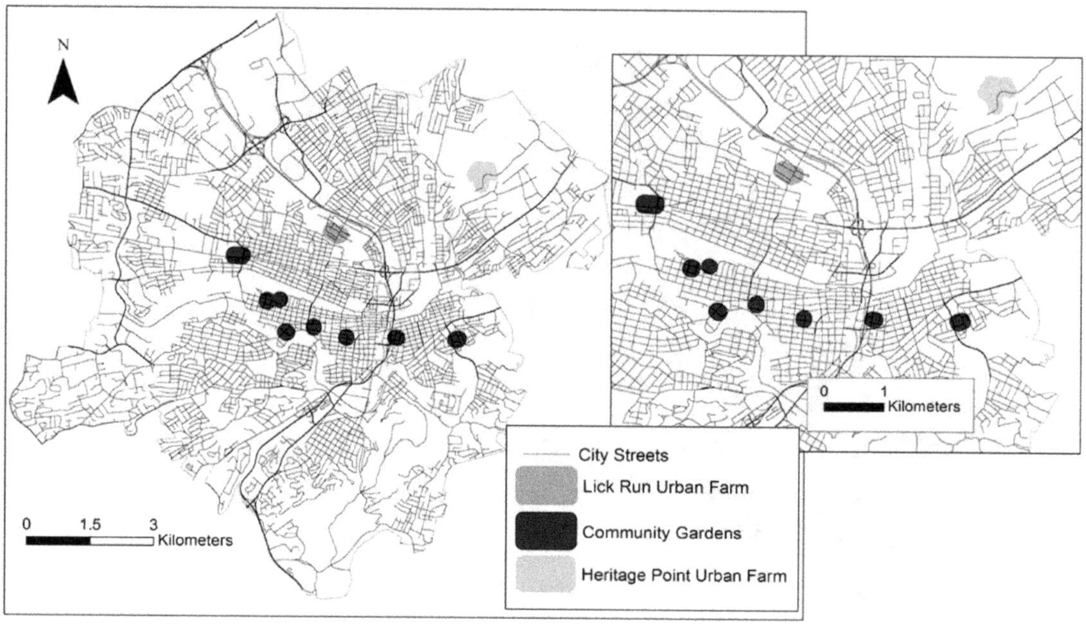

2013). The second urban farm is located at the site of a defunct nursery (almost center in Figure 3), and located within a residential neighborhood. This land was purchased by a private citizen in 2010 with the intention of starting an urban farm and farmer's market. This site was used as a dumping ground for trash and tires after the nursery closed in 2002. Since its purchase in 2010, the current owner has cleaned up the property and renovated the site's building (Nair, 2012).

The Roanoke Community Garden Association (RCGA), established in 2008, operates multiple locations across the city (Figure 3). RCGA owns some gardens, leases one from the city, and operates on land provided by local property owners - with one on land owned by Goodwill Industries of the Valleys. RCGA requires that members garden organically, and RCGA harvests rainwater at most of its locations. The two most recent gardens established include Mountain View Community Garden (2013), and Growing Goodwill Garden (2014). Growing Goodwill Garden, currently under development, embraces planting an orchard. Immigrants or refugees comprise 30% of RCGA gardeners (Powell, 2013). RCGA plans additional sites across the city (Powell, personal communication, 2015).

METHODS

This section describes our strategy for extracting those areas that are unsuitable or impractical for agricultural uses to produce our final inventory of land. Our goal is to distill, from existing information about Roanoke's landscape, those sites likely to form suitable opportunities for urban agriculture. In brief, our strategy systematically identifies, from varied sources outlined below, areas unsuitable for use as urban agriculture because of existing land cover and land uses, e.g., contamination, impervious surfaces, and other reasons, then eliminates these locations from consideration. The remaining areas identify prospective sites for the practice of urban agriculture.

Our data came from several sources (Table 3). Each source provided data downloading capability, in most cases in shapefile format, then we analyzed the data using Esri's ArcMap™ GIS software (Esri, 2015). The spatial extent of some files exceeded the boundaries of the City of Roanoke (those designated with an ••), so we clipped this data using the city's boundary shapefile. The US EPA and the National School Lunch Program provided data in spreadsheet files, which we spatially joined to specific shapefiles (further details are outlined below). In addition, the authors created one new dataset, discussed in more detail below.

Land Cover Analysis

The first step in a land inventory analysis is to identify the different types of land cover - impervious surfaces, tree canopy cover, water, and open areas (land not covered by the other three).

Impervious surfaces are a non-permeable land cover, usually man-made; examples include asphalt and concrete roads, walkways, parking lots, and buildings. A variety of remote sensing methods have been used to identify impervious surfaces (Parece & Campbell, 2013). For this study, we manually delineated impervious surfaces, starting with the city's shapefiles for buildings and parcels (parcels provide a basic outline for roads). We overlaid the shapefiles on the 2011 aerial photos via the Virginia Geographic Information Network's GIS server. In some instances, because of intense shadowing, either from buildings, trees, or the camera's field of view, we used 2008 aerial photos or Google Earth™ to clarify the land cover. Once completed, we employed this new shapefile to erase all impervious surfaces from the city's boundary shapefile, resulting in a new a shapefile representing all of the city's pervious surface cover.

We converted the tree canopy cover raster file (one-m^2 resolution) to a vector polygon shapefile and then erased the canopy cover from the pervious surface cover shapefile. With the city's stream shapefile, we added polygons for other areas of open water – ponds, and stormwater retention and detention areas. We likewise erased the water bodies from the pervious surface cover shapefile. The resulting shapefile contained one land cover - open areas, utilized in our land use analysis.

Table 3. Data and sources

Data	Source
Vector shapefiles for buildings, parcels, parks, greenways, streams, flood plains, city boundary, airport	City of Roanoke FTP site (2015)
Impervious Surfaces	Created by first author
Tree Canopy Cover, raster file••	Virginia Department of Forestry (2015d)
Aerial Photos, 2011, 15 cm resolution	Virginia Geospatial Information Network (VGIN) (2015e)
Vector shapefiles for Conservation Lands••	Virginia Department of Conservation and Recreation (2015a)
Brownfields, Superfund, RCRA, and other potentially polluted sites••	US EPA (2015a, 2015b, 2015c) and Virginia Department of Environmental Quality (2015c)
Poverty Rates	U.S. Census Bureau (2012)
National School Lunch Program	Virginia Department of Education (2015b)

Land Use Analysis

We identified land uses, not available as potential sites for urban agriculture greenspaces, from several files. From the city's shapefiles, we identified and eliminated locations for the Roanoke-Blacksburg Regional Airport, quarries, the wastewater treatment facility, and all railroad yards. The conservation land shapefiles included The Nature Conservancy lands, National Park Service (NPS) land, easements, National and State Forests, U.S. Fish and Wildlife areas, Department of Defense lands, and scenic rivers; the city only contained NPS and easements lands. We eliminated NPS lands, but not easements.

The city contains sixty-two parks within its boundaries, ranging in size from 0.1 to 247.4 hectares. We evaluated each park to determine if it was in active use either with sports fields, playground equipment, or walking trails. We looked closely at the larger city parks, examining the parks for open areas accessible for community garden sites.

We identified four golf course locations from the city parcel's file. Two of the golf courses are actively operating so we eliminated any open areas within them. The city designated the other two golf courses as vacant, and thus open areas within these two locations remained as potential urban agriculture sites.

Schoolyards are viable locations for urban agriculture, however, recognizing that school children need outside space for recreation, we decided not to include all of these open areas as potential urban agriculture sites. The open areas ranged from schoolyard to schoolyard, from 731 m^2 to 14.9 hectares. As such, for those schoolyards with more than 7.7 hectares of open area (n = 3), we designated 0.8 hectares (2 acres) as potential areas for urban agriculture. For those schoolyards of between 1.3 and 7.7 hectares of open areas (n = 14), we designated 0.4 hectares (1 acre) of potential land for urban agriculture. For the rest of the schoolyards (n = 12), we calculate 50% of the open area as available for urban agriculture. We exported the open schoolyard areas from the potential urban agriculture shapefile into its own shapefile named *schoolyard gardens*.

Other Land Use Considerations

We also evaluated prior land uses and potential soil contamination, most specifically, pollution sites noted by regulatory authorities (the US EPA and the Virginia Department of Environmental Quality). The data from the US EPA included the Toxic Relief Inventory (TRI), Superfund sites, and Resource Conservation and Recovery Act (RCRA) sites. Files from the Virginia Department of Environmental Quality contained shapefiles for solid waste facilities, petroleum release sites, and Virginia locations that qualify under the National Pollutant Discharge Elimination System Program (NPDES). Explanation for review of this data is provided as follows.

US EPA's Toxic Relief Inventory (TRI) "tracks the management of over 650 toxic chemicals that pose a threat to human health and the environment" (US EPA, 2015c). TRI data is available as a downloadable spreadsheet. US EPA Superfund identifies sites with uncontrolled hazardous waste; the US EPA evaluates the sites, identifies clean-up requirements and can place prohibitions and site restrictions before and after completion of clean-up activities (US EPA, 2015b). RCRA gives the federal government authority for involvement in regulation of hazardous solid waste disposal sites, including provisions for site evaluation and placement of use regulations if deemed hazardous to human health (US EPA, 2015a).

The Virginia Department of Environmental Quality (DEQ) tracks releases of petroleum products, the locations, and clean-up status. Virginia DEQ also documents locations for all active Virginia locations that qualify under the National Pollutant Discharge Elimination System Program (VPDES). The Clean Water Act (Section 402) established the National Pollutant Discharge Elimination System Program, which limits pollutant discharges into streams, rivers, and bays for industrial activities, municipal separate storm stormwater systems (MS4s), and construction activities qualifying as point sources of stormwater discharge to surface waters (Virginia, 2012).

For each of the above data categories, we identified locations (if any) within the city, through query of the parcel's shapefile or by joining the spreadsheet file with the parcels shapefile. Then, for each location, we reviewed the documentation from the US EPA or Virginia DEQ for the type of contaminant, status of clean-up, any use restrictions or regulations placed on the site, and the status of the land use (i.e. active or inactive). For any location that was designated as hazardous to human health, we erased those open areas from our potential urban agriculture sites.

The final areas requiring our consideration were floodways and floodplains identified by Federal Emergency Management Agency (FEMA), available as a shapefile from the city. FEMA defines a floodway as a stream and any adjacent land that must be reserved to discharge a base line flood; development on this adjacent land can be restricted (2014a). FEMA defines floodplains as all land areas with the potential to be overcome by floodwaters, regardless of the source (2014b). While development of land with these designations faces limitations, they offer viable opportunities for expanding the urban forest with fruit or nut bearing trees. We used the *select by location query* for those potential urban agriculture sites contained within the FEMA flood zone locations, then exported this selection into a new shapefile named *FEMA Orchards*. We then switched the selection and exported the remaining areas to a new potential urban agriculture open areas shapefile.

Urban Agriculture Form Analysis

Since urban agriculture takes multiple forms, the next step of the evaluation considered urban agriculture form. We located specific property descriptions within the city's parcels shapefile to consider in this part of the evaluation – vacant, commercial/industrial, multi-family, residential-non-living areas, and single family. We intersected the potential urban agriculture open areas shapefile with city's parcel shapefile to determine the property type description/land use categories for each parcel's open area.

However, intersecting with the parcel's shapefile excluded open areas not within any parcels, i.e. lands within highway cloverleaves, median strips, areas between sidewalks and streets. As such, we created another shapefile (erasing the parcel areas from the potential urban agriculture open areas shapefile), and named this shapefile *non-parcel open areas*.

We evaluated each of the previous two shapefiles (*open parcel areas* and *non-parcel open areas*) for location, property type, and size in meters^2 or hectares, and assigned the following urban agriculture form categories:

- **Too Small:** Less than 1.0 m^2 – for vacant parcels, non-parcel land or occupied commercial/industrial properties;
- **Urban Farm:** Greater than 0.4 hectares for vacant parcels, non-parcel land (with exceptions noted below), or residential non-living areas; all highway cloverleaves, and areas less than 0.4 hectares adjacent to highway cloverleaves;

- **Community Garden:** Areas equal to 1000 m^2 but less than 0.4 hectares for vacant parcels or non-parcel lands in non-residential neighborhoods, any non-parcel lands greater 0.4 hectares within residential neighborhoods, all multi-family residential and non-residential living areas (regardless of size), and all commercial/industrial property greater than 1.0 m^2;
- **Home Garden:** For occupied parcels that are single family residential and all-non-parcel lands within 6 meters of a residential parcel;
- **Schoolyard Garden:** As described previously; and
- **Orchards:** FEMA floodways and flood plains, highway or major road way median strips, non-parcel grassy areas next to roadways/streets in commercial/industrial areas greater than 1.0 m^2, all vacant areas greater than 1.0 m^2 but less than 1000 m^2.

Demographic Analysis

We joined the spreadsheet reporting the rate of qualification for the National School Lunch Program (NSLP) for each school to the corresponding school's point shapefile. We added a field to this shapefile for size of potential schoolyard gardens. We evaluated these two attributes, asking the question – can those schools with the highest NSLP qualification rates install a schoolyard garden and contribute to food security for its students? In addition, we completed a spatial analysis query to determine the proximity of each school to potential agriculture sites for urban farms, community gardens, and orchards.

Finally, we overlaid the urban agriculture form shapefiles (*open parcel areas, non-parcel areas, and FEMA Orchards*) with the shapefile for the 2010 Census data for poverty rate by block group, to identify potential urban agriculture sites within those block groups with high poverty rates (and therefore, at greatest risk for food insecurity). We defined high poverty as any rates above national level.

RESULTS AND DISCUSSION

Potential Urban Agriculture Sites

After the land cover analysis, i.e. elimination of impervious surfaces, tree canopy cover, and water, Roanoke contains 2,774.4 hectares of open areas (25% of total area) that may be suitable for urban agriculture.

As previously described in the land use analysis discussion, we erased any opens areas that were contained within the airport, quarries, National Park Service land, the two golf courses, and the wastewater treatment plant. For city parks, it was determined that the majority of the open areas within these parks are in active use, either with sports fields, playground equipment, or walking trails. The two largest parks – Mill Mountain Park (247.4 hectares) and Yellow Mountain Park (118.7 hectares) are adjacent to the National Park Service and tree canopy covers all but minute areas of these parks. Thus, we eliminated open areas within all city parks.

The US EPA Toxic Relief Inventory (TRI) file contained fifteen sites within the city boundaries and the city's parcel's shapefile noted all sites as active commercial or industrial locations. We considered each location inappropriate for food production.

The US EPA listed five Superfund sites for the city of Roanoke, we reviewed each location but only three were actually within the city boundaries:

- US EPA designated one site as a threat to public health (1998);
- US EPA found a second site with soil contaminated from multiple chemicals, and of immediate threat to public health, welfare and environment (1994); and
- US EPA noted the final site with elevated lead (in excess of 1,000 ppm) in soil, and because of its proximity to the Roanoke River, also deemed it a threat to aquatic life and water quality (2000).

We deemed any open areas within these three sites as inappropriate for agriculture and, thus, erased these areas from the potential urban agriculture sites shapefile. In addition, because of the notes in the report from the US EPA for the third site (i.e. the aquatic threat from runoff), we eliminated any non-parcel areas adjacent to this site and any locations with the *FEMA Orchards* shapefile as potential urban agriculture sites.

The US EPA identified four sites under the Resource Conservation and Recovery Act (RCRA) for the City of Roanoke; while evaluating these locations, we determined one site was outside of the city boundaries. For the other three sites, one site was already eliminated as part of the TRI evaluation. The US EPA noted that a second site should remain solely as industrial/commercial use and required that the company institute a management plan (US EPA, 2010). US EPA specifically restricted the third site from any residential use or as a public garden space (2009). We eliminated any open areas within these parcels as potential urban agriculture sites.

The Virginia DEQ solids waste facility shapefile noted two facility sites within the city, but both of these are offices not actual landfills, so those open areas were not erased. The shapefile for petroleum release sites designated all but three sites -- resolved and closed; none of the open incidents intersected with any potential urban agriculture location. The VPDES shapefile contained six such locations but did not provide any additional details if a specific pollutant was in the stormwater discharge. None of these locations required elimination from the open areas – four already eliminated and the final two noted as vacant properties by the city.

After the schoolyard evaluation, open areas on school properties totals 80.6 hectares. But, referring back to our methods on schoolyards, the total potential urban agriculture area on schoolyards equals 6.9 hectares. We did not identify the specific section of open area for agriculture as this would be a decision made by local school principals and teachers.

After categorization for urban agriculture form, excluding the *Too Small* category and reduction in schoolyard open areas, we identified 2,311.6 hectares (20.8% of total area) with potential for urban agriculture. Of these total hectares, we designated 179 potential sites for urban farms (278.5 hectares), 6,247 potential sites for community gardens (691.7 hectares), 986.9 hectares of potential home gardens, and 347.6 hectares for potential orchards (Figures 4 and 5).

Site Contribution to Food Security

While the city average rate for students qualifying for the National School Lunch Program (NSLP) is 74.8%, only two city public schools fall below national average qualification, one at 18% and one at 38% (note - some schools within the city boundaries are not part of the Roanoke City Public Schools system). All other city public schools qualify at 53% or greater - 17 schools exceed 75% and eight schools qualify at 90% or greater. After evaluating the schoolyard open areas with the percentage of students within each Roanoke City Public School that qualify for the NSLP, all schools have the potential to contribute to their students' food security by implementing a schoolyard garden. Two of the schools with potential

Figure 4. Major land cover classifications, including urban agriculture. Within the urban agriculture classification, the bar chart indicates amount of hectares for each urban agriculture form

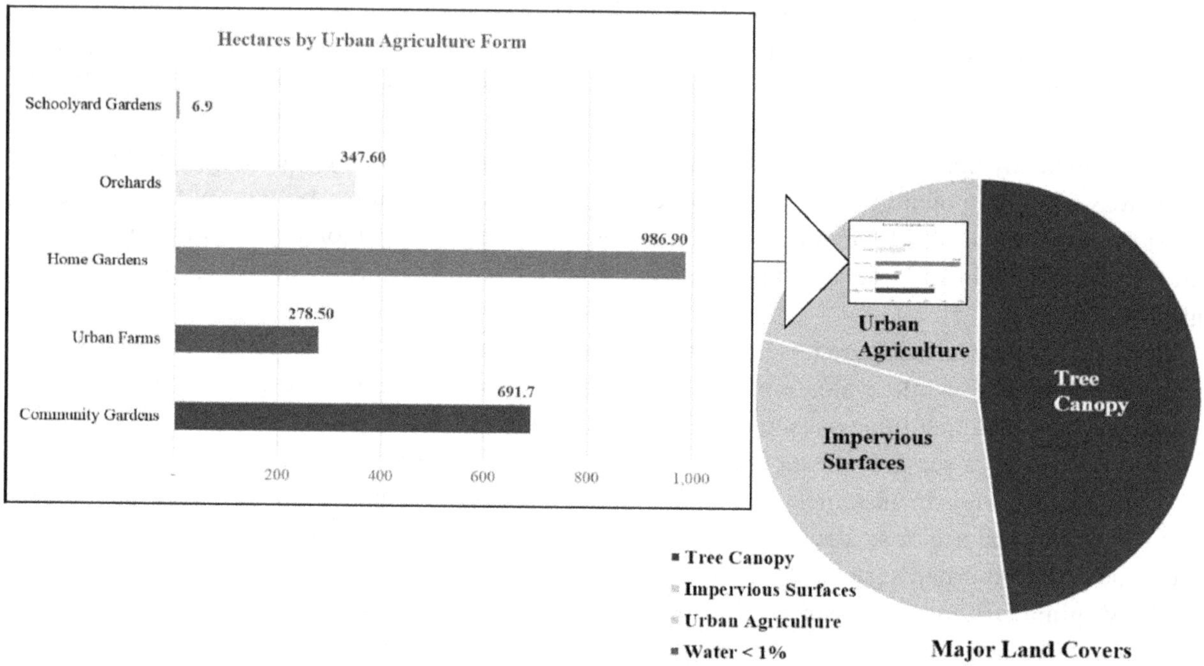

plots of 0.8 hectares, have NSLP rates of 77.5% and 90%. For those schools with potential plots of 0.4 hectares, nine have NSLP rates over 75% and all but one of these schools (a private school) have rates in excess of 53%. The eight schools, with 90% or greater rate of NSLP qualification, have enough open areas for potential urban agriculture plots from 364 m^2 to 0.8 hectares; four with potential plots of 0.4 hectares each.

All schools within the city have access to other potential urban agriculture sites. For potential community gardens sites, eighteen schools are within 100 meters, twenty-five schools are within 150 meters, and all schools but one are within 250 meters. For potential urban farm sites, three schools are within 100 meters, four schools within 150 meters, eight schools within 250 meters, thirteen schools within 500 meters, and twenty-three schools within 1 kilometer. For potential orchard sites, twenty schools are within 100 meters, twenty-six schools - 150 meters, and all schools are within 250 meters.

We identified the Household Federal Poverty Rate for each census block group (Figure 6), as a reminder the national average is 13.8%. For Roanoke, twenty block groups have high poverty rates, rates greater than 15% but less than 25%; 24,431 people live within these block groups. Ten block groups have very high poverty rates - greater than or equal to 25% but less than 35% (13,514 people). Eleven block groups have extremely high rates of poverty, 35% or more (15,178 people), and these rates range from 35.5% to 53%. As such, 55% of the city's population live within high poverty census block groups (97,032 people live in the city).

Potential urban agriculture sites exist within all block groups. For those block groups with high rates of poverty, five contain potential sites for urban farms and all twenty contain potential sites for community gardens and orchards. For those block groups with very high rates of poverty, three contain potential sites for urban farms and all contain potential sites for community gardens and orchards. For

Figure 5. Locations of potential urban farms, community gardens, backyard gardens, orchards, and schoolyard gardens within Roanoke, Virginia

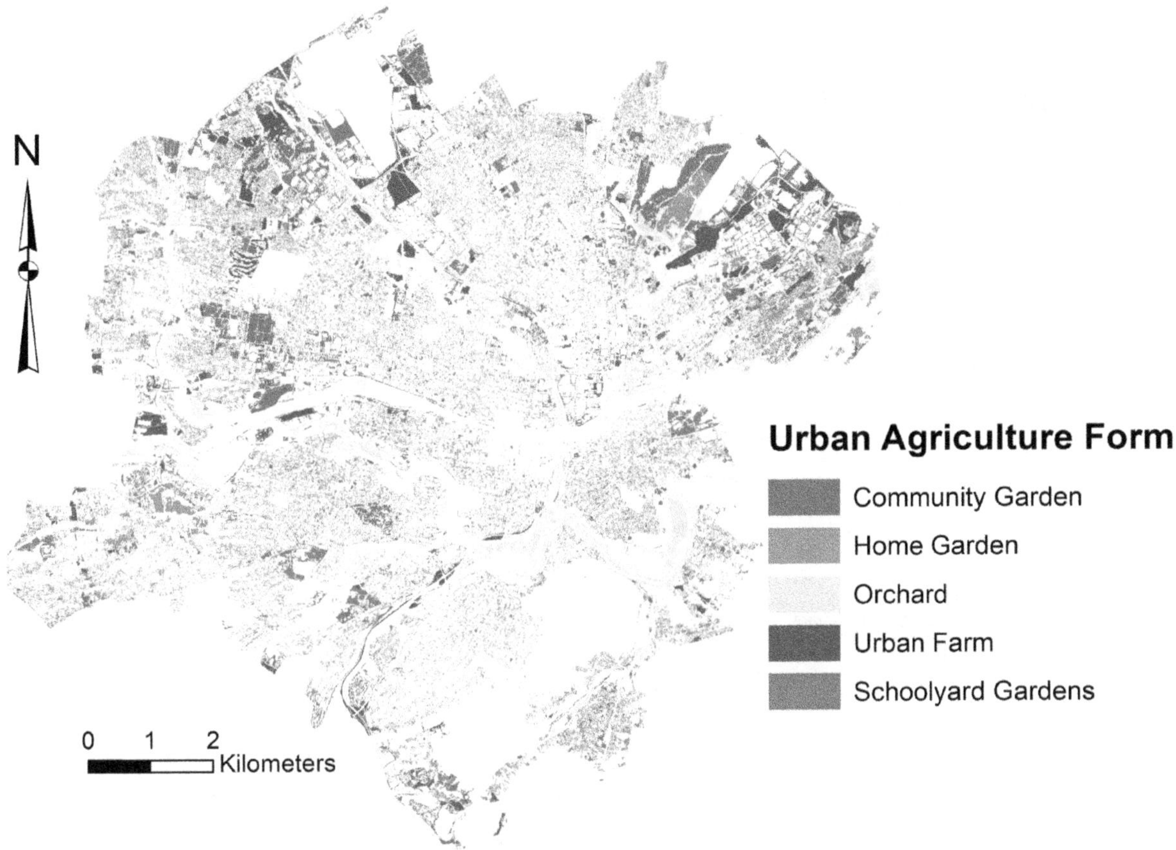

those block groups with extremely high rates of poverty (≥ 35%), all eleven contain potential locations for community gardens (108.9 hectares) and orchards (69.1 hectares), and six contain potential locations for urban farms (11.5 hectares) (Figure 7). Of note, the Roanoke Community Garden Association successfully sited six community gardens within these extremely high poverty rate block groups, and the privately owned urban farm is also located within one.

CONCLUSION

The premise for our evaluation is that we constructed our land inventory with a land cover analysis, followed by a land use assessment. Our procedure is contrary to most inventories that start with identification of vacant land uses and then evaluate these vacant lands for other characteristics – i.e. adjacency to existing agriculture, proximity to water, dense vegetation. This strategy ignores the basic fact that vacant lands are not the only locations that could support urban agriculture.

Our strategy eliminated areas that would not be available for new greenspaces without substantial alteration of surface cover – impervious surfaces, tree canopy cover and open water. We eliminated areas

Figure 6. Percent of households within each block group whose income was below the Federal Poverty Rate

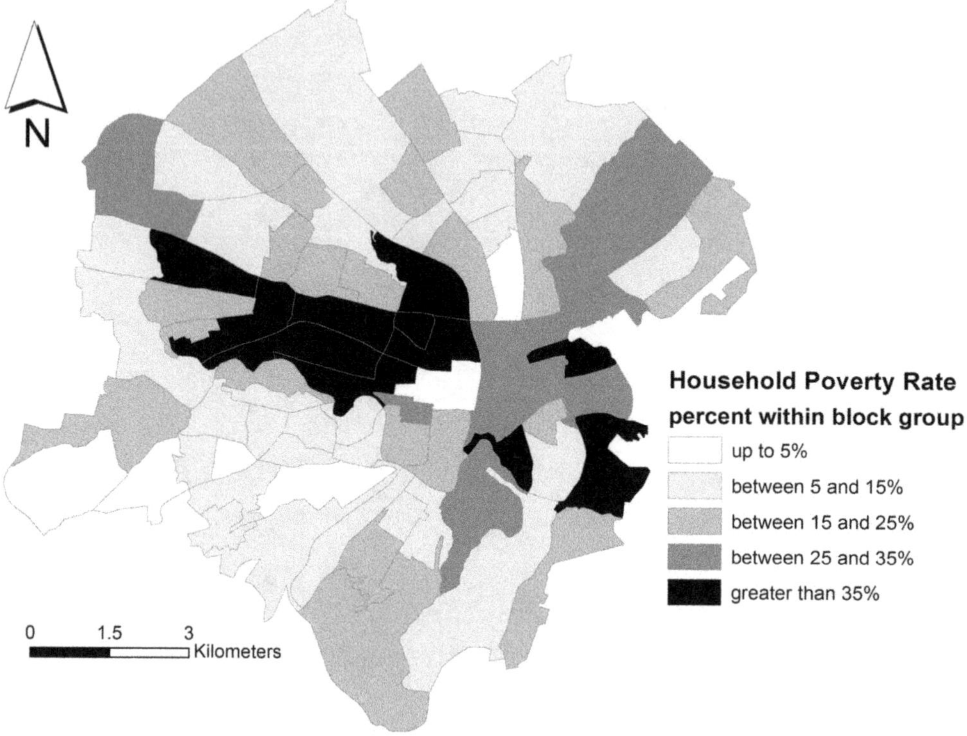

Figure 7. Block groups with extremely high rates of poverty (≥ 35%), black represents sites of potential urban farms and gray represents the locations of an existing urban farm and community gardens. The number within the block group shows the percent of households within that block group that exceed the Federal Poverty Rate

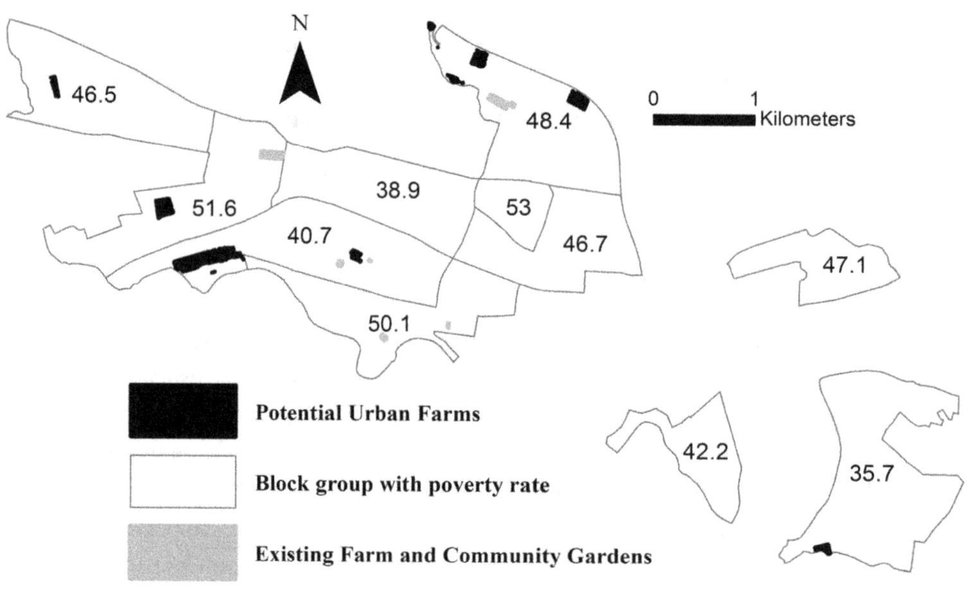

of national security concern (airport and railways) and areas where current land use is not amenable to agricultural production (wastewater treatment facilities, active recreational areas, National Park Service land, and quarries). While we eliminated all open areas with city parks, these sites could support gardens, provided that local residents have input. We included easements. We also included the final two VPDES sites that were designated by the city as vacant properties; although these locations are permitted pollutant discharge locations, this status does not especially mean that a human health toxic pollutant is located on the site. These two sites would require specific *in-situ* analysis, but should not be excluded as potential sites.

We also included FEMA floodways and flood plains. Some urban agriculture could be problematic within these areas, as during times of flood, plants could be washed away and pollutants, i.e. chemicals and sediments, could wash into the streams during normal rainfalls. However, these opens areas are amenable to trees. Although, rarely discussed as a form of urban agriculture, orchards form a viable source of food production for urban areas, and a form which requires discussion more often because most urban areas include expansion of urban forests in their sustainability plans.

We did not limit our analysis to the larger urban agriculture forms – community gardens and urban farms. We also included home gardens, schoolyard gardens and orchards. We did ignore rooftops as potential sites, although easily identifiable, these locales are problematic as their structural integrity requires assessment. We also did not place a lower size limit on all forms, as no site is too small for a home garden. As a result, our land inventory reveals that a significant portion of land (21% of total area) within the City of Roanoke is available as potential urban agriculture sites, with many potential sites available to the food insecure. Our analysis has only provided a brief and limited socio-economic analysis, other factors, e.g., locations of existing greenspaces and food deserts, should be included in identification of optimal sites for urban agriculture. Our analysis shows that the Roanoke Community Garden Association effectively chooses low-income neighborhoods for community garden sites.

These results report only hectares, and locations, of open areas available for urban agriculture, not production capability. Also, we do not comment on specific site qualities, such as soil properties, nor capabilities to support growth of specific crops. Each site would require an individual assessment (prior to cultivation), for soils, and for needed amendments, and length of growing season for plant species. Such evaluations are standard practice in any agriculture or development setting prior to land use implementation. We also recognize that, while we have identified 2,311.6 hectares as available for urban agriculture, this entire land area would not be cultivated as areas would be needed for site access, equipment, and other infrastructure, such as containers for storing harvested rainwater.

Although our analysis shows results for a specific municipality – the City of Roanoke, Virginia USA, our analysis can be implemented for any urban area, both in the Global North and Global South. While existing GIS data is not available for all characteristics for every urban area, shapefiles can be created using GIS software and remotely sensed images – both aerial photos and satellite imagery – ubiquitous for all land areas of the world. Furthermore, additional data can be acquired using unmanned aerial systems, which are low cost, becoming abundant, and in certain locales more easily flown than other aerial systems (Parece and Campbell, 2015).

Our inventory strategy provides a framework for understanding social and ecological contexts for urban agriculture. It offers an informational resource enabling examination of relationships of existing and prospective sites for contribution to ecosystem services, e.g., temperature and water regulation. Likewise, from a social perspective, an inventory allows communities to examine relationships between potential sites and their social settings; for example, between neighborhood demographics, proximity to schools, farmer's markets, parks, and walkways. With this information, communities are better prepared to employ less of a "target of opportunity" strategy for siting of community gardens, and instead undertake more extensive long-range planning that can consider a broader suite of informational resources, and involve a broader range of organizations and stakeholders. Because these data are already prepared in a geospatial context, they can seamlessly integrate with data used by planning agencies and community organizations.

ACKNOWLEDGMENT

We recognize the contributions of the *IJAGR* reviewers for their careful attention, comments, and insights that have strengthened this paper. This research was financially supported by The Virginia Space Grant Consortium 2014/2015 Graduate Research Fellowship. This publication was developed under STAR Fellowship Assistance Agreement No. FP-91769301-0 awarded by the U.S. Environmental Protection Agency (EPA). It has not been formally reviewed by U.S. EPA. The views expressed in this publication are solely those of Tammy Erlene Parece and U.S. EPA does not endorse any products or commercial services mentioned in this publication.

REFERENCES

Aitkenhead-Peterson, J. A., Steele, M. K., & Volder, A. (2010). Services in natural and human dominated ecosystems. In J. Aitkenhead-Peterson & A. Volder (Eds.), *Urban ecosystem ecology* (pp. 373–390). Madison, WI: American Society of Agronomy, Crop Science Society of America, Soil Science Society of America.

Andersson, E., Barthel, S., & Ahrné, K. (2007). Measuring social-ecological dynamics behind the generation of ecosystem services. *Ecological Applications*, *17*(5), 1267–1278. doi:10.1890/06-1116.1 PMID:17708207

Bader, M. D. M., Purciel, M., Yousefzadeh, P., & Neckerman, K. M. (2010). Disparities in neighborhood food environments: Implications of measurement strategies. *Economic Geography*, *86*(4), 409–430. doi:10.1111/j.1944-8287.2010.01084.x PMID:21117330

Balmer, K., Gill, J., Kaplinger, H., Miller, J., Peterson, M., Rhoads, A., . . . Wall, T. (2005). The diggable city: Making urban agriculture a planning priority. Nohad A. Toulan School of Urban Studies and Planning, Portland State University. Retrieved from http://community-wealth.org/sites/clone.community-wealth.org/files/downloads/report-balmer-et-al.pdf

Barton, H. (2009). Land use planning and health and well-being. *Land Use Policy*, 26(Suppl. 1), S115-S123.

Beer, A. R. (2010). Greenspaces, green structure and green infrastructure planning. In J. Aitkenhead-Peterson & A. Volder (Eds.), *Urban ecosystem ecology* (pp. 431–448). Madison, WI: American Society of Agronomy, Crop Science Society of America, Soil Science Society of America.

Bellows, A. C., Brown, K., & Smit, J. (2003). Health benefits of urban agriculture. Retrieved from http://www.co.fresno.ca.us/uploadedFiles/Departments/Behavioral_Health/MHSA/Health%20Benefits%20of%20Urban%20Agriculture%20%281-8%29.pdf

Bolund, P., & Hunhammar, S. (1999). Ecosystem services in urban areas. *Ecological Economics*, *29*(2), 293–301. doi:10.1016/S0921-8009(99)00013-0

Botkin, D. B., & Beveridge, C. E. (1997). Cities as Environments. *Urban Ecosystems*, *1*(1), 3–19. doi:10.1023/A:1014354923367

Brown, K. H., Bailkey, M., Meares-Cohen, A., Nasr, J., Smit, J., & Buchanan, T. (2002). Urban agriculture and community food security in the United States: Farming from the city center to the urban fringe. Retrieved from http://community-wealth.org/content/urban-agriculture-and-community-food-security-united-states-farming-city-center-urban-fringe

Census, U. S. (2012). Quick Facts. Retrieved from http://quickfacts.census.gov/qfd/states/51/51770.html

Colasanti, K. (2010). Assessing the local food supply capacity of Detroit, Michigan. *Journal of Agriculture, Food Systems, and Community Development*, *1*(2), 41–58. doi:10.5304/jafscd.2010.012.002

Coleman-Jensen, A., Nord, M., Andrews, M., & Carlson, S. (2011). Household food security in the United States in 2010. United States Department of Agriculture: Food Assistance & Nutrition Research Program. Retrieved from http://www.ers.usda.gov/publications/err-economic-research-report/err125.aspx

Deelstra, T., & Girardet, H. (2000). Urban agriculture and sustainable cities. In N. Bakker, M. Dubbeling, S. Gündel, U. Sabel-Koschella, & H. deZeeuw (Eds.), Growing cities, growing food: Urban agriculture on the policy agenda, a reader on urban agriculture (pp. 43-66). Feldafing, Germany: Deutsche Stiftung.

Dongus, S., & Drescher, A. W. (2006). First things first: Mapping the extent of urban agriculture. *ICT Update*, *33*, 4–6.

Doron, G. (2005). Urban agriculture: Small, medium, large. *Architectural Design*, *75*(3), 52–59. doi:10.1002/ad.76

Drescher, A. W., Holmer, R. J., & Iaquinta, D. L. (2006). Urban homegardens and allotment gardens for sustainable livelihoods: Management strategies and institutional environments. In B. M. Kumar & P. K. R. Nair (Eds.), *Tropical homegardens: A time-tested example of sustainable agroforestry* (pp. 317–338). The Netherlands: Springer. doi:10.1007/978-1-4020-4948-4_18

Dubbeling, M., & Merzthal, G. (2006). Sustaining urban agriculture requires the involvement of multiple stakeholders. In R. Van Veenhuizen (Ed.), Cities farming for the future: Urban agriculture for green and productive cities (pp. 19-40). The Netherlands: RUAF Foundation, IDRC, IIRR.

Dubbeling, M., van Veenhuizen, R., & de Zeeuw, H. (2010). *Cities, poverty and food: Multi-stakeholder policy and planning in urban agriculture*. Rugby, Warwickshire: Practical Action Publishing. doi:10.3362/9781780440545

Eisenhauer, E. (2001). In poor health: Supermarket redlining and urban nutrition. *GeoJournal*, *53*(2), 125–133. doi:10.1023/A:1015772503007

Esri. (2015). ArcMap 10.2.2. Retrieved from http://www.esri.com/software/arcgis

FAO (Food and Agriculture Organization of the United Nations). IFAD (International Fund for Agricultural Development of the United Nations), WFP (World Food Programme of the United Nations). (2014). The state of food insecurity in the world 2014. Strengthening the enabling environment for food security and nutrition. http://www.fao.org/publications/sofi/2014/en/

Feeding America. (2014). Map the Meal Gap, Food Insecurity in Your County. Retrieved from http://feedingamerica.org/hunger-in-america/hunger-studies/map-the-meal-gap.aspx

FEMA. (2014a). Floodway Definition/Description. Retrieved from https://www.fema.gov/floodplain-management/floodway

FEMA. (2014b). Definitions. Retrieved from https://www.fema.gov/national-flood-nsurance-program/definitions#F

Fuller, R. A., Irvine, K. N., Devine-Wright, P., Warren, P. H., & Gaston, K. J. (2007). Psychological benefits of greenspace increase with biodiversity. *Biology Letters*, *3*(4), 390–394. doi:10.1098/rsbl.2007.0149 PMID:17504734

Galvez, M. P., Morland, K., Raines, C., Kobil, J., Siskind, J., Godbold, J., & Brenner, B. (2008). Race and food store availability in an inner-city neighbourhood. *Public Health Nutrition*, *11*(6), 624–631. doi:10.1017/S1368980007001097 PMID:17935646

Grewal, S. S., & Grewal, P. S. (2012). Can cities become self-reliant in food? *Cities (London, England)*, *29*(1), 1–11. doi:10.1016/j.cities.2011.06.003

Hardman, M., & Larkham, P. J. (2014). Informal urban agriculture: The secret lives of guerrilla gardeners. In C. Aubry, E. Duchemin, & J. Nasr (Eds.), *Urban Agriculture* (Vol. 1). London: Springer. doi:10.1007/978-3-319-09534-9

Hofferth, S. L., & Curtin, S. (2005). Poverty, food programs, and childhood obesity. *Journal of Policy Analysis and Management*, *24*(4), 703–726. doi:10.1002/pam.20134 PMID:16201056

Horst, M. (2008). *Growing green: An inventory of public lands suitable for community gardening in Seattle, Washington* [Unpublished Masters' thesis]. University of Washington, Seattle, Washington.

Iaquinta, D. L., & Drescher, A. W. (2010). Urban agriculture: A comparative review of allotment and community gardens. In J. Aitkenhead-Peterson & A. Volder (Eds.), *Urban ecosystem ecology* (pp. 199–226). Madison, WI: American Society of Agronomy, Crop Science Society of America, Soil Science Society of America.

Kaethler, T. M. (2006). *Growing space: The potential for urban agriculture in the city of Vancouver* [Unpublished Masters' thesis]. University of British Columbia, Vancouver, British Columbia.

Kuo, F. E. (2010). *Parks and other green environments: Components of a healthy human habitat*. Ashburn, Virginia: National Parks and Recreation Association.

Lawson, L. J. (2005). *City bountiful: A century of community gardening in America*. Berkeley, CA: University of California Press.

Lehmann, S. (2012). Green urbanism: Formulating a series of holistic principles. In T. Haas (Ed.), *Sustainable urbanism and beyond* (pp. 24–30). New York, NY: Rizzoli International Publications Inc.

Lincoln Land Institute. (2015). Atlas of Urban Expansion. Retrieved from http://www.lincolninst.edu/subcenters/atlas-urban-expansion/

MacKenzie, M., & Cohen, S. (2007). *The Diggable City Phase III. Portland OR*. Portland: Multnomah Food Policy Council.

MacRae, R., Gallant, E., Patel, S., Michalak, M., Bunch, M., & Schaffner, S. (2010). Could Toronto provide 10% of its fresh vegetable requirements from within its own boundaries? Matching consumption requirements with growing spaces. *Journal of Agriculture, Food Systems, and Community Development, 1*(2), 105–127. doi:10.5304/jafscd.2010.012.008

Matsuo, E. (1998, August 2). Japanese perspectives of allotment and community gardens. XXV International Horticulture Conference Part 13. Brussels, Belgium. http://wwwlib.teiep.gr/images/stories/acta/Acta%20523/523_19.pdf

McClintock, N., Cooper, J., & Khandeshi, S. (2013). Assessing the potential contribution of vacant land to urban vegetable production and consumption in Oakland, California. *Landscape and Urban Planning, 111*, 46–58. doi:10.1016/j.landurbplan.2012.12.009

McGranahan, G., & Satterthwaite, D. (2003). Urban centers: An assessment of sustainability. *Annual Review of Environment and Resources, 28*(1), 243–274. doi:10.1146/annurev.energy.28.050302.105541

Meenar, M. R., & Hoover, B. M. (2012). Community food security via urban agriculture: Understanding people, place, economy, and accessibility from a food justice perspective. *Journal of Agriculture, Food Systems, and Community Development, 3*(1), 143–160. doi:10.5304/jafscd.2012.031.013

Mok, H., Williamson, V. G., Grove, J. R., Burry, K., Barker, S. F., & Hamilton, A. J. (2014). Strawberry fields forever? Urban agriculture in developed countries: A review. *Agronomy for Sustainable Development, 34*(1), 21–43. doi:10.100713593-013-0156-7

Moore, L. V., & Diez Roux, A. V. (2006). Associations of neighborhood characteristics with the location and type of food stores. *American Journal of Public Health, 96*(2), 325–331. doi:10.2105/AJPH.2004.058040 PMID:16380567

Moore, S. A., Wilson, J., Kelly-Richards, S., & Marston, S. A. (2015). School gardens as sites for forging progressive socioecological futures. *Annals of the Association of American Geographers, 105*(2), 407–415. doi:10.1080/00045608.2014.985627

Morland, K., Diez Roux, A. V., & Wing, S. (2006). Supermarkets, other food stores, and obesity: The atherosclerosis risk in communities study. *American Journal of Preventive Medicine, 30*(4), 333–339. doi:10.1016/j.amepre.2005.11.003 PMID:16530621

Mougeot, L. J. A. (2000). Urban agriculture: Definition, presence, potentials and risks. In N. Bakker, M. Dubbeling, S. Gündel, U. Sabel-Koschella, & H. DeZeeuw (Eds.), Growing cities, growing food: Urban agriculture on the policy agenda, A primer on urban agriculture (pp. 1-42). Feldafing, Germany: Deutsche Stiftung.

Nair, L. (2012, February 19). Cultivating Change. *Roanoke Times*. Retrieved from http://www.roanoke.com/extra/wb/304995

New York State Energy Research and Development Authority (NYSERDA). (2013). *Sustainable urban agriculture: Confirming viable scenarios for production.* Retrieved from http:\\nyserda.org

Nipen, A. (2009). *Assessing the available land area for urban agriculture on the Halifax Peninsula* [Unpublished Masters' thesis]. Dalhousie University, Halifax, Nova Scotia.

Parece, T. E., & Campbell, J. B. (2013). Comparing Urban Impervious Surface Identification Using Landsat and High Resolution Aerial Photography. *Remote Sensing*, *5*(10), 4942–4960. doi:10.3390/rs5104942

Parece, T. E., & Campbell, J. B. (2015). Land use/land cover monitoring and geospatial technologies: An overview. In T. Younos & T. E. Parece (Eds.), *Advances in watershed science and assessment* (pp. 1–32). Switzerland: Springer. doi:10.1007/978-3-319-14212-8_1

Parece, T. E., & Campbell, J. B. (forthcoming) A survey of urban gardeners across the United States. In A. WinklerPrins (Ed.), Global urban agriculture: Convergence of theory and practice between North and South. CABI.

Patel, S., & MacRae, R. (2012). Community supported agriculture in the city: The case of Toronto. *Journal of Agriculture, Food Systems, and Community Development*, *2*(4), 1–17.

Pearson, L., Pearson, L., & Pearson, C. (2010). Sustainable urban agriculture: Stocktake and opportunities. *International Journal of Agricultural Sustainability*, *8*(1/2), 7–19. doi:10.3763/ijas.2009.0468

Petts, J. (2001). *Urban agriculture in London.* WHO Regional Office for Europe, Copehagen Denmark. Retrieved from http://www.euro.who.int/__data/assets/pdf_file/0015/101625/E72421.pdf

Pickett, S. T. A., Cadenasso, M. L., Grove, J. M., Nilon, C. H., Pouyat, R. V., Zipperer, W. C., & Costanza, R. (2001). Urban ecological systems: Linking terrestrial ecological, physical, and socieoeconomic components of metropolitan areas. *Annual Review of Ecology and Systematics*, *32*(1), 127–157. doi:10.1146/annurev.ecolsys.32.081501.114012

Powell, M. (2013). *Growing Community Gardens. Presentation 1 November 2013.* Blacksburg, Virginia: Virginia Tech.

Pudup, M. B. (2008). It takes a garden: Cultivating citizen-subjects in organized garden projects. *Geoforum*, *39*(3), 1228–1240. doi:10.1016/j.geoforum.2007.06.012

Pugh, J. (2010). *A Report on the city of Roanoke's existing and possible urban tree canopy*. Charlottesville, Virginia: Virginia Department of Forestry; http://gep.frec.vt.edu/UTC/UTC_Report_RoanokeCity.pdf

Ress, D. (2013). Roanoke Natural Food Co-op farm fits in at industry park. *Roanoke Times*, *28*(May), 1.

Roanoke. (2015). GIS Shapefiles, FTP Site. Retrieved from ftp://ftp.roanokeva.gov/GIS/

Rose, D. (1999). Economic determinants and dietary consequences of food insecurity in the United States. *The Journal of Nutrition*, *129*(Suppl.), 517S–520S. PMID:10064321

Rose, D., & Richards, R. (2004). Food store access and household fruit and vegetable use among participants in the US Food Stamp Program. *Public Health Nutrition*, *7*(8), 1081–1088. doi:10.1079/PHN2004648 PMID:15548347

RUAF. (2010). *Cities, poverty and food: Multi-stakeholder policy and planning in urban agriculture*. Warwickshire, UK: Practical Action Publishing.

Seaman, P., Jones, R., & Ellaway, A. (2010). It's not just about the park, it's about integration too: Why people choose to use or not use urban greenspaces. *The International Journal of Behavioral Nutrition and Physical Activity*, *7*(1), 1–9. doi:10.1186/1479-5868-7-78 PMID:21029448

Smit, J., & Nasr, J. (1992). Urban agriculture for sustainable cities: Using wastes and idle land and water bodies as resources. *Environment and Urbanization*, *4*(2), 141–152. doi:10.1177/095624789200400214

Smit, J., Nasr, J., & Ratta, A. (2001). *Urban agriculture: Food, jobs and sustainable cities*. Canada: The Urban Agriculture Network Inc.

Taggart, M., Chaney, M., & Meaney, D. (2009). *Vacant land inventory for urban agriculture*. Cleveland, OH: Cleveland-Cuyahoga Food Policy Coalition. Retrieved from http://cccfoodpolicy.org/document/vacantland-inventory-urban-agriculture-report-urban-land-ecology-conference

United Nations. (2014). Department of Economic and Social Affairs, Population Division. World urbanization prospects: The 2014 Revision. New York.

University of Wisconsin. (2014). Population Health Institute County Health Rankings and Road Maps. Retrieved from http://www.countyhealthrankings.org/#app/virginia/2014/rankings/outcomes/overall

US EPA. (1998). Roanoke Drum Recycling, EPA ID# VA0001897289. Retrieved from http://www.epa.gov/reg3hwmd/super/sites/VA0001897289/index.htm

US EPA. (2000). Evans Chemical, EPA ID# VASFN0305570. Retrieved from http://www.epa.gov/reg3hwmd/super/sites/VASFN0305570/index.htm

US EPA. (2009). Public Works Service Center Roanoke City Virginia, EPA ID# VAD123725350. Retrieved from http://www.epa.gov/reg3wcmd/ca/va/webpages/vad123725350.html

US EPA. (2010). Ashland Distribution Co., EPA ID# VAD062373600. Retrieved from http://www.epa.gov/reg3wcmd/ca/va/webpages/vad062373600.html

US EPA. (2015a). RCRA Sites, Region 3. Retrieved from http://www.epa.gov/reg3wcmd/ca/va.htm

US EPA. (2015b). Superfund Sites, Region 3. Retrieved from http://www.epa.gov/reg3hwmd/super/va.htm

US EPA. (2015c). Toxic Relief Inventory. Retrieved from http://www.epa.gov/enviro/facts/tri/index.html

US EPA (US Environmental Protection Agency). (1994). Virginia Scrap Iron & Metal, EPA ID# VA0000807156. Retrieved from http://www.epa.gov/reg3hwmd/super/sites/VA0000807156/index.htm

USDA. (U.S. Department of Agriculture). (2013). National School Lunch Program Fact Sheet. Retrieved from http://www.fns.usda.gov/sites/default/files/NSLPFactSheet.pdf

USDA. (2014). Food Access Research Atlas. Retrieved from http://www.ers.usda.gov/data-products/food-access-research-atlas/go-to-the-atlas.aspx

USDA. (2015). Food Security in the U.S., Statistics. Retrieved from http://www.ers.usda.gov/topics/food-nutrition-assistance/food-security-in-the-us/key-statistics-graphics.aspx

van Leeuwen, E., Nijkamp, P., & de Noronha Vaz, T. (2010). The multifunctional use of urban greenspace. *International Journal of Agricultural Sustainability*, *8*(1/2), 20–25. doi:10.3763/ijas.2009.0466

Viljoen, A., Bohn, K., & Howe, J. (2005). More food with less space: Why bother? In A. Viljoen, K. Bohn, & J. Howe (Eds.), *Continuous productive urban landscapes (CPULs)* (pp. 20–31). Burlington, MA: Elsevier.

Virginia. (2012). Virginia Pollutant Discharge Elimination System Permit Program. Retrieved from http://www.deq.virginia.gov/Programs/Water/PermittingCompliance/PollutionDischargeElimination.aspx

Virginia. (2015a). Department of Conservation and Recreation, Downloads. Retrieved from http://wawww.dcr.virginia.gov/natural_heritage/cldownload.shtml

Virginia. (2015b). Department of Education School Nutrition Program Statistics and Reports: Eligibility Report for 2014/2015 School Year. Retrieved from http://www.doe.virginia.gov/support/nutrition/statistics/index.shtml

Virginia. (2015c). Department of Environmental Quality GIS Datasets. Retrieved from http://www.deq.virginia.gov/ConnectWithDEQ/VEGIS/VEGISDatasets.aspx

Virginia. (2015d). Department of Forestry, GIS Downloads. Retrieved from http://www.dof.virginia.gov/gis/dwnload/index.htm

Virginia. (2015e). Virginia Geographic Information Network (VGIN), Virginia Base Mapping Program GIS Server Connection. Retrieved from http://gismaps.vita.virginia.gov/arcgis/services

White, S. A. (2014). *Cultivating the city: Exploring the production of place and people through urban agriculture. Three studies from M'Bour, Senegal* [Unpublished doctoral dissertation]. Michigan State University, Michigan.

This research was previously published in the International Journal of Applied Geospatial Research (IJAGR), 8(1); edited by Donald Patrick Albert and Samuel Adu-Prah, pages 43-63, copyright year 2017 by IGI Publishing (an imprint of IGI Global).

Chapter 28
Urban Agriculture as Livelihood Strategy:
A Case Study in the Southern Fringe of Addis Ababa

Ashenafi Tilahun Hailie
Ministry of Construction, Ethiopia

ABSTRACT

This chapter assesses the benefits of urban agriculture in the study area. Data is generated through questionnaire, focus group discussions, interviews, and direct observations. Qualitative and quantitative analysis is made in line with sustainable livelihood framework. In the findings, sex indiscrimination, migrants' dominancy, poor educational status, and agricultural background made it as main stays. As assets, access to land, gentle slope, fertile soil, moderate climate, drainage, diversification, access to transportation, are promising. Skills, knowledge transfer, ability to labor, and good health are examined insufficient. Population explosion, absence of good governance, frequent indeterminable natural calamities, land tenure insecurity, and seasonality of prices are identified vulnerabilities. The strategies designed need intervention, organization, and persuasion effort. As outcomes, increasing well-being, reducing vulnerability, and improving food security, are substantial. Hence, incorporating a land use plan and promoting and providing support to the sector imply helping urban poor.

BACKGROUND OF THE STUDY

Over half of the absolute poor would be living in urban areas which are expected to surpass rural areas in population around the year 2005 (FAO, 1998). Africa gained an average 13 million additional urban dwellers per year in 2005-2010 and is expected to gain some 25 million per year in 2045-2050 (UN, 2011). Population growth is an important element in the growth of demand for food in the cities. Age structure of the population and income level also determine food demand. Most of the food in cities is purchased but poor urban consumers have also access to food through home production, bartering or food assistance programmes (Drescher et al. 1999).

DOI: 10.4018/978-1-5225-8063-8.ch028

"The urban food production for the well-being of producers in connection with nutrition or consumption, health, cash saving and income generation and employment is of paramount importance" (RUFA, 2001). "Finding the ways to provide food, shelter and basic services to the city residents; create "sustainable cities" are real challenges for many city authorities around the world especially in Sub-Saharan Africa" (Drescher; 2000).

The drive to integrate agriculture into urban landscapes has a long history and this drive has intensified since the Industrial Revolution. Garden cities and the benefits that early visionaries perceived were possible through them remain Utopian ideals. In the contemporary context of urban development, the possibilities of looking a new at agriculture relate more to implementing sustainability and addressing the structural changes brought about by globalization to communities, their food systems and quality of life for urbanites (Knowd et al, 2006).

Urban agriculture is a strategic resource and deserves a systematic approach to planning, one that recognizes the values of agriculture in economic, social and environmental terms. Decision-making processes that fail to consider UA risk denying current and future generations' choices in the way the economy, society and the environment are managed. These failures breach principles of sustainability and risk the loss of productive land, social capital and natural assets (Knowd et al, 2006).

Ethiopia is one of the Sub-Saharan African countries where, despite its existence and significant contribution, UA was not only unrecognized by researchers but also underestimated and given very little attention by urban development studies, urban planners and decision makers. Particularly in the city of Addis Ababa, where the practices of UA are highly prevalent, baseline data and information on the actual and potential roles of urban agriculture are scanty. (Mohammed, 2002).

Akaki Kaliti is one of the ten Sub-Cities of Addis Ababa. Geographically, it occupies the southern parts of Addis Ababa. From the city's 530.14 km^2 area, the Sub City covers a total area of 61.4km^2 consisting of eight Kebeles (UAT, 2008).

Economically, the Sub-City is an industrial zone of Addis Ababa where 60% of the industries are found. According to officials of the Sub City, there are more than 300 industries with estimated of 80,000 factory workers. The depth of workers problem can be revealed through the low wages they earn, large family size, being physical laborers or unskilled, structurally unemployed and insecurity in their work environment that forced to seek alternative means of subsistence (www.addisababacity.gov.et accessed 03/10/2009).

Demographically the population of Sub City as indicated by the Population Census Commission report (2008), is relatively with the lowest density of population (2951persons/km^2) accounting 181,202; 88,676 males and 92,526 females even as compared to the average density of the city (5,165.1/km^2).

The possible reasons for low density are the spread of numerous enterprises together with vast areal coverage, rangelands and reserved open spaces for investment. Many writers in this area of the study confirmed that even those portions of the population look for employment opportunities. The settlements are brought following the pulling effects of the stated establishments and open spaces. Besides, as alternative means of subsistence that fits with the existing assets, Urban Agriculture becomes one of the sectors for sustaining life in the study area. More specifically the purpose of the study is to: Survey the general characteristics of the farmers, Identify the major components, sort-out the relatively environmentally friendly and feasible ones, Assess its role as survival strategy and overall benefits, Indicate the challenges of Urban Agriculture, Survey stakeholders' involvement and comment on the prospect of Urban Agriculture.

THEORETICAL FRAMEWORK: THE SUSTAINABLE LIVELIHOOD SYSTEM APPROACH

A livelihood comprises the capabilities, assets (including both material and social resources) and activities required for a means of living. A livelihood is sustainable when it can cope with and recover from stresses and shocks and maintain or enhance its capabilities and assets both now and in the future, while not undermining the natural resource base. (Adapted from Chambers and Conway, 1992)

This working definition provided by Department For International Development (DFID) in the sustainable livelihoods guidance sheets(1999) is basically adopted in favour of the poor or people-centered approach to understanding the households strategies to survive, the impact of policy and institutional arrangements that influence the constraints people encountered and developing pro-poor policies(DFID,1999).

Moreover, the nature of approach's being holistic, attempting to identify the most pressing constraints faced by people everywhere, dynamicity to take account of positive patterns, mitigate negative patterns and analyze strengths rather than needs (DFID, 1999).

Hence, its stress on societies input, being modified open possibilities for urban livelihoods, its pro-poor in nature made it be more appropriate to change the livelihood status of the members of the cooperative in the study area. The framework can simplify the appropriateness through verifications of the stated components.

Figure 1. The Sustainable Livelihoods Systems Framework

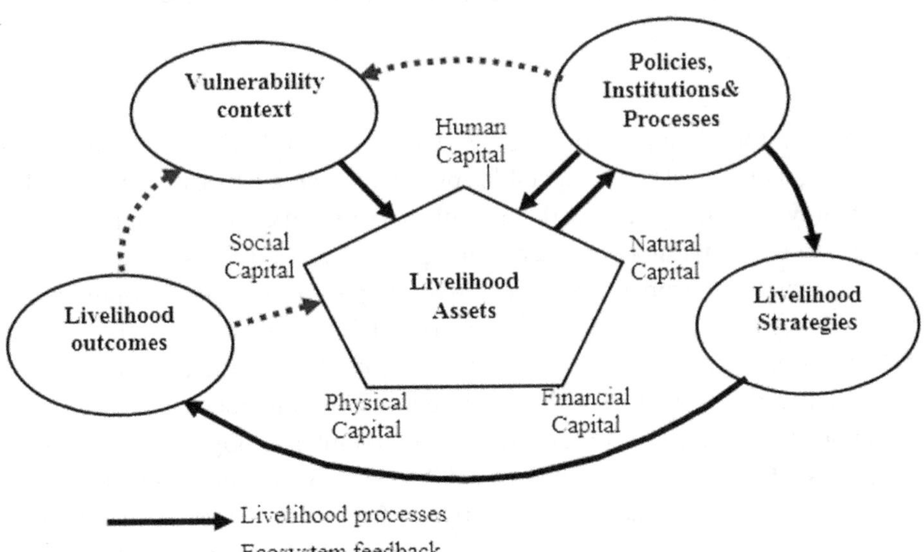

Adapted by Urban Harvest Program-designed by Thomas Zschoke from DFID, 2001, Sustainable Livelihoods Guidance Sheets

The relevancy of the framework to the study area in accordance with the basic components stated is the core of the approach, the issue of *assets*[1] stated in the definition, the vulnerability context, policies, institutions and processes since UA presupposes holistic approach.

Following the sub city's relative location it is better to justify from its proximity to natural resources (land, low slope, fertile soil, relatively cheap farmland for rent, moderate climate, and drainage). Of course, urban sprawl continues to consume land resources and make life difficult for urban commuters (Sernau, 2006). Being sub-urban; partial or entire dependence on urban farming (both crops and live-stock farming), from financial capital micro-finance (savings and credits) availability like Addis Credit Association, other Workers Associations, NGOs etc. and their potential to involve themselves in further production activities using their agricultural background.

On the *vulnerability*[2] context, vulnerability to market price instability, environmental pollution (both air and water) due to industrial concentration and poor waste disposal management hence, frequent flooding, plastics deposits, indeterminable natural catastrophes, internal conflicts over interest, structural unemployment, shocks and tensions as a result of unsecured environment and land tenure ship are the most prominent.

The involvement of stakeholders such as government, private sectors, civic societies and communities seen in a twilight form within the society has got concern in policies processes and structures part of the framework. The strategies designed are the inputs of the society which need only slight intervention, organization and persuasion to implement because almost all the cooperatives under study started production before a decade. The role of Cooperatives Organization& Development (COD) is to recognize and empower them. The outcomes forwarded in the framework are what the people dreamt.

Moreover, the strategies are designed for and revealed through the day to day struggle of the households for survival. In some cases, model households even had seen being succeeded in materializing the initial phase of the outcomes by the combined efforts of all household members who are committed to working.

METHODS OF SAMPLING, DATA COLLECTION AND PROCESSING

Methods of Sampling

According to the information obtained from the Sub City's Urban Agriculture team (UAT) extension workers in each *kebele*[3], there are about 6 vegetables producing *cooperatives*[4] with more than 200 members holding 1007 family size engaged at work currently. Document examination revealed that there are further cooperatives registered but not yet entered into production activities.

The necessary data generated from the representative samples of the survey population by stratifying into similar groups. The stratification is done on the basis of document examinations, interviewing key informants, field survey and pilot test. Out of six cooperatives; four of them namely *Biruh Tesfa*[5] Vegetables Producing Cooperative (BTVPC) with 52 members as melting pot of producers, *Cholo Enatoch*[6] Vegetables Producing Cooperative (CEVPC) of 42 women members cooperative, *Nur Aregawyan Abat Turetegnoch*[7] Vegetables Producing Cooperative (NAATVPC) 35 members of elders group and *Abu Chefie/Worku Sefer*[8] Vegetables Producing Cooperative (AC/WSVPC) with 21 members form spatially distant and unique feature (source of water for irrigation, working habit and cultural bond).

Taking the advantage of manageable size, available list of members and wide age range, socio-cultural diversifications in these four cooperatives, questionnaire are dispatched to all (145 members) to escape from wrong inference.

Out of 145 questionnaires, 17 of them were lost due to different inconveniences considered in the limitation session. So the remaining 128 members' responses are collected and analyzed accordingly. The specific sampling procedures initially followed were:

1. Selecting four vegetables producing cooperatives out of six by stratification.
2. Obtaining the list of members from the respective offices and leaders of the cooperatives.
3. Dispatching the questionnaire to the identified and available members.

Methods of Data Collection

The research is based on both qualitative and quantitative data. Information was gathered from the vegetable producing cooperatives, Urban Agriculture team, the Kebele Officers (extension workers, cooperatives organizers and other stakeholders) in the sub-city. These sources were supplemented by secondary data sources through library work, exploring the documents in governmental and non-governmental institutions using the following data-collection instruments.

- **Questionnaire Administration:** Both qualitative and quantitative questions were dispatched to the informant with the help of trained enumerators assigned in each cooperative. Two types of questionnaire were designed with varying details i.e. a bit summarized one for every member and the relatively detailed one for cooperative leaders.
- **Focus Group Discussion:** Conducted by forming flexible few sizes of small groups of selected informants from the survey population. It laid the ground for checking each other's opinion and to realize the accuracy of the responses. Moreover, the involvement of leaders and experienced members allowed the best chance for a gathering of valuable and detailed information from each cooperative.
- **Semi-Structured-Interview:** An extensive and qualitative interview conducted mainly with extension workers, cooperative officers and leaders of the four cooperatives on the overall performance, stakeholders' involvement and other related issues of the cooperatives. Besides, some others like Environmental Protection Authority, UA Team, Land Use Development and Administration in the areas of the cooperatives are also interviewed.
- **Direct Observations:** The researcher, along with the enumerators make frequent observations of the sites to check some of the responses of the respondents in practice and to perceive the feelings and day to day activities.

Data Processing

The collected data using all the stated collection methods were organized, analyzed and interpreted accordingly. Results of the quantitative session of the questionnaire were processed, refined and produced using descriptive statistical tools such as tables, bar graphs and pie charts whereas the qualitative contents were used as supplementary explanation inputs. Focus group discussions, semi-structured interviews and direct observations results were processed using logical interpretations and photographs. Satellite image to depict the current site of selected cooperatives and GIS to produce study area maps were also used.

DISCUSSION AND IMPLICATIONS

The General Characteristics of the Farmers

Age composition, gender dimension, educational background, residency status, religion, ethnicity, marital status, household size and mainstay of life are the most prominent factors that determine the overall performance of producers. The thematic approach of this research gives priority to the interpretation of these general characteristics to lay the ground for the rest of the objectives.

Adult age dominancy in the cooperatives justified by the fact that majority of them have entered into the cooperative on the basis of self-owned land using advantages of open spaces. The probability for youngsters to join is almost unthinkable unless they get an inheritance or other possible reasons such as social, economic or political. Many activities of developing countries are gender discrimination. The involvement of males and females in different activities is directly related to the attitude of people or the status of the activity in the eyes of the society. The gender data of producers in the sample cooperatives are

Figure 2. Map of the Study Area

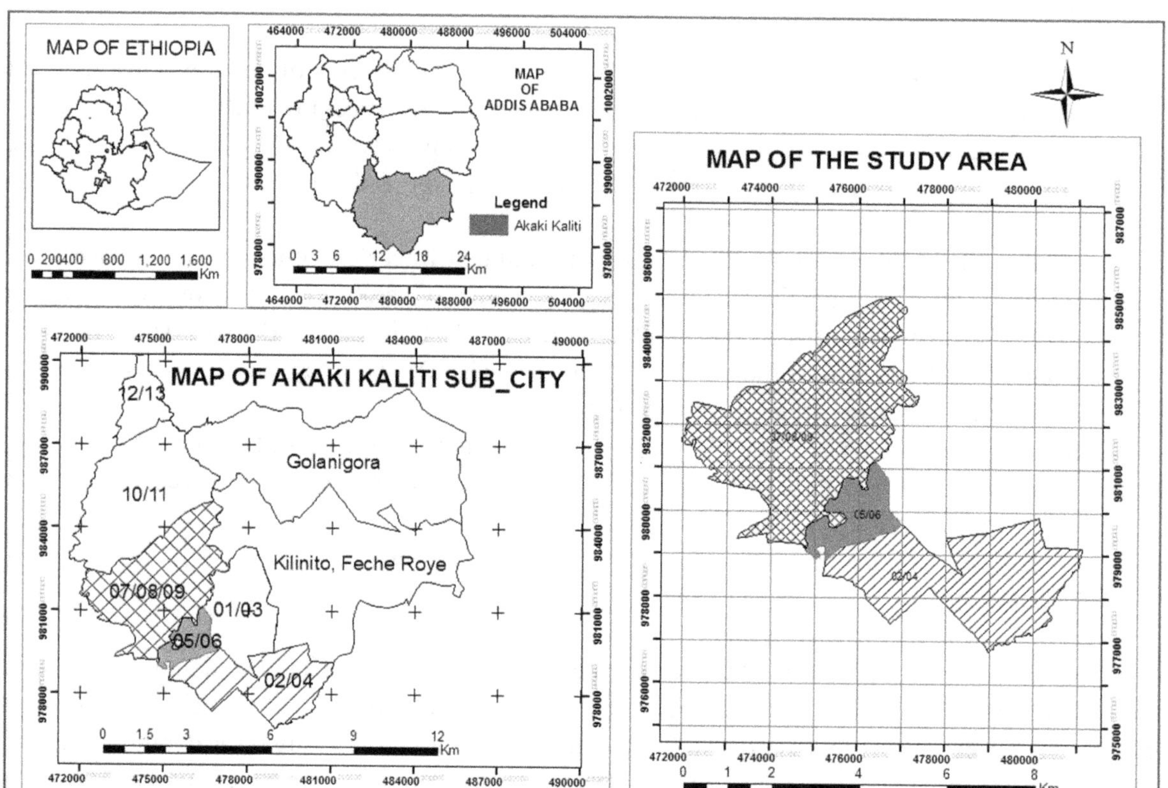

82 females: 46 males as examined from members' profile. Hence, UA has invited almost all age groups and both sexes indeed. Hence, it is an indispensable human capital. Many of the writings surveyed so far portray male dominance. Of course, gender involvement ratio in urban agricultural activities varies greatly from city to city, depending on cultural/religious context, the economic conjuncture, the economic activity, the production system, scale and areas involved (UNDP, 1996). In the *three*[9] cooperatives surveyed, men farmers were predominant and sole members of the cooperatives because the cooperatives primarily considered ownership of land/Physical capital/ in membership and the women did not have land on their own. Looking for better income and better survival options for themselves and their family members, it was only men who became tenants and waged farm laborers, and finally state-land occupiers. Given the stated few factors, the most important argument is that the sector is relatively open for both sexes. The bases of cooperatives matters. For instance, Biruh Tesfa cooperative invited both sexes following the indiscriminate nature of migration and unemployment by which the producers were victimized. In the case of Nur Aregawyan and Abu Chefie cooperatives, women and youth replaced *men*[10] due to death, illness, separation, inheritance, etc.

According to Table 1, out of the total 128 members, 68.8% is below 4th grade, shared almost evenly by 22.7% illiterates, 23.4% able to read and write, and 22.7% grade 1-4. Only 6.3% respondents have attained education level above grade 10. Hence, the cumulative 68.8% under the first cycle of primary education (1-4) need intervention in knowledge and skills empowerment. The results have two typical implications from the activity and producers composition point of view. Compared to other production sectors, UA requires fewer skills and level of sophistication. Hence, it holds physical labourers who have no other alternative jobs. Furthermore, the study witnessed this situation as: "a relatively easy industry to enter in terms of its educational and technical requirements, urban agriculture has given job opportunity to a large number of illiterate people (Mohammed, 2002). Besides, what is peculiar in this finding is that the educational background ranges of producers go to tertiary level. As a result, UA can resume as part-time job simultaneously with other professional engagements that are why first degree holders have got the chance to be seen frequently in the farmland at times of observation. Simultaneously, research work conducted in Peru where urban farmers' knowledge is transferred in various ways:

Table 1. Distribution of Respondents of Cooperatives by Educational Background.

Educational Background	Sample Cooperatives									
	Biruh Tesfa		Nur Aregawyan		Cholo Enatoch		Abu Chefie		Total	
	No.	%	No.	%	No.	%	No.	%	No.	%
Non-literate	16	34.8	13	48.2	.	.	.	.	29	22.7
Read& Write	12	26.2	6	22.2	.	.	12	63.2	30	23.4
1-4	3	6.5	8	29.6	18	50	.	.	29	22.7
5-8	7	15.2	.	.	6	16.7	.	.	13	10.1
9-10	2	4.3	.	.	12	33.3	2	10.5	16	12.5
11-12	2	4.3	.	.	.	.	.	.	2	1.6
12& above	1	2.2	.	.	.	.	5	26.3	6	4.7
*NS	3	6.5	.	.	.	.	.		3	2.3
Total	46	100	27	100	36	100	19	100	128	100

Source: Field Survey 2011 **NS: Not Stated*

from generation to generation, parents to children and by the knowledge acquired over time assured this reality (Villavicencio, 2008). Here lie the vast ranges of the *human capital* sounder. Generally, in all of the cooperatives, the need for intensive intervention in technical support and supervision is what their educational status is strongly in need of.

One of the population dynamics, migration, is paramount in its extent. As assessed from members profile and focus group discussions, 75% are *migrants*[11]. The major reasons for their migrations were the pulling effects of *natural and physical capital* such as access to open spaces, job opportunity and relatively lesser cost of living, lesser skill requirement. Ethnic-based social pulls are the other equally important factor of migration. The migrants' place of origin goes to the rural parts of the country. The three consecutive census results of CSA (1984, 1994& 2007) portrayed that the prime source of urban population explosion is city ward migration. Migrants from rural parts of the country with less skill, poor educational status and agricultural background were obliged to choose UA as the mainstay of livelihood. Even few migrants from regional urban areas are not as such to consider them unfamiliar to farming practices in a situation where the prime city, Addis Ababa is stuffed with UA. Importantly this mode of migration eases the technical and supervision provided in the contemporary areas of production. Ethnic and religious heterogeneity is more pronounced. This '*unity with diversity*' speaks of their tolerant, secular character and economic base that provide *human capital diversity* and experience exchange.

According to table 2, 93% of the total respondents are married, where 57.8% are living pairs (live together), 20.3% widowed, 9.4% divorced. The rest 7% are single. The implication of this statistics is that it is the responsibility of HHs to support significant numbers of family size. It has a direct correlation with the *HHs*[12] size result. Even the respondents living single asserted as they have the household size to support due to an extended family. Respondents of HHs size ascertained that 68% of the producers uphold 4-6 household size followed by 16.4% 7-10 and 11.7% with 1-3 family size (Figure 2). It is notable that more than 84% of the households have members greater than the average family size (4.1) of Addis Ababa (CSA, 2007). From the documentary analysis in the unpublished lists of UA cooperatives compiled by the sub city's UA team (2007) shows, 1007HHs were supported by 200 members of six cooperatives. These significant statistical results are symptoms to worry about the sustainability and productivity enhancement of the sector. Results of supplementary question "whether there are members of the family who have involvement in other activities?" assured that HH members help their parents

Table 2. Distribution of Respondents by Marital Status

Marital Status	Sample Cooperatives									
	Biruh Tesfa		Nur Aregawyan		Cholo Enatoch		Abu Chefie		Total	
	No.	%	No.	%	No.	%	No.	%	No.	%
Single	9	19.6	.	.	.	.	.	.	9	7
Pair intact	20	43.4	15	55.6	22	61.1	17	89.5	74	57.8
Widowed	7	15.2	7	25.9	10	27.8	2	10.5	26	20.3
Divorced	9	19.6	3	11.1	.	.	.	.	12	9.4
Separated	.	.	.		.	.	.	.	.	.
*NS	1	2.2	2	7.4	4	11.1	.	.	7	5.5
Total	46	100	27	100	36	100	19	100	128	100

Source: Field Survey 2011 ****NS: Not Stated***

in selected gender-based tasks stating"... even the university and high school students, graduates from different higher institutions who are short-term unemployed that look for vacancy and workers of other sector in time of their vacation have active participation in the activity..." hence, there is no gap in inter-generation work experience sharing. At cumulative level, 93% is entirely dependent on UA. Only a few members responded major occupation other than UA as a factory worker and government employee. The need for intensive technical and advisory service will be sounder since the lion shares of them have no alternative means of subsistence. Almost all of the issues discussed so far in the first objective imply UA is the ground where different assets are prevalent.

Major Production Types, Decision Makers and Products' Diversification Identification

Major Components of UA in the Selected Cooperatives

Identifying the major components of UA in the selected cooperatives is the basis for evaluating the role, challenge, stakeholders' involvement and prospect of UA in the study area. Names of all the cooperatives end up with "vegetable producers" phrase. But the pilot test ascertained that some of the cooperatives have included other components. Hence, the researcher finds it logical to add other components as follows. As evident from table 3, 74.2% of all the members of selected cooperatives are found to be vegetable producers and 18.7% are practising mixed farming i.e. livestock farming and vegetable farming. Crop cultivation is almost insignificant (only 3%). The dominance of vegetable production at the level of individual cooperatives is substantial. Besides 4.3% of Biruh Tesfa and 7.4% of Nur Aregawyan stated crop as additional products. 37% of Biruh Tesfa and 36.8% of Abu Chefie producers have simultaneous involvement in livestock and vegetables.

Who Determines What to Produce?

Table 4 is designed to crosscheck the consistency of products selection in the eyes of decision-making power. In principle decision making power is believed to be the function of all possible stakeholders'

Table 3. Distributions of Members by Activities Performed in Sample Cooperatives.

Activities	Biruh Tesfa		Nur Aregawyan		Cholo Enatoch		Abu Chefie		Total	
	No.	%	No.	%	No.	%	No.	%	No.	%
Dairy farming	.	.	.	.	.	.	.	.	.	.
Livestock fattening	.	.	.	.	.	.	.	.	.	.
Vegetables	25	54.4	24	88.9	34	94.4	12	63.2	95	74.2
Crops	2	4.3	2	7.4	.	.	.	.	4	3.1
Livestock & Vegetables	17	37	.	.	.	.	7	36.8	24	18.7
Other	.	.	.	.	.	.	.	.	.	.
*NS	2	4.3	1	3.7	2	5.6		.	5	4
Total	46	100	27	100	36	100	19	100	128	100

Source: Field Survey 2011 **NS: Not Stated*

Table 4. Distributions of Respondents on Decision Makers about Production items.

Decision Makers	Biruh Tesfa		Nur Aregawyan		Cholo Enatoch		Abu Chefie		Total	
	No.	%	No.	%	No.	%	No.	%	No.	%
Self/individual	6	13	.	.	.	.	5	26.3	11	8.6
Members together	18	39.2	12	44.4	8	22.2	14	73.7	52	40.6
AD. Extension worker	1	2.2	.	.	.	.	.	.	1	0.8
Members& Extension worker jointly	6	13	13	48.2	.	.	.	.	19	14.8
Government& NGOs	13	28.3	2	7.4	26	72.2	.	.	41	32
*NS	2	4.3	.	.	2	5.6	.	.	4	3.2
Total	46	100	27	100	36	100	19	100	128	100

Source: Field Survey 2011 *NS: Not Stated*

involvement. But the data shows deviation from this reality. At cumulative level, 40.6% of the respondents confirmed that decisions are confined among the members while in case of 32% there is the direct or indirect influence of *organizations*[13], and 14.8% by joint decision of members and extension workers. Only 8.6% members responded that they themselves were the decision makers. Abu Chefie Cooperative did not get the chance from other stakeholders for decision making. However, the majority of the discussants in the focus group discussion stated that there are no concerned bodies that frequently visit to provide technical and counselling services let alone to decide what to produce. Members and few organizations confinement on what to produce imply produced items in each season are subjected to the actual performance of these agents which in turn shadowed over the issue of sustainability.

Products' Diversification

Determination of average numbers of products produced in a single season is an input to have a sound figure in the most frequently produced items of UA in the sample cooperatives. Informants on average admitted having produced 4-5 items production per cropping season. Moreover, as pointed out by informants through discussions, interviews and observations as well, the status of diversification is the function of farm size, weather suitability, input accessibility and members' agreement. Beyond the identification of products, the assessments come up with another finding. Many of the producers are conscious of "what to produce when" by 'default' or using their lifelong try and error experience stating that:

...although we haven't received timely and sustainable support from anybody else, we have the know-how of 'what to produce when' in accordance with factors of production, nature of the products, the market demands... above all by following the footsteps of our forefathers and agricultural backgrounds of place of origin...

Almost all the cooperatives under consideration have similar justifications over the selection of items of production. Minor variations in the order of preferences were seen across cooperatives. Spring cabbage, Cabbage, Carrot, Potato, Lettuce, "Kosta", Tomato, green peppers etc. are usually produced vegetables. For all of the cooperatives potato, spring cabbage and carrot ranked as the top three preferred products. Bases of top three prioritizing among all others are these products adaptability to weather conditions, the requirement of relatively low cost of production inputs like labor, finance, land, water etc., high demands and lesser extent of the post-harvest damage. Besides, native cabbage and tomato are preferred for they are harvested every 15 days as results serve as the mainstay of home consumptions. Mixed farming to withstand challenges is also another area of diversification. In the stated cooperative there are cows, sheep and oxen reared using the advantages of the relatively vast open spaces as obtained from discussion, interviews and field observations. Besides, the cooperatives' preparation observed during the study is another area of diversification by planting *fruits*[14] and letting flood-prone areas free "*Qetiema*"[15] grass. The diversifications both in livestock and permanent fruits are the physical capitals of the producers that serve as means of capital accumulation and withstand challenges. Further exploration through discussions and field observations raised the need to categorize the cooperatives into two. Those who have commune and individually owned plots simultaneously and are biased to the selection of products for the commune from a profitability point of view, irrespective of the cost of production and compensating their daily consumption requirements flexibly from the individually owned ones. That is their *resilience*[16] from risks and shocks. On the other hand are those who joined the cooperatives with their entire plots under commune and prefer short season products to the rest stating:

We prefer short-term products such as cabbage, tomato, even green pepper because we don't have means of subsistence to tolerate, cover costs of production and to withstand seasonal catastrophes... hence we usually prefer shorter growing season products such as cabbage, tomato and green peppers against the long-term

Generally, the issue of "what to produce" is subject to multiple factors such as the nature of land tenure system, sex composition, age, economic status of the members, production season, etc. but producers are conscious and have their rational basis in selecting feasible components and withstand shocks. Simultaneously informality in land tenure, product diversification and shift to short-season vegetables is more prevalent following *vulnerability context*[17].

The Role of UA as Survival Strategy and Overall Benefits

Major Agro- Activities and Roles as Means of Subsistence

The relief gardens, vacant lot gardens, and subsistence gardens improved the health and spirit of participants by creating feelings of usefulness, productivity and importance while also providing opportunities for food and work (Tucker 1993). In many of African countries, UA could become an important component of urban development and make more food available to the urban poor besides providing employment. The common language in few kinds of literature appearing so far in this sector, in Addis Ababa, is significant in the magnitude of UA to the well-being of urban poor. Mohammed in his findings stated: "The urban vegetable production in Addis Ababa is making substantial contribution to improve the situation in the city letting vegetable products available in the market at large amount with relatively lower price that most low-income urban dwellers can afford" (2002).

The lion share from agro-activities identified in the study goes to vegetable production, followed distantly by dairy farming and livestock fattening. Typically they are registered as vegetable producing cooperatives. As a cooperative, the vegetable production activity is the main source for all of them but side by side they shift to mixed farming[18] to some extent. "Whether UA is the prime means of subsistence for producers?" responses 88.3% are under the umbrella of UA and 10.2% of them have alternative sources of income. All related questions discussed substantiate this reality with slight fluctuations. The major challenge of this study is a failure to explore the absolutely exact figure of subsistence for the producers were not free and willing to tell their revenue.

An Assessment of Production Seasons and Sources of Income Other Than UA

Urban Agriculture's economic importance has a direct relation to the number of the production season. More production seasons, mean more chance of income generation. In terms of production seasons, it is only one cooperative's farmers that reported a third growing season. The most frequently forwarded reasons across cooperatives in the focus group discussion laid on the cyclic problems encountered every year as factors of the production period. The factors of delay and confinement to one or two production seasons will be discussed in the challenges too UA session. The general concluding remark is that the sector has a significant positive role by means of subsistence. Sources of income other than UA cross-check questions are another support to the validity of the role because alternative sources like a government employee, petty traders and pensioned informants cumulatively accounted less than 20% (Figure 3). On the other hand, 96(75%) of the producers do not have other sources. To speak about income and expenditures with the urban farmers was quite a delicate subject because they are not open about this issue (Villavicencio, 2008). Given the difficulty to determine the actual income generated per producer, the contribution is of paramount significance. That is why some literature realizes this aspect and justify it as "Be it small or large, legal or illegal, informal or formally recognized, urban farmers around the world, especially in the developing countries, are producing competitive incomes through urban farming" (UNDP; 1996: 169).

Contribution to Annual Income and Status of Involvement

Following the respondents' feelings of unwillingness to tell their annual income, indirect approach to estimate the facts was followed. At cumulative level, UA as a major and partial contribution to annual income is proven sound. These successive data are stated in line with family size, efficiency and cost of production as grasped from the discussions, facial expressions and supportive sayings of the respondents. One of the widows willing to respond to the role of UA stating:

UA is everything for me. When I was displaced having lost all of my properties in Asmara[19], there was nothing to eat, even to purchase basic equipments. I and four of my children were waiting for wheat aids from rehabilitation program. Thanks to God! After getting organized my family's life has been changed. Today, four of my children are attending their education successfully, I afford up to 400birr housing rent, furnish my home, employ more than four workers and get an average 20,000 to 25,000 birr per year on self-owned land side by side to the commune.

Moreover, sample sayings like: "Something is better than nothing/involvement losing alternatives." Some of the informants, especially the elders and mothers groups responded also their involvement in UA "'insignificant' but waiting for the better tomorrow." Furnishing their home, covering social expenses like *idir*[20], *ikub*[21], shift to modern savings etc… are also areas of improvement as a result of income generation. As to the status of involvement; permanent members, seasonal employees and family members are there to fill the gaps in times when the cooperative members are not around the major actors.

Reasons and Attitudes Towards Getting Organized

Justifications on the reasons of getting organized by large are due to unemployment, production enhancement purpose and to withstand challenges. In addition to employment opportunity, "to withstand challenges" is the notion of Biruh Tesfa cooperatives, where the majorities of them were displaced from different parts of the *country*[22] and get organized as coping mechanisms. Production enhancement is the prime motive of them as well. Many of the respondents and focus group discussants common language "*unity is strength*" proves their strong and positive outlook towards their social bond. Further discussions and field survey assured that a significant number of employees joined the sector both permanently and seasonally. Producers' attitude towards getting organized is a bit subjective but many of them have an optimistic view. Reasons to perceive getting organized very well among all others are livelihood improvement; a chance to work in harmony, perform tasks within a short period of time and success in having different supports like aids, credits, training etc. Few of them who failed to judge is due to lack of consistency in the overall activity of the cooperative, loss of confidence in the *sustainability*[23] of UA as a sector, suspicion on land tenure security in the locality following the horizontal expansion of the city.

Producers Opinions on Livelihood Improvement

The attitude of respondents towards getting organized has a logical match for livelihood improvement. There necessitate the need to associate with the role of UA tri-modally to the producers, the community and the city's economy. The increment in the number of growing seasons has a direct correlation with annual income. Success in identifying UA as prime means of subsistence and lack of a significant alternative source of income (Figure 3) tends to show the benefit of UA. The overall performance, recognition, support provision and awareness of producers are in progress. As a survival strategy, the aforementioned features put the role of UA sounder even to the nearby community through the provision of job opportunity in the farmland, transportation of products for carts men, middlemen (marketing), permanent merchants, as means of subsistence for urban poor with relatively affordable price and fresh products provision. Besides, some of the outcomes frequently heard from the respondents are coverage of the locality with scenic beauty, environmental balance, grounds of mental excitement, prolong life (elders case), prevention of youngsters from undulating in bad habits such as addictions, sexual harassment, enhance working habits of generation, testing ground for urban born new generations by creating the image of rural Ethiopia. This study succeeded in exploring the essence of exercising life than what usually is told in theory. In the eyes of findings assessed so far, given the contemporary role of UA in the study area, it is logical to look for physical, natural, social, financial and human capital through sustainable technology and counselling support provision.

Challenges of Urban Agriculture

The practice of UA revolves around the fact that it is the most common survival strategy among the poor while there are no well laid down by-laws and policies to regulate the practice and the competition

Figure 3. UA is the prime means of subsistence for producers

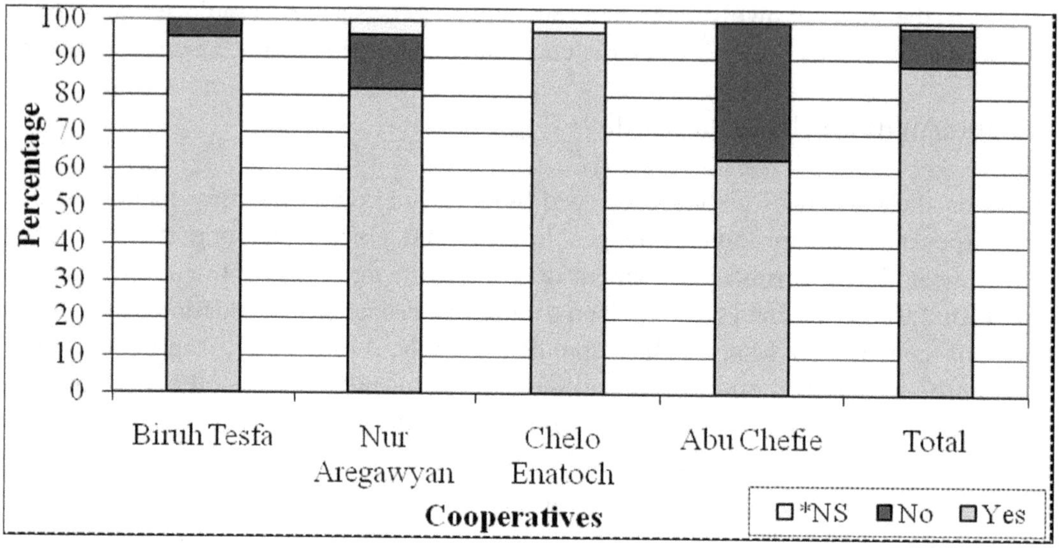

for access to land and water from other "developmental" projects such as housing and industry (Smith, 1992; Mensah 1996). The pilot test carried out prior to the actual implementation stage of this study enforced the scope of challenges to be confined to pre, during and post-production period challenges (vulnerabilities) with typical features, which are not the focus areas of this chapter.

Shock [24]Absorbing Mechanisms

Survey on the availability of short term and long term solutions was made. There are no long lasting attempts to alleviate problems encountered. All of the coping mechanisms seen are temporary in character.

Figure 4. Sample Photos of Challenges in Pre and during production stage

Shock absorbing mechanisms as explained by many of the respondents are their attempt at "struggling for survival." Few respondents also witnessed that they notify problems and receive timely support from stakeholders but the majority of them put this notion under question stating that: "It is one direction; we frequently notify our problems but we do not get a timely and appropriate answer. Our question faced complete ignorance or distortion by woreda[25] administrators" Hence, some informants hopelessly responded "Adaptation" justifying "no further attempt at all; it is beyond our capacity; we will struggle to the maximum effort, use the left season and harvest what is left from damage." Members' withdrawal, frustrations and despair are too proximate, especially in the old age groups. The frustrations originated from the merciless frequent damages by floods, plastics litters, market price fluctuation and information related challenges otherwise many of them are optimistic to better tomorrow.

Stakeholders Involvement

Urban agriculture takes place in a multi-sectoral environment, touches several urban management areas (e.g. land use planning, environmental and waste management, economic development, public health, social and community development), and involves numerous diverse systems and related actors (input provision, vegetable production, aquaculture, livestock production, processing and marketing) (Marielle et al., 2006). This session of the discussion considers the formal and informal reactions of possible stakeholders through interviews, focus group discussions, documentary analysis and observations besides the questionnaire.

Economic Actors/Stakeholders Involvement Assessment

According to Figure 6, Sources of Initial capital is one ground to evaluate the involvement of some of the stakeholders at the beginning. Top three sources of capital come from 33% aids from organizations. This is true of Biruh Tesfa cooperative which consists of displaced and unemployed HHs rehabilitated by FAO and for whom provision of the existing open space was made by government authorities. The efficiency and consistency of Cholo Enatoch cooperative also strongly linked with aids. One of the best evidence is the material and financial support by EDA. Following the robbery of cooperative's water

Figure 5. Shelterbelts as Copping Mechanisms

pumping generators and cultivating materials they were under the severe problem. It is EDA that gave immediate credit response. Personal savings take the second position accounting 30% followed by credit associations of 13% (Figure 6).

Asking alternative sources of capital during production to withstand challenges 51.6% members responded in favor of deposits or savings, 14.8% aids, 13.3% relatives and friends and 12.5% from credit associations (Table 5). These responses, at least, have implications on the involvement of stakeholders. Whatever the extent NGOs, credit associations, relatives, friends and community organizations like "Idirs" have initial roles to play as financial capitals.

Figure 6. Sources of Initial capital

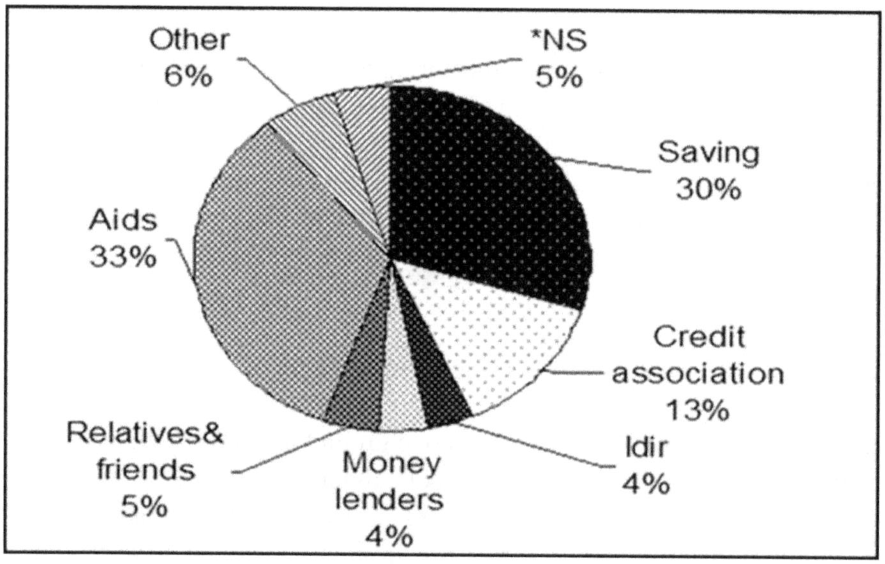

Table 5. Alternative Sources of Capital

Source of capital at times of adversities	Distribution of Responses By Cooperatives									
	Biruh Tesfa		Nur Aregawyan		Cholo Enatoch		Abu Chefie		Total	
	No.	%	No.	%	No.	%	No.	%	No.	%
Deposits/savings	11	23.9	17	62.9	25	69.4	13	68.4	66	51.6
Credit associations	16	34.8	.	.	.	.	.	.	16	12.5
Lenders	3	6.5	.	.	.	.	.	.	3	2.3
Relatives& friends	9	19.6	2	7.4	.	.	6	31.6	17	13.3
Aids	3	6.5	7	26	9	25	.	.	19	14.8
*NS	4	8.7	1	3.7	2	5.6	.	.	7	5.5
Total	46	100	27	100	36	100	19	100	128	100

Source: Field Survey 2011 *NS: Not Stated*

Overall Supports From Stakeholders Involvement

Question on whether there are provisions of any support or not (Table 6), 49.2% of them witnessed to materials support, 19.5% of technical support responded by 37% of Nur Aregawyan and 41.7% of Cholo Enatoch cooperatives stating agricultural extension worker the source of this support. Training/awareness provision is the cumulative responses of the cooperatives except for Abu Chefe cooperative that responded uniformly in the focus group discussion session as they are neglected from this service. Informants that notified the support came from relatives and friends had no formal stakeholders to mention rather they thank the extended family. Surprisingly the majority of the members get a single training on cooperatives perhaps in the time of establishment or using temporal advantages, and only a few of them got twice as confirmed in the focus group discussion session. Moreover, few leaders have got the chance to attend on different short-term workshops that are left theoretical due to inconsistency and lack of supervision.

Moreover, a checklist of possible institutions distributed to the leaders of the cooperative believing that the leaders have frequent exposure to external factors of the respective cooperative. Some of them responded they are familiar with but few of them are even not known as a concerned institution given the educational background and awareness of the producers. The potential stakeholders put on BPR manual interface and their performances crosschecked through the checklist by referring the *core processes model*.[26]

Comments on Prospect

Producers Prospect to Their Cooperatives

Cooperative leaders and permanent members are invited to guess rationally about the future of their cooperative. Hinted to consider both internal and external factors before prediction, hot focus group discussion was made. Hereby, their attitudes towards getting organized have a strong match with the cooperatives prospect description. Some of the sayings:

Table 6. Supports Obtained from Stakeholders

Supports Obtained So far	Distribution of Responses by Cooperatives								Total	
	Biruh Tesfa		Nur Aregawyan		Chelo Enatoch		Abu Chefie			
	No.	%	No.	%	No.	%	No.	%	No.	%
Materials	29	63	12	44.5	4	11.1	18	94.7	63	49.2
Technical	.	.	10	37	15	41.7	.	.	25	19.5
Financial	5	11	.	.	8	22.2	.	.	13	10.1
Training/ awareness	10	21.7	2	7.4	4	11.1	.	.	16	12.5
Others	.	.	2	7.4	4	11.1	.	.	6	4.7
*NS	2	4.3	1	3.7	1	2.8	1	5.3	5	4
Total	46	100	27	100	36	100	19	100	128	100

Source: Field Survey 2011 ****NS: Not Stated***

As compared to the past there are symptoms of improvement in living conditions, awareness change of members, local authorities and communities towards the cooperatives and the relative extent of visit of some stakeholders such as extension workers, urban agriculture team, cooperatives organizers, NGOs etc. are in a slightly increasing and promising trend.

However, this prediction is put into precondition. Many of the discussants concluded with an argument:

...we hopefully expect sustainability of all the promising internal and external improvements. If so, we hope the prospect of the cooperative will be good...'hope is the only beckon'...we hope...!

Prospect of UA in the Study Area

As for this study, the historical perspective of the city's master plan is taken as an input to have an intellectual guess for the future of UA. Addis Ababa has experienced several planning changes. Of which the latest Master Plan for Addis Ababa, meant to serve the city for 20 years, was prepared by the Addis Ababa Master Plan Project through Ethio-Italian technical cooperation during 1984-86. The anticipation of the socio-economic and spatial changes in the city's development and the transition toward political federalism and market-led economy necessitated revision (ORAAMP, 2009). The Master Plan Revision Office (ORAAMP) was set up and mandated primarily to develop an appropriate and viable city development plan for an effective urban management that promotes investment and facilitates stakeholder involvement. Sustaining Urban Agriculture Requires the Involvement of Stakeholders and consideration in the revision because, it touches on a large number of *urban management areas*[27] and involves a large diversity of systems and related actors (input provision, vegetable production, aquaculture, livestock production, processing and marketing) (Marielle et al., 2006). Its Multi stakeholder orientation made to face challenges. Jeong (2006) explained that "when material interests override ecological concerns, short-term economic gains are achieved at the expense of long-term environmental sustainability". Results from focus group discussion, interviews, document examination and field exposure ascertained this reality.

The more frequent argument from urban planning is that agriculture should be confined to rural areas, as it can interfere with more productive use/rent of land by other economic activities. They solely give attention to the accounting cost of land; tolerate the use of rivers for liquid disposal, industrialization without treatment plant and displacing the UA producers following the city's expansion by compensation but without consulting what to perform by ignoring the implicit role of the sector. This does not lead the city's healthy development an inch in a situation where the whole world is claiming the need to deliver environmentally friendly development approach.

The positive outlook of the city government is expressed by the recognition of its benefit and contribution to the city development effort. Inclusion as one of the land use component within the 10 years city structural plan (2001-2010) by reserving 7,175ha of land for UA representing 13.82% of the total land use within the city is initial promising ground for its prospect (Gete et al., 2007). Besides, the key informants from core process officers asserted the prospect of commenting transforming structures:

Its inclusion under Trade and Industry Development Bureau has both benefits and limitations. The benefit is that it can perform its task nearby with strong intra-sectoral linkages among MSEs Institutions.

On the contrary, UA is getting less emphasis, underestimation, ignorance of its basic requirements and empowerments both technically and budget allocation for it is neglected from its mother, Ministry of Agriculture and Rural Development.

Of course, the comparative advantage of its inclusion under Trade and Industry needs thorough assessment and exhaustive study. A respondent from extension workers forecasts the fate of UA with preconditions both from producers and government actors' perspective stating: "*I have no threat to the fate of UA because the producers are aware of the benefits of UA, as a result, they resume producing at the commune and individual level. But for its sustainability the attitude, commitment and vision of all government and nongovernmental actors needs improvement.*"

To be optimistic, the trends of UA in the city in general and the sub-city in particular, given the resolve of all the reviewed internal and external challenges, are promising.

CONCLUSION

Urban-specific poverty indices reveal that poverty had increased in the urban centers. Both the growth and re-distribution effect are dominant in escalating as well as reducing poverty. The poverty profile of the nation in general and Addis Ababa, in particular, reached to its indeterminable extent due to multiple factors. The vast and heterogeneous in nature of activities performed in the city made the assessment more blurred. Classifying economic activities into formal and informal and measuring the extent to which the informal economy generates employment opportunities are challenging tasks. According to the first National Labor Force Survey (1999) in Ethiopia, nearly three-quarters of the population were employed in the informal economy. Economic regression, adjustment policies, continued high rates of urbanization and population growth have led to the unpredicted expansion of informal sectors in the urban centers. Identification of basic and non-basic sectors in urban centers is a challenge. But this study ascertained that UA is one of the sectors where a significant number of the urban poor made their mainstay.

This research assessed the overall status of UA in the study area integrating with the Sustainable livelihoods framework and the Core Processes Interface. The selected cooperatives assets surveyed in the background and findings of the study, as to the natural capital access to natural resources (land free of tax, gentle slope, fertile soil, relatively cheap farmland for rent, moderate climate, drainage (Akaki River and groundwater) along the challenges are promising. Under physical capitals: shift to mixed farming, fruits, relative location proximity to Akaki market, access to roads and transportation are inviting grounds; the human capital in line with skills, knowledge transfer, ability to labor and good health is found to be the lowest status. Financial capital has got justifications in its relative context for the respondents are not open for this issue. The finding under Initial and alternative sources of capital find out personal savings, aids from organizations (both government and NGOs), micro-finance (savings and credits) availability like Addis Credit Association, other Workers Associations, NGOs etc. are symptoms of building assets, securing their basic needs requirements revealed through shift from destitution to affording capacity, a track to poverty reduction although frequently disrupted by catastrophes.

On the vulnerability context, the study area cooperatives producers livelihoods and assets vulnerability observed being affected by trends of (population, urbanization, resource utilization, economic crises and governance); Shocks due to frequent flooding, plastics deposits, indeterminable natural calamities, economic, internal conflicts over interest, Crop damages, insecurity in the working environment and land tenure ship; Seasonality of prices/inflation/, production, health, employment opportunities, structural unemployment, shocks and tensions as a result of are the most prominent.

Transforming structures (levels of government and private sectors) and processes (laws, policies, culture and institutions) that determining the gaps between assets and vulnerability context are reflected in the findings. Decentralization and privatization, the need for reconsideration and substantiation with clear working guides of tax-free cooperatives land tenure system, the existing cultural bonds and grounds of tolerance are instruments for enhancing livelihoods outcomes and reducing vulnerability. The involvements of stakeholders such as government, private sectors, civic societies and communities have been seen. The interface's implementation successions of the core process are the driving forces of livelihoods improvements. Besides, the strategies are designed for and revealed through the day to day struggle of the households for survival.

Considering livelihoods outcomes, UA as the mainstay of the urban poor, few cooperatives and households even had seen being succeeded in gaining income, increasing well-being, reducing vulnerability, improving food security, using the resource by mobilizing cooperative members and family size. Recognizing, promoting and providing relevant support to the sector by implication imply helping the poor.

This study has explored the general characteristics of UA actors; most frequently produced items in the eyes of diversification, explicit and implicit benefits, pre, during and post-production challenges, level of involvement of stakeholders and basic grounds for prediction of UA in the study area.

The prevalence and significance of Urban Agriculture, the challenges it poses to urban planners, decision-makers and residents are now widely recognized. This is proved by the growing research on UA and its inclusion in the urban land use plan. UA is also gaining momentum because it is increasingly seen as instrumental to furthering the sustainable development of human settlements.

Moreover, promoting Urban Agriculture seems an integral part of making cities more productive economically as well as environmentally appealing. Although Urban Agriculture is a traditional practice among city dwellers, it is a relatively new phenomenon in Urban Planning; there still exist some negative attitudes towards it. The notion that Urban Agriculture is an illegal activity or an expression of the "ruralization" of cities is not yet washed from the minds of stakeholders. This is what the research has proved in the assessment session of stakeholders' involvement in the study area. Unlike other researchers conducted so far, in this study multiple issues through comparative analysis across the sampled cooperatives is made. This approach will enable readers to justify the overall status of UA, actors' performances, concerned bodies actual performance, cooperatives fate in particular and the sector's prospect in general. Above all serving as a benchmark for newly coming researchers in the area especially on multiple open-ended features of the sector in the future have got beginning in this study.

ACKNOWLEDGMENT

My deepest gratitude goes to all of the people around me who have a positive outlook, substantial support and constructive comments on my effort.

Moreover, this chapter is dedicated to the memory of my lovely mother Tewabech Worku Askalemariyam.

REFERENCES

Chambers, R., & Conway, R. (1992). *Sustainable rural livelihoods: Practical concepts for the 21st C.* Institute of Development Studies. Discussion Paper No. 296.

CSA. (1999). The 1994 Population and Housing Census of Ethiopia. Results for Addis Ababa, Vol. II. Analytical Report. Addis Ababa: CSA.

CSA. (2008). *Summary and Statistical Report of the 2007 Population and Housing Census*. FDRE-PCC. Dec. 2008.AA.

DFID. (1999). *Sustainable Livelihoods Guidance Sheets*. DFID.

Drescher, A. W. (2000). *Urban Food Security: Urban Agriculture, a Response to Crisis*. Frankfurt: University of Freiburg.

Drescher, A. W., & Iaquinta, D. (1999). *Urban and peri-urban Agriculture: A new challenge for the UN Food and Agriculture Organization (FAO)*. FAO.

Dubbeling & Merzthal. (2006). *Sustaining Urban Agriculture Requires the Involvement of Multiple Stakeholders*. Academic Press.

FAO. (1998). *Majority of people lives in cities by 2005*. Retrieved from http//www.fao.org/NEWS/FACTFILE/FF9811-E.HTM

Getez, Trutman, & Aster. (Eds.). (2007). Fostering new development pathways: Harnessing Rural-Urban linkages to reduce poverty and improve environment in the highlands of Ethiopia. *Proceedings of a planning workshop on thematic research area of the global mountain program.*

IDRC. (2003). Urban Agriculture: Land Management and Physical Planning. Guidelines for municipal policymaking on urban agriculture. First Edition – March 2003, No. 3.

Jemal. (2002). *Urban Agriculture initiatives in A.A. with practical evidences from selected vegetable producing in the city*. Academic Press.

Jeong, H. (2006). *Globalization and the Physical Environment*. Philadelphia: CH P.

Knowd, I., Mason, D., & Docking, A. (2006). *Urban Agriculture: The New Frontier*. Planning For Food Seminar, Vancouver, Canada. Retrieved from www.agf.gov.bc.ca/resmgmt/sf/plan_food/Presentations/8_Mason.pdf

Mensah, O. K. (1996). Changes in official attitudes towards urban agriculture in Accra. *African Studies Quarterly, 6*(3). Retrieved from http://web.africa.ufl.edu/asq/v6i3a.htm

ORAAMP. (2009). *The Addis Ababa Master Plan: Historical Perspective*. ORAAMP.

Sernau, S. (2006). Global Problems: The search for equity, peace, and sustainability. Academic Press.

Smith, D. (1992). Strategies for meeting basic food needs in Harare. In J. Baker & P.O. Pedersen (Eds.), *The urban-rural interface*. Uppsala: Nordiska Afrikainsitutet.

Tucker. (1993). *Depression Relief Gardens.1929-39. Sprouts in the sidewalk*. Academic Press.

UNDP. (1996). *Urban Agriculture: Food, Jobs and Sustainable Cities*. United Nations Development Program.

United Nations. (2011). *Population Distribution, Urbanization, Internal Migration & Development: An International Perspective*. Author.

Villavicencio, L. M. (2008). Urban Agriculture as a Livelihood Strategy in Lima, Peru. Academic Press.

ENDNOTES

1 Material (Physical) and non-material (social) resources that people possess.

2 The probability of being exposed to risk.

3 The then lowest administrative structure in the city.

4 According to cooperatives organization offices of each kebele in the sub city, there are more number of cooperatives registered and entered to production activities during the research year. But these 6 cooperatives are relatively aged and can serve as input of stratification.

5 "Mekana – Biruh Tesfa" a merged phrase Literally means the "Barren land", which was named by an individual prior to 1986 and when they are given by 1986 added "Bright hope" or "Future Bright" to the site's previous name since many of them reached the current status from nothing as they explained.

6 Cholo Enatoch" literally mean "mothers" in Specific locality the so called "Cholo", which is established as women cooperation and organized by indigenous NGO to empower them.

7 Self explanatorily, the cooperative is an association of old aged pensioned people eager to live long.

8 Formerly it had its name called "Worku-Sefer" from its founder an individual called "Gash Worku". Still they preferred it to be known by that name but the office recognized it as "Abu Chefie" following its location in marshy area. Hence, throughout the discussion the second name/Abu Chefie/ is used.

9 Biruh Tesfa, Nur Aregawyan and Abu Chefie.

10 Their husbands or fathers.

11 More than two third of the migrant respondents left their places of origin before two decades.

12 Households.

13 Organizations like Emanuel Development Association (EDA), FAO & other rehabilitators.

14 Flood resistant tree-fruits such as: Apple, Grape, Coke, "Cazmire", "Prime", "Zeyitune".

15 Grass type highly demanded by the society in time of holidays, different social events, restaurants etc. to decorate houses.

16 The ability of households or individuals to recover from shocks and avoid a decline in wellbeing.

17 Vulnerability to market price instability, environmental pollution and poor waste disposal management hence, frequent flooding, plastics deposits, indeterminable natural catastrophes, internal conflicts over interest, structural unemployment, shocks and tensions as a result of unsecured environment, blurredness on land tenure ship.

18 Using the advantages of open space and vegetable residues; livestock fattening, diary and ship rearing took place.

19 Capital city of Eritrea.

20 ...voluntary association of nearby community to facilitate funeral ceremonies. Every member of the association is subjected to contribute some amount of money monthly. This money is mostly meant for mourning and related activities.

21 An old aged traditional but across generation and organized way of saving money every week, two weeks or once in a month based on the volunteer members nature of income that will be given to either of the member through the lottery method until member size based round completes.

22 Eritrea, Aseb, Massawa… and unemployed x-soldiers and civilian people from the rest part of the country and get organized in the locality.

23 The ability of a natural or human system to maintain output and productivity.

24 A sudden, unpredictable and traumatic event which leads to a marked decline in wellbeing.

25 Administrative hierarchy above Kebele.

26 Are: Urban Agriculture Extension professionals, Kebele Officers, Environmental protection Authority, MSEs development, NGOs involved in UA, Work& Urban Development Bureau, Cooperatives Organization& Development, Land use development& administration, Modern Marketing Agency, Credit Associations, Ministry of agriculture& rural development, Artificial livestock breeding center, fodder preparation factories, input production and distribution organization. The inter-sectoral linkages of these stakeholders both across offices and at field determine the efficiency and sustainable performance of UA.

27 Land use planning, environmental and waste management, economic development, public health, social and community development, etc.

This research was previously published in the Handbook of Research on Urban Governance and Management in the Developing World edited by Joshua Mugambwa and Mesharch W. Katusiimeh, pages 320-342, copyright year 2018 by Information Science Reference (an imprint of IGI Global).

Chapter 29
Challenges and Opportunities in Urban Farming:
A Study With Reference to Chennai

V. Hemanth Kumar
Sri Sairam Engineering College, India

K. Sentamilselvan
SRM Valliammai Engineering College, India

ABSTRACT

In the rapid growing urbanization, especially in the metros, it is not sure that it really leads to development of the society nor becoming a threat to it. The urbanization leads to shrinkage of the fertile and farming lands to concrete monuments for human to work in an artificial atmosphere. The vertical urbanization is very common in the cities than horizontal, due to shortage of lands. This leads to the vital resources of food and vegetables in scarce for the drastically overflowing population, ends up in fast foods and packed foods culture, that is far from organic foods, which is a clear path towards obesity, diabetes and high rate of heart disease among the youth. This study is an initiative to bring out the feasibility of urban farming that struggles to strengthen its roots in our nation. The research focuses on the challenges and opportunities of urban farming in the society. The outcomes shall bring out societal views on urban farming and the suggest ways to overcome the hindrances and facilitate the people towards utilizing the opportunities of urban farming with its benefits and 'Let the City Grow Green', which is the way to enrich health for their present and future generations.

INTRODUCTION

In the recent walks of urban life, the life is becoming intense. This also lead to the good fact of more popularity of urban farming in the last few years as concerns about the environment have combined with increased interest in health and community development. This gave raise to support for food systems in metro areas as an integral part of a sustainable development. However, they face many challenges, which accounts for the fact that a number of resources provided in this study pertaining to the recent data. The city's food system is governed by local, regional and global systems of production and consumption.

DOI: 10.4018/978-1-5225-8063-8.ch029

Urban farming is not a new concept; however, it is a rapidly growing, which developing economies are adopting. With more than 2 billion people, now residing in cities around the world it is increasingly important that we use forgotten spaces to adopt practices that will benefit both the environment and our society.

Urban farming is highly optimistic in the garden space and in the rooftops. The crops that are very much required for the day-to-day life are mostly grown in all climate and hence all the food crops of the city needs can be grown within city itself, provided the urban farming is spread completely. There are also legitimate health concerns surrounding urban farming, particularly in terms of recycling of wastes into fertilizers. The part of rainwater harvesting shall be directed for home based farms. This will also avoid the stagnation of water and other waste and helps in controlling the spread of diseases. Recently there is strong evidence, that these concerns are far outweighed by the current and potential benefits of urban farming. Particularly in developing countries like India, urban farming enhances household food security, which is free from chemicals. This currently draws the societal attention. Urban farming also plays a vital role in engaging the women and youth in gaining the fullest advantage of the available resources. It is increasingly recognized that, effectively managed, urban farming contributes to the eco-friendly environment.

Significance of Urban Farming

Urban farming is becoming a popular option among metro residents. Now-a-days lot of people are concerned over their health and quality of life. There are plenty of options for urban dwellers to have their own farm at home. These options make it easier for individuals and families to reap the benefits of growing their own vegetables. The following are some of the benefits of urban farming:

- Pollution free air.
- Reduces the risk of respiratory problems and headaches.
- Creates access to fresh vegetables at home.
- Reduced risk of chemical ingestion found in foods and vegetables.
- Increased awareness about the need for improving eco-friendly environment.
- Plants are safeguarded from pests and chemicals.

REVIEW OF LITERATURE

Jitendra Pandey and Ashima Singh (2012), states that Organic farming follows the principle of circular causation and has emerged in response to questions on health, environment and sustainability issues. In this review, we assess the status, opportunities and C- sequestration potentials of in India. We identify constraints that impede adoption of especially for small farm holders who constitute over 70% of farming community in India. With large land area and climate diversity, India has a considerable potential to contribute to C-sequestration.

According to Weiping Chen (2013), the aim is on three factors: to conceptualize a construct of the perceived value in CSA and its dimensions, to operationalize and validate the construct, and to empirically investigate the influence of perceived value in CSA on consumer satisfaction and loyalty. This research begins by reviewing relevant literature that frames the concept of perceived value in CSA and integrates this construct into a nomological network. Operational measures for each component of perceived value in CSA are then developed. The scale is validated, and then used to test the hypothesized model in a sample of 198 consumers of Beijing's five CSA farms using partial least squares (PLS) as an analytical tool. Perceived value in CSA is found to be a formative, multi-dimensional, third-order construct. Further, the results suggest that perceived value in CSA has both a direct positive effect on consumer loyalty and an indirect effect mediated through consumer satisfaction. Limitations and research directions refer to the need to increase the sample size and replicate research in other places, the possibility of conducting longitudinal research, the need to identify antecedents of perceived value, and the opportunity to assess perceived value across a number of CSA member characteristics. CSA farmers should recognize the importance of value creation to build and maintain long time relationships. This study suggests that CSA farmers can enhance value from two approaches. One approach is to build strategies to facilitate delivery of product benefits, emotional benefits, and social benefits. Another approach is to invest in efforts to reduce inconvenience and lower risk.

Poppy Arsil, Elton Li, Johan Bruwer, and Graham Lyons (2014), investigates how consumers from a developing country background such as Indonesia make local fresh food decisions for daily eating. The use of the means-end chain approach is utilized as a measure of attributes, consequences and values of locally produced products. For Javanese ethnic group in Indonesia, "save money" and "health benefits" are identified views that motivate consumers purchasing their local foods. Although investigating the largest ethnic groups in Indonesia, the results of this study cannot be generalized to all Indonesian consumers and a larger sample needs to be studied to generalize the results to the wider population of Indonesia. It is better for the Government to promote local food policies that is based on identified motivations of consumers. "Save money" and "health benefits" themes can be used as the central messages for the development of advertising strategies. This study identifies the Javanese motivations for buying local foods and examines the motivation differences between rural and urban locations.

Jan Falkowski (2015), reveals that, recent literature has considerably improved our understanding of vertical relations in the food chain. One area which has received relatively little attention however relates to the resilience of an agro-food supply chain, that is its ability to face and, if needed, to recover after a major disruption. This paper studies the factors, including characteristics of vertical links between upstream and downstream sectors, may make farmer-processor relationships more or less resilient to adverse shocks. To do so, the author uses a unique region-level data set on disruptions to dairy supply chain in Poland during the transition from a centrally planned economy to a market economy. More specifically, using between-region variation, the author investigates why in some regions supply chain disruptions, measured as the breakdown of relationships between farmers and processing industry, were smaller than in other regions. The findings suggest that the supply chain, which the author analysed, was less resilient to crisis, i.e. the author observed larger disruptions to supply relations, when the supply base was more fragmented and when farmers had better outside options to market their produce via direct sales to consumers. In addition, dairy supply chain in Poland seemed to be less resilient in regions with larger share of the state-owned land. The results come with several caveats. First, the empirical evidence comes from Poland during the specific period and thus it may not be easily generalized. Second, the results are based on historical correlations. Therefore, although they are robust across various specifications, which the author estimates, they may not establish causal relationship due to some omitted variables or potential indigeneity issue. Finally, what the author uses here are region-level data. One may argue therefore that farm-level data would give more fine-tuned focus for testing impacts and theories regarding supply chain resilience. To the best of the author's knowledge, this paper is the first to provide some evidence on agro-food supply chain from this perspective.

Jessica Cook, Kate Oviatt, Deborah S. Main, Harpreet Kaur, and John Brett (2015), in their study on Re-conceptualizing urban agriculture: an exploration of farming along the banks of the Yamuna River in Delhi, India, reveals that the proportion of the world's population living in urban areas is increasing rapidly, with the vast majority of this growth in developing countries. As growing populations in urban areas demand greater food supplies, coupled with a rise in rural to urban migration and the need to create livelihood options, there has been an increase in urban agriculture worldwide. Urban agriculture is commonly discussed as a sustainable solution for dealing with gaps in the local food system, and proponents often highlight the many social, environmental, and economic benefits. They argue that the sustainability of urban agriculture should not be assumed. There is a need for research to take a bottom-up approach, exploring the influence that city-level conditions have on livelihood decisions by farmers and how this shapes the practice of urban agriculture. This paper uses a case study for an in-depth look at urban agriculture in Delhi, India to understand from the farmers' perspective how urban agriculture is practiced and what factors influence farmers' livelihood decisions. Using a team-based, multi-method Rapid Assessment Process, data were collected through preliminary key informant interviews, field observations, semi-structured interviews with urban farmers, and geographic information systems mapping. This research provides an in-depth description of market-oriented urban agriculture in a developing country, explores how farmers' livelihood decisions are embedded in the urban context, and discusses the potential of urban agriculture as a sustainable city-system.

Elke Rogge, Eva Kerselaers, and Charlotte Prové (2016), In their study, focus on urban planning, peri-urban areas are often addressed with an urban-centric view on development, disregarding the multifunctional and dynamic opportunities that these spaces offer. Consequently, argument is that land use functions such as agriculture do not reach their full potential, despite the increasing enthusiasm for peri-urban and urban agriculture. The twofold structures include- first, it explores the opportunities and challenges for agriculture in peri-urban areas; and second, it studies success factors for envisioning processes promoting peri-urban agriculture in urban policy and planning. Through action research, we gather and compare data from two envisioning processes in the Flemish cities of Ghent and Kortrijk. Both processes were initiated by the local authorities, with the purpose of developing a spatial vision for agriculture in peri-urban areas. Results show that in both contexts, pressure on farmland is a key issue. In addition, it highlights that multi-functionality is rather complex, both in practice and from a governance perspective, but promising as a territorial concept in envisioning processes. Regarding the envisioning process itself, the analysis shows that clarity and consensus on the objectives of the process, delineation of the study area, policy support, clear leadership, and inserting sound and reliable data into the process are important success factors. It also provides insight into the visions, plans and strategies needed to embrace the potential of agriculture in peri-urban areas, through the exploration and valuation of participatory envisioning processes. Future research is needed to explore the implementation phases of envisioning processes in urban planning.

Need for the Study

The demand and supply of food and vegetables are constantly facing high challenges due to the rapid growth of urbanization, where the agriculture lands are converted into buildings. This requires some serious attention, and especially people living in the urban are facing the problem of junk food addiction. Now this study aims in creating the awareness about the healthy food habits that could be possible through urban farming and ascertains the level of interest among the various sects of the society towards it. The challenges and the opportunities are highlighted to encourage urban farming cities.

Objectives of the Study

- To identify the challenges and opportunities for urban farming.
- To ascertain the societal views towards junk food and create health awareness.
- To find out the level of acceptance towards the home-grown vegetables through urban farming.
- To suggest ways to overcome the hindrances and encourage urban farming at low cost.

RESEARCH METHODOLOGY

The primary purpose of Descriptive research is to provide an accurate description or picture of the status or characteristics of a situation or phenomenon and hence the same is adopted in this study. The Convenience sampling method is followed in-order to collect the primary data through the structured questionnaire (rated with 85% - Cronbach's alpha testing) from the respondents. The sample size is 160 respondents.

Analysis and Discussions

Pearson correlation coefficient was computed to assess the relationship between the Awareness of urban farming and planting vegetables at home. There was a positive correlation between the two variables, r = 0.414, n = 160, p = 0.001. Overall, there was a weak, positive correlation between Awareness of urban farming and planning vegetables at home.

Table 2 allows us to understand the different category of the income levels in terms of place of purchasing of vegetables. Out of the total population 50.1% people fall below 3 hundred thousand income, 37% fall between 3 – 6 hundred thousands, 20% fall between 6 – 9 hundred thousands, 2.5% fall between 9 – 12 hundred thousands and 4.4% fall above 12 thousands.

Table 1. Correlation coefficient: Awareness of urban farming vs. plant vegetables at home

Correlation			
		Aware of Urban Farming	**Plant Veg at Home**
Aware of urban farming	Pearson Correlation	1	.414**
	Sig. (2-tailed)		.000
	N	160	160
Plant veg at home	Pearson Correlation	.414**	1
	Sig. (2-tailed)	.000	
	N	160	160

**. Correlation is significant at the 0.01 level (2-tailed).

Table 2. Cross tabulation: income vs. place of purchase of vegetables (numbers in hundred thousands)

			Purchase Vegetables				Total
			Market	Super Market	Petty Shop	Street Vendors	
Income in Rs. (p.a.)	<3 lacs	Count	52	12	4	12	80
	3 - 6 lacs	Count	20	13	4	0	37
	6 - 9 lacs	Count	12	16	0	4	32
	9 - 12 lacs	Count	0	0	0	4	4
	> 12 lacs	Count	4	3	0	0	7
Total		Count	88	44	8	20	160

H0: There is no significant relationship between the level of education and purchase of organic vegetables.
H1: There is significant relationship between the level of education and purchase of organic vegetables.

Inference: The values chi square = 0.379, p = 0.945, denotes that since p-value is more than 0.05, it is concluded that, Null hypothesis is accepted and there is no significant relationship between education level and purchase of organic vegetables.

H0: There is no significant relationship between the Age group of the respondents and the frequency of consuming food outside.
H1: There is significant relationship between the Age group of the respondents and the frequency of consuming food outside.

The values chi square = 0.329, p = 0.001, denotes that since p-value is less than 0.05, it is concluded that, the Null hypothesis is rejected and there is significant relationship between the age group of the respondents and frequency of consuming food outside.

Table 3. Chi-square test: education level vs. purchase organic vegetables

	Value	df	Asymp. Sig. (2-Sided)
Pearson Chi-Square	.379[a]	3	.945
Likelihood Ratio	.378	3	.945
Linear-by-Linear Association	.120	1	.729
N of Valid Cases	160		

Table 4. Age group of the respondents vs. frequency of consuming food outside

	Value	df	Asymp. Sig. (2-Sided)
Pearson Chi-Square	.329[a]	9	.001
Likelihood Ratio	.053	9	.000
Linear-by-Linear Association	.159	1	.000
N of Valid Cases	160		

Table 5. Yule's co-efficient: gender vs. opinion towards planting vegetables at home

Gender	Yes (A)	No (α)	Total
Male (B)	30	62	92
Female (β)	38	30	68
Total	68	92	160

Yule's co-efficient of association Q = (AB) (α β) – (A β) (αB) / (AB) (α β) + (A β) (αB) = 316 / 2116 = 0.149

There is positive association between the Gender and Planting vegetables at home.

Summary of Findings

- Most of the respondents are in the age group of less than 35 years. Majority (58%) of the respondents are Male and the remaining are female (42%).
- All the respondents are educated and most (38%) have completed under graduation.
- Most (33%) of the respondents are salaried employees with (50%) of them drawing less than 3 lac p.a. and followed by (23%) respondents drawing between 3-6 lacs p.a.
- Majority (70%) of the respondents are staying on Rental accommodation and (30%) of the respondents own their houses.
- Majority (70%) of the respondents are in Chennai for more than 9 years. Majority (65%) are the native of Chennai and 70% of the respondents are feeling Good living in Chennai.
- Majority (65%) of the respondents have opined Good towards growing urbanization.
- Majority (70%) of the respondents comment Bad about converting agricultural land into buildings.
- People residing in Chennai are prone to consume food outside and 60% of them consume once in a week, 18% twice and 22% thrice a week. Moreover, all of them are aware of the adverse effect on health due to junk foods.
- 50% of the respondents purchase vegetables from vegetable market and 30% from super markets.
- Majority (63%) of the respondents found the quality of vegetables as Normal.
- 50% of the respondents feel the price of the vegetables is Normal and 43% feel its High.
- Majority (60%) of the respondents don't purchase organic vegetables, and 40% do purchase. Moreover 70% of them feel the price of organic vegetables are High compared to that of normal vegetables.
- Majority (55%) of the respondents feel that the availability of the organic vegetables is not adequate in the market place.
- Majority (62%) of the respondents do not have garden at home and 38% has the garden. The prime reason for not having the garden is the Lack of space at their residence.
- Among the surveyed respondents 53% are aware of Urban farming and the prime source of awareness is through Internet (38%).
- Majority (57%) of the respondents don't plant vegetables at home and 43% do plant at home by utilizing the terrace and open space.

- Majority (73%) of the respondents are interested to have their own farm at home and 65% of the respondents are willing to spend a nominal maintenance fee to take of their farm.
- Majority (68%) of the respondents would encourage the society to have their own farm at home to take care of their requirements.

Implications

- The respondents are mostly city based and are educated, in this regard, it becomes easy to reach out them through internet and campaigns that highlights the benefits of Urban farming.
- The main reason being lack of space to indulge in urban farming and most of them feel the land space and not the terrace, which can be utilized effectively for farming activities.
- Since the salaried group are high in numbers, the urban farming can be extended to their work place also in the aspect of community farming, so that it reaches out well.
- The people who consume food outside are very much aware about the adverse effect of junk foods and this shall be supplemented by inducing the interest towards urban farming, as the youth are much concerned about the community servicing.
- Moreover, banks shall give special loans for the urban farmers and organic farmers, which will motivate towards eco-friendly products in the metros. Once the supply is increased, the price obviously will be reasonable and quality/organic foods available affordable prices.
- The residents of Chennai are willing to have the farm of their own at home at a nominal price that also leads to ventures that promote urban farming in cities.

CONCLUSION

The study on challenges and opportunities in urban farming in Chennai, reveals the insights of the societal view towards urban farming, the level of awareness, concern towards the health & food, their consumption pattern and habits, their willingness to accept urban farming have given an optimistic outcome, that strengthens the feasibility of developing urban farming in the metros. The findings are arrived through appropriate analysis and suggestions are made based on the survey and observation. The suggestions if incorporated aptly by the people and authorities concerned, shall pave the way for successful urban farming in the cities undoubtedly in the near future.

REFERENCES

Arsil, P., Li, E., Bruwer, J., & Lyons, G. (2014). Exploring consumer motivations towards buying local fresh food products: A means-end chain approach. *British Food Journal*, *116*(10), 1533–1549. doi:10.1108/BFJ-04-2013-0083

Chen, W. (2013). Perceived value in community supported agriculture (CSA): A preliminary conceptualization, measurement, and nomological validity. *British Food Journal*, *115*(10), 1428–1453. doi:10.1108/BFJ-01-2011-0013

Cook, J., Oviatt, K., Main, D. S., Kaur, H., & Brett, J. (2015). Re-conceptualizing urban agriculture: An exploration of farming along the banks of the Yamuna River in Delhi, India. *Agriculture and Human Values, 32*(2), 265–279. doi:10.100710460-014-9545-z

Falkowski, J. (2015). Resilience of farmer-processor relationships to adverse shocks: The case of dairy sector in Poland. *British Food Journal, 117*(10), 2465–2483. doi:10.1108/BFJ-12-2014-0433

Pandey, J., & Singh, A. (2012). Opportunities and Constraints in Organic Farming: An Indian Perspective. *Journal of Scientific Research Banaras Hindu University, 56*, 47–72.

Rogge, E., Kerselaers, E., & Prové, C. (2016), Envisioning Opportunities for Agriculture in Peri-Urban Areas. In K. Andersson, S. Sjöblom, L. Granberg et al. (Eds.), *Metropolitan Ruralities (Research in Rural Sociology and Development* (Vol. 23, pp. 161-189). Emerald Group Publishing Limited.

This research was previously published in Start-Up Enterprises and Contemporary Innovation Strategies in the Global Marketplace edited by Rajagopal and Ramesh Behl, pages 28-36, copyright year 2018 by Business Science Reference (an imprint of IGI Global).

Index

A

B

C

D

E

F

G

H

I

K

L

M

N

O

P

R

S

T

U

V

W

www.ingramcontent.com/pod-product-compliance
Lightning Source LLC
LaVergne TN
LVHW080404011026
849161LV00009B/2347
* 9 7 8 1 5 2 2 5 8 0 6 3 8 *